Mustapha Hatti
Editor

AF801082

Smart Computing and Control Renewable Energy Systems

Advanced Computational Techniques for Wireless Network Optimization

Springer

Editor
Mustapha Hatti
Unité de Développement des Equipements Solaires,
EPST-CDER
Bou Ismail, Algeria

ISSN 2367-3370 ISSN 2367-3389 (electronic)
Lecture Notes in Networks and Systems
ISBN 978-3-031-80300-0 ISBN 978-3-031-80301-7 (eBook)
https://doi.org/10.1007/978-3-031-80301-7

© The Editor(s) (if applicable) and The Author(s), under exclusive license
to Springer Nature Switzerland AG 2025

This work is subject to copyright. All rights are solely and exclusively licensed by the Publisher, whether the whole or part of the material is concerned, specifically the rights of translation, reprinting, reuse of illustrations, recitation, broadcasting, reproduction on microfilms or in any other physical way, and transmission or information storage and retrieval, electronic adaptation, computer software, or by similar or dissimilar methodology now known or hereafter developed.
The use of general descriptive names, registered names, trademarks, service marks, etc. in this publication does not imply, even in the absence of a specific statement, that such names are exempt from the relevant protective laws and regulations and therefore free for general use.
The publisher, the authors and the editors are safe to assume that the advice and information in this book are believed to be true and accurate at the date of publication. Neither the publisher nor the authors or the editors give a warranty, expressed or implied, with respect to the material contained herein or for any errors or omissions that may have been made. The publisher remains neutral with regard to jurisdictional claims in published maps and institutional affiliations.

This Springer imprint is published by the registered company Springer Nature Switzerland AG
The registered company address is: Gewerbestrasse 11, 6330 Cham, Switzerland

If disposing of this product, please recycle the paper.

Lecture Notes in Networks and Systems 1238

Series Editor

Janusz Kacprzyk, *Systems Research Institute, Polish Academy of Sciences, Warsaw, Poland*

Advisory Editors

Fernando Gomide, *Department of Computer Engineering and Automation—DCA, School of Electrical and Computer Engineering—FEEC, University of Campinas—UNICAMP, São Paulo, Brazil*
Okyay Kaynak, *Department of Electrical and Electronic Engineering, Bogazici University, Istanbul, Türkiye*
Derong Liu, *Department of Electrical and Computer Engineering, University of Illinois at Chicago, Chicago, USA*
Institute of Automation, Chinese Academy of Sciences, Beijing, China
Witold Pedrycz, *Department of Electrical and Computer Engineering, University of Alberta, Alberta, Canada*
Systems Research Institute, Polish Academy of Sciences, Warsaw, Poland
Marios M. Polycarpou, *Department of Electrical and Computer Engineering, KIOS Research Center for Intelligent Systems and Networks, University of Cyprus, Nicosia, Cyprus*
Imre J. Rudas, *Óbuda University, Budapest, Hungary*
Jun Wang, *Department of Computer Science, City University of Hong Kong, Kowloon, Hong Kong*

The series "Lecture Notes in Networks and Systems" publishes the latest developments in Networks and Systems—quickly, informally and with high quality. Original research reported in proceedings and post-proceedings represents the core of LNNS.

Volumes published in LNNS embrace all aspects and subfields of, as well as new challenges in, Networks and Systems.

The series contains proceedings and edited volumes in systems and networks, spanning the areas of Cyber-Physical Systems, Autonomous Systems, Sensor Networks, Control Systems, Energy Systems, Automotive Systems, Biological Systems, Vehicular Networking and Connected Vehicles, Aerospace Systems, Automation, Manufacturing, Smart Grids, Nonlinear Systems, Power Systems, Robotics, Social Systems, Economic Systems and other. Of particular value to both the contributors and the readership are the short publication timeframe and the worldwide distribution and exposure which enable both a wide and rapid dissemination of research output.

The series covers the theory, applications, and perspectives on the state of the art and future developments relevant to systems and networks, decision making, control, complex processes and related areas, as embedded in the fields of interdisciplinary and applied sciences, engineering, computer science, physics, economics, social, and life sciences, as well as the paradigms and methodologies behind them.

Indexed by SCOPUS, EI Compendex, INSPEC, WTI Frankfurt eG, zbMATH, SCImago.

All books published in the series are submitted for consideration in Web of Science.

For proposals from Asia please contact Aninda Bose (aninda.bose@springer.com).

Contents

Energy Systems and Renewable Technologies

Numerical Investigation of a Novel Constant Volume Combustion Phase Strategy in a Diesel-Hydrogen Dual-Fuel Engine 3
Ghodbane Hassina, Fouad Khaldi, and Derradji Bahloul

Performance Analysis of Thermoelectric Generator with Sliding Mode Control 12
Asma Toualbia, Rachid Taleb, and Abdelkadir Belhadj Djilali

Passivity Based Control of Thermoelectric Generator System by Using a Boost Power Converter 18
Asma Toualbia, Rachid Taleb, and Abdelkadir Belhadj Djilali

Improvement of the Performance of a Proton Exchange Membrane Fuel Cell (PEMFC) System Designed to Propel an Electric Vehicle 27
Achouri Abdelhak and Abdi Hamid

Predictive Wind Turbine Power Analysis Based on SCADA Data and Machine Learning Algorithms 32
Zouhir Iourzikene, Fawzi Gougam, and Djamel Benazzouz

Applying an Optimized and Enhanced ANN-SMOOC for Wind Turbine Chains with DFIG Using PSO Algorithm 43
Lakhdar Saihi, Fateh Ferroudji, and Mebrouk Bellaoui

Parallel Photovoltaic Inverters Equipped Active Power Filters: Autonomous Mode and Grid Connection Mode Operations 51
Issam Abadlia, Fateh Abdoune, Mohamed Rida Bengourina, and Linda Hassaine

Comparative Analysis of MPPT Algorithms: P&O and Inc for Optimizing PV Systems 60
M. Belabed, S. Bechekir, M. Brahami, H. Bendaho, A. Brahimi, and I. S. Bousmaha

Enhanced Grid Integration of PV Systems Using Double-Stage Configuration and PUC7 Inverter Controlled by MPC 71
Soufian Khettab, Aissa Kheldoun, Rafik Bradai, and Intissar Hattabi

Fault Ride-Through Strategy for MMC-HVDC System Under Single Line-to-Ground Fault ... 78
Hicham Korzane, Abdeldjabar Benrabah, Farid Khoucha, and Khelifa Benmansour

Optimizing Crow Algorithm for Enhanced MPPT Control in Photovoltaic System Under Varying Illumination Conditions ... 86
Djihane Bougandoura, Sabrina Titri, and Cherif Larbes

Control Systems and Power Optimization

Design of Control Circuits for Dynamic Reconfiguration of Shaded Solar Panels ... 99
S. Arezki, A. Abdelghani, and M. Boudour

Optimal Design and Tuning of TID Power System Stabilizer for Multi-machine Power System Using Dandelion Optimizer ... 116
Intissar Hattabi, Aissa Kheldoun, Rafik Bradai, and Soufian Khettab

Integration of Energy Storage Systems in the Power System to Prevent Blackouts ... 124
L. Mizat, M. Lahdeb, Y. Oubbati, and Ahmed T. Hachemi

Improvement of Double Diode Model of Solar Cells Using Mountain Gazelle Optimization Algorithm and Based on Lambert's W Function ... 133
A. Abbadi, F. Hamidia, A. Morsli, M. R. Skender, and N. Douifi

Application of Artificial Intelligence to Reduce Conventional LFC Capacity in Utility Grid Using Multiple Interconnected Microgrids ... 141
N. Kouba, A. Boudoucha, A. Brik, M. A. Matouk, and M. Ammi

Design and Performance Investigation of Hierarchical Control Intended for Three-Phase Photovoltaic Microgrids ... 152
C. Ait Hammouda, R. Bradai, A. Kherbachi, A. Bendib, and R. Boukenoui

PD Controller Design for Three-Axis Attitude Stabilization of a Microsatellite Using Reaction Wheels ... 161
B. Khouane, J. E. Benmansour, I. Rached, and A. S. Zemouli

Stability and Performance Enhancement of a Step-Down Converter in Photovoltaic Systems Using an Optimization Algorithm-Based Fractional $PI^{\lambda}D^{\delta}$ Controller ... 170
Elouahab Bouguenna, Samir Ladaci, and Walid Merrouche

Fuzzy Logic Control of Hybrid PV-Wind-Battery On-Grid Systems 179
Zouak Fatma Zahra, Chaib Lakhdar, and Choucha Abdelghani

Optimizing MPPT Control of DC-DC Boost Converters Using Metaheuristic Search Algorithms and Nonlinear PID (N-PID) Controllers 187
L. Khellaf, A. Djellal, H. Mayache, and A. Sheta

Simplified Super-Twisting Sliding Mode Applied to Dc-Bus Voltage Regulator 196
Abdesselam Guerbouz, Imad Merzouk, Ahmed Hfaifa, and Amine Fihakhir

Computational Modeling and Simulation

Comparative Analysis of PV Module Models Using AI-Based Metaheuristic Optimization Techniques 207
El Mouatez Billah Messini, Yacine Bourek, and Chouaib Ammari

Experimental Investigation into the Effect of Building Orientation on Its Thermal Behavior 217
Abderrahmane Khelfaoui, Abdeldjalil Dahbi, Miloud Benmedjahed, and Boudjemaa Tidjar

Numerical Simulation of the Heat Exchange Surface Effect on the Performance of a Vortex Cooling Tower 226
A. L. Deghal Cheridi, A. Dahia, A. Bouam, A. Dadda Khorsi, A. Hadjam, and A. Kentouche

Modelling and Understanding the Lateral Displacement of the Bottom Hole Assembly (BHA) Under Bending Vibrations 235
Nora Benmir, Hamza Akroum, and Moahamed. Z. Doghman

Modeling of a Coupled Severe Axial-Torsional Vibrations in Drill String Petroleum System 244
S. Meddah, A. Idir, M. Z. Doghmane, S. A. Tadjer, and M. Kidouche

Numerical Study of an Aircraft Cargo Compartment Fire: The Impact of Fire Retardant Sheets 254
Houssam Eddine Nadir Hiber and Hadj Miloua

An Innovative Ionic Liquid for Use as an Electrolyte in PEM Fuel Cells: 1-(2-Hydroxyethyl)-3-Methylimidazolium Hydroxide 263
Saad Halloub, Abdelkader Benabdellah, EL-Habib Belarbi, Mansour Debdab, Yassine Chaker, Mustapha Hatti, and Abdelhak Guezam

Improving Power Quality by Using Active Power Filter Based on DQ Frame of Reference Theory 269
Benhaoua Larbi Daouli and Hassan Mana

An Overview of Analytical Methods Utilized for Microbiological and Physiochemical Characterization of Locally Grown Agricultural Crops During Solar Drying 276
Akil Loumani, Ahmed Mediani, and Meriama Foullanine

Emerging Technologies and Applications

OC Fault Detection in Single Phase NPC Inverter 285
H. Mechri, A. Kheldoun, and A. Ammar

An Improved-Detection System for Diagnosing Photovoltaic Power Plant Efficiency Degradation Using a UAV-Based System 293
N. Cheggaga, S. Tchoketch-Kebir, and A.-H. Kebir

Anomaly Detection and Classification of Solar Photovoltaic Modules Using Vision Transformers (ViT) 304
Yacine Boutana, Sofiane Haddad, and Ammar Soukkou

YOLO-Based Photovoltaic Panel Detection: A Comparative Study 312
C. Moussaoui, M. Bouzerdoum, N. Kellil, and A. Mellit

YOLOv8-Based Photovoltaic Module Detection Using Aerial Imagery 321
N. Kellil, C. Moussaoui, A. Mellit, and A. Boudiaf

Deep Learning-Based Segmentation for Oil Pipeline Leak Detection Using Quadcopter Drones 330
Dalila Cherifi, Sifeddine Mekroud, and Ali Boudaoud

QuantumSign: Online E-Signature Service Based on Blockchain and Post-Quantum Cryptography 342
Chaima Boumesaid, Yousra Guerfa, and Gueltoum Bendiab

Review and Comparative Analysis of Meta-Heuristic-MPPT Algorithms of Photovoltaic Panels Under PSC 354
Fethia Hamidia, Amel Abbadi, Ahmed Medjber, Abdelkader Morsli, Redha Skender Mohamed, and Madani Safaa

A New Strategy to Mitigate Overheating in Solar PV Systems Using Adaptive Solar Tracking 364
Khadidja Dahli, Karim Selam, and Nawal Cheggaga

Intelligent Adaptive Neural Network Controller for Speed Control of PMSM in Electric Vehicles 372
Slimani Amira, Bourek Amor, Ammar Abdelkarim, and Kakouche Khoudir

MTPA Flux Weakening Approach for IPMSM Motor Powertrain Used in Vehicle Propulsion System 383
Aouadj Norediene, Hartani Kada, Hammou Meriem, and Bouziane Yassmine

Environmental and Sustainable Solutions

Effect of Different Drying Modes on the Essential Oil Extract of the Tobacco Plant *(Nicotiana Tabacum L.)* Before Maturation 397
M. Foullanine, A. Mediani, W. Braham Chaouch, A. Djaber, A. A. Larbi, A. Loumani, Z. Belkacem, and C. Tigani

Optimizing Geothermal Heat Exchangers Performance for Energy-Efficient Poultry House Heating 405
Assia Meziane, Khaled Mansouri, Abdelmalek Atia, and Abdelbaki Mameri

Organic Waste Biomethanation Controlled by Artificial Intelligence in a Smart City Concept 416
L. Benhabyles and M. Saber

High-Performance Asymmetric Flexible Supercapacitor Based on Zinc Oxide/Copper Oxide 423
M. I. Khenissa, A. Benchettara, and S. E. Berrabah

Optimizing Annual Energy Output of a Residential Bifacial PV System Using Box-Behnken Design: A Case Study in Tamanrasset, Algeria 428
Khaoula Rahrah, Soufiane Taibi, and Hassen Bouzgou

Corrosion and Protection of Irrigation Pivots in Adrar Region, Algeria 437
Boukhlef Djedjiga, Belhi Fatima, Khaldi Manal, Saadi Zine, Boutadara Abdelkader, and Hadidi Abedelkader

Stick Slip and Bit Bounce: Survey of Bit-Rock Interaction Modeling 443
Samira Hadibi, Madjid Kidouche, and Mohamed Zinelabidine Doghmane

Optimized GA-Based PID Control Design for Stick-Slip Suppression in Drill String of Rotary Drilling System 453
Tellaa Mawloud, Aibeche Abderrezak, Doghmane Mohamed Zinelabidine, and Kidouche Madjid

Optimal Control Design for Stick-Slip Vibrations Mitigation in Rotary Drilling Systems 464
Mawloud Tellaa, Abderrezak Aibeche, Mohamed Zinelabidine Doghmane, and Madjid Kidouche

Sizing, Design and Simulation of Solar Water Pumping System as Sustainable Solution for El Oued Region, Algeria 475
M. Benmbarek, M. Daif, and A. Kheldoun

Optimized Energy Management of a Hybrid System with Storage Using Genetic Algorithm 483
Zoulikha Ouchefoun, Mourad Hasni, and Lakhdar Guenfaf

Advanced Algorithms and Machine Learning

Energy Aware and Stability Multipath Routing Protocol Based on the Cuckoo Search Algorithm for FANETs 493
Abdelkader Mekkaoui, Omar Smail, and Mohammed Rebbah

The Recent Global MPPT Methods: State-of-the-Art 504
N. Rouibah, A. Bouttout, S. Oukaci, A. Limam, and L. Derradji

Machine Learning Techniques for GDP Per Capita Prediction in Developing Countries: A Case Study of Sub-Saharan Africa 516
Hanane Zermane

Carbon Emission Forecasting Using Advanced Machine Learning Algorithms 525
Abdelghani Bouziane, Mohammed Bouziane, and Khatir Naima

Analysis and Prediction of Solar Power Plant Energy Production: A Comparative Study of Machine Learning Algorithms 536
Lakhdari Lahcen and Habbab Mohamed

Spatiotemporal Prediction of Solar Irradiance in Algeria: A CNN-Driven Approach for Solar Energy Mapping 544
Salah Eddine Megaache, Hassen Bouzgou, Amor Gama, and Naima Zerari

A Deep Reinforcement Learning Approach for Tracking the Maximum Power Point 554
S. Belarbi, N. Drir, and L. Barazane

Autoencoder, LSTM, RNN, and SVM for Anomaly Detection, Prediction, and Localization in Industrial Systems ... 563
Dalila Cherifi, Mohammed Amine Bedri, and Said Boutaghane

Multi-classification and Multi-labeling of UML Diagrams Using Transfer Learning and CNN ... 573
Abderrahim Lakehal, Adel Alti, Boubakeur Annane, Bensalah Imene, Smaani Anissa Afaf, and mami meriem

Determination of the Optimal Volume and Distribution of Learning Data for Solar Irradiance Forecasting Applications ... 584
Houssem Eddine Louchene, Hassen Bouzgou, and Christian Gueymard

Empowering MPPT Efficiency in Partial Shading with Machine Learning Driven Metaheuristic Strategies: Recent Developments ... 594
F. Belhachat, C. Larbes, and R. Bennia

Communication Systems and Optimization

IoT-Enhanced Photovoltaic Systems: Achieving Low LCOE and Improved Sustainability Through Advanced Energy Management ... 615
Mustapha Hatti

Up-Link OFDMA Hybrid Access for Enabling IoT-Based Remote Video Monitoring Over Wi-Fi 7 ... 623
Said Ould Amara, Mohand Yazid, and Souhila Mammeri

Double Threshold Energy Detection Algorithm in Cognitive Radio/mmWave Mobile Communications at 60 GHz ... 634
Amel Haouzi, Samir Kameche, and Haroun Errachid Adardour

TTC Antenna at S-band for the Drone to Geo Satellite Application ... 646
Z. Kourdi, I. Four, R. Mansour Bekhtache, and N. Addaoui

Efficient Predictive Dynamic Routing of IoT Communication Networks in Smart Farming: An Effective Fuzzy Q-Learning Solution ... 656
Abdesselem Beghriche, Boubekeur Annane, and Amina Abed

An Improved Energy-Efficient MAC Protocol for WSNs ... 672
Abdelmalek Bengheni

ALTEM: A New Lightweight Trust Evaluation Mechanism for Wireless Sensor Networks ... 684
Ahmed Saidi

Innovative Materials and Applications

Nanostructured Materials: Synthesis and Application 699
Henia Fraoucene, El hadi Khoumeri, and Rabea Cheggou

Optical Characteristics of Photonic Crystal Fiber Splitters: A Study on Refractive Index, Chromatic Dispersion, and Effective Frequency 708
Ahlem Assia Harrat, Mohammed Debbal, and Mohammed Chamse Eddine Ouadah

The Study of Cooling Systems for Superconducting Coils 716
Cheriet Rahmouna, Bensaad Burassia, Bouhadjela Fatiha, and Belhenini Soufyane

Estimate the Hardness of Deposited Thin Layers Using Fuzzy Logic 722
Nada Rihane Mechehat, Djilani Necib, Imane Rezgui, and Mondir Soualah

Improved TSR Based MPPT Control Using Hybrid Fuzzy Sliding Mode Control for Wind Turbines 731
Rania Kadri, Mohamed Abid, and Aissaoui Abdelghani

Evaluation of the Performance of a Savonius Micro Wind Turbine 741
Zohra Ouchiha, Hemiche Iddou, Tarek Legouini, and Abdelhakim Hamrit

Design and Development of Low-Power Output Vortex Tower Model 750
A. Bouam, A. L. Cheridi Deghal, A. Dadda Khorsi, A. Hadjam, A. Dahia, N. Koudiah, K. Attari, and L. Bouam

Ambient Temperature Effect on the Coriolis Force Produced in the Vortex Tower 758
A. Dadda Khorsi, A. L. Cheridi Deghal, A. Bouam, A. Dahia, A. Hadjam, and A. Kentouche

An Experimental Study of a Photovoltaic Stand-Alone House System in Adrar 766
Abdeldjalil Dahbi, Abderrahmane Khelfaoui, Miloud Benmedjahed, Ahmed Bouraiou, Tidjar Boudjema, Abdeldjalil Slimani, Nouar Aoun, and Messaoud Hamouda

Evaluation of Wind Energy Potential in the Dhaya Region, Sidi Bel Abbès 774
Fatima Zohra Bochra Amrani, Boutlilis Fatima, and Benaissa Bekkouche

A Systematic Review of AI-Powered Generative Design to Improve Efficiency and Thermal Comfort in Early-Stage Residential Design 783
Ibtissam Chekkal and Djamel Mestoul

Author Index 793

Short Biography

Dr. Mustapha Hatti, born in El-Asnam (now Chlef), Algeria, is a distinguished Research Director and IEEE Senior Member currently leading the "Tipasa Smart City" initiative. After completing his Electronics Engineering degree from USTHB Algiers and post-graduate studies at USTO-Oran, he held several key positions including Research Engineer at CDSE Ain Oussera, Scientist at Sonatrach CRD Hassi Messaoud, Research Engineer at CRNB Birine, and Senior Scientist at UDES/EPST-CDER Bou Ismail. His expertise spans smart sustainable energy systems, fuel cell technology, photovoltaic systems, optimization, intelligent embedded systems, and electric vehicles. He serves as the Chairman of ICAIRES (International Conference on Artificial Intelligence in Renewable Energy Systems). He has served as an expert examiner for numerous Master's and PhD theses and participated in multiple European and international research and development projects. As an accomplished researcher, he has authored numerous scientific papers and book chapters, contributing significantly to the advancement of sustainable energy technologies and smart city development.

Energy Systems and Renewable Technologies

Numerical Investigation of a Novel Constant Volume Combustion Phase Strategy in a Diesel-Hydrogen Dual-Fuel Engine

Ghodbane Hassina(✉), Fouad Khaldi, and Derradji Bahloul

LEREESI Laboratory, HNS-RE2SD, Batna, Algeria
{h.ghodbane,fouad.khaldi,derradji.bahloul}@hns-re2sd.dz

Abstract. Hydrogen presents a promising alternative in the realm of large-scale thermal power generation, particularly when integrated into internal combustion engines (ICEs). This research looks into whether it is possible and helpful to use a diesel-hydrogen dual-fuel engine that works with a new constant volume combustion phase (CVCP) strategy and is simulated using OpenModelica. The combustion model incorporates a double Wiebe function to characterize the combustion process comprehensively. Results indicate significant improvements in peak pressures and temperatures, alongside emission reductions, particularly NOx, with increasing CVCP duration. The study sheds light on the potential of hydrogen integration in ICEs for enhanced efficiency and reduced environmental impact.

Keywords: Hydrogen · Diesel · Dual-Fuel Engine · Combustion · OpenModelica

1 Introduction

The use of hydrogen fuel cell power generators in homes and cars is growing in popularity. Large-scale hydrogen-powered thermal power plants are greatly predicted to come into existence. Large-scale thermal power generators that use traditional fossil fuels consume a lot of fossil fuel; nevertheless, a significant amount of CO2 emissions can be avoided by substituting hydrogen for a portion of the fossil fuel. Making use of For the following reasons, Compared to hydrogen fuel cells (H2 FC), hydrogen used with thermal power generators, such as internal combustion engines (ICE), offers several advantages. One key benefit is cost savings, thanks to the availability of modern engine components and materials like rare-earth metals. Additionally, unlike hydrogen fuel cells (H2 FC), ICEs do not require highly purified hydrogen. This means that ICEs can efficiently use hydrogen produced from various processes, including steam reforming, water electrolysis, and the dehydrogenation of organic chemical hydroxides [1]. Combined heat and power (CHP) systems using ICEs are expected to achieve an overall efficiency of over 80%, and these systems are widely used today to meet diverse power and size demands. In the current energy landscape, carbon dioxide emissions and significant fossil fuel consumption are common. Therefore, integrating hydrogen as part

© The Author(s), under exclusive license to Springer Nature Switzerland AG 2025
M. Hatti (Ed.): IC-AIRES 2024, LNNS 1238, pp. 3–11, 2025.
https://doi.org/10.1007/978-3-031-80301-7_1

of a multi-fuel strategy in numerous existing ICE-powered CHP systems could lead to substantial reductions in CO2 emissions and other harmful pollutants generated from the combustion of carbon-based fuels. Multi-fuel combustion techniques in ICEs or thermal power systems are crucial for the efficient and environmentally friendly utilization of hydrogen. In addition, a sustainable, carbon-free, and renewable supply of hydrogen is necessary to achieve such a significant reduction in CO2 [2]. This study makes the assumption that renewable electricity from solar and wind power is used to manufacture hydrogen. Recently, hydrogen has become a practical large-scale energy storage option to help balance power grids impacted by the sporadic nature of solar and wind power-generated electricity. A notional schematic of the energy flow from renewable sources to hydrogen use in an internal combustion engine is shown in Fig. 1. Additionally, this diagram illustrates the process for storing hydrogen and organic chemical hydrides (OCH). Since ICEs are designed specifically for the combustion of hydrogen, it is ideal for them to run solely on hydrogen as fuel. Several automakers have already created an Otto engine, an ICE powered by hydrogen, and spark ignition. In the quest for improved engine efficiency, researchers have turned their attention to alternative fuel technologies, recognizing the inherent limitations of Otto cycle engines compared to their diesel counterparts in terms of thermal efficiency. One promising avenue of investigation involves the adaptation of hydrogen fuel to diesel engine architectures, leveraging the higher compression ratios characteristic of these systems. A notable study in this field was conducted by Antunes and colleagues, who examined the feasibility of hydrogen-fueled diesel engines. Their approach involved direct hydrogen injection into the combustion chamber, a method chosen to address potential pre-ignition issues associated with hydrogen fuel. The experimental setup utilized a single-cylinder engine modified to include a heated intake air system, allowing for precise control of intake conditions. The results of Antunes et al.'s investigation were encouraging, demonstrating a significant performance boost. Their hydrogen-fueled diesel engine achieved a 14% increase in maximum power output when compared to a standard diesel configuration. However, the researchers also identified areas requiring further development, particularly in enhancing the durability and reliability of hydrogen injection systems. Addressing another challenge inherent to hydrogen fuel, Fukuma and collaborators investigated novel ignition strategies [3]. Recognizing hydrogen's substantially higher auto-ignition temperature relative to conventional fossil fuels, they explored the efficacy of a hot surface ignition technique. This method was designed to initiate combustion in directly injected hydrogen jets within the combustion chamber. Fukuma's team also grappled with hydrogen's exceptionally high flame propagation speed. To mitigate the risk of severe combustion vibrations, they implemented precise adjustments to both ignition timing and hydrogen injection patterns. These refinements were crucial in achieving stable and controlled combustion processes. These studies collectively highlight the potential of hydrogen as an alternative fuel in diesel engine applications, while also underscoring the technical hurdles that must be overcome to fully realize this potential. However, in practical terms, hydrogen is not more cost-effective than traditional fuels derived from fossil fuels, and the supply chain for hydrogen has not yet been shortened. Thus, hydrogen would be introduced gradually, at least for the time being, and traditional fossil fuels would still be required. In light of this, multifuel internal combustion engines (ICEs) need to be further developed to integrate hydrogen as a fuel source in the near future. A promising method is diesel dual fuel (DDF) combustion, a multi-fuel technology that involves the use of two fuels: one serving as the primary fuel and the other acting as an ignition source. Figure 2 illustrates the schematic process of fuel supply and combustion in a DDF engine.

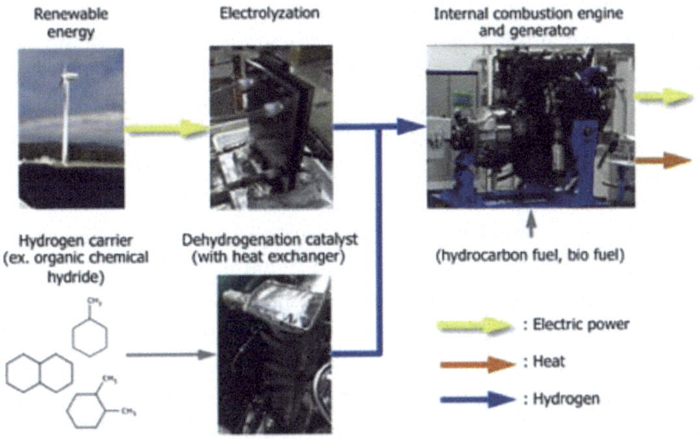

Fig. 1. The energy transfer from renewable energy sources to HICE use [14].

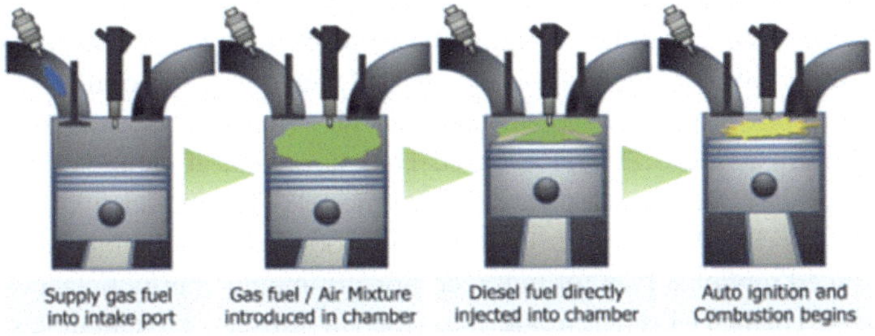

Fig. 2. Diagrammatic procedures for the fuel supply and combustion in Direct-Injected Dual-Fuel (DDF) engines [14].

In this study we use Open Modelica to similate Diesel-Hydrogen Dual-Fuel Engine Operating with a Novel Constant Volume Combustion Phase Strategy.

2 Combustion Model

2.1 Double Weibe Function

Semi-empirical functions, particularly Wiebe functions, play a crucial role in combustion modeling. These mathematical tools are extensively employed to estimate combustion processes at specific crank angles. Researchers have applied Wiebe functions across various domains, including studies on alternative fuels, advanced combustion techniques, and complex systems incorporating sequential turbocharging and variable geometry turbines in engines [4, 5].

The single Wiebe function is defined by three key parameters: combustion duration, shape factor, and combustion efficiency factor [6, 7]. Among these, the shape factor is particularly significant as it determines the combustion profile, also known as the mass fraction burned. In spark ignition (SI) gasoline engines, researchers typically use shape and efficiency factor values of 2 and 5, respectively [8].

Recent investigations have explored the versatility of Wiebe functions in diverse combustion scenarios. For example, Diaz et al. examined hydrogen-methane blend combustion in a 2 kW cooperative fuel research (CFR) engine, employing shape and efficiency factors of 2.8 and 2.4 to analyze knock occurrence crank angle (KOCA) [9]. In a separate study, Shivapuji and Dasappa demonstrated the necessity of modifying Wiebe coefficients to accurately model producer gas combustion in a 28 kW SI engine, accounting for variations in in-cylinder pressure [10].

Computational approaches have also been utilized to refine Wiebe function applications. Carrera et al. used computational fluid dynamics (CFD) simulations to derive Wiebe function coefficients for biogas combustion, considering variables such as compression ratio, engine speed, air-fuel ratio, spark timing, and CO_2 content [11]. Similarly, Rousseau et al. calculated Wiebe coefficients for natural gas combustion in lean-burn SI engines, factoring in parameters like air-fuel ratio, spark timing, and engine load [12]. However, their research highlighted limitations in using a single Wiebe model to fully capture experimental mass-burn fraction data, suggesting the need for further exploration of alternative mathematical functions.

The complexity of accurately modeling various combustion stages with a single Wiebe function has led researchers to adopt more sophisticated approaches, particularly for diesel engines. Double Wiebe functions (Eq. 2) have gained popularity as they can better represent both premixed and diffusion combustion phases. An example of this advanced approach is the 0-D Wiebe-based H-D combustion characterization model, which adapts the H-D heat release rate (HRR) model. This model fits the normalized rate of combustion (RCO) derived from the HRR model using a double-Wiebe function.

This advanced model solves for mass, composition, and energy balances, mirroring the approach used in the H-D HRR model described in Subsect. 1. It calculates in-cylinder temperature by solving Eq. 1 and determines the rate of combustion by differentiating the Wiebe-fitted normalized RCO. The model estimates in-cylinder pressure using the ideal gas law and computes heat loss using the Woschni correlation [13], similar to the H-D HRR model. Key outputs of this model include average in-cylinder pressures and temperatures at each crank angle, as well as quantifications of work done and heat released during the combustion process.

$$\mathrm{CRR} = \frac{U + W + Q_{loss}}{u_{comb-fuel}} = \frac{\mathrm{m\,C_v \frac{dT}{dt} + p\frac{dV}{dt} + Q_{loss}}}{\mathrm{u_{comb-fuel}}} \tag{1}$$

where:

- U: Internal energy (kJ/sec)
- W: Work (kJ/sec)
- Q_(loss): Heat loss (kJ/sec)
- dV/dt: Rate of volume change
- m: In-cylinder mass (kg)

- Cv: Specific heat at constant volume (kJ/kg·K)

$$RCO_{norm} = \mathrm{b}_1 * \left(1 - \mathrm{e}^{-\mathrm{a}_1(\mathrm{y}_1)^{(\mathrm{m}_1+1)}}\right) + \mathrm{b}_2 * \left(1 - \mathrm{e}^{-\mathrm{a}_2(\mathrm{y}_2)^{(\mathrm{m}_2+1)}}\right) \quad (2)$$

where,

$$y_1 = \frac{\theta - \theta_0}{\Delta\theta_1}, y_2 = \frac{\theta - \theta_0}{\Delta\theta_2}; a_1 = a_2 = -\log(1 - \eta_{comb}) \quad (3)$$

θ: Instantaneous crank-angle in degrees.

b_1, b_2Wiebe combustion weighing factors.

a_1, a_2 Wiebe combustion efficiency factors.

m_1, m_2 Wiebe shape parameters θ_0 Start of combustion.; $\Delta\theta_1$, $\Delta\theta_2$ Combustion duration for first and second Wiebe.

3 Numerical Methodology

The modeling in this study is carried out in the Open Modelica environment using the Modelica language. A high-level declarative language for describing mathematical behavior is called Modelica. It is frequently used in engineering systems and makes it simple to explain how various kinds of engineering components—such as springs, resistors, clutches, etc.—behave. Then, these parts can be joined to create systems, subsystems, or even entire architectures. A free and open-source platform for modeling, simulating, optimizing, and analyzing complicated dynamic systems, OpenModelica is built on the Modelica modeling language.

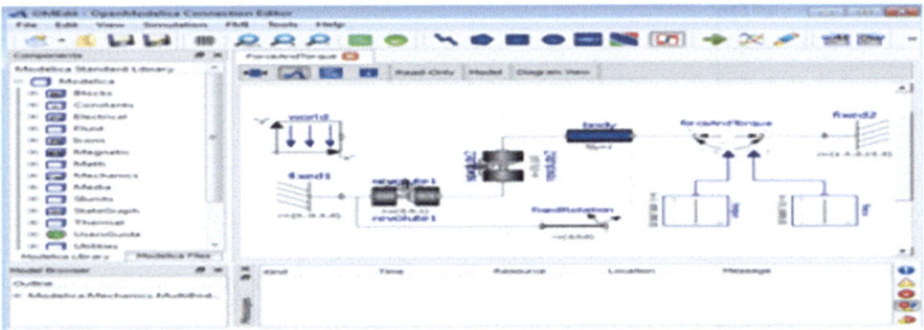

4 Results and Discussion

4.1 Combustion Characteristics

In Fig. 3, in-cylinder pressure measurements are shown for three distinct constant volume combustion process (CVCP) durations and are contrasted with high-load conventional engine test scenarios. Peak pressure in a typical engine under high load is 7.25 MPa. By comparison, the peak pressures of the CVCP configurations are much higher: 12.15 MPa for CV20, 14.86 MPa for CV30, and 15.51 MPa for CV40. The peak pressure of the traditional engine is 5.18 MPa under low load, whereas the peak pressures of the CV20, CV30, and CV40 are 6.92 MPa, 7.37 MPa, and 7.52 MPa, respectively.

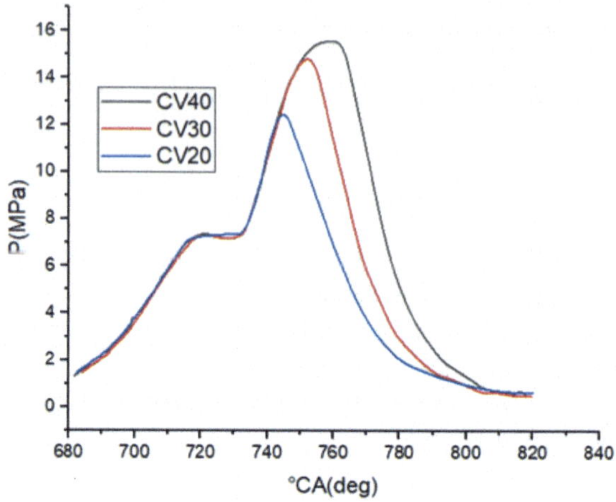

Fig. 3. Pressure at high load-CVCP duration.

The increased pressures in the CVCP cases are due to a larger portion of the combustion process occurring at constant volume, with the piston remaining at top dead center (TDC). Shorter CVCP durations result in lower peak pressures because of reduced time for pressure buildup and incomplete fuel combustion. These elevated pressures are maintained throughout much of the expansion stroke, continuing approximately 50 degrees beyond the end of each CVCP phase.

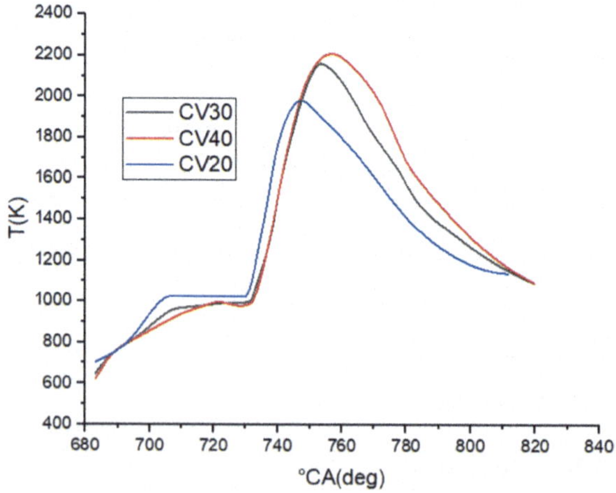

Fig. 4. Average temperature at high load-CVCP duration.

Figure 4 illustrates the average in-cylinder temperature profiles under high load conditions for various engine configurations. The graph reveals several key trends related to the Constant Volume Combustion Process (CVCP) duration:

- Temperature Peaks:

Cases with extended CVCP durations exhibit higher peak temperatures that persist for longer periods. This phenomenon can be attributed to the prolonged time spent at top dead center (TDC), allowing for greater accumulation of pressure and thermal energy.

- Initial Temperature Decline:

As the piston begins its descent from TDC, the temperature in CVCP cases decreases more rapidly compared to conventional engine designs. This accelerated cooling can be explained by three factors:

a) Enhanced heat transfer to the cylinder walls
b) Reduced availability of unburned fuel for ongoing combustion
c) The piston's faster movement away from the TDC position

- Late Expansion Stage:

During the latter part of the expansion stroke, all CVCP configurations display lower temperatures than their conventional counterparts. Notably, the cases with the longest CVCP durations demonstrate the lowest temperatures at the point of exhaust valve opening (EVO).

- Quantitative Temperature Increase:

Under high load conditions, each incremental increase in CVCP duration (in the range of 0 to 40 degrees) corresponds to an approximate rise of 30 K in the peak temperature.

5 Emissions

Figure 5 illustrates that nitrogen oxide (NOx) emissions increase with longer Constant Volume Combustion Process (CVCP) durations under both high and low load conditions. This is due to the prolonged periods of elevated temperatures as the CVCP duration

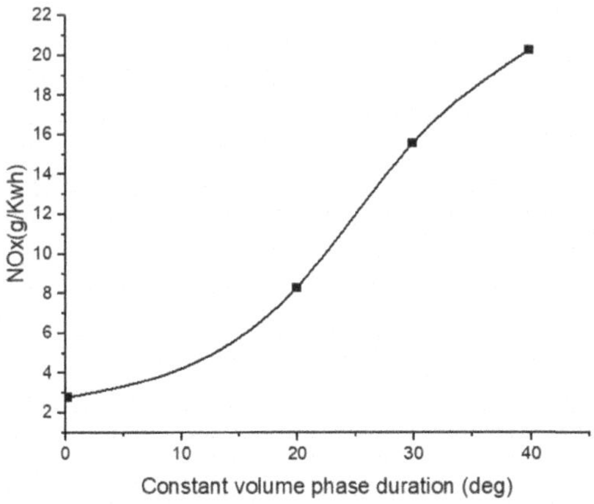

Fig. 5. NOx emissions at high loadCVCP duration.

extends. As observed in Fig. 4, the temperature profile increases with longer CVCP durations, leading to a corresponding increase in NOx emissions as higher temperatures promote the formation of nitrogen oxides.

6 Conclusion

In conclusion, the utilization of hydrogen in Diesel-Hydrogen Dual-Fuel Engines with innovative Constant Volume Combustion Phase strategies presents a compelling avenue for improving thermal power generation efficiency while mitigating environmental impacts. Through meticulous simulation using OpenModelica, it is evident that longer CVCP durations lead to notable enhancements in peak pressures and temperatures, attributing to more efficient combustion processes. Moreover, emissions, particularly NOx, exhibit a downward trend as CVCP duration increases, underscoring the environmental benefits of hydrogen integration. These findings underscore the potential of hydrogen as a multifaceted solution in large-scale power generation, urging further exploration and development in the pursuit of sustainable energy systems.

References

1. Okada, Y., Shimura, M.: Development of large-scale H2 storage and transportation technology with Liquid Organic Hydrogen Career (LOHC). Technical paper at Joint GCC-JAPAN Environment Symposia (2013)
2. Gomes Antunes, J.M., Mikalsen, R., Roskilly, A.P.: An experimental study of a direct injection compression ignition hydrogen engine. Int. J. Hydrogen Ener. **34**(15), 6516e22 (2009)
3. Fukuma, T., Fujita, T., Pichainarong, P., Furuhama, S.: Hydrogen combustion study in direct injection hot surface ignition engine. SAE Paper (1986). https://doi.org/10.4271/861579
4. Sun, Y., Wang, H., Yang, C., Wang, Y.: Development and validation of a marine sequential turbocharging diesel engine combustion model based on double Wiebe function and partial least squares method. Energy Convers. Manag. **151**, 481–495 (2017)
5. Baldi, F., Theotokatos, G., Andersson, K.: Development of a combined mean value-zero dimensional model and application for a large marine four-stroke Diesel engine simulation. Appl. Energy **154**, 402–415 (2015)
6. Ghojel, J.I.: Review of the development and applications of the Wiebe function: a tribute to the contribution of Ivan Wiebe to engine research. Int. J. Engine Res. **11**(4), 297–312 (2010)
7. Abbaszadehmosayebi, G., Ganippa, L.: Determination of specific heat ratio and error analysis for engine heat release calculations. Appl. Energy **122**, 143–150 (2014)
8. Yeliana, Y., Cooney, C., Worm, J., Michalek, D.J., Naber, J.D.: Estimation of double-Wiebe function parameters using least square method for burn durations of ethanol-gasoline blends in spark ignition engine over variable compression ratios and EGR levels. Appl. Therm. Eng. **31**(14–15), 2213–2220 (2011)
9. Amador Diaz, G.J., Gómez Montoya, J.P., Corredor Martinez, L.A., Olsen, D.B., Salazar Navarro, A. : Influence of engine operating conditions on combustion parameters in a spark ignited internal combustion engine fueled with blends of methane and hydrogen. Energy Convers. Manag. **181**, 414–424 (2019)
10. Carrera, J.L.E., Riesco, J.M.A., Martinez, S.M., Sanchez, F.A.C., Gallegos, A.M.: Numerical study on the combustion process of a biogas spark ignition engine. Therm. Sci. **17**(1), 241–254 (2013)

11. Shivapuji, A.M., Dasappa, S.: Experiments and zero D modeling studies using specific Wiebe coefficients for producer gas as fuel in spark-ignited engines. J. Mechan. Eng. Sci. **227**(3), 504–519 (2012)
12. Rousseau, S., Lemoult, B., Tazerout, M.: Combustion characterization of natural gas in a lean burn spark-ignition engine **213**(x), 481–490 (1999)
13. Torregrosa, A.J., Olmeda, P.C.: Revisng Engine Heat Transfer. J. Eng. Univ. Hunedoara 21 (2008)
14. Tsujimura, T., Suzuki, Y.: The utilization of hydrogen in hydrogen/diesel dual fuel engine. Int. J. Hydrogen Energy **42**(19), 14019–14029 (2017)

Performance Analysis of Thermoelectric Generator with Sliding Mode Control

Asma Toualbia[1], Rachid Taleb[1,2(✉)], and Abdelkadir Belhadj Djilali[1]

[1] Electrical Engineering Department, Laboratoire Génie Electrique Et Energies Renouvelables (LGEER), Hassiba Benbouali University of Chlef, Chlef, Algeria
rac.taleb@gmail.com

[2] Unité de Recherche en Systèmes Embarqués de Chlef, Centre de Recherche Sur L'Information Scientifique Et Technique (CERIST), Chlef, Algeria

Abstract. A thermoelectric generator (TEG) is capable of converting heat energy directly into electrical energy. Fluctuations in load and temperature can cause the TEG to operate at a voltage that is not optimal for generating maximum power for a given temperature gradient. This paper examines the electrical performance of a TEG system when the incremental conductance (INC-MPPT) control algorithm is used to optimize energy conversion efficiency. Additionally, the paper evaluates the system's performance under the sliding mode control (SMC- MPPT) algorithm to further enhance energy conversion efficiency.

Keywords: Thermoelectric generator · sliding mode control · MPPT

1 Introduction

Our world is currently facing significant energy crises, with environmental pollution on the rise. To address these issues, renewable energy sources have become essential. Thermoelectric energy is particularly important in this context as it is clean, inexhaustible, and widely available. A thermoelectric module (TEM) is a solid-state device that converts thermal energy into electrical energy, as described by Benhadouga (2022). Thermoelectric generation (TEG), which consists of multiple TEMs, has gained increased attention recently as an independent, clean, and renewable energy source, according to S. Benhadouga (2019).

Similar to photovoltaic systems, a DC-DC converter allows the power generated by the TEG to be injected into the DC grid. However, the electrical output of a TEG is highly dependent on temperature, leading to unstable power generation. As a result, maximum power point tracking (MPPT) techniques, such as perturb and observe (P&O) and incremental conductance (IC), are necessary to extract maximum power during dynamic changes in temperature and load, as noted by H. Zafar (2013).

This paper presents the validation of a novel technical control method—sliding mode control (SMC)—applied to the TEG for MPPT. This technique focuses on controlling a power boost converter that operates across multiple operating points. Originally developed for photovoltaic systems, these MPPT controls have been adapted for thermoelectric generation, as explained by A. Toualbia (2017).

© The Author(s), under exclusive license to Springer Nature Switzerland AG 2025
M. Hatti (Ed.): IC-AIRES 2024, LNNS 1238, pp. 12–17, 2025.
https://doi.org/10.1007/978-3-031-80301-7_2

2 MPPT System Modeling

The thermoelectric generation model comprises a thermoelectric module, a dc-to-dc boost converter, a maximum power point tracker (mppt), and a load, as illustrated in Fig. 1.

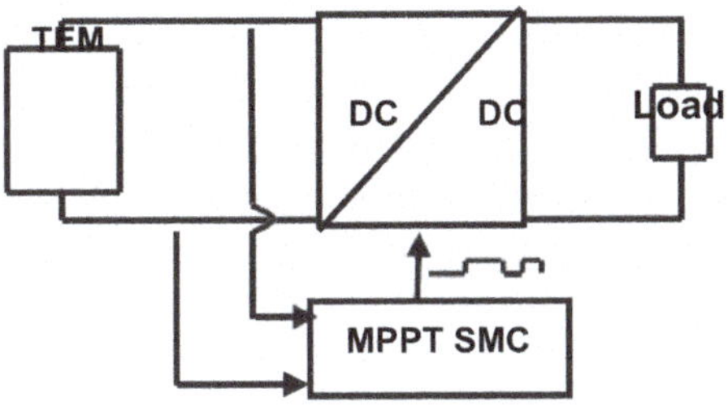

Fig. 1. Block diagram for proposed topology.

The voltage generated by the thermoelectric module (TEM) is influenced by the temperature difference (ΔT) across the module. The measured voltage and current from the TEM source are fed into the MPPT block to track the maximum power, as noted by Eakburanawat (2006). By utilizing the sliding mode control (SMC) algorithm for MPPT, a gating signal is produced to activate the switch of the DC-DC boost converter, ensuring the generation of maximum output power.

2.1 The Characteristics and Modeling of GTE

A GTE module can be modeled using an equivalent circuit, as illustrated in Fig. 2 by Xinyu (2022).

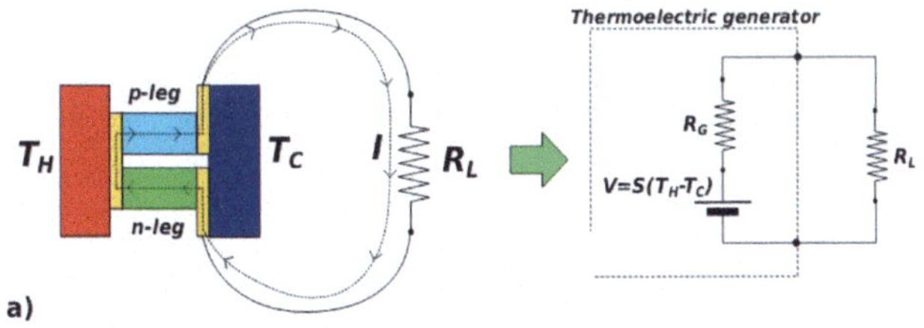

Fig. 2. Module thetrmoelectricand the electrical equivalent circuit of the thermoelectric generator.

The TEM voltage can be expressed as:

$$V = \alpha \Delta T - R_g I_g \tag{1}$$

2.2 Sliding Mode Control

Sliding mode control is a control strategy employed in systems to ensure that the output follows a desired reference signal, even in the presence of disturbances or uncertainties. This technique involves defining a sliding surface based on the system's state variables, which delineates the boundaries of desired and undesired behavior, as noted by Benhadouga (2019).

The goal of this control method is to characterize the surface σ(x) on which the control objectives are fulfilled. The main aim of sliding mode control is to design an appropriate surface defined by σ(x)=0, which confines the state trajectories to achieve the desired outcomes, including tracking, regulation, and stability, as explained by Farhat (2016).

$$u(t) = u(t)_{eq} + u(t)_n \tag{2}$$

To compute the two components of the sliding mode control (SMC) law, it is essential to first define an appropriate sliding surface σ(x). In this scenario, the sliding surface should be based on the power output of the thermoelectric generator (TEG). It is proposed in the following form:

$$\sigma = f(P_{TEG}) \tag{3}$$

2.3 Equivalent Circuit

This control is computed using the invariance condition, as described by the following equations, according to Labbaf Khaniki (2023):

$$\sigma = 0 \rightarrow \frac{d\sigma}{dt} = 0 \rightarrow (u = u_{eq}) \tag{4}$$

where the derivative of rrr represents the sliding surface.

$$\sigma = \frac{dP_{TEG}}{dt} = \frac{dV_{TEG}I}{dt} = I\frac{dV_{TEG}}{dt} + V_{TEG}\frac{dI}{dt} = 0 \tag{5}$$

where:

$$S = R(I) - r(I) \tag{6}$$

The derivative of Eq. (6) can be expressed as:

$$\frac{d\sigma}{dt} = \frac{dR(I)}{dt} - \frac{dr(I)}{dt} \tag{7}$$

While $I=I_L$

$$\frac{dI_L}{dt} = \left[-\frac{1-u_{eq}(t)}{L}V_o + \frac{1}{L}V_{GTE}\right] = 0 \tag{8}$$

The form of the equivalent control is provided by the following equation:

$$u_{eq}(t) = \left(1 - \frac{V_{GTE}}{V_o}\right) \tag{9}$$

The switching control un unt(t) in this case is given by:

$$u_n(t) = -Ksign(\sigma) \tag{10}$$

Finally, the sliding mode control law is given by:

$$u(t) = \left(1 - \frac{V_{GTE}}{V_o}\right) - Ksign(\sigma) \tag{11}$$

where K is a positive constant.

3 Resultats and Simulation

The performance results for output power, voltage, and current of the proposed sliding mode control (SMC) method are displayed in Fig. 3 (a, b, c).

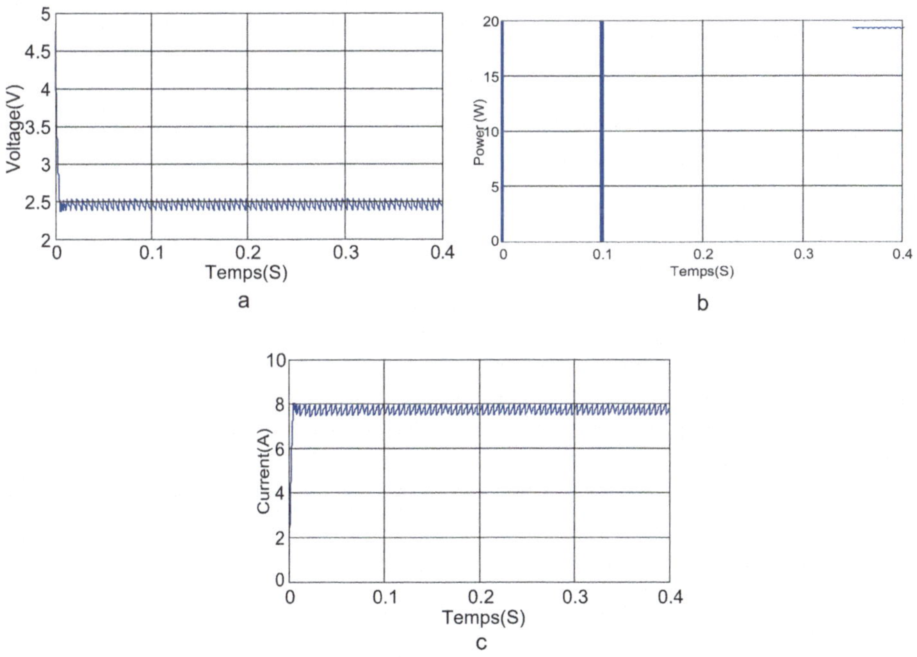

Fig. 3. Voltage, Power and Current of GTE

For the sliding mode control (SMC), the objective of achieving maximum power is reached in a shorter time, with a conversion efficiency of 99.56%.

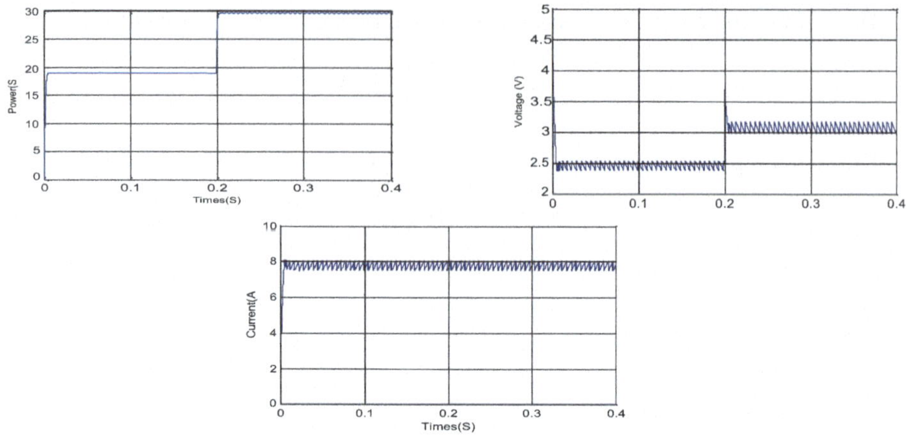

Fig. 4. Power, Voltage and Current of GTE

Figure 4 illustrates the simulation results for a sudden change in the temperature gradient from $\nabla T = 200$ °C to $\nabla T = 250$ °C at t=0.2 s

In this case, the GTE system is simulated for a sudden variation in temperature difference, using a resistive load of Rch = 3 Ω. The results indicate that the characteristic curves, particularly for power, show a faster tracking speed for the SMC MPPT. Moreover, when the temperature differential in the GTE system shifts rapidly from 200 °C to 250 °C at t=0.2 s, the SMC MPPT demonstrates excellent steady-state performance, adapting well to changes in the temperature gradient.

4 Conclusion

In this paper, a novel control method based on sliding mode control (SMC) was proposed for thermoelectric generator (TEG) energy harvesting systems. The simulation results demonstrate that the proposed TEG system, utilizing the SMC method, can accurately achieve the maximum output power of the TEG module with minimal fluctuation around the maximum power point (MPP), even during rapid temperature changes and under various operating conditions. The system also achieves a maximum efficiency of 98.6%, according to the simulation results.

References

Akhtar, N.: A novel MPPT controller based on mud ring optimization algorithm for centralized thermoelectric generator under dynamic thermal gradients. Applied Science 13 (2023)

Benhadouga, S., Meddad, M., Eddiai, M., Boukhetala, D., Khenferi, R.: Sliding Mode Control for MPPT of a Thermogenerator. Journal of Electronic Materials (2019)

Eakburanawat, J., Boonyaroonate, I.: Development of a thermoelectric battery-charge with microcontroller-based maximum power point tracking technique. Journal Applied Energy 687–704 (2006)

Farhat, M., Oscar, B., SbitA, L., Aymen, F.: A Robust MPP Tracker Based on Sliding Mode Control for a Photovoltaic Based Pumping System. International Journal of Automation and Computing, pp. 1–12 (2016)

Zafar, H., et al.: Modelling and simulation of solar thermoelectric generator. OSR Journal of Electrical and Electronics Engineering, 265–267 (2013)

Toualbia, A., Tadjine, M.: Perturb and observe maximum power point tracking based passivity approach for thermoelectric generator. the Mediterranean Journal of Measurement and Control, 355–365 (2017)

Xinyu, L., et al.: Theoretical and experimental research on control strategy ofmaximumpower point tracking formonolayer thermoelectricgenerator considering the degree of disturbance. Energy Reports, pp. 15124–15143 (2022)

Labbaf Khaniki, M.A., Manthouri, M., Ahmadieh Khanesar, M.: Adaptive non-singular fast terminal sliding mode control and synchronization of a chaotic system via interval type-2 fuzzy inference system with proportionate controller. Iranian Journal of Fuzzy Systems, pp. 171–185 (2023)

Benhadouga, S., et al.: Experimental Comparative Study MPPT between P&O and Sliding Control of a Small PV System. International Journal of renewable energy research, pp. 864–869 (2022)

Razman, A., et al.: Comparison of control strategies for thermoelectric generator emulator. International Journal of Power Electronics and Drive Systems (IJPEDS), pp. 2094–2106 (2023)

Passivity Based Control of Thermoelectric Generator System by Using a Boost Power Converter

Asma Toualbia[1], Rachid Taleb[1,2(✉)], and Abdelkadir Belhadj Djilali[1]

[1] Electrical Engineering Department, Laboratoire Génie Electrique et Energies Renouvelables (LGEER), Hassiba Benbouali University of Chlef, Chlef, Algeria
rac.taleb@gmail.com

[2] Unité de Recherche en Systèmes Embarqués de Chlef, Centre de Recherche Sur L'Information Scientifique et Technique (CERIST), Chlef, Algeria

Abstract. The thermoelectric module (TEM) is a solid-state device that converts thermal energy into electrical energy. A key feature of the TEM is that it delivers maximum power at only one specific operating point, defined by a certain localized voltage and current, referred to as the Maximum Power Point (MPP). Recently, much research has focused on improving the efficiency of power extraction from TEMs. Among the various Maximum Power Point Tracking (MPPT) methods proposed for thermoelectric systems, the perturb and observe (P&O) technique stands out for its low cost and simplicity of implementation. However, a limitation of the P&O method is that, under thermal equilibrium, the operating point oscillates around the MPP. Several enhancements have been suggested to reduce these oscillations. In this paper, a passivity-based P&O algorithm is introduced to address the shortcomings of the traditional P&O approach. To improve stability at the MPP, the Euler-Lagrange passivity technique is integrated with the standard P&O MPPT in a novel way. This combined approach, referred to as the P&O/EL-PBC algorithm, accelerates the tracking of maximum power and minimizes voltage fluctuations, showing excellent performance under varying load conditions.

Keywords: Thermoelectric generator · passivity based control · MPPT

1 Introduction

Our world is currently facing significant energy crises and escalating environmental pollution. To address these issues, renewable energy sources are being utilized, as noted by Anil (2013). Thermoelectric energy plays a crucial role in this context since it is clean, inexhaustible, and widely available. A thermoelectric generator (TEG) is a solid-state device that directly converts a temperature gradient into electricity and vice versa. TEGs are made up of numerous thermocouples connected electrically in series and thermally in parallel, as explained by Tang (2015). Given the low efficiency of thermoelectric systems, they must be operated at the maximum power point (MPP), which varies with

© The Author(s), under exclusive license to Springer Nature Switzerland AG 2025
M. Hatti (Ed.): IC-AIRES 2024, LNNS 1238, pp. 18–26, 2025.
https://doi.org/10.1007/978-3-031-80301-7_3

temperature or load changes. To address these fluctuations, a Maximum Power Point Tracker (MPPT) is used, as described by Eakburanawat (2006). This technique relies on an adaptive interface, typically a DC-DC converter, placed between the TEG module and the load. The perturbation and observation (P&O) method is an algorithm used for tracking the maximum power by periodically adjusting the operating point and monitoring the resulting power change. However, as Paraskevas (2016) points out, the drawback of this method is that the operating point of the TEM tends to oscillate around the MPP, leading to potential power losses. To mitigate this issue, a passivity-based perturb and observe (P&O) approach is proposed to maintain the system's power point at its maximum. This work will also provide a stability analysis of the P&O controller based on passivity techniques.

2 Thermoelectric Generator Characteristic

The electrical component of the thermoelectric module (tem) equivalent circuit consists of a voltage source and internal resistance, as described by kwan, t.h. (2016). The norton equivalent circuit of a tem with a resistive load is illustrated in Fig. 1. This representation simplifies the analysis of tem behavior by modeling the system with its corresponding electrical properties.

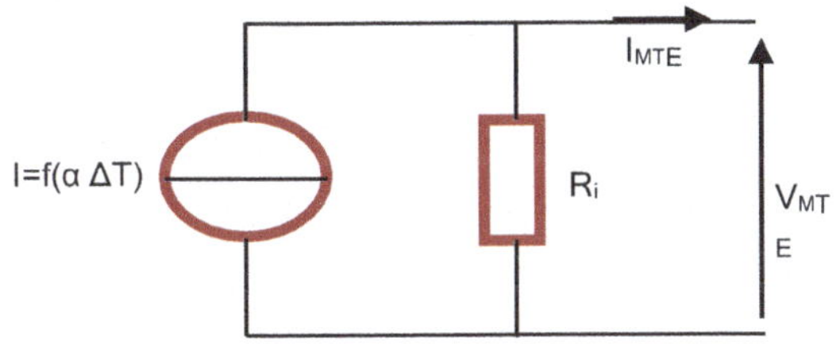

Fig. 1. TEM.

The TEM current can be expressed as:

$$I_{MTE} = \frac{N_S \cdot \alpha_{pn} \cdot \Delta T - V_{MTE}}{N_s R_i} \tag{1}$$

where, V_M, I_{MTE} and R_i are the voltage, current, and total electrical resistance of TEM.

3 MPPT Converter Configuration

Let us examine a boost converter connected to the thermoelectric module (TEM), as shown in Fig. 2 by Gao (2016).

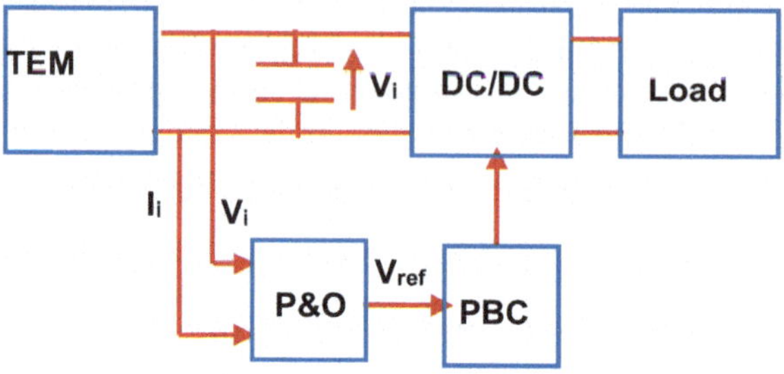

Fig. 2. Boost converter (BC).

As depicted in Fig. 2, the PBC controller functions to minimize the error between the reference (Vref), generated by the MPPT block, and the output voltage of the TEM by adjusting the duty cycle of the (BC) boost converter.

3.1 Perturb and Observe (P&O) MPPT

The most widely used MPPT technique is the Perturb and Observe (P&O) method. The P&O algorithm functions by periodically perturbing (either increasing or decreasing) the terminal voltage (VT) or current of the array and comparing Pout (output power) of the TEM with that from the previous perturbation cycle, as noted by Chermitti (2012).

A common issue with P&O algorithms is that the terminal voltage of the TEM is perturbed during each MPPT cycle; consequently, once MPP is reached, the output power tends to oscillate around this maximum, leading to power losses in the TEM system.

Incorporating an additional control loop with passivity enhances the efficiency of the P&O method, resulting in faster dynamic performance and better regulation of the TEM output voltage (Fig. 3).

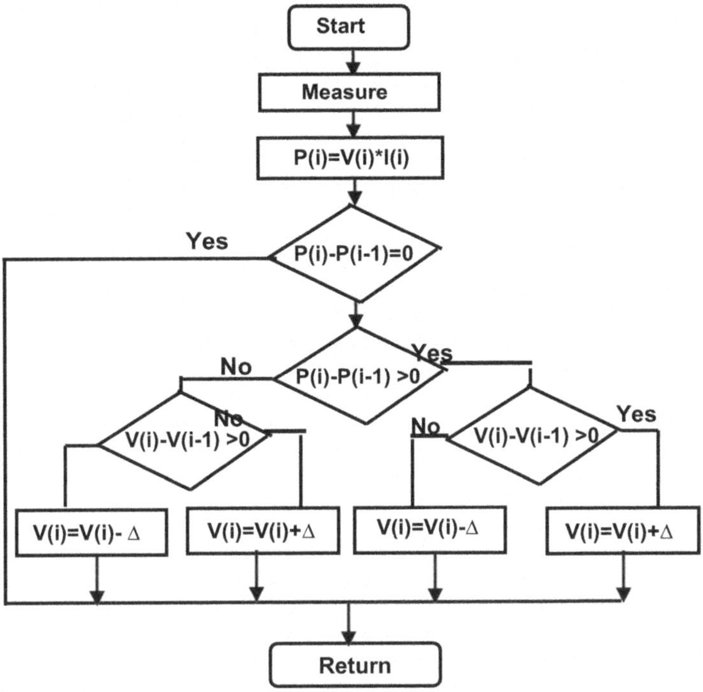

Fig. 3. P&O block diagram.

4 The Proposed Controller

Passivity-based controllers (PBC) have been utilized in various practical control applications, including DC/DC power converters, as noted by Yand (2004). Several researchers have introduced PBC methods to achieve robust output voltage regulation for DC/DC boost converters under varying load conditions. The fundamental concept involves introducing additional damping—known as injection damping—to attain asymptotic stability, as discussed by Lee (2004). This approach modifies the Rayleigh dissipation function to ensure that the new equilibrium point is globally asymptotically stable. This paper focuses on the Lagrangian modeling of the boost converter, with the basic circuit of the boost (step-up) illustrated in Fig. 2 by Gao (2002) (Fig. 4).

The Lagrangian $L(q, \dot{q})$, is defined as the difference between the magnetic co energy (MCE) of the circuit, denoted by $T(q, \dot{q})$ of the circuit and the electric field energy (EFE) of the circuit, denoted by $V(q)$.

$$L(\dot{q}, q) = T(\dot{q}, q) - V(q) \tag{2}$$

On the other hand, EL system can be expressed by:

$$\frac{d}{dt}\frac{\partial L(q, \dot{q})}{\partial \dot{q}} - \frac{\partial L(q, \dot{q})}{\partial q} = -\frac{\partial D(\dot{q})}{\partial \dot{q}} + F_q \tag{3}$$

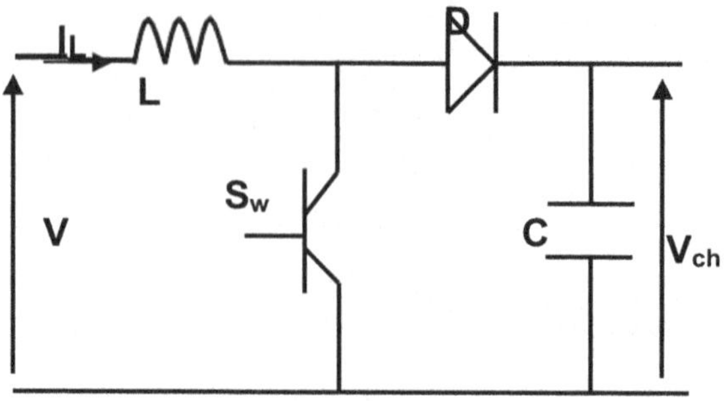

Fig. 4. Basic boost converter circuit.

To obtain the EL expression for the Boost converter, the following definitions are needed.

$$\begin{cases} T_\mu(\dot{q}) = \frac{1}{2}Lq_L^2 \\ V_\mu(q) = \frac{1}{2C}q_C^2 \\ F_\mu(\dot{q}) = \frac{1}{2}R_{ch}((1-\mu)\dot{q}_L - \dot{q}_C)^2 \\ Q_\mu = \begin{pmatrix} V \\ 0 \end{pmatrix} \end{cases} \tag{4}$$

where:

$D = \frac{1}{2}\dot{q}^T R\dot{q}$: The Rayleigh dissipation function.

F_q: are exogenous inputs to the system.

q_L : is the electrical charge on the inductor.

q_c: is the electrical charge on the capacitor.

Applying Eqs. (2) and (3), we get:

$$\begin{pmatrix} L & 0 \\ 0 & C_2 \end{pmatrix}\dot{x} = \left(\begin{pmatrix} 0 & -(1-\mu) \\ 1-\mu & 0 \end{pmatrix} - \begin{pmatrix} 0 & 0 \\ 0 & -\frac{1}{R_{ch}} \end{pmatrix}\right)x + \begin{pmatrix} V \\ 0 \end{pmatrix} \tag{5}$$

The dynamics as described as follows:

$$M\dot{x} + (1-\mu)Jx + Rx = E \tag{6}$$

where; $M = \begin{pmatrix} L & 0 \\ 0 & C_2 \end{pmatrix}$; $J = \begin{pmatrix} 0 & 1 \\ -1 & 0 \end{pmatrix}$; $R = \begin{pmatrix} 0 & 0 \\ 0 & \frac{1}{R_{ch}} \end{pmatrix}$; $E = \begin{pmatrix} V \\ 0 \end{pmatrix}$

In order to ensure the asymptotic stability error, we may introduce the damping injection function in the system.

$$R_b = \begin{pmatrix} R_1 & 0 \\ 0 & 0 \end{pmatrix} \tag{7}$$

$$M\dot{\tilde{x}} + (1-\mu)J\tilde{x} + R\tilde{x} = E - M\dot{x}_d - (1-\mu)Jx_d - Rx_d = \Phi \tag{8}$$

where:

Rd : the desired dissipation.

$\tilde{x}$: The error state vector

We can see that the error dynamic for Eq. (8) is asymptotically stable to zero. It can be expressed as:

$$M\dot{x}_d + (1-\mu)Jx_d + Rx_d - Rb\tilde{x} = E \tag{9}$$

According to Eq. (9), the scalar differential equations can be given:

$$\begin{cases} L\dot{x}_{1d} + (1-\mu)x_{2d} - R_1(x_1 - x_{1d}) = V \\ C\dot{x}_{2d} - (1-\mu)x_1 + \frac{1}{R_{ch}}x_{2d} = 0 \end{cases} \tag{10}$$

From Eq. (10), the dynamic duty ratio can be written by:

$$\dot{\mu} = \frac{(1-\mu)^2}{C[V + R_1(x_1 - x_{1d})]} * \left[x_{1d}(1-\mu) - \frac{V + R_1(x_1 - x_{1d})}{(1-\mu)R_{ch}} - \frac{C(R_{1b}x_1^*)}{1-\mu}\right] \tag{11}$$

5 Resultants and Simulation

To showcase the efficiency of the proposed system, simulations were performed using MATLAB Simulink. The thermoelectric module used for the simulation was the HZ-20.

5.1 Simulation Result for Temperature Gradient Constant

The simulation results are presented in Figs. 5, 6 and 7, depicting various temperature gradients.

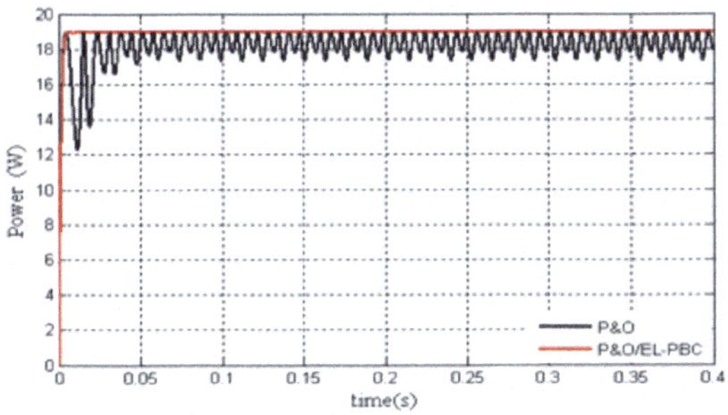

Fig. 5. Power of GTE

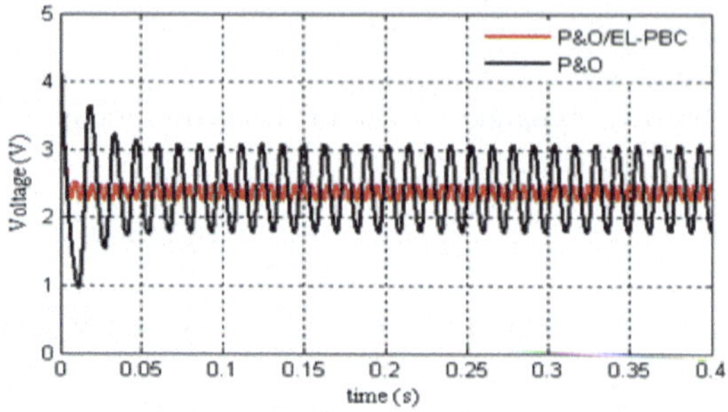

Fig. 6. Voltage of GTE

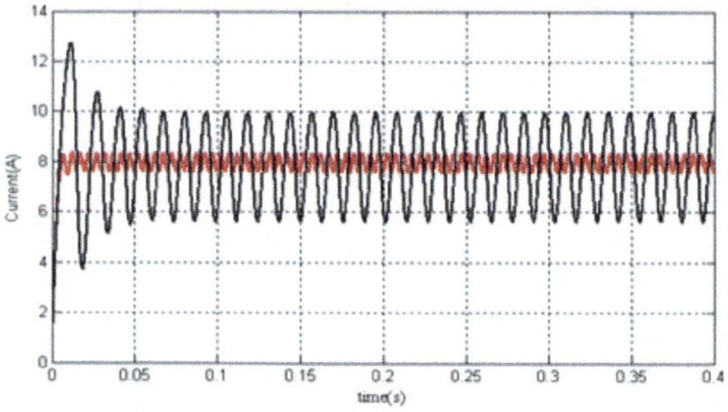

Fig. 7. Current of GTE

Simulation results distinctly indicate that the passivity-based P&O technique performs better than the conventional P&O method, offering faster settling times and fewer steady-state oscillations. Consequently, this leads to increased energy production by the thermoelectric module (TEM).

5.2 Simulation Result Fordifferent Temperature Gradient Variable

Figures 8, 9 and 10 present the simulation results following a sudden change in the temperature gradient ($\nabla T = 150$ °C to $\nabla T = 200$ °C) at t=0.2. The results clearly demonstrate that the passivity-based P&O method reaches the optimal maximum power more quickly and with fewer oscillations compared to the conventional technique. Additionally, passivity-based P&O exhibits a faster response to changes in temperature gradient conditions. The simulation indicates that the voltage and current for the conventional Perturb and Observe method are not stable. In contrast, the passivity-based P&O achieves a maximum output value of 99.74%, surpassing that of the Perturb and Observe controller.

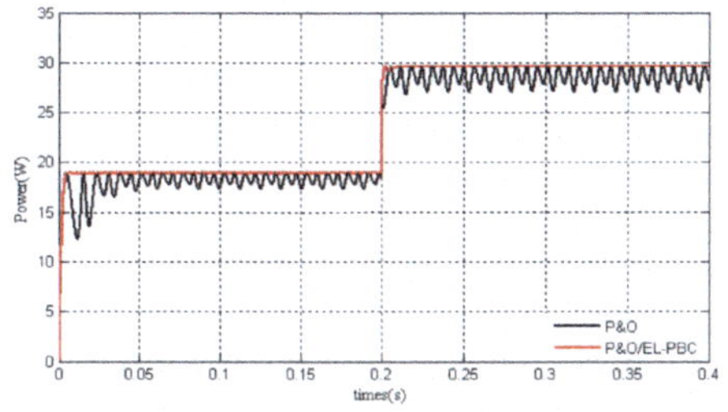

Fig. 8. Power of GTE.

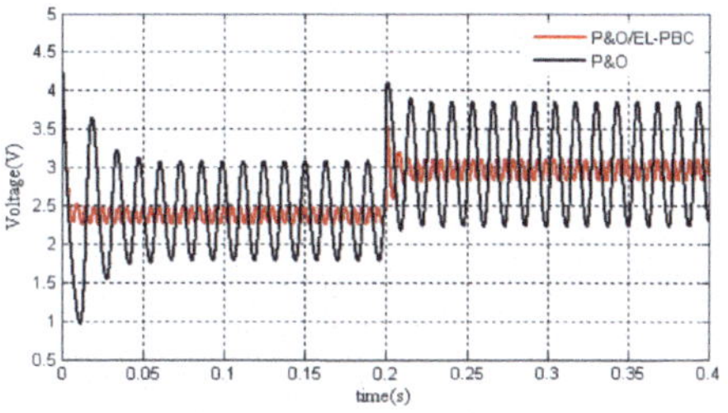

Fig. 9. Voltage of GTE.

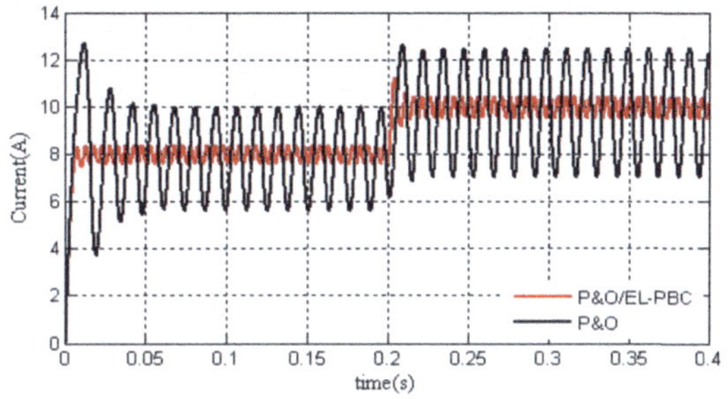

Fig. 10. Current of GTE.

6 Conclusion

In this paper, two MPPT techniques are compared based on both the transient and steady-state responses of the TEs. The simulation results indicate that the developed system demonstrates exceptional precision and speed in tracking the MPP, even in the presence of rapid temperature gradient changes. Once the system reaches the MPP, the efficiency of the P&O/EL-PBC MPPT exceeds 99%, surpassing the performance achieved by the conventional P&O method.

References

Anil, G., Murugan, N., Ubaid, M.: PI Controller based MPPT for a PV System. J. Elect. Electro. Eng. 265–267 (2013)

Tang, Z.B., Deng, Y.D., Su, C.Q., Shuani, W.W., Xie, C.J.: A research on thermoelectric generator's electrical performance under temperature mismatch conditions for automotive waste heat recovery system. Case Studies in Thermal Engineering, 143–150 (2015)

Eakburanawat, J., Boonyaroonate, I.: Development of a thermoelectric battery-charge with microcontroller- based maximum power point tracking technique. Journal Applied Energy, 687–704 (2006)

Paraskevas, A., Koutroulis, E.: A simple maximum power point tracker for thermoelectric generators. Energy Conversion and Management, 355–365 (2016)

Kwan, T.H., Wu, X.: Power and mass optimization of the hybrid solar panel and thermoelectric generators. Applied Energy, 297–307 (2016)

Gao, H.B., Huang, G.H., Li, H.J., Qu, Z.G., Zhang, Y.J.: Development of stove-powered thermoelectric generators: A review. Applied Thermal Engineering, 297–310 (2016)
Chermitti, A., Boukli-Hacene, O., Bencherif, M.: Improvement of the "Perturb and Observe" MPPT algorithm in a photovoltaic system under rapidly changing climatic conditions. International Journal of Computer Applications, 5–9 (2012)
Yand, Z., Wu, L.: A new passivity-based control method and simulation for DC/D converter. 5th world congress on intelligent control and automation, pp. 5582–5585 (2004)
Lee, T.S.: Lagrangian modeling and passivity-based control of three-phase AC/DC voltage-source converters. IEEE Transactions on industrial electronics, 892–902 (2004)
Gao, H.B., Huang, G.H., Li, H.J., Qu, Z.G., Zhang, Y.J.: Interconnection and damping assignment passivity- based control of port-controlled Hamiltonian systems. Automatica, pp. 585–596 (2002)

Improvement of the Performance of a Proton Exchange Membrane Fuel Cell (PEMFC) System Designed to Propel an Electric Vehicle

Achouri Abdelhak(✉) and Abdi Hamid

Department of Mechanics, Faculty of Technology, University of Blida 1, Blida, Algeria
achouriabdelhak1@gmail.com

Abstract. This study investigates the impact of inlet pressure and temperature of reactants, on the performance of an energy system powered by a proton exchange membrane fuel cell (PEMFC) for electricity production. A thorough energy analysis was conducted using thermodynamic modeling implemented in Matlab to assess the influence of these parameters on the overall system performance. Simulation results were carefully compared to those obtained in specialized literature, thereby reinforcing the validity of the conclusions. The parametric analysis reveals that increasing the inlet temperature of reactants from 303.15 K to 343.14 K results in a significant improvement in system efficiency (PEMFC) with an average increase of 8.96%. Conversely, raising the inlet pressure from 1.2 atm to 3.2 atm negatively affects the overall system performance (PEMFC), leading to an average decrease of 19.5%.

Keywords: PEMFC · thermodynamic modeling · inlet temperature · inlet pressure · parametric analysis · system efficiency

1 Introduction

The industrial boom and population growth have led to an exponential increase in energy needs, intensifying reliance on fossil fuels. This dependence raises major concerns such as air pollution, climate change, and global warming, threatening the future of our planet. The search for alternative energy sources, such as hydrogen, has therefore become crucial. This study stands out by examining the combined impact of inlet pressure and reactant temperature on the performance of PEMFC systems. It adopts an integrated approach that evaluates not only the intrinsic performance of the fuel cell but also the influence of operating conditions on auxiliary consumption and overall efficiency. By optimizing temperature and inlet pressure, the study proposes thermal management and gas flow strategies to maintain optimal performance. The analysis also includes an assessment of the long-term impacts on PEMFC durability, making this technology more efficient and viable for automotive applications.

© The Author(s), under exclusive license to Springer Nature Switzerland AG 2025
M. Hatti (Ed.): IC-AIRES 2024, LNNS 1238, pp. 27–31, 2025.
https://doi.org/10.1007/978-3-031-80301-7_4

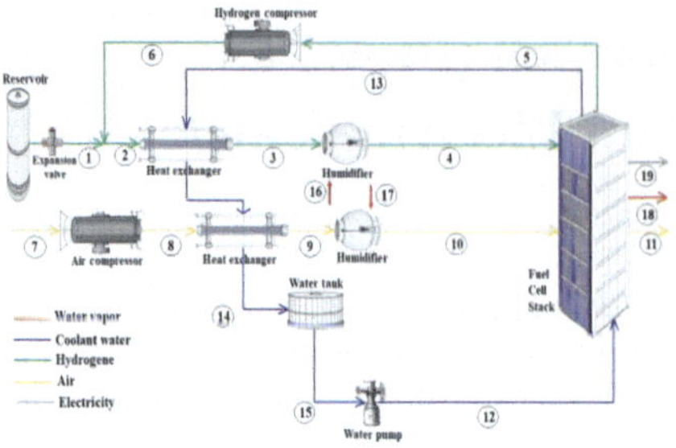

Fig. 1. System layout of the proposed PEMFC system.

Table 1. Input parameters of the PEMFC model

Parameters	Values
Number of cells in the stack	65
Active surface area	240 cm^2
Membrane's thickness	178 x 10^{-4} cm
Maximum current density	9375 (A cm^{-2})
Relative humidity	100%
Operating pressure	1.2 - 3.2 atm
Operating temperature	308.15 K - 358.15 K
Air compressor isentropic efficiency	55 %
Pump isentropic efficiency	43 %

2 Experimental/Theoretical Study

To better understand certain effects and phenomena in the fuel cell system, we employ thermodynamic modeling. This allows us to predict the consumption of various system accessories as well as the net power delivered. Additionally, by using the MATLAB simulation tool.

3 Results and Discussion

Among all the existing families of fuel cells, we are interested in low-temperature proton exchange membrane fuel cells (PEMFCs). Changes in operating conditions can have a beneficial or detrimental effect on the performance of the cell and the system. Predicting the behavior of the PEMFC system across its operating parameters is becoming increasingly essential for achieving widespread applications of this technology. Therefore, better knowledge and understanding of the behavior and operation of PEMFCs are necessary (Fig. 1 and Table 1).

Mathematical model of PEMFC system:

The cell voltage of the PEM fuel cell:	$Vstack = Ncell.\,(E - Vact - Vohm - Vconc)$
Stack electrical power:	$Pstack = Vstack.\,i$
Electrical efficiency of the PEMFC stack:	$\eta_{stack} = \frac{V_{cell}}{Sc.LHV}$
Power of all auxiliary:	$Wp = WHC + WAC + WAHE + WCHE + Wwp$
Net power of the power system:	$Wnet = Wstack - Wp$
Energy efficiency of the system:	$\eta_{sys} = \frac{W_{net}}{N_{cell}.i.V_{rev}}$

Increasing pressure has a positive impact on the power delivered by the fuel cell. For example, at a current density of 0.85 A/cm^2, the power densities are 32.16 W/cm^2 and 29.10 W/cm^2 for pressures of 1.2 atm and 3.2 atm, respectively. On average, the stack power is 8.28% higher for a pressure of 3.2 atm compared to 1.2 atm, primarily due to the improvement in stack voltage with pressure. However, operating at high-pressure results in increased load on the compressors and cooling water pump, leading to higher total consumption, with an average increase of 67.30% in total consumption when transitioning from a pressure of 1.2 atm to 3.2 atm. Consequently, the net energy production decreases with increasing pressure. It reveals a decrease in overall system efficiency with increasing inlet pressure of the reactive gases. For example, at a current density of 0.6 A/cm^2 and pressures of 1.2 atm and 3.2 atm, the system efficiency is 0.46 and 0.37, respectively. Thus, increasing pressure at the stack inlet has a negative effect on net power and overall efficiency of the PEMFC system. This is because the increase in reactant inlet pressure leads to a larger increase in energy consumption by the auxiliaries than that delivered by the fuel cell, resulting in an average decrease in overall efficiency of up to 19.50% (Figs. 2, 3, 4, 5, 6 and 7).

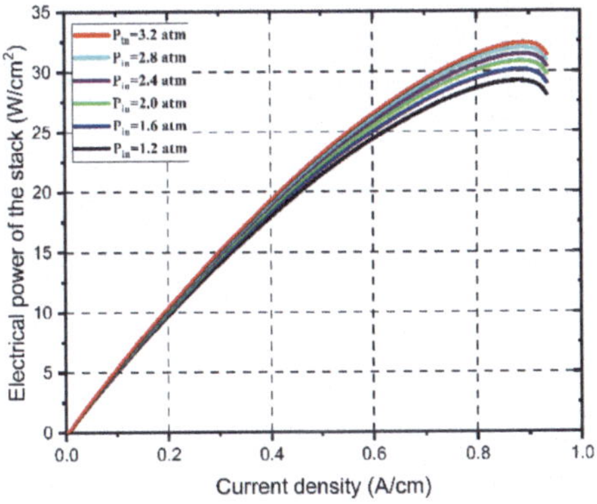

Fig. 2. Impact of pressure on the stack power.

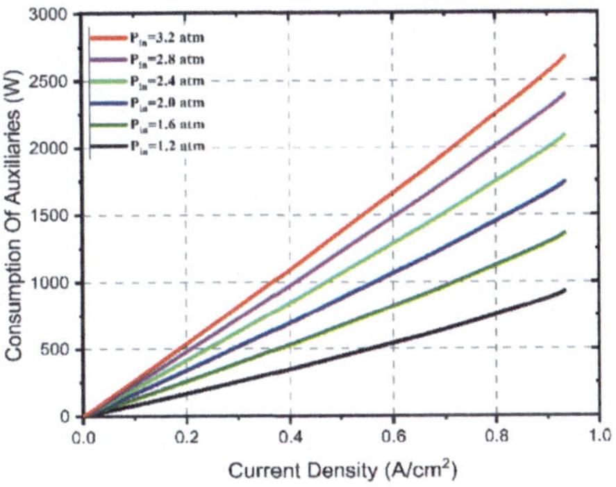

Fig. 3. Impact of pressure on the total power consumed.

Higher inlet gas temperatures promote the production of water vapor in the fuel cell, thereby increasing water activity and membrane water content. This results in an increase in membrane conductivity and a decrease in ohmic loss, despite an increase in activation overpotential. This overall improvement in performance leads to an increase in the power delivered by the fuel cell. Where we observe an increase in electrical power density with increasing temperature. However, increasing the inlet temperature of the reactive gases also leads to a significant increase in the total power consumed by the fuel cell accessories. This increased energy consumption is mainly due to increased demand on associated compressors and an increase in the flow rate of cooling water needed to maintain temperatures under control. Efficiencies increase with increasing operating temperature, with an average difference of 8.96%.

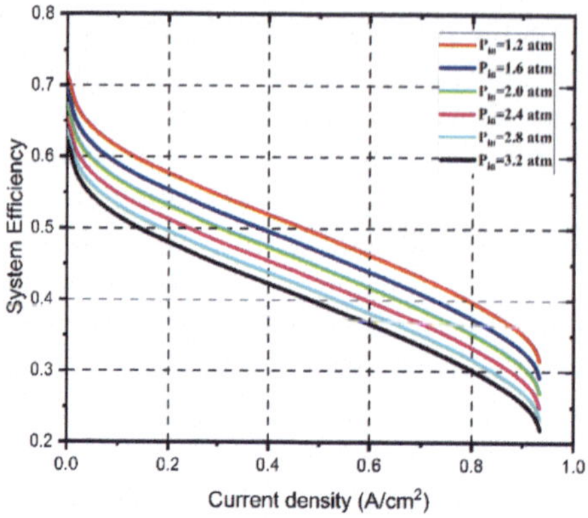

Fig. 4. Impact of pressure on system efficiency.

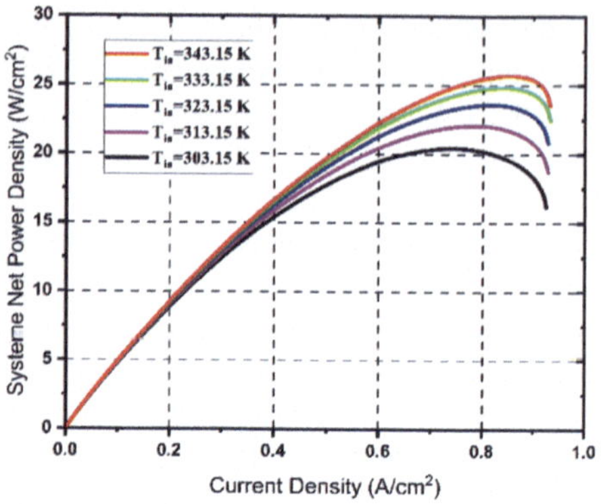

Fig. 5. Impact of temperature on the stack power.

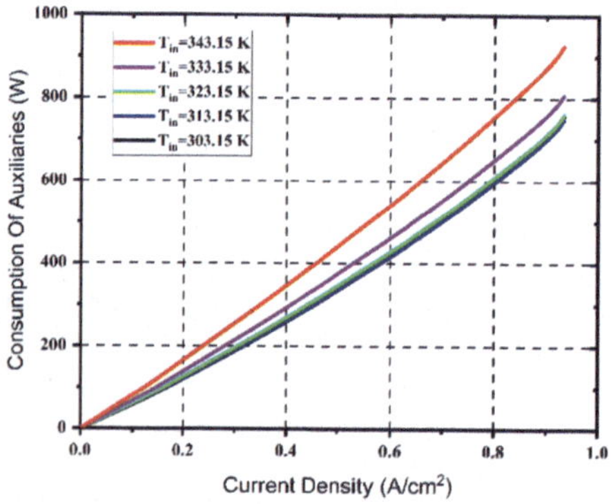

Fig. 6. Impact of Temperature on the total power consumed.

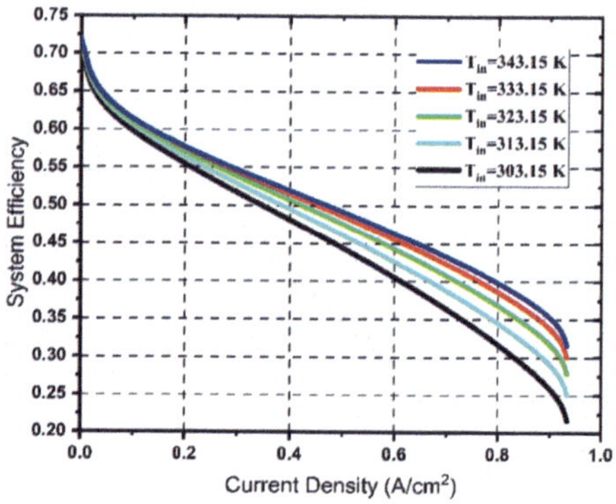

Fig. 7. Impact of Temperature on system efficiency.

4 Conclusions

This study highlights the significant impact of inlet temperature and pressure, as well as mole fraction of nitrogen in air, on proton exchange membrane fuel cell (PEMFC) system performance. Furthermore, it presents an in-depth analysis of the existing literature and a validation of the mathematical model used. The results obtained reveal the following points:

- Raising the cell inlet temperature leads to an overall improvement in system performance, notably due to the increase in reaction speed and mass transfer rate. Additionally, this increase in temperature helps reduce the ohmic resistance of the cell by promoting increased ionic conductivity of the electrolyte membrane.
- An increase in inlet pressure results in improved cell performance, by increasing the partial pressures of the reactive gases and reducing activation losses. However, a reduction in the overall performance of the system is observed, this phenomenon being explained by the increase in the consumption of the compression auxiliary.

References

1. Hussain, M.M., Baschuk, J.J., Li, X., Dincer, I.: Thermodynamic analysis of a PEM fuel cell power system International Journal of Thermal Sciences **44**, 903–911 (2005)
2. Mo, Z.J., Zhu, X.J., Wei, L.Y., Cao, G.Y.: Parameter optimization for a PEMFC model with a hybrid genetic algorithm. Int. J. Energy Res. **30**, 585–597 (2006)
3. El-Fergany, A.A.: Extracting optimal parameters of PEM fuel cells using Salp Swarm Optimizer. Renewable Energy **119**, 641–648 (2018)
4. Selem, S.I., Hasanien, H..M.., El-Fergany, A..A..: Parameters extraction of PEMFC's model using manta rays foraging optimizer. International Journal of Energy Researche **44**, 4629–4640 (2020)

Predictive Wind Turbine Power Analysis Based on SCADA Data and Machine Learning Algorithms

Zouhir Iourzikene(✉), Fawzi Gougam, and Djamel Benazzouz

Laboratoire de Mécanique des Solides et Systèmes (LMSS), Faculté de Technologie, Université M'Hamed BOUGARA de Boumerdes (UMBB), 35000 Boumerdes, Algeria
{z.iourzikene,f.gougam,d.benazzouz}@univ-boumerdes.dz

Abstract. In this study, we assess the performance of various regression models for predicting the theoretical power generated by a wind turbine using supervisory control and data acquisition. The tested models include linear regression, fine tree, support vector machine, rational quadratic Gaussian process regression, and the boosted trees optimizable. We use cross- validation to select the best-performing model. Our findings indicate that the Boosted Trees Optimizable model exhibits superior accuracy, indicated by its low root mean squared error value of 3.69 compared to other models. This high accuracy demonstrates the potential of the Boosted Trees Optimizable model for effective application in wind turbines management.

Keywords: Wind turbine · Scada Dataset · Machine Learning · Prediction · Regression

1 Introduction

In recent years' wind turbines have attracted increasing attention as an alternative source of renewable energy and a mature renewable energy technology, benefiting from technological innovation and political support. It has become an important component of the energy supply, on a par with solar energy [1–3]. Due to advancements in high-performance alternating current (AC) motor control theory, variable speed constant frequency technology, mainly defined by converters, plays a crucial role in harnessing wind power [4, 5]. Supervisory control and data acquisition (SCADA) systems play an essential role in infrastructure management, monitoring industrial facilities such as power grids, wind farms...ect. SCADA systems collect real-time data from various sensors and equipment in the field, which are subsequently transmitted to a control center [6, 7].

Machine learning (ML) models offer effective methods for capturing complex relationships among explanatory variables. Recently, reliability analysis based on ML has seen significant improvements in terms of performance and precision [8–10]. Various approaches have been developed to enhance computational efficiency. However, few studies focus on identifying the optimal theoretical strategy. Modern intelligent systems,

© The Author(s), under exclusive license to Springer Nature Switzerland AG 2025
M. Hatti (Ed.): IC-AIRES 2024, LNNS 1238, pp. 32–42, 2025.
https://doi.org/10.1007/978-3-031-80301-7_5

equipped with artificial intelligence (AI) capabilities, often rely on ML. This technology allows systems to learn from problem-specific data, enabling the automation of analytical model creation and task resolution [11, 12]. K-Fold Cross-validation (CV) is a commonly used method for assessing the predictive performance of statistical models. It involves dividing data into multiple subsets to conduct predictive evaluations [13]. CV technique is particularly valuable for model comparison and selection across various modeling techniques, whether for hypothesis testing or predictions samples. The flexibility of CV lies in its its capacity to utilize a diverse range of predictive metrics and data splitting strategies, allowing for consideration of data structure and management of computational costs and estimation biases. In the field of ML, classical methods such as dividing data into training, validation, and test sets, along with CV, are essential for reliably estimating model performance [14–17].

The prediction tool using time series data to estimate wind power and an exploration method for SCADA data, utilizing polar coordinate systems and scatter plots to better understand the relationship between wind and energy production, was recently designed. By incorporating features such as theoretical power, active power, wind direction, wind speed, month, and time, this approach employed various ML algorithms to predict the power generated by the turbine. Model performance was evaluated using several statistical criteria [18]. A model using a cloud-based digital twin infrastructure and a 5G network to monitor and predict wind turbine energy production, based on a temporal convolutional network and k- nearest neighbors (kNN) regression, can predict wind speed and energy production over periods ranging from one week to one month. The results demonstrate that this model outperforms classical models and can effectively assist in managing wind farms [19]. The use of K-means clustering and box plots can detect anomalies and erroneous behaviors in wind turbines, improving predictive maintenance without relying on labeled data [20]. Prediction models for wind energy forecasting use advanced ML techniques such as artificial neural networks (ANNs), recurrent neural networks (RNNs), convolutional neural networks (CNNs), and long short-term memory networks (LSTMs). The results have confirmed the effectiveness of these models in wind energy prediction [21]. A model using unsupervised SVM-KNN detection has significantly improved the accuracy of predictions for SCADA data related to pitch angle and rotor speed curves [22].

This document outlines the following key contributions for predicting the theoretical power output of a wind turbine, aiming to assess its efficiency in converting wind energy into electrical energy:

- The SCADA dataset from wind turbines operating and generating electricity in Turkey in 2018 was selected to train automatic prediction models for the theoretical power generated by the turbine. After preprocessing, this dataset contains the following information: date/time (in 10-min intervals), power generated by the turbine, wind speed at the turbine hub height, theoretical power generated by the turbine, and wind direction at the wind turbine hub height.
- CV method ($k = 5$) was used to divide the data into multiple subsets. The prediction model is then trained on a portion of the data and tested on the other parts repeatedly, in order to estimate the performance of the ML model on this dataset.

- Five ML models were evaluated for predicting the theoretical power of the turbine: linear regression (LR), fine tree (FT), support vector machine (SVM), rational quadratic gaussian-process regression (RQ-GPR), and boosted trees optimizable (BTO). The root mean squared error (RMSE) values obtained were 538.65, 8.84, 561.13, 5.31, and 3.69, respectively, with BTO achieving the best performance with an RMSE of 3.69.
- The following sections of this article are organized as follows: methodology, results, discussion, and conclusion.

2 Approach and Techniques Used

Figure 1 shows the key steps in the methodology used in this article to predict the theoretical power values generated by a turbine. Five important steps were followed. The first step consists in selecting the SCADA dataset of the wind turbines. The second step involves data pre-processing, carried out prior to training the regression models. The third step involves dividing the data into training and test sets using the CV method (k = 5). The fourth step involves selecting several regression models for analysis. Finally, the fifth step is dedicated to evaluating the performance of the different models and selecting the one that offers the best results in terms of predictive accuracy.

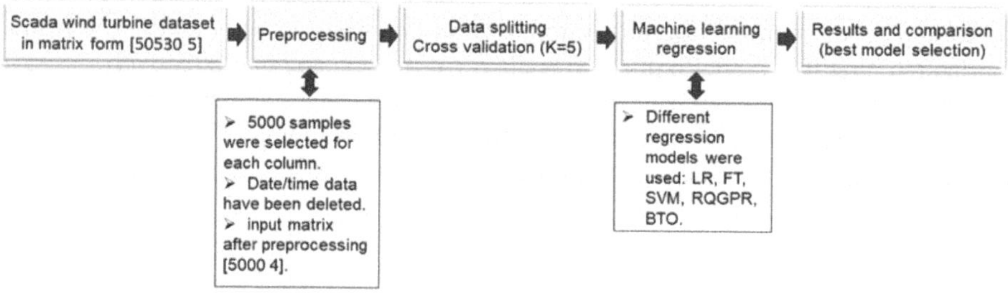

Fig. 1. Diagram of theoretical wind turbine power prediction methodology.

2.1 Database and Pre-processing

The 2018 SCADA dataset for wind turbines in Turkey used in this study is freely available on Kaggle [23]. Wind turbine SCADA systems monitor and record essential data such as wind speed and direction, as well as power output, at regular 10-minute intervals. This data set, presented as an excel file, comprises 5 columns and 50530 rows. The columns are as follows: date/time (at 10-minute intervals), active power, wind Speed, theoretical power curve and wind direction. The lines in this file represent the records of these characteristics at each 10-minute interval, covering the period from '01/01/2018, 00:00' to '31/12/2018, 23:50'.

Prior to learning the regression models, the dataset was pre-processed to ensure data quality and relevance. The first 5000 rows of the dataset were selected, covering the period from '01/01/2018, 00:00' to '09/02/2018, 05:00'. The column containing the dates was removed to simplify the data for analysis. This resulted in an input matrix of dimensions [5000, 4], used to predict theoretical power. This pre-processing step is crucial to eliminate superfluous temporal information and focus on variables directly linked to energy production.

2.2 Cross Validation

CV is used to evaluate the performance of a ML model. Unlike a simple division of data into training and test sets, CV divides the data set into several parts, called folds. It then trains the model on some folds and tests it on others. This process is repeated several times, alternating the folds used for training and testing. The advantage of CV lies in its ability to provide a more robust assessment of model performance. It helps avoid overtraining, as well as undertraining [24]. In this approach, the data set has been divided into five folds (K = 5). Figure 2 shows a description of the CV method.

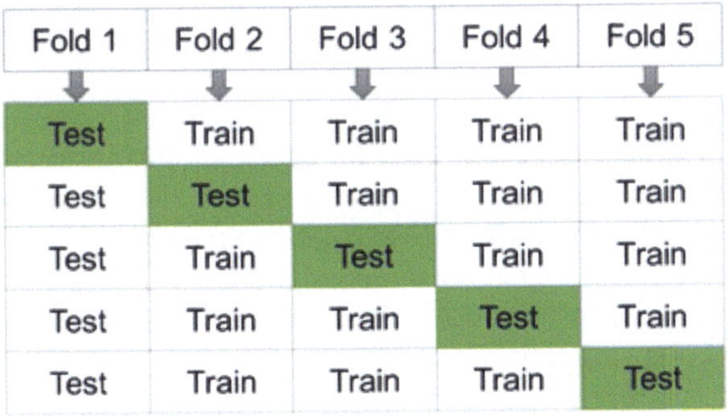

Fig. 2. Illustration of cross validation method.

2.3 Machine Learning Models

In our study, we explored the use of various regression models to predict the predictive power of a turbine. Figure 3 illustrates the prediction of the theoretical power generated by the turbine using various ML regression models. To do this, we exploited MATLAB ML toolbox, which offers a variety of regression algorithms. Here are the five models we chose:

Linear Regression

LR is a fundamental statistical technique used to model and predict a target variable as a function of known explanatory variables. It aims to determine a linear relationship between the variable to be predicted and the explanatory variables. Simple LR is used to predict a quantitative response 'Y' as a function of a single predictor variable 'X'. In other words, it establishes a linear relationship between these two variables to facilitate

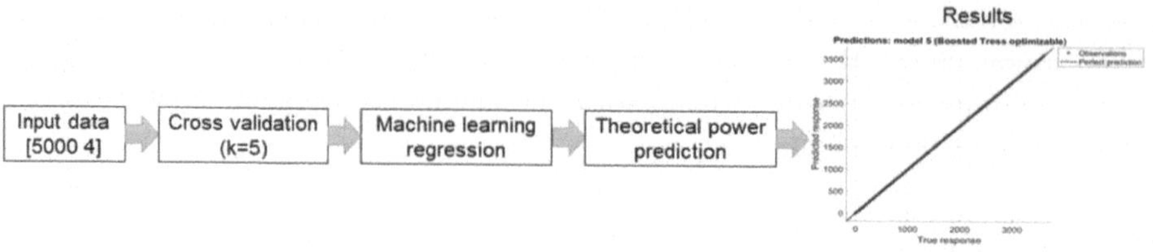

Fig. 3. Explanatory diagram of theoretical power detection using machine learning.

predictions [25]. Mathematically, we can express this linear relationship in the following Eq. 1:

$$Y = \beta_0 + \beta_1 X \tag{1}$$

The unknown coefficients β_0 and β_1 correspond to the y-intercept and slope of the linear model, respectively. Figure 4(a) shows the results of theoretical power predictions obtained by LR.

Fine Tree

A FT is a visual and analytical tool used for decision-making. It represents a series of choices and their consequences in the form of a tree structure, where each node corresponds to a decision to be made and each branch represents a possible choice leading to a final node or leaf. This hierarchical representation enables a complex problem to be broken down into a series of simpler decisions [26]. Mathematically, a FT can be expressed as a function $g(x)$ defined as in Eq. 2:

$$g(x) = \sum\nolimits_{n}^{N} P_n A(x \in B_n) \tag{2}$$

x : vector of explanatory variables; N: total number of leaves in the tree; P_n : value predicted by the sheet n; B_n : region of the explanatory variables space corresponding to leaf n; A : indicator function which is 1 if $x \in Bn$ and 0 otherwise.

Figure 4(b) shows the results of theoretical power predictions obtained by FT.

Support Vector Machine

SVM regression is a robust method for predicting continuous values, based on the principles of SVMs. This method aims to determine a function that maximizes the deviation from all target values, while remaining within a predefined margin [27]. The linear regression function can be expressed as in Eq. 3:

$$f(x) = Wx + b \tag{3}$$

where W is a weight vector and b is a bias.

Figure 4(c) shows the results of theoretical power predictions obtained by SVM.

Rational Quadratic Gaussian Process Regression

RQ-GPR is a probabilistic model based on non-parametric kernel functions. It uses a multivariate distribution to model a finite collection of random variables, guaranteeing a balanced distribution across all linear combinations. Unlike many recent methods, RQ-GPR requires few parameters to make predictions. It is based on the idea of the Gaussian distribution, as Eq. 4 shows, which enables it to capture complex relationships in the data while retaining a probabilistic interpretation of the predictions. This feature makes RQ-GPR particularly effective for regression tasks in contexts where relationships between variables can be complex and non-linear [28].

$$f(x, \mu, 6) = \frac{1}{6\sqrt{2\pi}} e^{-\frac{(x-\mu)^2}{26^2}} \quad (4)$$

x : random variable, μ: the average, Ϭ: standard deviation.

Figure 4(d) shows the results of theoretical power predictions obtained by RQ-GPR.

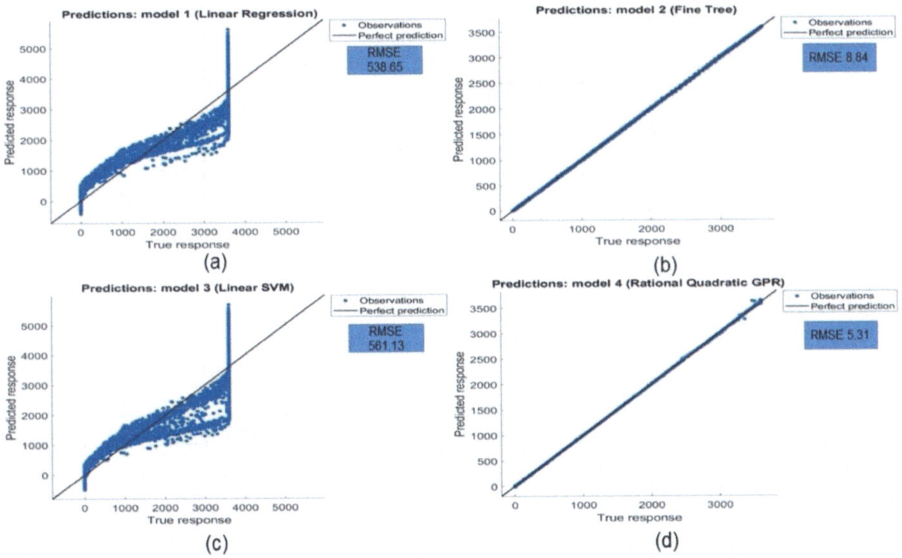

Fig. 4. Results of theoretical turbine power prediction using: (a) LR, (b) FT, (c) SVM, (d) RQ-GPR.

Boosted Tress Optimizable

This model combines several decision trees and the gradient boosting algorithm to improve performance can handle heterogeneous data [29].

In our study, we used the BTO model to predict the theoretical power values generated by a turbine. Table 1 shows how this model was optimized.

Figure 5(a) presents the results of theoretical power predictions obtained by BTO, 5(b) presents a graphical visualization that shows how the mean square error (MSE) varies as a function of the training variable to minimize the MSE in order to obtain the best model performance.

Table 1. BTO model optimization options and hyperparameters used.

Model hyperparameters	Optimization options
• The Bag method was chosen, which combines several decision trees to improve performance • Minimum leaf size is set to a value of 5, which controls tree complexity • Number of learners: 24 decision trees were used to build the model	• Bayesian optimization was used for the model • Expected improvement per second plus was selected as the acquisition function, which measures expected improvement as a function of time • 30 optimization iterations were performed to explore the space of hyperparameters

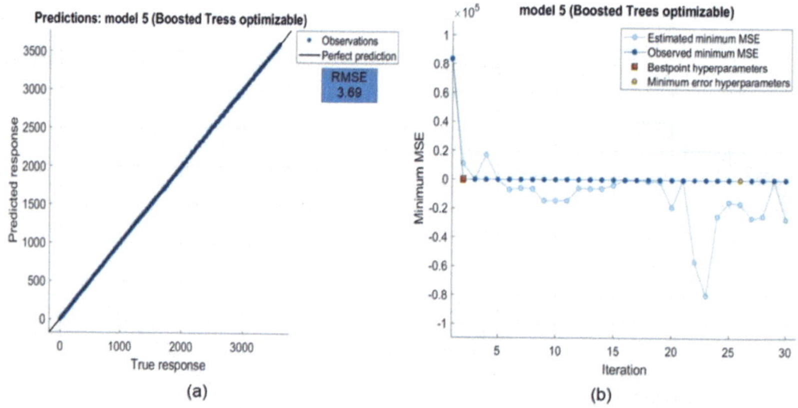

Fig. 5. Results of theoretical turbine power prediction using: (a) BTO, (b) Bayesian model optimization.

RMSE is used to assess the precision of regression models. It measures the differences between observed and predicted values (standard deviation.) A lower RMSE indicates more accurate predictions. The formula for RMSE is given in Eq. 5:

$$RMSE = \sqrt{\frac{1}{n}\sum\nolimits_{1}^{n}(f_n - \hat{f}_n)^2} \tag{5}$$

n: Number of data, fn: Observed value, $\hat{f}_n$: value predicted by the model.

3 Performance Results and Discussion

The results of predictions obtained from various regression models are illustrated in Figures 4 and 5. In an ideal situation, the points should align along the identity line (X = Y), indicating a perfect correspondence between predictions and actual values. The proximity of points to this line reflects the precision of predictions. A low dispersion of points around the identity line in the graphs demonstrates that the regression models used provide precise predictions.

The RMSE values for the five models used to predict the theoretical power of the turbine (LR, FT, SVM, RQ- GPR, BTO) are 538.65, 8.84, 561.13, 5.31, and 3.69, respectively. These values are illustrated in Figures 4 and 5.

These results show that the BTO model has the lowest RMSE value, indicating that it provides the most accurate predictions among the models tested in this article. Table 2 and Figure 6 show a comparison of the RMSE value for these models.

Table 2. Performance of the five prediction models used (RMSE values).

Models	RMSE
LR	538.65
FT	8.84
SVM	561.13
RQ-GPR	5.31
BTO	**3.69**

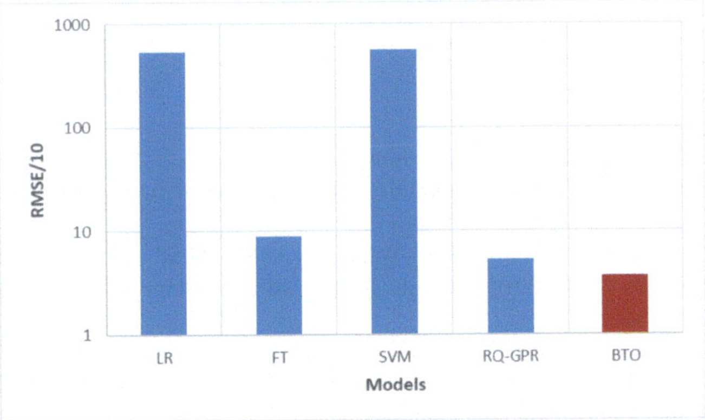

Fig. 6. Comparative histogram of the RMSE values of the five prediction models used.

The results of predicting the theoretical power of wind turbines, expressed by the RMSE values, show significant variations in the performance of the tested regression models. The BTO models stand out for their superior accuracy compared to other models.

These observations indicate that BTO models are particularly effective for this prediction task, thanks to their ability to capture complex nonlinear relationships present in the data.

The remarkable performance of BTO models suggests that they can be extremely useful for practical applications in the wind energy production sector. Their increased accuracy in predicting the theoretical power generated by wind turbines could enable more efficient management of wind farms, optimizing energy production and reducing operational costs.

However, this study has certain limitations. Data preprocessing, including the selection of the first 5000 rows and the removal of the date column, may have affected the results. Future research should explore the impact of different data preprocessing techniques and the selection of diverse datasets.

Furthermore, applying these models to larger and more diverse datasets could offer interesting perspectives and contribute to the generalization of the obtained results.

4 Conclusion

In this study, we evaluated the performance of different regression models to predict the theoretical power generated by a wind turbine using SCADA data. The tested models include LR, FT, SVM, RQ-GPR, and BTO.

Our results show that the BTO model stands out for its high precision, as indicated by its low RMSE value of 3.69 compared to the other models. This precision suggests that the BTO model can be effectively used for practical applications in wind farm management.

However, certain limitations of our study should be noted. Data preprocessing may have influenced our results. Future research should explore the impact of different preprocessing techniques and include larger and more diverse datasets to validate and generalize our findings.

References

1. Sid Amer, Y., Benammar, S., Tee, K.F., Iourzikene, Z.: Failure probability analysis of composite pressure tanks using subset simulation. Mechan. Advan. Compo. Struct. **11**(2), 363–374 (2024)
2. Amer, Y.S., Benammar, S., Tee, K.F., Wadi, M., Jouda, M.S.: A contribution to structural reliability analysis of composite high-pressure hydrogen storage tanks. Journal of KONBiN **53**(2), 33–44 (2023)
3. Sid Amer, Y., Benammar, S., Tee, K.F., Iourzikene, Z.: A study on the small failure probabilities of cylindrical composite hydrogen storage tanks using subset simulation. Proceedings of the Institution of Mechanical Engineers, Part O: Journal of Risk and Reliability, 1748006X241270735 (2024)
4. Zhang, X., Jia, J., Zheng, L., Yi, W., Zhang, Z.: Maximum power point tracking algorithms for wind power generation system: review, comparison and analysis. Ener. Sci. Eng. **11**(1), 430–444 (2023)
5. Liu, J., et al.: Life cycle cost modelling and economic analysis of wind power: A state of art review. Energy Convers. Manage. **277**(August), 116628 (2023)

6. Tidrea, A., Korodi, A.: Elliptic Curve Cryptography Considerations for Securing Automation and SCADA Systems. sensors. MDPI, 2686 (2023)
7. Basholli, F., Mema, B., Hyka, D., Basholli, A.: Analysis of security challenges in SCADA systems, a technical review on automated real-time systems. 8th Advanced Engineering Days (AED), pp. 18–22. Mersin, Türkiye (2023)
8. Gougam, F., Chemseddine, R., Benazzouz, D., Benaggoune, K., Zerhouni, N.: Fault prognostics of rolling element bearing based on feature extraction and supervised machine learning: Application to shaft wind turbine gearbox using vibration signal. Proc. Inst. Mech. Eng. C J. Mech. Eng. Sci. **235**(20), 5186–5197 (2021)
9. Afia, A., Gougam, F., Touzout, W., Rahmoune, C., Ouelmokhtar, H., Benazzouz, D.: Spectral proper orthogonal decomposition and machine learning algorithms for bearing fault diagnosis. J. Braz. Soc. Mech. Sci. Eng. **45**(10), 550 (2023)
10. Iourzikene, Z., Benazzouz, D., Gougam, F.: Edge detection of MRI brain images based on segmentation and classification using support vector machines and neural networks pattern recognition. In: International Conference on Artificial Intelligence in Renewable Energetic Systems, pp. 99–105. Cham: Springer International Publishing (2022)
11. Zhou, L., Luo, Z., Pan, X.: Machine learning-based system reliability analysis with Gaussian Process Regression (2024). arXiv preprint arXiv:2403.11125
12. Janiesch, C., Zschech, P., Heinrich, K.: Machine learning and deep learning. Electron. Mark. **31**, 685–695 (2021)
13. Iourzikene, Z., Gougam, F., Benazzouz, D.: Performance evaluation of feature extraction and SVM for brain tumor detection using MRI images. Traitement du Signal **41**(4), 1967–1979 (2024). https://doi.org/10.18280/ts.410426
14. Yates, L.A., Aandahl, Z., Richards, S.A., Brook, B.W.: Cross validation for model selection: a review with examples from ecology. Ecol. Monogr. **93**(1), 1–24 (2023)
15. Leinonen, T., et al.: Empirical investigation of multi-source cross- validation in clinical machine learning. arXiv, 1–14 (2024)
16. Gougam, F., Afia, A., Soualhi, A., Touzout, W., Rahmoune, C., Benazzouz, D.: Bearing faults classification using a new approach of signal processing combined with machine learning algorithms. J. Braz. Soc. Mech. Sci. Eng. **46**(2), 65 (2024)
17. Gougam, F., Afia, A., Aitchikh, M.A., Touzout, W., Rahmoune, C., Benazzouz, D.: Computer numerical control machine tool wear monitoring through a data-driven approach. Adv. Mech. Eng. **16**(2), 16878132241229314 (2024)
18. Singh, U., Rizwan, M.: SCADA system dataset exploration and machine learning based forecast for wind turbines. Results in Engineering **16**(August), 100640 (2022)
19. Fahim, M., Sharma, V., Cao, T.V., Canberk, B., Duong, T.Q.: Machine learning-based digital twin for predictive modeling in wind turbines. IEEE Access **10**, 14184–14194 (2022)
20. Rodriguez, P.C., et al.: Exploratory analysis of SCADA data from wind turbines using the K-means clustering algorithm for predictive maintenance purposes. Machines **11**(2) (2023)
21. Karaman, Ö.A.: Prediction of wind power with machine learning models. Appl. Sci. **13**(20), 11455 (2023)
22. Zhang, S., Robinson, E., Basu, M.: Wind turbine condition monitoring based on three fitted performance curves. Wind Energy (October 2022), 429-446 (2023)
23. Dataset website: https://www.kaggle.com/datasets/berkerisen/wind-turbine-scada-dataset?select=T1.csv
24. Kaliappan, J., Bagepalli, A.R., Almal, S., Mishra, R., Hu, Y.C., Srinivasan, K.: Impact of cross-validation on machine learning models for early detection of intrauterine fetal demise. Diagnostics **13**(10), 1–22 (2023)
25. James, G., Witten, D., Hastie, T., Tibshirani, R., Taylor, J.: Linear regression. In: An introduction to statistical learning: With applications in python, pp. 69–134 (2023)

26. Liu, J.H., Chen, J.C., Corbita, N.T., Jr.: Analysis and comparison of turbulence models on wind turbine performance using SCADA data and machine learning technique. Cogent Engineering **10**(1), 2167345 (2023)
27. Pandit, R., Xie, W.: Data-driven models for predicting remaining useful life of high-speed shaft bearings in wind turbines using vibration signal analysis and sparrow search algorithm. Energy Science & Engineering **11**(12), 4557–4569 (2023)
28. Yadav, A., Bareth, R., Kochar, M., Pazoki, M., Sehiemy, R.A.E.: Gaussian process regression-based load forecasting model. IET Gener. Transm. Distrib. **18**(5), 899–910 (2024)
29. Huang, H., Jia, R., Shi, X., Liang, J., Dang, J.: Feature selection and hyper parameters optimization for short-term wind power forecast. Appl/ied Intelligence **51**(10), 6752–6770 (2021)

Applying an Optimized and Enhanced ANN-SMOOC for Wind Turbine Chains with DFIG Using PSO Algorithm

Lakhdar Saihi(✉), Fateh Ferroudji, and Mebrouk Bellaoui

Unité de Recherche en Energies Renouvelables en Milieu Saharien (URERMS), Centre de Développement Des Energies Renouvelables (CDER), Adrar, Algeria
sadysaihi@live.fr

Abstract. This study explores an advanced Sliding Mode Order-One Controller with Artificial Neural Network (ANN) for wind turbine chain equipped with a Doubly Fed Induction Generator (DFIG). The control algorithm incorporates an optimization process, such as PSO, to determine optimal gain parameters. The DFIG is connected directly to the grid in the stator and is fed in the rotor by the Converter Side Rotor (CSR). The initial section of the paper presents the model of the wind turbine system, covering both the mechanical and electrical aspects. The subsequent section introduces a robust control strategy to regulate the active and reactive powers (stator powers) of the DFIG. The proposed control method combines Sliding Mode Order-One (SMOO) with Artificial Neural Network control, aiming to reduce the chattering effect and enhance quality of grid compared to conventional SMOO. This approach, referred to as ANN-SMOO, exhibits improved performance and robustness in DFIG control by mitigating chattering effects and improving energy quality. The paper concludes with the presentation and discussion of simulation results for the entire system.

Keywords: DFIG · Artificial Neural Network · sliding mode order-one · chattering phenomenon · ANN-SMOO-C

1 Introduction

Wind energy is recognized as a crucial player in the realm of renewable energy, providing a cleaner alternative to conventional energy sources that contribute to environmental pollution. The availability of wind, a free resource worldwide, positions it as a viable option for harnessing clean energy [1–3]. While many wind turbine systems operate at fixed speeds, the trend towards variable-speed turbines has gained momentum due to numerous advantages. DFIGs are extensively employed in wind turbine systems with variable-speed, offering several benefits in the wind power system [1, 2].

In the field of control systems, various methodologies have emerged to govern the behaviour of Doubly Fed Induction Generators (DFIGs), with one notable approach being the application of Sliding Mode Control of order-one. Recognized as an advanced

© The Author(s), under exclusive license to Springer Nature Switzerland AG 2025
M. Hatti (Ed.): IC-AIRES 2024, LNNS 1238, pp. 43–50, 2025.
https://doi.org/10.1007/978-3-031-80301-7_6

control technique for nonlinear systems [2–4], this method involves a systematic selection of the stability (Lyapunov) function to define the control law governing the system. The core principle revolves around the constructive formulation of the control law for the nonlinear system, treating specific state variables as virtual drivers and developing intermediary control laws [5, 6]. However, a significant drawback of SMOO is the introduction of ripples (chattering) induced by the discontinuous part of sliding mode. To address the issue of ripples and enhance the performance of this classical strategy, one approach is to hybridize it with other techniques, such as the ANN controller [7]. The ANN controller can be applied to reduce the ripple effect and improve the quality of energy in SMOO-C strategy [8]. Adding a layer of sophistication, the parameters of this suggested controller are carefully adjusted using the PSO algorithm [9].

The main goal of this paper is to emphasize the features and advantages of the ANN-SMOO-C for regulating the stator powers of a machine (DFIG) integrated into a horizontal-axis wind turbine. The structure of the study is outlined as follows: the second section presents the models for the mechanical part and the electrical part. The third part is dedicated to ANN-SMOO-C method for the DFIG, optimized through the PSO algorithm. Subsequently, results of simulation are illustrated, and the conclusions are outlined in the final section (Fig. 1).

2 Modeling a Wind Turbine System with DFIG

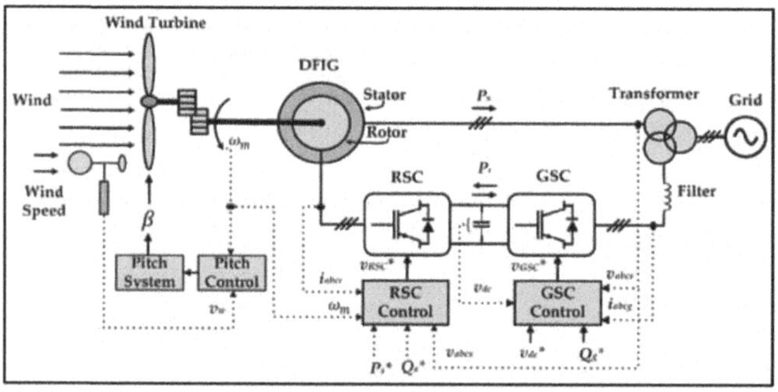

Fig. 1. Wind turbine system configuration.

The wind power is calculated using the following equation [3, 6]:

$$P_t = \frac{1}{2}\rho S v^3 C_p(\lambda, \beta) \tag{1}$$

with; $C_p(\lambda, \beta) = \frac{1}{2}\left(\frac{116}{\lambda_i} - 0.4\beta - 5\right)e^{-\left(\frac{21}{\lambda_i}\right)}$; $\lambda = \frac{r\Omega_t}{v}$.; ρ (kg.m^{-3}): density of air, $S = \pi r^2$, v: speed of wind (m.s^{-1}), $C_p(\lambda, \beta)$: coefficient of power, λ: ratio of speed, β: Pitch angle.

2.1 Mathematical Representation of DFIG

The mathematical representation of DFIG is expressed as [2, 5]:

$$\begin{cases} v_{ds} = R_s i_{ds} + \frac{d\varphi_{ds}}{dt} - \omega_s \varphi_{qs} \\ v_{qs} = R_s i_{qs} + \frac{d\varphi_{qs}}{dt} + \omega_s \varphi_{ds} \\ v_{dr} = R_r i_{dr} + \frac{d\varphi_{dr}}{dt} - (\omega_s - \omega_r)\varphi_{qr} \\ v_{qr} = R_r i_{qr} + \frac{d\varphi_{qr}}{dt} - (\omega_s - \omega_r)\varphi_{dr} \end{cases} \tag{2}$$

The two stator powers (reactive; active) [2, 5]:

$$\begin{cases} P_s = v_{ds} i_{ds} + v_{qs} i_{qs} \\ Q_s = v_{qs} i_{ds} - v_{ds} i_{qs} \end{cases} \tag{3}$$

The control of stator powers in DFIG is accomplished through the rotor currents regulation [4, 6]:

$$\begin{cases} v_{dr} = R_r i_{dr} + l_r \sigma \frac{di_{dr}}{dt} - \left(g\omega_s l_r \sigma i_{qr}\right) \\ v_{qr} = R_r i_{qr} + l_r \sigma \frac{di_{qr}}{dt} + g\omega_s l_r \sigma i_{dr} + g\frac{l_m v_s}{l_s} \end{cases} \tag{4}$$

3 SMOO-C (Siding Mode Order-One) Technique of DFIG

Sliding mode order-one control (SMOO-C) stands out as a highly intriguing nonlinear control approach. This method extends the fundamental SM concept by influencing the first-order time derivatives of the sliding surface. SMC typically involves three key steps [3, 4, 8]: -Selection of the sliding surface (SS), -Conditions for existence (Lyapunov function stability), -Control laws (SMC).

3.1 Control regulations (Laws)

The fundamental concept behind SM-C is to compel the system trajectories to reach a predefined surface, known as the sliding surface, and subsequently maintain their presence on that surface [3, 4, 8]. Designing this control involves three essential steps, starting with:

- **The selection of the SS:**

$$s(t) = \left(\lambda + \frac{d}{dt}\right)^{n-1} e(t) \tag{5}$$

we can formulate the next equations:

$$\begin{cases} e_{P_s} = P_s^{ref} - P_s \\ e_{Qs} = Q_s^{ref} - Q_s \end{cases} \begin{cases} \dot{e}_{P_s} = (\dot{P}_s^{ref} + v_s \frac{l_m}{l_s l_r \sigma}\left(\left(v_{qr}^{eq} + v_{qr}^{n}\right) - R_r \dot{i}_{qr}\right) \\ \dot{e}_{Q_s} = (\dot{Q}_s^{ref} + v_s \frac{l_m}{l_s l_r \sigma}\left(\left(v_{dr}^{eq} + v_{dr}^{n}\right) - R_r \dot{i}_{dr}\right) \end{cases} \tag{6}$$

By using the SM condition, we defined: $S(P) = 0, \dot{S}(P) = 0, v_{qr}^{n} = 0$ and $S(Q) = 0, \dot{S}(Q) = 0, v_{dr}^{n} = 0$. The equivalent and term of commutation control v_{qr}^{eq} and v_{dr}^{eq} (v_{qr}^{n} and v_{dr}^{n}) are get from the next expressions:

$$\begin{cases} v_{qr}^{eq} = -\dot{P}_{s}^{ref}\frac{l_s l_r \sigma}{v_s l_m} + R_r \dot{i}_{qr} \\ v_{dr}^{eq} = -\dot{Q}_{s}^{ref}\frac{l_s l_r \sigma}{v_s l_m} + R_r \dot{i}_{dr} \end{cases} \begin{cases} v_{qr}^{n} = K_{qr} v_{qr} sign(S(P)) \\ v_{dr}^{n} = K_{qr} v_{dr} sign(S(Q)) \end{cases} \quad (7)$$

where; Kv_{qr} and Kv_{dr} must be positive.

- **Conditions of stability** [7, 10]

$$V(x) = S\dot{S} < 0 \quad (8)$$

3.2 ANN-Sliding Mode Order-One Control

To enhance the DFIG performance in the realm of wind power generation, this article focuses on a novel control technique. The proposed technique is a hybrid method that combines ANN and SMOO-C. The structure of ANN-SMOO-C (Fig. 2) [8, 10]:

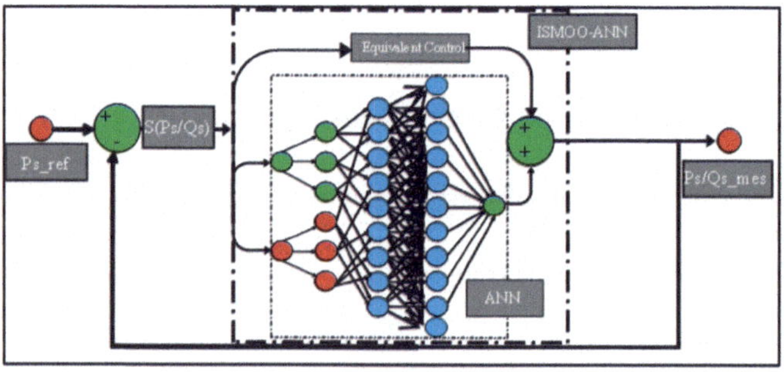

Fig. 2. ANN-SMOO-C diagram.

The ANN-SMOO-C represents a modification of the conventional sliding mode in this adaptation, the switching part, sign(S(x)), has been substituted by an ANN technique, as expressed below:

$$\begin{cases} v_{qr} = v_{qr}^{eq} + v_{qr-ANN}^{n} \\ v_{dr} = v_{dr}^{eq} + v_{qr-ANN}^{n} \end{cases} \quad (9)$$

In recent times, the ANN controller has emerged as one of the most captivating nonlinear control methodologies, offering increased robustness compared to classical nonlinear controllers. It stands as a crucial component within the realm of artificial intelligence, showcasing the ability to emulate human reasoning and holding a prominent position in contemporary research areas [11]. This strategy has gained significant dominance across various industrial domains, and notably, its implementation does not necessitate a mathematical model [12].

3.3 Algorithm of PSO

In 1995, Eberhart and Kennedy introduced PSO as an optimization technique based on populations, taking inspiration from the collective behaviour observed in birds and fish [9].

Note: PSO is a heuristic optimization algorithm that mimics the social behaviour of organisms in nature to discover optimal solutions. It finds applications in various fields for solving optimization problems [9] (Fig 3).

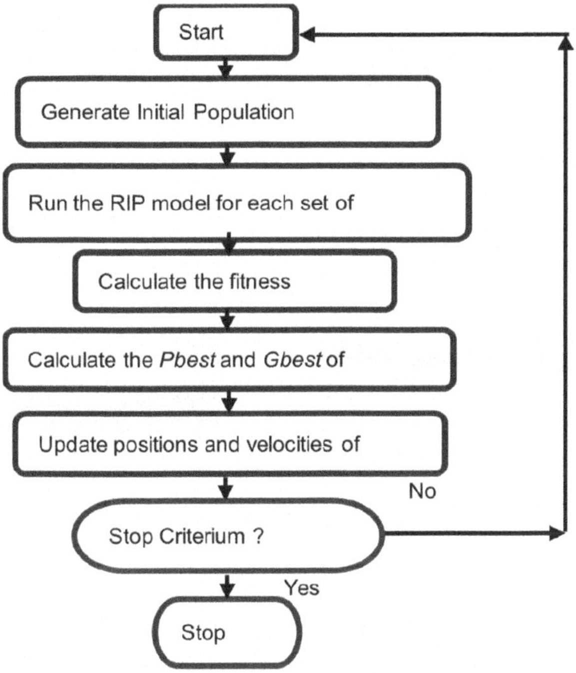

Fig. 3. PSO Flowchart for proposed controller.

4 Simulation Results

The control methodologies utilized to govern the stator powers of the DFIG consist of ANN-ISMOO-C and the traditional SMOO, both optimized through PSO algorithm. These control techniques are executed within the MATLAB-Simulink environment, providing a platform for the assessment and comparison of their performance.

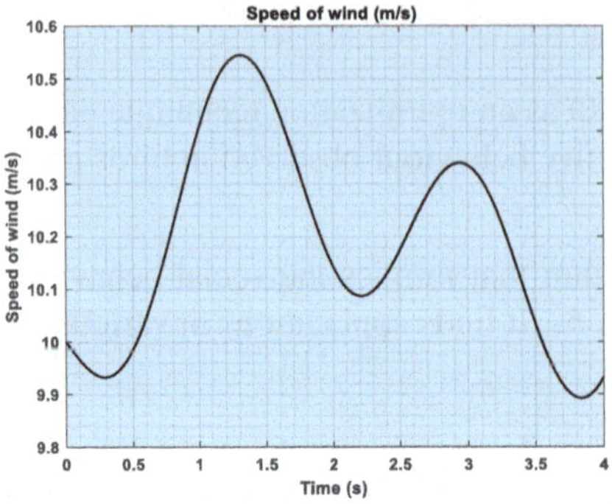

Fig. 4. Wind speed of DFIG.

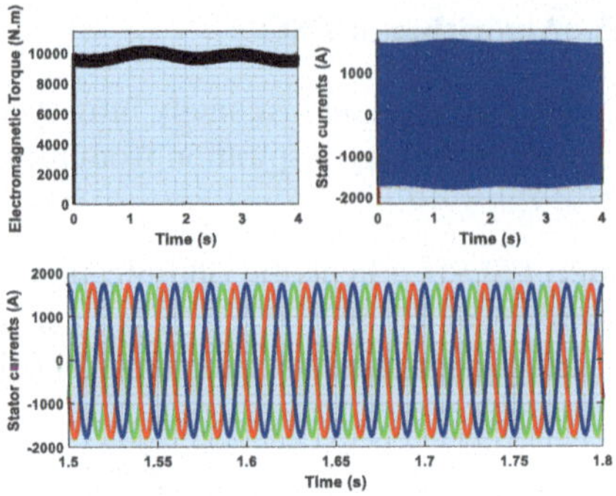

Fig. 5. Stator currents&electromagnetic torque of DFIG.

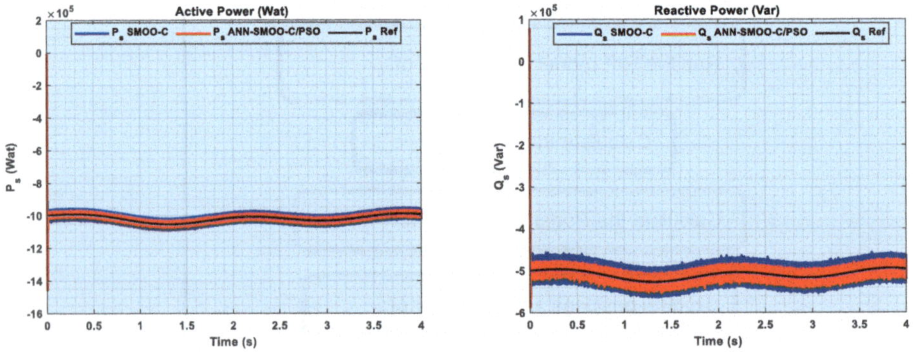

Fig. 6. Active power and reactive power.

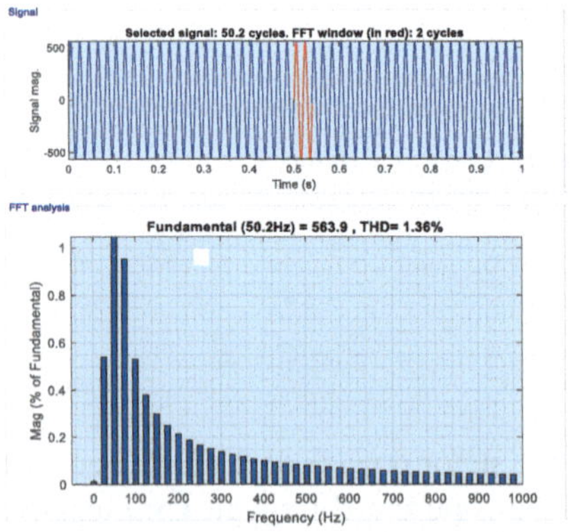

Fig. 7. THD of stator currents with SMOO-C.

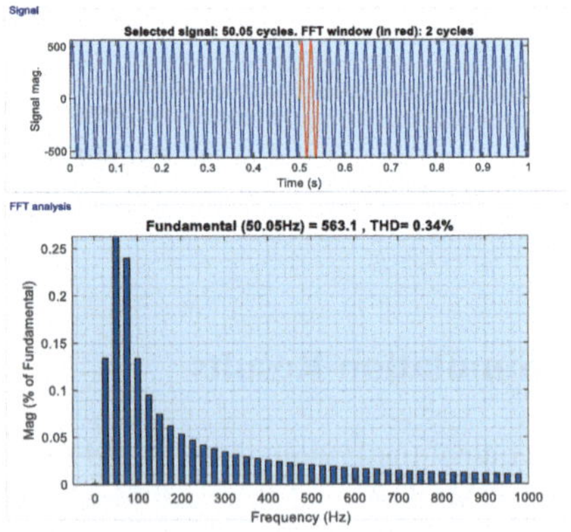

Fig. 8. THD of stator currents with ANN-SMOO-C.

Figure 4 displays the speed of wind in meters per second (m/s) within a range of ± 10 m/s, while Fig. 5 illustrates the behaviour of the currents of stator and electromagnetic torque of the machine under the ANN-SMOO control strategies. These variables are influenced by variations in wind speed. Furthermore Fig. 6 showcases the powers of stator of the machine regulated by using the proposed ANN-SMOO-C (red line) and SMOO-C (blue line), along with their references (black line). The obtained results demonstrate that the proposed ANN-SMOO-C controller exhibits superior effectiveness in terms of reference tracking (time response, overshoot), sensitivity to disturbances, and mitigation of ripple phenomena compared to the conventional SMOO-C, under the variations of reference. Additionally, Fig. 7 depicts a significantly improved THD value with SMOO-C (1.36%) compared to Fig. 8 illustrates THD of ANN-SMOO-C (0.34%), indicating an enhanced quality of energy generated by the wind turbine based on DFIG.

5 CONCLUSION

This research introduces the utilization of ANN-SMOO-C for a horizontal axis wind turbine system coupled with a DFIG. The DFIG is directly linked to the network through the stator and powered by the Converter Side Rotor (CSR), with its parameters optimized using the PSO algorithm. The success of this optimization relies on the careful selection of the algorithm and precise tuning of its parameters to achieve optimal practical outcomes.

The simulation results obtained with the novel ANN-SMOO-C were compared against the classical sliding mode technique. The comparison reveals that the ANN-SMOO technique showcases enhanced efficiency in tracking time-varying trajectories. Moreover, it demonstrates a significant reduction in chattering effects, leading to an improved quality of grid, especially during variable speed conditions.

References

1. Saihi, L., Berbaoui, B., Glaoui, H.: Robust control H∞ fuzzy of a doubly fed induction generator integrated to wind power system. Majlesi J. Elect. Eng. **14**(1), 59–69 (2020)
2. Bakou, Y., et a.: Design of robust control based on R∞ approach of DFIG for wind energy system. 1st Global Power, Energy and Communication Conference (GPECOM), pp. 337–341. IEEE (2019)
3. Saihi, L., et al.: Hybrid control based on sliding mode fuzzy of DFIG power associated WECS. In AIP Conference Proceedings **2123**(1), 030015 (2019)
4. Saihi, L., et al.: A comparative study between robust control sliding mode and backstepping of a DFIG integrated to wind power system. 2019 7th International Conference on Smart Grid (icSmartGrid), pp. 137–143 (2019)
5. Saihi, L., et al.: Super twisting high order sliding mode control of vertical axis wind turbine with direct attack based on doubly fed induction generators. In: Hatti, M., (eds.) Smart Energy Empowerment in Smart and Resilient Cities. ICAIRES 2019. Lecture Notes in Networks and Systems, vol 102. Springer, Cham (2020)
6. Echiheb, F., et al.: Robust Sliding-Backstepping Mode Control of a wind system based on the DFIG Generator (2022)

7. Djilali, L., Badillo-Olvera, A., Rios, Y.Y., López-Beltrán, H., Saihi, L.: Neural high order sliding mode control for doubly fed induction generator based wind turbines. IEEE Lat. Am. Trans. **20**(2), 223–232 (2021)
8. Zhou, M., Su, H., Liu, Y., Cai, W., Xu, W., Wang, D.: Full-order terminal sliding-mode control of brushless doubly fed induction generator for ship microgrids. Energies **14**(21), 7302 (2021)
9. Lakhdar, S., Ferroudji, F., Roummani, K., Koussa, K.: Enhanced sensor-less control of DFIG-based generators using hybrid MRAS-ANN observer and PSO parameter optimization. In: 2023 14th International Renewable Energy Congress (IREC), pp. 1–6. IEEE (2023)
10. Yaichi, I., Semmah, A., Wira, P.: Neuro-second order sliding mode control of a DFIG based wind turbine system. J. Electr. Electron. Eng. **13**(1), 63–68 (2020)
11. Saihi, L., Berbaoui, B., Glaoui, H., Djilali, L., Abdeldjalil, S.: Robust sliding mode H∞ controller of DFIG based on variable speed wind energy conversion system. Periodica Polytechnica Electrical Engineering and Computer Science (2019)
12. Owolabi, O.I., Madushele, N., Adedeji, P.A., Olatunji, O.O.: FEM and ANN approaches to wind turbine gearbox monitoring and diagnosis: a mini review. J. Reliable Intel. Environ. **9**(4), 399–419 (2023)

Parallel Photovoltaic Inverters Equipped Active Power Filters: Autonomous Mode and Grid Connection Mode Operations

Issam Abadlia(✉), Fateh Abdoune, Mohamed Rida Bengourina, and Linda Hassaine

Centre de Développement des Energies Renouvelables, CDERBP. 62 Route de l'Observatoire Bouzareah, 16340 Algies, Algeria
i.abadlia@cder.dz

Abstract. This paper proposes the study of a microgrid system based on photovoltaic sources capable of ensuring the operation in autonomous mode and grid connection mode considering the optimal power flow control using the Droop controller and the power quality improvement using an Active Power Filter. The proposed microgrid is composed of parallel individual PV inverters controlled in Droop mode. The second PV inverter is combined with an active power filter used to improve the energy quality. The simulation results, in different scenarios of operation of the system, indicate that the proposed system with the control strategies is an ideal microgrid that ensures optimal power flow control, the power sharing and is capable of improving energy quality.

Keywords: Parallel Photovoltaic Inverters · Active power filter · Microgrid · Power sharing · Droop control

1 Introduction

Distributed generation is a term commonly used to describe small producers of renewable energies (RE) located as close to users as possible. It continues to gain many applications in modern power system engineering. It thrives as advancements in distributed energy technologies enhance the subsequent implementation of distributed generation systems on Mini/micro grids [1–3]. On the other hand, harmonic pollution affects all domestic and industrial networks. No modern environment can escape this pollution, such as processers, servers, air conditioners, etc. This equipment generates harmonic currents, which cause a degradation of the electrical network [4]. The grid current and voltage quality are significantly degraded.

© The Author(s), under exclusive license to Springer Nature Switzerland AG 2025
M. Hatti (Ed.): IC-AIRES 2024, LNNS 1238, pp. 51–59, 2025.
https://doi.org/10.1007/978-3-031-80301-7_7

The combination of an Active Power Filter (APF) and a photovoltaic (PV) source, is not only a RE source, but also spotless, unlimited, and with a very low level of risk [5]. This combination gives us a RE sources and effectively improves the energy quality (EQ). The APF injects a current that opposes the harmonic current emitted by the nonlinear load to mitigate the effect of harmonic currents [6]. Thus, the current delivered by the power source remains sinusoidal with a Total Harmonic Distortion (THD) rate in accordance with the recommendation of the local grid code [7].

In this paper, a microgrid system composed of a parallel PV inverter integrated the APF is proposed. The microgrid system is capable to ensuring the operations of isolating and grid connected mods, the active (P) and reactive (Q) power sharing based on the Droop controller (DCL), the power flow control and the improvement of the energy quality.

2 Model and Topology of the Proposed Microgrid System

Figure 1 shows the model of the proposed microgrid system (MGs), executed in the Simulink interface of Matlab. The topology of the MG consists of two parallel inverter systems (PIs) controlled in Droop mode. Each inverter combines an individual PV source with a DC/DC Boost converter controlled via an incremental conductance INC-MPPT algorithm. The PIs is connected to the local loads devised into linear and non-linear loads using the LCL filters, capable also of integrating into the grid. The inverter system 2 is equipped with an APF in order to improving the EQ.

The proposed power system with their control strategies is capable of ensuring many functionalities, such:

1. Creation of an ideal microgrid, using the photovoltaic sources only and the parallel inverters, allowing:

   a)Extraction and injecting of maximum photovoltaic power;
   b)Operation in autonomous local mode and in the grid connection mode.

2. The P/Q sharing using the DCL.
3. Control and improvement of EQ by integrating the APF to eliminate unwanted harmonics.

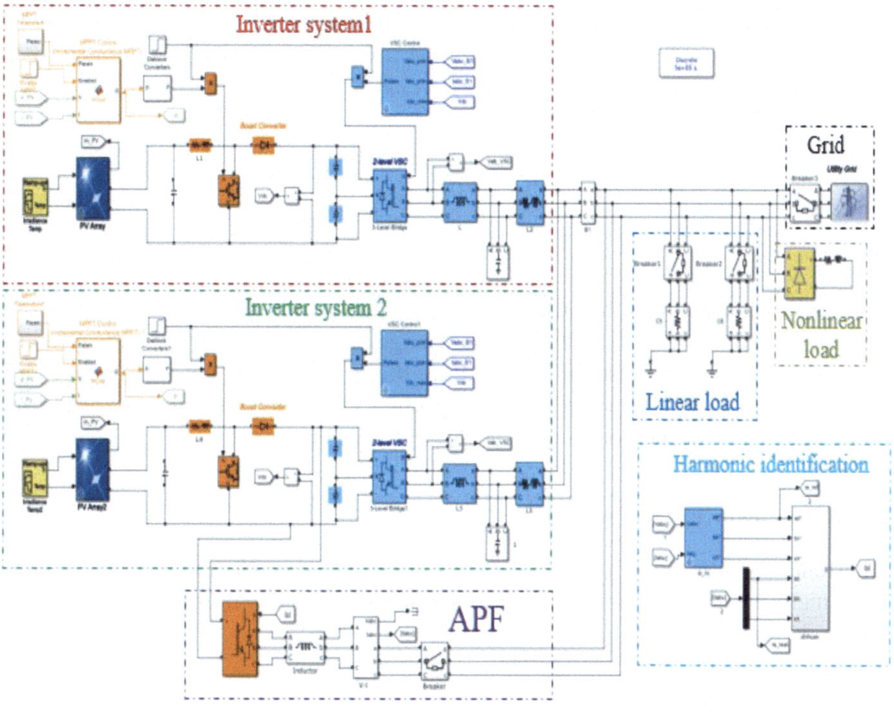

Fig. 1. Proposed microgrid model

3 Control Strategies of the Microgrid System

In this section, the control strategies of the MGs are presented. The production system is composed of a power PV source with a DC/DC Boost converter controlled using the INC-MPPT algorithm. The Droop control is investigated for the power system sharing. An APF is collected in parallel with the second inverter to improve the power quality.

3.1 Proposed Droop Controller of the Parallel Inverter

Parallel inverter systems are equipment that allows several UPSs to be placed in parallel in its installation or IT infrastructure to increase the availability and reliability of its equipment via taking over protection through protection redundancy. In the electrical MG, frequency is a global system variable, and the corresponding P of distributed generation can be allocated more precisely according to the droop curve. Since the output voltage is a local variable, the traditional droop controller [8] cannot accurately allocate the corresponding reactive power. It is therefore necessary to optimize the power allocation of the output. Conventional DCL has poor performance for MGs with complex impedance lines [9]. Which, voltage control by cascaded loops allows explicit limitation of the inverter output current. However, the saturation of the reference current renders the control of the voltage and power drop necessary for synchronization ineffective. As a result, the system loses its stability after a large transient. To improve the dynamic response and the power control of the MGs, some modified DCLs should be used. In this work, we proposed a droop adjusted by the pole placement method defined by the following P and Q equations:

$$f = f* - m(P - P*) + m_k \frac{dP}{dt} \tag{1}$$

$$U = U* - n(Q - Q*) + n_k \frac{dQ}{dt} \tag{2}$$

where: f is the frequency of the output voltage of the inverter system, f* is the reference frequency of the microgrid system, U is the voltage of the microgrid, U* is the reference voltage of the microgrid system, m and n are the droop coefficient of frequency and voltage droop coefficient in droop control respectively. Moreover, mk and nk are adjusted by the pole placement method to improve the control performance.

3.2 Active Power Filter Control

The current control strategy of the APF/PV system includes a harmonic current identification method and an inverter control loop with a DC bus control for the DC voltage input regulation.

3.2.1 Harmonic Current Identification Algorithm

The identification method allows us to calculate the disturbing currents, which will be injected by the inverter through an output filter, at the connection point in phase opposition. Therefore, the network side becomes clean, with sinusoidal current and voltage waves [10]. The method of instantaneous real and imaginary powers [11] is the most commonly used identification method. It offers the advantage of compensating harmonic currents, unbalanced currents and Q, precisely, quickly and easily. In addition, this identification method makes it very easy to integrate the INC algorithm when the DC voltage at the regulator output is the maximum power of the PV system (P-MPP) as shown in Fig. 2.

3.2.2 Current Control Algorithm

The current control loop is designed to ensure high precision tracking between the identified current reference and the injection current. Using the proposed structure and control strategy, a single PI controller is used to monitor between the injected and the reference currents although the APF/PV compensates for all current disturbances (up to 2500 Hz) including P generated at the fundamental frequency.

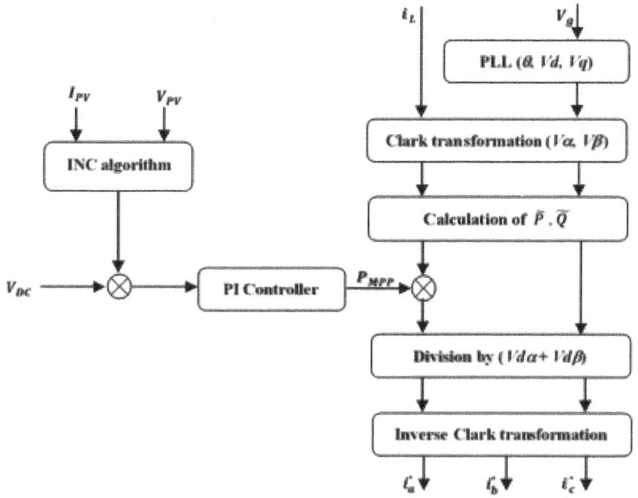

Fig. 2. Extended instantaneous power identification method.

3.2.3 DC Bus Control and Regulation

For the regulation of the DC bus voltage of the APF input, we will propose a proportional integral (PI) controller, with a pre-filtration action on the measured voltage Vdc of the second PV inverter system, this operation is carried out by a second order low pass filter as illustrated in the Fig. 4. For the regulation of the dc bus voltage of the APF input, we will propose a pi controller, with a pre-filtration action on the measured voltage Vdc of the second PV inverter system, this operation is carried-out by a second order low pass filter as illustrated in the Fig. 3.

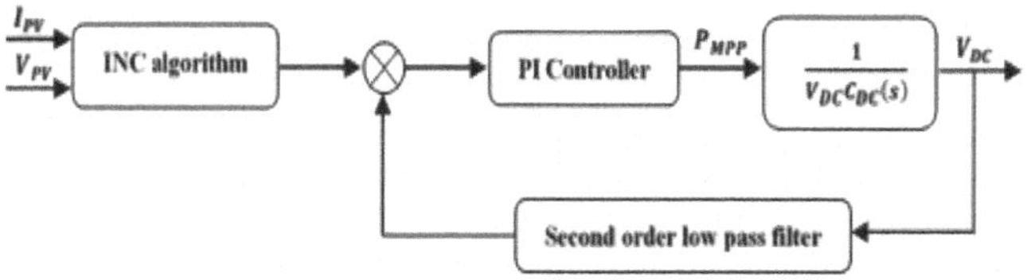

Fig. 3. DC bus regulation loop.

4 Simulations and Results

Table 1 gives the system simulation parameters used in the MGs.

Table 1. Parameters of the MGs simulation model.

System	Parameter and unit	Observations
PV inverter system 1	LCL filter (L1 = 0.1 mH, L2 = 0.2 mH, C = 25 μF)	Injected PV power Controlled on Droop
	Boost converter (L = 1 mH, C = 25 μF) VDC = 363 V, Ppvmax = 3 KW	
PV inverter system 2	LCL filter (L1 = 0.1 mH, L2 = 0.2 mH, C = 25 μF)	Injected PV power Controlled on Droop
	Boost converter (L = 1 mH, C = 25 μF) VDC = 363 V, Ppvmax = 3 KW	
APF	L = 2.5 mH	Harmonic identification + PWM control
Linear load	R1 = 10 Ω, R2 = 10 Ω	
Nonlinear load	L = 1 mH, R = 10 Ω + Rectifier	
Grid	V = 180 V (RMS), r = 0.05 Ω, l = 3 mH	
Droop coefficients	m = 0.005, n = 0.00038	
	mk = 0.00013, nk = 0.000021	
PI (DC-APF feedback regulation)	KP = 0.83, KI = 5.18	

The microgrid with the proposed control strategies is simulated in various scenarios of operations, in autonomous mode adopting nonlinear load without using APF and with using the APF, addition of the linear loads, and in the grid connection. Table 2 explains the scenarios of the operation microgrid system.

Table 2. Proposed scenarios operation of the microgrid system.

Time(s)	0→0.2	0.2→0.4	0.4→0.6	0.6→0.8	0.8→1
Mode of operation	Autonomous mode operation With only nonlinear load Without using the APF	Autonomous mode operation With only nonlinear load With using the APF	Autonomous mode operation With nonlinear load With linear load 1 With using the APF	Autonomous mode operation With nonlinear load With linear load 1 With linear load 2 With using the APF	Grid connection mode operation With nonlinear load With linear load 1 With linear load 2 With using the APF

Figure 4 shows the active and the reactive power sharing in the microgrid system. The first inverter system is participated with 67% of the total power delivery to the loads and the second inverter participate with the 33%. The adopted Droop controller strategy prove the efficiency of the power sharing.

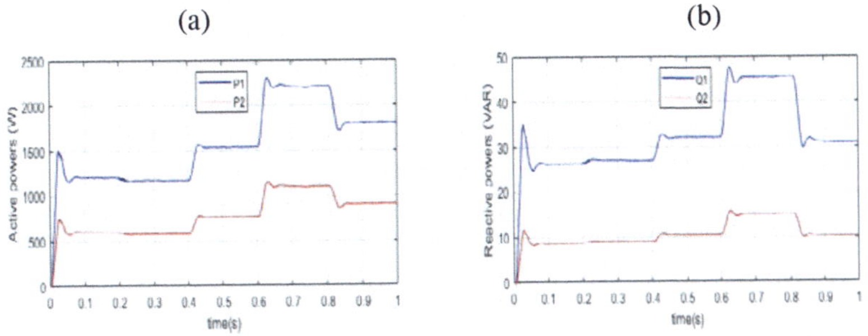

Fig. 4. (a): system active powers sharing, (b): system reactive powers sharing.

Figure 5 shows the line voltage in the common coupling point PCC of the inverters. It observe that a drop in voltage caused by the insuring of the load but this drop is <5% it is very satisfied to the local grid norms applications.

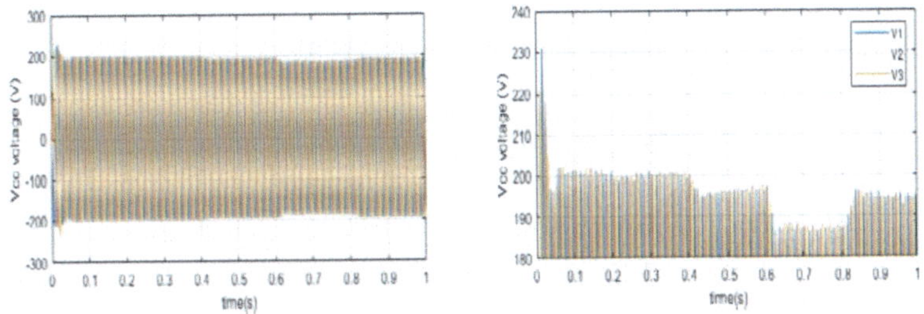

Fig. 5. Voltage on the PCC.

Figure 6 shows the line current. That prove the improvement of the energy quality with using the APF in all scenarios of the system operation.

Table 1 indicate the THD values of these scenarios. It observe that the potential of the APF. Without using the APF, the THD is equal to 21.30% and with using the APF the THD<4% educate and very satisfied with the recommended local THD (Table 3).

Figure 7 shows a good follows of the DC bus voltage of the APF with good regulation and compensation of any perturbations.

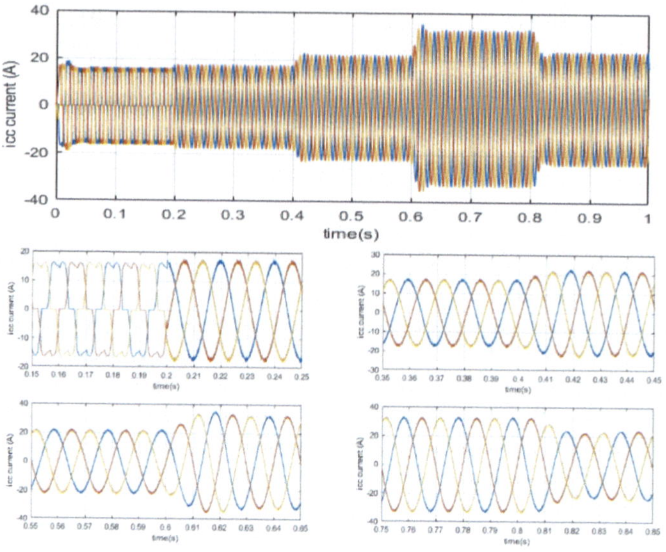

Fig. 6. Line current on the PCC.

Table 3. THD of the line current.

Time(s)	0→0.2	0.2→0.4	0.4→0.6	0.6→0.8	0.8→1
THD (%)	21.30	3.49	2.88	1.73	2.65

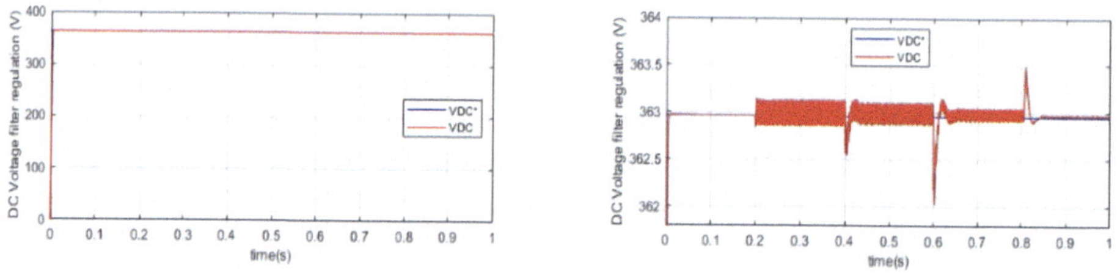

Fig. 7. DC bus regulation of the APF.

5 Conclusion

In this paper, a DCL based on pole placement is introduced to exploit the flexibility and fast dynamics of parallel inverters. In the proposed microgrid scheme, an APF is meeting in parallel with the second inverter system to improve the power quality. The simulation results verify the accuracy and feasibility of the proposed control strategy for power sharing and energy quality improvement.

References

1. Kumar, A., Bhadu, M.: A comprehensive study of wide-area damping controller requirements through real-time evaluation with operational uncertainties in modern power systems. IETE **69**(11), 8382–8403 (2023)
2. Pijarski, P., Kacejko, P., Miller, P.: Advanced Optimisation and Forecasting Methods in Power Engineering—Special Issue. Energies **16**(6), 2804 (2023)
3. Abdoune, F., et al.: Analysis and design of voltage-controller based on single-state feedback active damping for grid-forming converters. PPEECS **68**(1), 54–63 (2024)
4. Gopalapillai, A., Fernandez, F.M.: Quantification of harmonic pollution using non-fundamental apparent power. Electrical Engineering, 1–16 (2024)
5. Eslami, A., et al.: Review of AI applications in harmonic analysis in power systems. Renewable and Sustainable Energy Reviews **154**, 111897 (2022)
6. Taghvaie, A., et al.: A comprehensive review of harmonic issues and estimation techniques in power system networks based on traditional and artificial intelligence/machine learning. Ieee Access **11**, 31417–31442 (2023)
7. Beddar, A., et al.: Investigation and implementation of a nonlinear decoupled power control for a single-phase grid- connected via an LCL filter. JER **11**(1), 100010 (2023)
8. Tayab, U.B., et al.: A review of droop control techniques for microgrid. Renewable and Sustainable Energy Reviews **76**, 717–727 (2017)
9. Xiao, J., et al.: An inertial droop control based on comparisons between virtual synchronous generator. Energy Reports **6**, 104–112 (2020)
10. Alali, M.A.E., et al.: Grid-connected shunt active photovoltaic filter. 2020 IEEE 29th International Symposium on Industrial Electronics (ISIE). IEEE (2020)
11. Bitoleanu, A., Popescu, M.: Shunt active power filter overview on the reference current methods calculation and their implementation. 2013 4th (ISEEE). IEEE (2013)

Comparative Analysis of MPPT Algorithms: P&O and Inc for Optimizing PV Systems

M. Belabed[1(✉)], S. Bechekir[1], M. Brahami[1], H. Bendaho[2], A. Brahimi[2], and I. S. Bousmaha[3]

[1] ICEPS Laboratory, Djillali Liabes University of Sidi Bel Abbès, Sidi Bel Abbès, Algeria
meriem.belabed@univ-sba.dz
[2] Salhi Ahmed University Center of Naâma, Naama, Algeria
[3] N ESSA-Tlemcen, Higher School of Applied Sciences of Tlemcen, BP 165 RP Bel Horizon, 13000 Tlemcen, Algeria

Abstract. Photovoltaic energy is a key component of the field of renewable energy. Methods for Maximum Power Point Tracking (MPPT) are critical for optimizing photovoltaic energy generation, which relies on meteorological conditions. This paper focusses on Two main Maximum Power Point Tracking methods approaches: Observe and Perturb (P&O) MPPT control Incremental conductance (InC) in electrical systems. An analysis of the P&O control system reveals that it is sluggish and struggles to accurately monitor maximum power point (MPP). Furthermore, it can distort the point of maximum power (MPP) during sudden fluctuations in air conditions. The INC algorithm demonstrates exceptional efficiency in monitoring several operating sites. The objective of this essay is to examine and contrast two distinct strategies of tracking of Maximum Power Points (MPPT). So as to optimise power extraction in photovoltaic (PV) systems, employment of maximum power point tracking is essential. We evaluate and understand the simulation results of these approaches using the Matlab Simulink environment.

Keywords: Photovoltaic energy · Tracking of Maximum Power Points (MPPT) methods such as Observe and Perturb and Incremental Conductance · implemented in Matlab/Simulink

1 Introduction

Photovoltaic technology transforms light energy into electricity to tackle challenges in energy generation using photovoltaic devices. [1–3] Research continues on solar energy-harvesting cells and systems. Solar cells (PV) semiconductors are materials that transform visible electrical conversion of light, often termed solar energy generators. [4, 5] The performance of a photovoltaic system depends on factors like solar irradiance, ambient temperature, and electrical demand. [6, 7]

MPPT algorithms optimize power output by continuously adjusting up to the point of maximum power of solar modules given different circumstances. [8] Factors such as sunlight intensity, temperature, and cell age directly affect solar cell performance, while the

© The Author(s), under exclusive license to Springer Nature Switzerland AG 2025
M. Hatti (Ed.): IC-AIRES 2024, LNNS 1238, pp. 60–70, 2025.
https://doi.org/10.1007/978-3-031-80301-7_8

connected load determines the system's operating point. The choice of MPPT algorithm significantly impacts the enhancement of efficiency and performance of solar systems. [6, 9] Researchers have developed various MPPT Methods of control to enhance efficiency and reduce costs, but challenges remain, including slow tracking, MPP oscillations, and suboptimal efficiency, Given that Perturb and Observe (P&O) and incremental conductance are the prevailing techniques. [10–12] To improve efficiency, the objective of this effort is to optimise the monitoring of the application of Maximum Power Point Tracking (MPPT) in photovoltaic generators (GPV). It focuses on employing MPPT techniques like Perturbation-Observation (P&O) technique and Incremental Conductance (Inc) theory, comparing these commonly used approaches in solar systems. The study will detail the simulation of the photovoltaic System implemented with Matlab/Simulink, allowing for easy modification of parameters such as meteorological conditions and visualization of the control input on the photovoltaic generator. [13, 14]

The work is organised by a clear as follows: after this introduction, I shall undertake. First show the PV cell model, the photovoltaic module, and the typical MPPT control algorithms P&O and INC. Subsequently, a presentation of simulation findings will be provided, followed by a discussion.

2 Photovoltaic Systems

Before achieving Symmetrical control of an optimal system, it is crucial to establish its mathematical formulation. In this particular section, we have duplicated the solar energy source for the purpose of production. [15].

2.1 Solar Cell Equivalent Model

Modeling solar cells requires the utmost precision and meticulousness. For an accurate simulation of solar cells, it is critical to carefully select the appropriate electrical equivalent circuits. To faithfully reproduce the solar circuit diagram cell, a profound comprehension in the physical arrangement of its components and the electrical properties of each constituent is essential, while considering various levels of precision. Various mathematical models have been devised to depict very nonlinear systems.

By virtue of its simplicity and accuracy in computing voltage and current in a solar cell, this study uses the mono-diode model for the circuit [16, 17] (Fig. 1).

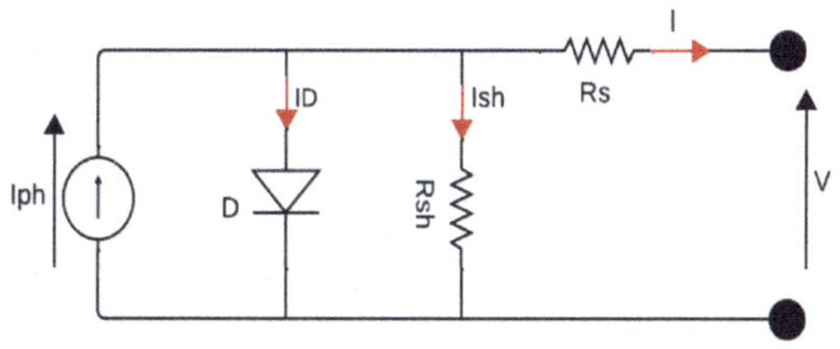

Fig. 1. Equivalent model of a PV cell

Therefore, the equation representing the curve I = f(V), which defines the solar cell, is:

$$I = I_{ph} - I_s \left[e^{\left(\frac{q(V+I.R_{s)}}{A.K..T} \right)} - 1 \right] - \frac{V + I.R_s}{R_{sh}} \quad (1)$$

The cell's photocurrent is generated by the saturation current of the light, where A is the diode's ideality factor. Constant k represents absolute Boltzmann constant (1.3806531 × 0–23 J/K^{-1}), q denotes the absolute values magnitude measurement of the electron's charge (1.60217646 × 10–19 C) and T stands for the junction temperature. This thermal condition is typically considered to be in close proximity to the cell's own temperature. Nevertheless. The thermal voltage (VT) is designated as the ratio kT/q. [18].

2.2 Synoptic Diagram

Figure 2 displays the synoptic visual representation of the PV system fitted with the MPPT controls. The system consists of many blocks:

- The monocrystalline silicon PV panel generates 135.3 W of power, with a current of 4.1 A and a voltage of 33 V, when tested under standard conditions (STC).
- A power block formed by a Boost converter serves as a DC/DC energy application-specific converter that require a voltage above the ideal voltage of a photovoltaic panel.
- An MPPT controller facilitates the monitoring of the maxima power point of the solar panel, which is affected by changes in weather conditions and load. [19]

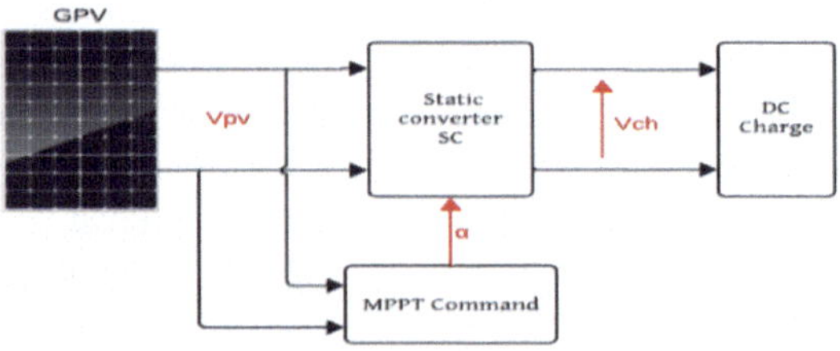

Fig. 2. Synoptic diagram of the system studied.

2.3 DC-DC Static Converter

The boost converter, sometimes referred to as the step-up converter, is the optimal DC-DC converter design for this particular application owing to its simplicity, cost-effectiveness, and higher performance. A precise equation specifies the duty cycle of the switch, which determines the correlation among the voltages of input and output of the well-designed boost converter:

$$V_s = \frac{1}{1-\alpha} V_e \tag{2}$$

Since $\alpha \in [0, 1]$, the boost converter consistently delivers an output voltage of higher magnitude than its input voltage. [20] .

3 Power Optimization Method

Precision control of Maximum Power Point Tracking (MPPT) is essential for optimizing effic in solar systems. This method automatically adjusts the duty cycle to achieve the optimal power generation from photovoltaic (PV) panels. The following sections will present and analyze these control techniques. [21].

3.1 Perturb and Observe "P&O"

Often utilized in the commercial goods industry, the P&O method is the fundamental basis for other elaborate algorithms detailed in academic writing. It is often used in practical applications. The P&O method operates intermittently through perturbation the operating voltage (V) and monitoring the power fluctuation to determine the adjustment needed for the reference voltage Vref. [22] It performs well with gradual changes in irradiation, but the P&O approach is ineffective in tracking the PPM when the radiation abruptly shifts due to a sluggish dynamic response. [23] This approach relies on calculating the photovoltaic output Power and monitoring power fluctuations are measured by sampling the current and PV voltage. [24] This command fonctionnes by altering the α cyclic ratio to induce disturbances and then analysing the impact on the power output of the GPV, as shown in Fig. 3. [25].

Initially, the voltage V and current I are measured to determine the power P(K). Compare the current P(K) value to the power value acquired during the previous measurement, $P(K^{-1})$. If the power output from the panel has increased since the last measurement, the cyclical ratio α will either continue to rise or fall in the same orientation as the preceding thematic cycle, as indicated by the sur dV test. [26].

3.2 Incremental Conductance "INC"

This approach is based on knowing the conductivity value, $G = I/V$, and the change in conductance (dG) to identify the spatial location of the point of operation relative to PPM. [27] Fig. 4 displays the organizational structure illustrating the functionality of this algorithm. The process starts by measuring the current voltage and current of the GPVs. Subsequently, incremental variations, dI and dV, are computed based on the current value and the prior and current voltage values. The primary control is executed using the power derivative equation relationships. If the rise in conductivity (dG) exceeds the negative conductance (-G), the cyclic ratio decreases. If the increased conductivity is smaller than the inverse of the conductance, the cyclical ratio will grow.

This procedure is utilized until the PPM is reached. At the end of the cycle, it records calculate the voltage and current values for the upcoming cycle [28].

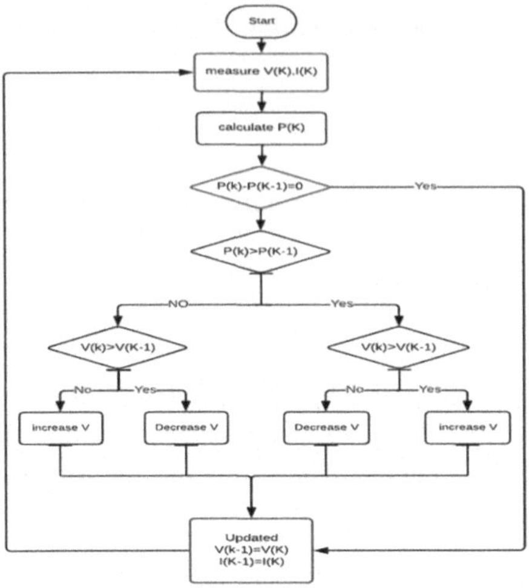

Fig. 3. The flowchart of the P&O algorithm.

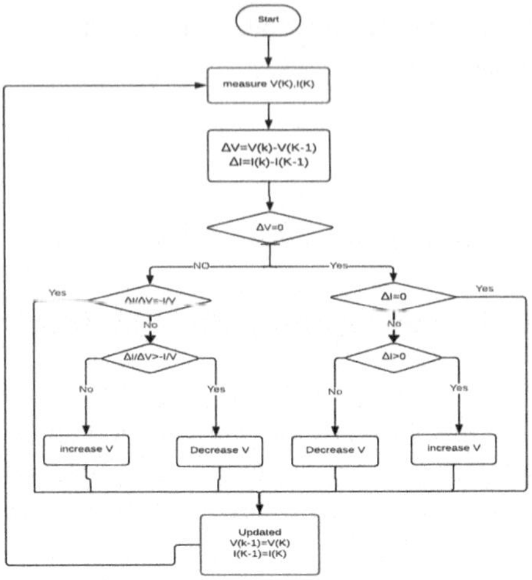

Fig. 4. The flowchart of the INC algorithm.

4 Simulation Results

Various types of DC-DC systems have utilized a range of tracking algorithms. This study focused on the techniques of "perturb and observe" and "incremental conductance". Selecting the appropriate converter is a crucial first step for effective Max Power Point Tracking (MPPT) analysis management. The boost-type DC-DC converter facilitates consistent tracking of the Power Profile Modulation (PPM), enabling easy adjustments of system parameters, such as meteorological conditions, and allowing visualization of control inputs on the photovoltaic generator. [6].

To simulate the photovoltaic system, we used the software Matlab/Simulink and the behaviour exhibited by the GPV is based on the mathematical model.

4.1 Simulation of the Photovoltaic System

Figure 5 illustrates the blocking scheme of our photovoltaic system, including the simulated floors: GPV, DC-DC, and MPPT. The simulation utilized a PVL136 solar module and a Boost cutting machine with specific component values: C1 = 3000e-6F, C2 = 2e-3F, L = 0.01 H, R = 0.01 Ω, and a charge resistor of 100 Ω.

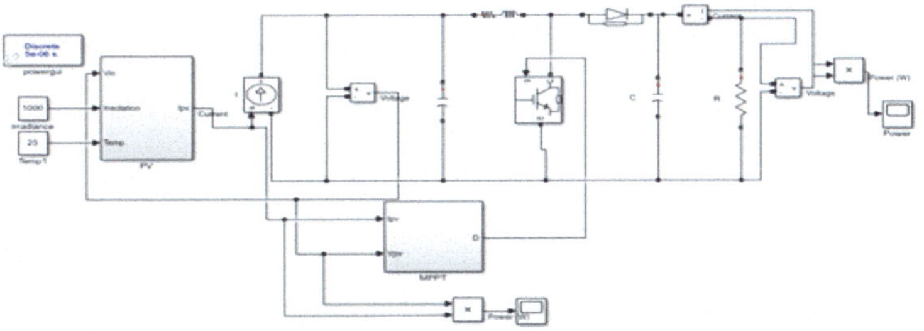

Fig. 5. Simulink block diagram of the overall system.

4.2 Simulation of Known MPPT Algorithms

The simulation models for the P&O and Inc. Algorithms are displayed as follows in the MATLAB software (Figs. 6 and 7).

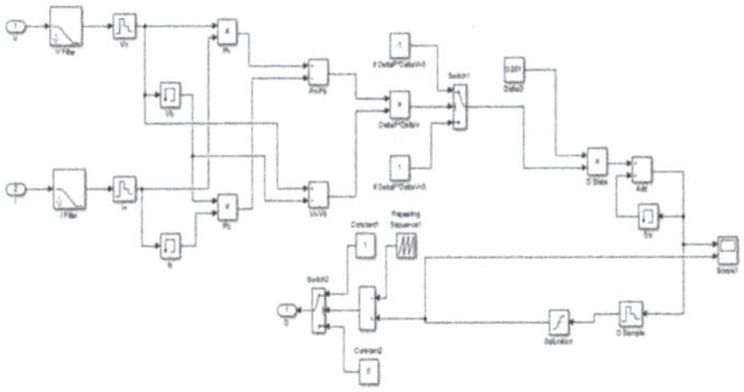

Fig. 6. Perturbation algorithm and observation simulation scheme (P&O).

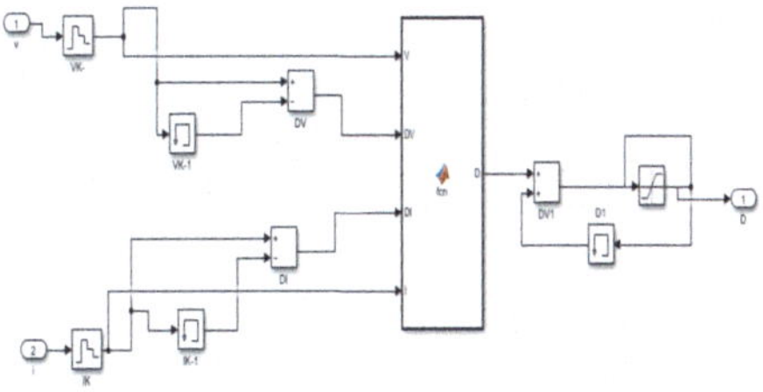

Fig. 7. Command INC. Block scheme.

4.3 Command INC. Block Scheme

1. Algorithme P&O

The figures display the power and current of a photovoltaic system (PVL136 with a power output of 135.3 W) regulated by a Perturb and Observe (P&O) controller under consistent conditions of 25 °C temperature and 1000 W/m^2 light (Figs. 8 and 9).

a. Temperature variation and Irradiation

To validate the P&O algorithm, we tested our controller for various irradiation and temperature values depending on the perturbation step. Figure 10 shows a comparison of power curves given a consistent irradiation of 1000 W/m^2 and a variable temperature [0 25 50] °C.

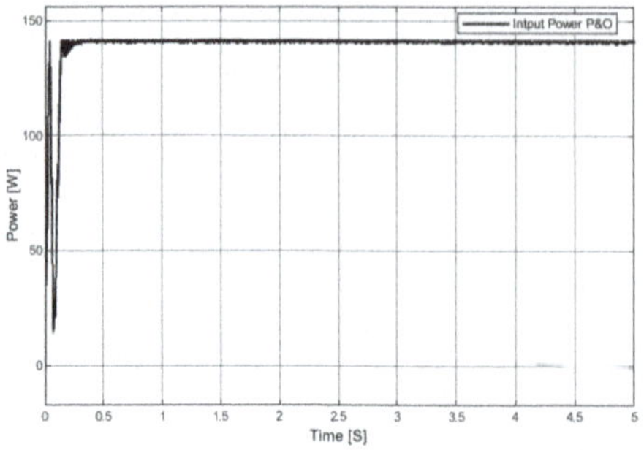

Fig. 8. Input power curve depending on time for (T = 25 °C, G = 1000 W/m^2).

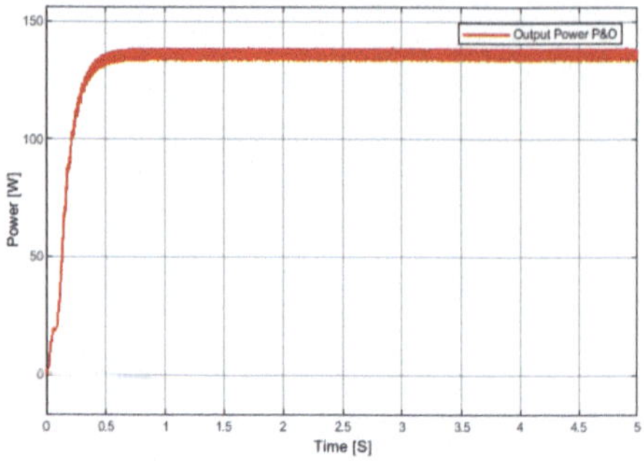

Fig. 9. Output power curve depending on time for (T = 25 °C, G = 1000 W/m^2).

Figure 11 compares the power output curves under conditions of constant temperature 25 °C and variable irradiance [1000 800 600] W/m^2.

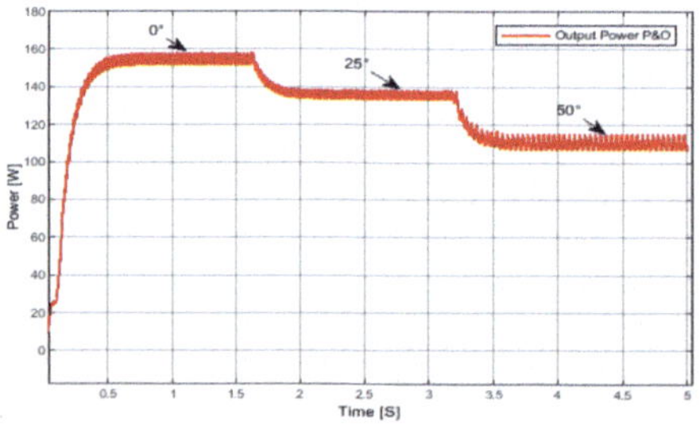

Fig. 10. Power curve in case of temperature change.

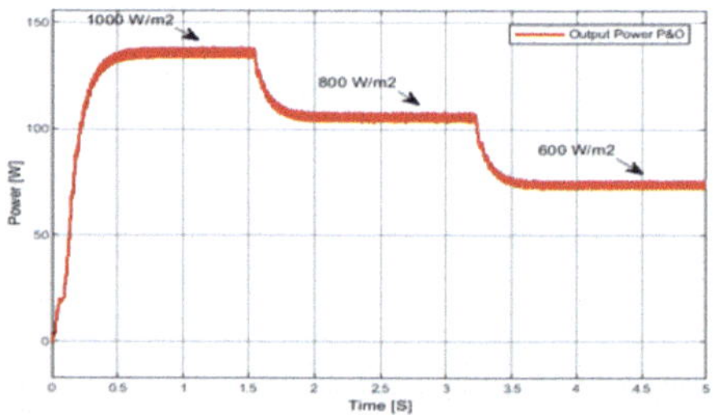

Fig. 11. Power curve in case of irradiation change.

The data shows that power levels vary in response to changes in illumination, consistently reaching the maximum power output. The P&O approach is directly implementable. Nevertheless, the limitations consist of oscillations around PPM in permanent mode, Sluggish reactive speed, and inaccurate tracking in dynamically shifting atmospheric conditions.

2. Algorithme Inc

The following figures show the power features of a photovoltaic system (PVL136 of power 135.3 W) controlled by a controller Inc at a constant temperature of 25 °C and constant illuminations of 1000 W/m^2 (Figs. 12 and 13):

a. Temperature variation and Irradiation

To validate the Inc algorithm, we tested our controller for various irradiation and temperature values depending on the perturbation step. Figure 14 shows a comparison of power curves for a constant irradiation of 1000 W/m^2 and a variable temperature [0 25 50] °C.

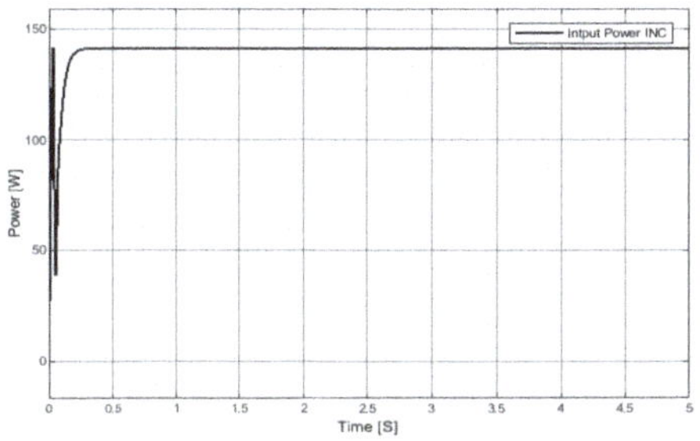

Fig. 12. Input power curve depending on time for (T = 25 °C, 1000 W/m^2).

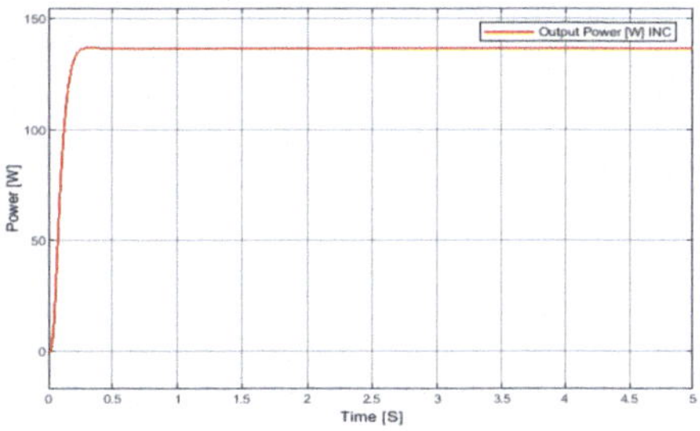

Fig. 13. Output power curve depending on time for (T = 25 °C, 1000 W/m^2).

Figure 15 illustrates the power curves with a reference temperature of 25 °C and varying levels of irradiation [1000 800 600] W/m^2.

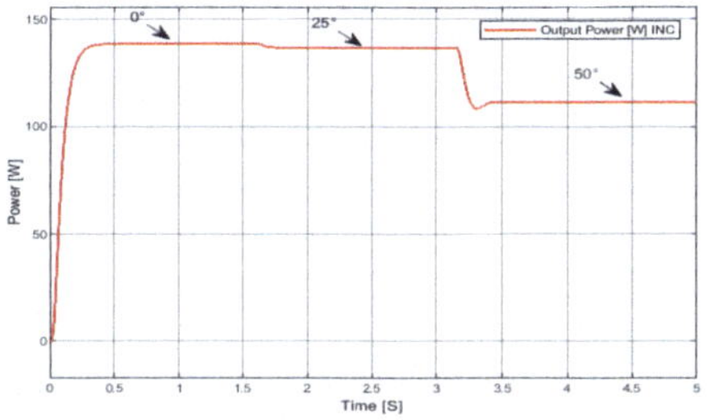

Fig. 14. Power curve in case of temperature change.

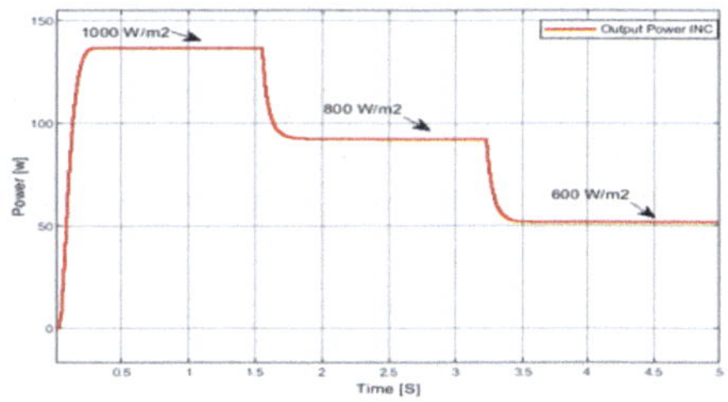

Fig. 15. Power curve in case of irradiation change.

The Inc function seems to be an enhancement of the P&O method. In fact, demonstrates superior behavior when rapidly changing metrological conditions, nevertheless, the voltage wave is important in the latter case. Nevertheless, the current algorithm is more intricate than its predecessor.

4.4 Comparison Between the Two Commands

For a better judgment between previous orders; we did a comparative study and obtained the following simulation results.

Figure 16 shows, for comparison, the GPV output power simulation results for perturbation and observation (P&O) and (Inc) MPPT. We can clearly see the command Inc reduces the response time of the PV system (Zoom1). It is obviously noted that the system with conventional P&O MPPT has a large loss of energy in continuous power (Zoom2).

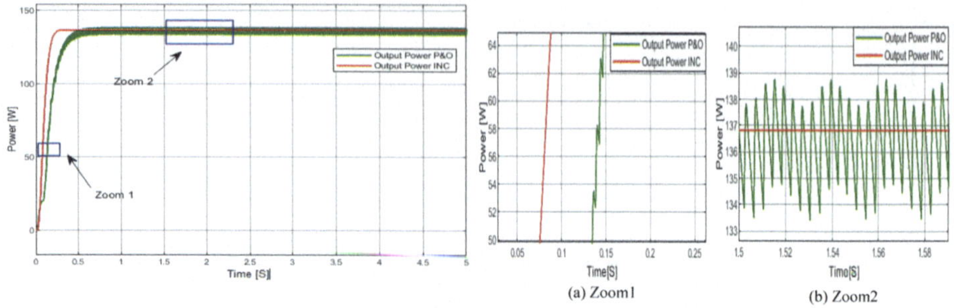

Fig. 16. P&O- Inc output power; (a) Zoom1, (b) zoom2

5 Conclusion

This paper outlines the primary components of the PV system. We then review the fundamental principles of two widely used MPPT algorithms and ultimately create modelling the two algorithms. The simulations of both algorithms indicate that the 'P&O' technique is often utilized, while the 'INC' algorithm demonstrates superior behavior and performance compared to other methods. By merging these two algorithms, we want to expand the range of MPPT methods, discover new ones, and improve their suitability in different situations.

The simulation results, obtained using the MATLAB/Simulink software, illustrate the performance of the PV side and grid side as time and irradiance change. In contrast to the P&O technique, the INC algorithm provides more precise tracking of rapidly changing irradiation circumstances. When employing the P&O approach, the voltage exhibits fluctuations around the maximum power point (MPP) instead of consistently attaining a precise number. Thus, the INC approach achieves the MPP more quickly and effectively than P&O because it doesn't have a drifting issue and is most effective in situations that change quickly.

References

1. Stéphane, P.: Nouvelles architectures distribuées de gestion et de conversion de l'énergie pour les applications photovoltaïques. In: French, Doctorat Thesis, de l'Université de Toulouse. Toulouse, France (2009)
2. Castillo-Calzadilla, T., Macarulla, A.M., KamaraEsteban, O., lBorges, C.E.: Analysis and assessment of an off-grid services building through the usage of a DC photovoltaic microgrid. Sustainable Cities and Society **38**, 405–419 (2018)
3. Castillo-Calzadilla, T., Macarulla, A.M., Borges, C.E.: Design of sizing algorithms for a direct current off-grid photovoltaic installation. IEEE Latin America Transactions **16**(8), 2168–2176 (2018)
4. Meriem, B., et al.: Modeling of PV systems and identification of electrical systems. In: 2024 2nd International Conference on Electrical Engineering and Automatic Control (ICEEAC), pp. 1–6. IEEE (2024)
5. Nicola, E., Petrone, G.: Optimization of perturb and observe maximum power point tracking method. IEEE Trans. Power Electron **20**, 963–973 (2005)

6. Bennis, E., Mohammed, K., Lagrioui, A.: Comparison of the MPPT command based on the PSO, P&O, and IncCond algorithm implemented in Dspace, pp. 2769–2507 (2024)
7. Nasreddine, B., Fethi, B., Yassine, H., Imane, H., Riyadh, B.: Optimal sizing and placement of distributed generation with short-circuit analysis using a combined technique based on modified PSO and ETAP. In: 2024 2nd International Conference on Electrical Engineering and Automatic Control (ICEEAC), pp. 1–6. IEEE (2024)
8. Bechekir, S., Bousmaha, I.S., Brahami, M., Abdeslam, D.O.: Realization of an inverter with PWM command for photovoltaic system. In: 2017 5th International Conference on Electrical Engineering-Boumerdes (ICEE-B), pp. 1–6. IEEE (2017)
9. Baba, A.O., Liu, G., Chen, X.: Classification and evaluation review of maximum power point tracking methods. Sustain. Futur. **2**, 100020 (2020)
10. Alphousseyni, N., Mactar, F., et al.: Experimental validation of PSO and neuro-fuzzy soft-computing methods for power optimization of PV installations. In: 2020 8th International Conference on Smart Grid (icSmartGrid), pp. 189–197. IEEE (2020)
11. Femia, N., Petrone, Spagnuolo, G., et al.: Optimisation de la méthode de suivi du point de puissance maximale par Perturbation et Observation. IEEE Trans. Électron de puissance (2005)
12. Yuansheng, X., Suxiang, Q., Jianming, X.: Research on constant voltage with incremental conductance MPPT method. Asia-Pacific Power and Energy Engineering Conference (APPEEC), pp. 27–29 (2012)
13. Dkhichi, E., Tabaa, M., Alami, K., et al.: Comparison between artificial neural network and its combination with perturb & observe as a MPPT control method of photovoltaic system. In: Proceedings of the 1st International Conference on Electronic Engineering and Renewable Energy: ICEERE 2018, 15–17 April 2018, Saidia, Morocco 1, pp. 454–460. Springer Singapore (2019)
14. Jie, L., Yunda, L., Xufang, B.: The research of maximum power point tracking method for photovoltaic system. In: Future Wireless Networks and Information Systems: Volume 2, pp. 629–636. Springer Berlin Heidelberg (2012)
15. Oumar Amadou, C.A., Alphousseyni, N., Amadou, B. A., et al.: Optimization of the P&O-MPPT controller by the adaptive method (Ad-P&O) for stand-alone PV systems. In: 2023 11th International Conference on Smart Grid (icSmartGrid), p. 1–8. IEEE (2023)
16. Khalis, M., Mir, Y., Hemine, J., et al.: Extraction of equivalent circuit parameters of solar cell: influence of temperature. The European Physical Journal-Applied Physics **54**(1), 10102 (2011)
17. El Hammedi, E., Chrouta, J., Khaterchi, H., et al.: Comparative study of MPPT algorithms: P&O, INC, and PSO for PV system optimization." In: 2023 9th International Conference on Control, Decision and Information Technologies (CoDIT), pp. 2683–2688. IEEE (2023)
18. Araújo, N., Sousa, F.J.P., Costa, F.B.: Equivalent models for photovoltaic cell–a review. Revista de Engenharia Térmica **19**(2), 77–98 (2020)
19. Baghaz., C., Melhaoui, M., Yaden, F., et al.: Design, realization and optimization of the photovoltaic systems equipped with analog and digital MPPT commands. Energy Procedia **42**, 270–279 (2013)
20. Roman, C., Alonso, R., Ibañez, P., Elorduizapatarietxe, S., et al.: Intelligent PV module for grid-connected PV systems. IEEE Transactions on Industrial Electronics **53**(4), 1066–1073 (2006)
21. Abbes, E., Abid, H., Loukil, K., Toumi, A., et al.: Etude comparative de cinq algorithmes de commande MPPT pour un système photovoltaïque. Journal of Renewable Energies **17**(3), 435–445 (2014)
22. Bechekir, S., Bousmaha, I.S., Brahami, M., Abdeslam, D.O.: Realization of an inverter with PWM command for photovoltaic system. In: 2017 5th International Conference on Electrical

Engineering-Boumerdes (ICEE-B), pp. 1–6. IEEE. Young, The Technical Writer's Handbook. Mill Valley, CA: University Science, 1989 (2017)
23. Jung, Y., So, J., Yu, G., Choi' J.: Improved perturbation and observation method (ip80) of mppt control for photovoltaic power systems. IEEE **1788**(1), 1788–1791 (2005). 0-7803-8707-4/051$20.00 02005
24. Azab, M.: A new maximum power point tracking for photovoltaic systems. 2002, 702–705 (2009)
25. Ayache, Z., et al.: Commande MPPT et contrôle d ' un système photovoltaïque par la logique floue, pp. 3–8
26. Tan, D.W., Green, T.C., a Hernandez-aramburo, C.: A Current-Mode Controlled Maximum Power Point Tracking Converter for Building Integrated Photovoltaics, pp. 1–10 (2010)
27. Oi, A.: Design and simulation of photovoltaic water pumping system. A Thesis Presented to the Faculty of California Polytechnic State University. San Luis Obispo (2005)
28. Sourov, M.R.: A high performance maximum power point tracker for photovoltaic power system using dc-dc boost converter. IOSR J. Eng. **02**(12), 12–20 (2012)

Enhanced Grid Integration of PV Systems Using Double-Stage Configuration and PUC7 Inverter Controlled by MPC

Soufian Khettab[1](✉), Aissa Kheldoun[2], Rafik Bradai[3], and Intissar Hattabi[1]

[1] SET Laboratory, Blida 1 University, Faculty of Technology, Automation and Electrical Engineering Department, 09000 Blida, Algeria
khettab_soufian@univ-blida.dz

[2] Laboratory of Signals and Systems, Institute of Electrical and Electronic Engineering, University M'hamed Bougara, 35000 Boumerdes, Algeria

[3] LATSI, Blida 1 University, Faculty of Technology, Electronics Department, 09000 Blida, Algeria

Abstract. The paper introduces a double-stage, single-phase photovoltaic (PV) system connected to the grid using a packed U-cell seven-level (PUC7) inverter, governed by Model Predictive Control (MPC). This system aims to optimize grid integration by achieving enhanced current injection with minimal Total Harmonic Distortion (THD), reactive power compensation, and efficient extraction and injection of maximum power from the PV source. Simulation results underscore the efficacy of the proposed control strategy in enhancing power quality, responsiveness, and accuracy across varying weather conditions, thereby validating the feasibility and robustness of the proposed PV system.

Keywords: PV system · double -stage configuration · PUC7 inverter · MPC · power quality · reactive power compensation · MPPT

1 Introduction

In recent years, the power system has grappled with surging energy demand due to increased consumer numbers and thriving businesses. Thanks to advancements in semiconductor technology, bulky transformers are gradually giving way to power electronics equipment. Renewable energy conversion systems now heavily rely on power inverters to deliver eco-friendly electricity to clients. These inverters, equipped with cost-effective power switches, are gaining ground in the market. Moreover, low switching frequency multilevel inverters are swiftly replacing traditional two-level converters, which suffer from significant power losses and harmonic content. The addition of extra switches to the inverter design no longer significantly escalates costs [1, 2]. The incorporation of renewable energy supplies into the electrical grid has grown in importance in recent years [3]. Solar panels are becoming widely used in several nations to address local energy needs. Inverters with power electronics are crucial interfaces for supplying renewable

© The Author(s), under exclusive license to Springer Nature Switzerland AG 2025
M. Hatti (Ed.): IC-AIRES 2024, LNNS 1238, pp. 71–77, 2025.
https://doi.org/10.1007/978-3-031-80301-7_9

energy to loads and the grid. As several publications have indicated, increasing efficiency and reducing power losses in these interfaces are the main goals of efforts to improve environmental sustainability [4, 5]. In this context, single-phase multilevel inverters are essential because they transform the DC voltage from solar systems into a smooth AC waveform that is appropriate for loads and the grid. These inverters have great efficiency and less harmonic filters to accomplish this. Furthermore, a potential alternative that offers benefits like fewer components and the capacity to produce various voltage levels at the output is the seven-level PUC inverter [6, 7].

This study presents the development of a single-phase grid-tied photovoltaic system utilizing a PUC7 multilevel inverter. To regulate the single-phase PUC7 inverter while feeding solar electricity into the grid, a modified Model Predictive Control (MPC) approach is employed. The key objectives of this technique include injecting actual power generated by a PV array with minimal Total Harmonic Distortion (THD), improving reactive power management, mitigating harmonic current content, and addressing the impact of a polluted load with a low power factor. Notably, the proposed design continues to provide services even when PV power fluctuates due to changes in irradiance. The MPC approach ensures high power factor correction, limits switching frequency, minimizes computational overhead, and maintains balance in the DC-link capacitor voltage. Detailed simulations in the MATLAB/Simulink environment are used to verify the performance of this setup.

2 System Describing

Figure 1 shows the suggested MPC controller and related control system, as well as the proposed architecture of the double-stage, single-phase PUC7 inverter. The AC source and PUC7 inverter, which are linked in series to the grid via a line inductor (Lf), make up the system under study. A boost converter, which can elevate the voltage of the PV array to a high DC-link voltage, is used to connect the PV system to the single DC-link capacitor of the PUC7 inverter. The effective integration of produced PV electricity into the electrical grid while maintaining high-quality grid current are the primary goals of this particular design.

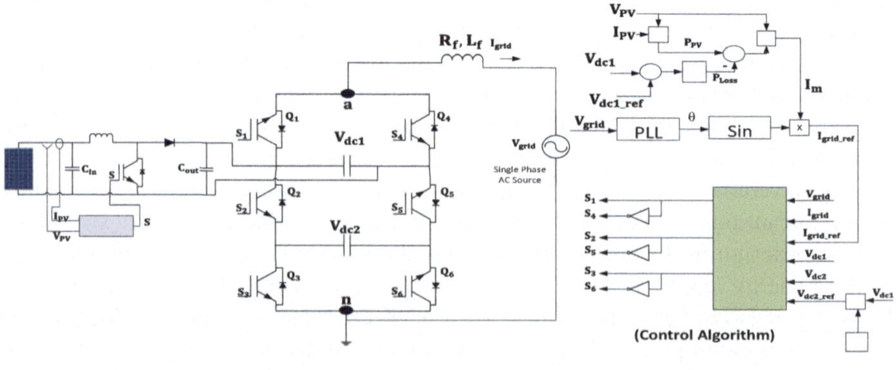

Fig. 1. The double-stage, single-phase PUC7 inverter's suggested setup.

2.1 PUC7 Inverter

The PUC contains the same amount of parts as previous systems, but because of switches and gate drives, it consumes less power and costs less money. The output voltage levels are shown in Table 1, where switches S4, S5, and S6 are connected to switches S1, S2, and S3, respectively. It is crucial to remember that no two switches (S1, S4), (S2, S5), and (S3,S6) can be turned on at the same time.

With the use of two capacitors, the PUC inverter can provide seven different output voltage levels. To achieve the desired output voltage levels of 0, $\pm V_{dc_2}$, $\pm 2\ V_{dc_2}$, and $\pm 3\ V_{dc_2}$ the capacitor's voltage (V_{dc2}) should be one-third of the DC bus voltage (V_{dc1}). The PUC inverter's inability to produce an output voltage higher than the DC bus voltage, however, is its main limitation. One advantage of using the PUC inverter is that it may reduce load voltage harmonics by dividing the DC bus voltage into many levels. Consequently, this reduces the need for large filters at the output of the inverter. The following equation shows the single-phase grid-connected system with the PUC7 inverter represented mathematically:

$$\frac{dI_{grid}(t)}{dt} = -\frac{R_f}{L_f} I_{grid} + \frac{1}{L_f}\left(V_{an} - V_{grid}\right) \tag{1}$$

$$\frac{dV_{dc}(t)}{dt} = \frac{1}{C} I_{grid} \tag{2}$$

Table 1. PUC7 inverter state switching.

State Switching (x)	s_1	s_2	s_3	s_4	s_5	s_6	V_{an}
State 1	1	0	0	0	1	1	Vdc1
State 2	1	0	1	0	1	0	V_{dc1}-V_{dc2}
State 3	1	1	0	0	0	1	V_{dc2}
State 4	1	1	1	0	0	0	0
State 5	0	0	0	1	1	1	0
State 6	0	0	1	1	1	0	-V_{dc2}
State 7	0	1	0	1	0	1	Vdc2-Vdc1
State 8	0	1	1	1	0	0	-V_{dc1}

2.2 Design of MPC

MPC is an advanced and useful method of regulating power converters. It makes use of the mathematical model of the system under study to predict future behavior of the controllable variables. To do this, a cost function must be defined and assessed in order to identify the optimal course of action. Power converter performance and efficiency may be increased by using MPC, which is simply a powerful tool that forecasts power converter behavior and takes appropriate action in advance. The flowchart shown in Fig. 2 illustrates the functionality of the suggested MPC.

One way to generate a discrete-time equation to predict the future value of injected current is to use the forward Euler approximation to create a discrete-time system representation. The discrete-time model may be obtained using the basic derivative approximation. As a result, predictions of the output current for different values of the voltage vector V (k) at time (k + 1) may be derived.

$$\frac{dI_{grid}(t)}{dt} = \frac{I_{grid}(k+1) - I_{grid}(k)}{T_s} \tag{3}$$

For where the sample time is *Ts*.

An formula that forecasts the future injected current at time (k + 1) for each of the eight values of the voltage vector V_an (k) produced by the inverter is substituted for Eq. 1.

This phrase is provided as follows:

$$I_{grid\_ref}(k+1) = \left(1 - \frac{R_f{}^*T_s}{L_f}\right)^* I_{grid}(k) + \frac{T_s}{L_f} * \left(V_{an}(k) - V_{grid}(k)\right) \tag{4}$$

The expected values of the inverter current (Iref_cal) and the capacitor voltage (Vdc2_cal) are determined for each of the eight switching states using the observed values of the grid current (Igrid), grid voltage (Vgrid), and capacitor voltages (V_{dc1}, V_{dc2}) (Table 1). Then, using a cost function (g), the optimal switching state that correlates to its lowest value is selected.

2.3 MPPT Controller

Under a range of climatic conditions, a solar system's power production can be increased by a technique known as maximum power point tracking (MPPT). MPPT controllers are used to synchronize the PV array's maximum power point (MPP) with the PV system's operating point. Perturb and Observe (P&O), an MPPT algorithm, is one of the most used ones. It works by gradually adjusting the PV array's voltage or current and measuring the resulting power.

If the power increases, the MPPT controller keeps adjusting the voltage or current in the same direction. It changes in the other direction if the power is reduced. To satisfy the requirement that the power fluctuation with regard to the voltage or current variation is null, this procedure is done as many times as necessary. The MPP of the "P-V" characteristic is this point. The P&O method flowchart that is shown in Fig. 3 summarizes these processes. The flowchart makes clear that the P&O algorithm is straightforward, simple to use, and broadly applicable.

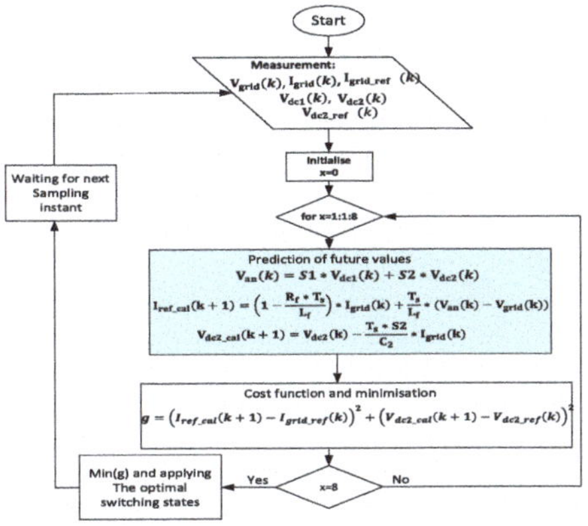

Fig. 2. Flow chart of MPC.

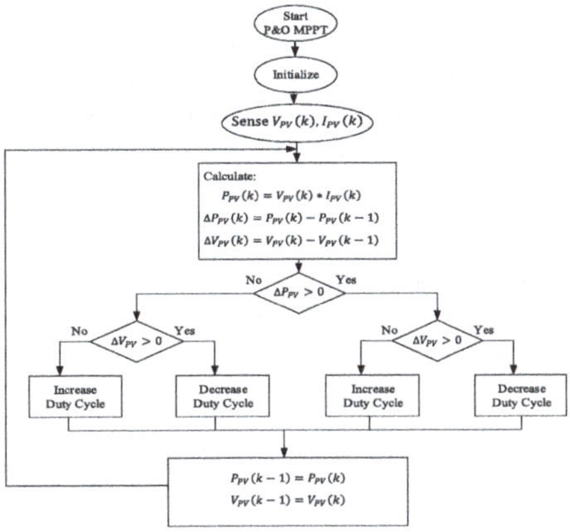

Fig. 3. The P&O method flowchart

3 Results of the Grid-Connected System's Simulation

To demonstrate the performance of the double-stage single-phase PUC7 inverter in transferring power from PV arrays to the grid, we conducted a MATLAB simulation using the configuration shown in Fig. 1. The PV panel specifications and system parameters are detailed in Table 2. It should be mentioned that the a sampling frequency of 20 kHz.

Table 2. System components specifications

Grid	Grid voltage, V_{grid} (RMS)	63 V
	Grid frequency, F_s	50 Hz
Filter	Filter inductance, L_f, R_f	10 mH, 0.25 mΩ
PUC7 inverter	DC capacitor, $C_1 = C_2$	1500 μF
	DC voltage reference	120 V
Boost converter	Input capacitor, C_{in}	47 μF
	Output capacitor, C_{out}	390 μF
	Inductance, L	5 mH
PV array (STC)	Maximum power (P_{max})	600 W
	Open circuit voltage (V_{oc})	90.5 V
	Short circuit current (I_{sc})	8.49 A
	Maximum power voltage (V_{mp})	74.6 V
	Maximum power current (I_{mp})	8.04 A

Figure 4c illustrates the variation of grid current and its reference. Notably, during different irradiances as depicted in Fig. 4a, the injected grid current corresponds to the minimum total harmonic distortion (THD). The THD remains below 5% across all scenarios, adhering to the IEEE 519-2022 standard [8]. The highest THD occurs during power injection, reaching 3.03%. The DC-link voltage, denoted as V_dc, is maintained at the reference value of 120 V (as shown in Fig. 4e).Additionally, precise voltage balance between the two capacitors, V_dc1 and V_dc2, is achieved. The voltage across the second capacitor remains constant and equals one-third of the voltage across the first cell capacitor (40 V). This voltage constraint is essential for the PUC7 inverter to deliver a seven-level single-phase voltage waveform, as depicted in Fig. 4(f). Figure 4(d) displays the PV power output for various irradiance levels. In summary, this proposed

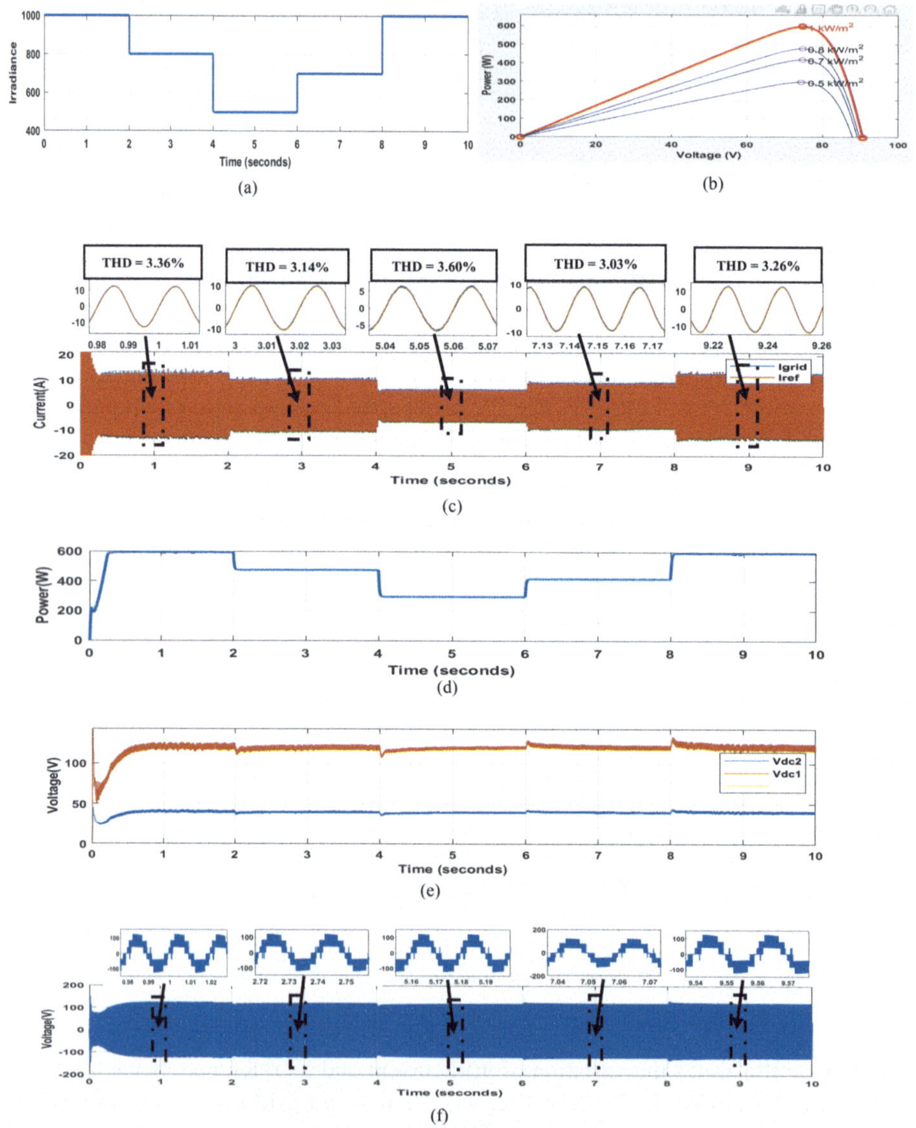

Fig. 4. (a) Irradiance (b) P-V curve (c) Grid current and its reference,(d) Power (e) DC-link capacitor voltages (f) Voltage of the PUC7.

control system effectively manages power distribution between the grid and the PV source while maintaining THD at its minimum level.

4 Conclusion

This research paper introduces a comprehensive control strategy for a dual-stage, single-phase PUC7 inverter used in PV applications. The inverter operates in grid-connected mode, aiming to achieve high-quality power injection while maintaining low grid current THD and a stable dc-link voltage. Notably, the proposed MPC strategy ensures consistent voltage profiles, even under varying irradiance conditions. This stability is evident through sustained voltage levels across capacitors, maintaining the desired proportionality. Additionally, the P&O algorithm effectively tracks and delivers maximum PV power. Overall, this innovative topology demonstrates remarkable adaptability to changing irradiance levels, including power factor adjustments, reactive power compensation, and active power injection.

References

1. Kadhim, I.J., Hasan, M.J.: Enhancing power stability and efficiency with multilevel inverter technology based on renewable energy sources. Electr. Power Syst. Res. **231**, 110290 (2024). https://doi.org/10.1016/j.epsr.2024.110290
2. Lu, M., et al.: Loss optimization for bottom IGBTs of half-bridge submodules in a high-power hybrid modular multilevel converter. Int. J. Electr. Power Energy Syst. **158**, 109910 (2024)
3. Louiza, A.M., et al.: Seasonal forecasting of global horizontal irradiance for grid-connected PV plants: a combined CNN-BiGRU approach. In: 2024 3rd International conference on Power Electronics and IoT Applications in Renewable Energy and its Control (PARC). IEEE (2024)
4. Saleem, M.I., Saha, S., Izhar, U., Ang, L.: Stability assessment of inverter-based renewable energy sources integrated to weak grids. IET Energy Syst. Integr. **6**(4), 525–544 (2024)
5. Aljarrah, R., et al.: Issues and challenges of grid-following converters interfacing renewable energy sources in low inertia systems: a review. IEEE Access **12**, 5534–5561 (2024)
6. Sifat, Z., et al.: Selective harmonic elimination in PUC-5 multilevel inverter using hybrid IGWO-DE algorithm, p. e12883 (2024)
7. Khettab, S., et al.: Performance evaluation of PUC7-based multifunction single-phase solar active filter in real outdoor environments: Experimental insights. IET Renew. Power Gen. **18**(11), 1740–1757 (2024)
8. Committee, T.D.: IEEE Standard for Harmonic Control in Electric Power Systems. IEEE Power and Energy Society (2022)

Fault Ride-Through Strategy for MMC-HVDC System Under Single Line-to-Ground Fault

Hicham Korzane[1(✉)], Abdeldjabar Benrabah[1], Farid Khoucha[1], and Khelifa Benmansour[2]

[1] Department of Electrical Engineering, Ecole Militaire Polytechnique, Algiers, Algeria
korzanehicham@gmail.com

[2] Department of Electrical Engineering, University of Medea, Medea, Algeria

Abstract. This paper investigates the single-line-to-ground (SLG) faults that can occur on the valve side of modular multilevel converter (MMC) stations used for integrating offshore wind farms. The grounding faults at the onshore converter installations result in significant threats that could compromise the stable performance of the power system. Furthermore, the paper also explores how SLG faults influence the DC link voltage and impact the wind farm operation. To address this issue, a control strategy using zero-sequence voltage modulation is used to reduce overvoltage in the DC voltage from such faults. Finally, MATLAB/Simulink simulations are carried out to verify the mathematical model of valve-side SLG faults and validate the effectiveness of the proposed control strategy.

Keywords: single-line-to ground fault · MMC-HVDC · offshore wind · onshore converter station

1 Introduction

The modular multilevel converter (MMC) plays a crucial role in integrating offshore wind farms into power grids, offering exceptional power quality, enhanced controllability, and the ability to withstand grid faults [1]. Currently, many offshore wind power projects have been already installed all over the world such as the Rudong project in China and the DolWin project in Germany [2].

The transformers are normally installed outside the onshore MMC station. The bushings on the valve side of these transformers are extended through the station wall to establish a connection with the MMC ac side. However, the degradation of electrical insulation of these bushings, may cause a single line-to-ground (SLG) fault [3, 4].

To address the above problem, various solutions have been suggested in the literature. The control method based on nonideal proportional resonant controllers proposed in [5], requires a period time to effectively mitigate the oscillation in the dc transmission line voltage. These delays could potentially result in a damage to the MMC equipment. The authors in [6] introduce in parallel double-thyristors with all the sub-modules of the lower arm in the MMC converter. By activating these thyristors, an SLG fault is transformed into a three-phase short circuit, which generates current zero crossing through the AC

© The Author(s), under exclusive license to Springer Nature Switzerland AG 2025
M. Hatti (Ed.): IC-AIRES 2024, LNNS 1238, pp. 78–85, 2025.
https://doi.org/10.1007/978-3-031-80301-7_10

circuit breaker for fault protection. However, this method introduces additional hardware and cost in the installation. In this article, an overvoltage reducing control method is used for limiting the effect of SLG fault without any additional system cost and increasing the dynamic response after the SLG occurs.

This paper is divided into the following sections. In Sect. 2, we introduce the basic architecture of the MMC-HVDC system and the impact of the valve-side single- line-to-ground (VSSLG) fault in the MMC station. Then, an overvoltage reducing strategies in MMC-HVDC structure during SLG fault is provided in Sect. 3. Moreover, we present the numerical results in Sect. 4, to illustrate effectiveness of the control strategies used in this paper. Finally, conclusion is drawn in Sect. 5.

2 The MMC-HVDC System Under SLG

2.1 General Structure of an MMC-HVDC System

The main configuration of the MMC-HVDC system used to establish a connection between offshore wind energy and the grid is illustrated in Fig. 1. This configuration includes an offshore wind turbine, wind-farm side MMC (WF-MMC), DC transmission line, grid-side MMC (GS-MMC) and an AC grid. The transformer in both MMC station are connected in delta/star configuration, in order to suppresses the zero-sequence current. In additional, a half-bridge MMC (HBMMC) is used in both WF-MMC and GS-MMC [7, 8].

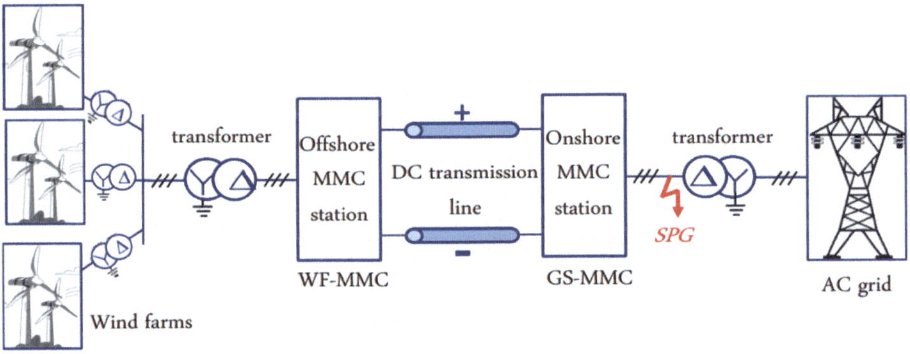

Fig. 1. Main configuration of MMC-HVDC system.

2.2 Mathematical Model of Valve-Side SLG Faults

Figure 2 presents the symmetrical monopole MMC connected to the grid, where U_{vj} $(j = a, b, c)$ denotes the phase-to ground voltage, I_{vj} the phase current, U_{pj} U_{nj}, represents the voltage of the upper and lower arm respectively. Similarly, I_{jp} and I_{jn} are the arm current, I_{cir} represents the circulating current, while the pole-to-ground voltage is indicated by

U_{dcp} and U_{dcn} [9]. In the normal operation the upper and lower arm voltage can be expressed as follows [3]:

$$\begin{aligned} U_{pj} &= U_j^{max} \times sin(\omega t + \theta_j) - \tfrac{1}{2}U_{dc} \\ U_{nj} &= U_j^{max} \times sin(\omega t + \theta_j) + \tfrac{1}{2}U_{dc} \end{aligned} \quad (1)$$

where U_j^{max} and θ_j are the phase voltage amplitude and the phase angle of phase j respectively. If we neglect the voltage drops across arm resistors (R_{arm}) and reactors (L_{arm}), the DC pole voltages can be derived as follows:

$$\begin{aligned} V_{dcp} &= U_{pj} + U_j \\ V_{dcp} &= -U_{pj} + U_j \end{aligned} \quad (2)$$

During a VSSLG fault, the voltage of the faulty phase drops to 0, while the voltage in the unaffected phases increases to the line voltage ($U_{va} = 0, U_{vb} = -U_{vab}$ *and* $U_{vc} = U_{vac}$). Moreover, the SLG will introduce an oscillation Δu onto the DC pole voltages, with an amplitude and opposite direction of the faulted phase.

$$\begin{aligned} V_{dcp} = U_{pj} &= \tfrac{1}{2}U_{dc} - U_j^{max} \times sin(\omega t + \theta_j) \\ V_{dcp} = -U_{nj} &= -\left[\tfrac{1}{2}U_{dc} + U_j^{max} \times sin(\omega t + \theta_j)\right] \end{aligned} \quad (3)$$

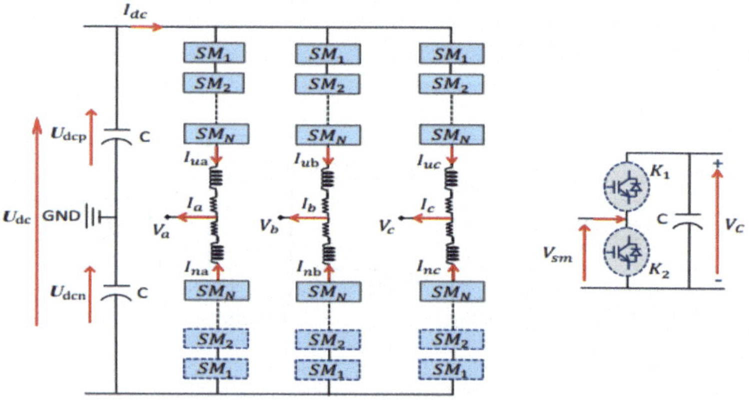

Fig. 2. Topology of MMC station.

Figure 3 shows that during VSSLG fault, the valve-side line voltage's amplitude remains unchanged. Furthermore, both the AC voltage on the onshore grid side and the grid-connected voltage of the wind farm are unaffected by a VSSLG failure due to the Δ/*Y* configuration of the converter transformer.

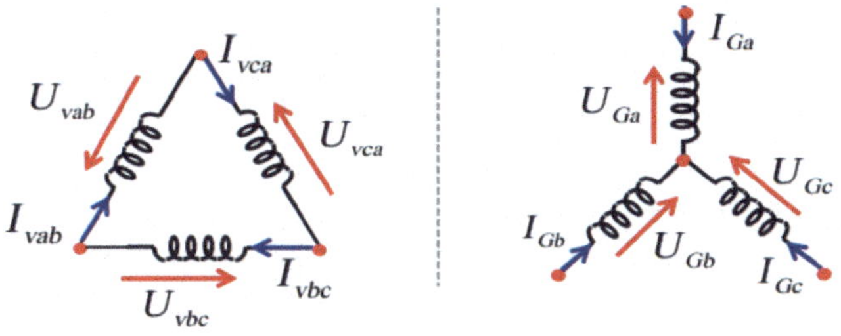

Fig. 3. Topology of MMC station.

3 Overvoltage Reducing Strategies in MMC-HVDC Under VSSLG

As we have seen in the previous section, a VSSLG fault may introduce an overvoltage in the dc pole to ground voltage.

During this fault, the MMC-HVDC station will automatically block operation to limit potential damage in the system if the overvoltage exceeds the threshold value expressed in Table 1.

Table 1. The capacity for tolerate dc overvoltage in an mmc-hvdc project.

Operating Scenarios	Maximum time limit
(1,02 – – 1,15) Udc/2	Extended tolerance period
(1,15 – – 1,40) Udc/2	> 5min
1,70 Udc/2	Acceptable for 500ms

Since the SLG fault appears, the oscillation Δu in the DC voltage is equal to the valve-side zero-sequence voltages U_{Gv}^{0}. Based on the Eq. 3, the overvoltage in the DC side can be limited by compensating the magnitude of the oscillation Δu.

The blue area in Fig. 4 presents the control method of the GS-MMC under a normal operation, which utilizes the DC voltage control method. In this control strategy, the D-axis is employed to stabilize the DC voltage, while the Q axis is utilized to control reactive power. However, the red area illustrates that the overvoltage fast reducing strategy is based on multiplying the zero-sequence voltages by a compensation number k and feeding the resulting value into the phase-shifted pulse-width-modulation (PS-PWM) [10] block, where the value of k ranges between 0 and 1. Moreover, the remainder of the scheme presents the circulating current suppression control (CCSC) and the voltage balancing control (VBC) method for the capacitor voltage of HB-MMC. From the above paragraph, it's clear that for fully eliminating the oscillation in the DC side voltage, we must set the k factor value to 1. However, in this case, an overmodulation to the MMC station is introduced. According to the Table 1, it's only satisfying to select the appropriate k factor value in order to reduce the DC overvoltage voltage below than $1.5U_{dc}$.

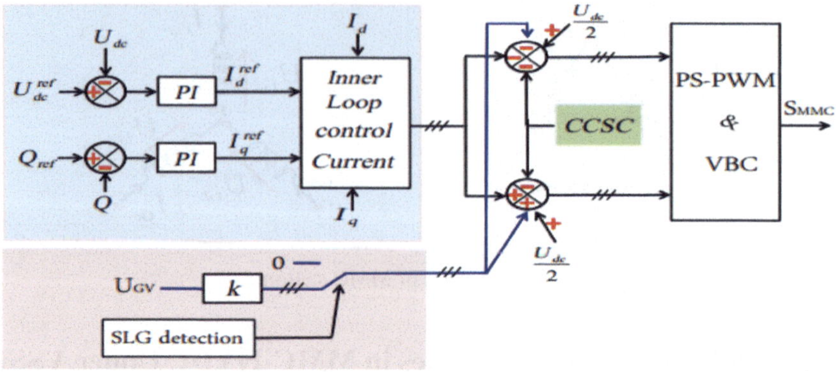

Fig. 4. Overvoltage reducing strategy.

4 Performance Evaluation Results

In order to validate the analysis and the control strategy discussed in earlier sections, a simulation model of the MMC-HVDC system has been implemented in MATLAB/Simulink. The system parameters employed in the simulation are detailed in Table 2.

Table 2. Parameters for mmc-HVDC system.

Parameters	Symbol	Value
Rated power	*Pnom*	5 MW
Nominal DC voltage	*Udc*	6 kV
Nominal DC pole to ground voltage	*Udcp/Udcn*	3 kV
Nominal AC voltage grid side	*Ug*	32 kV
Nominal AC voltage valve side	*Uvg*	2 kV
Grid frequency	*f*	50 Hz
Number of SMs in each arm	*N*	10

4.1 SLG Mathematical Model Validation

Simulation setup: at 5.0 s, a VSSLG fault occurs in the valve-side of the GS-MMC.

The phase to ground voltage on the valve-side of the GSMMC and the phase-to-phase voltage on AC grid are presented in Fig. 5.

According to Fig. 5, after the occurrence of the SLG fault, the voltage of the affected phase decreases to 0, while the voltage of the non-fault phases increases to the line voltage. Additionally, the phase-to-phase voltage of the AC side remains unaffected by the SLG fault, due to the configuration of the transformer winding in a delta/star setup.

Figure 6 illustrates the DC pole-to-ground voltage and the DC pole-to-pole voltage of the HVDC system during VSSLG fault.

Figure 6 illustrates that the oscillation voltage is introduced into the DC pole-to-ground voltage following the occurrence of the SLG fault. Additionally, this oscillation is equal to the valve-side zero-sequence voltages. While, the DC pole-to-pole voltage remains constant and unaffected by the SLG fault.

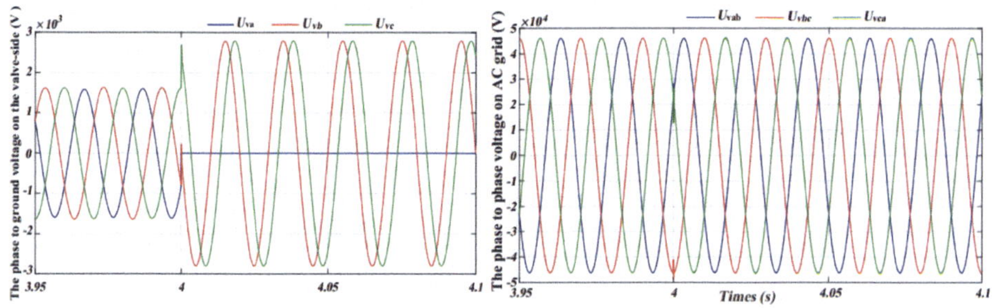

Fig. 5. Phase voltage on the valve-side of the GSMMC and grid side under SLG.

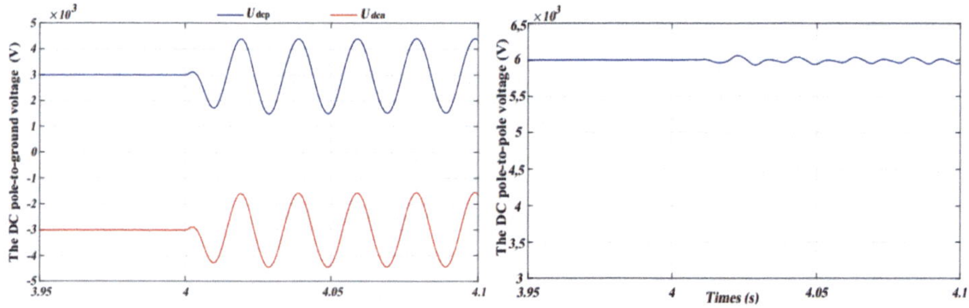

Fig. 6. HVDC DC voltage evolution system during VSSLG fault.

Figure 7 illustrates the phase to ground voltage on the valve-side of the WF-MMC and the phase-to-phase voltage on wind side.

Referring to Fig. 7, during the SLG failure, the oscillation voltage causes the two phases voltages on the valve-side of the WF-MMC to increase while the other phase voltage decreases. However, due to the delta/star configuration of the transformer winding, the phase-to-phase voltage on wind side remains unaffected by the SLG failure.

4.2 Validation of Overvoltage Reducing Control Methods

Using the suggested overvoltage reducing control methods, Fig. 8 presents the DC pole-to-ground voltages U_{dcp} and U_{dcn} for multiple values of k.

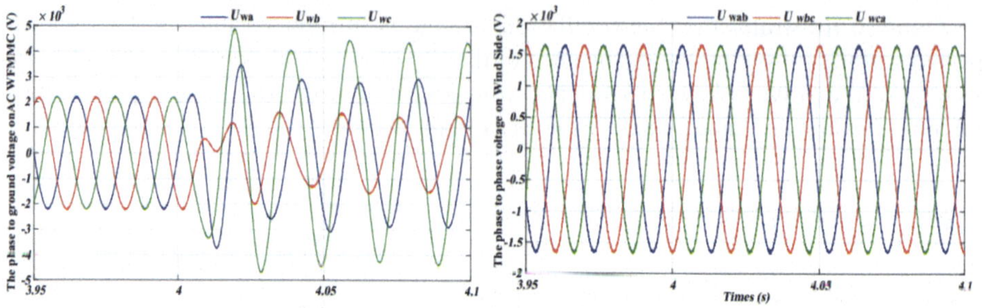

Fig. 7. Phase voltage waveform on the valve-side of the WF-MMC and wind side during SLG.

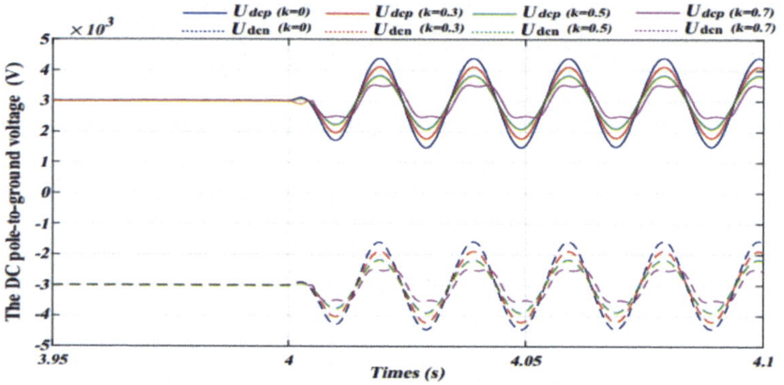

Fig. 8. The impact of overvoltage reduction strategies on HVDC DC voltage

According to Fig. 8, the suggested control method effectively mitigates overvoltage up to a certain threshold when k is non-zero. Increasing the value of k enhances the reduction of overvoltage. Nevertheless, elevating the value of k may introduce overmodulation to the MMC station, as discussed in Sect. 3.

5 Conclusion

This paper examines the impact of single-line-to-ground (SLG) faults on onshore MMC stations that connect offshore wind farms. The analysis focuses on how such faults affect various components, including HVDC transmission lines and the offshore converter station. It has been observed that valve-side SLG faults can lead to overvoltage issues in the DC voltage and AC overvoltage at the valve-side of the offshore converter station.

To mitigate such problems, this paper proposes an overvoltage reduction strategy that uses zero-sequence voltage modulation to limit damage to the HVDC system. The mathematical model of valve-side SLG faults is validated through simulations on the MATLAB/Simulink platform. The simulations also demonstrate the effectiveness of the proposed control strategy for mitigating overvoltage during such faults.

References

1. Collins, C.T., Clemow, P.R., Green, T.C.: Experimental study on the influence of MMC control design on the propagation of AC grid-unbalance to a DC network. IET Conf. Proc. **2023**(1), 115–121 (2023). https://doi.org/10.1049/icp.2023.1317
2. Zhou, H., Yao, W., Sun, K., Ai, X., Wen, J., Cheng, S.: Dynamic reactive current optimization based onshore ac fault ride-through strategy for mmc-HVDC integrated offshore wind farms. IEEE Trans. Sustainable Energy **15**(2), 735–746 (2024)
3. Li, G., Liang, J., Ma, F., Ugalde-Loo, C.E., Liang, H.: Analysis of single-phase-to-ground faults at the valve-side of hb-mmcs in hvdc systems. IEEE Trans. Ind. Electron. **66**(3), 2444–2453 (2019)
4. Liu, X., Li, G., Zheng, T., Yang, X., Guerrero, J.M.: Thyristor-pair- and damping-submodule-based protection against valve-side single-phase-to-ground faults in MMC-MTDC systems. IEEE Trans. Power Delivery **37**(4), 3257–3269 (2022)
5. Li, S., Wang, X., Yao, Z., Li, T., Peng, Z.: Circulating current suppressing strategy for MMC-HVDC based on nonideal proportional resonant controllers under unbalanced grid conditions. IEEE Trans. Power Electron. **30**(1), 387–397 (2014)
6. Li, G., Liu, W., Joseph, T., Liang, J., Song, Z.: Doublethyristor-based protection for valve-side single-phase-to-ground faults in HB-MMC-based bipolar HVDC systems. IEEE Trans. Industr. Electron. **67**(7), 5810–5815 (2019)
7. Zhou, H., Yao, W., Sun, K., Ai, X., Wen, J., Cheng, S.: Characteristic investigation and overvoltage suppression of mmc-hvdc integrated offshore wind farms under onshore valve-side spg fault. IEEE Trans. Power Syst. **39**(1), 1346–1359 (2024). https://doi.org/10.1109/TPWRS.2023.3236095
8. Zhou, H., Yao, W., Sun, K., Zhao, H., Zong, Q., Wen, J.: Two-level suppressing controller of offshore dc overvoltage using zero-sequence voltage and gfl-gfm switching. IEEE Trans. Power Deliv. **38**(6), 3991–4003 (2023)
9. Bensiali, H., Khoucha, F., Benrabah, A., Benbouzid, M.: Virtual synchronous generator control of grid connected modular multilevel converters with an improved capacitor voltage balancing method. Available at SSRN 4587288 (2017)
10. Bekhouche, R. Khoucha, F., Benrabah, A., Benmansour, K ., Benbouzid, M.E.H.: Comparison of PWM techniques for modular multilevel converter: a comparison based on different voltage level waveforms. In: 2020 1st International Conference on Communications, Control Systems and Signal Processing (CCSSP), pp. 460–465. IEEE (2020)

Optimizing Crow Algorithm for Enhanced MPPT Control in Photovoltaic System Under Varying Illumination Conditions

Djihane Bougandoura[1](✉), Sabrina Titri[2], and Cherif Larbes[1]

[1] Laboratoire Des Dispositifs de Communication et de Conversion Photovoltaïque LDCCP, Département D'Electronique, ENP, El-Harrach, Algeria
djihane.bougandoura@g.enp.edu.dz

[2] Division Micro-Électroniques et Nanotechnologie, Centre de Développement Des Technologies Avancées CDTA, Baba Hassen, Algeria

ABSTRACT. This article presents a study of the performance of the crow algorithm, applied to the control of the Maximum Power Point (MPPT) of photovoltaic panels. The performance of this algorithm was optimized by relying on a meticulous search approach in the promising area. This strategy consists of refining the selection of parameters and the number of search agents and their initial positions to maximize the efficiency of the controller. The proposed algorithm was evaluated by comparing it with the PSO algorithm in terms of efficiency, convergence time and reliability under homogeneous illumination and partial shading conditions. To evaluate the reliability of the controllers, the tests were repeated 60 times, and the standard deviation was used as a measurement indicator. The simulation results in all scenarios (homogeneous illumination and partial shading) validate the superior performance of CSA compared to the PSO technique. This superiority is evident in terms of efficiency, convergence time and reliability.

Keywords: Photovoltaic System · Maximum Power Point Tracking (MPPT) · Bio-Inspired Algorithms · Crow Search Algorithm · Particle Swarm Algorithm

1 Introduction

The environmental and socio-economic crises caused by our reliance on fossil and fissile resources have highlighted the need to reduce their share in energy supply and replace them with renewable sources like sunlight, wind, rain, tides, and geothermal heat [1]. Solar energy stands out as a clean, abundant alternative. Technologies like solar thermal energy and photovoltaic (PV) solar energy convert sunlight into usable energy. A PV generator comprises solar cells made of semiconductor materials, typically silicon, with N- type and P-type layers. Sunlight hitting the cells creates an electric field at the junction, generating an electric current. PV generators exhibit non-linear current-voltage and power-voltage characteristics, which vary with meteorological conditions such as temperature and solar radiation. Uniform solar exposure results in a unimodal power-voltage

© The Author(s), under exclusive license to Springer Nature Switzerland AG 2025
M. Hatti (Ed.): IC-AIRES 2024, LNNS 1238, pp. 86–95, 2025.
https://doi.org/10.1007/978-3-031-80301-7_11

characteristic with a single maximum power point, whereas partial shading creates multiple peaks, complicating power extraction. Given the high cost and low efficiency of PV modules, optimizing energy use is crucial. Partial shading, in particular, can significantly reduce power output and efficiency. Consequently, research has focused on mitigating these losses through system architectures, converter topologies, PV array configurations, and maximum power point tracking (MPPT) techniques. The first algorithms used in maximum power point tracking (MPPT) for photovoltaic systems were mainly simple methods based on observing the current-voltage (I-V) characteristics of the solar panel [2]. The most famous of these techniques are P&O and IC. These techniques are simple but struggle with non-uniform irradiance and can cause steady-state oscillations and high power consumption. Bio-inspired algorithms have emerged as a solution to these issues, handling nonlinear, multimodal optimization problems effectively and requiring minimal complex calculations [3]. Examples include Ant Colony Algorithm (ACO) [4], Particle Swarm Optimization (PSO) [5], Bat Swarm Optimization (BSO) [6], Artificial Bee Colony [7], Firefly Algorithm [8], and Crow Search Algorithm (CSA) [9]. This paper proposes an MPPT controller based on the Crow Search Algorithm proposed by [10]. We enhance the algorithm to target areas rich in maximum power on the PV characteristic curve. The proposed controller's performance is compared with the PSO technique regarding accuracy, convergence time, and reliability under various illumination conditions. The paper is organized as follows: Sect. 2 presents the PV system used, Sect. 3 demonstrates the CSA and its MPPT implementation, Sect. 4 provides simulation and evaluation results, and Sect. 5 concludes the study.

2 Photovoltaic System Overview

The proposed photovoltaic system architecture is shown in Fig. 1 and simulated using Matlab/Simulink. It comprises a photovoltaic generator with 8 MSX60 PV modules, a buck-boost converter with an MPPT controller, and a resistive load. Each module has 36 series-connected cells protected by an antiparallel bypass diode. Specifications of the MSX60 module are in Table 1.

The model of the photovoltaic cell considered in this work is the model of two diodes [11]. The mathematical equation that models the current-voltage characteristic (I-V) of a photovoltaic module based on this model of cell is represented as follows:

$$I = I_{PV} N_{pp} - I_{01} N_{pp} \left[\exp\left(\frac{v + IR_S N_S \left(\frac{N_{ss}}{N_{pp}} \right)}{a_1 V_T N_S N_{SS}} \right) - 1 \right] - I_{02} N_{pp} \left[\exp\left(\frac{V + IR_S N_S \left(\frac{N_{ss}}{N_{pp}} \right)}{a_2 V_T N_s N_{ss}} \right) \right] - \left(\frac{V + IR_S N_S \left(\frac{N_{SS}}{N_{pp}} \right)}{R_p N_s \left(\frac{N_{ss}}{N_{pp}} \right)} \right) \tag{1}$$

$$I_{PV} = I_{sc\_STC} + K_I (T - T_{STC}) \tag{2}$$

$$I_{01} = I_{02} = \frac{I_{sc\_STC} + K_I (T - T_{STC})}{\exp\left(\frac{V_{OC\_STC} + K_V (T - T_{STC})}{N_S V_T} \right)} \tag{3}$$

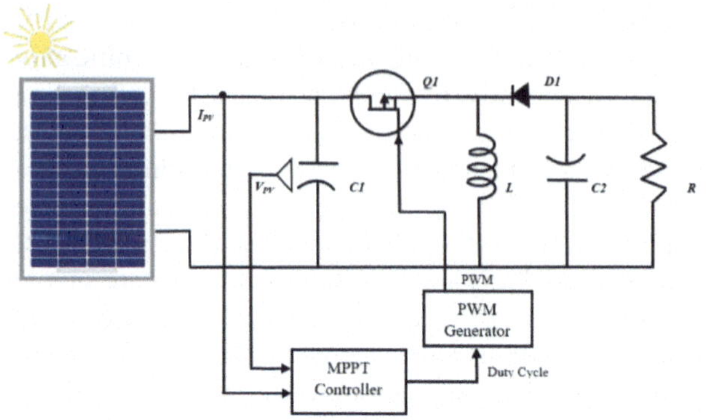

Fig. 1. The electrical diagram of the proposed photovoltaic system.

Table 1. The specifications of MSX60 module system.

Parameters	Value
Maximum Power	60 W
Short circuit current (Isc)	3.8 A
Open circuit voltage (Voc)	21.1 V
Maximum power current (Impp)	3.5 A
Maximum power voltage (Vmpp)	17.1 V
Temperature coefficient of Isc (KI)	0.003 A/°C
Temperature coefficient of Voc (KV)	−0.08 V/°C
Number of series cells in the module (Ns)	36
Parallel resistor, Rp	176.4
Series resistor, Rs	0.35Ω

K: Boltzmann's constant
V_{oc_STC}: Open circuit voltage in standard test condition (STC)
I_{sc_STC}: Short circuit current in standard test condition (STC)
V_T : $\frac{N_s KT}{q}$: The thermal voltage of the PV having N_s connected in series
q: Electron charge, 1: 6021 × 10^{-19} C
R_S: Series resistor
R_P: Shunt resistor
K_V: Short circuit voltage coefficient
K_I: Short circuit current coefficient
KV: Short circuit voltage coefficient
T_{STC}: Temperature in Standard Test Condition

3 The Suggested Crow Search Algorithm Based Mppt

3.1 Overview of Crow Search Algorithm

The Crow Search Algorithm (CSA), introduced by Alireza [10], is inspired by the intelligence behavior of crows. Known for their problem-solving skills, memory, and tool use, crows can track and steal from others while also using deception to protect their own food. The CSA emulates this behavior with these principles:

- Crows live and search for food in groups.
- They remember food hiding locations.
- They secretly follow others to find hidden food.
- They protect their food from thieves.

Like all other swarm algorithms. It is supposed that there is a search space with d dimension contains N number of crows (swarm size). Each crow i at time iter is defined by a position vector *xi,iter* ($i = 1, \ldots, N$; $iter = 1, \ldots, itermax$) with $x_{i,iter} = [x^1_{i,iter}, x^2_{i,iter}, x^3_{i,iter}, \ldots, x^d_{i,iter}]$, *itermax* is maximum number of iterations. The hiding location of each crow i is retained in its memory. This location presents the best position that crow i achieved so far and it is presented by a memory vector $m_{i,iter} = [m^1_{i,iter}, m^2_{i,iter}, m^3_{i,iter}, \ldots, m^d_{i,iter}]$.

Crows search for food by tracking others to discover their hiding locations. At time t=iter, if crow j goes to its hiding spot, crow i may follow to find the food. Two situations arise based on crow j's awareness:

Status 1: $raw \geq AP$

Crow j is unaware of being followed by crow i. This is expressed by $raw \geq AP$ where *raw* presents a random number uniformly distributed between 0 and 1, and *AP* is the awareness probability of crow j. Crow i then updates its position towards crow j's hiding location using the equation:

$$x_{i,iter} = x_{i,iter-1} + r_i * f_l * (m_{j,iter-1} - x_{i,iter-1}) \tag{4}$$

Where *ri* is a random variable uniformly distributed between 0 and 1, and *fl* is the flight length. Small *fl* values ($fl << 1$) make the search process local, near $xi,iter-1$, while larger *fl* values ($fl > 1$) guide the algorithm towards a global search, far from $xi,iter-1$.

As shown in Fig. 2, when *fl* is less than one, the next position of $xi,iter-1$ will be between $xi,iter-1$ and $mj,iter-1$. When *fl* is greater than one, the next step may skip $mj,iter-1$.

Status 2: $raw < AP$

If crow j is aware of being followed by crow i, it will protect its hidden food by moving to a random location to deceive crow i, causing crow i to update its position accordingly.

We can combine the two statuses with the following mathematical expression:

$$\{x_{i,iter} = x_{i,iter-1} + r_i * f_l * (m_{j,iter-1} - x_{i,iter-1}) \qquad r_{aw} > AP \tag{5}$$

$$\{x_{i,iter} = r_j \quad otherwise \tag{6}$$

Intensification and diversification are key in metaheuristic algorithms, requiring balance for optimal global search. In CSA, the awareness probability (*AP*) parameter controls this balance. Increasing *AP* promotes global exploration by generating diverse solutions, while decreasing *AP* focuses the search locally, exploiting current good solutions.

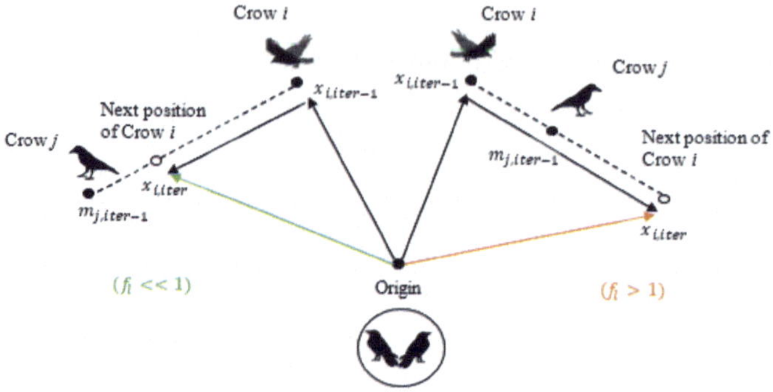

Fig. 2. How *fl* affects the position update.

3.2 CSA Implementation for MPPT

This controller automatically finds the photovoltaic generator's maximum power point (MPP) by varying the converter's duty cycle (D) based on climatic conditions. The optimization problem is represented as:

$$P_{pv} = f(d), d \in [d_{\min}, d_{\max}]$$

Where the PV system's power output is the objective function, and the duty cycle (d) represents the positions of crows. The MPP search involves:

1) Creating a population of N particles uniformly distributed in the range [0.02, 0.98].
2) Determining the best duty cycle by comparing fitness values.
3) Updating the duty cycle using the formulas (5) and (6).
4) Outputting the best value once the convergence criterion is met.
5) Restarting the process if climatic conditions change.

The flowchart depicted in Fig. 3 illustrates the sequential steps involved in the CSA controller.

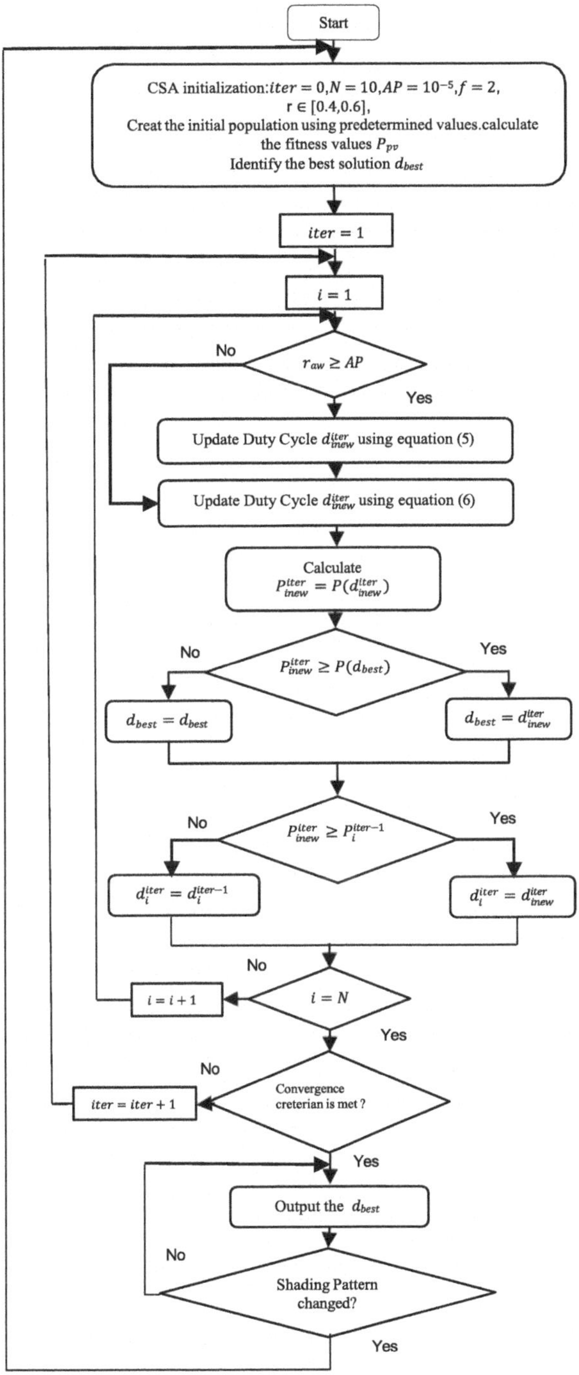

Fig. 3. CSA Flowchart

4 Simulation and Evaluation Results

4.1 Selction of CSA Parameters

The selection of the number of the particles and their positioning is very important to ensure the good performance of the MPPT controller. Different bio-inspired algorithms have different requirements for the number of agents. The selection of the number of agents depend on how a specific bio-inspired algorithm works. Some algorithms are more efficient with larger populations; while others can achieve good results with smaller populations. This number must be chosen to guarantee a good tracking accuracy and reduced convergence time. In our research, ten particles are considered. Their initial positions are uniformly distributed in the search space [0.02,0.98].The aim of this choice is to explore the different parts of the search space and determine the area which seems promising in terms of optimization.

The idea is to guide the algorithm towards potentially better solutions by searching in a promising area. This is achieved by adjusting the parameters of the algorithm to favor the exploration of the region where the probability of finding the point of maximum power is highest. In order to serve this, the parameters of the algorithm are chosen as follows:

The values assigned to parameters AP and fl are 0.0001 and 2 respectively.

To guide the algorithm towards the search in the promising zone, The random number r is limited to the range [0.4, 0.6]. These speeds up the convergence of the algorithm by avoiding searching in less promising regions.

4.2 Performance Evaluation Tests.

To evaluate the proposed algorithm, a comparative study with the PSO algorithm in terms of efficiency, convergence time and reliability is carried out. The controllers are examined according two tests. In the first test, the PV panels are exposed to three different patterns of homogenous illumination. While in the second test, the PV panels are exposed to three different patterns of partial shading conditions. The change from one pattern to another is surprising. The period of each pattern is 5 second, Figs. 4 and 5 shows the different P-V characteristics corresponding to these patterns. Throughout the duration of the tests, the temperature is maintained at 25°C. The tests were repeated 60 times in order to study the reliability of the controllers.

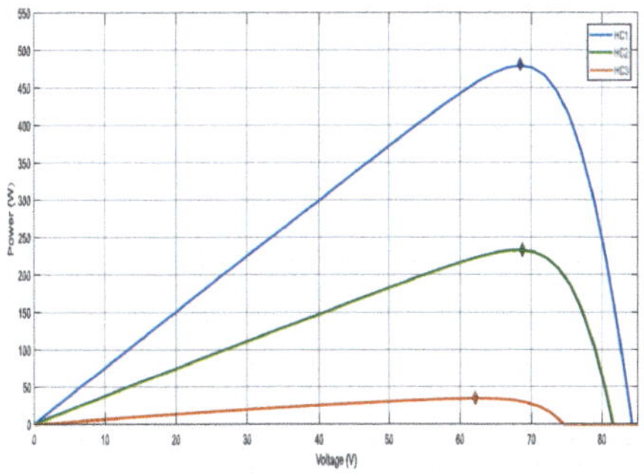

Fig. 4. The P-V characteristics of the photovoltaic generator corresponding to the first test configurations.

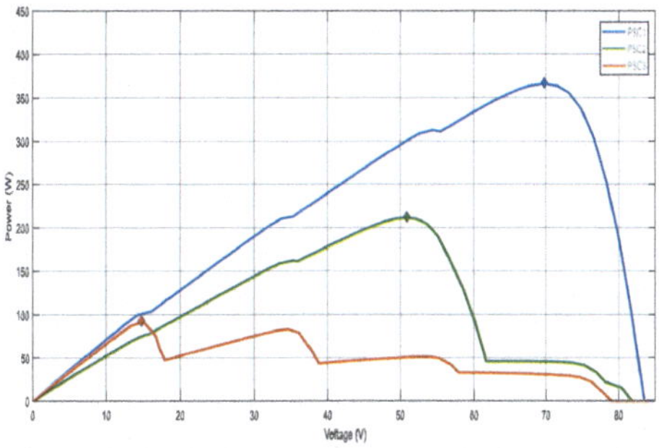

Fig. 5. The P-V characteristics of the photovoltaic generator corresponding to the second test configurations.

In the first test, the crows algorithm (CSA) demonstrated a superior ability to reach the maximum power point (MPP) quickly. The convergence times for CSA were consistently 1.5 s across all patterns, while PSO took 2.35 s, 1.6 s, and 2 s. These results are illustrated in Figs. 6 and 7. Similarly, the second test, involving partial shading conditions, highlighted CSA's efficiency, with convergence times of 1.5 s, 1.5 s, and 1 s, compared to PSO's 2 s, 2 s, and 1.8 s. These results are presented in Figs. 8 and 9. Repeating these tests 60 times, we used standard deviation to evaluate reliability. The results confirmed that the proposed algorithm is stable and efficient in all configurations. Compared to PSO, CSA proved more reliable with less result dispersion, indicating its potential to improve the stability of maximum power point tracking systems in varying solar environments. The results of the two tests are assembled and presented in Table 2.

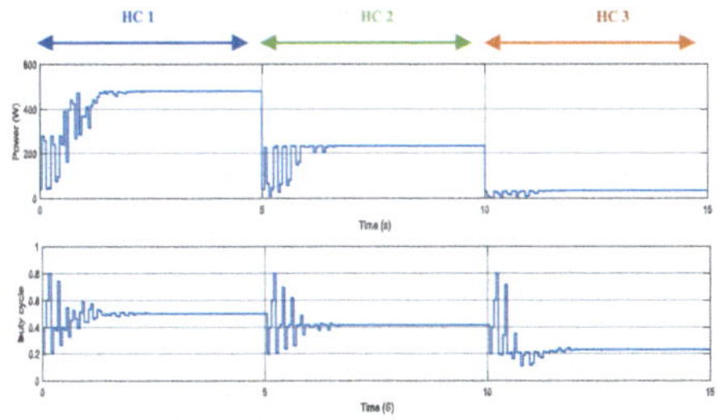

Fig. 6. Variation of PV power and duty cycle during MPP tracking using PSO algorithm under homogenous conditions.

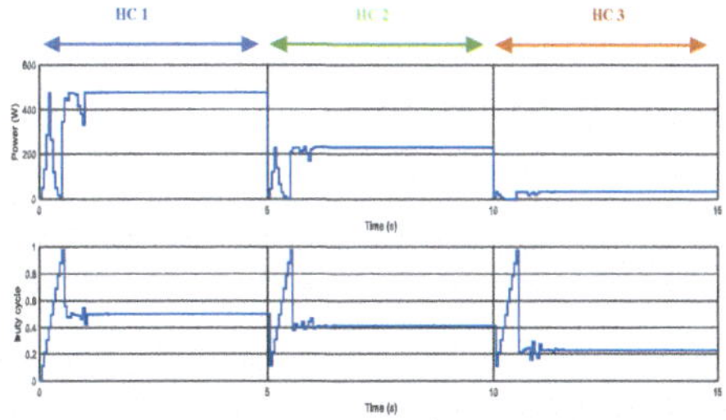

Fig. 7. Variation of PV power and duty cycle during MPP tracking using CSA algorithm under homogenous conditions.

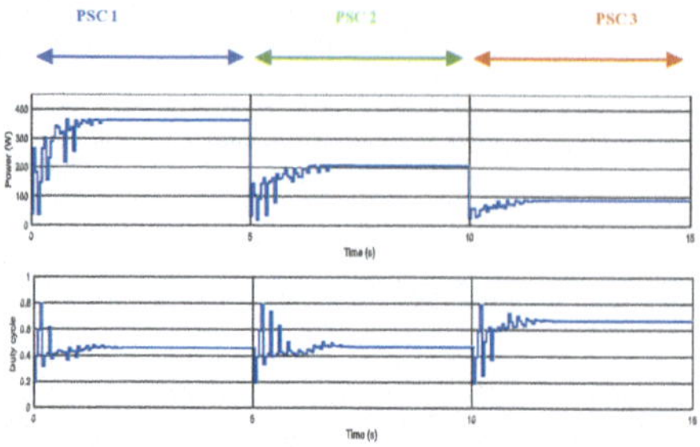

Fig. 8. Variation of PV power and duty cycle during MPP tracking using PSO algorithm under partial shading conditions.

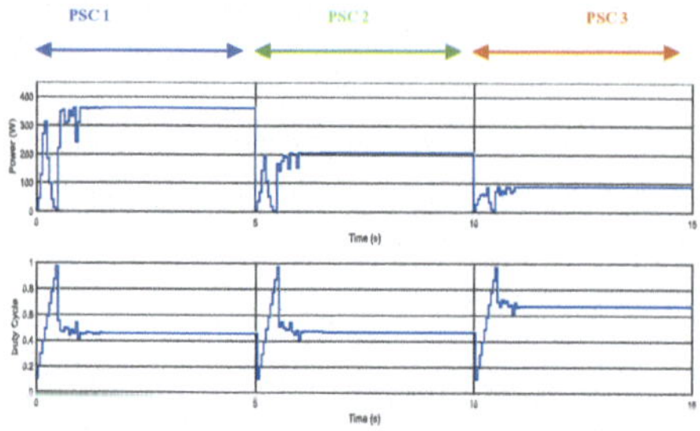

Fig. 9. Variation of PV power and duty cycle during MPP tracking using CSA algorithm under partial shading conditions.

Table 2. Outcomes of simulating tests 60 times.

Illumination Condition	Algorithm	Maximum Power (W)	Average Best Power (W)	Convergence Time (s)	Static efficacity (%)	Sttandard deviation
HC 1 [1000 1000 1000 1000 1000 1000 1000 1000]	CSA	479.2795	479.2793	1.5	99.99	0.3647
	PSO		479.2792	2.35	99.99	1.7654
HC 2 [500 500 500 500 500 500 500 500]	CSA	232.9005	232.9003	1.5	99.99	1.2184
	PSO		232.8458	1.6	99.97	1.2654
HC 3 [100 100 100 100 100 100 100 100]	CSA	34.9184	34.9142	1.5	99.98	0.4198
	PSO		34.9026	2	99.95	1.1189
PSC 1 [1000 1000 1000 1000 900 700 500 600]	CSA	365.9393	365.9382	1.5	99.99	1.6762
	PSO		365.9151	2	99.99	2.9924
PSC 2 [100 200 300 400 100 1000 1000 1000]	CSA	211.4555	211.4539	1.5	99.99	1.2788
	PSO		211.3957	2	99.97	1.58697
PSC 3 [50 100 400 800 100 200 300 1000]	CSA	92.5410	92.5390	1	99.99	0.7091
	PSO		92.5139	1.8	99.97	1.6762

5 Conclusion

In this study, we applied the CSA algorithm to develop an MPPT controller for finding the maximum power point in a photovoltaic system. We enhanced the algorithm with a 'promising zone' strategy. We analyzed the algorithm's precision, convergence time, and reliability, comparing it with the PSO algorithm under homogeneous illumination and partial shading conditions. Simulations in MATLAB/Simulink confirmed the proposed algorithm's superior accuracy, convergence time, and reliability. In future work, we plan to deploy the proposed algorithm on FPGA hardware and evaluate its performance under real temperature and irradiance conditions.

References

1. Mercedes, M., Cantarero, V.: Of renewable energy, energy democracy, and sustainable development: a roadmap to accelerate the energy transition in developing countries. Energy Res. Soc. Sci. **70**, 101716 (2020). https://doi.org/10.1016/j.erss.2020.101716
2. Abdel-Salam, M., El-Mohandes, M., Goda, M.: History of maximum power point tracking. In: Eltamaly, A.M., Abdelaziz, A.Y. (eds.) Modern Maximum Power Point Tracking Techniques for Photovoltaic Energy Systems, pp. 1–29. Springer, Switzerland (2020)
3. Belhachat, F., Larbes, C.: Comprehensive review on global maximum power point tracking techniques for PV systems subjected to partial shading conditions. Solar Energy **183**, 476–500 (2019). https://doi.org/10.1016/j.solener.2019.03.045
4. Titri, S., Larbes, C., Toumi, K.Y., Benatchba, K.: A new MPPT controller based on the Ant colony optimization algorithm for Photovoltaic systems under partial shading conditions. Appl. Soft Comput. **58**, 465–479 (2017)
5. Sangrody, R., Taheri, S., Cretu, A.-M., Pouresmaeil, E.: An Improved PSO-based MPPT technique using stability and steady state analyses under partial shading conditions. IEEE Transa. Sustain. Energy **15**(1), 136–145 (2024)
6. Kaced, K., Larbes, C., Ramzan, N., Bounabi, M., Dahmane, Z.E.: Bat algorithm based maximum power point tracking for photovoltaic system under partial shading conditions. Solar Energy **158**, 490–503 (2017)
7. Benyoucef, A.S., Chouder, A., Kara, K., Silvestre, S., Ait Sahed, O.: Artificial bee colony based algorithm for maximum power point tracking (MPPT) for PV systems operating under partial shaded conditions. Appl. Soft Comput. **32**, 38–48 (2015)
8. Abo-Khalil, A.G., Alharbi, W., Al-Qawasmi, A.R., Alobaid, M., Alarifi, I.M.: Maximum power point tracking of PV systems under partial shading conditions based on opposition-based learning firefly algorithm. Sustainability **13**, 1–18 (2021)
9. Houam, Y., Terki, A., Bouarroudj, N.: An efficient metaheuristic technique to control the maximum power point of a partially shaded photovoltaic system using crow search algorithm (CSA). J. Electr. Eng. Technol. **16**(1), 381–402 (2021)
10. Askarzadeh, A.: A novel metaheuristic method for solving constrained engineering optimization problems: crow search algorithm. Comput. Struct. **169**, 1–12 (2016)
11. Ishaque, K., Salam, Z., Taheri, H.: Simple, fast and accurate two-diode model for photovoltaic modules. Solar Energy Mater. Solar Cells **95**(2), 586–594 (2011). https://doi.org/10.1016/j.solmat.2010.09.023

Control Systems and Power Optimization

Design of Control Circuits for Dynamic Reconfiguration of Shaded Solar Panels

S. Arezki[1](✉), A. Abdelghani[2], and M. Boudour[1]

[1] Electrical and Industrial System Laboratory (LSEI), University of Science and Technology Houari Boumediene Bab Ezzouar, Bab Ezzouar, Algeria
arezki.saliha@gmail.com

[2] Department of Electrical Engineering, University of Tahri Mohamed Bechar, Bechar, Algeria

Abstract. This study highlights the critical importance of reconfiguring photovoltaic (PV) arrays to optimize energy production despite various forms of shading. The research shows that reconfiguration, whether static or dynamic, has been widely explored in different PV array topologies. A detailed modeling of PV arrays is first presented, including evaluation criteria and behavioral analysis. This preliminary study establishes the necessary foundation for a comprehensive understanding of the system and the exploration of advanced optimization methods. Two optimization methods through reconfiguration are then proposed. The first method, called modified SP, involves transitioning from the series-parallel (SP) circuit to the Total Cross-Tie (TCT) circuit, known as SP-TCT. The second method, SP_RECFG, allows for the redistribution of solar modules affected by shading. Simulations show that these approaches optimize the power output of PV arrays even in complex shading scenarios. A comparative study also demonstrates the superior efficiency of the SP_RECFG method. Additionally, the theoretical approaches have been experimentally validated. The SP-TCT method was successfully implemented, minimizing power losses due to partial shading. This research demonstrates that reconfiguring PV arrays is essential for maximizing solar energy production and contributing to a sustainable energy transition.

Keywords: Dynamic reconfiguration · Photovoltaic (PV) array · Partial shading · SP-TCT method · $SP_{\_RECFG}$ reconfiguration method · Experimental validation

1 Introduction

Photovoltaic (PV) system reconfiguration is essential for optimizing the performance of solar installations, especially under partial shading conditions. This approach aims to minimize power losses by adjusting the connections of PV modules. The evolution of reconfiguration techniques, particularly series-parallel (SP) and Total Cross-Tied (TCT) configurations, has been extensively studied in recent literature.

TCT and SP configurations are among the most studied techniques for PV system reconfiguration. Research has shown that TCT configurations can offer significant benefits in terms of energy efficiency under complex shading conditions. Ali and Kumar

© The Author(s), under exclusive license to Springer Nature Switzerland AG 2025
M. Hatti (Ed.): IC-AIRES 2024, LNNS 1238, pp. 99–115, 2025.
https://doi.org/10.1007/978-3-031-80301-7_12

(2023) explored optimized reconfiguration strategies for PV networks, highlighting that the TCT configuration often performs better in the presence of shading dispersion [1]. Similarly, Chen, Yang, and Zhang (2022) combined TCT and SP strategies, demonstrating that hybrid approaches can enhance PV system performance in dynamic shading scenarios [2].

Desai and Mikkili (2019) conducted a detailed analysis of PV configurations, including TCT and SP. Their study modeled and analyzed various configurations such as TCT, series-parallel, and others to extract maximum power under partial shading conditions. Results revealed the specific advantages of each configuration based on shading conditions [3]. Meanwhile, Pendem and Mikkili (2018) examined the performance of series, series-parallel, and honeycomb PV configurations. Their research indicated that although the series-parallel configuration is advantageous in some cases, TCT configurations remain competitive for more complex shading situations [4]. The Firefly algorithm was introduced by Nazeri et al. (2021) for reconfiguring PV networks. This method optimizes power extraction under mismatch conditions, showing that bio-inspired algorithms can significantly improve PV system performance under variable shading conditions [5]. Rao and Kumar (2023) also investigated the impact of reconfiguration strategies on PV system performance, highlighting the effectiveness of advanced techniques in dynamic shading scenarios [6]. Boghdady et al. (2022) conducted a comparative study of optimal PV configurations and Maximum Power Point Tracking (MPPT) techniques under partial shading with fast dynamic changes in hybrid loads. Their research demonstrated that optimal configurations combined with improved MPPT techniques can significantly enhance PV system performance [7]. Khan and Ali (2022) explored dynamic reconfiguration approaches by comparing TCT and SP configurations, finding that dynamic methods offer significant benefits for improving PV system performance in fluctuating shading conditions [8].

Recent research indicates a trend towards optimizing photovoltaic configurations through advanced reconfiguration approaches, including TCT and SP configurations. Studies have shown that these techniques can significantly improve the performance of PV systems under partial shading. Optimization algorithms and hybrid approaches continue to evolve, providing increasingly effective solutions for maximizing solar energy production. For example, the study by Liu et al. (2023) proposed a dynamic reconfiguration method based on intelligent sensor networks to mitigate the effects of partial shading, demonstrating a significant improvement in energy production [9]. Similarly, Zhang et al. (2022) developed a hybrid optimization algorithm combining physical reassignment and dynamic reconfiguration techniques, allowing for better management of temperature variations [10]. Another study by Kim et al. (2021) highlighted the use of advanced Maximum Power Point Tracking (MPPT) systems to improve the overall efficiency of PV systems under varied shading conditions [11].

However, it should be noted that dynamic reconfiguration methods require a substantial number of sensors, switches, and a complex control algorithm [12]. Concurrently, physical reassignment methods may present limitations in terms of efficiency and wiring complexity [12]. In this context, it becomes essential to formulate a simple, cost-effective, and efficient algorithm to achieve better energy harvesting in PV systems.

This study aims to address this challenge by proposing an innovative and rigorous approach to optimize the energy production of PV systems under unfavorable conditions [12]. Through extensive research, experimental tests, and simulations, we will develop and evaluate various optimization strategies, focusing on efficiency. The results of this study will contribute to enhancing the efficiency, reliability, and cost- effectiveness of PV systems, thereby promoting their widespread adoption [12]

2 Analysis of SP-TCT and Reconfigured SP Configurations for Optimal Shading Management

In this section, we will present new configurations inspired by the basic SP configuration, as well as reconfiguration mechanisms between panels to reduce power losses due to shading.

2.1 Series-Parallel Configuration (SP)

In a series-parallel configuration (Fig. 1), photovoltaic modules are connected both in series and in parallel. This configuration is the most commonly used in photovoltaic fields due to its numerous advantages. It optimizes the system voltage, provides redundancy in case of solar module failure, allows for design flexibility, facilitates shade management, and offers scalability.

$$P_{champ\_SP} = V_{SP} \times I_{SP} \tag{1}$$

$$V_{SP} = \sum V_i = V_1 + V_2 \tag{2}$$

$$I_{SP} = \sum I_j = I_1 + I_2 \tag{3}$$

where $p_{\text{champ-SP}}$, V_{SP} , I_{SP} respectively denote the total power of the array, the voltage, and the current of the PV array.

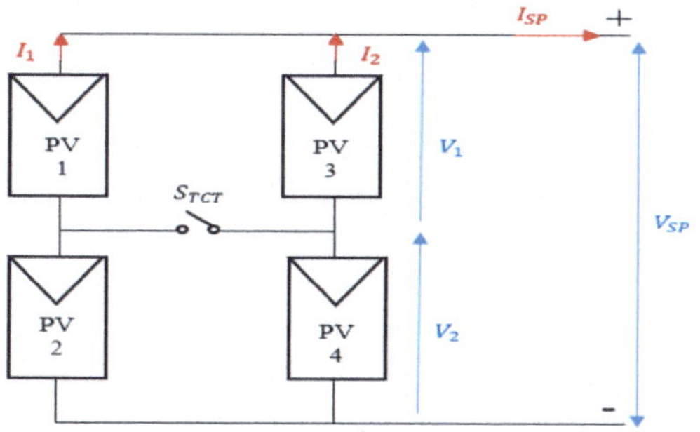

Fig. 1. Serie-parallel SP configuration.

2.2 Improvement of the SP Configuration (SP-TCT)

The Total-Cross-Tied (TCT) configuration is a specific enhancement of the series-parallel (SP) configuration, where solar modules are interconnected with adjacent modules. This configuration improves shading tolerance and the overall system performance. However, it should be noted that the TCT configuration requires additional cables, resulting in higher production costs. Considering the benefits of this configuration while minimizing the number of cables, we propose introducing a relay (STCT switch) between the two branches. The control of this relay will be based on a well-defined algorithm, which will be presented below. For the modeling of this configuration, the STCT switch is considered closed (Fig. 2).

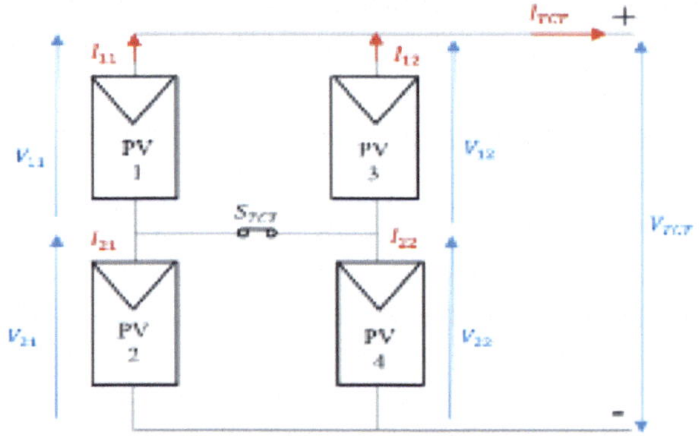

Fig. 2. Reconfiguration Total-Cross-Tied TCT.

For any PV module in the jth row and kth column, the current generated by the PV module at the irradiation level "g" is given by: [3, 13]

$$I_{jk} = g_{jk} I_m \tag{4}$$

where Im is the current generated by the module at standard irradiation $G_{STC} = 1000\ W//m^2$: $g_{jk} = G_{jk}/G_{STC}$

$$I_{11} = I_{21};\ I_{12} = I_{22} \tag{5}$$

$$I_{TCT} = I_{11} + I_{12} = I_{21} + I_{22} \tag{6}$$

$$V_{11} = V_{12};\ V_{21} = V_{22} \tag{7}$$

$$V_{TCT} = V_{11} + V_{21} = V_{12} + V_{22} \tag{8}$$

Similarly, for "m" rows and "n" columns, the equation is as follows:

$$\left.\begin{array}{l} I_{11} = I_{21} = I_{31} = I_{m1} \\ I_{12} = I_{22} = I_{32} = I_{m2} \\ \quad\quad\quad\quad \ldots \\ I_{1n} = I_{2n} = I_{3n} = I_{mn} \end{array}\right\} \tag{9}$$

And

$$\left.\begin{aligned} I_{TCT} &= I_{11} + I_{12} + I_{13} \ldots I_{1n} \\ &= I_{21} + I_{22} + I_{23} \ldots I_{2n} \\ &\quad \ldots \\ &= I_{m1} + I_{m2} + I_{m3} \ldots I_{mn} \end{aligned}\right\} \tag{10}$$

$$\left.\begin{aligned} V_{11} &= V_{12} = V_{13} \ldots V_{1n} \\ V_{21} &= V_{22} = V_{23} \ldots V_{2n} \\ &\quad \ldots \\ V_{m1} &= V_{m2} = V_{m3} \ldots V_{mn} \end{aligned}\right\} \tag{11}$$

And

$$\left.\begin{aligned} V_{TCT} &= V_{11} + V_{21} + V_{31} \ldots V_{m1} \\ &= V_{12} + V_{22} + V_{32} \ldots V_{m2} \\ &\quad \ldots \\ &= V_{1n} + V_{2n} + V_{3n} \ldots V_{mn} \end{aligned}\right\} \tag{12}$$

The reconfiguration strategy algorithm is given below (Fig. 3).

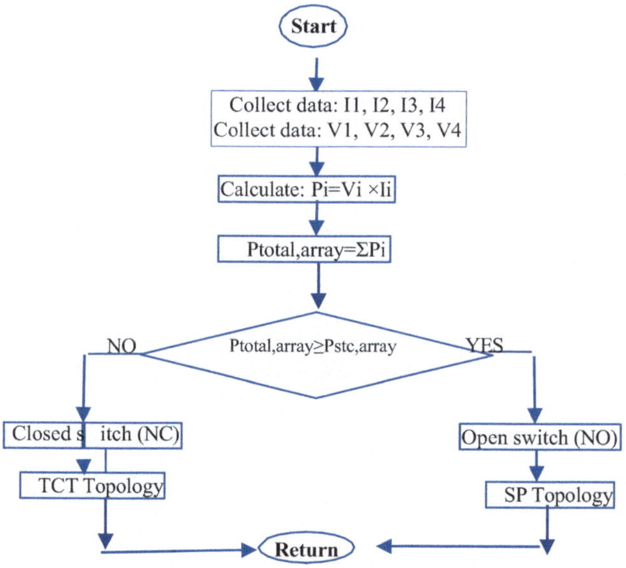

Fig. 3. Flowchart of the SP-TCT reconfiguration (Modified SP).

2.3 Proposed Configuration: SP_RECFG

When shading occurs, it has been observed that the current is significantly affected, while the voltage can be influenced by other factors, such as temperature. This study focuses on power, which is calculated by multiplying voltage by current (P = V × I). To find an optimal solution, current is considered a crucial parameter. Therefore, reconfiguration will be performed by grouping the lines with the lowest currents onto the same branch. This approach aims to improve the output power of the array by minimizing shading losses. Rearranging the lines in this manner can reduce the negative effects of shading on current and increase solar energy production in the overall array configuration.

The mechanism of the proposed reconfiguration is illustrated in Fig. 4.

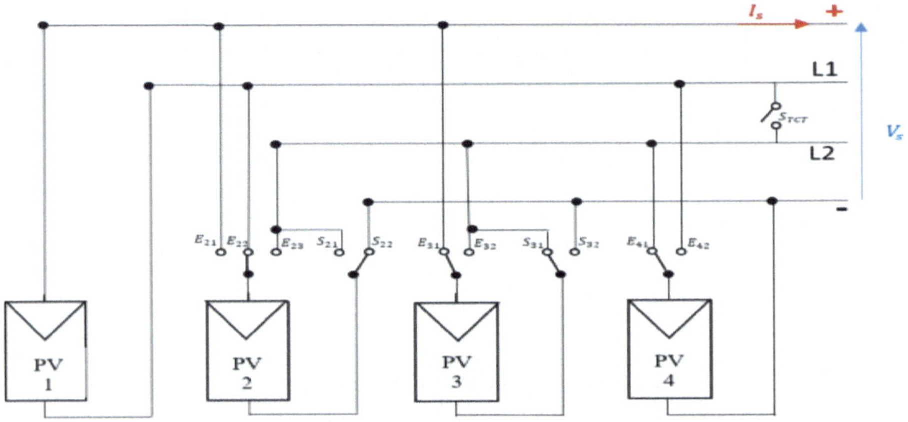

Fig. 4. Schematic of the modified SP-TCT circuit for a 22 PV array.

The photovoltaic array studied is 2 × 2 in dimensions. The specifications of the photovoltaic installation are summarized in Table 1.

Table 1. The specifications of the 80 W monocrystalline panel used.

Characteristic	Value
Maximum Power Pmax	80 Wp
Power Tolerance	−5/ + 10%
Open Circuit Voltage Voc	22.5 V
Maximum Power Point Voltage Vmpp	18.5 V
Short Circuit Current Isc	4.66 A
Maximum Power Point Current Impp	4.35 A

2.3.1 SP$_{\text{RECFG}}$ Configuration Principle

To study different scenarios and analyze shading losses resulting from this new reconfiguration, an initial setup is proposed that enables switching between the rows of the array. This setup allows for the adjustment of solar panel positions using the relays E_{21},E_{22},E_{23},S_{21},S_{22},E_{31},E_{32},S_{31},S_{32},E_{41},E_{42} (Figs. 5, 6 and 7).

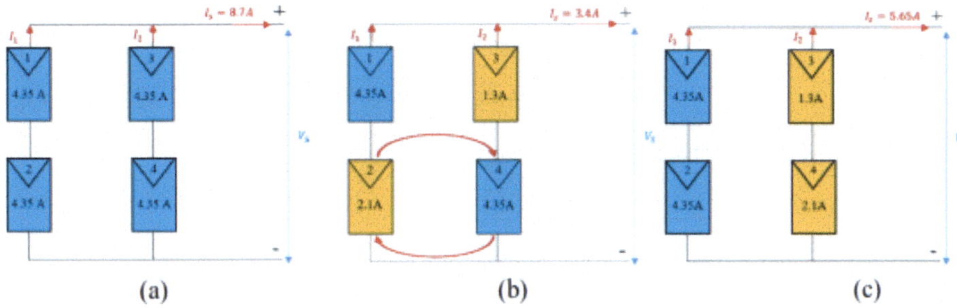

Fig. 5. Mechanism of the reconfiguration: (a) healthy state, (b) under shading, and (c) after reconfiguration.

The reconfiguration procedure is summarized as follows:

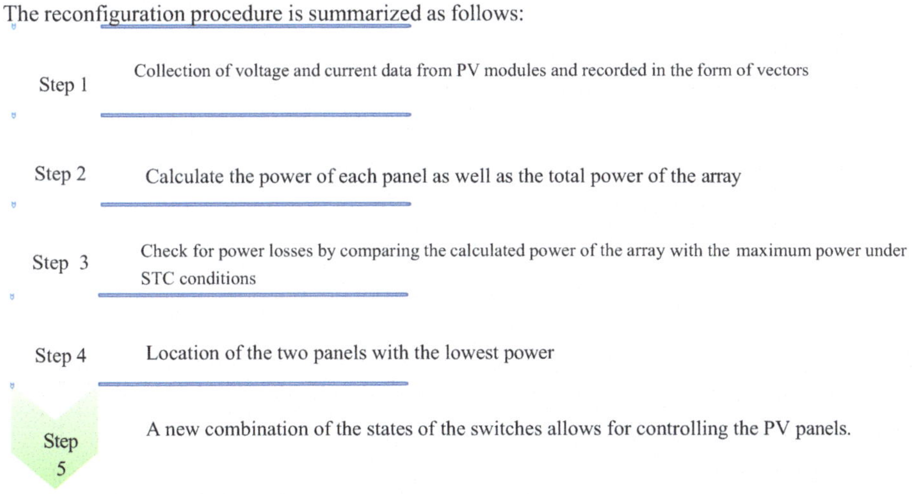

Fig. 6. Reconfiguration Procedures.

The following flowchart illustrates the procedure for the reconfiguration mechanism:

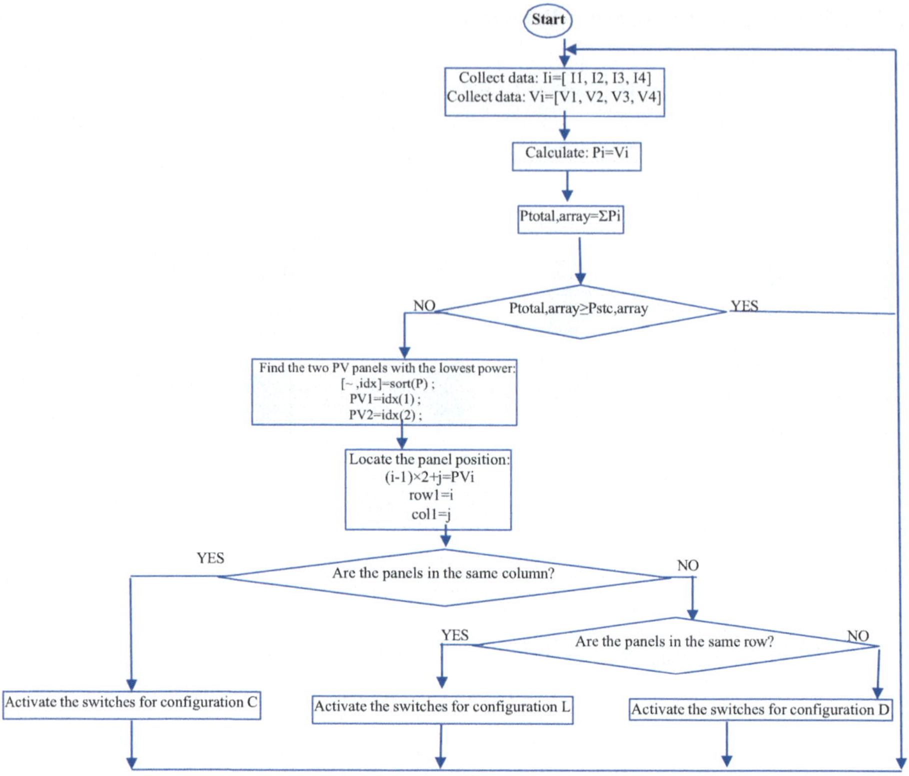

Fig. 7. Flowchart of the Reconfiguration ***SP_RECFG***.

Based on the identified shading scenarios, the different states of the relays are developed and summarized in Table 2.

Table 2. Relay States of the Circuit ***SP_RECFG.***

State of Relays	E_{21}	E_{22}	E_{23}	E_{21}	E_{22}	E_{31}	E_{32}	E_{31}	E_{32}	E_{41}	E_{42}	S_{TCT}
Reconfiguration C	1	0	0	1	0	0	1	0	1	1	0	0
Reconfiguration L	0	1	0	0	1	1	0	1	0	1	0	0
Reconfiguration D	0	0	1	0	1	1	0	1	0	0	1	0

3 Simulation Performance for Configuration Evaluation

To evaluate the performance of the proposed configurations, simulations were conducted at a temperature of 25 °C and for various shading scenarios described below (Fig. 8):

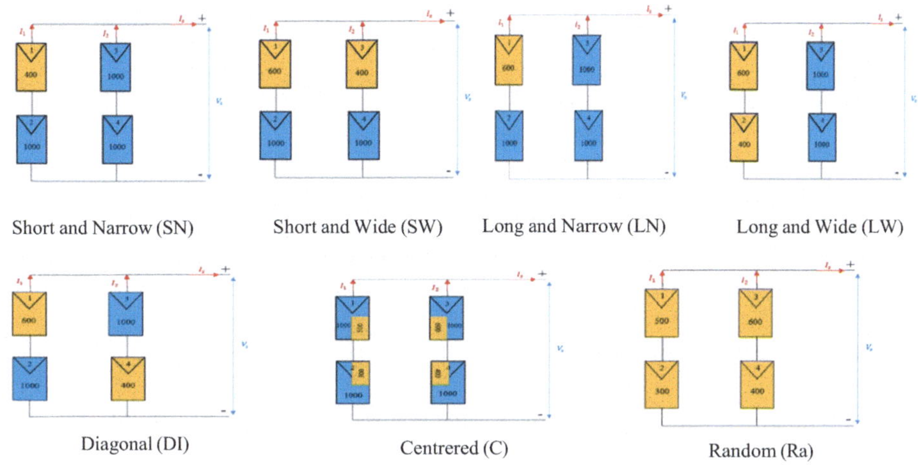

Fig. 8. The Different Shading Scenarios Studied

The results of the simulation for the six shading scenarios for the SP and SP-TCT configurations are presented in a Table 3, summarizing the key findings from the simulated I-V and P-V characteristics.

Table 3. Simulation Results under Various Partial Shading Conditions.

Topologies	Shading types	Voc(V)	Isc(A)	Pmax(W)	Vm(V)	Im(A)	Losses(%)	FF (%)	η (%)
SP	SN/LN	44.47	9.32	231.16	37.9	6.1	29	55.8	71.1
	SW	44	9.32	172.5	38.6	4.47	47	52.4	53.1
	LW	44.16	7.46	228.766	37.45	6.11	30	69.4	70.4
	Ra	42.7	5.13	115.44	37.16	3.1	64	52.7	35.5
	C	43.8	9.32	143.74	28.83	4.99	56	35.2	44.2
	DI	44	9.32	172.53	38.31	4.5	47	44.1	53.1
TCT	SN/LN	44.47	9.32	240.47	38.18	6.3%	26	58	83.2
	SW	44	9.32	172.74	38.62	4.52	47	42.1	53.2
	LW	44.13	7.46	232.4	37.11	6.26	29	70.6	71.5
	Ra	42.7	5.13	115.44	37.16	3.1	64	52.7	35.5
	C	43.8	9.32	143.8	28.83	4.99	56	35.2	44.2
	DI	44.16	7.56	232.34	37.17	6.25	29	69.6	71.5

Based on the I-V and P-V characteristics results, the following observations are made:

- **In the series-parallel configuration:**
  - Power losses are low in the short-narrow and long-wide shading cases (28.9% and 30%, respectively), with efficiencies of 70.1% and 70%.
  - Power losses are significantly higher in the short-wide, diagonal, centered, and random shading cases (46.9%, 47%, 55.8%, and 64%, respectively), resulting in lower efficiencies (53.1%, 53%, 44.2%, and 35.5%).
- **In the TCT reconfiguration:**
  - For the short-narrow shading case, relatively low power losses of 26% are observed, with an efficiency of 73.2%.
  - In contrast, the short-wide, centered, and random shading cases show higher power losses (46.8%, 55.8%, and 64%, respectively), which correspond to lower efficiencies (53.2%, 44.2%, and 35.5%).
  - The long-wide and diagonal shading cases exhibit similar power losses around 28.5%, with an efficiency of 71.5%.

For the SP configuration, the results before and after the $SP_{\_RECFG}$ reconfiguration can be compared to assess the improvements achieved (Fig. 9).

The summary and comparison of the results for the various scenarios studied, both before and after reconfiguration, are presented in the following Table 4.

The results of all the studied scenarios are summarized in the diagram shown in Fig. 10.

The analysis of photovoltaic system performance under various configurations shows that for**:**

Performance under Different Configurations

- Modified reconfigurations (SP-modified and TCT-modified) provide notable improvements over standard configurations.
- The TCT reconfiguration tends to deliver slightly higher maximum power and better overall efficiency compared to the SP reconfiguration.

Impact of Topologies

- Certain topologies, such as LL and DI, respond better to reconfigurations, showing significant increases in power and efficiency.
- Other topologies, like AL, exhibit weaker performance, indicating greater sensitivity to shading effects.

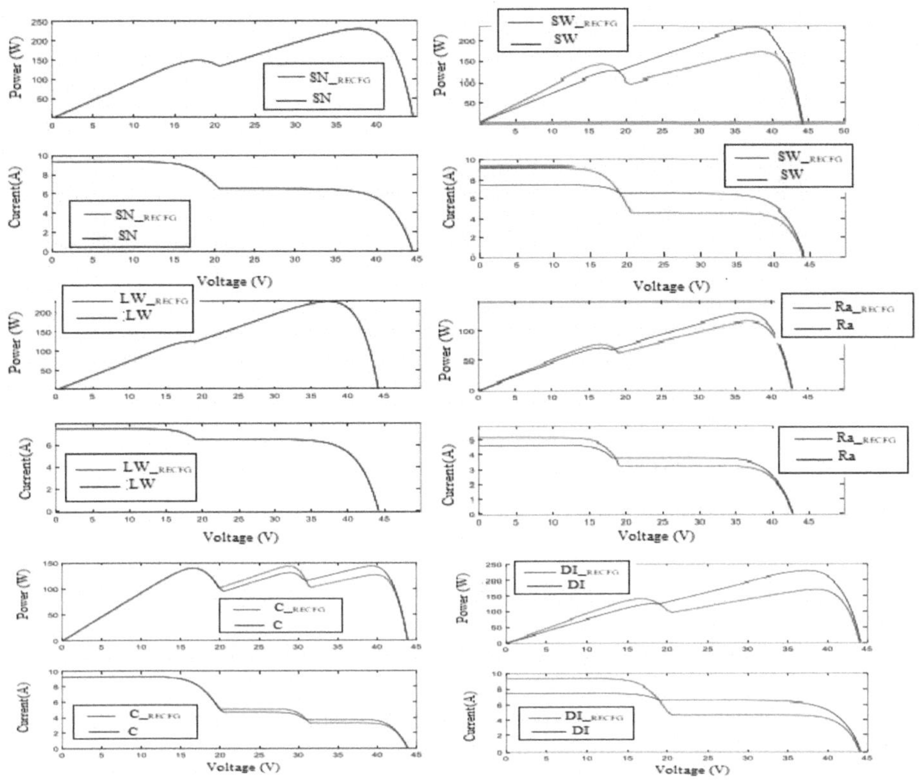

Fig. 9. P-V and I-V Characteristics of SP and SP$_{_RECFG}$.

Table 4. Simulation Results for Various Partial Shading Conditions after Reconfiguration.

Topologies	Type of shading	Voc(V)	Isc (A)	Pm (W)	Vm(V)	Im (A)	Losses (%)	FF (%)	η (%)
Reconfiguration with SP (SP_RECFG)	SN_RECFG/LN_RECFG	44.47	9.32	231.16	37.9	6.1	29	0.56	71
	SW_RECFG	44.16	7.46	228.766	37.45	6.11	30	0.42	70.4
	LW_RECFG	44.16	7.46	228.766	37.45	6.11	30	0.69	70.4
	Ra_RECFG	42.7	4.66	129.67	36.34	3.57	60	0.48	39.8
	C_RECFG	43.8	9.32	143.64	28.83	4.99	56	0.41	44
	DI_RECFG	44.16	7.46	228.766	37.45	6.11	30	0.42	70.4
Reconfiguration with TCT (TCT_RECFG)	SN_RECFG/LN_RECFG	44.47	9.32	240.47	38.18	6.3%	26	0.58	74
	SW_RECFG	44.13	7.46	232.4	37.11	6.26	29	0.42	71.5
	LW_RECFG	44.13	7.46	232.4	37.11	6.26	29	0.71	71.5
	Ra_RECFG	42.7	4.66	129.67	37.34	3.56	10	0.48	39.8
	C_RECFG	43.8	9.32	143.8	28.83	4.99	56	0.42	44
	DI_RECFG	44.13	7.46	232.4	37.11	6.26	29	0.7	71.5

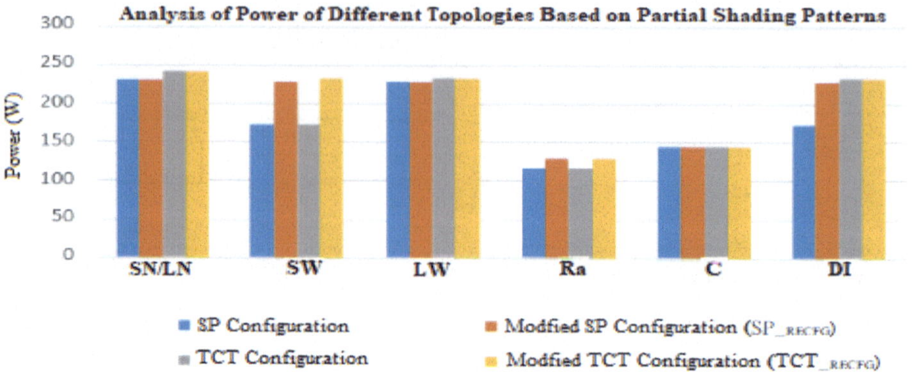

Fig. 10. Analysis Diagram Between Different Reconfigurations Under Various Shading Distributions.

Optimization Recommendations

- Adopting modified TCT configurations can maximize power and efficiency in partially shaded environments.
- However, this requires additional wiring and monitoring systems, which can increase costs.
- The TCT configuration offers good results but is still outperformed by the TCT-modified reconfiguration, which reduces losses and costs.

It is interesting to note that while the TCT configuration offers good results compared to the SP configuration, it is outperformed by the TCT-modified reconfiguration. The latter can be considered a viable solution offering advantages such as reduced losses and costs. Therefore, for photovoltaic systems located in partially shaded areas, investing in a TCT-modified reconfiguration could prove profitable in the long term due to increased efficiency and maximum power output.

4 Experimental Validation

4.1 Presentation of the Schematics of the Realized Circuits: Circuit "SP-TCT"

Figure 11 describes the realized SP-TCT circuit with its components. The overall system is divided into three main parts. The first part involves the configuration of photovoltaic panels arranged in a 2 $\times$ 2 topology, with each panel having a nominal power of 80 V. The second part concerns the control and measurement unit, which is housed in a cabinet and plays a crucial role in the reconfiguration of the SP-TCT field interconnection. Finally, the last part includes the load, which in our case consists of five variable resistors connected in series with a rating of 3.3 Ω–10 A.

Fig. 11. Realized SP-TCT Circuit.

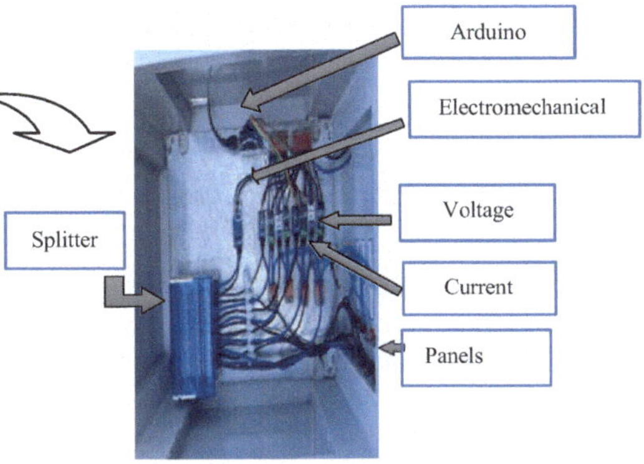

Fig. 12. Cabinet of the Implemented SP-TCT Circuit.

4.2 Description of the Control System

The control section of the system is integrated into the cabinet (Fig. 12). It includes several key components that are crucial for the operation of the SP-TCT circuit. These components are:

- **4 Voltage Measurement Sensors (25 Vdc)**: Each sensor monitors the voltage of each panel in real-time, with a maximum range of 25 Vdc. They provide precise voltage level information for monitoring and ensure compatibility with the Arduino.
- **4 Current Sensors (20 A)**: Each sensor is specifically designed for use with the Arduino and measures the current in real-time from the panel, with a maximum range of 20 A. They offer accurate readings of current levels.
- **1 Electromechanical Relay**: This component plays a critical role in the system. It functions as a switch controlled by the Arduino Nano, enabling real-time adjustment of the interconnections between the photovoltaic modules.
- **Arduino Nano**: The Arduino Nano is a programmable development board typically featuring 14 digital pins (6 of which can be used for PWM outputs) and 8 analog pins for analog I/O. It is essential for the system, as it reads real-time data from the voltage and current sensors and controls the relay based on a pre-loaded program.

4.3 Operation of the SP-TCT System

The operation of the SP-TCT system is based on continuous monitoring of the photovoltaic modules' performance and dynamic real-time reconfiguration to optimize output power and minimize losses caused by various types of faults, such as shading. The program, developed from the previous algorithm, enables dynamic reconfiguration of the PV module interconnections when the power generated by the array decreases in real time. The algorithm analyzes voltage and current measurements from the modules, identifies shaded areas or degraded modules, and determines the optimal configuration for connecting the modules. When shading variations are detected, the relay is controlled by a signal from the Arduino to switch from series-parallel connection mode to TCT (Total Cross-Tie) mode, thereby enhancing the overall output power.

5 Results of the Experimentation and Interpretations

Several test scenarios under different shading conditions were conducted to validate the proposed idea for reconfiguration while minimizing power losses. The results for a 2 * 2 SP PV array under E = 990 W/m^2, T = 30 °C (Bab Ezzouar, June) were recorded under various shading fault conditions:

- Shading fault SN with one panel shaded at 50%
- Shading fault SW with two panels shaded at 50% and 25%
- Shading fault LW with two panels shaded at 50% and 25%
- Shading fault Ra with four panels shaded at 0%, 50%, 25%, and 75%
- Shading fault DI with two panels shaded at 50% and 25%
- Shading fault C with four panels shaded at 25%

Below, the I-V and P-V characteristics corresponding to these results are presented in Figs. 13, 14, 15, 16, 17 and 18.

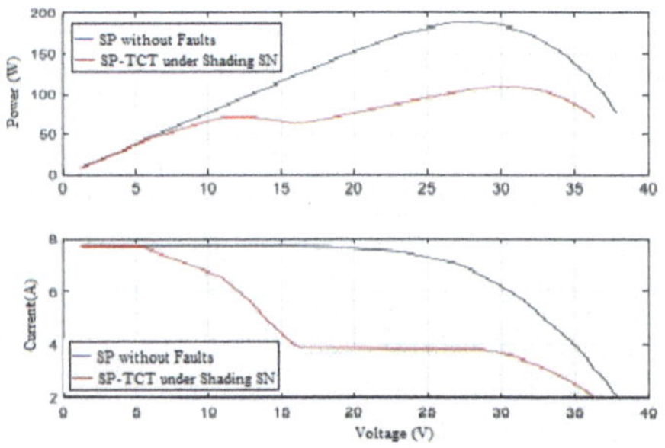

Fig. 13. I-V and P-V Characteristics of an SP-TCT Array with Shading Fault SN.

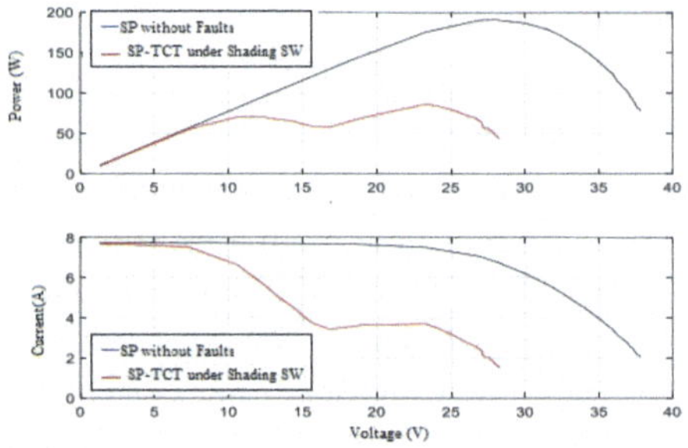

Fig. 14. I-V and P-V Characteristics of an SP-TCT Array with Shading Fault SW.

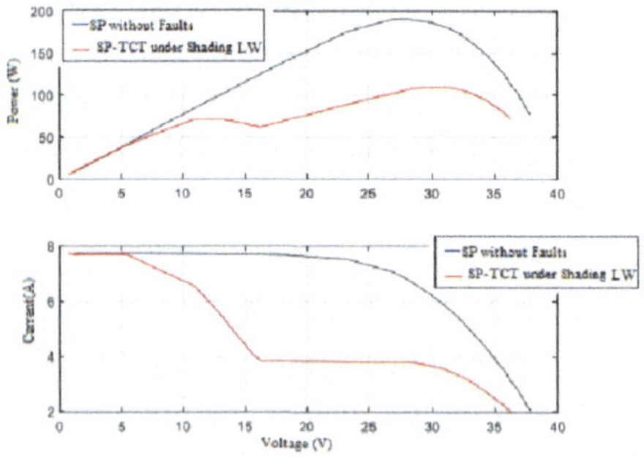

Fig. 15. I-V and P-V Characteristics of an SP-TCT Array with Shading Fault LW.

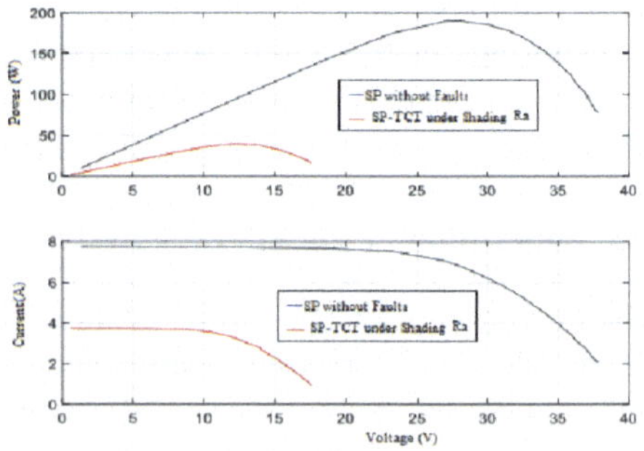

Fig. 16. I-V and P-V Characteristics of an SP-TCT Array with Shading Fault Ra.

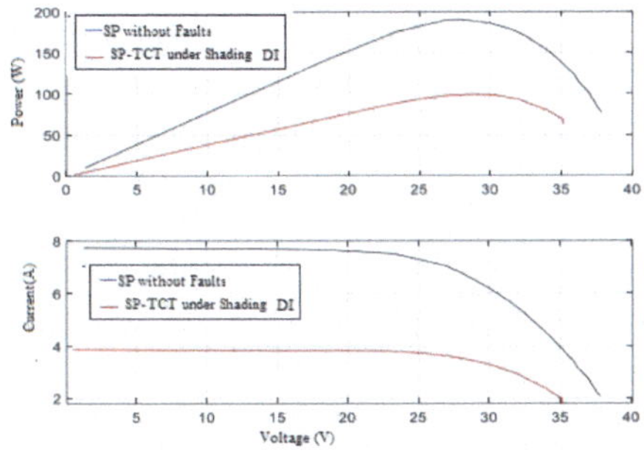

Fig. 17. I-V and P-V Characteristics of an SP-TCT Array with Shading Fault DI.

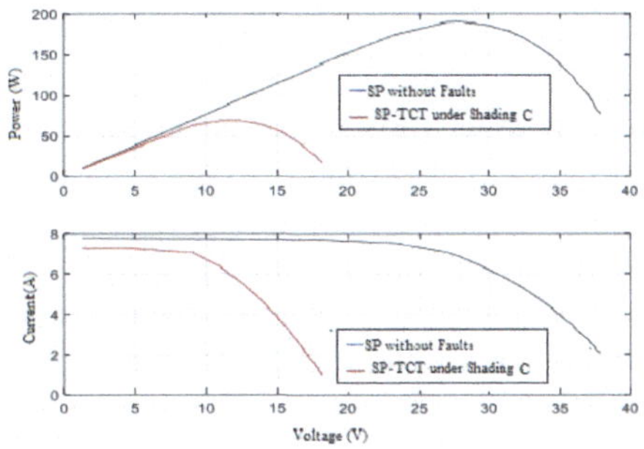

Fig. 18. I-V and P-V Characteristics of an SP-TCT Array with Shading Fault C.

Based on the I-V and P-V characteristics results, we observe the following:

- Power losses are very high in the cases of short-wide, centered, and random shading (54.7%, 63%, 79%), resulting in low efficiencies (45.3%, 36.5%, 20.9%).
- Power losses in the case of centered shading are lower compared to the others (42%), with an efficiency of 57.8%.
- Power losses in the cases of long-wide and diagonal shading are almost the same (49%, 47.6%), with efficiencies of 51% and 52.4%, respectively.

The table below summarizes the effectiveness of the SP-TCT technique under different shading conditions. This effectiveness is reflected by the parameters of losses, fill factor, and efficiency (Table 5).

Table 5. Performance of the SP-TCT technique under shading effects.

Topologie	Voc(V)	Isc (A)	Pm(W)	Vm(V)	Im(A)	Losses (%)	FF (%)	η(%)
SP-TCT	37.8	7.66	190.4	28	6.8	0	66.76	100
SN	36.3	7.72	110.05	31	3.55	42	39.3	57.8
LW	35.2	3.87	97.1	28.3	3.43	49	71.3	51
SW	28.2	7.66	86.21	23.3	3.7	54.7	39.9	45.3
Ra	17.61	3.75	39.76	12.31	3.23	79	60.2	20.9
C	18.14	7.26	69.52	12.37	5.62	63	52.79	36.5
DI	35.1	3.88	99.7	28.9	3.45	47.6	73.21	52.4

6 Conclusion

This paper proposed two main branching approaches for photovoltaic (PV) arrays to address the challenge of power losses caused by shading. The first approach introduces the series-parallel (SP) circuit, which can be reconfigured into a Total Cross-Tie (TCT) circuit, termed SP-TCT. This reconfiguration is triggered in the event of power degradation due to shading. Extensive tests were conducted to analyze the performance of this method under various shading conditions. The second approach involves relocating and rearranging low-power PV panels to maximize overall efficiency by primarily utilizing the most efficient modules. The results demonstrate that the SP-TCT reconfiguration effectively optimizes output power, even in complex shading scenarios, showing superior performance compared to the physical rearrangement of panels. These methods were validated through practical testing, confirming their ability to reduce power losses and enhance the efficiency of PV systems. In conclusion, these innovative reconfiguration approaches offer significant solutions for improving the performance of PV arrays under shading conditions, providing valuable tools to maximize energy production.

References

1. Ali, M., Kumar, P.: Optimized reconfiguration strategies for PV networks. J. Solar Energy **79**, 123–135 (2023). https://doi.org/10.1016/j.solener.2023.01.045
2. Chen, W., Yang, S., Zhang, X.: Enhanced photovoltaic array reconfiguration for partial shading using hybrid TCT and series-parallel strategies. Sol. Energy **242**, 148–159 (2022). https://doi.org/10.1016/j.solener.2022.09.007
3. Khan, M.S., Ali, J.: Dynamic reconfiguration of PV arrays based on TCT and SP configurations for improved performance in shaded conditions. Energy Rep. **8**, 118–131 (2022). https://doi.org/10.1016/j.egyr.2022.09.004
4. Kim, J., et al.: Advanced MPPT systems for PV efficiency. Renew. Power Gener. J. **15**(4), 531–546 (2021). https://doi.org/10.1049/rpg2.12039
5. Zhang, Y., et al.: Hybrid optimization algorithm for PV systems. Int. J. Energy Res. **46**(7), 8250–8265 (2022). https://doi.org/10.1002/er.5650
6. Liu, H., et al.: Dynamic reconfiguration method based on intelligent sensor networks. Sol. Energy Mater. Sol. Cells **258**, 112361 (2023). https://doi.org/10.1016/j.solmat.2023.112361
7. Rao, P.V., Kumar, P.: Impact of reconfiguration strategies on PV array performance under dynamic partial shading conditions. J. Sol. Energy Eng. **145**(2), 021010 (2023). https://doi.org/10.1115/1.4056078
8. Nazeri, M.N.R., Tajuddin, M.F.N., Babu, T.S., Azmi, A., Malvoni, M., Kumar, N.M.: Firefly algorithm-based photovoltaic array reconfiguration for maximum power extraction during mismatch conditions. Sustainability **13**(6), 3206 (2021). https://doi.org/10.3390/su13063206
9. Desai, A.A., Mikkili, S.: Modelling and analysis of PV configurations (alternate TCT-BL, total cross-tied, series, series- parallel, bridge-linked and honeycomb) to extract maximum power under partial shading conditions. CSEE J. Power Energy Syst. **99**, 1670–1683 (2019). https://doi.org/10.17775/CSEEJPES.2020.00900
10. Boghdady, T.A., Kotb, Y.E., Aljumah, A., Sayed, M.M.: Comparative study of optimal PV array configurations and MPPT under partial shading with fast dynamical change of hybrid load. Electrical Power and Machines Department, Faculty of Engineering, Cairo University, Giza 12613, Egypt (2022)
11. Pendem, S.R., Mikkili, S.: Simulation and performance analysis of solar PV array configurations (series, series-parallel, and honey-comb) to extract maximum power under partial shading conditions. Energy Reports **4**, 274–287 (2018). https://doi.org/10.1016/j.egyr.2018.03.003
12. Wang, Y.-J., Hsu, P.-C.: An investigation on partial shading of PV modules with different connection configurations of PV cells. Energy **36**(6), 3692–3701 (2011). https://doi.org/10.1016/j.energy.2011.03.053

13. Chowdhury, S., Yadav, A., Das, S.: Dynamic performance and cost analysis of a grid-tied photovoltaic system. IEEE Trans. Sustain. Energy **10**(3), 1393–1402 (2019). https://doi.org/10.1109/TSTE.2018.2865974

Optimal Design and Tuning of TID Power System Stabilizer for Multi-machine Power System Using Dandelion Optimizer

Intissar Hattabi[1](✉), Aissa Kheldoun[2], Rafik Bradai[3], and Soufian Khettab[1]

[1] SET Laboratory, Blida 1 University, Faculty of Technology, Electrical and Control Department, Blida, Algeria
hattabi_intissar@univ-blida.dz

[2] Laboratory of Signals and Systems, Institute of Electrical and Electronic Engineering, University M'hamed Bougara, 35000 Boumerdes, Algeria

[3] LATSI, Blida 1 University, Faculty of Technology, Electronics Department, 09000 Blida, Algeria

Abstract. In this study, we explore the enhancement of stability in a four-machine power system through the optimized tuning of Tilt-Integral-Derivative Power System Stabilizer (TID-PSS) parameters. We introduce a novel application of the Dandelion Optimizer (DO) algorithm, designed to provide robust damping and stability under fault conditions. To improve the effectiveness of the DO algorithm, we compare its performance with several established optimization techniques, including Particle Swarm Optimization (PSO), HGS (Hunger Games Search), HBA (Honey Badger Algorithm), EO (Equilibrium Optimizer), and AVOA (African Vulture Optimization algorithm.

Keywords: Tilt-Integral-Derivative (TID) · Dandelion Optimizer (DO) · Power System Stabilizer (PSS)

1 Introduction

Ensuring the reliability and security of electrical power systems requires the elimination of poorly damped oscillations. This condition is crucial in modern power grids, particularly as the interconnection of power systems continues to expand. Low-frequency oscillations can arise in weakly interconnected large power systems, even in response to minor disturbances. These oscillations necessitate appropriate damping forces to prevent interruptions and maintain stable power system operation [1]. According to the literature, Conventional PSSs (CPSS with fixed parameters) have been widely used as effective damping controllers in power systems due to their simplicity and satisfactory performance [2, 3]. Unfortunately, CPSS designs struggle when dealing with the wide range of operating conditions found in non-linear power systems. That's why many studies have incorporated more advanced controllers, such as proportional-integral (PI), proportional-integral-derivative (PID) [4], fractional-order PID (FOPID) [5] and Fractional order fuzzy PID (FOFPID) [6], controllers, to enhance damping performance

© The Author(s), under exclusive license to Springer Nature Switzerland AG 2025
M. Hatti (Ed.): IC-AIRES 2024, LNNS 1238, pp. 116–123, 2025.
https://doi.org/10.1007/978-3-031-80301-7_13

and adapt to varying system dynamics. Additionally, a number of metaheuristic algorithms have been proposed for effectively adjusting the PSS parameters in order to improve system stability, including differential evolution (DE) [7], genetic algorithm (GA) [8], particle swarm optimization (PSO) [9], cuckoo search algorithm [10], and the firefly algorithm (FA) [11]. In this paper, an optimal Tilt-Integral-Derivative (TID)-based Power System Stabilizer (PSS) is developed to enhance the transient stability of a multi-machine power system (MMPS) under a three-phase fault condition. The parameters of the TID-based PSS are adjusted using the Dandelion Optimizer (DO), with a fitness function aimed at shifting the eigenvalues to the left side of a D-shaped region. A comparative study is conducted where the TID-based PSS optimized using DO is compared with those optimized using Particle Swarm Optimization (PSO), Hunger Games Search (HGS), Honey Badger Algorithm (HBA), Equilibrium Optimizer (EO), and African Vulture Optimization Algorithm (AVOA). The results demonstrate the superior accuracy and effectiveness of the proposed algorithm and controller.

2 Problem Mathematical Modelling

2.1 Four Machine Power System

The four-machine system in Fig. 1 is commonly studied in the power system community (Kundur,1994). The system contains 4 generators G1-G4, 11 buses, 4 transformers, 8 transmission lines, 2 loads (indicated by arrows) and 2 shunt capacitances.

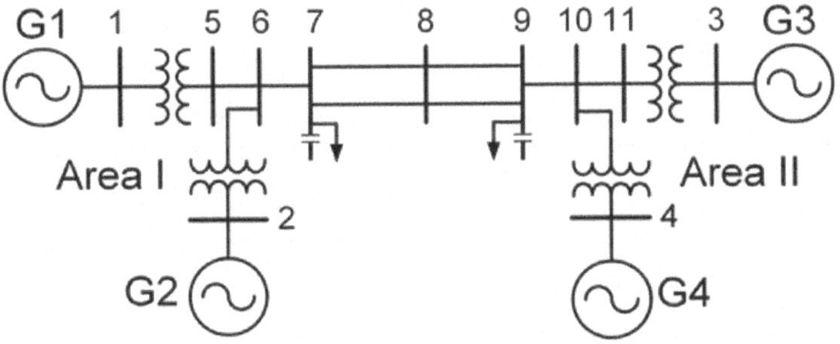

Fig. 1. A four-machine, two-area test system (Kundur,1994).

This section provides all the differential algebraic equations required for a time domain simulation of a synchronous machine. The equations capture the subtransient dynamic behaviour of a synchronous generator.

$$\frac{d\delta}{dt} = \omega - \omega_s \tag{1}$$

Equation (1) captures the change in generator rotor angle, due to a mismatch between the synchronous speed of the network and the rotor angular speed.

$$\frac{2H}{\omega_s}\frac{d\omega}{dt} = T_M - T_e - D(\omega - \omega_s) \tag{2}$$

Equation (2) is needed to describe the change in the rotor angular speed due to a mismatch of torques in the mechanical system. If the input mechanical torque is greater than the electrical torque,while accounting for damping, the rotor will speed up.If the electrical torque is greater than the mechanical torque,considering damping,the rotor willslow down. The rate at which the rotor speeds up or slows down is determined by the inertia of the generator.

The mechanical torque is an input quantity for the synchronous machine simulation.The electrical torque can be calculated from the electrical equations of the machine,as shown in Eq. (3).

$$T_e = \frac{X_d''-X_{ls}}{X_d'-X_{ls}}E_q' I_q + \frac{X_d'-X_d''}{X_d'-X_{ls}}\psi_{1d} I_q + \frac{X_d''-X_{ls}}{X_d'-X_{ls}}E_q' I_d - \frac{X_d'-X_d''}{X_d'-X_{ls}}\psi_{2q} I_d - \left(X_d'' - X_q''\right) I_d I_q \tag{3}$$

Synchronous machines use governors to regulate the system frequency during changes in the generation -load balance, by adjusting the input torque to the generator. A simple governor can be represented in Eq. (4)

$$\frac{dT_m}{dt} = \frac{1}{T_g}\left(T_{m2} - T_m - \frac{\omega_r-\omega_s}{R_{gov}}\right) \tag{4}$$

Synchronous machines use excitation systems to regulate the generator bus voltage, by adjusting the generator field voltage. A simple static excitation system can be represented in Eq. (5)

$$\frac{dE_{fd}}{dt} = \frac{1}{T_a}\left(K_a E_t - K_a V_{ref} - E_{fd}\right) \tag{5}$$

2.2 Structure of TID

The proposed TID controller, which is a sort of FOC that uses FO calculus for its design. The TID structure is similar to that of the PID controller, but the proportional component is tilted and has a transfer function, which is the only change. The FOCs are presented based on the order of $1/s^{\frac{1}{n}}$ derivative and integral terms is not an integer. The concept of FO depends on expanding integer order integration and differentiation to a fractional order operator.

The transfer function of TID can be formulated as:

$$TID = \frac{K_t}{S^{\frac{1}{n}}} + \frac{K_I}{S} + K_D S \tag{6}$$

where K_I, K_D and K_t express the proportion, integral, and differential tilt gain of the proportional component. These gains can be tuned in the range of [0.1,50]. Whereas the FO operator can be tuned in the range of [0, 1]. This component simplifies the tuning process, improves the system's ability in the face of the disturbance, and enhances the robustness in facing the uncertainty of system parameters. Thus, more robustness for the TID controller is provided by the tilt component in comparison to the PID one. In this study, the proposed TID controller is designed for the LFC issue of modern power systems. The proposed TID controller has the advantage of being simple, ease of use, and the ability to improve the controller's transient response, particularly the overshoot time, without influencing the other parameters.

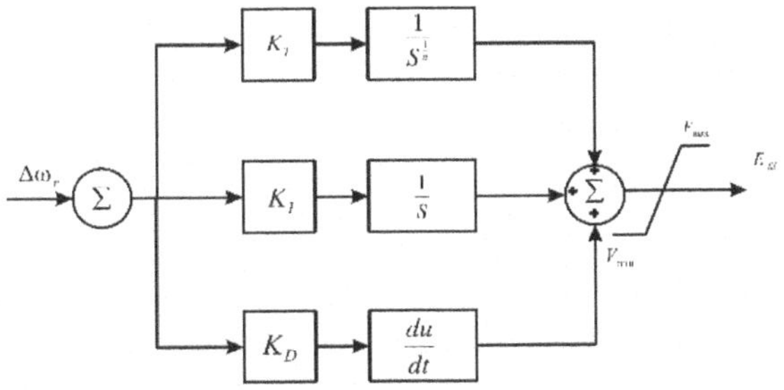

Fig. 2. Structure of TID controller

2.3 Fitness Function

To minimize the error signal, specifically Low-Frequency Oscillations (LFOs), a cost function has been adopted from [11] for comprehensive analysis. This formulation involve considering the eigenvalues of the system, which are essentially the "natural frequencies" of its dynamic response. By adjusting crucial parameters such as damping ratios, damping factors parameters, this cost functions aim to achieve a specific outcome: shifting the eigenvalues to the left half of the complex s-plane, as shown in Fig. 2. This displacement signifies enhanced stability, as complex numbers in the left half plane decay over time, indicating a damped and stable system response.

$$F1 = \sum_{i=1}^{N}(\sigma_0 - \sigma_i)^2 + \sum_{i=1}^{N}(\xi_0 - \xi_i)^2 \quad (7)$$
$$with\ \sigma_i \geq \sigma_0\ and\ \xi_i \leq \xi_0$$

The first part of this objective function focuses on maximizing the real part of the eigenvalue, while the second part aims to maximize the damping ratio. This strategy shifts the eigenvalues towards the left half of the S-plane, enhancing stability by ensuring larger negative real parts with higher damping ratios.

3 Dandelion Optimization Algorithm

DO is a metaheuristic algorithm, inspired by the long-distance flight dynamics of dandelion seeds, which rely on wind to travel and disperse. The algorithm emulates this natural process through three distinct stages as described in [12].

4 Results and Discussions

In this section, we employed the DO (Dandelion Optimization) algorithm to effectively tune the parameters of the proposed TID-PSS, aiming to enhance the performance of a four-machine power system. The primary goal was to improve the damping characteristics under a symmetrical fault condition at bus 8, a critical scenario for assessing system stability. To evaluate the effectiveness of the proposed DO-TID, we performed a comparative analysis using several other optimization algorithms: PSO (Particle Swarm Optimization), HGS (Hunger Games Search), HBA (Honey Badger Algorithm), EO (Equilibrium Optimizer), and AVOA (African Vulture Optimization alghorithm).

Figure 3 illustrates that without the Power System Stabilizer (PSS), the system lacks adequate damping, which hinders its ability to settle. It is clear that when speed of generators G1 and G2 at area 1 is increasing while speed of generators G3 and G4 at area 2 is reducing and vice versa. Following the three phase fault at 1s, an oscillation is triggered in the system in which generator from different parts or areas in the system participate. Hence, it is called interarea oscillations.

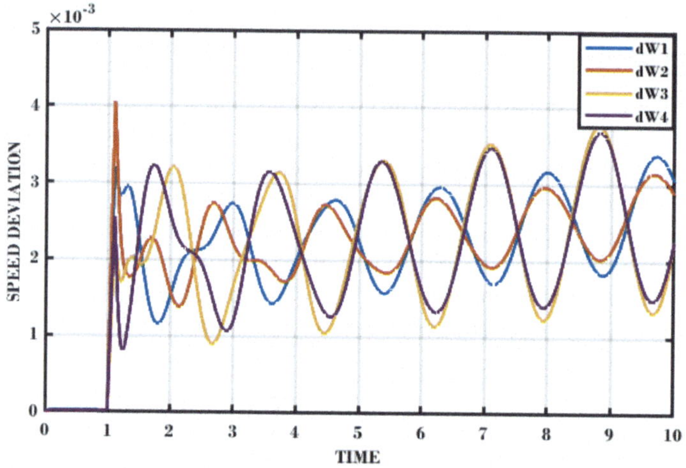

Fig. 3. Two-area system simulation - speed of generators without PSS.

The stability analysis based on Eigenvalues and the damping of the electromechanical modes is presented in Table 1, it is seen that one-pair of eigenvalues without-PSS fall in the unstable region of the s-plane and have a negative damping ratio (-0.0099).

Table 1. Complex eigenvalues of the two-area test system without PSS

	Eigenvalues	Damping ratio	Frequency
Local area eigenvalues	$-14.7148 \pm 16.7249i$	0.6605	2.6619
	$-0.5678 \pm 7.0682i$	0.0801	1.1249
	$-0.5828 \pm 7.1239i$	0.0815	1.1338
Inter-area eigenvalues	$-17.3557 \pm 5.6146i$	0.9515	0.8936
	$-17.4152 \pm 4.9938i$	0.9613	0.7948
	$0.0362 \pm 3.6547i$	-0.0099	0.5817

The incorporation of TID-PSS enhances the system's damping. The performance of the DO-TID, as demonstrated in Fig. 4 (a)–(d), is noteworthy. It maintains satisfactory performance indices in the four generator, less oscillation and minimum settling time. In comparison to the TID optimized by PSO/HBA/HGS/AVOA and EO.

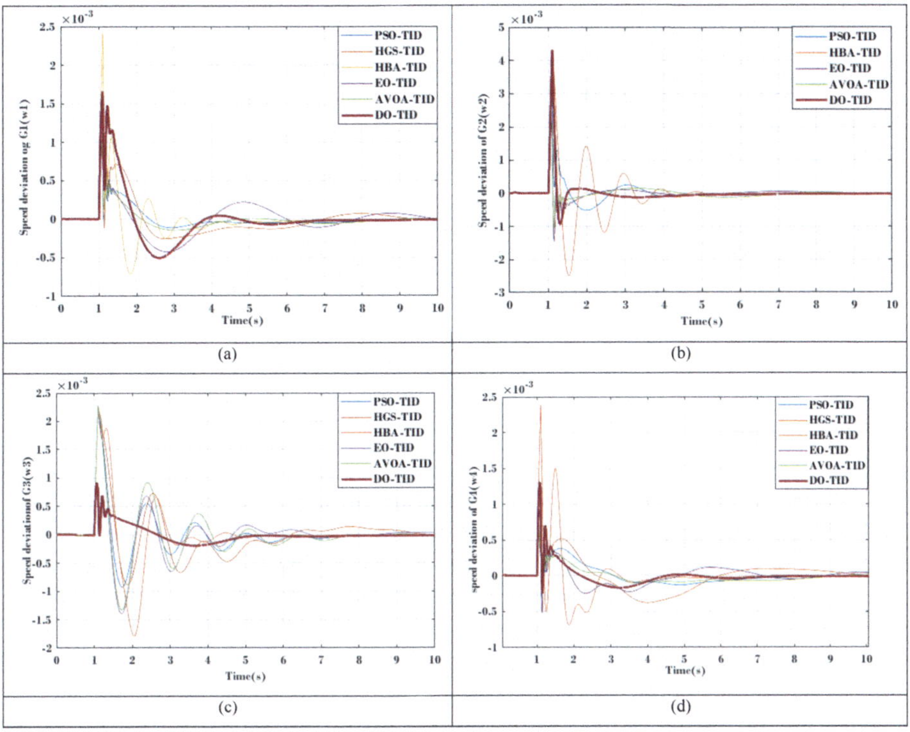

Fig. 4. Time domain simulation for speed deviation in areas 1 and 2 for generators G1, G2, G3 and G4 after TID-PSS integration. (a): w1, (b): w2, (c): w3 and (d): w4.

As depicted in Table 2, the eigenvalues obtained from the DO-TID approach consistently exhibit more negative real parts across both local and inter-area modes, indicating superior damping and system stability. For instance, in the local modes, the DO-TID method achieved eigenvalues such as $-12.649 \pm 17.71i$ and $-13.880 \pm 14.64i$, which are significantly more negative compared to those obtained from HGS-TID (e.g., $-0.5395 \pm 8.889i$).

Furthermore, in the inter-area modes, the DO-TID approach demonstrated the most negative real parts, with eigenvalues such as $-25.554 \pm 0.404i$, indicating a higher degree of stability and damping. This is contrasted with the results from other algorithms, where, for example, the HGS-TID method yielded eigenvalues like $-19.195 \pm 3.835i$. These findings highlight that DO-TID is more effective in mitigating oscillations between different areas of the power system, thus enhancing overall system robustness.

Overall, the comparative analysis of the six optimization techniques underscores the superiority of the DO algorithm. It consistently achieves more stable performance, as evidenced by the lower eigenvalue real parts in both local and inter-area modes. This robustness in providing effective damping and stability is crucial for maintaining power system reliability under various operating conditions and disturbances.

Table 2. Complex eigenvalues of the two-area test system after TID-PSS integration..

Eigen-values	PSO-TID	HGS-TID	HBA-TID	EO-TID	AVOA-TID	DO-TID
Local area	−12.785 ± 17.8i −14.30 ± 14.58i −2.134 ± 13.01i −2.473 ± 9.956i	−14.668 ± 16.17i −0.5395 ± 8.889i −0.7825 ± 8.827i	−12.816 ± 17.69i −14.484 ± 14.56i −2.186 ± 3.867i −2.5792 ± 10.84i	−12.888 ± 18.05i −14.64 ± 15.491i −1.826 ± 12.396i −1.988 ± 9.1748i	−12.726 ± 17.75i 14.071 ± 14.23i −1.9200 ± 13.86 −2.331 ± 9.698i	−12.649 ± 17.71i −13.880 ± 14.64i −2.081 ± 11.628i −1.9050 ± 11.08i
Inter-area	−16.279 ± 0.01i −2.267 ± 5.049i −2.141 ± 3.992i − 3.689 ± 0.642i	−19.195 ± 3.835i −18.287 ± 4.459i −0.334 ± 4.158i −3.699 ± 0.0370i	−1.1981 ± 6.010i −1.6351 ± 3.61i −3.7355 ± 0.35i	−1.821 ± 4.998i −1.821 ± 4.998i −1.8454 v3.746i −1.8454 ± 3.746i	−16.28 ± 0.024i −1.9202 ± 5.27i −1.9202 ± 5.276i −1.9199 ± 3.263i	−25.554 ± 0.404i −7.0127 ± 0.782i −2.599 ± 4.1485i −2.6325 ± 3.516i

5 Conclusion

In this study, we have developed and evaluated a Tilt-Integral-Derivative (TID)-based Power System Stabilizer (PSS) to improve the transient stability of a Two area four machines under three-phase fault conditions. By employing the Dandelion Optimizer (DO) for parameter tuning, we demonstrated significant improvements in system stability compared to other optimization techniques, including PSO, HGS, HBA, EO, and AVOA. The proposed DO-TID PSS not only achieved better eigenvalue placement but also exhibited superior damping performance, highlighting its potential as a robust solution for enhancing the reliability and security of modern power systems.

References

1. Kundur, P.: Power system stability. Power Syst. Stab. Control **10**, 7–1 (2007)
2. Demello, F.P., Concordia, C.: Concepts of synchronous machine stability as affected by excitation control. IEEE Trans. Power Appar. Syst. **88**(4), 316–329 (1969)
3. Kundur, P., et al.: Definition and classification of power system stability IEEE/CIGRE joint task force on stability terms and definitions. IEEE Trans. Power Syst. **19**(3), 1387–1401 (2004)
4. Sambariya, D.K., Prasad, R.: Design of optimal proportional integral derivative based power system stabilizer using bat algorithm. Appl. Comput. Intelli. Soft Comput. **2016**, 1–22 (2016)
5. Majidabad, S.S., Shandiz, H.T., Hajizadeh, A.: Nonlinear fractional-order power system stabilizer for multi-machine power systems based on sliding mode technique. Int. J. Robust Nonlinear Control **25**(10), 1548–1568 (2015)
6. Das, S., Pan, I., Das, S.: Performance comparison of optimal fractional order hybrid fuzzy PID controllers for handling oscillatory fractional order processes with dead time. ISA Trans. **52**(4), 550–566 (2013)
7. Sun, Z., Wang, N., Srinivasan, D., Bi, Y.: Optimal tunning of type-2 fuzzy logic power system stabilizer based on differential evolution algorithm. Int. J. Electr. Power Energy Syst. **62**, 19–28 (2014)

8. Mustapha, H., Buhari, M., Ahmad, A.S.: An improved genetic algorithm based power system stabilizer for power system stabilization. In: 2019 IEEE AFRICON, pp. 1–5. IEEE (2019)
9. Labdelaoui, H., Boudjema, F., Boukhetala, D.: A multiobjective tuning approach of power system stabilizers using particle swarm optimization. Turk. J. Electr. Eng. Comput. Sci. **24**(5), 3898–3909 (2016)
10. Chitara, D., Niazi, K.R., Swarnkar, A., Gupta, N.: Cuckoo search optimization algorithm for designing of a multimachine power system stabilizer. IEEE Trans. Ind. Appl. **54**(4), 3056–3065 (2018)
11. Ray, P.K., et al.: A hybrid firefly-swarm optimized fractional order interval type-2 fuzzy PID-PSS for transient stability improvement. IEEE Trans. Ind. Appl. **55**(6), 6486–6498 (2019). https://doi.org/10.1109/TIA.2019.2938473
12. Zhao, S., Zhang, T., Ma, S., Chen, M.: Dandelion optimizer: a nature-inspired metaheuristic algorithm for engineering applications. Eng. Appl. Artif. Intell. **114**, 105075 (2022)

Integration of Energy Storage Systems in the Power System to Prevent Blackouts

L. Mizat[1(✉)], M. Lahdeb[1], Y. Oubbati[1], and Ahmed T. Hachemi[1,2]

[1] LACoSERE Laboratory, University of Amar Telidji, 03000 Laghouat, Algeria
{l.mizate,m.lahdeb,y.oubbati}@lagh-univ.dz,
hachemi.ahmedtidjani@univ-ouargla.dz

[2] Electrical Engineering Laboratory, University of Kasdi Merbah Ouargla, Ouargla, Algeria

Abstract. The penetration of distributed generation (DG) has led to the development of smart grids that incorporate Renewable Energy Sources (RES), thus improving the reliability of Distribution Networks (DN). The stability of such networks can be further increased by integrating Energy Storage Systems (ESSs), a promising technology that can reduce environmental impact. This paper presents an effective method for designing ESS units to enhance distribution system performance. One of these benefits is the ability to increase system reliability through efficient islanding operations. This work proposes an approach to improving system reliability in distribution networks at minimal cost by allocating energy storage systems to avoid blackouts in power systems in the presence of RES (wind turbines). The willingness of consumers to pay to prevent power outages serves as a proxy for reliability value, which optimises the cost of installing the energy storage system. The main objective of this study will be to identify the most appropriate combination of storage units to be installed and loads to be shed to address all conceivable scenarios successfully. Therefore, a probabilistic method that will be used is the genetic algorithm (GA), which considers the stochastic nature of the system components.

Keywords: Distributed Generation (DG) · Distribution Network (DN) · Energy Storage System (ESS) · Blackouts · Load Shedding · Renewable Energy Sources (RES)

1 Introduction

Various factors contribute to power outages. One of the most difficult challenges is closely linked to the problem of transient stability. An in-depth study of transient stability is needed because it involves electrical energy systems that are not linear and change quickly after big shocks. Consequently, it is necessary to control the power system, and one method to avoid blackouts is to implement an ESS [1, 2].

The conventional regulated power grid has been replaced with a deregulated structure, allowing independent generators direct connection to DN. ESSs have the potential to facilitate islanding and support the integration of RESs. They are considered a promising technology for enabling the adoption of smart grids, also known as distributed storage

© The Author(s), under exclusive license to Springer Nature Switzerland AG 2025
M. Hatti (Ed.): IC-AIRES 2024, LNNS 1238, pp. 124–132, 2025.
https://doi.org/10.1007/978-3-031-80301-7_14

systems (DS) [3]. The main challenge in developing ESS is the high installation cost associated with non-volatile DN. Distribution utilities must strategically choose the most advantageous location and appropriate capacity of ESS units to minimise installation expenses while enhancing dependability [4–6]. Results from the research indicate that significant efforts have been made to examine the reliability of ESSs in DNs and to evaluate the suitability of power networks with high degrees of wind energy integration [7, 8]. Research suggests that the problem of improving system reliability by identifying the most efficient and economical placement and dimensions of ESS units in distribution systems remains unresolved. Distribution utilities prioritise discretionary targets for their dependability indices instead of mandated standards. Typically, distribution utilities lack a system of rewards or penalties to incentivize them to reach certain levels of reliability. The following are the main contributions of this paper:

The proposed method involves choosing specific places for load shedding in emergencies, which decreases the total costs of interruptions by enhancing the likelihood of successful islanding operations. This research presents a novel approach that differs from previous studies. Previous studies focused on using time series patterns to optimise the size of ESS units. Instead, this study introduces a probabilistic way to consider the unreliability of system components.

The organization of this paper is as outlined below: Sect. 2 addresses the formulation of the issue while Sect. 3 provides more details on the methods used. Section 4 is focused explicitly on presenting the outcomes of the simulations. The document closes in Sect. 5 summarizing the study's findings.

2 Problem Description

The main objective of this research was to determine the most efficient configuration of ESS units for installation and load shedding to enhance the reliability of the DN while minimising costs. Hence, the proposed challenge involves strategic planning elements such as determining the dimensions and placements of the ESS units to be installed, as well as making informed judgments during emergencies, such as selecting the most efficient load shedding strategy. The primary objective of the contingency plan is to optimise the probability of a successful islanding operation and minimise the total financial impact of any interruption. The customer's classification which can be residential, industrial, or commercial and the interruption's characteristics, such as its duration and frequency, frequently determine the cost of the interruption.

3 Methodology

This section describes the overall technique employed for the work reported in this paper, using two models in conjunction with a GA to minimise the objective function under consideration. Each solution (chromosome) has integer variables that show the discrete sizes of ESS units and the discrete loads that can be shed as decision factors at each bus.

At first, an N-1 contingency analysis will be performed to consider the possibility of distribution system line failure. The protective mechanism separates the broken part to protect the rest of the system, which could mean setting up separate areas with DG and/or ESS units that meet the power needs of those areas. Figure 1 presents an overview of the flowchart for the first phase. It is important to note that the power values estimated in Eqs. (1) and (2) may be either lower or larger than the size of the ESS power, which may determine the success or failure of the islanding operation.

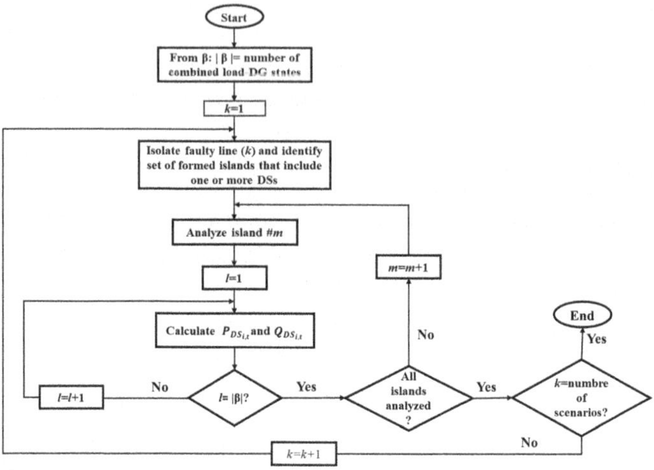

Fig. 1. Flowchart of calculating the size of ESS

The power flow equations of ESS are placed on the bus i during contingency k and state l, with N being the total number of system buses.

$$P_{ESS_{i,k,l}} = \sum_{j=1}^{N} V_{i,k,l} \times V_{j,k,l} \times Y_{ij,k} \times \cos(\theta_{ij,k} + \delta_{j,k,l} - \delta_{i,k,l}) - \left[P_{G_{i,l}} - P_{D_{i,l}}(1 - P_{SH_i})\right] \quad (1)$$

$$Q_{ESS_{i,k,l}} = -\sum_{j=1}^{N} V_{i,k,l} \times V_{j,k,l} \times Y_{ij,k} \times \sin(\theta_{ij,k} + \delta_{j,k,l} - \delta_{i,k,l}) - \left[Q_{G_{i,l}} - Q_{D_{i,l}}(1 - Q_{SH_i})\right] \quad (2)$$

where the P_{ESS} and Q_{ESS} are the active and reactive powers of ESS respectively; P_D and Q_D are the demand active and reactive powers, respectively; P_G and Q_G are the generated active and reactive powers, respectively; P_{SH} is the portion of the load to be shed if the load is totally shed it will be equal to 1; V and δ are the magnitudes of the voltage and the phase angle, respectively; Y and θ represent the bus admittance matrix element magnitude and angle, respectively.

Furthermore, the electricity needed from each ESS unit throughout each hour is obtained using a contingency analysis. When the energy saved is not enough or the power needed is more than the ESS's rated power, the islanding function fails, and all loads need to be disconnected. Inadequate total generation may fail to fulfil the load demand plus losses in a certain scenario.

The superscripts indicate the charging and discharging processes and the ESS's round-trip efficiency. The approach is iterated several times (using samples) until the variation in quality between the dependability index of interest and the sample mean of the same index meets a predetermined tolerance. This technique is succinctly outlined. This process is summarized in Fig. 2.

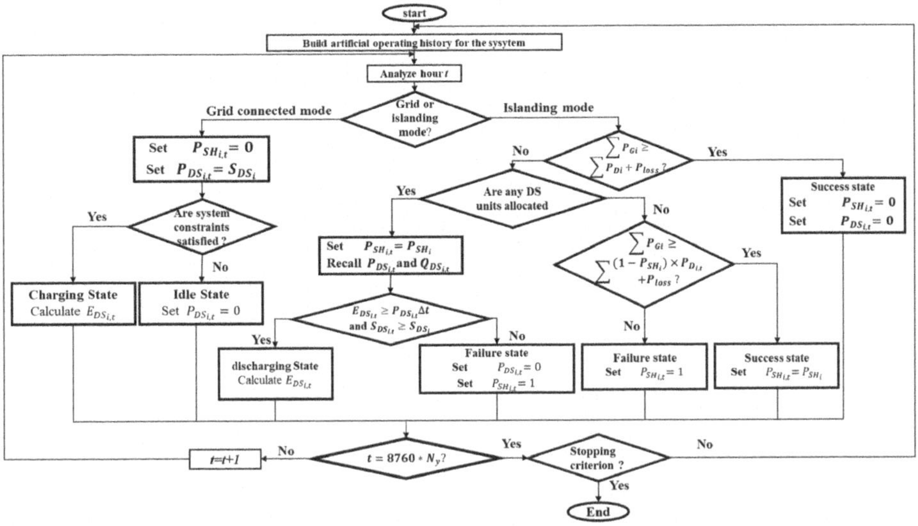

Fig. 2. Flowchart of calculating the power size of ESS and its location during contingency

The annual Energy Not Supplied (ENS) is determined by summing the amount of energy disrupted at all nodes throughout the research, as estimated in (3). The yearly interruption costs (COST) are determined in (4) by using the Customer Damage Function (CDFN) shown in Fig. 3 [5]. This function mostly depends on the length of the interruption.

The anticipated values for each index, the Expected Energy Not Supplied (EENS) and the Expected Cost (ECOST), are determined by averaging the ENS and COST over the entire period. The ECOST is then used to assess the objective function in (5), with the S_{ESS} and E_{ESS} representing the power and energy size of ESS, respectively.

$$ENS = \frac{1}{N_y} \sum_{t=1}^{8760*N_y} \sum_{i=1}^{N} P_{SH_{i,t}} \times P_{D_{i,t}} \tag{3}$$

$$COST = \sum_{\forall d} CDFN(f_d) \times ENS \tag{4}$$

$$Obj = \sum_{i=1}^{N} \left[(C_p + C_M) \times S_{ESS_i} + C_E \times E_{ESS_i} \right] + ECOST \tag{5}$$

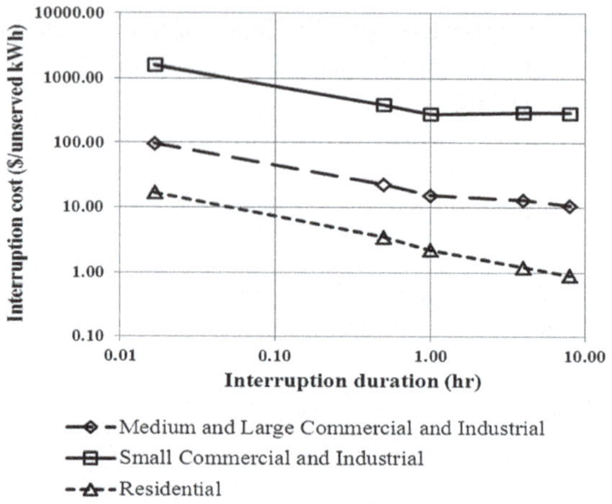

Fig. 3. CDFN for a variety of customers

4 Simulation and Results

4.1 Case Study

Figure 4 depicts the 33 bus test radial distribution system, which includes DGs connected at buses 18 and 33. These DGs are designed to contribute 30% of the system's total active power (Pg) and 21% of the total reactive power (Qg). Figure 5 compares the voltage profiles with and without the inclusion of DG units.

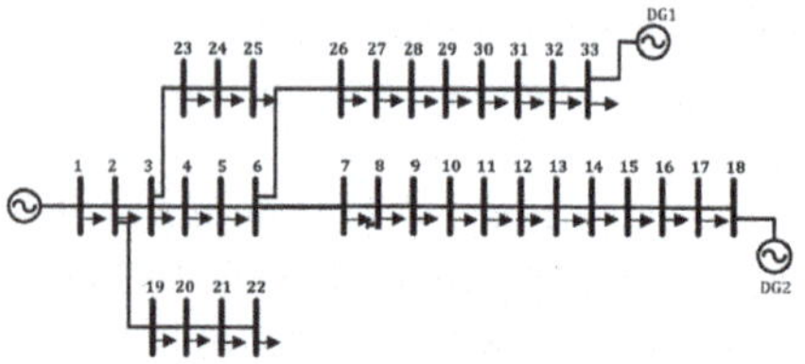

Fig. 4. 33-Bus test distribution system [9]

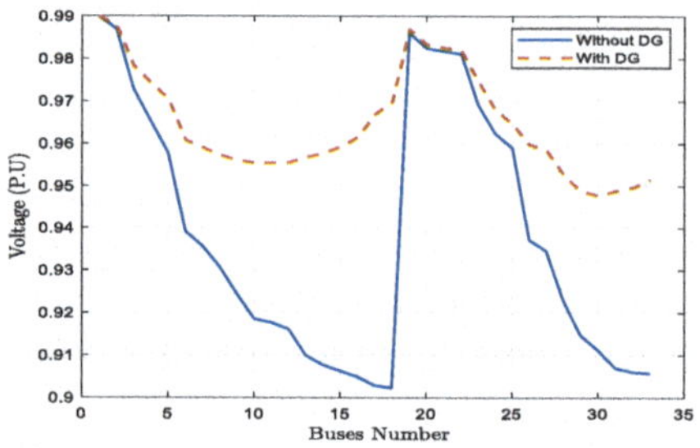

Fig. 5. Voltage profiles of 33 bus test distribution system

Tables 1 lists all case study financial assumptions, including interest and inflation rates. About 70% of DN clients are residential, and 30% are small commercial and industrial consumers. This proportion, shown by CDFNs in Fig. 2, estimates the customer base. The capital power cost CP of the rotating machine/converter interface (in $/kVA), the capital energy cost CE (in $/kWh), and the annual fixed O&M expenditures (in $/kVA) are similarly divided into three portions. Because they can provide adequate power and time to drain, we considered lead-acid (LA), compressed air in vessels (CAS), sodium-sulfur (Na/S), and vanadium redox (VR) batteries [10, 11]. These storage systems are 100 kVA/kWh discrete sizes and have been around for 30 years. Table 1 and [12, 13] show the four potential technologies' yearly capital and maintenance costs. Like other metaheuristics, GA control parameters require adjustment. After conducting many experiments, we determined the optimal settings for the GA as follows: **Population size** = 40; **Number of generations** = 100; **Crossover probability** = 0.6; **Probability of mutation** = 0.01.

Table 1. ESS technologies' annual capital and maintenance costs

	LA	CAS	Na/S	VR
Capital and maintenance costs				
Capital power cost ($/kVA)	175	1000	1000	See note
Capital energy cost ($/kWh)	305	250	500	740
Annual O&M cost ($/kVA)	15	28	20	20
Annual capital cost, IR=5%, F=1% and n=30years				
Annual capital power cost	10.07	57.55	57.77	-
Annual capital energy cost	17.55	14.39	28.78	42.59
Capital power cost for VR battery is included in capital energy cost				

4.2 Simulation Results

This part presents the research results, which determine the optimal location of ESS units to enhance system reliability, considering all possible island configurations. In this case, a load shedding technology called dual load shedding was used. It is essential to mention that GA functions optimally when the choice factors are binary. We will analyse the outcomes as instances when there is a connection between buses 6 and 7. The average grid in Fig. 6 supplies electricity to the final consumers (30% Commercial, 70% Residential).

Contingency Exists Without Load Shedding: When we first have a contingency exit but do not have load shedding at any bus, as shown in Table 2, we notice that ESS_1 is 300 kW and located on bus 9, and ESS_2 is 300 kW and located on bus 16, which is the optimal ESS unit.

Contingency Exist, Load Shed at any two buses: Existing contingency, load shedding on any two nodes.

- The first load to be shed is situated at bus 9.
- The second load to be shed is located at bus15.

The results are summarized in Table 3 and include the optimum ESS size, position, and load shedding for buses 9 and 15.

In Fig. 7, the optimal response of ESS and the necessary load shedding due to a contingency between buses 6 and 7 are depicted.

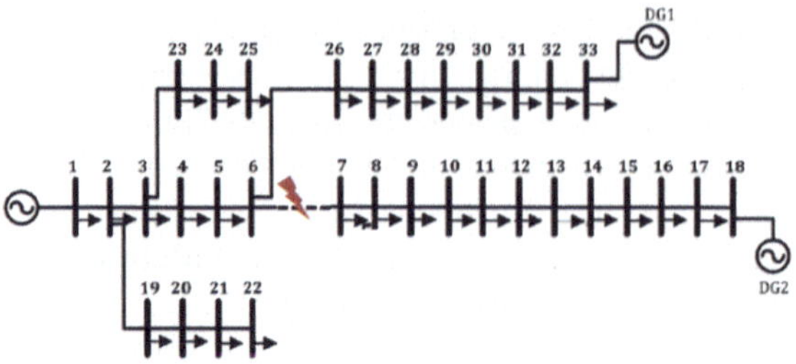

Fig. 6. System under study with contingency between nodes 6 and 7

Table 2. Optimal ESS size and location without load shed

	Size (kW)	Location
ESS1	300	9
ESS2	300	16

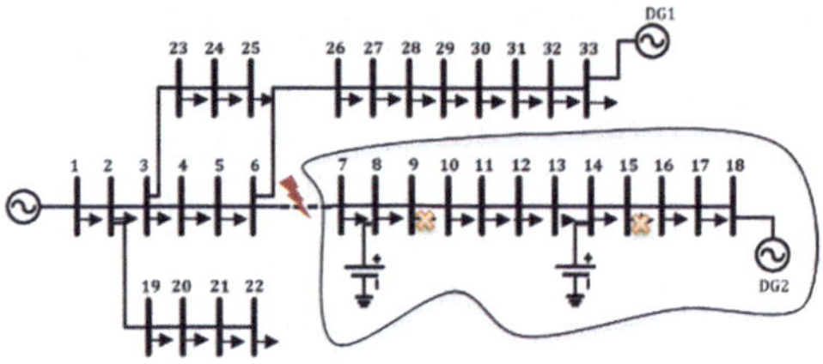

Fig. 7. Graphical representation of ESS location and load shedding during a contingency on the line between bus 6 and 7

Table 3. Optimal ESS size and location with load shed at bus 9 and 15

	Size (kW)	Location
ESS1	200	8
ESS2	200	14

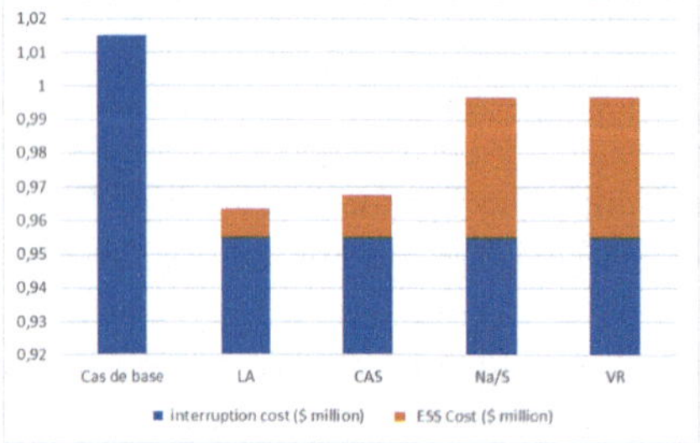

Fig.8. Total annual cost (million $) of base case and batteries technology

Table 4. Total annual cost (million $) of base case and batteries technology (case.b)

	Base case	LA	CAS	Na/S	VR
Interruption cost (million $)	1.02	0.95	0.95	0.95	0.95
ESS cost (million $)	0	0.0087	0.012	0.042	0.042

- **Case 1** assumes the contingency lasts for 8 h, matching the supply duration of SSE technologies. The total annual costs for the base case and four different battery technologies are presented in Table 4 for this scenario. A review of Fig. 8 reveals that, when comparing the base case and the other three technologies, the LA battery emerges as the most cost-effective solution in terms of total annual expenses.
- **In Case 2**, we examine the Expected Energy Not Supplied (EENS) for both the base case and the different battery technologies. As shown in Table 5, the results detail the EENS comparisons between the base case and the battery technologies. Figure 9 illustrates the reliability levels, measured in EENS, for each storage technology with the base case. Since the loads to be shed are identical across all technologies, Fig. 9 reflects consistent EENS values for each scenario.

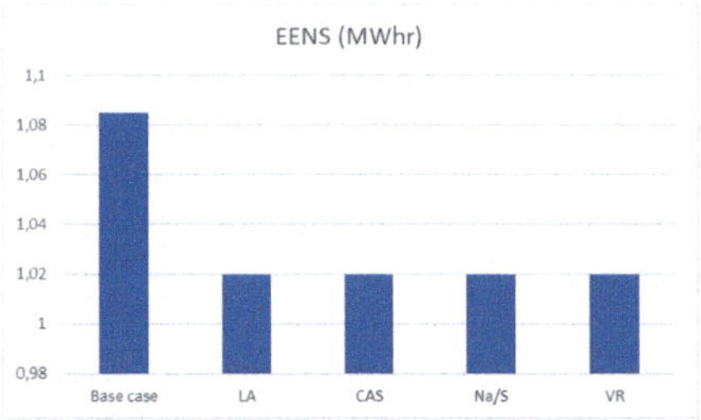

Fig. 9. EENS for the different storage technologies and the base case

Table 5. EENS for the base case and the different storage technologies

ESS Types	Base case	LA	CAS	Na/S	VR
EENS (MWhr)	1.09	1.02	1.02	1.02	1.02

After examining the impact of each technology on annual costs, a further case study was conducted to assess the allocation of different combinations of the four potential ESS technologies. This analysis was performed by introducing an integer-choice variable to represent the selection of ESS units to be installed. Findings indicate that LA batteries are the preferred choice for ESS units. However, it is important to note that this conclusion is contingent upon specific system configurations and the varying costs of ESS units.

5 Conclusion

This paper presents a study that discusses the selection of optimal ESS units based on location, size, and load-shedding strategies to enhance the efficiency of the DN. Additionally, it proposes a cost-effective, two-stage methodology for allocating ESS units in DN to boost system reliability. This approach utilizes a value-based reliability strategy that accounts for customer benefits as an efficiency advantage in enhancing system dependability. A probabilistic method determines how much power the given ESS units will need, considering how random all system parts are, such as loads and existing DGs. The paper includes a case study in which four storage strategies are analysed and compared to a baseline scenario without ESS.

References

1. Lahdeb, M., Oubbati, Y.: Actions préventives relatives aux pannes d'électricité (Preventive Actions for Blackouts) (Full text in French), vol. 65, no. 4 (2017)
2. Demetriou, P., Quirós-Tortós, J., Kyriakides, E.: When to island for blackout prevention. IEEE Syst. J. **13**(3), 3326–3336 (2019). https://doi.org/10.1109/JSYST.2018.2866937
3. IEEE Guide for Design, Operation, and Integration of Distributed Resource Island Systems with Electric Power Systems. IEEE Std 1547.4-2011, pp. 1–54 (2011). https://doi.org/10.1109/IEEESTD.2011.5960751
4. Banerjee, B., Islam, S.M.: Reliability based optimum location of distributed generation. Int. J. Electr. Power Energy Syst. **33**(8), 1470–1478 (2011). https://doi.org/10.1016/J.IJEPES.2011.06.029
5. Awad, A.S.A., El-Fouly, T.H.M., Salama, M.M.A.: Optimal ESS allocation and load shedding for improving distribution system reliability. IEEE Trans Smart Grid **5**(5), 2339–2349 (2014). https://doi.org/10.1109/TSG.2014.2316197

6. Devineni, G.K., Ganesh, A., Naga Malleswara Rao, D.S., Saravanan, S.: Optimal sizing and placement of DGs to reduce the fuel cost and T D losses by using GA PSO optimization algorithms. In: 2021 International Conference on Sustainable Energy and Future Electric Transportation, SeFet 2021 (2021). https://doi.org/10.1109/SeFet48154.2021.9375701
7. Pregelj, A., Begovic, M., Rohatgi, A.: Recloser allocation for improved reliability of DG-enhanced distribution networks. IEEE Trans. Power Syst. **21**(3), 1442–1449 (2006). https://doi.org/10.1109/TPWRS.2006.876649
8. Chowdhury, A.A., Mielnik, T.C., Lawton, L.E., Sullivan, M.J., Katz, A., Koval, D.: System reliability worth assessment using the customer survey approach. IEEE Trans. Ind. Appl. **4**, 2162–2167 (2024). https://doi.org/10.1109/IAS.2004.1348776
9. Jasmon, G.B., Lee, L.H.C.C.: Distribution network reduction for voltage stability analysis and loadflow calculations. Int. J. Electr. Power Energy Syst. **13**(1), 9 (1991). https://doi.org/10.1016/0142-0615(91)90011-J
10. Sullivan, M.J., Mercurio, M., Schellenberg, J., Freeman, M.A: Estimated value of service reliability for electric utility customers in the United States. Methodology (2009)
11. Das, C.K., Bass, O., Kothapalli, G., Mahmoud, T.S., Habibi, D.: Overview of energy storage systems in distribution networks: placement, sizing, operation, and power quality. Renew. Sustain. Energy Rev. **91**, 1205–1230 (2018). https://doi.org/10.1016/j.rser.2018.03.068
12. Schoenung, S.M., Burns, C.: Utility energy storage applications studies. IEEE Trans. Energy Convers. **11**(3), 658 (1996). https://doi.org/10.1109/60.537039
13. Poonpun, P., Jewell, W.T.: Analysis of the cost per kilowatt hour to store electricity. IEEE Trans. Energy Convers. **23**(2), 529 (2008). https://doi.org/10.1109/TEC.2007.914157

Improvement of Double Diode Model of Solar Cells Using Mountain Gazelle Optimization Algorithm and Based on Lambert's W Function

A. Abbadi[1](✉), F. Hamidia[1], A. Morsli[1], M. R. Skender[2], and N. Douifi[3]

[1] Electrical Engineering Department, Research Laboratory in Electrical Engineering and Automatic (LREA), Faculty of Technology, University of Médéa, Médéa, Algeria
amel.abbadi@yahoo.fr

[2] Electrical Engineering Department, Renewable Energy and Materials Laboratory (REML), University of Médéa, Médéa, Algeria

[3] Electrical Engineering Department, Laboratory of Advanced Electronic Systems (LSEA), Faculty of Technology, University of Médéa, Médéa, Algeria

Abstract. In this paper, we aim to extract the model parameters of the double diode photovoltaic cell. Double Diode Models (DDM) make it possible to understand the behavioral characteristics of a photovoltaic cell. The mountain gazelle optimizer carries out parameter estimation. This estimation is performed by minimizing the root mean square of the difference between the experimental and estimated current values. The mathematical expression for the current-voltage relationship of this circuit is derived based on Lambert's W function. The estimation accuracy of the proposed algorithm is tested on a standard RTC France solar cell. Based on all the results obtained, it was demonstrated that the mountain gazelle optimizer is effective and outperforms other algorithms in terms of accuracy.

Keywords: Solar cell · Double diode model · Lambert W-function · Parameter extraction · Mountain gazelle optimizer

1 Introduction

Energy production presents a significant challenge in the coming years. As the energy demands of both industrialized and developing societies continue to grow, this increased consumption leads to higher greenhouse gas emissions, further intensifying pollution levels.

The use of non-polluting energy then appeared to be the ultimate solution to this problem [1]. Renewable energy refers to sources derived from the sun, wind, geothermal heat, water, and biomass. Unlike fossil fuels, renewable energies are energies with unlimited resources [2].

Photovoltaic solar energy has experienced rapid growth due to its significant potential as a renewable energy source which makes it possible to produce electricity by transforming part of solar radiation using a photovoltaic cell [2]. These photovoltaic

© The Author(s), under exclusive license to Springer Nature Switzerland AG 2025
M. Hatti (Ed.): IC-AIRES 2024, LNNS 1238, pp. 133–140, 2025.
https://doi.org/10.1007/978-3-031-80301-7_15

systems for electricity generation can be applied in a variety of areas: electrification of isolated sites, installation in buildings, or direct connection to the electric network [3].

To enhance the performance of photovoltaic systems, it is essential to identify the electrical characteristics of their fundamental component, the photovoltaic cell.One of the most used models to characterize the behavior of solar cells is the double diode model [4].

The efficiency of the photovoltaic system can be improved if the model parameters are estimated and identified accurately. Most researchers optimize PV model parameters using three methodologies: deterministic, analytical, or metaheuristic [5]. Analytical approaches use mathematical equations to identify the unknown parameters [6]. Assumptions are made in advance in these methods, and therefore the findings are not accurate [7]. The deterministic strategy generates PV model parameters using gradient calculation [8]. As a result, this strategy is inefficient if the beginning points are the best local ones but are far from the global optima [4]. The metaheuristic algorithms deal with the aforementioned problems while avoiding local optimal problems which allows us to say that these methods generate satisfactory results regarding the estimation of the parameters of photovoltaic solar cells compared to other methods [4, 5].

In this work, a novel metaheuristic optimization algorithm is used to estimate the parameters of solar PV models, called the gazelle optimization algorithm [9]. The objective function used to estimate the parameters is the Root Mean Square Error (RMSE) between the estimated current and the measured PV current. In our case we have opted for the Lambert- W function, since it is a nonlinear estimation problem.

This paper is structured in the following manner: Section 2 provides an overview of the double diode model under analysis. Section 3 outlines the proposed optimization algorithm. Section 4 assesses the parameter estimation of DDM, and Section 6 closes this paper.

2 Double Diode Model and Objective Function

2.1 Double Diode Mathematical Model of Solar Photovoltaic Cell

The use of equivalent circuits has made the modeling of photovoltaic cells possible. The simplest model of photovoltaic cell is called a single diode model. It consists of an ideal current source connected in parallel with a diode, a shunt resistor (Rsh) and a series resistor (Rs) [10]. Nevertheless, to better represent the photovoltaic cell model, another model called the Double Diode Model (DDM) is often used [11–14].

The load current generated from the equivalent circuit illustrated in Fig. 1 is defined as follows:

$$I_L = I_{ph} - I_{d1} - I_{d2} - I_{sh} \quad (1)$$

where I_{ph} represents the photocurrent and I_L denotes the output current of the PV cell, I_{se1} is current of the first diode 1, I_{se2} is current of the second diode and I_{sh} is the shunt resistance current [11].

This current can also be represented by the following expression:

$$I_L = I_{ph} - I_{sd1}\left(e^{\frac{V_L+I_LR_s}{n_1V_{th}}} - 1\right) - I_{sd2}\left(e^{\frac{V_L+I_LR_s}{n_2V_{th}}} - 1\right)^{\frac{V_L+I_LR_s}{R_{sh}}} \quad (2)$$

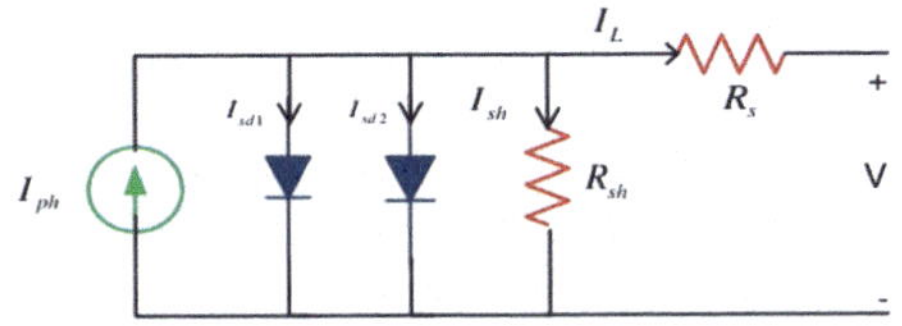

Fig. 1. Equivalent circuit of the double diode model [11]

In this equation, I_{se1}, and I_{se2} represent the reverse saturation current of the two diodes, respectively. n1, n2, represent the ideality factors of the diodes, respectively, and V_{th} is the thermal voltage, which equals K_BT/q, where K_B is the Boltzmann constant, q is the charge of the electron, and T is the temperature in Kelvin.

The DDM requires the estimation of seven unknown parameters (I_{ph}, I_{se1}, I_{se2}, R_{sh}, R_s, $k1$, $k2$).

2.2 Objective Function Formulation

The main objective in identifying the unknown parameters of solar photovoltaic models is to define the objective function. This function aims to minimize the root mean square error (RMSE) for the differences between the estimated values and the actual experimental values.

$$RMSE = \sqrt{\frac{1}{N_p}\sum_{i=1}^{N_p}\left(I_i^{meas} - I_i^{calc}\right)^2} \quad (3)$$

where I_i^{meas} is the experimental current and I_i^{calc} is the calculated current. Np is the number of data samples.

In our case, we use the Lambert W function to extract the unknown parameters of the solar PV model. The Lambert W function is used to reduce the voltage-current coupling represented by the Eq. (2) [15]. The estimated current expression for the DDM using the Lambert W function is presented as follows:

$$F_{obj} = \sqrt{\frac{1}{N_p}\sum_{i=1}^{N_p}\left(I_i^{meas} - \left\{\frac{R_{sh}\left(I_{ph} + I_{sd1} + I_{sd2}\right) - V_i^{meas}}{R_s + R_{sh}} - \frac{n_1V_{th}}{R_s}W(\beta_1) - \frac{n_2V_{th}}{R_s}W(\beta_2)\right\}\right)^2} \quad (4)$$

where

$$W(\beta_i) = \frac{I_{di}R_{sh}R_s}{n_iV_{th}(R_{sh} + R_s)}\exp\left(\frac{R_{sh}\left(R_sI_{ph} + R_sI_{sdi} + V_i^{meas}\right)}{n_iV_{th}(R_{sh} + R_s)}\right); i = 1, 2.$$

And Np is the number of measured points.

3 Mountain Gazelle Optimizer

The mountain gazelle or Edmi gazelle (Gazella gazella) is a species of gazelle. Gazelles are small, slender antelopes, agile, lively and very fast in running. The mountain gazelle is a species that spends the majority of its time awake during the day and sleeping during the night. They generally remain organized in groups of three to eight individuals. There are three main types of herds in the mountain gazelle community, namely mother-offspring herds, young male herds, and single male herds. In this community, the solitary males, the oldest, patrol and demarcate territories. The MGO optimization process is based on four major aspects of mountain gazelle life: maternity herds, single male herds, solitary and territorial males, and foraging movements [9].

3.1 Territorial Solitary Males

Once male mountain gazelles reach adulthood and are strong enough, they establish solitary territories that are widely separated. Competition among adult males arises over these territories and the possession of females

The territory of the adult male is expressed by:

$$TSM = male_{gazelle} - |(ri_1 \times BH - ri_2 \times X(T)) \times F| \times Cof_r \tag{5}$$

where the young male herd coefficient vector (BH), random coefficient F and the random coefficient vector (Cof_i), are formulated as

$$TBH = X_{ra} \times [r_1] + M_{pr} \times [r_2], ra = \left\{\left[\frac{N}{3} \ldots N\right]\right\}; \quad F = N_1(D) \times \exp\left(2 - Iter \times \left(\frac{2}{MaxIter}\right)\right)$$
$$Cof = \begin{cases} (a+1) + r_3, \\ a \times N_2(D), \\ r_4(D), \\ N_3(D) \times N_4(D)^2 \times \cos((r_4 \times 2) \times N_3(D)), \end{cases} \quad a = -1 + Iter \times \left(\frac{-1}{MaxIter}\right) \tag{6}$$

$male_{gazelle}$ represents the position vector of the optimal global solution, corresponding to an adult male gazelle. The parameters ri_x (x=1,..6) are random integers that can either be 1 or 2. *Xra* denotes a random solution, akin to a young male gazelle, within the interval of *ra*.

M_{pr} indicates the average number of search agents [N/3] which were randomly selected. Here, N is the total number of gazelles, while r_1 and r_2 are random values ranging between 0 and 1. r_3, r_4, and rand are random numbers in the range of 0 and 1. N_2, N_3 and N_4 are random numbers in the normal range and the dimensions of the problem. Max_{Iter} represents the maximum number of iterations, and *Iter* denotes the current iterations number.

3.2 Maternity Herds

Maternity herds are vital to the mountain gazelle life cycle because they give birth to male mountain gazelles. Male gazelles may also play a role in this cycle since it is these young males who attempt to possess females. This attitude is expressed as

$$MH = (BH + Cof_{1,r}) + (ri_3 \times male_{gazelle} - ri_4 \times X_{rand}) \times Cof_{1,r} \tag{7}$$

where *Cof2,r* and *Cof3,r* are randomly selected coefficient vectors. *Xrand* represents the position vector of a gazelle randomly chosen from the entire population.

3.3 Bachelor Male Herds

As the young male gazelles grow up, they will try to create their own territory and take over the female gazelles by entering into battles with the adult male gazelles. This behavior is mathematically formulated as

$$BMH = (X(t) - D) + \left(ri_5 \times male_{gazelle} - ri_6 \times BH\right) \times Cof_r$$
$$D = \left(|X(t)| + \left|male_{gazelle}\right|\right) \times (2 \times r_6 - 1) \quad (8)$$

where $X(t)$ denotes the position vector of the gazelle during the current iteration.

3.4 Migration in Search of Food

Mountain gazelles are continually foraging for food and cover long distances. Additionally, they possess high running speeds and impressive jumping ability. This behavior is mathematically represented as

$$MSF = (ub - lb) \times r_7 + lb \quad (9)$$

ub and lb represent the upper and lower limits of the problem, respectively.

Gazelles that are deemed old or weak are removed from the population. The best gazelle is typically considered to be the mature male that owns the territory (Fig. 2).

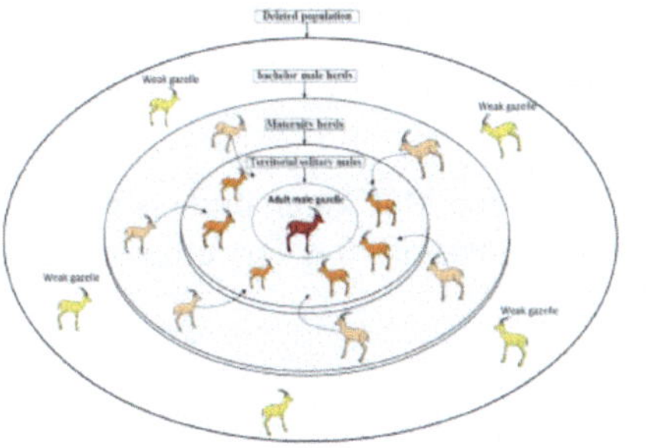

a. Optimization procedure

```
Pseudo-code of MGO.
% MGO setting
Inputs: The population size N and maximum number of iterations T
Outputs: Gazelle's location and fitness potential
% initialization
Create a random population using X_i(i = 1, 2, ..., N)
Calculate Gazelle's fitness levels.
While (stopping condition is not met) do
for (each Gazelle (X_i)) do
%Alone male realm
Calculate TSM using Eq.(5)
% Mother and child herd
Calculate MH using Eq.(7)
% Young male herd
Calculate BMH using Eq.(8)
% Migration to search for food
Calculate MSF using Eq.(9)
Calculate the fitness values of TSM, MH, BMH, and MSF, Then add them to the habitat
end for
Sort the entire population in ascending order
Update best_Gazelle
Save the N Best Gazelles in the Max number of population
end while
Return X_BestGazelle, best Fitness
```

b. Pseudo-code

Fig. 2. Mountain gazelle optimizer [9]

4 Finding and Analysis

In order to prove the effectiveness of the proposed algorithm for identifying the parameters of solar cell models, a commercial silicon solar cell of 57 mm diameter (R.T.C. France) was taken into account. The experimental data were obtained from the system under irradiation of 1000 W/m^2 at 33 °C [16].

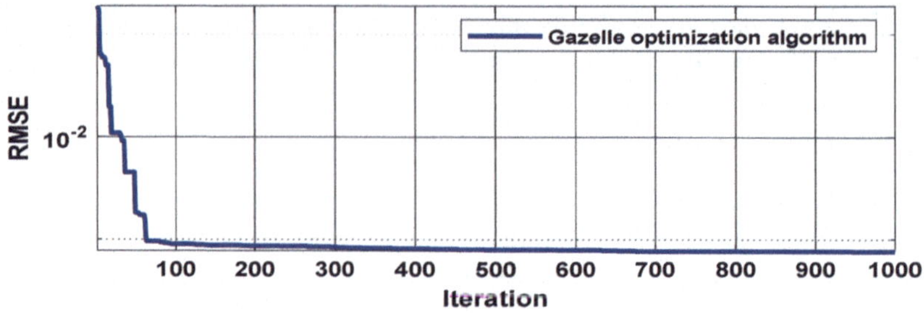

Fig. 3. Convergence curve of the proposed optimization strategy

The Fig. 3 represents the convergence curve of the proposed technique.

The I–V and P-V characteristics curve of the RTC France silicon solar cell model are shown in Fig. 4.

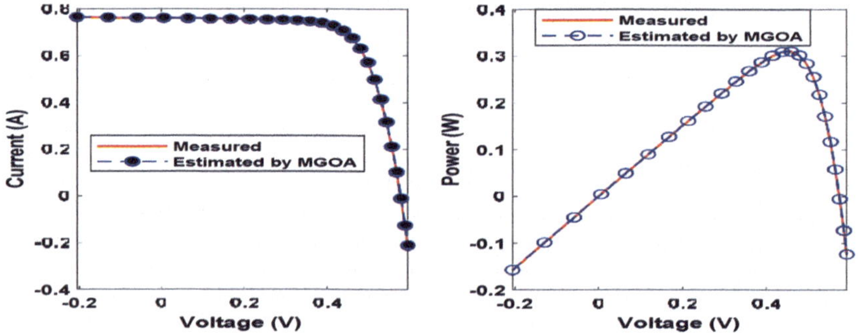

Fig. 4. I–V and P-V characteristics curve of the RTC France silicon solar cell model.

It is clear from the Fig. 4 that the data generated by the MGO technique is virtually identical to the experimental data, proving that the MGO algorithm is very efficient when estimating the PV parameters of the model.

Table 1. Optimal values of parameters estimated by four optimization algorithms of the RTC France silicon solar cell double diode model with their corresponding RMSE.

Optimization technique	*Iph*	I_{se1}	I_{se2}	*Rsh*	*Rs*	*n1*	*n2*	RMSE
Search range	[0, 1]	[0, 1]	[0,1]	[0,100]	[0,0.5]	[1, 2]	[1, 2]	-
GOA	0.760802	0.249925	0.001033	55.004443	0.039641	1.937321	1.463721	**7.829577e-04**
AHO	0.760848	0.2818422	0.019603	58.268813	0.038772	1.481895	1.965029	8.962918e-04
PSO	0.760635	0.273452	0.0006585	63.113509	0.041040	1.477948	1.8694352	0.001000617
HHO	0.7633340	0.76531978089	0.90044874182	90.501712	0.041869	1.851947	1.775601	0.005587289

The lower and upper limits of the seven parameters $\{I_{ph}, I_{se1}, I_{se2}, R_{sh}, R_s, k_1, k_2\}$ of the DDM are shown in Table 1. Table 1 also lists the parameters extracted from four metaheuristics algorithms namely the proposed MGOA, Harris Hawks Optimization (HHO) [17], Particle Swarm Optimization (PSO) [18], Artificial Humming bird Optimization (AHO) algorithms [19]. Parameter estimation with the proposed algorithm gave better results compared to other techniques (RMSE = 7.829577e-04).

The best statistical results for the DDM estimation of RTC France, presented in Table 2, are obtained by the MGO algorithm.

Table 2. Statistical results obtained during the identification of DDM of the RTC France.

Algorithm	RMSEmin	RMSEmean	RMSEmax	STD
HHO	0.001414570928660	0.006569	0.019487	0.004464364592597
PSO	7.898122e-04	0.001024	0.001506	2.318178e-04
AHO	7.585859e-04	9.914470e-04	0.001659	2.855819e-04
Proposed MGO	7.616596e-04	**8.412428e-04**	**0.001018**	**5.901777e-05**

5 Conclusion

This article proposes a metaheuristic optimization method inspired by the behavior of mountain gazelles to identify the parameters of photovoltaic solar cell models. The parameters of the RTC France solar cell are extracted using the Lambert W function. The cost function used for the parameter estimation is the Root Mean Square Error (RMSE). The performances of the MGO algorithm for extracting the seven parameters of the DDM are compared to those of three other metaheuristics algorithms: Harris hawks optimization, particle swarm optimization and artificial humming bird optimization algorithms. The results undoubtedly show that the model obtained by the proposed MGO algorithm demonstrated better accuracy and efficiency than the other techniques.

References

1. Luque, A., Hegedus, S.: Handbook of Photovoltaic Science and Engineering, 2nd edn. John Wiley & Sons Ltd., West Sussex, England (2003)
2. Twidell, J., Weir, T.: Renewable Energy Resources, 2nd edn. Taylor & Francis, New York, USA (2006)
3. Akarslan, F.: Photovoltaic Systems and Applications, Chapters. In: Sencan, A. (ed.) Modeling and Optimization of Renewable Energy Systems, IntechOpen, pp. 21–52 (2012)
4. Abdel-Basset, M., El-Shahat, D., Chakrabortty, R.K., Ryan, M.: Parameter estimation of photovoltaic models using an improved marine predators algorithm. Energy Convers. Manage. **227**, 113491 (2021)

5. Mostafa, M., Rezk, H., Aly, M., Ahmed, E.M.: A new strategy based on slime mould algorithm to extract the optimal model parameters of solar PV panel. Sustain. Energy Technol. Assessments **42**, 100849 (2020)
6. Sakthivel, S.S., Gokulakrishnan, L., Arunachaleashwer, S., Lokesh, V., Venkadesan, A.: Parameter Estimation of a Single Diode Model of a PV Panel using Gauss-Seidel Method: An Experimental Validation. 2023 International Conference for Advancement in Technology (ICONAT), pp. 1–6. Goa, India (2023)
7. Wang, D., Ding, F.: Parameter estimation algorithms for multivariable Hammerstein CARMA systems. Inf. Sci. (Ny). **355–356**, 237–248 (2016)
8. Ismaeel, A.A.K., Houssein, E.H., Oliva, D., Said, M.: Gradient-based optimizer for parameter extraction in photovoltaic models. IEEE Access **9**, 13403–13416 (2021)
9. Abdollahzadeh, B., Gharehchopogh, F.S., Khodadadi, N., Mirjalili, S.: Mountain gazelle optimizer: a new nature-inspired metaheuristic algorithm for global optimization problems. Adv. Eng. Softw. **174**, 103282 (2022)
10. Lidaighbi, S., Elyaqouti, M., Ben Hmamou, D., Saadaoui, D. Assalaou, K. Arjdal, E.: A new hybrid method to estimate the single-diode model parameters of solar photovoltaic panel. Energy Conversion and Management: X **15**, 100234 (2022)
11. Xu, S., Wang, Y.: Parameter estimation of photovoltaic modules using a hybrid flower pollination algorithm. Energy Convers. Manag. **144**, 53–68 (2017)
12. Gude, S., Jana, K.C.: Parameter extraction of photovoltaic cell using an improved cuckoo search optimization. Sol. Energy **204**, 280–293 (2020)
13. Premkumar, M., Babu, T.S., Umashankar, S., Sowmya, R.: A new metaphor-less algorithms for the photovoltaic cell parameter estimation. Optik **208**, 164559 (2020)
14. Yu, K., et al.: A performance-guided JAYA algorithm for parameters identification of photovoltaic cell and module. Appl. Energy **237**, 241–257 (2019)
15. Patel, S.: The single diode model of I-V and P-V characteristics using the lambert w function. Int. J. Innovative Res. Sci. Eng. Technol. **5**(5), 7034–7039 (2016)
16. Easwarakhanthan, T., Bottin, J., Bouhouch, I., Boutrit, C.: Nonlinear minimization algorithm for determining the solar cell parameters with microcomputers. Int. J. Solar Ener. **4**, 1–12 (1986)
17. Ramadan, A., Kamel, S., Korashy, A., et al.: An enhanced Harris Hawk optimization algorithm for parameter estimation of single, double and triple diode photovoltaic models. Soft Comput **26**, 7233–7257 (2022)
18. Khanna, V., Das, B.K., Bisht, D., Vandana, Singh, P.K.: Estimation of photovoltaic cells model parameters using particle swarm optimization. In: Jain, V., Verma, A. (eds.) Physics of Semiconductor Devices. Environmental Science and Engineering(), pp. 391–394. Springer, Cham (2014)
19. Ayyarao, T.S.L.V., Kishore, G.I.: Parameter estimation of solar PV models with artificial humming bird optimization algorithm using various objective functions. Soft Comput **28**, 3371–3392 (2024)

Application of Artificial Intelligence to Reduce Conventional LFC Capacity in Utility Grid Using Multiple Interconnected Microgrids

N. Kouba(✉), A. Boudoucha, A. Brik, M. A. Matouk, and M. Ammi

Faculty of Electrical Engineering, University of Sciences and Technology Houari Boumediene, Algiers, Algeria
koubanour@gmail.com

Abstract. Operating reserves, with the particular subclass of secondary reserves, are one type of ancillary services and fundamental in dealing with disturbances of power system. Operating reserves are an essential component of utility grid to keep electric network stability and reliability. This paper deals with the design of an optimal active power management and Load Frequency Control (LFC) in microgrid including distributed generations (DGs) and hybrid energy storage system (HESS). The investigated microgrid was operated in isolated and grid-connected modes. The proposed strategy involves a multi-stage regulation scheme with optimal combined cascade Proportional-Derivative-Fuzzy-Proportional-Integral-Derivative controller with filter (PD-Fuzzy-PIDN). At the first stage, only the load frequency loop is used to cope with fluctuations, then at the second stage, the hybrid storage system will be activated to support the LFC loop in coordination with the Tie-Line power flow transmission control. The HESS storage system was employed to ensure an optimal power management in presences of DGs for reducing the conventional LFC ability. In doing so, a novel bio-inspired meta-heuristic optimization approach called Artificial Hummingbird Algorithm (AHA) has been used to fine tune the controller parameters. To promote the use of renewable energy sources (RESs) and minimize the use of fuel by reducing the LFC capacity, a novel method have been proposed by using the surplus power of the distributed generations in the microgrid. To prove the effectiveness and superiority of the proposed strategy, various scenarios have been performed and compared.

Keywords: Microgrid · Renewable Energy · Load Frequency Control · Storage System · Fuzzy Control · PIDN Controller · Artificial Hummingbird Algorithm (AHA)

1 Introduction

Integrating renewable energies into utility grid requires adapting infrastructures and power system management. Smart grid technologies will make it possible to develop network flexibility and thus manage the intermittency and variability of renewable energies. The control of microgrids integrating storage systems and distributed generations

© The Author(s), under exclusive license to Springer Nature Switzerland AG 2025
M. Hatti (Ed.): IC-AIRES 2024, LNNS 1238, pp. 141–151, 2025.
https://doi.org/10.1007/978-3-031-80301-7_16

stands at the forefront of the evolving global energy landscape. Multi-microgrids, combining diverse renewable energy sources, play a pivotal role in the transition toward more sustainable energy production. In the last decades, many published articles have presented significant milestones in the field of controlling and managing multi-microgrids while integrating energy storage systems. Their diverse and complementary contributions enrich our understanding of the challenges and opportunities in this rapidly evolving field. These insights are essential to guide our own research and contribute to the advancement of sustainable and distributed energy systems. The review of relevant articles demonstrates significant advancements and insights into managing these complex systems. In essence, multi-microgrid control lies at the heart of the global energy transition. Previous research lays a solid foundation, and my contribution intends to provide practical recommendations to enhance grid reliability, emphasizing renewable energy integration and energy quality optimization. The future of multi-microgrid control looks promising, leading toward a more sustainable and resilient electric grid [1–6]. In reference [7], author's have addresses the crucial issue of stability in microgrids incorporating diverse wind energy sources. As one of the most dynamic renewable energy sources, wind energy can exhibit significant fluctuations. This article explores control strategies to ensure the operational stability of microgrids while leveraging these diversified wind resources. In reference [8], author's have proposed an optimal strategy for multi-microgrid with Hydrogen Storage. In [9], author's have used the model of predictive control for dynamic management of isolated microgrids. This is a significant advancement in improving the management of energy resources at the community level. In [10], author's have analyzed the challenges and solutions related to coordinated control and energy sharing among multiple microgrids. The insights from this research are essential for our understanding of managing distributed energy systems. Microgrid research would be incomplete without considering the context of smart grids and the use of artificial intelligence. This published work in [11] addresses a wide range of topics, including microgrid stability, operational optimization, and demand participation in a smart grid environment. It provides a comprehensive overview of the challenges and opportunities in microgrid control within the context of a smart grid. Managing load frequency is a crucial component of microgrid control. The published work in [12] propose control strategies to ensure load frequency stability using energy storage systems. Maintaining power quality and effective power sharing in unbalanced multi-microgrid scenarios is addressed in [13]. Moreover, a lot of research works have been published on the scope of using divers storage systems to enhance system frequency operation and control [14–17].

In this paper, the author's contribution will focus on an evaluation of the connection of multiple multi-microgrid to the conventional interconnected transmission network. A multi-stage active power management and frequency control strategy was proposed using a recently bio-inspired meta-heuristic algorithm to cope with system fluctuations and increase the installation of uncontrollable distributed generation units such as wind and solar power generation into the utility grid, which can minimize the use of the conventional LFC ability of the classical power plants. In addition, this paper underscores the significance and contribution of hybrid energy storage system in islanded and connected microgrid to reduce the conventional frequency regulation. In addition, the optimal design of the automatic generation controller allows to minimize the frequency deviation and maintain the set the scheduled tie line power, which allow to analyze the power transfer between conventional networks and multi-microgrid including storage system. In order to perform optimization tasks and test the effectiveness of the optimal secondary LFC loop in isolated and connected microgrid to the utility grid, an effective combined cascade PD-Fuzzy-PIDN controller was designed and tested. The rest of the paper is organized as follows. Section 2 presents the analytic model of the investigated system. Section 3 was devoted to presents the multi-stage power management and optimal frequency control method Artificial Hummingbird Algorithm (AHA). The obtained simulation results are presented in Sect. 4. Finally, Sect. 5 concludes the paper.

2 Analytical Model

A isolated power system or microgrid is a small scale electrical network that can includes various distributed generation units incorporate conventional Diesel generator, solar photovoltaic (PV) plant, wind turbines and storage devices such as batteries. Microgrids (MG) are often designed to reduce carbon emissions and enhance power system resilience [3, 4]. The MG have the ability to be operated in islanded mode provides the ideally platform to study the impact of integration various kinds of distributed generation. Also, the MG can be operated in ON-grid oar connected mode to support the utility grid with power supply and reduce the conventional LFC capacity with the large integration of renewable energies sources.

2.1 Microgrid: OFF-Grid Model

During many years, the integration of green power sources in power system is significantly increasing [16]. The microgrid is a system that have the capability to be operated on two modes: isolated or connected. This system presents a perfect platform to analyze the integration impact of various kind of distributed units. In this part, hybrid microgrid (MG) model sketched in Fig. 1 was used for simulation. The assumed microgrid consists of Diesel, Wind, PV units, Solar-thermal power generator (CSP), Hybrid Energy Storage devices (HESS) which are: Superconducting Magnetic Energy Storage (SMES), Redox Flow Batteries (RFBs), and Fuel Cells (FCs) and Load. This MG was used to show the contribution of sharing the capacity of distributed generations to support utility grid to face with load changes and minimize the conventional LFC ability.

2.2 Power System Model

Interconnected electrical network can includes a huge number of multi-sources power stations in parallel to supply various loads. The primary task of power system is to give an uninterrupted continuous power supply and satisfy the loads with good power quality with ensured by the operation and control system (OS). In this part, a multi-machines multi-sources power system was considered for the frequency assessment in presence of multi-microgrid integrated renewable energies (RESs) and faced to load changes as shown in Fig. 2. The investigated system incorporate Nuclear, Thermal and Gas units. In addition, a battery energy storage system (BESS) was installed to support the LFC regulation loop in case of large scale green power integration. Over normal operating conditions, the utility grid is continually subjected to various disturbances. To regulate the speed variation following the load disturbances, a centralized LFC control signal was inputted to the governor of each unit, even the storage system.

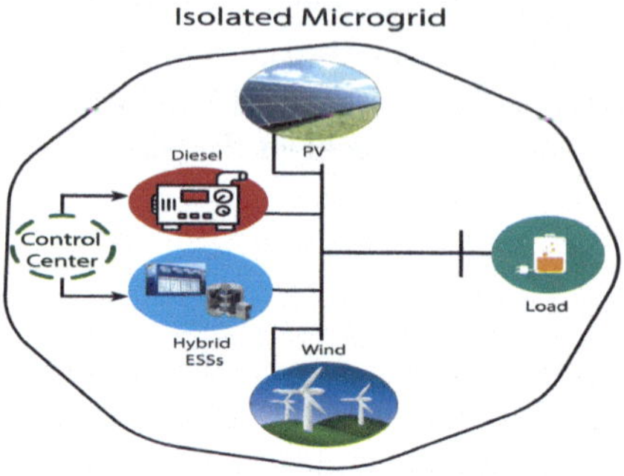

Fig. 1. Islanded Microgrid Model.

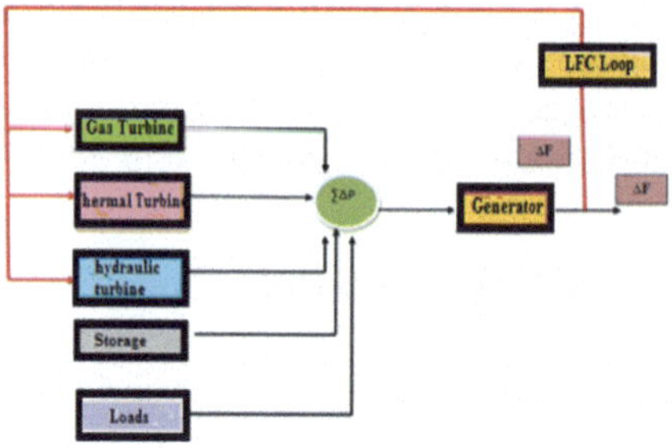

Fig. 2. Interconnected Multi-sources Power System Model.

2.3 Microgrid Connected Utility Grid Model

In this part, the main grid was connected to multiple microgrid in the aim to reduce the conventional LFC capacity and enhance frequency regulation such presented in Fig. 3. The multi-microgrids have a physical connection to the utility grid via a switching mechanism at the point of common coupling (PCC), but they also can disconnect into island mode and reconnect back to the main grid as needed. Various distributed generation types and hybrid storage systems have been used for each microgrid. Figure 4 presents the schematic blocks of the applied control strategy of the multi-microgrid connected to the utility grid. The LFC signal was dispatched between the classical power generation units of the main grid and the microgrid. Using this method microgrid can support the frequency regulation and power system stability in the utility grid by using various kinds of distributed generations units and storage systems. In addition, using smart power management algorithm can minimize the conventional power generation cost and at the same time can reduce the CO2 gas emission by increasing the use of green sources.

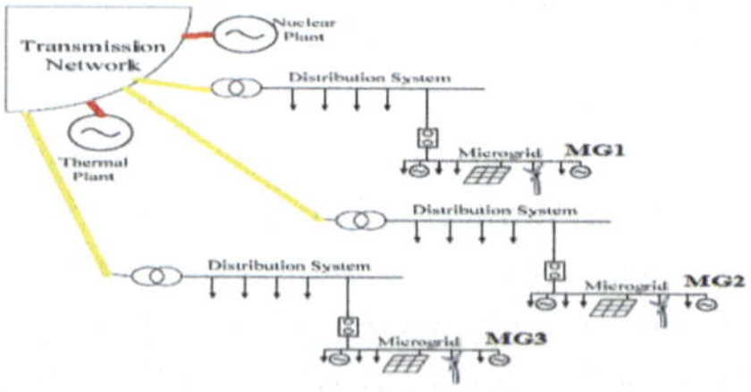

Fig. 3. Topology of Multi-Microgrid Connected to the Utility Grid.

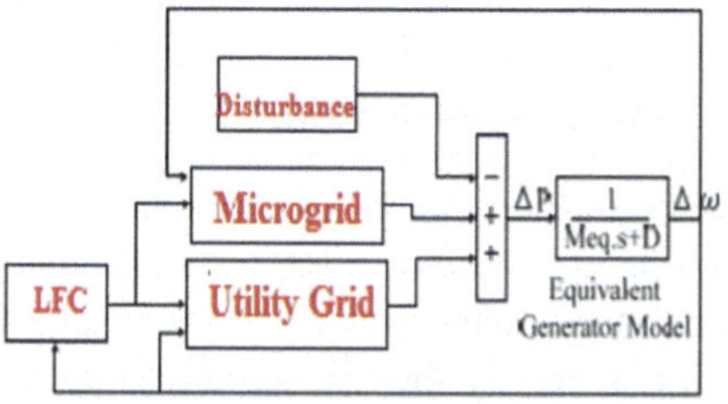

Fig. 4. Proposed Contribution of Microrid Connected into the Main Grid.

The analytical model used for dynamic frequency stability and control analysis is based on the linear model shown in Fig. 4 [14, 15]. This model is simply the transfer function model based on the deviation from a given operating point. More details are available in our previous published work, see references [3, 4, 14, 15, 21, 22].

3 Multi-stage Power Management and Control

3.1 Optimal Power Management and Frequency Control

In conventional power system, Governor -Turbine regulation system named also primary or Droop control is used for the load fluctuation with the period of tens of seconds to several minutes; Where the secondary automatic frequency control LFC is used for the fluctuation with the period of several minutes to about thirty minutes. Finally, the tertiary control action named economic dispatch control (EDC) is used for the fluctuation with the period of more than about thirty minutes because the fluctuation is large and can be predicted. In presence of large penetration of RESs, the power system operation and control face more challenges in view of voltage and frequency stability. In this context, and such presented in Fig. 5, this part present a novel coordinated optimal active power management and frequency control using artificial intelligence based Fuzzy and PID controller. An optimal LFC scheme was designed to cope with frequency fluctuations, where, an effective optimal active power management strategy was applied in the aim to support the main grid power supply and reduce the conventional LFC capacity and free the primary operating reserve, which can minimize both of generation coast and CO2 emission. Focusing mainly on the power system operation and control, the power management system of the microgrid and the LFC loop were modeled separately as shown in Figs. 6 and 7 respectively. The LFC signal dispatching control technique was proposed in this work to achieve the contribution of the integrated distributed generation units into the microgrid connected to the main grid in the aim of minimizing the LFC capacity. In the main grid, and in presence of microgrid connection, the LFC signal is divided between the conventional power generation units and the connected microgrid as presented in Fig. 7 [23]. The variable x presents the dispatching ratio. The proposed LFC control scheme is shown in Fig. 8.

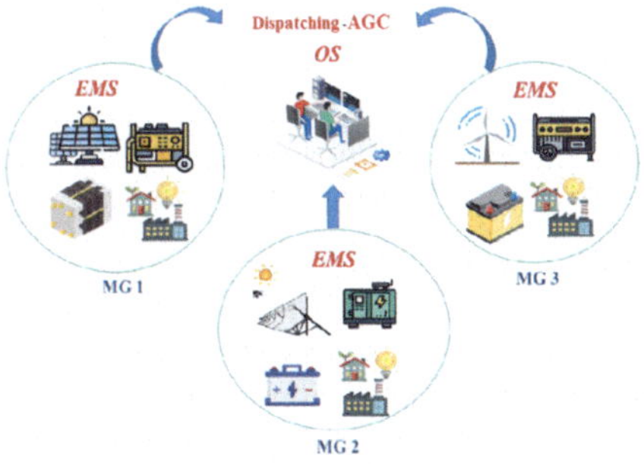

Fig. 5. Proposed Active Power Management and Frequency Control.

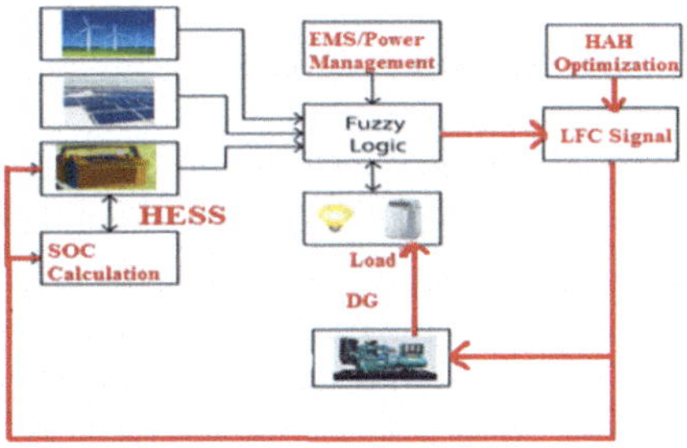

Fig. 6. Microgrid Active Power Management and Frequency Control.

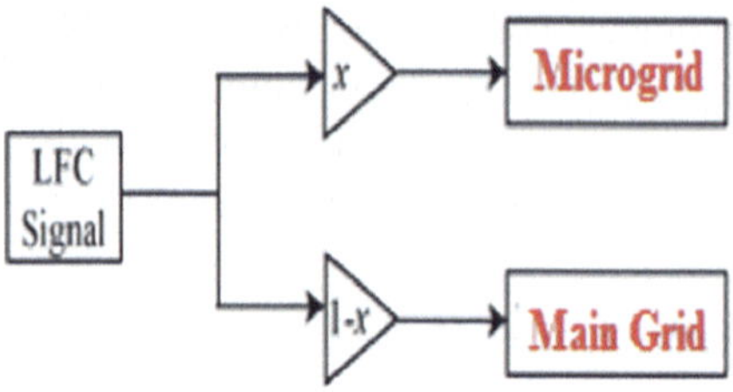

Fig. 7. Dispatching ratio x.

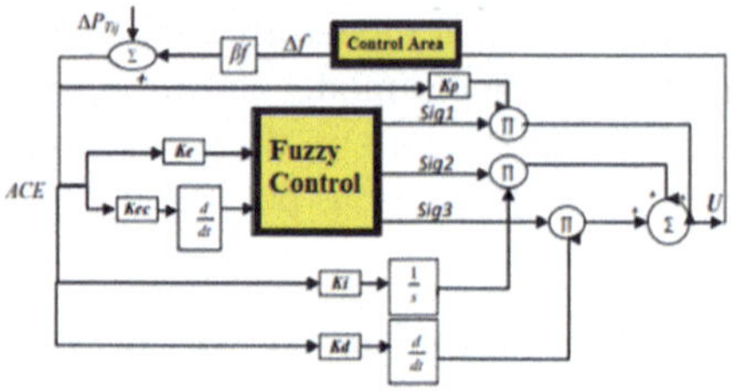

Fig. 8. Proposed LFC Model.

3.2 Artificial Hummingbird Algorithm (AHA)

Over the last decades, Artificial Intelligence (AI) has been a very important scope in almost engineering fields, which leads to create a variety of smart and robust control actions that have shown great promise in a large number of applications especially in power system operation and control. Nature-inspired, Bio-inspired and metaheuristics optimization algorithms are a set of approaches that are illumed by the behavior of natural situations. Those approaches have been inspired by animal and insects behaviors, biology, chemical reactions, etc. Metaheuristics a algorithms are widely used to solve many optimization problems. They are often used when the problem is too complex for classical methods, or when the traditional methods are too slow [18–22]. In this part, a recently optimization algorithm named Artificial Hummingbird Algorithm (AHA) was introduced to be used to design an optimal power management and control of interconnected multi-microgrid connected to the utility grid. In general, the algorithmic framework of AHA is simple, and easy to implement to solve optimization issues with a small-size population and only a few control parameters. AHA is the newest bio-inspired evolutionary optimization technique and algorithm which was recently developed by Weiguo Zhao in 2021 [24]. It illustrates the action of small hummingbirds to wisely search for food in their territory to gather the maximum amount of nectar. The entire phenomenon has been exercised by hummingbirds for the foraging of food sources with the highest level of nectar, resulting in a close-optimal solution [24–26]. The pseudo code of AHA algorithm is shown in Fig. 9 [24–26]. More detailed descriptions of all AHA algorithm steps are available in [24].

4 Simulation Results

In this section, the designed coordinated and enhanced active power management and frequency control LFC loop based on cascade PD-Fuzzy-PIDN controller based AHA algorithm was applied and test on the investigated interconnected multi-microgrid shown in Fig. 3. Three microgrid including wind farms, Solar PV and CSP power generators were connected to the power system shown in Fig. 2. In order to show the validity of the proposed strategy five case studies have been simulated and presented in this section, including the contribution of the hybrid energy storage system, the connection of the microgrid to the utility grid, and the application of artificial intelligence.

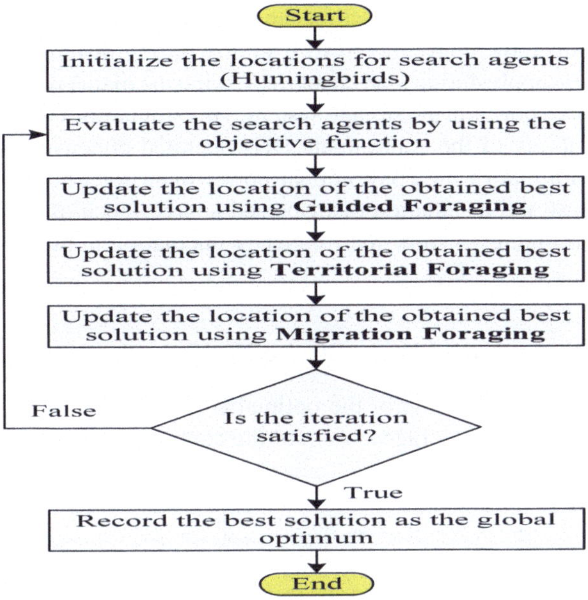

Fig. 9. Schematic flowchart of AHA [24–26].

4.1 Islanded Microgrid Without HESS

In this part, the microgrid was investigated for the simulation in isolated operation mode using the LFC method as shown in Fig. 1. The microgrid was simulated during load disturbances and renewable energies integration without insertion of hybrid storage system. It is clear from the obtained results in Fig. 10 that LFC loop can be used to keep system frequency at scheduled value.

4.2 Islanded Microgrid with HESS

In this part, a hybrid storage system was installed in the isolated microgrid to support the LFC loop. It is clear from the obtained results in Fig. 11 that the storage system participate in microgrid frequency regulation and can be used to support the LFC loop in presence of RESs.

4.3 Utility Grid Without Microgrid

In this scenario, the microgrid was not considered for the simulation, only the main grid was simulated. Focusing mainly on conventional power system frequency control, it can be observed from the obtained results that the designed controller has proven its efficiency to regulate the power system frequency as shown in Fig. 12.

4.4 Utility Grid Connected Microgrid Without HESS

In this scenario, the microgrid was connected to the main grid and the simulation was considered as on-grid operation mode. The purpose of this connection was mainly for two objectives: firstly to use the MG regulation capacity to enhance the main grid frequency regulation and reduce the conventional LFC ability during disturbances. Secondly, to support MG power supply during load changes by using the main grid Primary Operating Reserve or the stored power in the storage system. This simulation can be helpful to show how much LFC ability of the controllable DGs operated effectively in the microgrid in cooperation with that of the conventional power plant for LFC in the main power system can increase the installation of dispersed sources such as wind and solar PV power generation. Figure 13 show the frequency fluctuation in presence of connected microgrid.

4.5 Utility Grid Connected Microgrid with HESS

Figure 14 show the effectiveness of the optimal coordinated control strategy of LFC loop with the hybrid energy storage system of the connected microgrid.

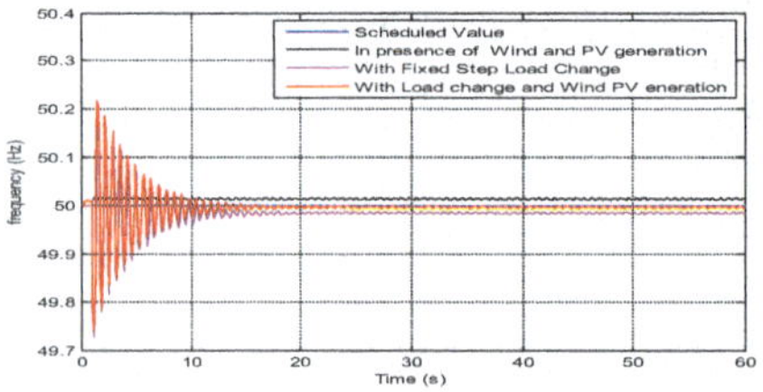

Fig. 10. Islanded Micro-Grid Frequency Fluctuation Without HESS.

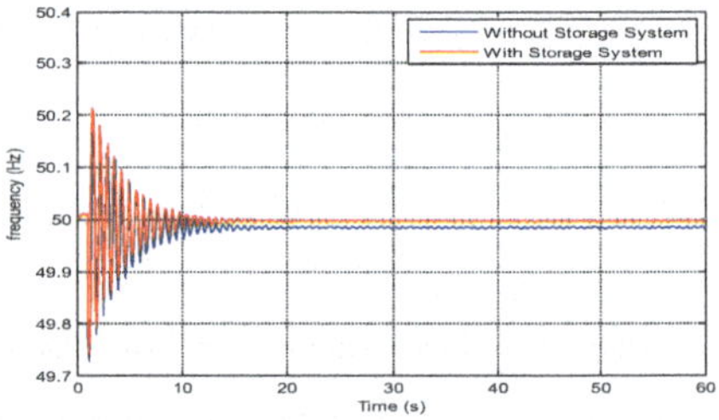

Fig. 11. Islanded Micro-Grid Frequency Fluctuation With HESS.

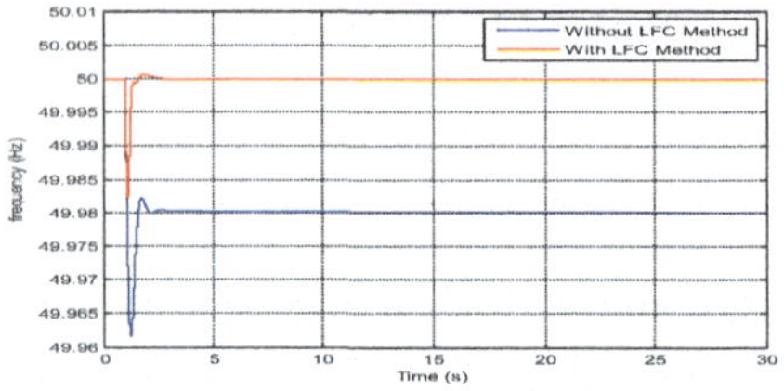

Fig. 12. Main Grid Frequency Fluctuation Before Micro-Grid connection.

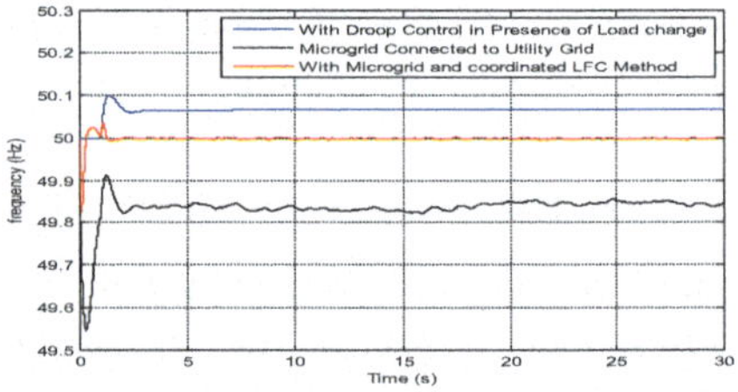

Fig. 13. Main Grid Frequency Fluctuation After Micro-Grid Connection.

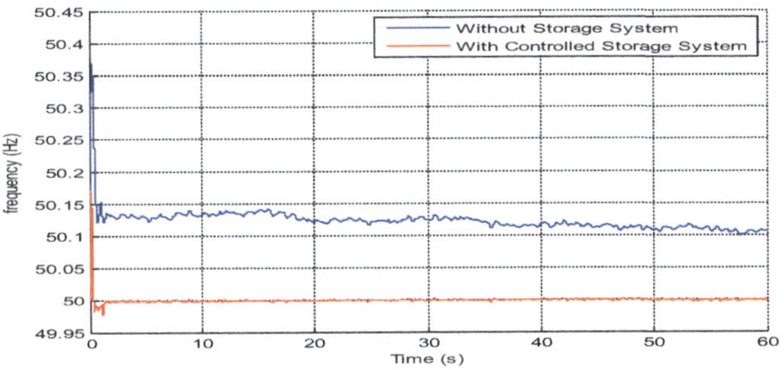

Fig. 14. Main Grid Frequency Fluctuation With Micro-Grid and HESS.

5 Conclusion

This paper has presented a novel effective optimal active power management and control of interconnected multi-microgrid connected to the utility grid using a recently developed bio-inspired optimization algorithm called Artificial Hummingbird Algorithm (AHA). A multi-stage coordinated operation and control strategy based on combined PD-Fuzzy-PIDN controller, HVDC transmission link, and hybrid energy storage system involving SMES, RFB and FC have been used in order to perform optimization tasks and reduce the conventional LFC capacity in the interconnected power system by using various sources of distributed generation such wind and solar power. Also, the developed optimal AGC scheme was employed to solve dynamic grid and microgrid frequency regulation issue in presence of load disturbances. Several scenarios have been simulated and presented to show the validity of the suggested method. The application of artificial intelligence algorithm in isolated grid for active power management of renewable energies and storage systems achieve significant improvement in the dynamic performance of the utility grid as well as the microgrid. Furthermore, the storage system can support the dynamic frequency regulation and enhance the LFC loop to keep system stability at scheduled values. In summary, the presented work reveals that the coordination between optimal LFC based AHA optimization algorithm, HVDC link and hybrid storage system can significantly reduce frequency deviations and quickly drive the dynamic system responses to steady state under disturbance. Also, this strategy can control faster system frequency and can increase the installation of uncontrollable DGs such as wind farm and solar PV array into the classical electric network.

References

1. Kumar, R.P., Karthikeyan, G.: A multi-objective optimization solution for distributed generation energy management in microgrids with hybrid energy sources and battery storage system. J. Energy Storage **75**, 109702 (2024)
2. Tooryan, F., Hassanzadehfard, H., Collins, E.R., et al.: Optimization and energy management of distributed energy resources for a hybrid residential microgrid. J. Energy Storage **30**, 101556 (2020)

3. El Yakine Kouba, N., Amrane, Y.: Application of artificial intelligence to enhance marine microgrid frequency control coordinated with battery energy storage system and HVDC transmission link. In: Innovations in Machine Intelligence, pp. 1–10 (2023)
4. El Yakine Kouba, N., Sadoudi, S.: Application of soft computing for optimal control of AC/DC interconnected multi-microgrid coordinated with fuel cell storage system. In: The 1st National Conference on Electronics, Electrical Engineering, Telecommunications, & Computer Vision: Boumerdes, Algeria (2023)
5. Abdelghany, M.B., Al-Durra, A., Gao, F.: A coordinated optimal operation of a grid-connected wind-solar microgrid incorporating hybrid energy storage management systems. IEEE Trans. Sustain. Energy **15**, 39 (2023)
6. Roy, N.B., Das, D.: Probabilistic optimal power allocation of dispatchable DGs and energy storage units in a reconfigurable grid-connected CCHP microgrid considering demand response. J. Energy Storage **72**, 108207 (2023)
7. Naz, M.N., Samad, A., Das, N., Imtiaz, S., Altaf, M.W.: Effective stability improvement of multi-microgrids assisted diversified wind energy resources. In: 5th International Conference on Energy Conservation and Efficiency, pp. 1–6. IEEE, Pakistan (2022)
8. Chen, X., Chen, X., Zhang, L., Zhou, Z., Chen, J., Guo, Z.: A stackelberg game strategy for multi-microgrid operation considering seasonal hydrogen storage. In: 5th Asia Energy and Electrical Engineering Symposium, pp. 1023–1030. IEEE, China (2023)
9. Zhang, Y., Zhang, T., Wang, R., Liu, Y., Guo, B., Zhang, T.: Dynamic dispatch of isolated neighboring multi-microgrids based on model predictive control. In: International Conference on Smart Grid and Clean Energy Technologies, pp. 50–55. IEEE, China (2016)
10. Sahoo, S.K., Kishore, N.K.: Coordinated control and operation of a multi-microgrid system. In: 2017 7th International Conference on Power Systems (ICPS), pp. 283–288. IEEE, Pune (2017)
11. Sofla, M.A., King, R.: Control method for multi-microgrid systems in smart grid environment—stability, optimization and smart demand participation. In: 2012 IEEE PES Innovative Smart Grid Technologies (ISGT), pp. 1–5. IEEE, Washington (2012)
12. Chowdhury, A.H., Asaduz-Zaman, M.: Load frequency control of multi-microgrid using energy storage system. In: 8th International Conference on Electrical and Computer Engineering, p. 548–551. IEEE, Dhaka (2014)
13. Akdogan, M.E., Ahmed, S.: Improving power quality and power sharing in unbalanced multi-microgrids using energy storage system. In: 2021 IEEE Applied Power Electronics Conference and Exposition, pp. 2733-2739. IEEE, Phoenix (2021)
14. El Yakine Kouba, N., Brik, A.: Islanded and grid-connected control in microgrid using artificial intelligence and storage system. In: The 1st National Conference on Emergent Technologies in Electrical Engineering (NCETEE'23), Setif, Algeria (2023)
15. El Yakine Kouba, N., Sadoudi, S.: Islanded microgrid frequency control in presence of HVDC-connected marine power plant coordinated with optimal fuzzy-PID and hybrid energy storage devices. In: 3rd IEEE International Conference on Signal, Control and Communication, pp. 1–6 (2023)
16. Ibrahim, M.M., Hasanien, H.M., Farag, H.E.Z., et al.: Energy management of multi-area islanded hybrid microgrids: a stochastic approach. IEEE Access **11**, 101409 (2023)
17. Yousri, D., Farag, H.E.Z., et al.: Integrated model for optimal energy management and demand response of microgrids considering hybrid hydrogen-battery storage systems. Energy Convers. Manag. **280**, 116809 (2023)
18. Ali, Z.M., Calasan, M., et al.: Applications of energy storage systems in enhancing energy management and access in microgrids: a review. Energies **16**(16), 5930 (2023)
19. Behera, P.K., Pattnaik, M.: Coordinated power management of a laboratory scale wind energy assisted LVDC microgrid with hybrid energy storage system. IEEE Trans. Consumer Electron. **69**, 467 (2023)

20. Cavus, M., Allahham, A., Adhikari, K., et al.: Energy management of grid-connected microgrids using an optimal systems approach. IEEE Access **11**, 9907–9919 (2023)
21. El Yakine Kouba, N., Sadoudi, S., Youssouf, A.: Optimal multi-microgrid active power management and frequency control coordinated with hybrid energy storage system and HVDC link. In: International Conference on Solar Energy and Hybrid Systems (icsehs), Laghouat, Algeria, pp.1–6 (2023)
22. El Yakine Kouba, N., Sadoudi, S.: Power system stability analysis of the Algerian isolated grid in Salah-Adrar-Timimoun in presence of renewable energy sources. In: Conference: 1er Séminaire National en Matériaux et Environnement (SNME) At: Tamanrasset, Algeria (2023)
23. Yoshihara, T., et al.: Contribution of distributed generations in microgrid to reduction of LFC capacity in utility grid. In: The International Conference on Electrical Engineering, pp.1–6 (2009)
24. Zhao, W., Wang, L., Mirjalili, S.: Artificial hummingbird algorithm: a new bio-inspired optimizer with its engineering applications. Comput. Methods Appl. Mech. Eng. **388**, 11419 (2022)
25. Waleed, U., Haseeb, A., Ashraf, M.M., Siddiq, F., Rafiq, M., Shafique, M.: A multiobjective artificial-hummingbird-algorithm-based framework for optimal reactive power dispatch considering renewable energy sources. Energies **15**, 9250 (2022)
26. Hu, G., Zhong, J., Zhao, C., et al.: LCAHA: a hybrid artificial hummingbird algorithm with multi-strategy for engineering applications. Comput. Methods Appl. Mech. Eng.. Methods Appl. Mech. Eng. **415**, 116238 (2023)

Design and Performance Investigation of Hierarchical Control Intended for Three-Phase Photovoltaic Microgrids

C. Ait Hammouda[1](✉), R. Bradai[1], A. Kherbachi[2], A. Bendib[3], and R. Boukenoui[3]

[1] LATSI Laboratory, Department of Electronics, Blida 1 University, Blida, Algeria
aithammouda_camelia@univ-blida.dz

[2] Wind Energy Division, Center of Renewable Energy Development, Algeries, Algeria

[3] SET Laboratory, Department of Electronics, Blida 1 University, Blida, Algeria

Abstract. The present paper focuses on investigating the performance of a hierarchical control designed based on DESOGI for three-phase Photovoltaic Microgrid (PVMG) systems. The main objective of this paper is to ensure effective and optimal control of the PVMG. In this paper, a comprehensive description of the PVMG system and the corresponding control strategy are provided. The design of the primary, secondary, and tertiary controls based on MESOGI is established. In addition, the paper covers the adaptation stage, involving the DC-DC converter with a Maximum Power Point Tracker (MPPT) controller intended for the PV generator (PVG). Furthermore, it investigates the performance of the designed hierarchical control through simulation tests conducted in MATLAB. The results show how well the hierarchical control that was created was able to provide precise power sharing, adequate power flow, smooth synchronization, and proper restoration of frequency and amplitude under a variety of operating situations.

Keywords: Three-Phase microgrid (MG) · Photovoltaic (PV) system · Hierarchical control

1 Introduction

A Microgrid (MG) is commonly described as a distribution network operating at low or medium voltage levels, comprising various Distributed Generation (DG) sources, energy storage elements, and controllable loads. The majority of DG sources rely on Renewable Energy Sources (RESs), including microturbines, solar systems, and energy storage systems, which need to be connected to the MGs and the utility grid at the Point of Common Coupling (PCC) through Voltage-Source Inverter (VSI) enabling the seamless operation of MGs in both grid-connected and islanded modes as needed [1, 2]. They serve as a significant component in converting DC power generated by RESs to AC power suitable for local loads. However, it is imperative to design an efficient and robust control system for the VSI to avoid possible MG problems [3], such as power quality, stability, and reliability as well as achieving accurate power sharing and power flow among all kinds of sources. The design of control schemes for three-phase inverters in

© The Author(s), under exclusive license to Springer Nature Switzerland AG 2025
M. Hatti (Ed.): IC-AIRES 2024, LNNS 1238, pp. 152–160, 2025.
https://doi.org/10.1007/978-3-031-80301-7_17

islanded or grid-connected AC MG systems has been the subject of extensive research. Most published works adopt a hierarchical control strategy, which consists of three control levels: primary, secondary, and tertiary. This strategy strikes a balance between decentralized, centralized, and distributed control structures [4]. In the realm of scientific literature within this field, the predominant focus has been on the techno-economic aspects, largely overlooking other crucial dimensions. While some studies have delved into the integration of renewable energies, they often confine their models to specific geographic locations and rural areas. Furthermore, hierarchical control mechanisms have gained attention in MG control, yet their implementation has predominantly centered on control strategies alone. However, it's pertinent to note that while numerous papers claim their proposed controllers are tailored for integrating renewable energies, none have substantiated this assertion through empirical testing. From here, the true litmus test lies in the application of the MGs control strategies incorporating RESs, an area that remains largely unexplored in validating the efficacy of such approaches. Hence, the issue at hand is whether PVMG can operate steadily and robustly with improved performance using the proposed primary, secondary, and tertiary control schemes.

The present paper focuses on the design of control schemes within a hierarchical control based on DESOGI intended for a three-phase PVMG. It begins by giving an overview of the system under study, introducing the structure of the PVG, and discussing the different used components in Sect. 2. Following that, Sect. 3 provides a comprehensive description of a hierarchical control approach including the three control levels designed based on DESOGI. Moving forward to Sect. 4 the simulation results and discussions pertaining to the entire PV system, encompassing two PVGs controlled by an MPPT-controlled boost converter, interconnected to the PCC through a three-phase voltage source inverter (VSI) controlled by a hierarchical control. Finally, Sect. 5 summarizes the main conclusions of this work.

2 Description of the PVMG System Under Study

In solar PV conversion, solar energy is converted into electricity through PV panels. These serve as a voltage generator to power a load. However, the problem here is that the voltage supplied by these PVGs remains lower, which is why an adaptation stage (regulation) is required. This converter is used to adapt the energy drawn from the sun to suit the load and meet all the desired requirements. In itself, needs a switching controller. To regulate the overall output power to the load, the PVGs must be able to communicate with each other.

To this end, the system considered in this study is made up of two PV arrays or two PVGs. As illustrated in Fig. 1. Each PV array output power is processed by a Perturb and Observe (P&O) MPPT-based DC-DC converter. Additionally, the connection to the PCC is performed by an LC-filtered three-phase VSI and a line impedance according to the grid restrictions. The inverter is controlled by the hierarchical control developed in the next section.

3 Hierarchical Control Design Based on DESOGI

One common method for monitoring the operation of developed AC MGs is a hierarchical control structure. It serves as a compromise between fully decentralized and fully centralized approaches. This control structure consists of three levels: primary, secondary, and tertiary. Figure 2 illustrates the multilevel hierarchical control with brief descriptions of the key responsibilities at each control level.

The hierarchical control system, as depicted in Fig. 3, includes the three control levels for AC MGs with parallel-connected VSIs, operating in grid-connected and islanded modes. The primary control layer is designed for decentralized operation, ensuring power sharing among DG units and MG stability. It often relies on a droop control approach to achieve fast power balance in a decentralized manner. However, the droop-based primary control has its limitations, resulting in variations in voltage and frequency as well as inaccuracies in the transfer of reactive power between DG units [5].

To address these issues, the secondary control layer, whether centralized or distributed, is applied to return the MG voltage frequency and amplitude to their nominal values, and hence improve power quality. Additionally, the secondary control also seeks to the synchronization of the MG with the main utility grid before connecting them [6]. The last controller is the slowest one, it is usually centralized and only activates when the MG is connected to the main grid. In addition, this level is also responsible for islanding detection and managing bidirectional power flow between the MG and the main grid[7]. It's worth noting that the secondary and tertiary control levels are included in the term "energy management system." The ensuing subsections will give a thorough explanation of each control level's functions as well as an outline of its basic architecture.

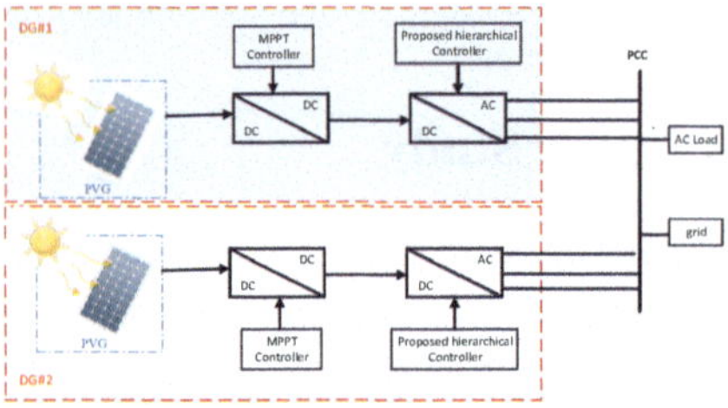

Fig. 1. Block diagram of the PVMG system under study

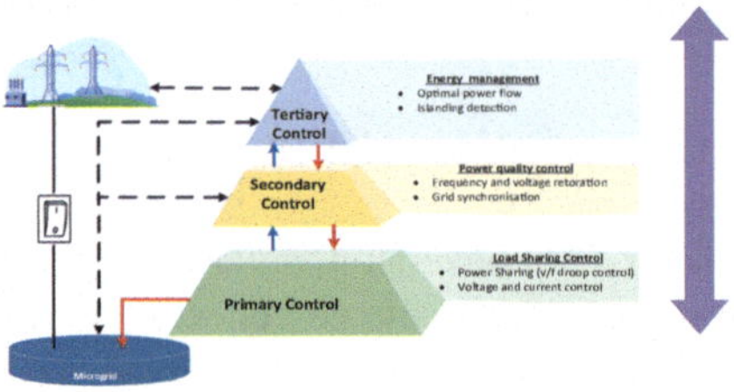

Fig. 2. Block diagram of hierarchical control

3.1 Primary Control Level

In an MG system, the primary control level is the lowest in the hierarchy. Its main role is to manage any sudden changes in the power load [8], whether those loads are linear or nonlinear. This control is local and decentralized, meaning it relies solely on local measurements with each inverter having its dedicated control stage, enabling each DG unit to function independently, without the need for communication during normal operation. It plays a crucial role in guaranteeing the reliability of the MG and enhancing its overall performance and stability [3]. It's responsible for adjusting the frequency and voltage output of inverters to serve as a reference for the inner multiloop controller. This helps reduce circulating currents that can occur when multiple converters are connected in parallel. Additionally, it ensures the precise sharing of active and reactive power among these converters. It can function in both grid-connected and autonomous modes, with a seamless transition between them.

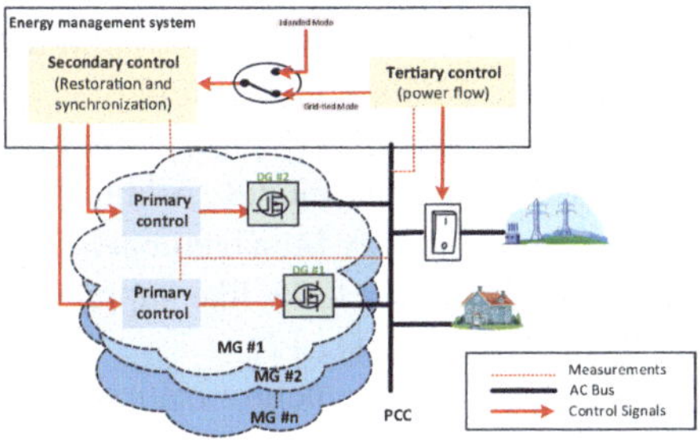

Fig. 3. Hierarchical control structure

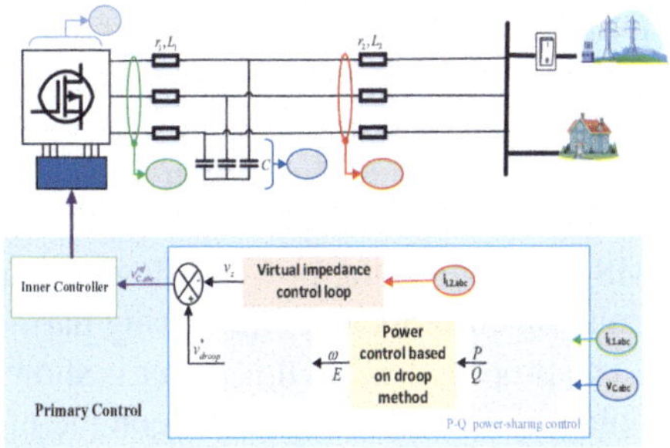

Fig. 4. Primary control level schematic diagram

As shown in Fig. 4, several blocks can be further distinguished within the primary control:

- The inner control also referred to as zero level, includes an external voltage control loop (VCL) loop and an internal current control loop (CCL), using proportional-integrator (PI) and P regulators, tasked with maintaining the power stage voltage and frequency within desired references. More details are given in [9].
- Power-sharing control, based on the droop technique, is responsible for coordinating the DG units feeding a common load within an MG. One of the major advantages of this control strategy is the ability to enable communication-less power-sharing among parallel-connected DG units. In this control level, a power calculation unit based on DESOGI is included. Thus, the active and reactive power is obtained by using the direct and quadrature components estimated by the DESOGI. The droop controlled adopts these calculated powers as inputs and produces the output voltage reference of the inverter. The operation principle of the droop method is based on adjusting the frequency (f), and voltage (E), of an inverter in order to control the power balance. This concept is inspired by synchronous generators, which drop their frequency to inject more active power.
- Virtual impedance: by using the droop method, the inverters' outputs and line impedances can affect the power-sharing accuracy. Also, feeding nonlinear loads creates current harmonics, therefore, a virtual impedance control loop (VICL) is proposed. This later is introduced to improve the accuracy of the power-sharing and allow the share of current harmonics by normalizing the output impedance of the VSIs. This control block is also based on the DESOGI, in which the estimated direct and quadrature components from the inverter output currents by the DESOGI are used to generate an output voltage. This voltage is then subtracted from the droop voltage reference, as proposed in [10]. During this stage, the MG's key parameters, such as frequency and amplitude, may deviate; hence, the role of the secondary level

is to restore them to their nominal values, facilitating synchronization with the grid. The tertiary, being the last level, plays a crucial role in regulating power flow between the MG and the main grid.

3.2 Secondary Control Level

The secondary controller (SC) assumes a critical role in rectifying the frequency and amplitude deviations that may occur due to the primary control's droop-based approach. This helps power quality enhancement by returning the MG's voltage amplitude and frequency to their nominal values. Moreover, it is in charge of MG's synchronization with the main grid during the switch from islanded to grid-connected modes. Additionally, during islanded mode operation, the SC serves as the MG's Energy Management System (EMS), managing the power's flow and quality, which includes tasks like suppressing circulating currents and eliminating harmonics within the MG [11, 12].

The proposed SC in this paper is shown in Fig. 5. During the synchronization phase, the phase angle difference between the utility grid and the MG is determined using the orthogonal components of the PCC and grid voltages that are extracted, using DESOGI-FLL with Positive Sequence Detector (PSD), then transmitted through a P controller to achieve alignment. The restoration control loop updates its frequency reference using this generated control signal.

As we proceed to the restoration stage, the frequency and amplitude $\widehat{\omega}_{MG}$, $\widehat{E}_{MG}$ estimated are compared to their intended nominal values ω^*_{MG}, E^*_{MG} and then, PI controllers process these values and provide the needed modifications for the actual frequency and amplitude. A low bandwidth communication channel is then used to send the resulting control signals, which are designated as $\delta\omega_{res}$ and δE_{res}, to the primary control level. This makes it easier for the intended nominal values to be recovered inside this later.

3.3 Tertiary Control Level

The MG can operate both independently and in conjunction with the main grid, as was previously noted. In grid-tied mode, the MG's power generation can exceed local consumption, particularly when renewable energy sources, such as energy storage devices, are producing at their peak. As a result, excess active and reactive power is sent via an inverter and line impedance to the main AC grid. On the other hand, the main grid supplies additional power to meet demand in the event that the MG's supply is insufficient.

As a centralized controller, the tertiary control level is responsible for managing the bidirectional power transfer between the distributed generation units in the MG and the main grid at the PCC by controlling the flow of both active and reactive power. The distributed generation units' voltage amplitudes and frequency can be changed to regulate this power exchange. It's crucial to remember that, in comparison to the secondary and primary controls at lower levels, this control level often displays a slower dynamic reaction.

The suggested tertiary controller, which is based on the DESOGI-FLL scheme, and the power stage of a three-phase MG running in grid-tied mode are shown in Fig. 6. The load in this configuration is linked to the AC common bus, where the DG units share the load power demand equally. The suggested tertiary control scheme comprises a PI controller and a DESOGI-FLL-based power calculation block, as shown in Fig. 6. The DESOGI-FLL is used to estimate the direct and quadrature components of the grid voltage and current. Based on the computed powers P_g and Q_g, the MG's active and reactive power references are modified by the PI controller. The corrected references for the MG represented as f_{MG} and E_{MG}, respectively, are then transmitted to the secondary control level by the tertiary control along with the updated frequency and voltage amplitude values.

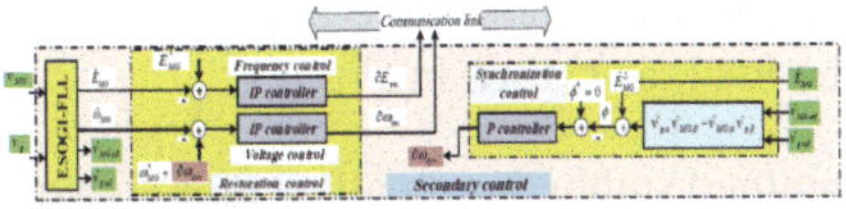

Fig. 5. Secondary control level schematic diagram

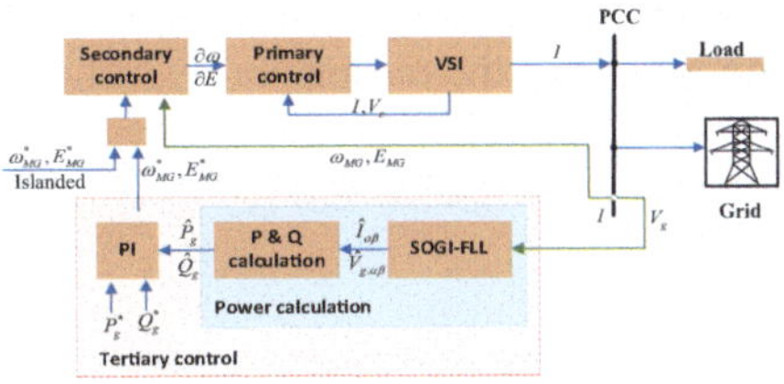

Fig. 6. Tertiary control diagram for a grid-tied MG

4 Simulation

The suggested scheme is introduced within the context of an AC PVMG's hierarchical control structure, as shown in Fig. 1. It is composed of two 2 kW rated power parallel PV arrays using three-phase VSIs, considering primary, secondary, and tertiary controls. According to this scheme, each PV array output power is processed by a P&O MPPT-based DC-DC converter, in addition to a three-phase VSI associated with its own local controller. The whole system is linked to the utility grid and a common load via a common AC bus according to the grid restrictions.

The studied PVMG along with the designed controller, is tested considering the case studies listed below, note that they all start with no load, and then a common load is added at t = 0.1s , where only the primary control is activated. The secondary control is activated later at t = 0.5 s.

Case 01: Uniform Condition: here, the two PVGs are exposed to the same irradiation level set to 850 W/m^2. Meaning that both of them generate the same power rate.

Case 02: Shading Condition, In this case, PVG#02 is shaded, which means that the irradiation level drops from 850 W/m^2 to 500W/m^2, hence, the generated power from the two units is not the same.

Case 03: Generator disconnection, In this case, a failure occurs in the second PV array at t = 0.7 s, which leads to the disconnection of this later.

Case 04: On-grid mode, In this case, the tertiary control is activated, and the two PVGs are not exposed to irradiation (i.e. night), the load is then fed from the grid.

Figure 7, 8, 9, 10, 11, and 12 show the obtained results of the performed simulation tests, where the active power, voltage, and current of the PVGs, grid, and at the PCC are illustrated. In Fig. 7, the active power is shared equally between the two generators when they are exposed to the same irradiation level. At t = 0.5 s, when the secondary control was activated, the frequency and amplitude are restored to their nominal values. Meaning that the proposed controller is working as required with the PVMG as shown in Fig. 8.

In the case of the shading condition, the load power is shared between PVGs but with different rates, so PVG#01 generates more than PVG#02 as illustrated in Fig. 9.

Figure 10 represents the active power generated in the case of the disconnection of PVG#02 at t = 0.8 s. It can be seen that this failure didn't affect the operation of the controller, where the PVG#01 continued generating the required power to the common load.

In the last scenario, it is considered that the PVGs are not lighted, in this case, the main grid takes the responsibility of feeding the load, as shown in Fig. 11.

In addition, the voltage and current are maintained stable during all tests as shown in Fig. 12.

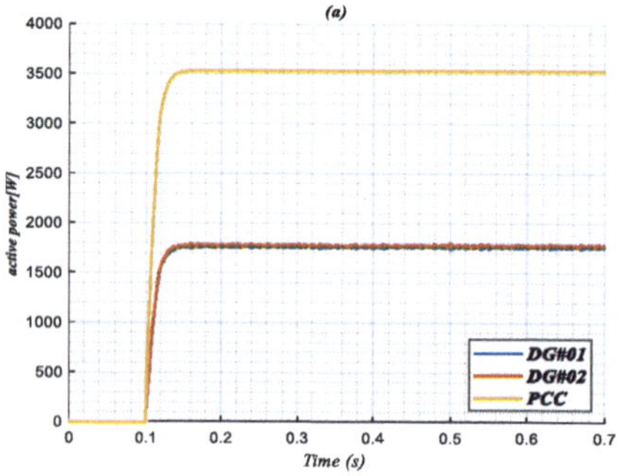

Fig. 7. Obtained generated power in response to case 01

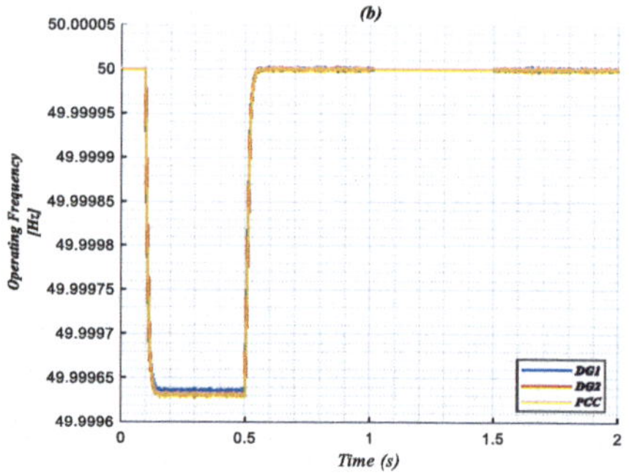

Fig. 8. Frequency restoration

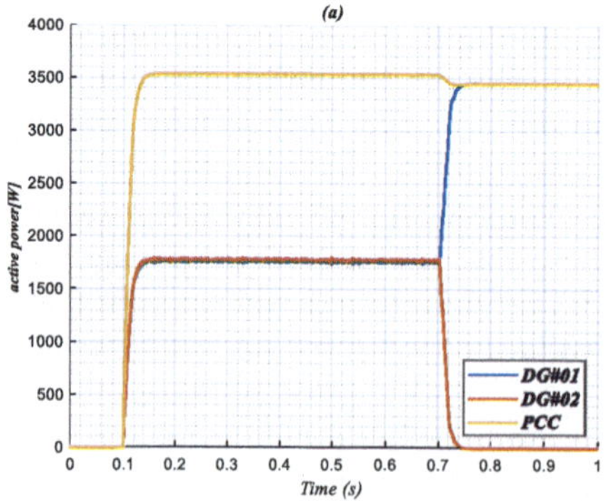

Fig. 9. Obtained generated power in response to case 02

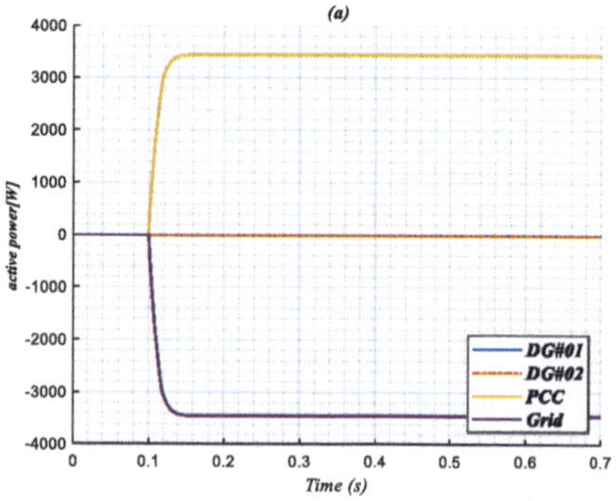

Fig. 10. Obtained generated power in response to case 03

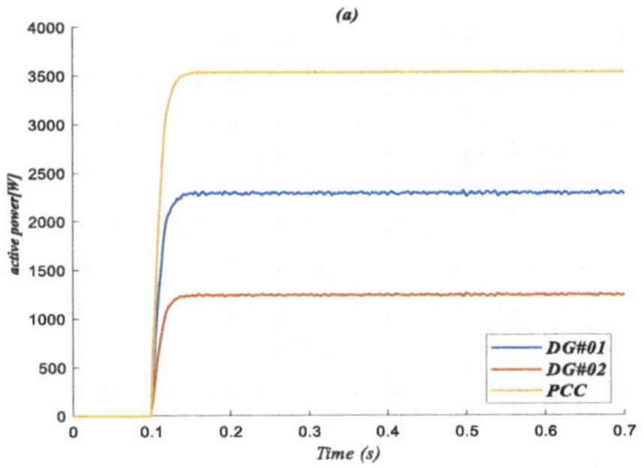

Fig. 11. Obtained generated power in response to case 04

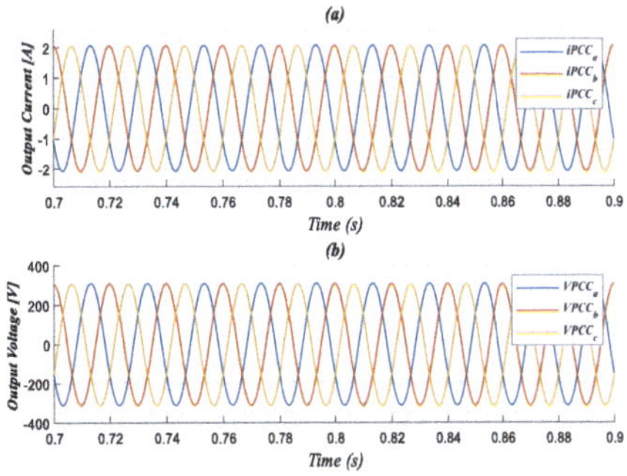

Fig. 12. Current and voltage waveforms

5 Conclusion

In this paper, the design and performance investigation of a hierarchical control for grid-connected and islanded operations of three-phase droop-controlled VSIs-interfaced PVMG. The controlled PVMG system was simulated and tested under various conditions to validate the proposed control approach using the MATLAB/Simulink software. Firstly, the designed primary control successfully achieved its objectives of attaining the desired references as well as sharing the load between the different PV arrays. In addition, the outcomes have demonstrated the secondary control's resilience in efficiently restoring the nominal values of frequency and amplitude with strong transient and steady-state responses, even under a variety of operating scenarios. Finally, the power flow regulation of the PVMG running in grid-supporting mode has yielded data that show how well the planned controller works to provide a system that can react quickly to changes in load. Overall, the results in this research have successfully proven that the primary, secondary, and tertiary schemes of the hierarchical control proposed for a three-phase PVMG are effective.

References

1. Aryanezhad, M.: Management and coordination of LTC, SVR, shunt capacitor and energy storage with high PV penetration in power distribution system for voltage regulation and power loss minimization. Int. J. Electr. Power Energy Syst. **100**, 178–192 (2018)
2. Mehrizi-Sani, A., Iravani, R.: Potential-function based control of a microgrid in islanded and grid-connected modes. IEEE Trans. Power Syst. **25**(4), 1883–1891 (2010)
3. Choudhury, S.: A comprehensive review on issues, investigations, control and protection trends, technical challenges and future directions for microgrid technology. Int. Trans. Electr. Energy Syst. **30**(9), e12446 (2020)
4. Sen, S., Kumar, V.: Microgrid control: a comprehensive survey. Ann. Rev. Control **45**, 118–151 (2018)
5. Guerrero, J.M., Vasquez, J.C., Matas, J., Vicuna, LGd., Castilla, M.: Hierarchical control of droop-controlled AC and DC microgrids—a general approach toward standardization. IEEE Trans. Ind. Electron. **58**(1), 158–172 (2011)

6. Micallef, A., Apap, M., Spiteri-Staines, C., Guerrero, J.M.: Single-phase microgrid with seamless transition capabilities between modes of operation. IEEE Trans. Smart Grid **6**(6), 2736–2745 (2015)
7. Meng, L., et al.: Flexible system integration and advanced hierarchical control architectures in the microgrid research laboratory of Aalborg University. IEEE Trans. Ind. Appl. **52**(2), 1736–1749 (2016)
8. Anwar, M., Marei, M.I., El-Sattar, A.A.: Generalized droop-based control for an islanded microgrid. pp. 717–722
9. Hammouda, C.A., Bradai, R., Kherbachi, A., Bendib, A., Boukenoui, R., Kara, K.: ESOGI-Based Virtual Impedance Control Scheme with Performance Improvement for Droop-Operated Three-Phase VSIs in Islanded AC Microgrid, pp. 1–6
10. Hammouda, C.A., Bradai, R., Boukenoui, R., Kherbachi, A., Bendib, A.: Modeling and Design of Primary Control's Inner Loops for Droop-Controlled Three-Phase Inverters within a Microgrid System, pp. 1–6
11. Wang, X., Guerrero, J.M., Blaabjerg, F., Chen, Z.: Secondary voltage control for harmonics suppression in islanded microgrids, pp. 1–8
12. Savaghebi, M., Jalilian, A., Vasquez, J.C., Guerrero, J.M., Lee, T.L.: Voltage harmonic compensation of a microgrid operating in islanded and grid-connected modes, p. 1

PD Controller Design for Three-Axis Attitude Stabilization of a Microsatellite Using Reaction Wheels

B. Khouane[1(✉)], J. E. Benmansour[1], I. Rached[2], and A. S. Zemouli[2]

[1] Satellite Development Center, Algerian Space Agency, Ibn Rochd USTO, BP 4065, 31130 Oran, Algeria
b.khouane@gmail.com

[2] Department of Electrical Engineering, National Polytechnic School of Oran – Maurice Audin, 31000 Oran, Algeria

Abstract. This work presents an attitude control strategy for a small satellite in a low Earth orbit (LEO) equipped with solar panels on all sides except the underside, where the payload is located. The proposed approach employs reaction wheels as actuators to produce torque along the three axes of the satellite, configured in a typical way to ensure three-axis attitude control in a low earth orbit. A controller PD is employed to distribute three-dimensional torque and commands optimally. The simulated attitude control system was subjected to external disturbances caused by gravity gradient and magnetic torques affecting its performance in open-loop control. This controller uses unit quaternions to achieve accurate stabilization of the three-axis attitude. Numerical simulations are performed to assess the effectiveness of the closed-loop system, taking advantage of the proposed attitude control strategy.

Keywords: Attitude control · PD control · LEO Small Satellite · Reaction wheels · allocation control systems

1 Introduction

The Attitude Determination and Control Subsystem (ADCS) is an essential component of satellite systems. It plays a crucial role in the success of space missions, including Earth observation, scientific experiments, and gravity measurements[1, 2]. Conventional ADCS designs commonly include multiple components to achieve precise control of spacecraft orientation. These components include actuators such as reaction wheels, magnetic torque coils, and thrusters, as well as sensors like gyroscopes and star trackers. The main part of attitude control often relies on Reaction Wheels (RWs), which are flywheels with high inertia driven by brushless DC motors. These RWs provide torque directly proportional to the electric current supplied, enabling exact modifications to the satellite's attitude. Several companies design and deliver RWs for space applications, ranging from very small (0.1 Nms) to large ones (20 Nms or higher) [3]. The

© The Author(s), under exclusive license to Springer Nature Switzerland AG 2025
M. Hatti (Ed.): IC-AIRES 2024, LNNS 1238, pp. 161–169, 2025.
https://doi.org/10.1007/978-3-031-80301-7_18

torque generated by the wheel is directly dependent on the motor current. These companies cater to various satellite sizes and mission requirements, from CubeSats to large geostationary satellites. Most spacecraft rely on multiple Reaction RWs for three-axis stabilization [4, 5].Various configurations have been proposed to enhance the efficiency of attitude control systems. A key advantage of RWs is that their maximum torque is nearly constant along their fixed spin axis. This characteristic simplifies spacecraft attitude control and allows for precise maneuvering. When these wheels, commonly known as Reaction Wheel Assembly (RWA), are combined with their motors and electronics, they frequently need to be redundant. This means including supplementary wheels that can serve as backups in the case of a failure in the main wheel. Advanced configurations often incorporate redundant reaction wheels to improve the system's reliability. The additional wheels function as a backup if the primary wheel fails, ensuring continuous control of the spacecraft's orientation [6, 7]. To assess the robustness and effectiveness of these control systems in the presence of multiple disturbances. As a scenario, we have applied the configuration of AlSat-1, the first Algerian microsatellite [8], as our case study, focusing on attitude control. Considering the satellite's configuration when designing control systems, particularly its kinematic constraints and dynamic model, are crucial, as it significantly impacts performance. The Orthogonal 3-wheel Configuration is frequently employed for smaller satellites due to its simplicity and cost-effectiveness. This configuration places three reaction wheels along orthogonal axes, providing control around all three rotational axes. As mentioned in [4, 9], satellites often use three reaction wheels positioned orthogonally to control their attitude accurately in all three axes. This structure provides efficient spacecraft maneuverability and stabilization in the space environment. While various configurations and control allocation strategies exist for spacecraft attitude control using reaction wheels, focusing on optimal torque distribution isn't our primary concern in this case. Our research uses the specific parameters of the Algerian spacecraft AlSat-1 to demonstrate a practical implementation of this technology.

In this paper, we introduce a model to control the attitude of a small satellite using only reaction wheels as actuators in a circular orbit. The efficiency and accuracy of reaction wheels are verified, and the simulation results highlight the performance and limits of this configuration. Hence, the stability of the satellite is guaranteed based on PD control adopted using an orthogonal configuration of reaction wheels. The rest of the paper is structured as follows: Sect. 2 outlines the mathematical model, including dynamics and kinematics. Section 3 examines the external forces acting on the spacecraft and its mathematical model in orbit, which informs the satellite mission design. Section 4 describes the configuration model of the reaction wheels designed for it. Section 5 details the control law used in the satellite. Section 6 showcases the simulation results, and finally, Sect. 7 offers conclusions and proposes future work.

2 Spacecraft Mathematical Model

2.1 Attitude Dynamics

The following nonlinear differential equation can describe the attitude dynamics of a rigid spacecraft controlled by reaction wheels [10, 11]:

$$I\dot{\omega}_B|_I^B + \omega_B|_I^B \times I\,\omega_B|_I^B = -\omega_B|_I^B \times h_{rw} - \dot{h}_{rw} + \tau_d|^B \tag{1}$$

where I is the inertia matrix of the spacecraft, ω denotes the angular velocity of the spacecraft, τ_d represents the external torques action the satellite, h_{rw} represents the entire angular momentum of the wheels. A torque control generated by the three wheels to get the desired satellite attitude can be represented by:

$$u = -\dot{h}_{rw} \tag{2}$$

2.2 Kinematics Equations

Quaternions are used in this section to represent attitude. The quaternion kinematic equation can be used to determine the time evolution of the Euler parameters. This method allows the ongoing updating of the attitude representation.

$$\dot{q} = \begin{bmatrix} q \\ \dot{q}_4 \end{bmatrix} = \begin{cases} q = \frac{1}{2}(q_0 I_3 + q^\times)\,\omega_B|_O^B = \frac{1}{2}\Lambda(q)\,\omega_B|_O^B \\ \dot{q}_4 = -\frac{1}{2} q\,\omega_B|_O^B \end{cases} \tag{3}$$

where,

$$\Lambda(q) = \begin{bmatrix} q_4 I_3 + [q^\times] \\ -q^T \end{bmatrix} = \begin{bmatrix} q_4 & -q_3 & q_2 \\ q_3 & q_4 & -q_1 \\ -q_2 & q_1 & q_4 \\ -q_1 & -q_2 & -q_3 \end{bmatrix} \tag{4}$$

Note that the quaternion q, expresses the rotation from the inertial frame to the body frame in the following form:

$$q = \begin{bmatrix} q_1 \; q_2 \; q_3 \; q_4 \end{bmatrix}^T \tag{5}$$

A quaternion q is composed of a scalar part denoted by q_4 and a vector part denoted by $q_{13} = \begin{bmatrix} q_1 \; q_2 \; q_3 \end{bmatrix}^T$. Also, q needs a unit norm in order to describe a pure rotation, as shown in the following equation.

$$\text{Norm}(q) = \|q\| = \sqrt{q_1^2 + q_2^2 + q_3^2 + q_4^2} \tag{6}$$

Knowing that:

$$\omega_B|_O^B = \omega_B|_I^B - A|_O^B\,\omega_O|_I^O \tag{7}$$

where

$$\omega_0|_I^O = \left[0\ -\omega_0\ 0\right]^T \tag{8}$$

$\omega_B|_O^B$ is the vector that represents the angular velocity of the satellite body in relation to the orbital frame within the body.

$A|_O^B$ is the rotation matrix from the orbital frame to the body frame.

ω_0 represents the satellite's orbital angular velocity.

3 External Disturbances Acting on the Spacecraft

Various external torques affect spacecraft in the Low Earth Orbit. For our case study, the main disturbances are caused by gravity gradient torque. τg and magnetic field torque τ_m, defined in [12, 13]. The gravity gradient torque, expressed in the body frame, is as follows:

$$\tau_g|^B = 3\omega_0^2 c_3|^B \times I c_3|^B \tag{9}$$

where

$$\omega_0^2 = \frac{\mu}{r_c^3} \tag{10}$$

And,

$$c_3 = C(q)\begin{bmatrix} 0 \\ 0 \\ 1 \end{bmatrix} = \begin{bmatrix} 2(q_1q_3 - q_2q_4) \\ 2(q_2q_3 + q_1q_4) \\ 1 - 2\left(q_1^2 - q_2^2\right) \end{bmatrix} \tag{11}$$

Next, the magnetic torque is represented through the following expression:

$$\tau_m|^B = m|^B \times B|^B = \begin{bmatrix} B_z|^B m_y - B_y|^B m_z \\ B_x|^B m_z - B_z|^B m_x \\ B_y|^B m_z - B_x|^B m_y \end{bmatrix} \tag{12}$$

We define the geomagnetic field vector $B|^B$, relative to the body frame, as follows:

$$B|^B = \frac{\mu}{R_0^3} = \begin{bmatrix} \cos nt \sin i \\ -\sin i \\ 2 \sin nt \sin i \end{bmatrix} \tag{13}$$

where,

$$n = \frac{2\pi}{T} = 2\pi\sqrt{\frac{R_0^3}{\mu}} \tag{14}$$

where, μ is the gravitational coefficient of the earth, with a value of $3.986004418 \times 10^{14}\ \text{m}^3/\text{s}^2$. r_c represents the distance from the center of the Earth. $C(q)$ represents the attitude matrix. c_3 is the Nadir- pointing unit vector.

4 Three Reaction Wheels Spacecraft Model

The orthogonal 3-wheel configuration is used. It is relatively simple and cost-effective to implement. We simulate a small satellite using this reaction wheel distribution (as shown in Figs. 1 and 2) to examine the impact of reaction wheels on the system.

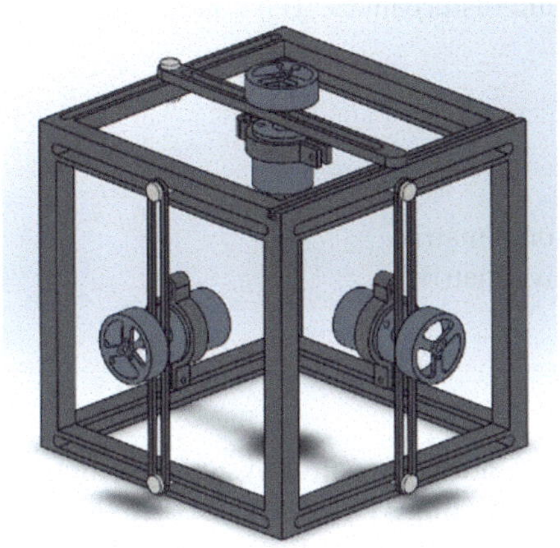

Fig. 1. Small satellite structure with reaction wheels

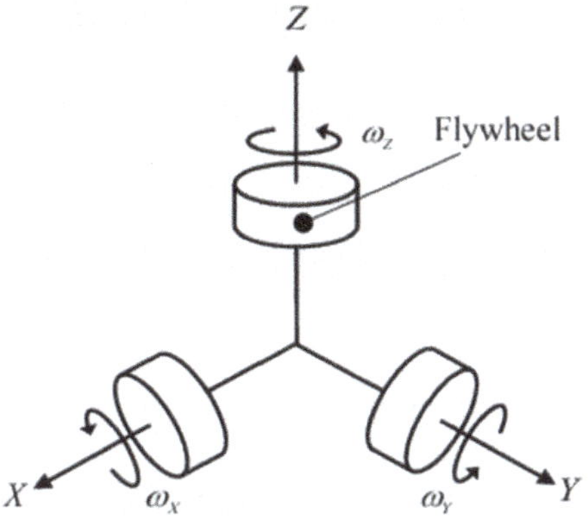

Fig. 2. Distribution of RWs

The distribution matrix L consists of n columns corresponding to n reaction wheels. Each column vector illustrates the distribution of torques from the reaction wheels across the satellite's rotational axes. Then,

$$T_{x,y,z} = L_{3\times n} T_{1,\ldots,n_{rws}} \tag{15}$$

where $T_{x,y,z} = \begin{bmatrix} T_x \; T_y \; T_z \end{bmatrix}^T$ are the moments acted on the spacecraft. $T_{1,\ldots,n_{rws}}$ Represent the torques generated by each reaction wheel.

The distribution matrix of the orthogonal 3-wheel configuration can be defined as:

$$L_{3\times 3} = \begin{bmatrix} 1\,0\,0 \\ 0\,1\,0 \\ 0\,0\,1 \end{bmatrix} \tag{16}$$

The symbol n_{rws} denotes the number of wheels, which in this case is 3. Then,

$$\begin{cases} T_x = T_1 \\ T_y = T_2 \\ T_z = T_3 \end{cases} \tag{17}$$

The reaction wheel dynamics can be expressed as follows:

$$\tau_{rw}|^B = \frac{dh_{rw}}{dt} + \omega_B|_I^B \times h_{rw} \tag{18}$$

5 Attitude Control Law for the Spacecraft Stabilization

This study applies a proportional-derivative control to the reaction wheels to manage the spacecraft's attitude control. The primary objective is to design a controller capable of stabilizing the spacecraft by driving its initial attitude (q) to the desired attitude (qd), thereby satisfying the kinematic equation. This approach aims to ensure precise orientation control while minimizing errors and disturbances. The control law applied to the system is

$$u = P.q_e + K.\omega_B^o \tag{19}$$

where,

$K_p = diag\left(\left[K_{p1}\ K_{p2}\ K_{p3}\right]\right)$ is the proportional matrix.
$K_D = diag([K_{D1}\ K_{D2}\ K_{D3}])$ is the derivative matrix.

6 Numerical Simulations

Simulation results verified the efficiency of the attitude control system. We obtained the results using an attitude propagator simulation using MATLAB/SIMULINK, which describes the satellite's dynamics and kinematics with external disturbances. As demonstrated in [14], which describes a satellite's initial parameters, these parameters are crucial for simulating the behavior and performance of the satellite in one orbit (6000 s). First, the external disturbances, as shown below:

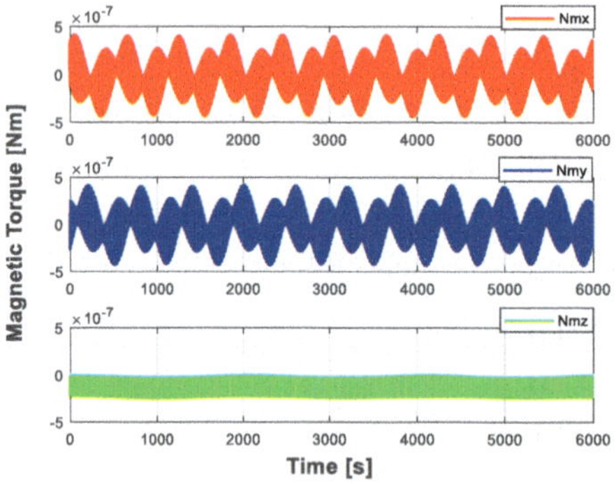

Fig. 3. Magnetic disturbance torques

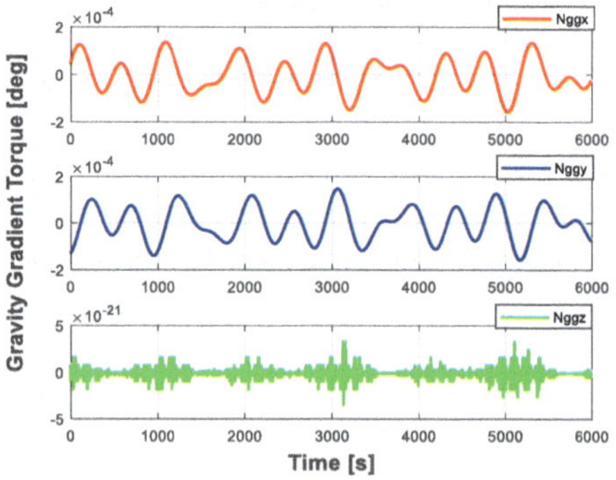

Fig. 4. Gravity gradient disturbance torques

Figures 3 and 4 demonstrate how the magnetic and gravity gradient torques fluctuate over time and impact the satellite. We can see that the gravity gradient torques are the primary and most influential perturbing torques. These figures are essential for designing the satellite's attitude control system because they help ensure that the system can compensate for these disturbances and maintain the desired satellite orientation.

Figures 5 and 6 demonstrate the quaternion angle and velocity without a controller when disturbance torques act on the satellite. We can see clearly that the system is not stable. Next, we will see the attitude control system stabilized using a PD controller.

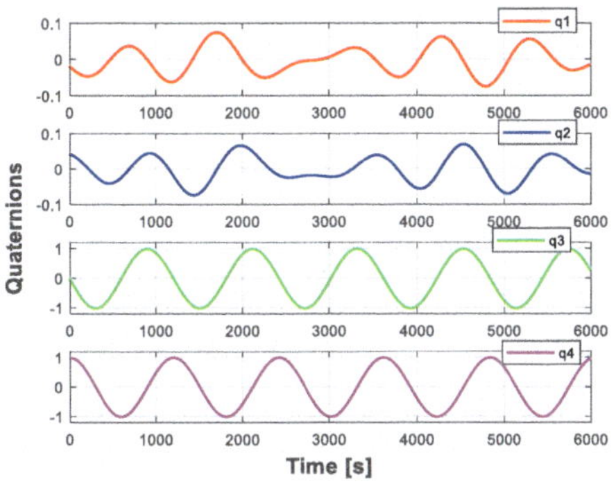

Fig. 5. Quaternions <<Without controller>>

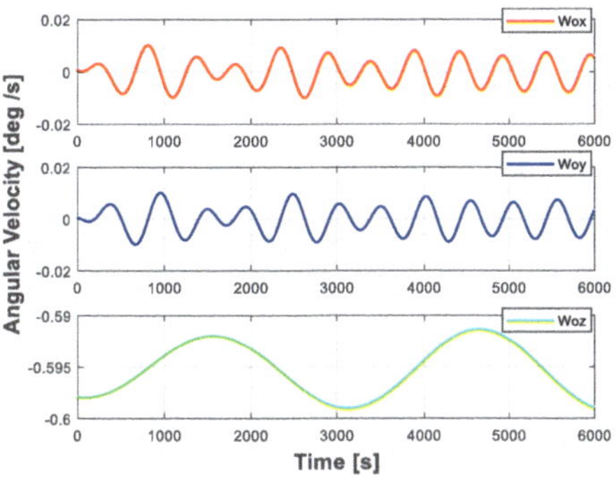

Fig. 6. Orbital angular velocity <<Without controller>>

Figures 7 and 8 illustrate the evolution of quaternions and the orbital angular velocity of the satellite under PD control. We observe that all three angles follow the reference trajectory excellently. Figure 10 clearly demonstrates the convergence of errors to zero. Figure 9 illustrates the applied control torque to the satellite, showing that the attitude control system remains stable under this control. Next, in the simulation, we can observe the torque the configuration wheels exert on the satellite.

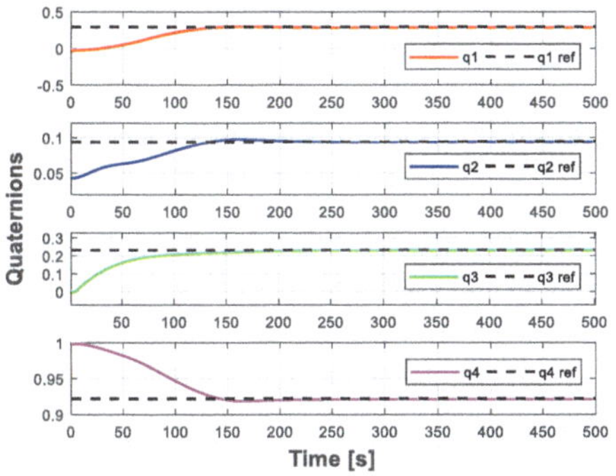

Fig. 7. Quaternions "PD-Control"

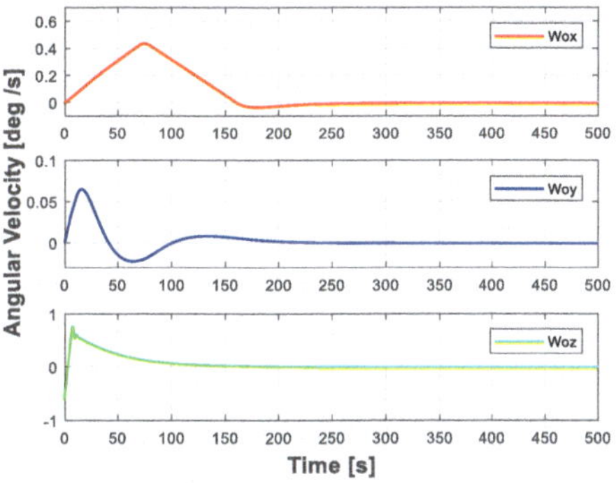

Fig. 8. Orbital angular velocity "PD-Control"

Figures 11 and 12 represent how the angular velocity of reaction wheels and wheel torque vary over time during the attitude control process. Initially, the rotational speed increases until the maximum orbital angular velocity of $\langle 0.4387\ 0.0658\ 0.7724 \rangle$deg/s, which helps bring the attitude trajectory closer to the desired reference. The maximum saturation value for wheel torque is ± 0.015 Nm. Table 1 shows the RMSE for the PD controller in terms of roll, pitch, and yaw axes, power consumption, and settling time. To enhance comprehension, we present the results in degrees.

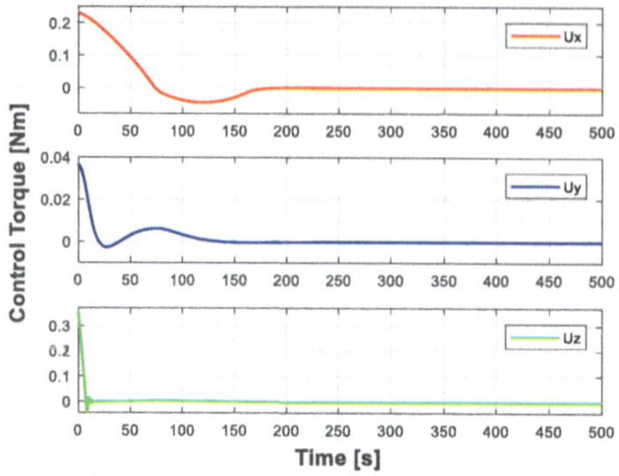

Fig. 9. Control torque PD-Control

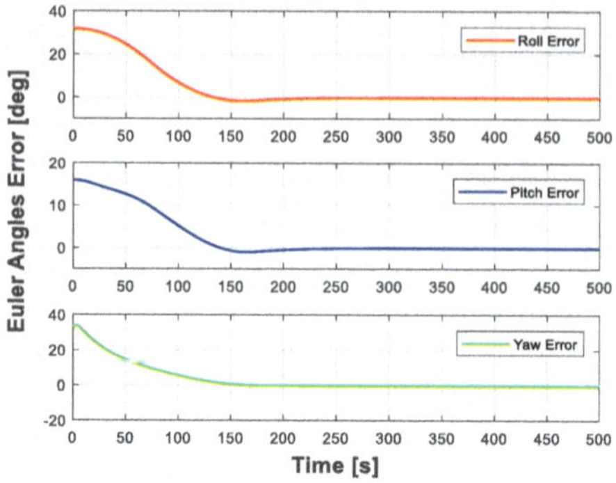

Fig. 10. Euler angles error PD

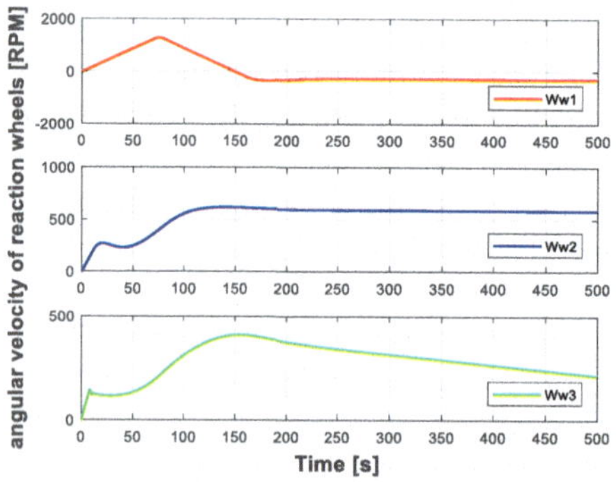

Fig. 11. Angular velocity of reaction wheels PD

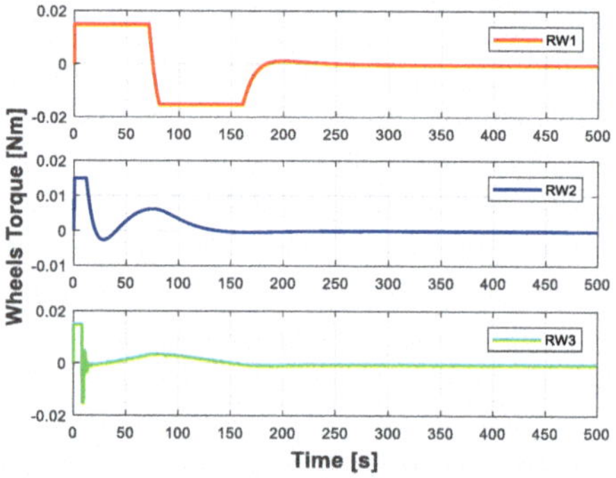

Fig. 12. Wheels Torque PD with saturation

Table 1. The performance of the PD controller results

Roll angle	Pitch angle	Yaw angle	Power consumption	Settling time
0.0542°	0.0240°	0.0352°	7.6849W	210.77s

7 Conclusions

The attitude control system based on the reaction wheels of a small satellite is presented. The control laws are defined to manage the attitude of a small satellite, which is AlSat-1 in our case, by using reaction wheels in orthogonal configurations as actuators. The attitude system is simulated with and without a controller, and the simulation results demonstrate our approach. The root mean square method shows the system's performance. In the standard configuration, the reaction wheels consumed 7.6849 W of power, which reached their maximum. Although this is satisfactory, further optimization is required to improve battery life and ensure the health of the reaction wheels. In future research, we aim to test the efficiency of the pyramid configuration and assess whether it can provide a solution.

References

1. Saadat Dehghan, A.: Design and Test Optimizations for Spacecraft Attitude Determination and Control Subsystems (2022)
2. Ibrahim, B.: Study and Analyze the Attitude Determination and Control Subsystem ADCS of a Small Remote Sensing Satellite (2020)
3. Carrara, V., Siqueira, R.H., Oliveira, D.: Speed and current control mode strategy comparison in satellite attitude control with reaction wheels. In: Proceedings of the 21st Brazilian Congress of Mechanical Engineering (COBEM'11) (2011)
4. Nudehi, S.S., et al.: Satellite attitude control using three reaction wheels. In: 2008 American Control Conference. IEEE (2008)
5. De Simone, M.C., et al.: Attitude controller design for micro-satellites. In: International Conference "New Technologies, Development and Applications". Springer (2021)
6. Jin, J., Ko, S., Ryoo, C.-K.: Fault tolerant control for satellites with four reaction wheels. Control. Eng. Pract. **16**(10), 1250–1258 (2008)
7. Ismail, Z., Varatharajoo, R.: A study of reaction wheel configurations for a 3-axis satellite attitude control. Adv. Space Res. **45**(6), 750–759 (2010)
8. Bellar, A., Mohammed, M.A.S., Adnane, A.: Minimum power consumption of the microsatellite attitude control using pyramidal reaction wheel configuration. In: 2016 8th International Conference on Modelling, Identification and Control (ICMIC). IEEE (2016)
9. Kim, B., Lee, H., Choi, S.: Three-axis reaction wheel attitude control system for Kitsat-3 microsatellite. Space Technology **5**(16), 291–296 (1996)
10. Bang, H., Tahk, M.-J., Choi, H.-D.: Large angle attitude control of spacecraft with actuator saturation. Control. Eng. Pract. **11**(9), 989–997 (2003)
11. Wie, B.: Space Vehicle Dynamics and Control. Aiaa (1998)
12. Forbes, J.R.: Fundamentals of spacecraft attitude determination and control [bookshelf]. IEEE Control. Syst. Mag. **35**(4), 56–58 (2015)
13. Markley, F.L., et al.: Correction to: Fundamentals of Spacecraft Attitude Determination and Control. Springer (2014)
14. Adnane, A., et al.: Adaptive attitude controller design for a rigid satellite with feedback linearization approach. J. Astronaut. Sci. **67**(4), 1445–1469 (2020)

Stability and Performance Enhancement of a Step-Down Converter in Photovoltaic Systems Using an Optimization Algorithm-Based Fractional $PI^{\lambda}D^{\delta}$ Controller

Elouahab Bouguenna[1,2](✉), Samir Ladaci[2], and Walid Merrouche[1]

[1] Renewable Energy Development Center, CDER, 16340 Bouzareah, Algiers, Algeria
a.bouguenna@cder.dz

[2] Electronic Department, Polytechnical National School, Algiers, Algeria

Abstract. Recent advancements in renewable energy, particularly the widespread deployment of photovoltaic systems within power sectors, have introduced notable challenges in terms of their operational intricacies, protection, and control mechanisms. This paper aims to enhance the stability and performance of a non-isolated step-down converter by implementing a Fractional–Order Proportional–Integral–Derivative (FOPID or $PI^{\lambda}D^{\delta}$) controller. A novel Artificial Intelligence (AI) tuning method is specifically designed to optimize the FOPID controller parameters, utilizing the Artificial Hummingbird Algorithm (AHA) optimizer. A comparative analysis with the conventional PID controller is conducted. Simulation results demonstrate a significant enhancement in the transient performance of the DC–DC buck converter dynamics when employing the AHA-optimized FOPID controller.

Keywords: Fractional PID controller · DC–DC Buck converter · Artificial Hummingbird Algorithm · Control System

1 Introduction

In recent years, the quest for energy derived from clean, efficient, and environmentally sustainable sources has emerged as a significant challenge for engineers and scientists [1]. Solar power systems have gained considerable attention among the array of renewable energy sources, offering a compelling opportunity to generate electricity while simultaneously mitigating greenhouse gas emissions [1, 2]. Despite these advantages, it is important to note that solar power systems, unfortunately, do not exhibit the desired level of efficiency [3]. Due to the inherently nonlinear dynamics in photovoltaic (PV) systems, effectively controlling them using conventional regulators poses a significant challenge. Conventional approaches require multiple simplifications and linearization around the equilibrium point of the system, often deviating from the complex realities of actual system dynamics. In the context of standalone PV systems, (i.e., not grid-connected), batteries play a crucial role in storing excess energy and providing power during periods of insufficient sunlight. However, these batteries are susceptible to issues such as

© The Author(s), under exclusive license to Springer Nature Switzerland AG 2025
M. Hatti (Ed.): IC-AIRES 2024, LNNS 1238, pp. 170–178, 2025.
https://doi.org/10.1007/978-3-031-80301-7_19

overcharging, deep discharges, temperature fluctuations, and current drifts [4]. Consequently, the integration of a regulator becomes essential to safeguard their operation. The importance of a charge controller in such standalone PV systems is undisputed, requiring careful implementation to achieve an optimal balance between cost, simplicity, and reliability [5]. The challenge is to design robust compensators that simultaneously meet the criteria for transient and frequency responses, while ensuring high tracking accuracy. Unidirectional DC–DC power converters were based on control strategies using linear models. However, their performance tends to degrade when subjected to disturbances and process uncertainties. Many power converters widely adopt PID controllers owing to their flexibility. The latter find extensive application in renewable energy systems (RESs) for managing power electronic converters [6]. These controllers are straightforward to tune, implement a simpler control law, and therefore, prove highly adaptable to industry [7]. In practical scenarios, variations in parameters and loads, along with non-linearities, are inevitable. In such circumstances, the performance of the PID controller tends to be suboptimal, necessitating the use of more sophisticated controllers [8]. This paper explores a powerful solution: the Fractional PID controller (FOPID or $PI^{\lambda}D^{\delta}$) which introduce two extra parameters, λ and δ. These additional degrees of freedom empower for more precise control and enhanced effectiveness compared to classical PID controllers. Optimizing the FOPID parameters is crucial. Numerical techniques work by finding the ideal settings for the FOPID controller based on a predetermined objective function. This function typically prioritizes time-domain characteristics like settling time, overshoot and rise time, ensuring the converter swiftly tracks the reference signal and maintains stability. While established metaheuristic algorithms (MA) like Particle Swarm Optimization (PSO) [9], Genetic Algorithm (GA) [10], and PID-based Search Algorithm [11], have been prevalent, this study introduces an innovative MA: the Artificial Hummingbird Algorithm (AHA) [12]. Inspired by the remarkable hovering and foraging behaviors of hummingbirds, AHA presents a promising optimization technique for fine-tuning FOPID controllers. By combining the power of FOPID control with the innovative optimization capabilities of AHA, this paper proposes a significant advancement in the stability and dynamical performance of buck converters, particularly within solar photovoltaic systems.

The next sections of this paper are further distributed as. A comprehensive analysis and modeling of the step-down converter, along with some preliminaries on Fractional Calculus (FC) are presented in Sect. 2. AHA modelling and control design are detailed in Sect. 3. Section 4 illustrates the numerical results of this work. Finally, the conclusions of Sect. 5 close this paper.

2 System Overview and Theoretical Background

2.1 Modelling of the DC-DC Step-Down Converter

This system serves as a crucial switch mode compensator employed to step down DC voltage for low power applications. This conversion is achieved by toggling a high-power device, such as a MOSFET, switching *on* and *off* at elevated frequency. The pulse-width modulation (PWM) is employed to control the on-time T_{on} of the switching device, thereby enabling the adjustment of the output voltage. The basic buck converter configuration is illustrated in Fig. 1. The power converter steps down voltage from its input v_{in} to its output v_o at the terminals of the load R using a combination of switching power device Q, an inductor L, a diode D, and a capacitor C.

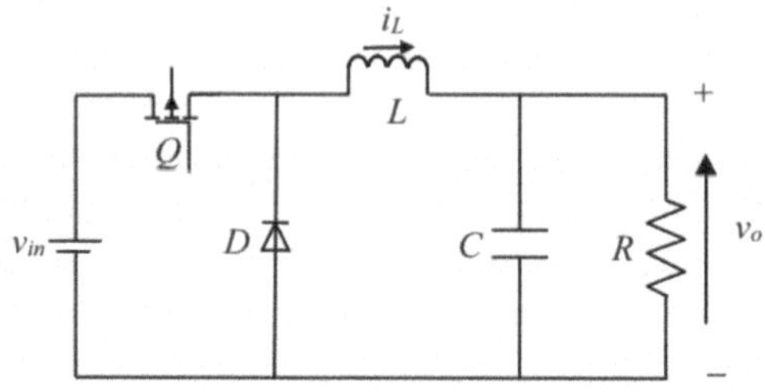

Fig. 1. Buck converter diagram.

From the inductor volt-balance equations, $v_o = dv_{in}$ where $d = T_{on}/T$ is the duty ratio and T is the period of switching. The design parameters of the step-down converter can be determined through [8]. The inductor and capacitor have reach minimum values at, $L_{\min} = (v_{in} - v_o)d/\Delta i_L f$ and $C_{\min} = (1 - d)v_o/8Lf^2 v_o$. Where, f represents the frequency of switching. Δi_L and Δv_o are the inductor current ripple and the output voltage ripple, respectively. The step-down converter proposed in this study operates in continuous conduction mode (CCM), and the averaged transfer function has the form,

$$G(s) = \frac{v_{in}}{as^2 + bs + c} \tag{1}$$

where the coefficients a, b and c have the values, $a = LC$, $b = \frac{L}{R} + R_L C$, $c = 1 + \frac{R_L}{R}$. Table 1 presents the parameters of the proposed step-down converter.

Table 1. Specifications of step-down Converter Parameters.

Parameters	Input Voltage, $v_{in}(V)$	Reference Output Voltage, $v_o(V)$	Load Resistance, $R(\Omega)$	Inductor Resistance, $R_L(\Omega)$	Inductance, $L(mH)$	Capacitance, $C(\mu F)$
Value	207	55	10	0.52	692	3.125

Implementing the PID controller to the buck converter offers several advantages. The primary objective is to eliminate the steady-state error, to improve the dynamic response, and to provide tracking performance of the desired reference signal. The transfer function of the PID controller denoted $G_{PID}(s)$, is expressed by the following equation,

$$G_{PID}(s) = K_P + \frac{K_I}{s} + K_D s \tag{2}$$

Here K_P,K_I **and** K_D are the controller **gains ans** s **is the laplace coefficient.**

2.2 Elements of Fractional Calculus and Approximation Technique

The concept of fractional calculus dates back to the early development of traditional (integer-order) calculus, with the earliest reference likely associated with Leibniz and L'Hôpital in 1695, where the notion of a half-order derivative was introduced. FC further extends the principles of differentiation/integration to non-integer orders [13]. An integro-differential operator, combining both differentiation and integration, can be expressed as ${}_aD_t^{\alpha}$,

$$ {}_aD_t^{\alpha} = \begin{cases} \Re(\alpha) > 0 \rightarrow d^{\alpha}/dt^{\alpha} \text{ FO derivative} \\ \Re(\alpha) < 0 \rightarrow \int_a^t dt^{\alpha} \text{ FO integral} \\ \Re(\alpha) = 0 \rightarrow 1 \end{cases} \quad (3) $$

where $0 < \alpha < 1$ is a complex factor and the bounds are a and t. Numerous definitions of fractional calculus exist, among which, the Riemann–Liouville (RL) definition being widely adopted [13]. The RL definition for function $f(t) \in C^{n+1}([0, +\infty), R)$, within a to t limits is given by,

$$ {}_a^{RL}D_t^{\alpha} = \frac{1}{\Gamma(n-\alpha)}\left(\frac{d}{dt}\right)^n \int_a^t \frac{f(\tau)}{(t-\tau)^{\alpha-n+1}} d\tau \quad (4) $$

where $\Gamma(x)$ is the Gamma function, represented as,

$$ \Gamma(x) = \int_0^{\infty} t^{x-1}e^{-t}dt \quad (5) $$

For convenience, the Laplace transform serves as a widely utilized tool in the analysis of control systems. With null initial conditions, the Laplace transform of the RL fractional derivative for order α can be expressed as follows,

$$ L\{D^{\alpha}f(t)\} = s^{\alpha}F(s) \quad (6) $$

A commonly employed approach for obtaining an approximated rational function of s^{α} is the Oustaloup Recursive Approximation (ORA) [14], which is expressed as follows,

$$ s^{\alpha} = c\prod_{i=1}^{N} \frac{1 + s/\omega_{z,i}}{1 + s/\omega_{p,i}} \quad (7) $$

This approach employs N poles and zeros, laying alternately across the negative real ($\mathbb{R}^-$) axis within a restricted frequency bandwidth $\{\omega_l, \omega_h\}$, where N is the approximation order.

$$ \omega_{z,i} = \omega_l(\omega_h/\omega_l)^{(2i-1-\alpha)/2N} \quad (8) $$

$$ \omega_{p,i} = \omega_l(\omega_h/\omega_l)^{(2i-1+\alpha)/2N} \quad (9) $$

In this paper, the implemented approximation is based on ($N = 5$) with the frequency $\omega \in [10^{-3}, 10^3]$ rad/s.

2.3 Fractional PID Control

Fractional-order control represents a major application of fractional calculus, extending the controller parameters to fractional orders. The PID regulator features three adjustable gains, whereas the FOPID boasts a more complex structure characterized by five distinct tunable parameters, including three gains and two additional real numbers λ and δ which can vary from 0 to 1 [8], improving its adaptability and degree of freedom, providing a more flexible control solution [13]. The FOPID controller transfer function is written as,

$$C(s) = K_P + K_I s^{-\lambda} + K_D s^{\delta} \tag{10}$$

3 AHA Mathematical Model and Design Process

3.1 AHA Mathematical Model

The Artificial Hummingbird Algorithm (AHA), introduced in [12], is a novel bio-inspired optimization technique considering the remarkable flight abilities of hummingbirds when foraging and prey hunting. In the AHA optimizer, hummingbirds demonstrate three distinct foraging behaviors: guided, territorial, and migrating.

- Guided foraging

  Three hummingbird flight techniques, encompassing omni-directional, axial, and diagonal skills, are harnessed to emulate the proposed optimizer. The flight by axis is mathematically expressed as [12],

$$D_i = \begin{cases} 1, \; i = randi([1, d]) \\ 0, \; \text{otherwise} \end{cases} \; i = 1, 2, ..., d \tag{11}$$

  Diagonal flights are expressed by,

$$D_i = \begin{cases} 1, \; i = P(j), j \in [1, k] \\ 0, \; \text{otherwise} \end{cases} \tag{12}$$

$$P = randperm(k), k \in [2, [r_1(d-2)] + 1] \tag{13}$$

  Besides, omni-directional flights are given as follows,

$$D_i = 1, \; i = 1, 2, ..., d \tag{14}$$

  Here,$randi([1, d])$ is a random vector computed between1 and d, while $randperm(k)$ gives randomly permutations between1 and k. Moreover, r_1 denotes a random number from 0 to 1. The model of this foraging behavior is,

$$v_i^{t+1} = x_{i,tar}^t + aD\left(x_i^t - x_{i,tar}^t\right), \; a \sim N(0, 1) \tag{15}$$

where at time t, x_i^t symbolizes the ith food source position, $x_{i,tar}^t$ indicates the objective location, while a is modeled by a Gaussian distribution $N(\mu, \sigma) = N(0, 1)$ is a guided factor. The ith food source novel position becomes,

$$x_i^{t+1} = \begin{cases} x_i^t f(x_i^t) \le f(v_i^{t+1}) \\ v_i^{t+1} f(x_i^t) > f(v_i^{t+1}) \end{cases} \tag{16}$$

Here, f denotes the cost function value.

- Territorial Foraging

  The local exploration behavior of hummingbirds during this phase, targeting a particular food source, is expressed by,

$$v_i^{t+1} = x_i^t + bD(x_i^t),\ b \sim N(0, 1) \tag{17}$$

  The territorial factor, denoted by b, corresponds to a Gaussian distribution $N(0, 1)$.

- Migration Foraging

  The hummingbird has a migration foraging, moving from the food source associated with the poorest nectar to a randomly generated one, is described as,

$$x_{wor}(t+1) = L + r(U - L) \tag{18}$$

  The variable x_{wor} denotes the food source in the population with the worst nectar-refilling rate.

3.2 Design Process of AHA-Based Control

For the optimum synthesis of the FOPID regulator by the fine tuning of its five parameters, with a minimizing cost function, Fig. 2 depicts the design technique of the proposed AHA-based Buck converter feedback control loop.

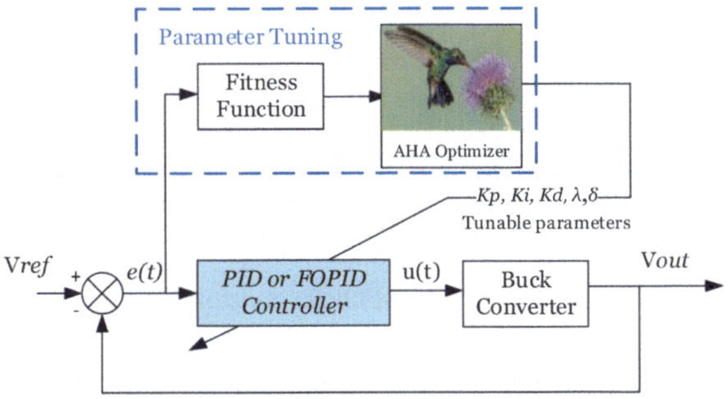

Fig. 2. AHA-based Buck converter optimization.

In this work, the Integral of the Absolute Error (IAE) is used as the cost function for the optimization issue. The voltage error signal is employed to calculate the objective function iteratively, and the AHA optimizer aims to pinpoint the optimal parameters for the system. The voltage error signal and the IAE are given by,

$$e(t) = V_{ref}(t) - V_{out}(t) \tag{19}$$

$$IAE = \int |e(t)|dt \tag{20}$$

The control signals for both PID and FOPID controllers are given in the time domain, respectively, as follows,

$$u_{PID}(t) = K_P e(t) + K_I \int e(t)dt + K_D de(t)/dt \tag{21}$$

$$u_{FOPID}(t) = K_P e(t) + K_I D^{-\lambda} e(t) + K_D D^{\delta} e(t) \tag{22}$$

4 Results and Discussion

The comparison of dynamic performance between the PID and AHA-FOPID controllers for the designed step-down converter is depicted in Fig. 3. It is clear that the AHA-FOPID controller improves the terminal voltage response compared to the PID controller, particularly in terms of overshoot, rise time, and settling time. The optimal parameters and performance metrics for both controllers are provided in Table 2.

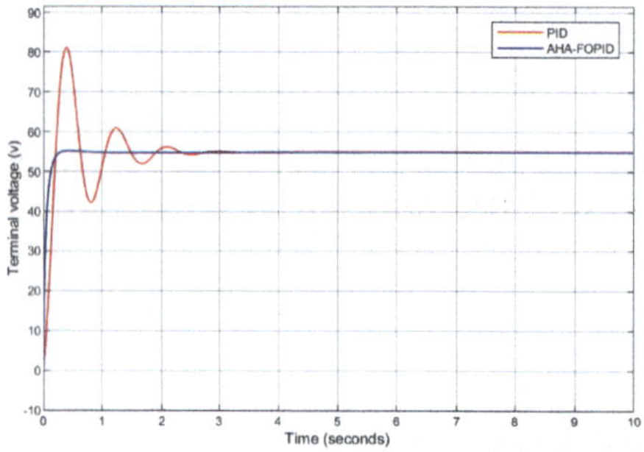

Fig. 3. Response Comparison of PID and AHA-FOPID

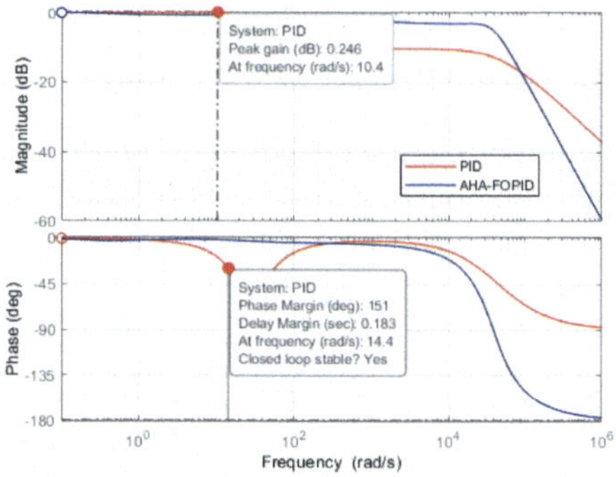

Fig. 4. Bode plot of the closed loop buck converter with different controllers.

Table 2. Comparison of PID and AHA-FOPID controllers dynamic performances.

Controller	K_P	K_I	K_D	λ	δ	$t_r(s)$	t_s (s)	Overshoot (%)	Peak time t_p (s)	Peak value (v)
PID	0.0067	0.2055	0.000147	1	1	0.23	3.012	26	0.45	81
AHA-FOPID	0.0567	0.1205	0.005147	0.85	0.55	0.11	0.19	0.01	0.38	55.01

The system performance criteria, such as IAE and ISE, are obtained using the adjusted controller parameters. For the PID controller, the IAE and ISE are 19.57 and 421.3, respectively, while for the AHA-FOPID controller, they are 3.602 and 58.63, respectively. Based on the optimal parameters outlined in Table 2, a thorough analysis entails assessing the Bode diagrams of the buck converter, as depicted in Fig. 4. The assessment of Bode plot parameters was carried out to verify system stability. The analysis shows that both the gain margin (GM) and the phase margin (PM) of the closed-loop feedback system are positive, indicating enhanced stability of the buck converter when using AHA-FOPID controller. Using the IAE performance criteria, Figs. 5 and 6 depict the error and control signals, respectively, for both the PID and AHA-FOPID controllers. The comparative analysis reveals that both the error signal and the control signal are significantly higher in the PID scenario compared to the FOPID case. This demonstrates the precision and efficacy of the AHA-optimized FOPID controller in improving the output trajectory of the buck converter by significantly reducing the tracking error between the reference and measured voltages. The AHA-tuned FOPID controller exhibits superior performance in terms of both steady-state error reduction and transient response, resulting in improved setpoint tracking and system stability. The enhanced control action is attributed to the fractional-order nature of the FOPID controller, which offers greater flexibility in shaping the system dynamics, particularly in minimizing overshoot, reducing rise and settling times, and achieving robust stability margins. This underscores the controller's ability to handle nonlinearity and system disturbances more effectively compared to conventional PID control strategies.

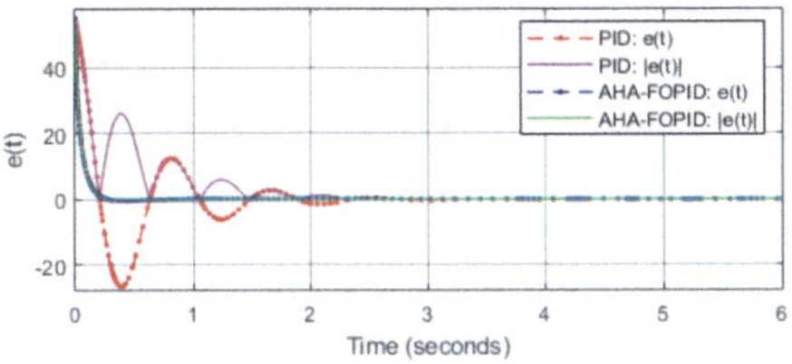

Fig. 5. Error signals as reported in Table 2

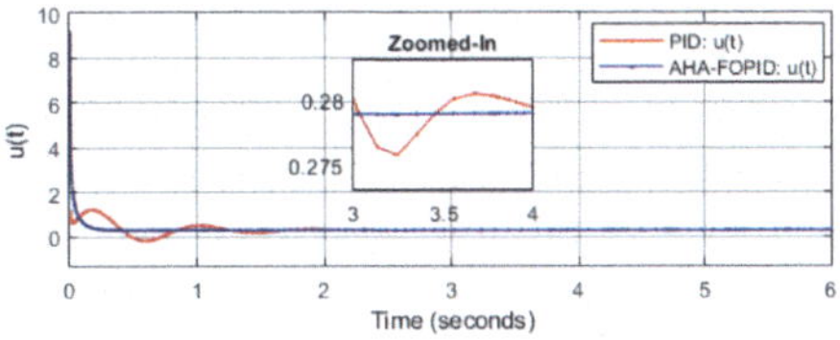

Fig. 6. Control signals as reported in Table 2 (Zoomed-In View).

5 Conclusion

In conclusion, our investigation into the AHA metaheuristic optimization approach for regulating a DC-DC buck converter with its nonsmooth dynamics, particularly when applied to a fractional PID controller, has effectively identified optimal parameters. This represents a significant advancement in refining control strategies for such converters. The AHA-based FOPID controller, guided by a flexible cost function, exhibits substantial improvements in both frequency-domain and time-domain performance metrics, including stability margins, overshoot, settling time, rise time, and steady-state error. The integration of the AHA optimizer with the FOPID controller underscores the controller adaptability and dynamism, essential for addressing the complex challenges of buck converter stability and regulation. This work contributes to the field by enhancing the controller ability to cope with uncertainties and non-linearities, thereby paving the way for more robust and efficient control solutions in power electronics.

References

1. Wai, R., Member, S., Wang, W., Lin, C.: High-performance stand-alone photovoltaic generation system. IEEE Trans. Ind. Electron. **55**(1), 240–250 (2008)
2. Mutoh, N., Member, S., Inoue, T.: A control method to charge series-connected ultraelectric double-layer capacitors suitable for photovoltaic generation systems combining mppt control method. IEEE Trans. Ind. Electron. **54**(1), 2–3 (2007)
3. Faranda, R., Leva, S., Maugeri, V.: MPPT techniques for PV systems : energetic and cost comparison, pp. 1–6 (2008)
4. Merrouche, W., Lekouaghet, B., Bouguenna, E., Himeur, Y.: Parameter estimation of ECM model for Li-Ion battery using the weighted mean of vectors algorithm. J. Energy Storage **76**, 109891 (2024). https://doi.org/10.1016/j.est.2023.109891
5. Messikh, L., Chikhi, S., Chergui, F.C.T.: Mise au point d ' un régulateur de charge / décharge de batterie avec seuils adaptatifs de tension pour les applications photovoltaïques. Revue **11**, 281–290 (2008)
6. Padula, F., Visioli, A.: Tuning rules for optimal PID and fractional-order PID controllers. J. Process. Control. **21**(1), 69–81 (2011). https://doi.org/10.1016/j.jprocont.2010.10.006
7. Zhang, G., Li, Z., Zhang, B., Halang, W.A.: Power electronics converters: past, present and future. Renew. Sustain. Energy Rev. **81**, 2028–2044 (2018). https://doi.org/10.1016/j.rser.2017.05.290
8. Warrier, P., Shah, P.: Fractional order control of power electronic converters in industrial drives and renewable energy systems: a review. IEEE Access (2021). https://doi.org/10.1109/ACCESS.2021.3073033
9. Maiti, D., Acharya, A., Chakraborty, M., Konar, A., Janarthanan, R.: Tuning pid and piλdδ controllers using the integral time absolute error criterion. In: Proceedings 2008 4th International Conference on Information and Automation for Sustainability ICIAFS 2008, pp. 457–462 (2008). https://doi.org/10.1109/ICIAFS.2008.4783932
10. Cao, J.Y., Liang, J., Cao, B.G.: Optimization of fractional order PID controllers based on genetic algorithms. In: 2005 International Conference on Machine Learning and Cybernetics ICMLC 2005, no. August, pp. 5686–5689 (2005). https://doi.org/10.1109/icmlc.2005.1527950
11. Bouguenna, E., Lekouaghet, B., Haddad, M.: Improved fractional P Iα Dβ controller for AVR system via a new optimization algorithm. In: 2024 2nd International Conference on Electrical Engineering and Automatic Control ICEEAC 2024 (2024). https://doi.org/10.1109/ICEEAC61226.2024.10576215
12. Zhao, W., Wang, L., Mirjalili, S.: Artificial hummingbird algorithm: a new bio-inspired optimizer with its engineering applications. Comput. Methods Appl. Mech. Eng. **388**, 114194 (2022). https://doi.org/10.1016/j.cma.2021.114194
13. Shah, P., Agashe, S.: Review of fractional PID controller. Mechatronics **38**, 29–41 (2016). https://doi.org/10.1016/j.mechatronics.2016.06.005
14. Oustaloup, A., Levron, F., Mathieu, B., Nanot, F.M.: Frequency-band complex noninteger differentiator: characterization and synthesis. IEEE Trans. Circuits Syst. I Fundam. Theory Appl. **47**(1), 25–39 (2000). https://doi.org/10.1109/81.817385

Fuzzy Logic Control of Hybrid PV-Wind-Battery On-Grid Systems

Zouak Fatma Zahra(✉), Chaib Lakhdar, and Choucha Abdelghani

Energy and Materials Laboratory, Faculty of Sciences and Technology, University of Tamanghasset, 11000 Tamanrasset, Algeria
Zouakftzahra@gmail.com

Abstract. This paper proposes a novel approach utilizing fuzzy logic control (FLC) to optimize the performance of hybrid photovoltaic (PV) and wind energy systems with energy storage. By dynamically adjusting the operation of PV and wind components, FLC ensures efficient power generation and storage utilization while maintaining system stability. Simulation results demonstrate significant improvements in system efficiency and reliability compared to traditional control methods. This research advances the integration of renewable energy by providing a promising approach to improve the performance of hybrid PV-wind systems with energy storage.

Keywords: Hybrid renewable energy systems · Photovoltaic (PV) · Wind energy · Fuzzy logic control (FLC) · Energy storage · System optimization · Renewable energy integration · Sustainability · Power generation · System stability

1 Introduction

The increasing global focus on discovering sustainable methods for power generation has emerged as an important development in many nations [1]. The incorporation of energy from renewable sources, particularly photovoltaic (PV) and wind technologies, has gained important value in the last years for its potential to address energy sustainability and climate change challenges. This paper proposes an innovative method to enhance the efficiency of hybrid photovoltaic-wind systems by incorporating fuzzy logic control (FLC) and energy storage. FLC dynamically adjusts the operation of PV and wind components, maximizing power generation while ensuring system stability. Additionally, energy storage enhances system reliability by mitigating intermittency and variability issues associated with renewable energy sources. This study seeks to assist to the advancement of renewable energy integration, providing a viable route for improving the efficiency and economy of hybrid PV-wind systems.

© The Author(s), under exclusive license to Springer Nature Switzerland AG 2025
M. Hatti (Ed.): IC-AIRES 2024, LNNS 1238, pp. 179–186, 2025.
https://doi.org/10.1007/978-3-031-80301-7_20

2 System Modelling

2.1 Photovoltaic System Mathematical Modelling

The PV system, a leading clean energy solution, uses the single-diode model to represent its electrical properties, characterized by a current source with a shunt diode and resistors, as illustrated in Fig. 1.

$$I = Iph - Io\left(e^{\frac{V+Irs}{nVt}} - 1\right) - \frac{V + Irs}{Rsh} \tag{1}$$

where:

I is the PV cell current, V is the pv cell voltage, Iph is the photo current, Io is the reseverse saturation current, n is the diode ideality factor, Vt is the thermal voltage, Rs is the series resistance, Rsh is the shunt resistance.

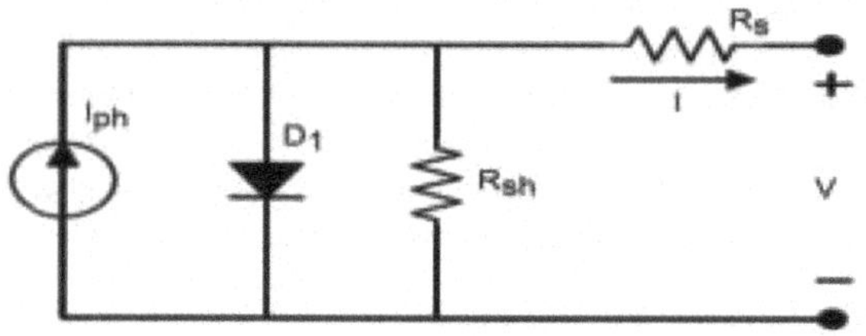

Fig. 1. Structure of PV Cell [3].

2.2 Wind Sytem Modelling

Wind energy is among the rapidly expanding renewable energy sources that thrive annually in several places [4]. Modelling the wind turbine's dynamic response to variations in wind speed is crucial for understanding its performance.

The mechanical power of the wind turbine is [5]:

$$P = \frac{1}{2}\rho ACpV\omega^3 \tag{2}$$

- Where: P represents the mechanical power produced of the wind turbine, and ρ denotes the air density. The rotor's swept area. Vw: the velocity of the wind, Cp: the power coefficient of the wind blades of a turbine.

The power coefficient Cp is associated with the tip speed ratio λ and the blade pitch angle β, given by [6]:

$$Cp(\lambda, \beta) = 0.5176\left(\frac{116}{\lambda i} - 0.4\lambda i - 5e^{\frac{21}{\lambda i}} - 0.0068\lambda\right) \tag{3}$$

where:

$$\lambda i = \left(\frac{1}{\lambda + 0.08\beta} - \frac{0.0035}{\beta^3 + 1}\right) \tag{4}$$

2.3 Model of Permanent Magnet Synchronous Generator

The dynamic model of PMSG is formulated in the d–q axis reference frame. The stator voltages in the d–q axis can be described as [7]:

$$Vqs = (Rsiqs + \omega\psi ds + P\psi qs) \tag{5}$$

$$Vds = (Rsids + \omega\psi qs + P\psi ds) \tag{6}$$

Where:

- *Vqs*, *Vds*: the stator voltages in the *d–q* axis, *iqs*, *ids*: the stator currents in the *d–q* axis, *Rs*: the stator resistance, *ψds*, *ψqs*: the stator fluxes in the *d–q* axis, *ωr*: the rotor speed.

$$\psi ds = (\psi ds + \lambda m) \tag{7}$$

$$\psi qs = (\psi qs) \tag{8}$$

where: *λm* is the rotor flux linkage,

The electromagnetic torque in the rotor is showed [8]:

$$Te = 1.5x\frac{P}{2}[\psi pisq + is\ disq(Lsd\ Lsq)] \tag{9}$$

2.4 Modelling of the DC-DC Booster

A DC-DC A Boost Converter elevates the input voltage to a superior output voltage. While maintaining constant overall power and minimal power consumption, effectively compensating for voltage differences [9].

The mathematical equations that describe the operation ofa DC-DC Boost Converter are as follows:

The inductance (L) of the boost converter's inductor iscalculated using the desired output voltage, input voltage range, and switching frequency (f):

$$L = \frac{Vin(\,Vout\ -\ Vin)}{\Delta ILfsVout} \tag{10}$$

Vin is the minimum input voltage, *ΔIL* is the ripple in inductor current from peak to peak, *fs* is the switching frequency, Vout is the output voltage.

To lessen output voltage ripple and stabilize the output voltage, the output capacitor Cout is selected. The minimum required capacitance is showed in the following equation:

$$C = \left(\frac{Iout_{\max\left(1-\frac{Vin}{Vout}\right)}}{\Delta\ Ioutfs}\right) \tag{11}$$

The maps, f The duty cycle (D) of the boost converter is calculated as:

$$D = \left(\frac{Vout - Vin}{Vout} \right) \tag{12}$$

2.5 Sizing of the Battery Bank

Solar or wind energy storage is essential. Proper battery sizing ensures efficient use of surplus energy and reliable power during low generation periods [10]. The number of battery cells in series is determined by the DC bus voltage [11]:

$$Ns = \frac{Vdc}{Vbn} \tag{13}$$

where: Ns the number of batteries connected in series, Vbn the nominal voltage of the batteries (V),}Vdc the voltage of the DC bus (V).

2.6 Control of the Full Brigde Converter

Synchronous reference frame control, commonly known as dq-control [12]. Synchronous reference frame control in Full Bridge Inverters synchronizes with grid voltage, employing PIcontrollers for efficient power flow regulation and synchronization.

2.7 The Fuzzy Logic Control

The FLC consists of three main parts [13]. Fuzzification, the inference system, and defuzzification.

1) *FLC process in The PV system:* Fuzzification determinesthe degree of membership function [14], within a fuzzy logic controller (FLC) framework. Within a photovoltaic (PV) system,The Fuzzy Logic Controller (FLC) captures the voltage and current values from the PV array output. These valuesare then used to determine the PV power, which is fed intothe controller. The controller receives two inputs: The error (E) wich is defined as the change in power relative to the change in voltage and CE change in error, and it is represented by the following equation [15]:

$$E(k) = \frac{P(k) - P(k-1)}{V(k) - V(k-1)} \tag{14}$$

$$CE(k) = E(k) - E(k-1) \tag{15}$$

Equation (14) calculates the error E(k) the change in power divided by the change in voltage between consecutive steps. Equation (15) computes the change in error ΔE(k) by subtracting the previous error from the current one.

FLC process in the wind system: FLC accurately tracks changes in wind speed. Fuzzified inputs ‘e’ and ‘ec’ are fed to the controller, resulting in fuzzy sets. Fuzzy rules are applied by the inference system to produce suitable fuzzy output. After defuzzification, The fuzzy output is transformed into a duty cycle. To control the boost converter's switching element for MPPT [16]. The following equations define the inputs and outputs:

$$\delta(p) = \frac{P(n) - P(n-1)}{\omega(n) - \omega(n-1)} \tag{16}$$

$$CE = En - E(n-1) \tag{17}$$

Equations (16) and (17) calculate changes in power and error over time, essential for efficient fuzzy logic MPPT control in PV and wind systems.

These variables E, CE, and D utilize linguistic labels (NB, NS, Z0, PB, PM, PS) through fundamental fuzzy subsets. Figures 2, 3, and 4 depict the fuzzy memberships for the PV systemand the wind system and Table 1 presents the fuzzy rules used in our system.

Table 1. Fuzzy logic rules

CE/E	NB	NM	NS	Z	PS	PM	PB
NB	Z	Z	NS	NM	PS	PM	PB
NM	Z	Z	Z	NS	PM	PM	PB
NS	Z	NS	Z	Z	PM	PM	PB
Z	NB	Z	Z	Z	PS	Z	Z
PS	NB	NM	NS	Z	Z	Z	Z
PM	NB	NM	NS	PS	Z	Z	Z
PB	NM	NM	NS	PM	Z	Z	Z

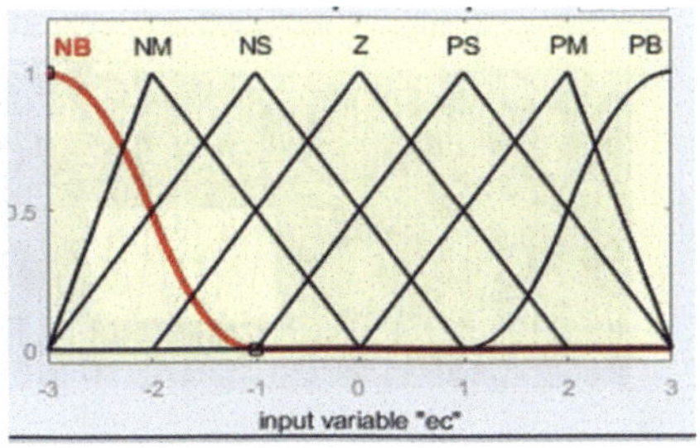

Fig. 2. Input variable ce

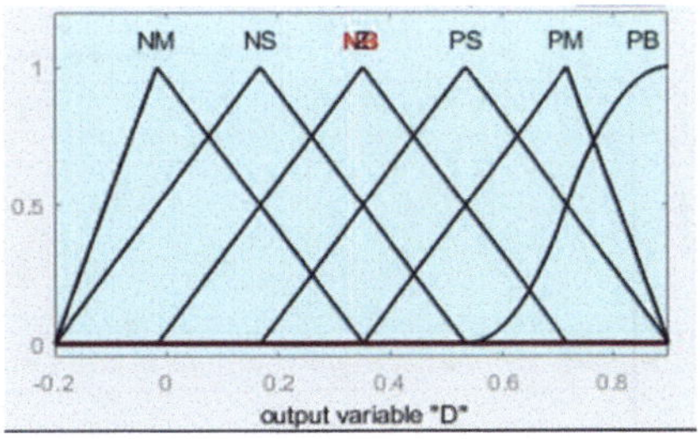

Fig. 3. Output variable D

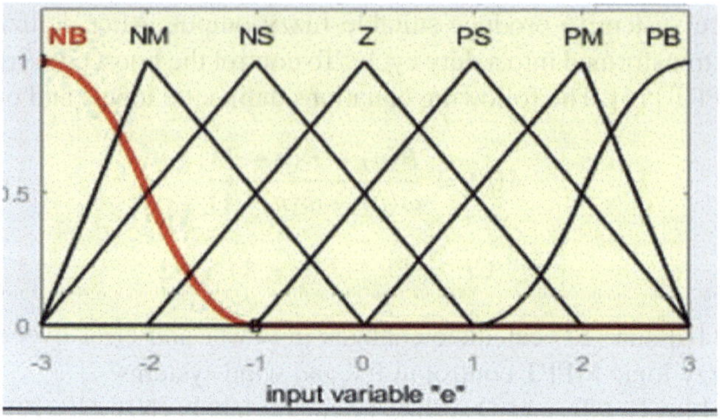

Fig. 4. Input variable e

3 Simulation Results and Discussions

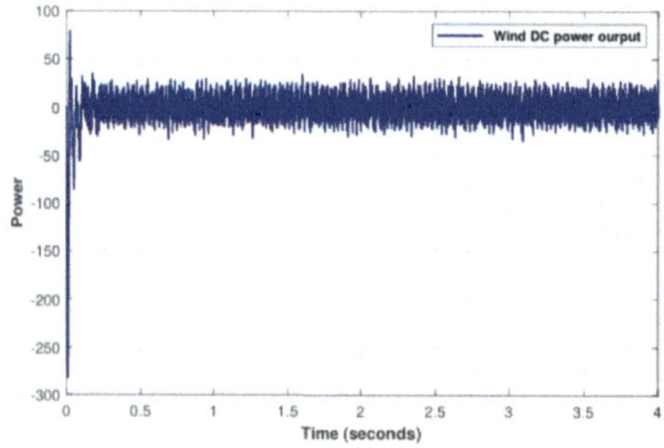

Fig. 5. Output of the DC-DC booster for the wind system.

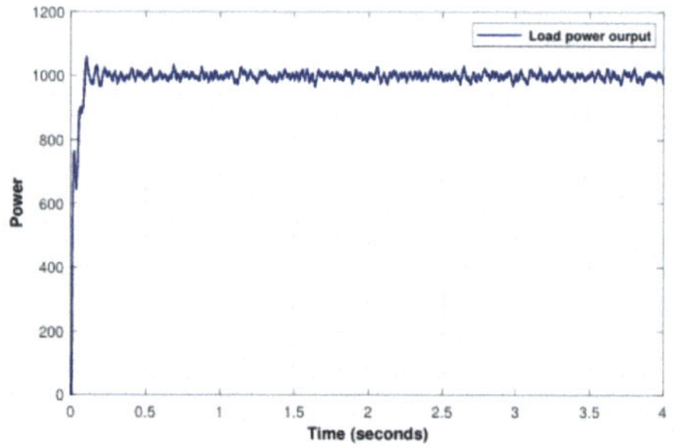

Fig. 6. Power output of the PV system's load.

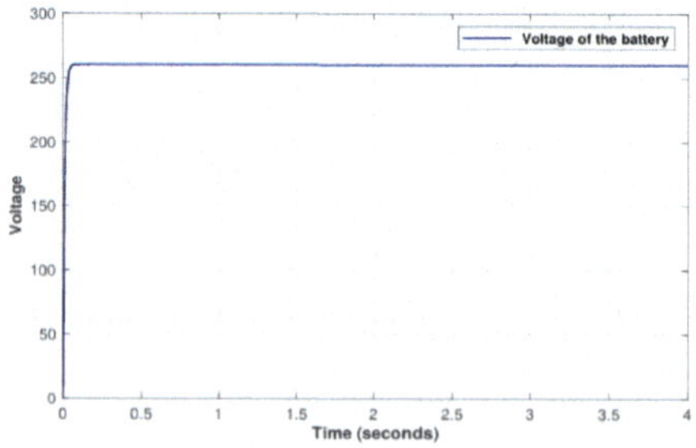

Fig. 7. Volatge of the battery

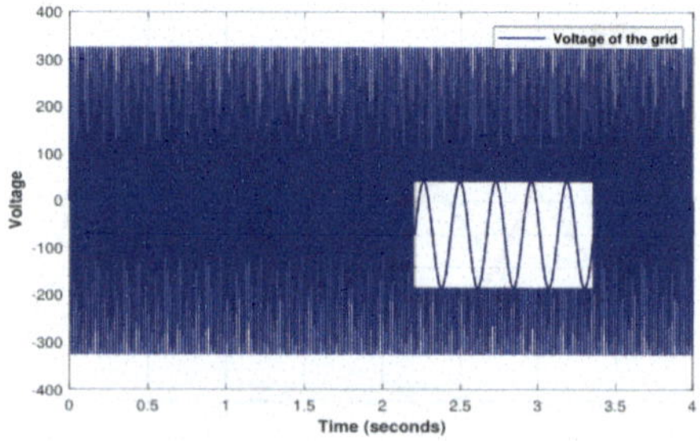

Fig. 8. Grid voltage

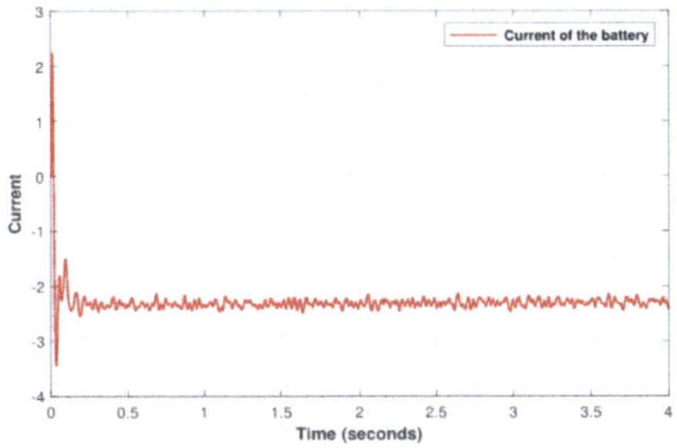

Fig. 9. The battery current.

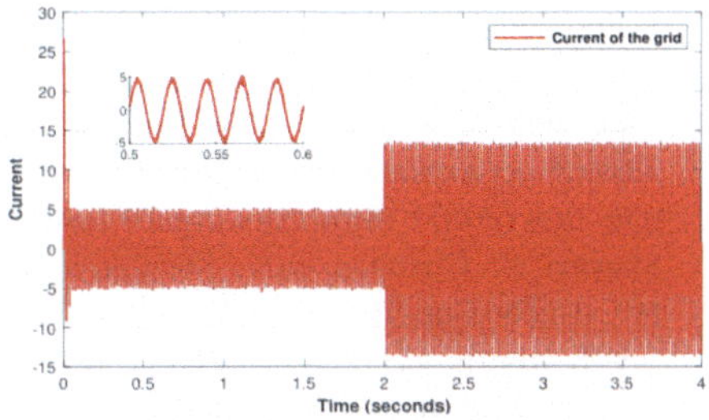

Fig. 10. Current of the grid.

In Fig. 5, the power output of the DC booster under variable wind conditions shows the Fuzzy Logic Controller (FLC) fluctuating and peaking at around 3000 W before stabilizing.

Figure 6 depicts load power variations, where the FLC initially achieves higher outputs with multiple peaks, stabilizing at 1000 W and reaching 800 W between 0.1 and 0.2 s.

Figures 7 and 9 show battery voltage around 260 V and effective current regulation by the FLC, especially between 0.1 to 0.3 s and 0.9 to 2 s.

Figure 8 presents grid voltage fluctuations ranging from −300 V to 300 V, managed effectively by the FLC.

Figure 10 shows grid current variations, with the FLC initially near 0 A but exhibiting wider fluctuations (10 A to −10 A) after 2 s, indicating responsiveness to changing conditions.

4 Conclusion

The integration of Fuzzy Logic Controllers (FLC) in renewable energy systems, like wind and PV DC systems, demonstrates its effectiveness in optimizing power extraction and managing output fluctuations. By ensuring stable operation, FLC enhances the reliability and performance of grid-connected converters. Continued advancements in FLC technology will further improve renewable energy integration, supporting sustainable and efficient energy solutions for the future.

References

1. Mourabit, Y.E., et al.: Enhanced performance in PMSG-based wind turbine systems: experimental validation of Adaptive Backstepping control design. Energies **16**(22), 7481 (2023). https://doi.org/10.3390/en16227481
2. Maroua, B., Zaid, L., Benbouhenni, H., Fateh, M., Debdouche, N., Colak, I.: Robust type 2 fuzzy logic control microgrid-connected photovoltaic system with battery energy storage through multi-functional voltage source inverter using direct power control. Energy Rep. **11**, 3117–3134 (2024). https://doi.org/10.1016/j.egyr.2024.02.047

3. Rasheed, M., Mohammed, O.Y., Shihab, S., Al-Adili, A.: A comparative analysis of PV cell mathematical model. J. Phys. Conf. Ser. **1795**(1), 012042 (2021). https://doi.org/10.1088/1742-6596/1795/1/012042
4. Yaramasu, V., Dekka, A., Durán, M.J., Kouro, S., Wu, B.: PMSG- based wind energy conversion systems: survey on power converters and controls. IET Electr. Power Appl. **11**(6), 956–968 (2017). https://doi.org/10.1049/iet-epa.2016.0799
5. Sharma, S., Chauhan, B.K., Saxena, N.K.: Artificial neural network grid-connected MPPT-based techniques for hybrid PV-WIND with battery energy storage system. J. Inst. Eng. (India) Ser. B **104**(6), 1217–1226 (2023). https://doi.org/10.1007/s40031-023-00922-y
6. Aloo, L.A., Kihato, P.K., Kamau, S.I., Orenge, R.S.: Modeling and control of a photovoltaic-wind hybrid microgrid system using GA-ANFIS. Heliyon **9**(4), e14678 (2023). https://doi.org/10.1016/j.heliyon.2023.e14678
7. Wang, C.N., Lin, W.C., Le, X.K.: Modelling of a PMSG wind turbine with autonomous control. Math. Probl. Eng. (2014). https://doi.org/10.1155/2014/856173
8. Hichem, L., Amar, O., Leila, M.: Optimized ANN-fuzzy MPPT controller for a stand-alone PV system under fast-changing atmospheric conditions. Bull. Electr. Eng. Inform. **12**(4), 1960–1981 (2023). https://doi.org/10.11591/eei.v12i4.5099
9. Anshory, I., et al.: Optimization DC-DC boost converter of BLDC motor drive by solar panel using PID and firefly algorithm. Res. Eng. **21**, 101727 (2024). https://doi.org/10.1016/j.rineng.2023.101727
10. Rekioua, D.: Energy storage systems for photovoltaic and wind systems: a review. Energies **16**(9), 3893 (2023). https://doi.org/10.3390/en16093893
11. Attik, N.: Contribution to the modeling and control of renewable energy systems, application to a hybrid PV/wind/storage system. PhD thesis, University Ferhat Abbas Setif-1, Faculty of Technology, Sciences and Technology, Electrical Engineering, Automatic & systems (2023). Available: http://dspace.univ-setif.dz:8888/jspui/handle/123456789/4094
12. Jia, J., Yang, G., Nielsen, A.H., Rønne-Hansen, P.: Study of control strategies of power electronics during faults in microgrids. In: Hybrid-renewable energy systems in microgrids, pp. 109–146. Woodhead Publishing (2018). https://doi.org/10.1016/B978-0-08-102493-5.00007-8
13. Dehghani, M., Taghipour, M., Gharehpetian, G.B., Abedi, M.: Optimized fuzzy controller for MPPT of grid-connected PV systems in rapidly changing atmospheric conditions. J. Mod. Power Syst. Clean Energy **9**(2), 376–383 (2020). https://doi.org/10.35833/MPCE.2019.000086
14. Agoub, A.A.R., Hançerlioğullari, A., Rahebi, J., Lopez-Guede, J.M.: Battery charge control in solar photovoltaic systems based on fuzzy logic and jellyfish optimization algorithm. Appl. Sci. **13**(20), 11409 (2023). https://doi.org/10.3390/app132011409
15. Ali, U.S.: Z-source DC-DC converter with fuzzy logic MPPT control for photovoltaic applications. Energy Procedia **90**, 163–170 (2016). https://doi.org/10.1016/j.egypro.2016.11.181
16. Pande, J., Nasikkar, P., Kotecha, K., Abraham, A.: An ingenious technique to track the maximum power point for a wind energy system. IEEE Access **12**, 10160 (2024). https://doi.org/10.1109/ACCESS.2024.3354708

Optimizing MPPT Control of DC-DC Boost Converters Using Metaheuristic Search Algorithms and Nonlinear PID (N-PID) Controllers

L. Khellaf[1](✉), A. Djellal[1], H. Mayache[1], and A. Sheta[2]

[1] Department of E.E.A, National Higher School of Technology and Engineering, 23005 Annaba, Algeria
l.khellaf@etu.ensti-annaba.dz, {a.djellal, h.mayache}@ensti-annaba.dz

[2] Computer Science Department, Southern Connecticut State University, New Haven, CT 06515, USA
shetaa1@southernct.edu

Abstract. Photovoltaic systems have proven to be one of the most widely used renewable energies and the best replacement for conventional energy. Yet, their non-linear nature remains a challenge when it comes to extracting maximum power from photovoltaic modules. Therefore, in this work, a nonlinear PID controller has been used to meet the requirements of the photovoltaic system. In addition, to improve system performance and response, metaheuristic search algorithms were introduced into the tuning process of both the NPID controller and conventional PID controller parameters in order to compare them. The use of Artificial Intelligence to fine-tune the controller parameters will enable the optimum values of proportional, integral, derivative and nonlinear gains to be determined as system condition change. Finally, a comparison between the algorithms applied is conducted in terms of efficiency, rise time, settling time and overshoot as well as the overall system stability.

Keywords: Machine Learning · Control · Swarm Intelligence Algorithm · Evolutionary Algorithm

1 Introduction

Today, renewable energies are the world's most common and modern alternative for generating electricity. That said, photovoltaic solar energy has experienced impressive growth, and is now considered the main source of clean energy for the coming decade [1].

The PID controller has become very popular and the most widely used in many industries, due to its simple structure and ease of implementation. However, the strong non-linearities of photovoltaic systems make the task of tracking the maximum power point with sudden changes and variations in environmental conditions, i.e., irradiation

© The Author(s), under exclusive license to Springer Nature Switzerland AG 2025
M. Hatti (Ed.): IC-AIRES 2024, LNNS 1238, pp. 187–195, 2025.
https://doi.org/10.1007/978-3-031-80301-7_21

and temperature, even more tricky [2]. In order to cope with these non-linearities, Nonlinear PID controllers can be a great alternative. Nonlinear PID controllers are known for their high robustness and stability in noisy, non-linear systems, as well as for their ease of implementation, which make them compatible with photovoltaic systems. However, in nonlinear systems, finding the optimum parameters for PID and NPID controllers can be a difficult task, so in order to further contribute to their efficiency, metaheuristic search algorithms are introduced into the controller gain setting process. Among the many existing algorithms, Particle Swarm Optimization (PSO) and Genetic Algorithm (GA) techniques have been remarkably successful in various applications [3, 4].

In this work, Nonlinear PID controller's optimal gains values are obtained through a tuning process using different learning algorithms to enhance its performance, i.e., Particle Swarm Optimization (PSO) and Genetic Algorithm (GA). As for the MPPT Algorithm, conventional Perturb & Observe (P&O) algorithm is used to extract the maximum power point and reach the maximum voltage, which then serves as the voltage reference for the NPID controller. This paper is organized as follows: Section 2 presents a brief review of previous related work. Section 3 deals with the design of non-linear PID controller, as well as metaheuristics algorithms used and the implementation of the controller in the system. Results and performance comparisons are presented in Sect. 4. Finally, Conclusion is given in Sect. 5.

2 Background and Related Work

Mohamed Issa et al. in [5] used meta-heuristic algorithms to tune PID controller parameters. The particle swarm optimization algorithm, the genetic algorithm and the sinusoidal optimization algorithm were chosen in the study and compared the results obtained with the results of the traditional method of Ziegler and Nicolas. The overall results showed that the optimization methods outperformed the ZN method, with a faster and more stable transient response of the control system.

Layachi Zaghba et al. in [6], discussed the optimization of MPPT PI controller via the GA technique in a grid connected PV system. Simulation results showed improved performance with efficient tracking response of maximum power point with a 25% in energy efficiency.

An algorithm based on particle swarm optimization to tune the parameters of the PID controller in [7], Deqing Yang was designed. Results showed system stability with reduced overshoot, faster response time and good robustness compared to the trial method.

Dhruv Kler et al. in [8], the N-PID controller structure was proposed for the MPPT of the PV system using the Teaching Learning-Based Optimization method (TLBO) to fine-tune the parameters of the PID and N-PID controllers. The performance results are promising in terms of undershoot, settling time and ripple when comparing the response of the PID and N-PID systems.

3 Materials and Methods

A. N-PID Controller

Nonlinear PID controller has been developed to overcome shortcomings of classic PID controller, which are mainly in the lack of adaptability to changes in the system's environmental operating conditions.

Usually, conventional PID controller is expressed in classical form, used in [9, 10], incorporating the filter coefficient N and with K_p, K_i, K_d be the proportional, integral and derivative gains is given as follows:

$$U_c(t) = \left(K_p E(t) + K_i \int E(t)dt + K_d \frac{dE(t)}{dt} Ne^{-Nt}\right) \tag{1}$$

Hence, nonlinear PID controller is obtained by incorporating a nonlinear gain in cascade with the conventional PID structure [11, 12]. Figures 1, 2, depict control block of conventional PID and NPID respectively.

The design of the Nonlinear gain is based on hyper stability criterion ensuring robust, global stability [13]. Therefore, the output of NPID controller is given as:

$$U_c(t) = (K(e))\left(K_p E(t) + K_i \int E(t)dt + K_d \frac{dE(t)}{dt} Ne^{-Nt}\right) \tag{2}$$

$$U_c(t) = (K(e)E(t))\left(K_p + K_i \int dt + K_d \frac{d}{dt} Ne^{-Nt}\right) \tag{3}$$

$$\psi(e) = (K(e)E(t)) \tag{4}$$

$$U_c(t) = (\psi(e))\left(K_p + K_i \int dt + K_d \frac{d}{dt} Ne^{-Nt}\right) \tag{5}$$

where $K(e)$ represents nonlinear gain, $\psi(e)$ represents nonlinear function of error, which is formulated as follows:

$$K(e) = K_n \frac{exp(ae) + exp(-ae)}{2} = K_n cosh(ae) \tag{6}$$

where:

$$e = \begin{cases} e, & |e| \le e_{max} \\ e_{max} sign(e), & |e| > e_{max} \end{cases}$$

$$K(e) = \begin{cases} K_n, & e = 0 \\ K_n \cosh(ae_{max}), & |e| > e_{max} \end{cases}$$

B. Learning Algorithms.

In the realm of optimization and control, learning algorithms have emerged as powerful tools for solving complex problems that are otherwise challenging to address using traditional methods. These algorithms are designed to iteratively improve solutions by learning from the data and the environment, making them particularly suitable for dynamic

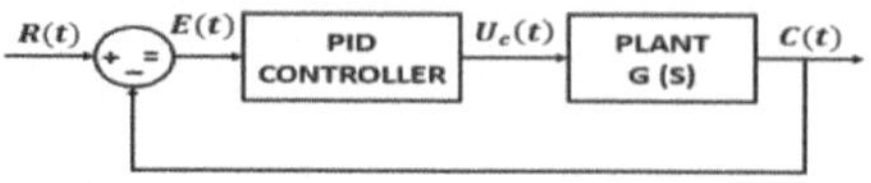

Fig. 1. Linear PID control block

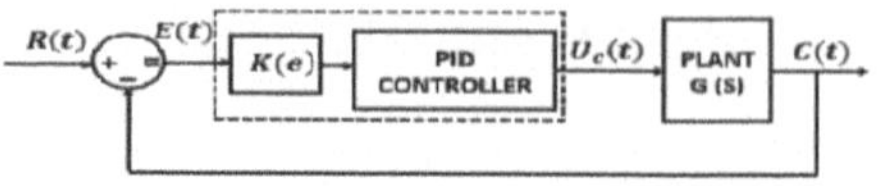

Fig. 2. Nonlinear PID (NPID) control block

and nonlinear systems. Among the myriad of learning algorithms available, Particle Swarm Optimization (PSO), Genetic Algorithm (GA)stand out due to their robustness and efficacy in various applications.

Several studies have demonstrated the effectiveness of these learning algorithms. For instance, PSO has been successfully applied to optimize the tuning of PID controllers in dynamic systems, as seen in the work by Kennedy and Eberhart (1995) [14]. Genetic Algorithms have been used extensively in structural optimization and feature selection in machine learning, with notable applications highlighted by Holland (1992) [15].

1) Genetic Algorithm (GA):

The genetic algorithm is based on a random search technique and is known for solving complex problems and is very well suited to non-linear systems, mimicking the nature of the behavior of the process of natural evolution. The algorithm relies on bio-inspired operators such as reproduction, crossover and mutation to find high-quality solutions, which has led to its implementation in control engineering, particularly in the tuning of PID controllers, receiving much attention and recognition in recent times [1617].

2) Particle Swarm Optimization (PSO):

The PSO algorithm was created in 1995 by Kennedy and Eberhart [14], capable of finding solutions to mathematical problems encountered in engineering. Known for its robustness and speed of convergence, it is in fact inspired by the collective behavior of social animals in search of food, such as flocks of birds, shoals of fish, ants, bees and fireflies.

Their strategy involves global communication among swarming particles, where the success of some influences others to follow the fitter particles, with the aim reaching the best solution. Each particle adjusts its movement based on global and personal data until all converge on the optimal solution guided by velocity and position updating equations. Equation (7) represent the mathematical model of the PSO algorithm used in [18].

$$V_i = \omega V_{i-1} + c_1 r_1 (P_{best} - X_{i-1}) + c_2 r_2 (G_{best} - X_{i-1}) \tag{7}$$

$$X_i = X_{i-1} + V_i$$

where is:

V_i: particle's speed; X_i: particle's position; P_{best}: best individual value of a particle; G_{best}: best global value of the entire population; ω: weights for the function; c_1, c_2: acceleration coefficients; r_1, r_2:: random numbers between (0–1).

C. Problem formulation

Finding the best parameter values for the PID controller can be a difficult task, especially in photovoltaic systems which are known for their non-linear nature due to changing environmental conditions. However, with the help of AI and metaheuristic algorithms, it is now possible to find optimal parameter values, which will improve efficiency by eliminating oscillations and overshoots. Based on the above motivations:

- A nonlinear PID controller is implemented for optimized MPPT control.
- To further improve control performance, metaheuristic algorithms have been used to fine-tune the values of the controller parameters.
- In addition, the fitness function plays a crucial role in obtaining optimized parameters. The mean square error was chosen as the fitness function. Control system design is depicted in Fig. 3

D. Fitness Function

In optimization problems, the fitness function plays a crucial role by directing the algorithm toward finding the best solution. It evaluates the performance of the system and quantifies how closely a proposed solution matches the desired setpoint, considering metrics like settling time and rise time. Our research focuses on fitness functions such as rise time, settling time, overshoot and Mean Square Error (MSE). MSE, in particular calculates the average squared difference between the setpoint and the process variable.

$$MSE = \frac{1}{n}\sum_{i=1}^{n}(y(i) - w(i))^2$$

where $y(i)$ and $w(i)$ are system output and desired output, respectively.

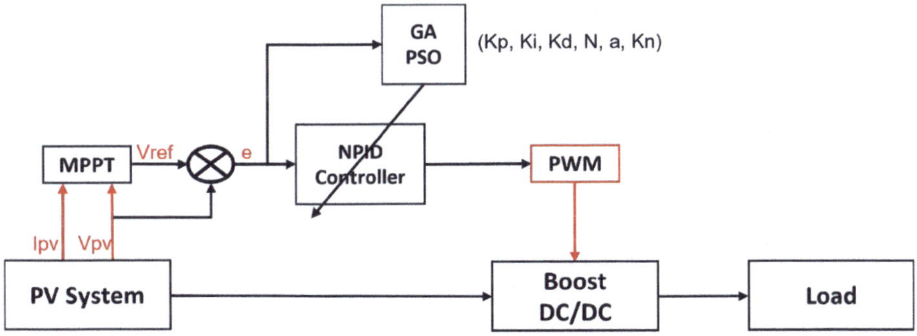

Fig. 3. Control System Sturture

4 Experimental Results

To refine the PID controller's parameters, a trial-and-error method was utilized to identify the most appropriate parameters for Genetic Algorithms (GAs), Particle Swarm Optimization (PSO). The outcomes are presented in the Tables 1 and 2, respectively.

Table 1. GA Parameters Settings

Parameter	Value
Population size	30
Mutation Fraction	0.3
Crossover Fraction	0.7
Lower Bound	0
Upper Bound	1
Iteration	100

Table 2. PSO Parameters Settings

Parameter	Value
Swarm size	30
Inertia Weight (ω)	0.5
Cognitive Parameter (c_1)	1
Social Parameter (c_2)	2
Lower Bound	0
Upper Bound	1
Iteration	100

System results were obtained according to a varying Irradiance, Fig. 4 depicts the variation of Irradiance (W/m2) in Time. Analysis and performance comparison is contributed in terms of Rise time, Settling time, Overshoot, and Efficiency. System efficiency is calculated during steady state according to Equation:

$$\boldsymbol{Efficiency}(\eta) = \frac{\boldsymbol{ExperimentPower}}{\boldsymbol{ActualPower}} \times 100 \tag{8}$$

The convergence graph for GA and PSO of the PID and NPID controllers is shown in Fig. 5. As can be seen, with the nonlinear PID controller, the optimum values of the gains are reached more quickly than with the PID controller and with a lower number of iterations, which improves the response time of the system. Table 3 represents the tuned gains of PID and NPID Controllers.

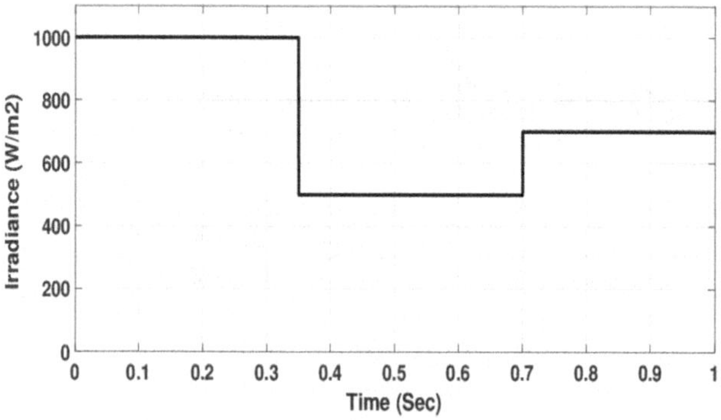

Fig. 4. Variation of Irradiance

A. Tuning of Controllers

The corresponding results are presented in Figs. 6, 7, the application of the PID and NPID

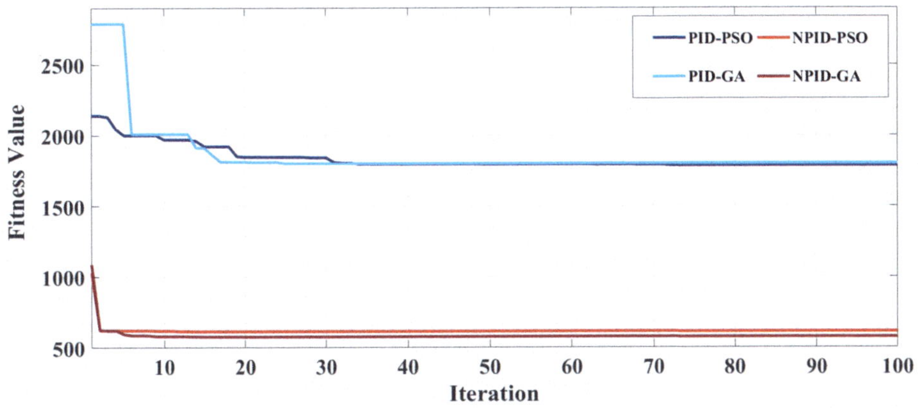

Fig. 5. Variation of fitness value with respect to iterations

Table 3. Tuned gains of PID and NPID Controllers

Controller		Kp	Ki	Kd	N	a	Kn
PID	GA	0.0073	0	0.5815	0	-	-
	PSO	0.0007	0.9992	0.2443	0	-	-
NPID	GA	0	0.2508	0.7471	0	0.2666	0.6593
	PSO	0.9071	0.4312	0.0458	2.34e-06	0.3788	0.8778

controllers tuned by GA and PSO algorithms respectively. From Figs. 6 and 7, it can clearly be observed that the NPID controller has better efficiency during steady state in comparison with PID controller which means it perfectly reaches the Maximum Power Point (MPP) of the system and power loss is very minimal. Howerver, it is networthy to mention that with PSO-tuned PID controller efficiency is still better than with GA-PID.

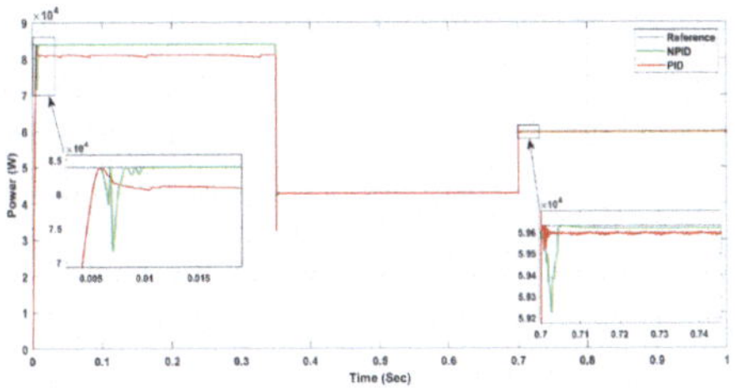

Fig. 6. Application of GA-tuned Controllers

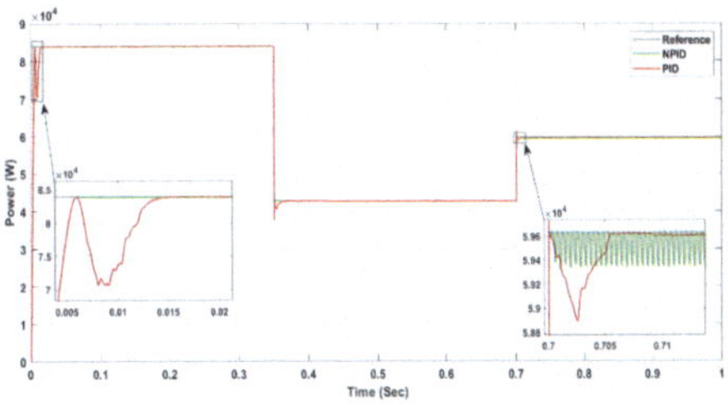

Fig. 7. Application of PSO-tuned Controllers

Performance of controllers is depicted in Tables 4 and 5, which indicates that PSO outperformed GA algorithm in terms of efficiency as it allowed to reach the maximum power point under different Irradiance conditions and also has less overshoot in the PID controller. Meanwhile, PID-GA has lower settling time and rise time but couldn't reach

Table 4. Results of PID Controller

Algorithm	GA	PSO
Rise Time (ms)	1.204	1.306
Settling Time (ms)	1.793	6.482
Overshoot (%)	8.849	5.759
Efficiency (%)	98.28	99.95

Table 5. Results of NPID Controller

Algorithm	GA	PSO
Rise Time (ms)	1.3063	1.3068
Settling Time (ms)	3.666	1.462
Overshoot (%)	5.759	5.759
Efficiency (%)	99.98	99.98

the MPP at 1000 W/m^2 & 700 W/m^2 irradiance level, hence, it has lower efficiency percentage. As for the NPID controller, PSO has lower settling time compared to GA algorithm and very close rise time and overshoot.

5 Conclusion

In this paper, an optimized MPPT control has been proposed using metaheuristic search algorithms and nonlinear PID controller for PV system. GA and PSO algorithms are applied to fine-tune the gains of the controller to obtain the optimum values and enhance system performance. To investigate the robustness of the proposed control, MATLAB/Simulink simulations are done under different Irradiance levels. NPID controller showed better results compared to PID controller with higher efficiency. Furthermore, in comparison between algorithms, results showed that PSO outperformed GA in PID controller with higher percentage of efficiency and lower overshoot. As for Nonlinear PID (NPID), results showed very close values of rise time and overshoot, however, PSO algorithm provides less settling time as compared to GA algorithm.

References

1. Chala, G.T., Al Alshaikh, S.M.: Solar photovoltaic energy as a promising enhanced share of clean energy sources in the future—A comprehensive review energies **16**(24), 7919 (2023)
2. Borase, R.P., Maghade, D.K., Sondkar, S.Y., Pawar, S.N.: A review of PID control, tuning methods and applications. International Journal of Dynamics and Control **9**, 818–827 (2021)
3. Pathak, D., Sagar, G., Gaur, P.: An application of intelligent non-linear discrete-PID controller for MPPT of PV system. Procedia Comput. Sci. **167**, 1574–1583 (2020)

4. Su, Y., Sun, D., Duan, B.: Design of an enhanced nonlinear PID controller. Mechatronics **15**(8), 1005–1024 (2005)
5. Issa, M., Elbaset, A.A., Hassanien, A.E., Ziedan, I.: PID controller tuning parameters using meta-heuristics algorithms: comparative analysis. In: Machine Learning Paradigms: Theory and Application, pp. 413–430 (2019)
6. Zaghba, L., et al.: A genetic algorithm based improve P&O-PI MPPT controller for stationary and tracking grid-connected photovoltaic system. In:2019 7th International Renewable and Sustainable Energy Conference (IRSEC), pp. 1–6 (2019)
7. Yang, D.: Tuning of PID parameters based on particle swarm optimization. In: IOP Conference Series: Materials Science and Engineering, vol. 452, IOP Publishing, p. 042179 (2018)
8. Kler, D., Rana, K., Kumar, V.: A nonlinear PID controller based novel maximum power point tracker for PV systems. J. Franklin Inst. **355**(16), 7827–7864 (2018)
9. Djellal, A., Lakel, R.: Adapted reference input to control PID-based active suspension system. Journal Européen des Systèmes Automatisés **51**(1–3), 7–23 (2018)
10. Eski I., Yildirim, S.: Vibration control of vehicle active suspension system using a new robust neural network control system. Simulat. Model. Pract. Theory (2009)
11. Maddi, A., Guessoum, A., Berkani, D.: Design of nonlinear PID controllers based on hyper-stability criteria. In: 2014 15th International Conference on Sciences and Techniques of Automatic Control and Computer Engineering (STA), pp. 736–741 (2014)
12. Malarvili, S., Mageshwari, S.: Nonlinear PID (N-PID) controller for SSSP grid connected inverter control of photovoltaic systems. Electric Power Syst. Res. **211**, 108175 (2022)
13. Dobra P, Tru M (2004) Robust stability of time-varying nonlinear systems using Popov criterion. IFAC Proc. Vol. 37(13):1213–1216
14. Kennedy, J.: Particle swarm optimization. Encyclopedia of Mach. learn. Springer **760**, 766 (2011)
15. Holland, J.H.: Genetic algorithms. Sci. Am. **267**(1), 66–73 (1992)
16. Joseph, S.B., Dada, E.G., Abidemi, A., Oyewola D.O., Khammas, B.M.: Metaheuristic algorithms for PID controller parameters tuning: review, approaches and open problems. Heliyon **8**(5) (2022)
17. Nasir, M., Saloumi, M., Nassif, A.B.: Review of various metaheuristics Techniques for tuning parameters of PID/FOPID controllers. ITM Web Conf. **43**, 01002 (2022)
18. Nanping, L., Kewen, X., Juan, Z., Chun, G.: Numerical simulation on transistor with CQPSO algorithm. In: 2009 4th IEEE Conference on Industrial Electronics and Applications, pp. 3732–3736 (2009)

Simplified Super-Twisting Sliding Mode Applied to Dc-Bus Voltage Regulator

Abdesselam Guerbouz[1(✉)], Imad Merzouk[1], Ahmed Hfaifa[1], and Amine Fihakhir[2]

[1] Laadi Laboratory, Faculty of Science and Technology, Djelfa University, Djelfa, Algeria
abdesselam.guerbouz@univ-djelfa.dz

[2] Faculty of Science and Technology, University of Ghardaia, Bounoura, Algeria

Abstract. To perform the dc-link voltage stability and its robustness of a photovoltaic (PV) grid- connected system, simplified super twisting sliding mode (SSTC) strategy is presented in this work. Where the tracking superiority of the SSTC control is clearful in dc-link bus voltage regulation with fast transient response, low oscillations and the reducing of current total Harmonic Distortion (THD) under any weather conditions;to show the superiority of the between the proposed controller(SSTC) comprehensive comparisons are done with the conventional Proportional–Integral (PI), using MATLAB/SIMULINK.

Keywords: PV · Dc-Link regulator · Super twisting control · THD

1 Introduction

Fossil fuel-based electrical energy generation is expensive and contributes to 40% of global CO2 emissions [1]. Renewable energy sources like solar energy are being used to replace fossil fuels and meet increasing energy demand at lower costs. Photovoltaic electricity production and grid-connected inverters are crucial for converting sunlight into usable electricity. Proper regulation of DC-link voltage is essential for high-performance injection into the grid [2]. The DC-Link Voltage Regulator controls the DC voltage level, ensuring efficient and safe operation. Monitoring and adjusting output of photovoltaic arrays are crucial for maximizing power generation; several methodes have been explored to regulate the voltage of dc-bus. Linear regulators like PID [3]. Also, Various non- linear control methods have been developed to regulate the dc-link voltage of Grid connected PV systems, and sliding mode control technique [5], model pridectif control [8] and AI methods like fuzzy logic controller [4], fuzzy neural network [9] and ANFIS [10].

In this paper, a dc-link capacitor voltage control based on the simplified super twisting control (SSTC) is proposed. We have organized our paper as follows: Sect. 2 introduce a description about the studied system. A brief introduction about the super twisting sliding mode control is given in Sect. 3, next we have describe developement of the proposed controller in Sect. 4 a. Simulation results are shown in Sect. 5, and finally Sect. 6 represent the conclusions of the work.

© The Author(s), under exclusive license to Springer Nature Switzerland AG 2025
M. Hatti (Ed.): IC-AIRES 2024, LNNS 1238, pp. 196–203, 2025.
https://doi.org/10.1007/978-3-031-80301-7_22

2 Proposed System Design

The proposed system is shown in Fig 1. The MPPT and voltage boost conversion is assuered by the DC-DC boost converter in the front of dc-link stage where the dc/ac inverter ensure the regulation of the active injected current on the grid side in synchronization with it; AC power is filtered out by the low pass filter (R, L).

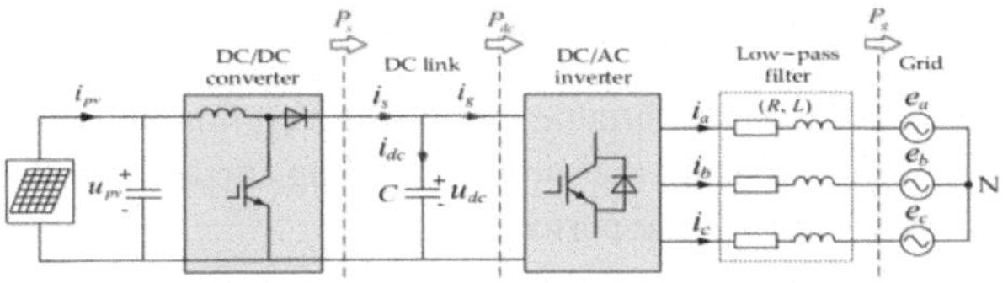

Fig. 1. Proposed two stage PV grid-connected system.

The given system operate by the following equations:

$$C\frac{du_{dc}}{dt} = i_s - i_g \tag{1}$$

$$U_{abc} - e_{abc} = Ri_{abc} + L\frac{di_{abc}}{dt} \tag{2}$$

where U_{abc}, i_{abc} represents the output voltage and current of the inverter respectively and e_{abc} represents the grid voltage; with the Park transformation we get in the dq reference system:

$$L\frac{di_d}{dt} = -Ri_d + wLi_q - e_d + V_d \tag{3}$$

$$L\frac{di_q}{dt} = -Ri_q + wLi_d - e_q + V_q \tag{4}$$

3 Super Twisting Sliding Mode Control

The non-linear sliding-mode-control (SMC) is a known technique which has significant advantages in its robustness against model uncertainties or several classes of perturbations; But its significant drawbacks is chattering phenomena; one of the best solution reduce SMC drawback is the use of novel super twisting control; which is a control method that has There multiple benefits in terms of monitoring accuracy and the ability to handle disruptions and resilience, including [6]:

- Robustness: STC is renowned for its ability to withstand fluctuations, uncertainties, and nonlinearities in the model. Because of its ability to efficiently manage both external disturbances and parameter fluctuations, it is well suited for situations where exact modeling is challenging or impracticable.

- Chattering reduction: Chattering, or high-frequency oscillations surrounding the sliding surface, is one of the primary problems with classic sliding mode control. Chattering might lead to unfavorable control behaviors. However, STC addresses this issue.

The single input first order system differential equations can be described as follow:

$$\begin{cases} \frac{dx(t)}{dt} = Ax + Bu \\ y = \psi(x) \end{cases} \tag{5}$$

We consider the x as the state variable and u is the command input; the goal of the sliding mode controller is to ensure that both the sliding surface and its derivative must be null and reach zero within a finite period of time:

$$\psi_w(x) = 0; \dot{\psi}_w(x) = 0 \tag{6}$$

We consider that C, λ_1, λ_2, w and α as positive constant the we get

$$\begin{cases} |\dot{A}| + W|\dot{B}| \le C \\ 0 \le \lambda_1 \le B(t, x) \le \lambda_2 \\ \left|\frac{A}{B}\right| < \alpha W; 0 < \alpha < 1 \end{cases} \tag{7}$$

If all of the following conditions are fulfilled, the STC controller can be represented as follows [7]:

$$U_{sstc} = -k\sqrt{|\psi_w|}\text{sign}(\psi_w) \tag{8}$$

The SSTC algorithm is introduced as a simplified version of the standard supert twisiting sliding mode control approach. Therefore, a more streamlined approach that can be readily created and implemented for any system, regardless of its intricacy, is devised where equations (9), (10) represents the SSTC algorithm:

$$U_{sstc} = -k\sqrt{|\psi_w|}\text{sign}(\psi_w) \tag{9}$$

$$U = B(x)^{-1}(U_{\text{sstc}} - A(x)) \tag{10}$$

where ψ_w is the sliding surface, k_1, k_2 and k are positive constant selected as mentioned in [7].

4 The Proposed SSTC Dc-Link Voltage Regulator

As already stated, the aim of DC-Link control is estimation of the reference d-axis grid current to regulate the DC- Link voltage to be as its reference value whatever the produced PV power, the proposed sliding surface is selected as:

$$\psi_{dc} = e_{dc} \tag{11}$$

where, e_{dc} is the voltage error between the DC-Link voltage u_{dc} and its reference value u^{*}_{dc}, the time derivative of the previous equation is:

$$\dot{\psi}_{dc} = \dot{e}_{dc} \tag{12}$$

where $e_{dc} = u^{*}_{dc} - u_{dc}$ and $\dot{e}_{dc} = -\dot{u}_{dc}$.

By considering energy conservation principle we can provide that

$$P_s = P_{dc} = P_g \tag{13}$$

where, P_s, P_{dc} and P_g are the PV panel power, Dc link side power and the injected power to the grid, respectively.

In synchronous d-q frame the injected power into grid as follow:

$$P_g = \frac{3}{2}\left(V_d i_d + V_q i_q\right) \tag{14}$$

For injecting only the produced PV power, the q-axis grid current *iqg* must be null then we can rewrite the Eq. 14 as

$$P_g = \frac{3}{2}(V_d i_d) \tag{15}$$

The inverter input power is:

$$P_{dc} = u_{dc} i_g \tag{16}$$

By applying Kirchhoff current law at the DC-Link between the DC-DC stage and DC-AC stage we get:

$$i_{dc} = C\frac{du_{dc}}{dt} = i_s - i_g \xrightarrow{\text{yields}} \dot{u}_{dc} = \frac{i_s - i_g}{C} \tag{17}$$

By substituting Eqs. (16) and (17) in Eq. (14) we get

$$u_{dc} i_g = \frac{3}{2}(V_d i_d) \xrightarrow{\text{yields}} i_g = \frac{3}{2}\frac{V_d}{u_{dc}} i_d \tag{18}$$

We appreciate that the derivative of the DC-Link voltage u_{dc} approaches zero in the case of DC-Link voltage is appropriately managed so

$$i_s = i_g \tag{19}$$

Then

$$P_{dc} = u_{dc}\, i_g = P_s \tag{20}$$

The current i_s can be defined as:

$$i_s = \frac{P_s}{u_{dc}} \tag{21}$$

We obtain By replacing Eqs. (18) and (21) in Eq. (18)

$$\dot{u}_{dc} = \frac{1}{C}\left(\frac{P_s}{u_{dc}} - \frac{3}{2}\frac{V_d}{u_{dc}} i_d\right) \tag{22}$$

From equation (9) the desired control law is defined as

$$i_d^* = -\frac{2}{3}\frac{Cu_{dc}}{V_d}\left[k\sqrt{|\psi_{dc}|}\mathrm{sign}(\psi_{dc}) - \frac{P_s}{Cu_{dc}}\right] \tag{23}$$

5 Simulation and Results

The method is simulated using Simulink in a 100 kW PV array connected to the grid. Perturb and observe are used to extract MPP and DC-DC converter duty cycle, with the proposed control controlling DC-link capacitor voltage and d-q grid currents.

5.1 Test 1: Stable Conditions

The study focuses on a scenario with temperature and irradiation set at 25 °C and 1000 W/m2, and a desired DC- Link voltage reference of 800 V. Results show the proposed SSTC outperforms PI control in Dc-link voltage regulation, with fast tracking of desired voltage in 0.035 s and better PI output Dc-link voltage tracking in 0.1 s with first overshoot and noticeable DC-link voltage fluctuation (Fig. 2).

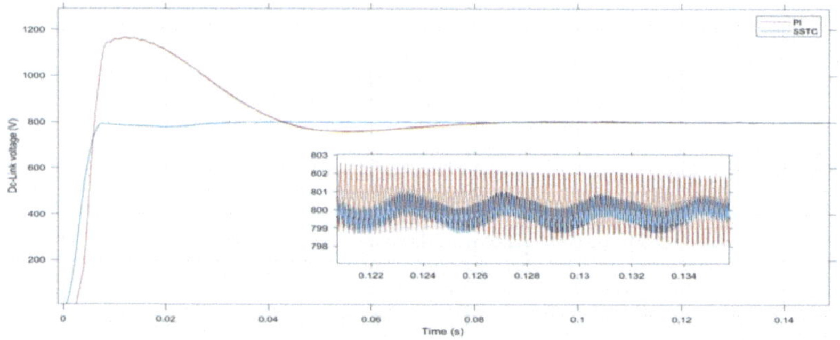

Fig. 2. Dc-link voltage.

5.2 Test 2: Variable Irradiation

The test at 25 °C with variable irradiation showed the superiority of the proposed SSTC control over PI control in Dc-link voltage regulation and fast tracking of desired dc-link voltage. The proposed FSTC system provided quicker reaching time with less overshoot and undershoot, injecting a current with less THD (Figs. 3, 4 and Table 1).

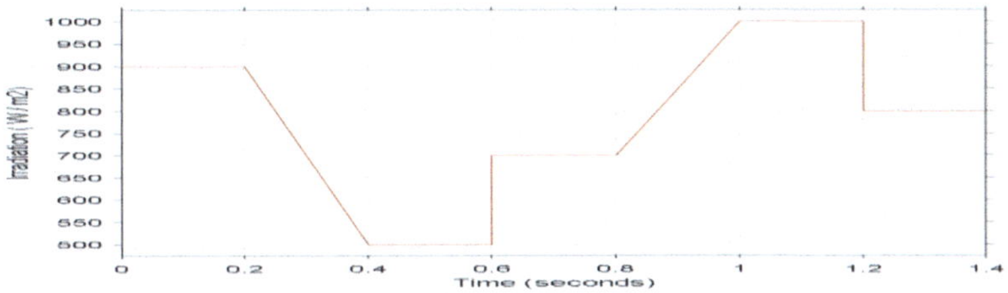

Fig. 3. Variable irradiance.

Table 1. Analysis for DC bus voltage umder variable irradiance

	900 → 500 W/m^2	500 → 700 W/m^2		700 → 1000 W/m^2	1000 → 800 W/m^2	
Parameter	Error (V)	Time of Settlin g (s)	Overshoot/Undershoot (V)	error (V)	Time of Settling (s)	Under-shoot
PI	14	0.12	30	12	0.18	30
SSTC	0.3	0.022	11	0.9	0.012	20

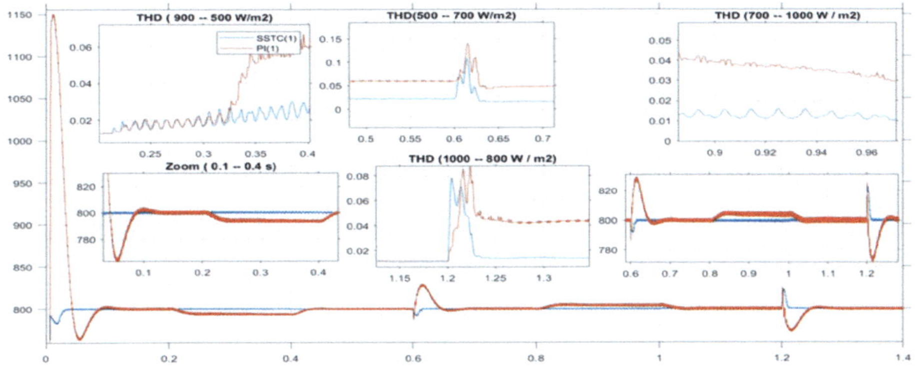

Fig. 4. Dc-link voltage and THD value under irradiance change.

5.3 Test 3: Robustness Test

The proposed SSTC was tested for robustness by changing the DC-Link voltage reference and irradiance as shown in following figures. Results showed that the control achieved the desired voltage level quickly and effectively, with no overshoot. The SSTC reached the desired voltage after 0.015 s, while the PI reached it in 0.07 s. The second variation in DC-Link voltage (from 896 V to 800 V) showed superiority and effectiveness, with a noticeable undershoot (Figs. 5 and 6).

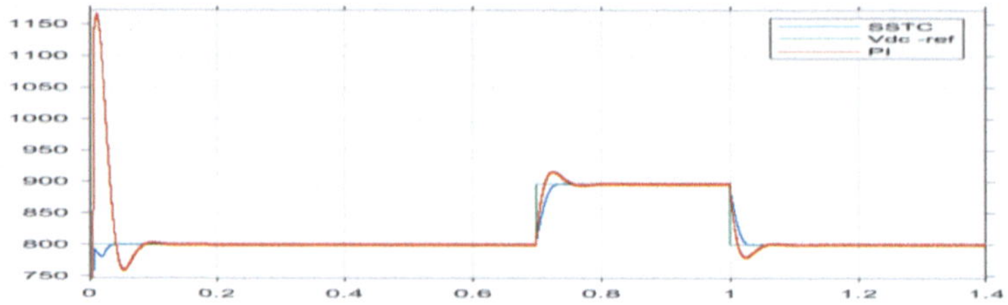

Fig. 5. Dc-Link voltage- robustness test 1 -

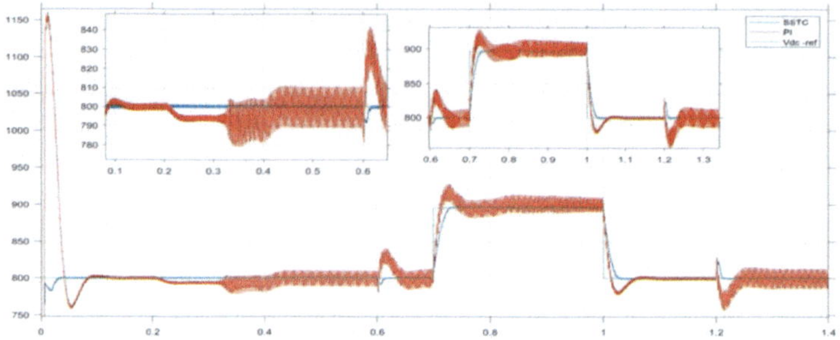

Fig. 6. Dc-Link voltage- robustness test 2-

6 Conclusion

The paper introduces a simplified supert twisting method regulation strategy for grid connected PV system inverters, improving dc-link voltage control performance. Simulation results show fast transient response, low oscillations, and reduced total harmonic distortion.

References

1. Ahmed et, J., Salam, Z.: An improved perturb and observe (p & o) maximum power point tracking (mppt) algorithm for higher efficiency. Appl. Energy t. 150, 97–108 (2015). ISSN: 0306-2619. https://doi.org/10.1016/j.apenergy.2015.04.006
2. Chinnappan, R., Logamani, P., Ramasubbu, R.: Fixed frequency integral sliding-mode current-controlled MPPT boost converter for two-stage PV generation system. IET Circ. Devices Syst. **13**, 793–805 (2019)
3. Krama, A., Zellouma, L., Rabhi, B., Laib, A., Fuzzy logic controller for improving DC side of PV connected shunt active filter based on MPPT sliding mode control. In: International Conference in Artificial Intelligence in Renewable Energetic Systems, pp. 224–235. Springer, Cham (2017)
4. H. Suryanarayana, H., Mishra, M.K.: Fuzzy logic based supervision of DC link PI control in a DSTATCOM. In: 2008 Annual IEEE India Conference, vol. 2, pp. 453–458 (2008)
5. Harnefors, L., Yepes, A.G., Vidal, A., Doval- Gandoy, J.: Passivity based controller design of grid-connected vscs for prevention of electrical resonance instability. IEEE Trans. Ind. Electron. **62**(2), 702–710 (2015). https://doi.org/10.1109/TIE.2014.2336632

6. Djennoune, S., Bettayeb, M.: Optimal synergetic control for fractional-order systems. Automatica **49**(7), 2243–2249 (2013)
7. Benbouhenni, H., Bizon, N., Colak, I., Thounthong, P., Takorabet, N.: Simplified super twisting sliding mode approaches of the double-powered induction generator-based multi-rotor wind turbine system. Sustainability **14**, 5014 (2022). https://doi.org/10.3390/su14095014
8. Ma, M., Liu, X., Lee, K.Y.: Maximum power point tracking and voltage regulation of two-stage grid-tied PV system based on model predictive control. Energies **13**, 1304 (2020)
9. Lin, F.-J., Lu, K.-C., Ke, T.H.: Probabilistic wavelet fuzzy neural network based reactive power control for grid-connected three-phase PV system during grid faults. Renew. Energy **92**, 437–449 (2016)
10. Mahmud, N., Zahedi, A., Mahmud, A.: A cooperative operation of novel PV inverter control scheme and storage energy management system based on ANFIS for voltage regulation of grid-tied PV system. IEEE Trans. Ind. Inf. **13**, 2657–2668 (2017)

Computational Modeling and Simulation

Comparative Analysis of PV Module Models Using AI-Based Metaheuristic Optimization Techniques

El Mouatez Billah Messini[1,2,3](✉), Yacine Bourek[2], and Chouaib Ammari[4]

[1] LAGE Laboratory, Faculty of Applied Sciences, University of Ouargla, 30000 Ouargla, Algeria
messini.elmouatezbillah@univ-ouargla.dz

[2] Electrical Engineering Department, Faculty of Applied Sciences, Ouargla University, Road Ghardaia, 30000 Ouargla, Algeria

[3] Sonatrach - Institut Algérien du Pétrole - École de Hassi Messaoud, BP 54 – Hassi Messaoud, 30001 Ouargla, Algeria

[4] Department of renewable energy, Faculty of Hydrocarbons, Renewable Energy, Science, Earth and Universe, University Kasdi Merbah-Ouargla, Ouargla, Algeria

Abstract. For successful modelling of PV system characteristics, an accurate PV module model is necessary. This research, which investigates and compares three PV module models: the single diode model (SDM), the double diode model (DDM), and the triple diode model (TDM), employs modern AI-based methods. Metaheuristic optimization techniques have solved the difficulty of obtaining parameters for the equivalent circuit of a solar module. Two artificial intelligence methods, Genetic Algorithm (GA) and Particle Swarm Optimization (PSO), are employed to achieve the optimal solution for error minimization between measured and estimated data using the three models. The optimization process involves minimizing the Root Mean Squared Error (RMSE) between measured and estimated data. A convergence comparison between GA and PSO after 50 runs demonstrates that both techniques effectively extract parameters for all three models. The optimization findings demonstrate accurate modelling with RMSE results as follows: for the SDM, 5.49% (GA) and 1.02% (PSO); for the DDM, 3.91% (GA) and 0.79% (PSO); and for the TDM, 4.94% (GA) and 0.94% (PSO). These results validate the effectiveness of AI-based methods for parameter extraction, with the DDM with PSO method showing superior accuracy and timing.

Keywords: Artificial Intelligence Optimization · Genetic Algorithm (GA) · Particle Swarm Optimization (PSO) · Renewable Energy · parameters estimation · PV module modelling

1 Introduction

The steady depletion of fossil fuels has been a driving force behind the growing importance of renewable energy sources over the last few decades [1, 2]. The number of photovoltaic systems (PVS) that are being installed is increasing, as indicated by estimates provided by the International Energy Agency (IEA) [3]. As a result, researchers

© The Author(s), under exclusive license to Springer Nature Switzerland AG 2025
M. Hatti (Ed.): IC-AIRES 2024, LNNS 1238, pp. 207–216, 2025.
https://doi.org/10.1007/978-3-031-80301-7_23

are focusing their attention on the complex and challenging issues concerning the design and appropriate application of this type of energy [4]. The single diode model (SDM) [5], the double diode model (DDM) [6], and the triple diode model (TDM) [3] are the three most common types of models that are used to depict a photovoltaic (PV) module. In order to attain the desired level of efficiency with this model, the manufacturer is unable to offer the necessary settings.

Furthermore, due to the nonlinear properties of photovoltaics, it is challenging to generate an accurate photovoltaic model; hence, it is vital to have precise modelling of the PVS components [7]. The three models (SDM, DDM, and TDM) offer a realistic behavior that imitates the characteristics of a solar photovoltaic (PV) cell or module under a variety of situations [8, 9]. As a result of the fact that many academics [9–11] are concentrating on developing various photovoltaic modelling methodologies, one of the most critical steps in the process of establishing a stable model is selecting acceptable and correct model parameters.

At this juncture, it's crucial to note the significant strides made in developing accurate PV models. The application of metaheuristic algorithms has revolutionized the use of a wide array of performance optimization tools [9, 10], thereby streamlining optimization the process of establishing photovoltaic parameters [3]. This research is a testament to the continuous evolution in the field of photovoltaics and renewable energy.

The field of PV model development is rich with diversity, as evidenced by the inclusion of a variety of optimization algorithms. The imperial competitive algorithm [12], the improved moth search algorithm (EMSA) [13], the random search optimization algorithm fractal (SFS) [14], the chaotic whale optimization algorithm [15], and the Harris hawk optimizer [16] are just a few examples of the many optimizations that fall under this category. This diversity is a testament to the complexity and depth of our research in this field.

This paper is at the forefront of research, employing advanced techniques such as Artificial intelligence- based optimization algorithms. Specifically, the Genetic Algorithm (GA [17, 18] and the Particle Swarm Optimization (PSO) [19, 20] algorithm are used. These heuristic methods, being evolutionary-based and swarm-based algorithms, respectively, are instrumental in accurately estimating the equivalent circuit coefficients. Their primary goal is to reduce the root mean square error (RMSE) between the experimental and computed I-V data, thereby enhancing the accuracy of the PV models.

2 Materials and Methods

The main objective of this research is to identify the parameters of the three (SDM, DDM and TDM) model for the PV modules by minimizing the RMSE between measured and estimated current using GA and PSO algorithm.

2.1 Single Diode Model

The equivalent circuit for the SDM is presented in Fig. 1,

The estimated current from the SDM is calculated based on the equivalent circuit presented in Fig. 1 and is given using Eq. (1)

$$I = f_1(X_1) = I_{ph} - I_{sd} - I_{sh} \tag{1}$$

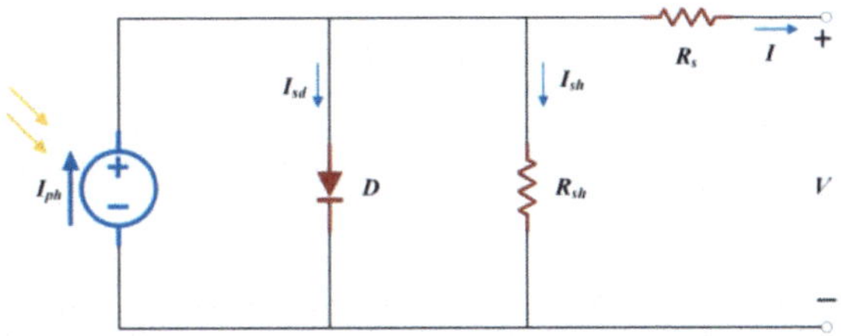

Fig. 1. Equivalent Circuit of Single Diode Model (SDM) for the solar PV module.

I_{ph}, I_{sd} and I_{sh} are photon current, diode current and shunt current respectively. I_{sd} and I_{sh} are expressed as follows:

$$I_{\text{sd}} = I_{s1} \cdot \left(e^{\frac{q(V+R_s \cdot I)}{N_s \cdot n_1 \cdot K \cdot T}} - 1 \right) \tag{2}$$

$$I_{\text{sh}} = \frac{(V + R_s \cdot I)}{R_p} \tag{3}$$

$$I = f_1(X_1) = I_{ph} - I_{s1} \cdot \left(e^{\frac{q(V+R_s \cdot I)}{N_s \cdot n_1 \cdot K \cdot T}} - 1 \right) - \frac{(V + R_s \cdot I)}{R_{sh}} \tag{4}$$

where $X_1 = \left[I_{ph}, I_{s1}, n_1, R_s, R_{sh} \right]^T$ is the vector of the 05 unknown parameters to be identified.

2.2 Double Diode Model (DDM)

The equivalent circuit for the SDM is presented in Fig. 2,

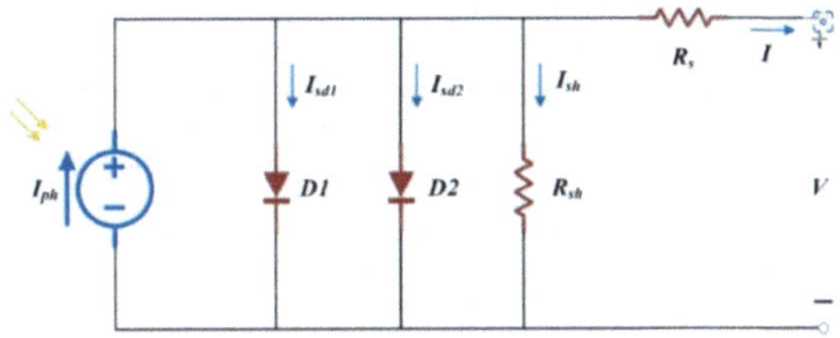

Fig. 2. Equivalent Circuit of Double Diode Model (DDM) for the solar PV module.

The estimated current from the SDM is calculated based on the equivalent circuit presented in Fig. 2 and is given using Eq. (5)

$$I = f_2(X_2) = I_{ph} - I_{sd1} - I_{sd2} - I_{sh} \tag{5}$$

I_{ph}, I_{sd1}, I_{sd2} and I_{sh} are photon current, 1st diode current, 2nd diode current and shunt current respectively. I_{sd1}, I_{sd2} are expressed as follows:

$$I_{\text{sd1}} = I_{s1} \cdot \left(e^{\frac{q(V+R_s \cdot I)}{N_s \cdot n_1 \cdot K \cdot T}} - 1 \right) \tag{6}$$

$$I_{\text{sd2}} = I_{s2} \cdot \left(e^{\frac{q(V+R_S \cdot I)}{N_s \cdot n_2 \cdot K \cdot T}} - 1 \right) \tag{7}$$

whereas I_{sh} is given in Eq. (3), Finally the estimated current for the DDM is expressed as follows (Eq. (8)):

$$I = f_2(X_2) = I_{ph} - I_{s1} \cdot \left(e^{\frac{q(V+R_S \cdot I)}{N_s \cdot n_1 \cdot \cdot T}} - 1 \right) - I_{s2} \cdot \left(e^{\frac{q(V+R_S \cdot I)}{N_s \cdot n_2 \cdot K \cdot T}} - 1 \right) - \frac{(V + R_s \cdot I)}{R_{sh}} \tag{8}$$

where $X_2 = \left[I_{ph}, I_{s1}, I_{s2}, n_1, n_2, R_s, R_{sh} \right]^T$ is the vector of the 07 unknown parameters to be identified for the DDM.

2.3 Triple Diode Model (TDM)

The equivalent circuit for the SDM is presented in Fig. 3,

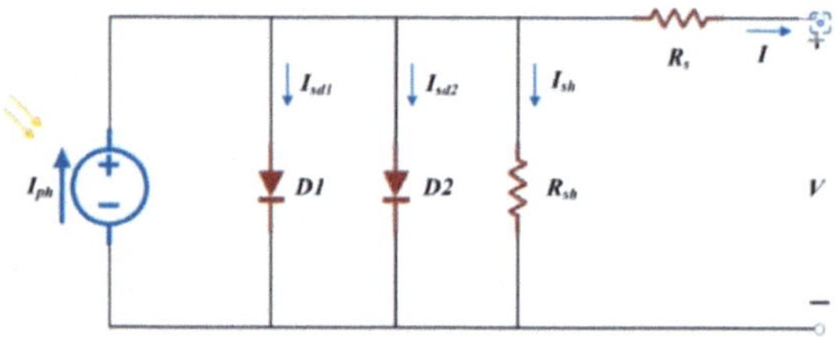

Fig. 3. Equivalent Circuit of Triple Diode Model (TDM) for the solar PV module.

The estimated current from the TDM is calculated based on the equivalent circuit presented in Fig. 3 and is given using Eq. (9)

$$I = f_3(X_3) = I_{ph} - I_{sd1} - I_{sd2} - I_{sd3} - I_{sh} \tag{9}$$

Iph, *Isd*1, *Isd*2, *Isd*3 and *Ish* are photon current, 1st diode current, 2nd diode current, 3rd diode current and shunt current respectively. *Isd*1, *Isd*2 and *Isd*3 are expressed as follows:

$$I_{\text{sd1}} = I_{s1} \cdot \left(e^{\frac{q(V+R_S \cdot I)}{N_S \cdot n_1 \cdot K \cdot T}} - 1 \right) \tag{10}$$

$$I_{\text{sd2}} = I_{s2} \cdot \left(e^{\frac{q(V+R_S \cdot I)}{N_s \cdot n_2 \cdot K \cdot T}} - 1 \right) \tag{11}$$

$$I_{\text{sd3}} = I_{s3} \cdot \left(e^{\frac{q(V+R_S \cdot I)}{N_s \cdot n_3 \cdot K \cdot T}} - 1 \right) \tag{12}$$

where as *Ish* is given in Eq. (3), Finally the estimated current for the DDM is expressed as follows (Eq. (13)):

$$I = f(X) = I_{ph} - I_{s1} \cdot \left(e^{\frac{q(V+R_S \cdot I)}{N_s \cdot n_1 \cdot K \cdot T}} - 1 \right) - I_{s2} \cdot \left(e^{\frac{q(V+R_S \cdot I)}{N_s \cdot n_2 \cdot K \cdot T}} - 1 \right) - I_{s3} \cdot \left(e^{\frac{q(V+R_S \cdot I)}{N_s \cdot n_3 \cdot K \cdot T}} - 1 \right) - \frac{(V + R_s \cdot I)}{R_{sh}} \tag{13}$$

where $X_3 = \left[I_{ph}, I_{s1}, I_{s2}, I_{s3}, n_1, n_2, n_3, R_s, R_{sh} \right]^T$ is the vector of the 09 unknown parameters to be identified for the TDM.

2.4 Objective Function

The objective function to be optimized is considered to be the Root Mean Square Error (RMSE) between the measured current I_{mes} from lab bench PV module and the estimated current from the 03 models developed above. The Table 1 recap the Objective functions for all 03 models.

The optimization problem is expressed as follow: $\underset{subject\ to\ X}{\text{minimize}}\ RMSE\ (X)$

The upper and the lower bounds for the decision variable are presented in Table 2

I. SOLUTION METHODS:

The minimization of the objective function is conducted using genetic algorithm (GA) and Particle swarm optimization (PSO) algorithms. The flow chart of the optimization process for GA is presented in Fig. 4(a), whereas the flowchart for the PSO algorithm is presented in Fig. 4(b).

Table 1. Objective functions

Model	Objective function
SDM	$\mathrm{RMSE}_1 = \sqrt{\sum_{i=1}^{N} (I_{mes}(i) - f_1(X_1)(i))^2}$ (14)
DDM	$\mathrm{RMSE}_{21} = \sqrt{\sum_{i=1}^{N} (I_{mes}(i) - f_3(X_3)(i))^2}$ (15)
TDM	$\mathrm{RMSE}_3 = \sqrt{\sum_{i=1}^{N} (I_{mes}(i) - f_3(X_3)(i))^2}$ (16)

Table 2. Lower and upper bounds for the decision variables

Parameter	Lower bound	Upper bound
$Rs(\Omega)$	0.0	1.00
$Rsh(\mathrm{k}\Omega)$	0.10	10.00
$Iph(A)$	3.6	6.00
$Is1(\mu A)$	10^{-3}	1000.00
$Is2\ (\mu A)$	10^{-3}	1000.00
$Is3\ (\mu A)$	10^{-3}	1000.00
$n1$	1.00	3.00
$n2$	1.00	3.00
$n3$	1.00	3.00

3 Results and Discussions

The optimization process for each model and for each algorithm has been run 50 times in order to get a statistical behaviour of both optimization techniques when applied to each model. Best, worst, mean and standard deviation for RMSE and elapsed time are recorded and presented in Table 3

Another insightful information after 50 runs of the optimization algorithms are presented in Fig. 5, this presentation shows that PSO surpass the GA in getting lower RMSE, when it comes to duration of the optimization process GA can give an acceptable results (RMSE<5%) with minimum time.

The three models parameters estimation with both optimization methods were compared using measured and estimated IV-curves (Fig. 6), and the results for the root mean square error (RMSE) were as follows: 5.49% (GA) and 1.02% (PSO) for the SDM, 3.91% (GA) and 0.79% (PSO) for the second model, and 4.94% (GA) and 0.94% (PSO) for the third model. The results show that the DDM using the PSO approach obtained greater accuracy in parameter extraction for PV module modelling. This is demonstrated by the average deviation of the estimated values from the actual measured values, confirming the efficiency of the method.

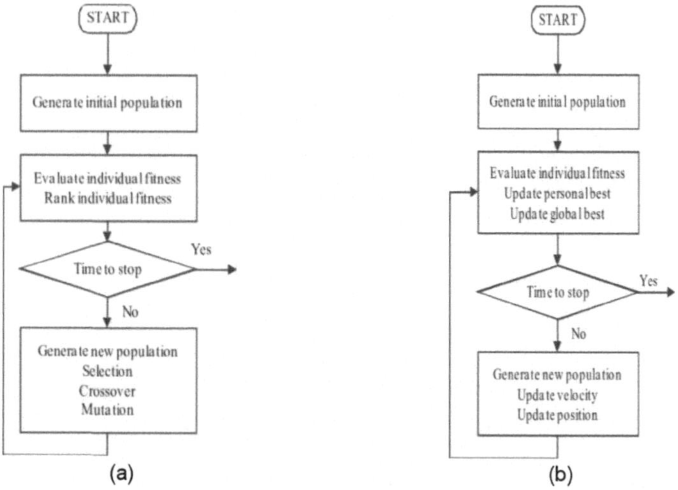

Fig. 4. (a) GA Flowchart [21] and (b) PSO method.

Table 3. Recap results after 50-runs optimization

		GA			PSO		
		SDM	DDM	TDM	SDM	DDM	TDM
$R_s(\Omega)$	For best RMSE value	0.0145	0.0394		0.3856	0.4421	
$R_{sh}(\mathbf{k\Omega})$		9.3665	6.1465		10.00	1.49	
$I_{ph}(A)$		3.9066	3.8720	3.9073	3.8432	3.8478	3.8439
$I_{s1}(\mu A)$		33.0680	10.8185	10.1159	0.2660	0.0030	0.2709
I_{s2} (μA)		-	12.2589	0.2324	-	0.2631	0.4262
I_{s3} (μA)		-	-	278.8820	-	-	0.1296
n_1		1.7099	1.5689	1.5592	1.2129	1.00	1.2143
n_2		-	2.0047	1.8124	-	1.2491	2.7581
n_3		-	-	2.9807	-	-	2.8257
RMSE	Best	5.4881	3.9114	4.9377	1.0164	0.7895	0.9356
	worst	20.8571	20.0861	17.9127	5.6924	6.7600	5.0558
	mean	14.1685	12.2437	10.8767	2.0549	1.7738	1.9830
	Std	3.9678	3.6492	3.8247	0.9871	0.9357	0.9524
Time	Best	0.8650	83.6727	130.6713	0.4837	94.78	76.8896
	worst	9.3010	1151.2	1732.0	16.5894	2636.7	1987.5
	mean	1.9384	254.6078	431.317	6.0922	938.22	719.6401
	Std	1.8298	232.8563	305.1968	3.6855	562.85	424.1028

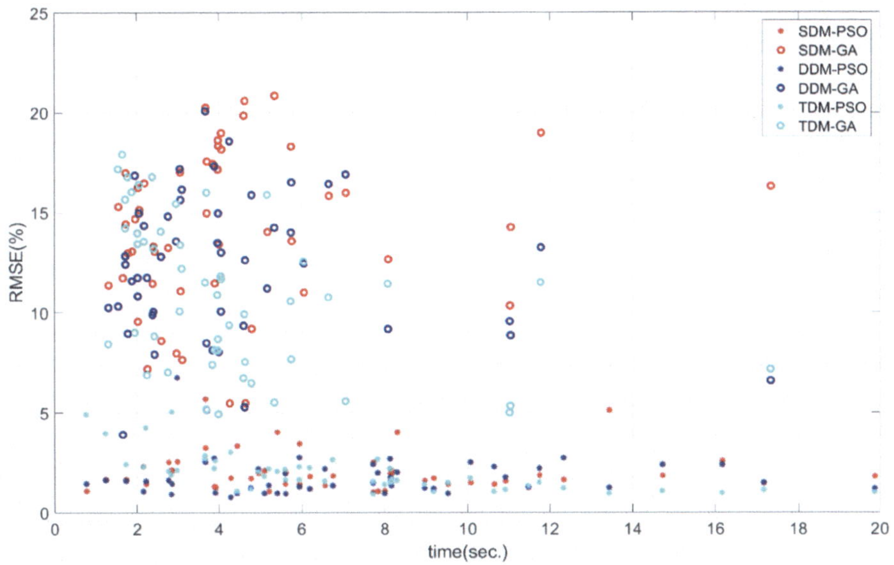

Fig. 5. 50 runs solutions.

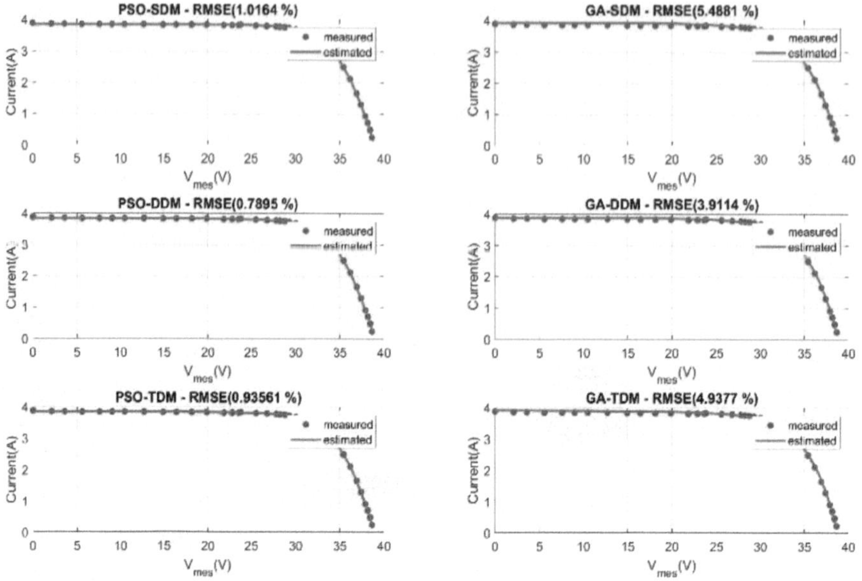

Fig. 6. Measured and estimated currents.

4 Conclusion

This study has made a significant stride in the field of photovoltaic systems by successfully employing Artificial Intelligence-based optimization methods to model PV modules using three different models: the single diode model (SDM), the double diode model (DDM), and the triple diode model (TDM). Both Genetic Algorithm (GA) and Particle Swarm Optimization (PSO) techniques were utilized to estimate the parameters for these models, aiming to minimise the Root Mean Squared Error (RMSE) between measured and calculated data.

The comparison of the three models using both optimization methods yielded the following RMSE results: 5.49% (GA) and 1.02% (PSO) for the SDM, 3.91% (GA) and 0.79% (PSO), and 4.94% (GA) and 0.94% (PSO) for the TDM. These results, which indicate the average deviation of the estimated values from the actual measured values, demonstrate that the DDM with the PSO method achieved superior accuracy, indicating its effectiveness in parameter extraction for PV module modelling.

The GA and PSO optimization techniques provided robust solutions, with PSO consistently outperforming GA in terms of lower RMSE values and faster convergence. This convergence comparison after 50 runs underscores the efficiency of PSO in handling the non-linear characteristics of photovoltaic systems.

The results validate the efficacy of AI-based optimization methods in PV module parameter estimation, with significant implications for improving the accuracy and reliability of PV system models. The superior performance of the PSO method, especially with the DDM, suggests its potential for broader application in photovoltaic research and development.

In summary, this research highlights the importance of selecting appropriate PV models and optimization techniques to enhance the accuracy of PV system simulations. Future work will focus on further refining these models and exploring additional AI-based methods to achieve even greater precision in PV module parameter estimation. This study paves the way for more efficient and reliable photovoltaic system designs, contributing to the advancement of renewable energy technologies.

5 List of Symbols

Symbol	Nomenclature
Rs	Series resistance (Ω)
Rsh	Shunt resistance (Ω)
Iph	Photon current (A)
$Is1, Is2, Is2$	Diodes saturation current (A)
$n1, n2, n3$	Ideality factor for diodes
K	Boltzmann constant (1.3806×10^{-23} J/K)
q	electron charge (1.6022×10^{-19} C).
T	absolute temperature

References

1. Mustafa, R.J., Gomaa, M.R., Al-Dhaifallah, M., Rezk, H.: Environmental impacts on the performance of solar photovoltaic systems. Sustainability **12**, 608 (2020). https://doi.org/10.3390/su12020608
2. Messini, E.M.B., Bourek, Y., Ammari, C., Pesyridis, A.: The integration of solar-hydrogen hybrid renewable energy systems in oil and gas industries for energy efficiency: optimal sizing using Fick's Law optimisation Algorithm. Energy Convers Manag **308**, 118372 (2024). https://doi.org/10.1016/j.enconman.2024.118372
3. Rezk, H., Abdelkareem, M.A.: Optimal parameter identification of triple diode model for solar photovoltaic panel and cells. Energy Rep. **8**, 1179–88 (2022). https://doi.org/10.1016/j.egyr.2021.11.179
4. Navarro, M.A., Oliva, D., Ramos-Michel, A., Haro, E.H.: An analysis on the performance of metaheuristic algorithms for the estimation of parameters in solar cell models. Energy Convers. Manag. **276**, 116523 (2023). https://doi.org/10.1016/j.enconman.2022.116523
5. Azzouzi, M., Popescu, D., Bouchahdane, M.: Modeling of electrical characteristics of photovoltaic cell considering single-diode model. J. Clean Energy Technol. **4**, 414–20 (2016)
6. Danoune, M.B., Djafour, A., Gougui, A, Khelfaoui, N., Boutli, H.: Characterization of photovoltaic panel using single diode and double diode models a comparative study with experimental validation. In: The 5th International Seminar on New and Renewable Energies, Gharda\"\ia–Algeria, vol. 24, p. 25 (2018)
7. Khanna, V., Das, B.K., Bisht, D., Singh, P.K.: A three diode model for industrial solar cells and estimation of solar cell parameters using PSO algorithm. Renew. Energy **78**, 105–13 (2015). https://doi.org/10.1016/j.renene.2014.12.072

8. Ibrahim, I.A., Hossain, M.J., Duck, B.C., Nadarajah, M.: An improved wind driven optimization algorithm for parameters identification of a triple-diode photovoltaic cell model. Energy Convers. Manag. **213**, 112872 (2020). https://doi.org/10.1016/j.enconman.2020.112872
9. Gu, Z., Xiong, G., Fu, X.: Parameter extraction of solar photovoltaic cell and module models with metaheuristic algorithms: a review. Sustainability **15**, 3312 (2023). https://doi.org/10.3390/su15043312
10. Merchaoui, M., Sakly, A., Mimouni, M.F.: Particle swarm optimisation with adaptive mutation strategy for photovoltaic solar cell/module parameter extraction. Energy Convers. Manag. **175**, 151–63 (2018). https://doi.org/10.1016/j.enconman.2018.08.081
11. Ridha, H.M., Heidari, A.A., Wang, M., Chen, H.: Boosted mutation-based Harris hawks optimizer for parameters identification of single-diode solar cell models. Energy Convers. Manag. **209**, 112660 (2020). https://doi.org/10.1016/j.enconman.2020.112660
12. Fathy, A., Rezk, H.: Parameter estimation of photovoltaic system using imperialist competitive algorithm. Renew. Energy **111**, 307–20 (2017). https://doi.org/10.1016/j.renene.2017.04.014
13. Fathy, A., Elaziz, M.A., Sayed, E.T., Olabi, A.G., Rezk, H.: Optimal parameter identification of triple-junction photovoltaic panel based on enhanced moth search algorithm. Energy **188**, 116025 (2019). https://doi.org/10.1016/j.energy.2019.116025
14. Rezk, H., Babu, T.S., Al-Dhaifallah, M., Ziedan, H.A.: A robust parameter estimation approach based on stochastic fractal search optimization algorithm applied to solar PV parameters. Energy Rep. **7**, 620–40 (2021). https://doi.org/10.1016/j.egyr.2021.01.024
15. Oliva, D., Abd El Aziz, M., Ella Hassanien, A.: Parameter estimation of photovoltaic cells using an improved chaotic whale optimization algorithm. Appl. Energy **200**, 141–54 (2017). https://doi.org/10.1016/j.apenergy.2017.05.029.
16. Qais, M.H., Hasanien, H.M., Alghuwainem, S.: Parameters extraction of three-diode photovoltaic model using computation and Harris Hawks optimization. Energy **195**, 117040 (2020). https://doi.org/10.1016/j.energy.2020.117040
17. Mirjalili, S.: Genetic Algorithm, p. 43–55 (2019). https://doi.org/10.1007/978-3-319-93025-1_4
18. Holland, J.H.: Genetic algorithms. Sci. Am. **267**, 66–72 (1992). https://doi.org/10.1038/scientificamerican0792-66
19. Kennedy, J., Eberhart, R.: Particle swarm optimization. In: Proceedings of ICNN 1995 - International Conference on Neural Networks, IEEE; n.d., pp. 1942–1948 (1945). https://doi.org/10.1109/ICNN.1995.488968.
20. Mirjalili, S.: Particle swarm optimisation, pp. 15–31 (2019). https://doi.org/10.1007/978-3-319-93025-1_2.
21. Kachitvichyanukul, V.: Comparison of three evolutionary algorithms: GA, PSO, and DE. Industr. Eng. Manag. Syst. **11**, 215–23 (2012). https://doi.org/10.7232/iems.2012.11.3.215

Experimental Investigation into the Effect of Building Orientation on Its Thermal Behavior

Abderrahmane Khelfaoui[1(✉)], Abdeldjalil Dahbi[1,2], Miloud Benmedjahed[1], and Boudjemaa Tidjar[1]

[1] Unité de recherche en énergies renouvelables en milieu saharien (URERMS), centre de développement des énergies renouvelables (CDER), 01000 Adrar, Algeria
dihe.khelfaoui59@gmail.com

[2] Laboratoire de Développement Durable et Informatique (LDDI), Adrar University, Adrar, Algeria

Abstract. Energy consumption in the building sector, which is an energy-intensive sector, leads us to deploy a lot of effort in energy efficiency and the rationalization of energy through the use of renewable energies. Several parameters come into play in the thermal behavior of the building, such as insulation, ventilation, orientation, etc. This work concerns the study of the effect of the building orientation on its thermal behavior; a field investigation (temperature reading, solar radiation) is undertaken to evaluate the overall quantitative response. To synthesize the numerical results of the experiment we use the mathematical entity which is the standard deviation of the temperature difference across the walls of the building such as the temperature gradient across the walls, roof and floor. The best location and orientation provide fundamental benefits to building designers in terms of energy efficiency and thermal comfort.

Keywords: Solar radiation · energy efficiency · thermal insulation · building thermal · thermal comfort

1 Introduction

The Due to extensive industrialization, energy consumption has substantially expanded during the previous century. Energy demand forecasts for the foreseeable future confirm and even magnify this tendency, particularly in light of population shifts and the development of some geographical regions, particularly in Asia. On the other hand, traditional energy resource deposits, which are primarily of fossil origin, can only be utilized for a limited period of time, signaling an impending worldwide energy deficit. However, nuclear power plant waste raises additional issues related to radioactive waste contamination, the impending decommissioning of ageing power stations, and industrial dangers [1–3]. Finding the right answers and diversifying them are required to meet Algeria's energy needs. There are now two actionable strategies. The first is to raise the productivity of power plants while decreasing the consumption of energy receivers by boosting each one's efficiency. The second method is to look for and create new energy

© The Author(s), under exclusive license to Springer Nature Switzerland AG 2025
M. Hatti (Ed.): IC-AIRES 2024, LNNS 1238, pp. 217–225, 2025.
https://doi.org/10.1007/978-3-031-80301-7_24

sources [4]. We will soon have access to cleaner and more accessible renewable energy sources like solar and wind energy [5]. But given that these technologies have long been disregarded, further study and development are needed to improve their dependability, reduce their costs (of production, usage, and recycling), and boost their energy efficiency [6–10]. Guidance for solar panels must take into account declining measurement. Many works have addressed the influence of orientation, construction materials in the building [11–13] and the performance of Trombe walls [14, 15]. In this article we first propose to study experimentally the effect of the orientation of the building on its thermal behavior before proceeding with its insulation to improve its comfort.

2 Metodology

The trailer was subject to a programmed move which allowed us to study the effect of orientation on its thermal behavior. An investigation of measuring interior and exterior temperatures in the initial and final position of the trailer was undertaken to evaluate the results. We used the standard deviation of the temperature differences of each wall as a comparison criterion. The standard deviation of each entity is calculated according to the following Eq. 1:

$$\sigma = \sqrt{\frac{\sum_{i=1}^{n} |x_i - \mu|^2}{n}} \tag{1}$$

$\sum$: means "sum of" xi: designates a value from statistical series

μ: Arithmetic mean of the series, n: number of statistical series

The standard deviation is used to determine the dispersion of the data in a statistical series compared to the mean. Figure 1 below represents a general view of the trailer studied with its actual dimensions.

3 Result and Mesures

3.1 Initial Orientation of the Trailer

During the preparation of the test bench of the trailer subject to improvement of its thermal comfort, its position was identified in relation to the Earth's poles as shown in Figure 2 below:

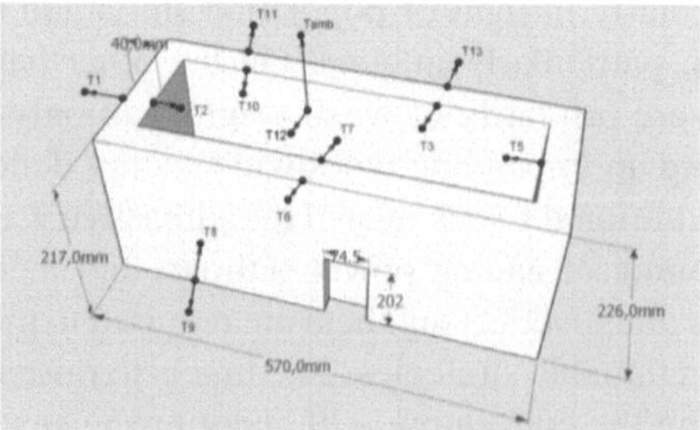

Fig. 1. General view of the trailer object of study

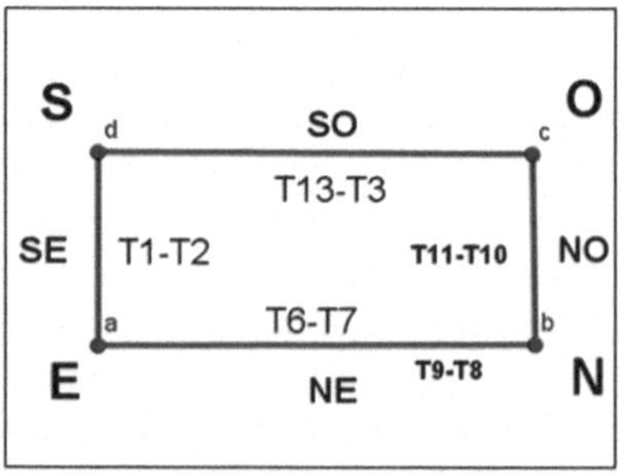

Fig. 2. Initial orientation of the trailer

The results of measuring the exterior and interior temperatures of the different walls constituting the trailer (walls, floor and roof) during the period from 20/02/2023 to 01/03/2023 were thus collected; we summarize the typical ecarts of the quantities T1-T2, T6-T7, T9-T8, T10-T11 and T13-T3 in Table 1 below:

Table 1. Standard deviation of temperature differences between walls in the trailer's initial position

Temperature difference	T1-T2	T6-T7	T9-T8	T11-T10	T13-T3	T5-T5′
Standard deviation 1	6.85	4.36	3.03	14.41	5.96	/
Orientation	SE	NE	FLOOR	ROOF	SO	NO
Rank, Orientation	2, SE	4, NE	5, F	1, R	3, SO	/

N/B: The wall corresponding to number 5 in Fig. 1 was not measured because the adjacent wall is air-conditioned. The heat flow per unit area transmitted by conduction through a wall is given by (Fourier's law):

$$\varphi = \frac{\lambda}{e} \mathrm{x}\,(\mathrm{T1} - \mathrm{T2}) \tag{2}$$

where the quantities λ, e, T1 and T2 designate the thermal conductivity, the thickness of the wall, the hot and cold temperatures respectively.

The surface heat flux is therefore proportional to the temperature difference through a wall and the thickness of the wall being kept constant.

In what follows, we study the temperature difference through the walls as a function of local time and their standard deviations as shown in Fig. 3 below.

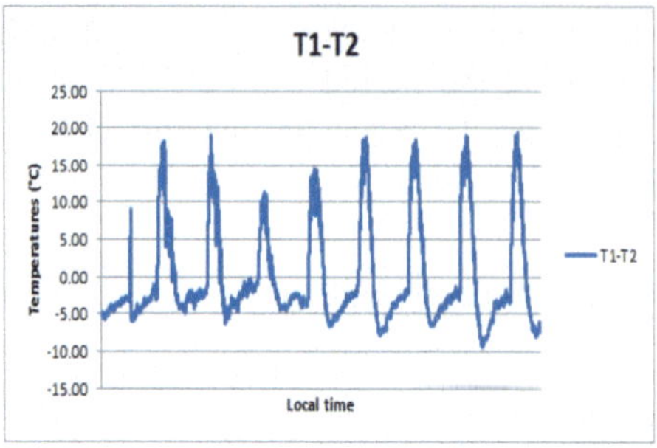

(a1)

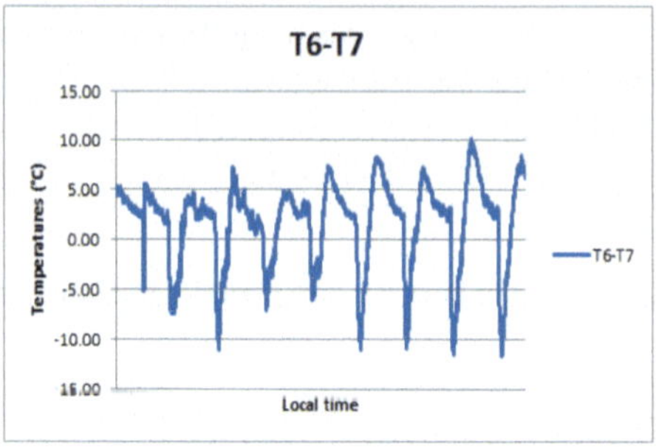

(b1)

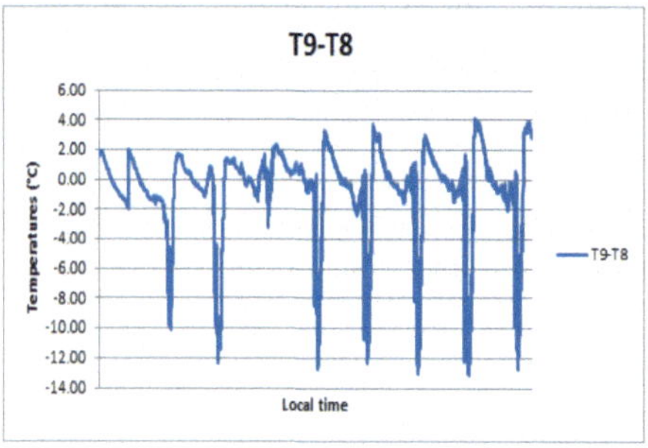

(c1)

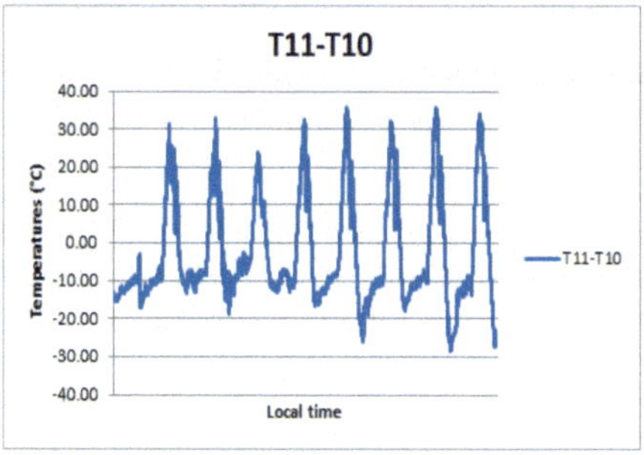

(d1)

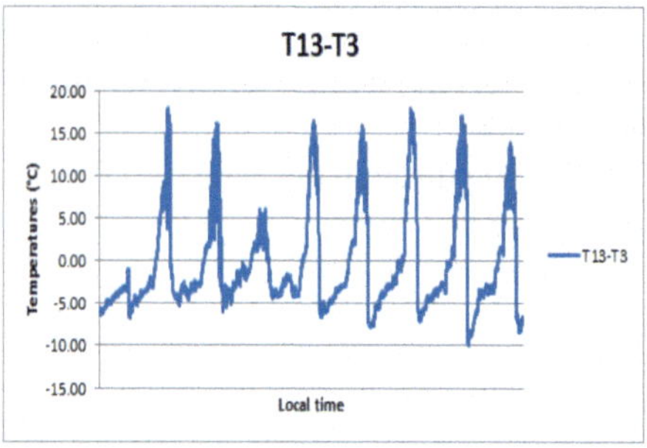

(e1)

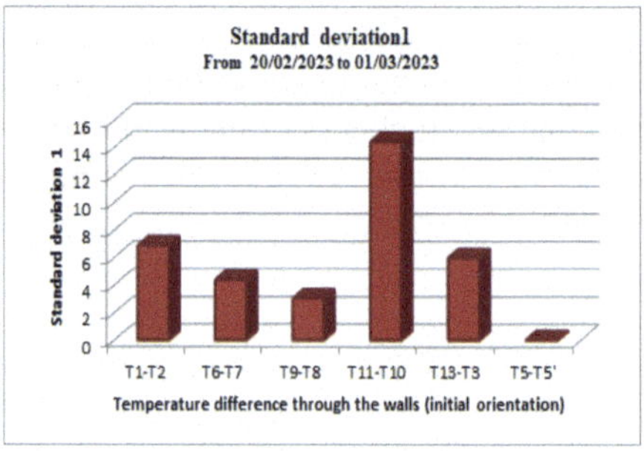

(f1)

Fig. 3. Wall temperature differences (a1, b1, c1, d1, e1) and standard deviations (f1) at initial orientation.

3.2 Final Orientation of the Trailer

When moving the trailer with a rotation of 90° counterclockwise (final position), the same operation was thus repeated from the date of 03/04/2023 to 13/ 04/2023 Layout (Fig. 4).

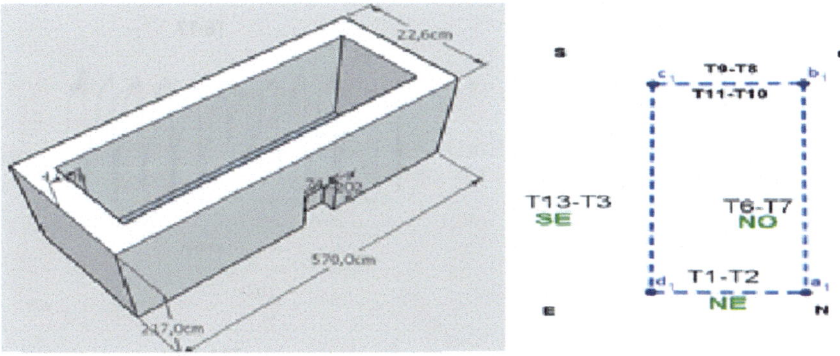

Fig. 4. Final orientation of the trailer

The results are collected and summarized in Table 2 below:

Table 2. Standard deviation of temperature differences between walls in the trailer's final position

Temperature difference	T1-T2	T6-T7	T9-T8	T11- T10	T13- T3	T5-T5'
Orientation	NE	NO	FLOOR	ROOF	SE	SO
Standard deviation **2**	3.61	4.59	1.24	7.95	5.53	/
Rank, Orientation	4, NE	3, NO	5, F	1, R	2, SE	/

The temperature difference between walls as a function of time and their standard deviations are shown in Fig. 5.

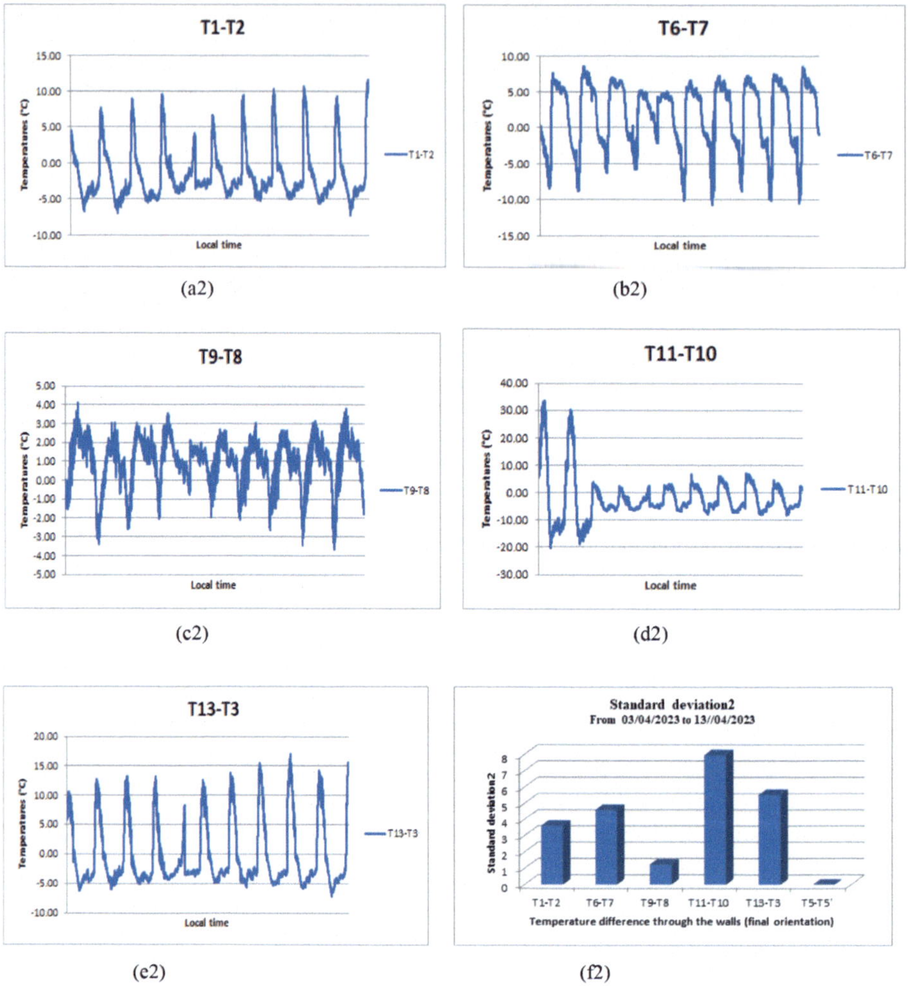

Fig. 5. Wall temperature differences (a2, b2, c2, d2, e2) and standard deviations (f2) at final orientation

3.3 Comparison of Final and Initial Orientations

The standard deviation of the following quantities T1-T2, T6-T7, T9-T8, T11-T10, T13-T3is given in Table 3 below:

Table 3. Standard deviation of temperature differences between walls in the trailer's at initial and final position.

Temperature Difference	T1-T2	T6-T7	T9-T8	T11-T10	T13-T3	T5-T5'
Standard deviation 1	6.85	4.36	3.03	14.41	5.96	/
Standard deviation 2	3.61	4.59	1.24	7.95	5.53	/

We give a graphical representation of Table 3 in Fig. 6 below in the form of histogram.

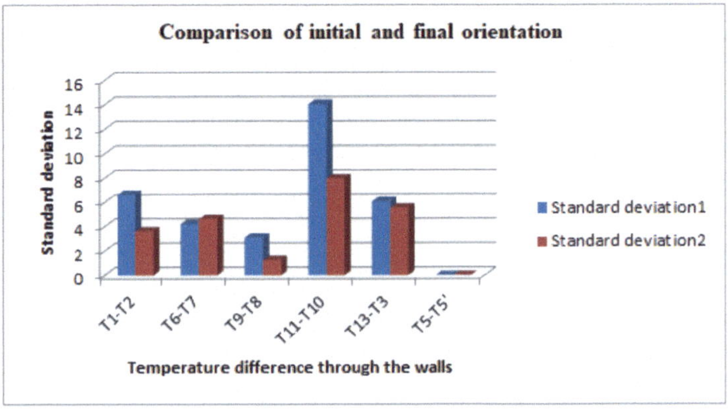

Fig. 6. Standard deviation at initial and final position of the trailer

4 Discussion

Analysis of the results of the standard deviation of temperature differences through the walls (see Table 1), led us to conclude that the surface heat flux is much more evident and in the following decreasing order: roof (T11-T10), south-east (T1-T2), south-west (T13-T3), north-east (T6-T7) and floor (T9-T8) respectively with a standard deviation of 14.41; 6.85; 5.96; 4.36 and 3.03; this is due to the way these walls are exposed to the sun's rays.

Likewise in the final orientation of the trailer (see Table 2) the surface heat flow is much more obvious and in the following descending order: roof (T11-T10), south-east (T13-T3), north-west (T6-T7), north-east (T1-T2) and floor (T9-T8) respectively with a standard deviation of 7.95; 5.53; 4.59; 3.61 and 1.24.

The order in which the walls of the building receive the heat flow by conduction is seen hierarchically: T (roof), SE (south-east), SO (south-west), NE (north-east), and F (floor).

5 Conclusion

- From a thermal point of view, the orientation of the building plays a key role in the supply of energy from the sun's rays; therefore the south orientation is the most suitable if we want to benefit from more free energy.
- This experience, justified by real measurements taken through the walls of the building, allowed us to know which walls are the most exposed to the sun's rays in order to intervene judiciously in the phase of improving the thermal comfort of this building.
- The results show that taking into account the orientation criterion makes the building participate in a more thermally efficient and more energy efficient design.

Acknowledgements. With the collaboration of the ENERGARID simullA laboratory of Tahri Mohamed Bechar University (UTMB) This work was carried out in the thermal and thermodynamic conversion division, "Building-Integrated Solar Thermal Systems in Sahara Medium" section of the research unit in renewable energies in the Saharan environment Adrar, Algeria attached to the CDER renewable energy development center.

References

1. Benmedjahed, M.: Choix du site et optimisation du dimensionnement d'une installation éolienne dans le nord Algérien et son impact sur l'environnement. (PhD Thesis) (2014)
2. Benmedjahed, M., Dahbi, A., Khelfaoui, A., Hadidi, A., Mouhadjer, S., Djaafri, O.: Statistical examination of wind energy's contribution to carbon dioxide emissions reduction*: *Study: Adrar, located in Southern Algeria. In: 2023 International Conference on Advances in Electronics, Control and Communication Systems (ICAECCS) (pp. 1–5) (2023). https://doi.org/10.1109/icaeccs56710.2023.10105015
3. Dahbi, A., et al.: Sizing of a solar parking system connected to the grid in Adrar. In: Smart Energy Empowerment in Smart and Resilient Cities: Renewable Energy for Smart and Sustainable Cities (pp. 506–514). Springer (2020). https://doi.org/10.1007/978-3-030-37207-1_54
4. Dahbi, A., et al.: Control of permanent magnet synchronous machine using speed estimation. In: Advanced Computational Techniques for Renewable Energy Systems (pp. 814–819). Springer (2023). https://doi.org/10.1007/978-3-031-21216-1_83
5. Abderrahmane, K., Mohammed, T., Houari, A., Abdeldjalil, D.: Design and Realization of an Air Solar Heater and Thermal Measurement, Instrumentation Mesure Métrologie, vol. 18, no. 6, December, 2019, pp. 595–602, https://doi.org/10.18280/i2m.180612
6. Benmedjahed, M., Dahbi, A., Hadidi, A., Mouhadjer, S.: Temperature and wind distribution effects on wind energy production in Adrar Region (Southern Algeria). Planning **16**(8), 1473–1477 (2021). https://doi.org/10.18280/ijsdp.160808
7. Boudghene Stambouli, A.: Algerian renewable energy assessment: The challenge of sustainability. Energy Policy, At the Crossroads: Pathways Renew. Nucl. Energy Policy North Africa **39**(8), 4507–4519 (2011) https://doi.org/10.1016/j.enpol.2010.10.005
8. Dahbi, A., et al.: MPPT fuzzy logic control of a variable speed wind turbine. J. Renew. Energ. 73–80 (2022)
9. Hadidi, A., Saba, D., Benmedjahed, M., Blal, M.: An overview on the development of the irrigation system in the province of Adrar (Algeria). Arab. J. Geosci. **15**(9), 854. Springer (2022). https://doi.org/10.1007/s12517-022-10103-8

10. Sadok, M., Benyoucef, B., Benmedjahed, M.: Assessment of PV modules degradation based on performances and visual inspection in Algerian Sahara. Int. J. Renew. Energy Res. (IJRER) **6**(1), 106–116 (2016)
11. Hamdani, M., Bekkouche, S.M.E.A., Benouaz, T., Cherier, M.: K. A new modelling approach of a multizone building to assess the influence of building orientation in Saharan climate. February 2014.Thermal Science 2015, vol. 19, no. suppl. 2, pp. 591–601 (2015). https://doi.org/10.2298/TSCI131217026H
12. Cherier, M.K., Bekkouche, S.M.A., Benouaz, T., Hamdani, M., Benamrane, N., Halloufi, O.: Studies and choice of localbuilding materials for improving interior temperatures of a building located in Ghardaia region. Renew. Energy Sources Clean Technol. 235–242 (2013). https://doi.org/10.5593/sgem2013/bd4/s17.030
13. Bekkouche, S.M.A., Benouaz, T., Cherier, M.K., Hamdani, M., Benamrane, N., Yaiche, M.R.: Thermal resistances of local building materials and their effect upon the interior temperatures case of a building located in Ghardaïa region. Constr. Build. Mater. **52**, 59–70 (2014)
14. Cherier, M.K., Benouaz, T., Bekkouche et S.M.A., Hamdani, M.: Etude des Performances Thermiques d'un Mur Trombe ", 15èmes Journées Internationales de Thermique, Tlemcen les 24, 25 et 26 Septembre 2011
15. Imessad et, K., Belhamel, M.: Evaluation des Performances d'un Mur Trombe, Revue Des Energies Renouvelables: Valorisation 195–198 (1999)

Numerical Simulation of the Heat Exchange Surface Effect on the Performance of a Vortex Cooling Tower

A. L. Deghal Cheridi(✉), A. Dahia, A. Bouam, A. Dadda Khorsi, A. Hadjam, and A. Kentouche

Nuclear Research Centre of Birine, Djelfa, Algeria
al_cheridi@yahoo.fr, a-l.deghal@crnb.dz

Abstract. The Atmospheric Vortex Motor is a cutting-edge technology designed to meet our growing need for renewable and sustainable energy; it could be a promising future solution for large-scale production of clean and low-cost energy. This research aims to analyze and forecast vortex tower behavior using the Relap5 system code. A parametric study is also conducted to assess the impact of the heat exchange area on the vortex tower performance, aiming to identify optimal operating conditions and design by exploring a heat area ratio ranging from 0.1 to 5. The objective of this device is to enhance electrical energy generation by recovering energy previously released to the environment during cooling processes. The key data sought is the ideal location in the air stream with a suitable velocity for turbine installation. The tower model has been developed and validated through both numerical and experimental results, and it is well accepted. Findings show that this vortex tower model can produce air flow with a maximum speed of 5.5411 m/s at a height of 0.560 m from the base. As a result, a turbine can be installed in this location to optimize kinetic energy. Furthermore, the simulation demonstrated the importance of geometrical factors in improving vortex tower performance, revealing that increasing the heat surface ratio to 5 boosted the air velocity to a maximum of 5.6991 m/s. Hence, the Vortex Generator System could be a promising technique for generating electrical power.

Keywords: Vortex generation · Modeling and simulation · Relap5 · Performance analysis · Parametric study · Heat exchange area

1 Introduction

The rapid growth in population and economic development lead to a rising demand for energy to support a sustainable and modern lifestyle. Energy sources, suchas fossil fuels and renewable resources, are the raw materials or natural processes used to generate energy [1]. Fossil energy reserves, like oil and gas, are often hard to reach (e.g., deep beneath the oceans) and poorly distributed on the surface of our planet, and the pollution issues generated by the utilization of these sources. Additionally, the current reliance on nuclear energy requires advanced technology and can be dangerous, hence, this

© The Author(s), under exclusive license to Springer Nature Switzerland AG 2025
M. Hatti (Ed.): IC-AIRES 2024, LNNS 1238, pp. 226–234, 2025.
https://doi.org/10.1007/978-3-031-80301-7_25

reasons lead to finding other sources of energy [2]. In response to energy challenges and global ecological threats, renewable energy is becoming increasingly significant. Several methods exist for harnessing energy, the atmospheric vortex motor (AVM) is one of the latest technologies which uses natural forces to provide renewable and sustainable energy [3]. The AVM is a power generator that utilizes a heat source, such as waste heat from industries or thermal and nuclear plants, to increase the air internal energy flowing through the system, thereby transforming the useful gain of the system into flow kinetic energy, which can be converted into electrical energy via an appropriate turbine [4]. Physically, the AVM is a complex system where different phenomena occur, with natural convection being the most important mechanism [5]. The operating principle of the AVM is based on the concept that air is heated from below and cooled from above, producing more mechanical energy in a manner similar to a solar chimney. The AVM concept was first introduced by Louis Michaud [6]. He stated that hot air works when it is transported to high altitude by convection, and that some of the hot air works by cooling in the during the ascending direction [7, 8]. In addition to experimental studies, computer simulations canbe used to model and simulate the installation's global behavior [9, 10]. The aim of this framework is to employ Relap5 system code to simulate a vortex tower and assess its viability for generating airflow. The tower under investigation is situated at the Birine Nuclear Research Center. By determining and observing the maximum air velocity, the optimal turbine placement, where kinetic energy is highest, can be identified. A parametric study is presented to assess how the heat transfer surface affects the vortex tower performance.

2 Vortex Tower Presentation

The investigated vortex tower is shown in Fig. 1. The study assesses the air behavior as it transfers heat from a hot source to the stack outlet. The tower is positioned away from obstacles to allow air to penetrate through the inlet openings. A cylindrical model was used in the initial test; it was suggested to add to a conventional tower an inlet that introduces a pre-rotating airflow after heat evacuation toward the chimney, aiming to accelerate the air. The vortex tower consists oftwo sections: a lower section with eight peripheral air inlet openings equipped with 08 vanes to artificially generate a convective vortex, and an upper section serves as a chimney and houses the speed anemometer support. The base is flared with raised concrete openings to ensure stability and allow air to be drawn in. Further details of the tower can be found in reference [11].

The density difference creates the air flow required to generate the vortex. This airflow moves across the hot source at the base of the tower, enters the second level tangentially, and continues toward the chimney. In our case, the warm spring is a concentric reservoir filled with water placed on three supports and warmed by electrical resistors maintained at a constant temperature. The spring contains three pipes: one for overflow, one for supply, and one for draining the tank. The heat is removed from the hot spring in two ways: firstly, by the passing the air from the bottom to the chimney outlet, and secondly, by eight inlets located around the tower base.

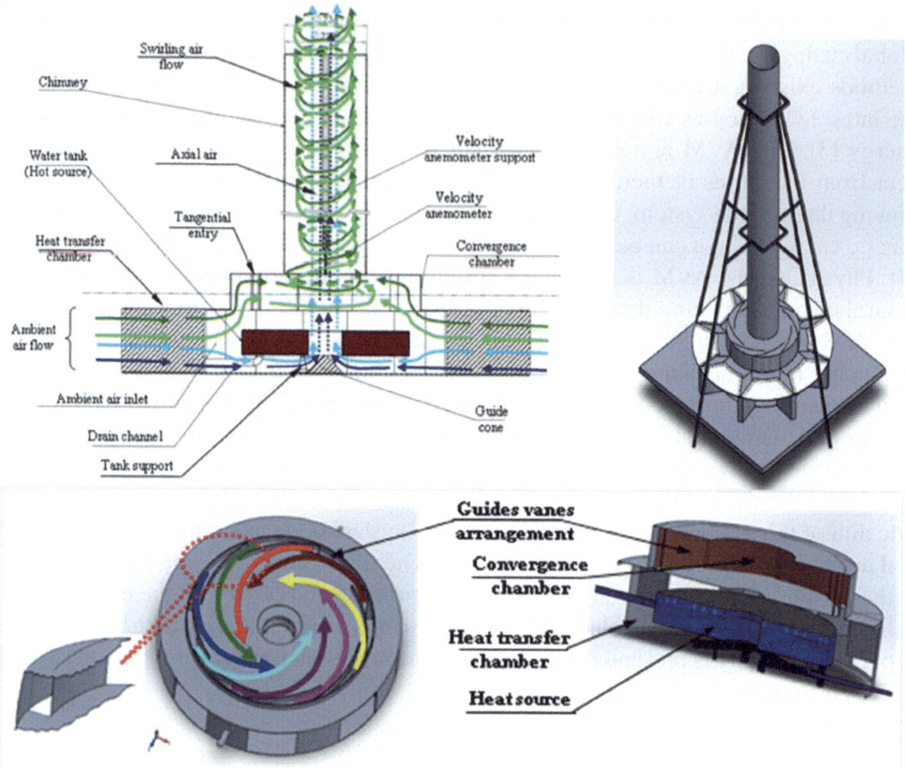

Fig. 1. Design diagram and airflow through the vortex tower model.

3 Vortex Tower Modeling and Nodalization

The vortex tower was modeled by subdividing itinto control volumes linked by flow connections [12], as shown in Fig 2. The air inlets are modeled by 08 Single-Vol components (501–508) which are connected to the Time- Dependent-Vols (401-408) to impose the weather conditions. The Brunch component 100 models the heat exchange chamber, while the Brunch component 200 models the vortex chamber. To prevent the air backflow, the 08 blades are modeled by 08 Single-Vol 601, 602, 603, 604, 605, 606, 607 and 608, which are connected to the vortex creation chamber by Check Valve components 201, 202, 203, 204, 205, 207 and 208.To model the chimney, the Pipe 300 is used and connected to the Time-Dependent-Vol component 500 to impose the weather conditions at the chimney outlet. The Brunch 700 represents the water tank. The thermal structures in the Relap5 code model were utilized to simulate the metal components thermal behavior making up the tower, specifically focusing on the heat exchange between air and water. These structures are connected to components 700 and 100.

The boundary conditions required to simulate airflow along the tower are shown in Table 1. To excite the natural circulation phenomenon, an initial air velocity was imposed at eight inlets at the tower's base. Ambient temperature and atmospheric pressure were also imposed at the tower inlet. The same conditions were applied downstream of the

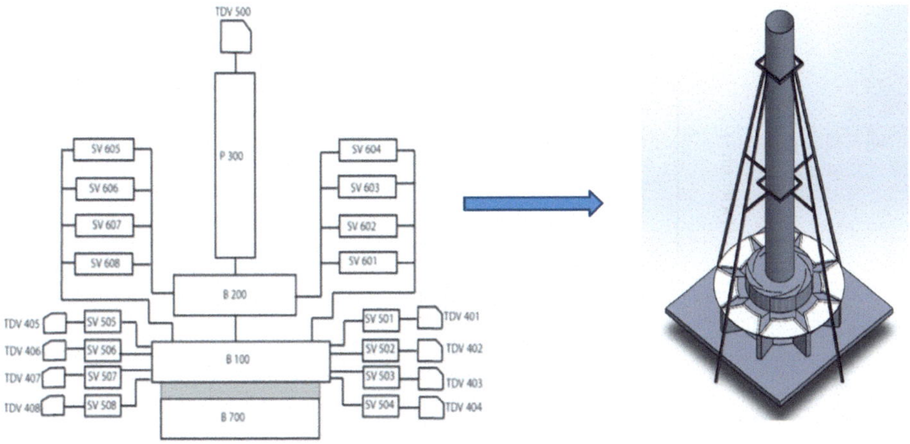

Fig. 2. Nodalization scheme of the vortex tower.

stack. The water temperature in the hot spring was set at 60°C. Adiabatic boundary conditions were used to eliminate heat exchange between the tower walls and the outside and inside air.

Table 1. Boundary conditions for simulation.

Parameters	Unit	Value
Inlet velocity: (Vi)	m/s	0.1
Heat source temperature: (TRe)	°C	60
Inlet air temperature: (Ti)	°C	25
Air inlet pressure: (Pi)	bar	Atmospheric pressure

4 Results and Discussion

The simulation aim objective is to study the tower kinematic behavior by examining the evolution of the air velocity from the tower inlet to the stack outlet. In addition, the impact of the heat transfer area on the tower's performance will be determined. Avortex tower model was developed, and the Relap5 code took 3000 seconds to reach a steady state. To visualize the steady state, we presented the evolution of the air velocity at different locations in the vortex tower. As shown in Fig. 3, the steady state is well established.

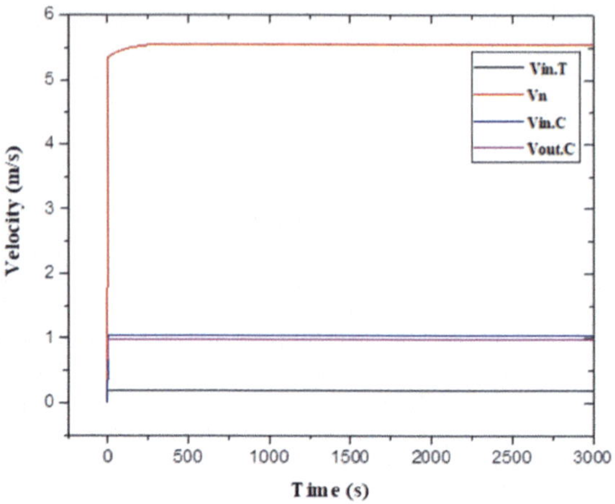

Fig. 3. Air velocity along the vortex tower in steady condition.

To validate the developed tower model, we have compared our numerical results with both the numerical and experimental findings from references [15, 16, 18], and [19]. Table 2 demonstrates that the Relap5 results align well with those reported in the literature.

Table 2. Validation of the elaborated tower model.

Reference	M.Ahmed& al. [15]	R.Balijepali& al [16]	Pritam Das & al [18]	Hadjam& al. [19]	Relap5
Max Velocity (m/s)	4.8	5.5	4.5	4.38	5.5411

4.1 Analysis of Vortex Tower Flow Parameters

Figure 4 illustrates the air velocity evolution as a function of the height ofthe vortex measured at several points: 0.28 m at the tower entrance, 0.56 m at the vortex creation chamber entrance, 0.863 m at the chimney entrance, and further along at 1.623, 2.383, 3.143, 3.903, 4.663, 5.423, 6.183, 6.943, 7.703, and 8.463 m, up to the chimney exit. The results indicate that the velocity increases rapidly from the tower inlet, rising to 5.5411 m/s at 0.56 m between the heat transfer chamber and the chamber creating the vortex. The airflow from the tower base to the stack shrinks, with a significant increase in velocity followed by a gradual decrease as the stack exit is approached. This change is induced by the conversion of kinetic energy into mechanical energy. Thus, at a height of 0.560 m, the maximum velocity of 5.5411 m/s has a velocity ratio of 55 with the inlet velocity of 0.1 m/s. As a result, this height is the optimum and most suitable location for the installation of a turbine.

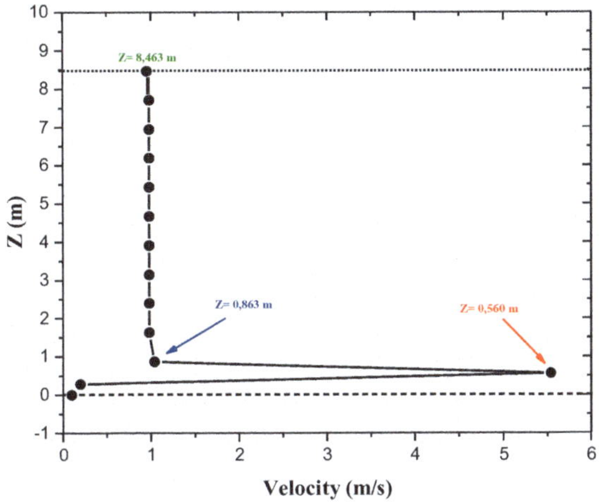

Fig. 4. Velocity variation along the tower.

Table 3 gives the numerical results of some parameters at the steady state.

Table 3. Relap5 steady state results.

Parameters	Unit	Value
Tower inlet temperature	°C	25.08
Chimney inlet temperature	°C	25.39
Chimney middle temperature	°C	25.38
Chimney outlet temperature	°C	25.38
Narrowing temperature	°C	25.39
Tank temperature	°C	35.64
Tower inlet velocity	m/s	0.198
Velocity at narrowing	m/s	5.5411
Chimney inlet velocity	m/s	1.039
Chimney outlet velocity	m/s	0.956

4.2 Effect of Heat Transfer Area

Air velocity is the key parameter that reflects the rotational flow [17]. In this section, we studied and examined the impact of the heat transfer area on tower performance, considering heat area ratios ranging from 0.1 to 5. Table 4 provides the data for the parameters used in this parametric investigation, along with the working conditions.

Figure 5 illustrates the impact of the heat exchange surface on air velocity at various locations. Numerical simulations were conducted for heat surface ratios of 0.1, 0.25, 0.5,

Table 4. Parametric study data.

T (°C)	D (m)	Vin (m/s)	*S ratio	Nb
25	0.720	0.1	0.1, 0.25, 0.5, 1, 1.5, 2, 3, 5	8

* Area ratio to total area

1, 1.5, 2, 3, and 5. The same inlet velocity of 0.1 m/s was applied to all eight (08) tower air inlets, while other parameters remained constant (see Table 4). The simulation aimed to assess the velocity profile. The results revealed that variation in the heat exchange surface affects the velocity, with the maximum observed at the same point, the chimney base (0.560 m). It was also observed that the heat exchange area significantly affects velocity distribution within the studied area, with air velocity increasing as the heated surface area inside the tower increases, resulting in increased heat transfer to the air circulating inside.

For surface ratios of 0.1, 0.25, 0.5, 1, 1.5, 2, 3, and 5, the highest velocity values are 4.3474 m/s, 4.8242 m/s, 5.3529 m/s, 5.5411 m/s, 5.5909 m/s, 5.6252 m/s, 5.6539 m/s, and 5.6991 m/s, respectively. As the heat transfer surface ratio increases from 0.1 to 5, the velocity at the chimney exit rises from 0.95589 m/s to 0.97048 m/s.

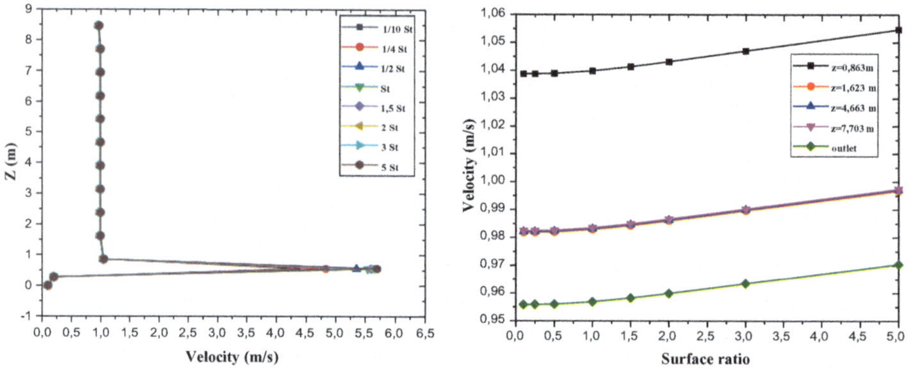

a. Velocity along the tower. b. Velocity at different points.

Fig. 5. Effect of heat transfer area on air velocity at several positions.

5 Conclusion

In this work, we employed the Relap5 code to conduct a steady-state investigation of the thermal-hydraulic characteristics of air flow within a vortex tower prototype. The tower model was developed and validated against both experimental and numerical data from several references, and a satisfactory agreement was foud. Flow parameters such as velocity and temperature were calculated and analyzed at various points along the tower during the simulation. The main objective of this device is to identify the highest velocity to determine the optimal location for turbine installation to generate energy. The results analysis revealed that the current vortex tower device is capable of producing an air flow, with a velocity ratio of 55 obtained between the maximum speed reached at the narrowing point at a height of 0.560 m and the speed at the entrance to the tower. Therefore, this position allows efficient conversion of kinetic energy into mechanical energy, which is subsequently converted into electrical energy. Furthermore, the effect of the heat exchange area on the vortex tower kinematic behaviour was investigated using the developed vortex tower model. This model provides an excellent starting point for other designs. The results showed that the heat transfer surface has a direct impact on the air velocity inside the tower. In fact, increasing the heat exchange surface improves the air velocity. The maximum air velocity increases from 4.3474 to 5.6991 m/s when the heat exchange surface ratio increases from 0.1 to 5.

References

1. Dhahri, A., Omri, A., Orfi, J.: Numerical Study of a solar chimney power plant. Res. J. Appl. Sci. Eng. Technol. **8**(8), 1953–1965 (2014)
2. Zuo, L., et al.: A vortex-type solar updraft power desalination integrated system. Energy Convers. Manag. **222**, 113216 (2020)
3. Semmani, K.: Simulation numériqued'unecheminéesolaire. Master thesis in Mechanical Engineering, Mouloud MAMMERI University, Tizi-Ouzou, Algeria (2013)
4. Michaud, L.M.: On the energy and control of atmospheric vortices. Journal de RecherchesAtmospheriques (1977)
5. Church, C.R., Snow, J.T., Baker, G.L., Agee, E.M.: Characteristics of tornado like vortices as a function of swirl ratio", A laboratory investigation, J. Atmosph. Sci. (1979)
6. Michaud, L.: Proposal for the use of a controlled tornado-like vortex to capture the mechanical energy produced in the atmosphere from solar energy. Bull. Amer. Meteor. Soc. **56**, 530–534 (1975)
7. Michaud, L.: Thermodynamic Cycle of the Atmospheric Upward Heat Convection Process. Meteor. Atmos. Phys. **72**, 29–46 (2000)
8. Michaud, L.: Heat to work convection during upward heat convection Part I: carnot engine method. Atmos. Res. **39**, 157–178 (1995)
9. DeghalCheridi, A.L., et al.: Modeling and simulation of a natural circulation water-tube steam boiler. SN Appl. Sci. A Springer Nat. J. 1, 1405 (2019). https://doi.org/10.1007/s42452-019-1452-x
10. Juslin, K.: A Companion Model Approach to Modelling and Simulation of Industrial Processes, Ph.D Thesis, Helsinki University of Technology, Finland, 23 (2005)
11. Deghal Cheridi, A.L.: Numerical investigation of a novel cooling vortex tower using Relap5 computer code. Nuclear Eng. Des. **391**, 111730 (2022)
12. IAEA-TECDOC-1395, Inter comparison and validation of computer codes for thermal-hydraulic safety analysis of heavy water reactors (2004)
13. RELAP5/Mo3.2 code manual volume: code structure, system models, and solution methods (1998)
14. RELAP5/MOD3.2 code manual volume: User's guidelines (1998)
15. Ahmed, M.R., Pate, S.K.: Computational and experimental studies on solar chimney power plants for power generation in pacific island countries. Energy Convers. Manage. **149**, 61–78 (2017)

16. Balijepali, R., Chandramohan, V.P., Kirankumar, K.: Ddevelopment of a small scale plant for a solar chimney power plant: a detailed fabrication procedure, experiments and performance parameters Renew. Energy (2019)https://doi.org/10.1016/j.renene.2019.12.001
17. Das, P., Chandramohan, V.P.: Estimation of flow parameters and power potential of solar vortex engine by varying its geometrical configuration: a numerical study. Energy Conver. Manag. **223**, 113272 (2020)
18. Das, P., Chandramohan, V.P.: CFD analysis on flow and performance parameters estimation of solar updraft tower (SUT) plant varying its geometrical configurations. Energy Sources Parts A: Recov. Utiliz. Environ. Effects (2018). https://doi.org/10.1080/15567036.2018.1477881
19. Hadjam, A., et al.: Numerical analysis of a small-scale novel vortex tower integrated with heat source for Nuclear Application. Nucl. Eng. Des. **414**, 112660 (2023)

Modelling and Understanding the Lateral Displacement of the Bottom Hole Assembly (BHA) Under Bending Vibrations

Nora Benmir[1], Hamza Akroum[1,2](✉), and Moahamed. Z. Doghman[3,4]

[1] Automation Applied Laboratory, Faculty of Hydrocarbons and Chemistry, M'Hamed Bougara University, Boumerdes, Algeria
{n.benmir,akroum}@univ-boumerdes.dz

[2] Systems Engineering and Telecommunications Laboratory, Faculty of Technology, M'Hamed Bougara University, Boumerdes, Algeria

[3] Exploration Operation Department: Exploration Division, SONATRACH, Hassi Messaoud, Ouargla, Algeria

[4] Bob L. Herd Department of Petroleum Engineering, Texas Tech University, Lubbock 79401, Texas, USA

Abstract. This technical article explores the behavior of drill string under bending vibration conditions, specifically targeting the unique challenges faced by the Bottom Hole Assembly (BHA). Undesirable oscillations during drilling operations, including axial, lateral, and torsional vibrations, can lead to complex and dynamic issues. The lateral vibration mode, with its multifaceted dynamics involving bending, buckling, and transverse motion, presents the most intricate challenges. Our research focuses on presenting three fundamental models to describe and predict drill string dynamics under bending effects. By simplifying the BHA as a cylindrical beam with fixed or rolling periodic touch points. In this article, we differentiate between three distinct cases: first, a static beam with fixed periodic touch points, excluding axial load influence; second, the inclusion of rolling touch points; and finally, fixed touch points under axial load, which will be the subject of further validation. Through simulations, we validate the novel contribution of this work, an ancient SLB expert's mathematical tool to effectively describe BHA dynamics under bending phenomena.

Keywords: Lateral vibrations · bending phenomenon · basics models · periodic touch points · vertical deflection · axial load

1 Introduction

Drilling operations are fraught with challenges, complexities can limits progress, and increase costs. The drilling system itself is intricate, with the drilling rig, including the mast and the top drive forming the heart of the operation. Land rigs though similar, have slightly different requirements, as they do not need to be as self-contained. The power and machinery involved in moving and rotating the drill string, as well as pumping fluid

© The Author(s), under exclusive license to Springer Nature Switzerland AG 2025
M. Hatti (Ed.): IC-AIRES 2024, LNNS 1238, pp. 235–243, 2025.
https://doi.org/10.1007/978-3-031-80301-7_26

are significant. Among key challenges faced by drilling operations is the emergence of unwanted vibrations specially, torsional, axial, and lateral modes. Lateral vibrations are the least understood and the most challenging to address. This is due two main reasons: the absence of visible signs on the surface and the complex dynamic of drill string operating with six degree of freedom, four of them are consumed by lateral motion. According to analysis effected by a former expert at Schlumberger, lateral movements of drill string can categorize in three phenomenon which are: bending, buckling and lateral dynamics [1]. This study focus on analyze the effect of the bending phenomenon on drill string, and aims to describe the dynamics of drill string using a modelling strategy based on the periodic contact between drill string and the walls of drilling well. By using a simple strategy of assuming touching points along the drill string, we can gain insights into how the lower part of the drill string behaves. In reality, there are no visible touch points, but this mathematical analysis technique allows to simplify the complex lateral dynamics. Three cases are explored: The effect of fixed touch points without axial compression or static beam, the impact of periodic rolling touch points, and finally the combined effect of touch points and axial load. The literature reviewing in first part of this study will discuss previous work on modeling, suppressing and mitigate bending vibrations in drilling systems. This is will be followed by a detailed analysis and modeling of the lower part of the drill sting using the effect of periodic touch points.

2 Mathematical Modeling of BHA Under Bending Effect: A Review

The bending behavior of drill strings or Bottom Hole Assemblies (BHA) has been a subject of extensive research, with various mathematical models proposed to explain and predict this complex dynamic phenomenon. Early field data from Shell Oil Company in 1984 confirmed the existence of parametric coupling between bending and axial vibration modes, sparking further investigation into this area. Another one notable contribution was made by Rong-Juin Shu in 1989 [2]. Shu examined the mechanisms causing bending vibrations in the BHA and proposed a finite difference scheme model that considered the effects of linear varying compression, mud, added mass, and damping forces. The model successfully predicted forward whirling and bending-stick-slip coupled vibrations but fell short in explaining backward whirling and the influence of axial loads on bending behavior. In 2012, Moradi et al. Investigated the suppression of transverse vibrations using a tunable vibration absorber. They employed the Euler-Bernoulli beam equation to describe the behavior of the drill string and proposed a control strategy to mitigate torsional-transverse coupled vibrations. While their approach showed effectiveness in this regard, it had less success in addressing axial- lateral coupled vibrations, indicating the need for more versatile modeling and control techniques [3]. Fesmi Abdul Majeed and colleagues conducted a detailed analysis of lateral vibrations using an unbalanced drill bit model [3]. Their novel approach successfully validated the bit whirl phenomenon, providing valuable insights into this particular aspect of BHA dynamics. However, their model's applicability to coupled vibration modes remained limited. More recently, in 2022, Valeriy M. Pestrenin et al. they employed a classical beam model to study the causes of bending vibrations in inclined and horizontal drilling wells. They considered the effects of rotation speed, gravity, and damping forces, providing a comprehensive understanding of BHA behavior in these challenging drilling scenarios [4].

3 Drilling Direction Mathematical Modeling: Bending Deflection Periodic Touch Points

The model used here is mathematically represented as a cylinder with either fixed or restricted periodic points.

The tool joints in drill collars with periodic stabilizers or drill pipes are shown by this simplified (Fig. 1). Although there are obviously no point contacts in reality, this is a very useful simplification for understanding lateral deflection. Three situations involving bending deflection and periodic touch points exist. First, in the situation when they are fixed and the pipe is not under any compression, or axial load. This translates to the pipe's vertical deflection on the low side of the hole. Next, rolling touch points also uncompressed pipe (horizontal deflection), and finally fixed periodic touch points with compressed pipe (axial load). We concentrated on the first two scenarios and examples in this section of the study [5].

3.1 Fixed Touch Points, Uncompressed Pipe (Vertical Deflection)

We started with the simplest case, which involved a pipe with fixed periodic touch points propagating on all of the cylinder beams and a constant cross-section. An idealized illustration of drill pipe resting on the side of a borehole at tool joints or BHA cutters with periodic touch contact points with wellbore walls represents in (Fig. 1).

The Lateral displacement is denoted by y, the deflection is expressed as the third derivative of Lateral displacement, which is disappeared in the touching points of this simple case, and the displacement between the contact sites, or pivots, follows a third-order polynomial:

$$y = a\left(1 - x^2\right) + \mathrm{bx}\left(1 - x^2\right) \tag{1}$$

y: lateral dicplacement. x: the distance.

Proposing that the contact points are in -1 and $+1$. At these, touch points, the displacement y concerning the coordinate x must disappear. Just as with propagating waves in periodic structures, we could write all solutions as a sum of periodic or exponential solutions, which we find by imposing these boundary conditions:

$$y'(-1) = \lambda.y'(1) \tag{2}$$

$$y''(-1) = \lambda.y''(1) \tag{3}$$

The solutions are exponential, not periodic, and involve a scale factor. This scale factor is represented by lambda (λ), which is expressed by:

$$\lambda = -2 \pm \surd 3 \tag{4}$$

The systeme has two degree of freedom, and there are two solutions one is inverse to author, and any solution (static deviation) can be built from these two solution.

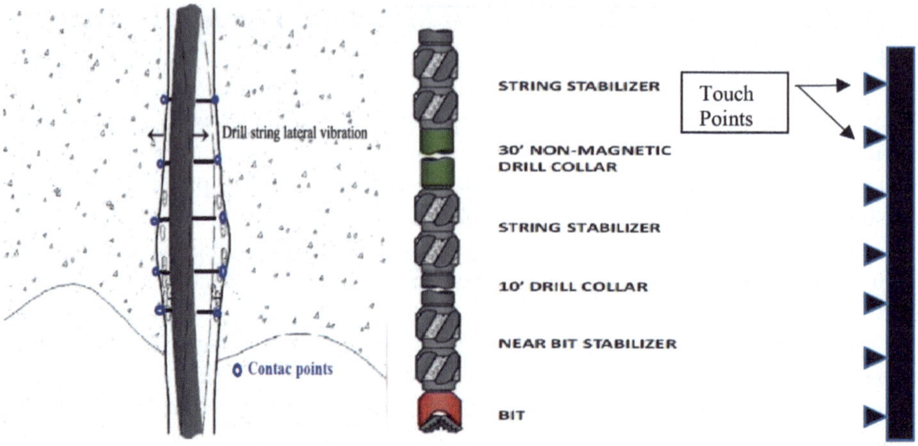

Fig. 1. Schematic of the contact between the drill-string and the wellbore wall (left, middle), Idealized illustration model of drill string (right).

Example and Simulation Validation (Vertical Deflection)

In this situation, we assumed the bending deflection at a point below the drill string and assigned it the value(−1), and the touched points are propagated by a step of (1).

We developed a program that replicates the evolution of lateral displacement beginning at the designated location. The end outcome was the Fig. 2.

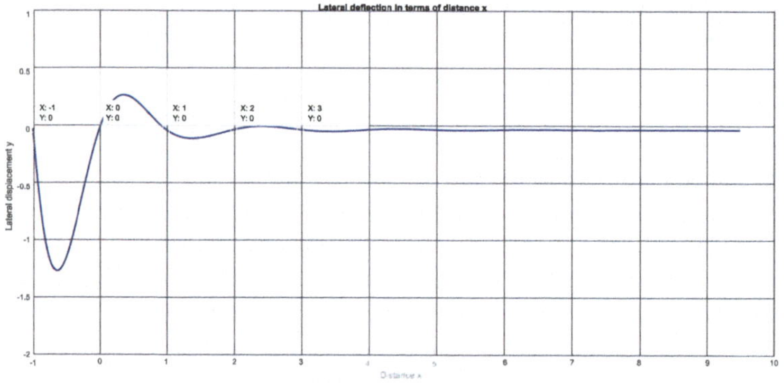

Fig. 2. Lateral deflection in terms of x (vertical deflection, no compression)

Discussion

The beam has been objected to a deflection at point $x = -1$, and the fixed touch points are chosen at x = (−1) and by a step equal to one. This choice of the specific periodic touch point location (−1, + 1) where the lateral deflection disappears is due of the periodicity in BHA structure (periodic stabilizers or drill pipes). The lateral deflection changes its sign from point to another one. This sign changement of lateral displacement may explained by the change in lateral direction waves direction during its traveling from the bottom of the hole into the surface. The deflection was reduced from point to another point by factor of $\frac{1}{(2+\sqrt{3})} \approx 4$. After three touch points, the lateral deflection almost disappeared. The observed reduction in lateral displacement, as shown in the previous figures, can be largely explained by the rule of three point [6]. This help us understand why the lateral vibrations mod does not manifest at the surface, due to the rapid attenuation of these vibrations as they travel from the bottom to the surface of the well prevents their detection at the surface.

3.2 Rolling Touch Points, Uncompressed Pipe (Horizontal Deflection)

When considering the vertical and horizontal movements of a pipe within a wellbore, the influence of gravity becomes a crucial factor. Vertically, for small forces, gravity guides the pipe to rest on the bottom of the hole. However, in the horizontal direction, the pipe exhibits a different behavior – it can roll as shown in (Fig. 3).

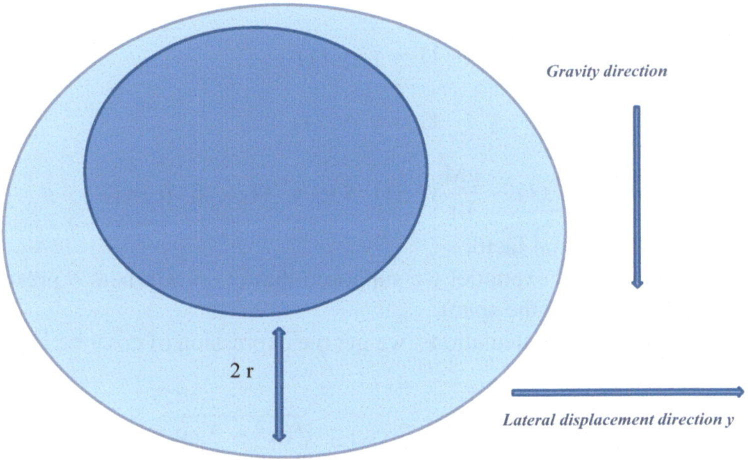

Fig. 3. Drill joint/ pipe horizontal cross-section

Considering the pipe's ability to roll around the hole, we introduced the rolling radius$r$, which refers to the path traced by the pipe during its rolling. The tool joint can modeled as point masses objected to a restoring force F: here is the steps of modelling:

The gravitational load exerted on the pipe span is concentrated at the joints, resulting in point force [7]. This concentration of force is defined as step change in the third derivative of the deflection as following:

$$F = -B\Delta \dddot{y} \tag{5}$$

The restoring force responsible for counteracting this deflection is directly proportional to the sine of the normalized deflection:

$$F = gM \sin\left(\frac{y}{r}\right) \approx gM\frac{y}{r} \tag{6}$$

By applying Linearization technique [8], we can apply this relationship as following equation:

$$F \approx gM\frac{y}{r} \tag{7}$$

As a result, the difference in the third derivative of the deflection y is directly proportional to the magnitude of the deflection itself as following:

$$\Delta y''' = \frac{gM}{Br}y \tag{9}$$

F: Total lateral force on the joint.
g: Gravity component.
M: Mass of each span.
r: Difference in radius between hole and joint.
B: Bending moment between joints (unite of length).
In order to solve the equation in 9, the following initial conditions are proposed:

$$\mathrm{y}(-1) = \gamma \cdot \mathrm{y}(1) \tag{10}$$

$$\mathrm{y}'(-1) = \gamma \cdot \mathrm{y}'(1) \tag{11}$$

$$\mathrm{y}''(-1) = \gamma \cdot \mathrm{y}''(1) \tag{12}$$

$$\mathrm{y}'''^{(-1)} = \gamma \cdot \mathrm{y}'''(1) - \frac{\mathrm{gM}}{\mathrm{Br}}\mathrm{y}(-1) = \gamma \cdot \mathrm{y}'''(1) - 6 \cdot \mu \cdot \mathrm{y}(-1) \tag{13}$$

γ: Scale factor, μ: Proportional factor.

The solutions will be an expontiel we supposed that $e^{i\theta} = \gamma$, where θ presents the angle of deflection (rolling of the span).

After applying analytical calculations, we get the expression of $\cos\theta = \frac{1}{2}(\gamma + \gamma^{-1})$ in terms of μ as following:

$$\frac{1}{2}(\gamma + \gamma^{-1}) = (1 - 2 \cdot \mu) \mp \sqrt{(\mu^2 - 3 \cdot \mu)} \tag{14}$$

Regarding of the value of μ this expression involved four solutions it means four degree of freedom.

Simulation and Validation

We developed a program that calculate the amplitude and the phase of $\frac{1}{2}(\gamma + \gamma^{-1})$ in terms of $\frac{\mu}{1+\mu}$, Here are the resulting Figs. 4 and 5.

Discussion

At the point $\mu = 0$ the phase is equal zero in the real case explained that there is no deflection at this location, unconstrained pipe or joints.

When μ takes a large values, the periodic touch points retune to the case Fixed touch points where phase $= -2$ where there is no deflection(Static bending), for the second case θ tends to the infinity so this explained by an infinite deflection from spans to another. For $\mu = 0.5$, cos $\theta$is pure imaginary, and $\theta = \frac{\pi}{2}$, i.e. 90-degree phase shift in deflection from one span to the next one.

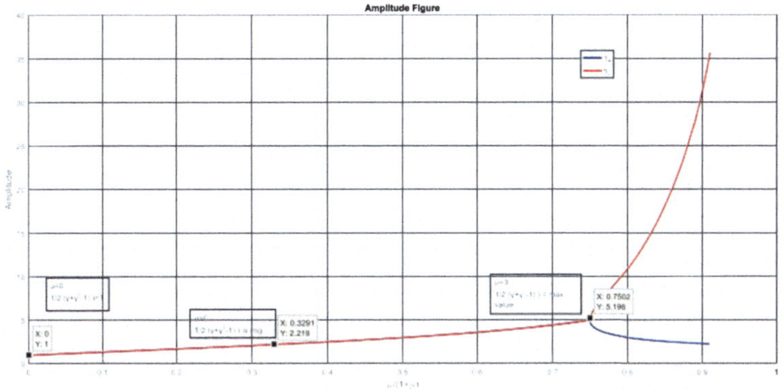

Fig. 4. Amplitude figure of the lateral deflection in the case of rolling touch points

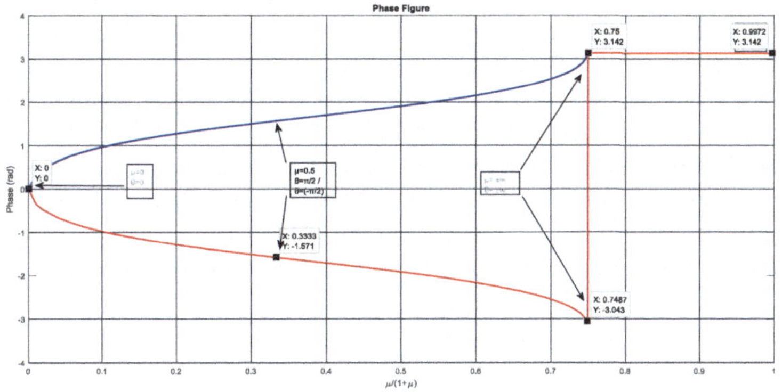

Fig. 5. The phase figure of the lateral deflection in the case of rolling touch points

3.3 Comparison Results Between Periodic Fixed and Periodic Rolling Points (Table 1)

Table 1. A Comparative Analysis of Beam Deflection: Fixed versus Rolling Touch Points

Cases/ results and conditions	Periodicity	Contact with the beam	Gravity factor	Axial load	Deflection decay
Fixed touch points	Periodic points	Fixed on the beam	Not considered	No axial load	Significant
Rolling touch points	Periodic points	Rotating with the beam	Considered	No axial load	Depends on various factors

The previous table summarizes and compiles the results and conclusions between the two studied cases:fixed and rolling periodic touch points taken into account the gravity and the axial load factors effects on the obtained results.

4 Conclusion

This study focuses on understanding and modeling the lateral dynamics and bending vibrations of a Bottom Hole Assembly (BHA) in drill strings. The lateral mode of vibration in drilling operations is particularly challenging due to its complex dynamics and potential for damage. By treating the BHA as a simple cylindrical beam with periodic touch points, three distinct cases are analyzed to describe and predict the drill string dynamics under bending effects. The novel mathematical approach validates the effectiveness of this method in describing BHA dynamics, under bending phenomena. The choice of touch points is critical in controlling lateral deflection, and periodicity in the BHA structure plays a significant role. The study also highlights the rapid attenuation of lateral vibrations as they travel from the bottom to the surface of the well, making their detection challenging. Additionally, the phase relationship between deflection and the touch point periodicity provides insights into the behavior of the BHA, with different phases indicating varying degrees of deflection or bending. Our analysis reveals a significant reduction in vertical displacements between touch points, with a decrease of almost four times from one span to the next. The behavior of horizontal displacements is more complex and dependent on various parameters, including the angle and size of the hole.

References

1. Osogba, O.J., Otamere, B., Abdullahi, G.S.: Analysis of vibrations in drilling systems. Accessed: Aug. 07, 2024. [Online]. Available: https://doi.org/10.2118/193397-MS
2. Shyu, R.-J.: Bending vibration of rotating drill strings. Thesis, Massachusetts Institute of Technology, 1989. Accessed: May 07, 2024. [Online]. Available: https://dspace.mit.edu/handle/1721.1/14205
3. Moradi, H., Moradi, H., Bakhtiari-Nejad, F., Bakhtiari-Nejad, F., Sadighi, M., Sadighi, M.: Suppression of bending vibration of drill-strings via an adjustable vibration absorber. Int. J. Acoust. Vib., **17**(3) (Jan 2012). https://doi.org/10.20855/ijav.2012.17.3308
4. Majeed, F.A., Majeed, F.A., Majeed, F.A., Majeed, F.A.: Analysis and control of rotary drilling rigs (Jan 2013)
5. Pestrenin, V.M., Pestrenina, I.V., Perelman, O.M., Fadeykin, A.S., Derkach, N.D.: Bending vibrations of a heavy horizontal composite rotating shaft limited by a cylindrical baffle. Vestn. Tomsk. Gos. Univ. Mat. Mekhanika **79**, 135–151 (2022). https://doi.org/10.17223/19988621/79/12
6. "Greatest Hits," Benjaminnovation. Accessed: May 28, 2024. [Online]. Available: https://www.benjaminnovation.com/work/greatest-hits
7. Keshavarzzadeh, V., Kirby, R.M., Narayan, A.: Generation of nested quadrature rules for generic weight functions via numerical optimization: application to sparse grids. J. Comput. Phys. **400**, 1089791 (2020). https://doi.org/10.1016/j.jcp.2019.108979

8. Capecelatro, J., Desjardins, O.: Eulerian–Lagrangian modeling of turbulent liquid–solid slurries in horizontal pipes. Int. J. Multiph. Flow, **55**, 64–79 (2013). https://doi.org/10.1016/j.ijmultiphaseflow.2013.04.006
9. Meleshko, S.V.: Methods for Constructing Exact Solutions of Partial Differential Equations: Mathematical and Analytical Techniques with Applications to Engineering. Springer Science and Business Media (2006)

Modeling of a Coupled Severe Axial-Torsional Vibrations in Drill String Petroleum System

S. Meddah[1,2](✉), A. Idir[1,3], M. Z. Doghmane[1,4], S. A. Tadjer[5], and M. Kidouche[1]

[1] Applied Automation Laboratory, F.H.C, University of Boumerdes, 35000 Boumerdes, Algeria
s.meddah@univ-boumerdes.com

[2] Institute of Electrical and Electronic Engineering (IGEE), University of Boumerdes, Boumerdes, Algeria

[3] Electrical Engineering Department, Mohamed Boudiaf University of M'sila, M'sila, Algeria

[4] Reservoirs Evaluation Department, OPE, Exploration Division, SONATRACH, 30500 Hassi Messaoud, Algeria

[5] Laboratory for the Electrification of Industrial Enterprises, University of Boumerdes, Boumerdes, Algeria

Abstract. The drilling industry faces a complex challenge in modeling dynamic vibrations during drilling because it directly affects drilling efficiency, tool longevity, and overall operational safety. The emphasis on creating reliable control systems that can either completely eliminate or significantly reduce these vibrations' negative effects has increased as a result of the necessity to understand and manage them. Thus, developing a thorough grasp of the drill bit's response to these dynamic forces is an essential first step in creating efficient control schemes. Numerous studies in the literature have previously examined the intricacies of the drill string's behavior when subjected to axial-torsional vibrations. Numerous models have been put forth in these studies to capture the complex interaction between axial and torsional forces. But incorporating such control mechanisms calls for a more thorough investigation of the ways in which the drill string's dynamic behavior and the control systems interact, especially in relation to the axial-torsional effects. Within this framework, our work aims to advance current understanding by providing a thorough model of our system that takes into account the particular subtleties of torsional and axial vibrations. Our main goal is to understand the complex interrelationship between these two vibration types and how they affect the drilling process as a whole. By dissecting this relationship, we hope to reduce the harmful effects of the coupled vibrations and increase the control system's robustness and effectiveness. We think that this research has a lot of potential to improve our knowledge of and ability to control dynamic vibrations that occur during drilling operations, which will ultimately increase the overall effectiveness and safety of drilling procedures.

Keywords: Dynamic vibrations · Drill string behavior · Axial-torsional effects · Control mechanisms · Drilling efficiency · Operational safety

© The Author(s), under exclusive license to Springer Nature Switzerland AG 2025
M. Hatti (Ed.): IC-AIRES 2024, LNNS 1238, pp. 244–253, 2025.
https://doi.org/10.1007/978-3-031-80301-7_27

1 Introduction

Current energy sources include oil and gas, which are typically extracted from subterranean reservoirs through targeted drilling techniques. The primary method used in this process is rotary drilling, which consists of a rotating mechanism driven by a top drive and connected to a bottom hole drill bit via a drill string [1].

Numerous academics have studied the different kinds of vibrations that come with drilling oil wells. Axial vibration first surfaced as a major issue, lining up with the drilling penetration axis and having a substantial impact on the Rate of Penetration (ROP) [2]. Intermittent contact between the drill bit and the rock formation being drilled is another serious problem that is referred to as bit-bounce [3]. Axial vibrations may be exacerbated by lateral vibrations, which mainly appear at the Bottom Hole Assembly (BHA) level and produce the swirling effect without transmitting axially to the surface. In addition, torsional vibrations caused by the elasticity of the drill string and the complex torque on bit interaction frequently cause the stick-slip phenomenon. This creates a series of problems in the surface and subsurface parts of rotary drilling systems that ultimately raise drilling costs and reduce ROP.

Prior research examined vibrations in rotary drilling systems in great detail and provided real-time control measures, but it did not address the effect of axial vibrations on stick-slip severity [5, 6]. Furthermore, little study has been done on the important interaction between axial-torsional vibrations and the robustness of the controllers designed to lessen these vibrations [7].

2 Drill String Model

2.1 System Design

One of the most commonly used methods in the petroleum industry is the rotary drilling system, which is mainly used to drill holes from the surface to reach reservoirs. The drill bit is connected to the top drive, which controls the drilling tool's rotation, through a drill string, which is a network of interconnected drill pipes. This system is composed of two main components [8]: the surface-based drilling rig, which has a derrick or mast for hoisting equipment, and a floor that makes it easier to manipulate rods, thread tools, adjust them, and replace them. Contrary, the underground portion includes the drill string, which is divided into two parts: a lower part that includes the Bottom Hole Assembly (BHA) and an upper part that consists of a series of rods [9]. The drilling tool, the drill collar, and the stabilizers are the three main components of the BHA itself [10].

When drilling, different kinds of vibrations can occur one after the other or separately. In either case, the equipment will deteriorate and may malfunction. Axial, lateral, and torsional vibrations are the three main categories into which these vibrations are usually divided according to the directions in which they propagate [11].

Moreover, these vibrations could occur in combined modes like lateral-axial-torsional, axial-torsional, or torsional-lateral modes [12]. Given their frequency and the significant damage they can cause to rotary drilling systems, special attention is paid to the interconnected axial-torsional vibrations for the purposes of this study [13, 14].

2.2 Axial-Torsional Vibrations

An example of how to create a 3-degree-of-freedom (3-DOF) mass-damping model for coupled axial-torsional vibrations is shown in Fig. 1.

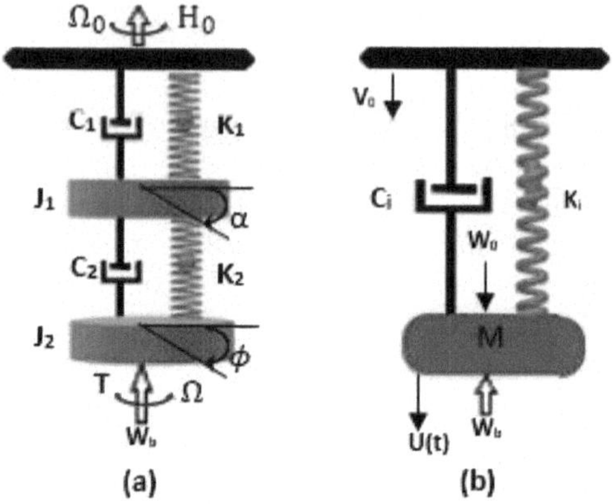

Fig. 1. Rotary drilling system equivalent dynamic for: a) Torsional Vibrations,b) Axial Vibrations.

In this case, torsional and axial vibrations are shown in (a) and (b), respectively [14]. A drillstring with an axial displacement U, a torsional angle, and a bending angle make up the system [15, 16]. The system's equations of motion can be obtained as follows eq. (1):

$$\begin{cases} J_1\ddot{\alpha} + c_1(\dot{\alpha} - \Omega_0) + k_1(\alpha - \Omega_0 t) + c_2(\dot{\alpha} - \dot{\phi}) + k_2(\alpha - \phi) = 0 \\ J_2\ddot{\phi} + c_2(\dot{\phi} - \dot{\alpha}) + k_2(\phi - \alpha) = -T_b \\ M\ddot{U} + c_i\dot{U} + k_i(U - V_0 t) = w_0 - w_b \end{cases} \tag{1}$$

The system parameters are displayed in the Table 1 below:

Table 1. List of parameters

Symbol	Description
V_0	constant axial velocity
Ω_0	constant torsional velocity
J_1	The moment of inertia of the drill string
J_2	The sum of the moment of inertia of the drill string and the moment of inertia of the drill bit
c_i	Axial damping
c_1, c_2	torsional damping

(continued)

Table 1. *(continued)*

Symbol	Description
k_i	axial stiffness
k_1, k_2	the torsional stiffness
M	the mass of drill string including BHA
w_0	the nominal drilling pressure
w_b	Force on bit
T_b	Torque on bit

The expression of w_b and T_b are:

$$w_b = \begin{pmatrix} k_C\left(\frac{w_0}{k_C} + U - V_0 t - s_0 \sin(n_b\phi)\right) if \left(\frac{w_0}{k_C} + U - V_0 t\right) \geq s_0 \sin(n_b\phi) \\ 0 \; if \left(\frac{w_0}{k_C} + U - V_0 t\right) < s_0 \sin(n_b\phi) \end{pmatrix} \quad (2)$$

$$T_b = w_b\left(\frac{2}{3} r_b \mu(\dot{\phi}) + \xi\sqrt{r_b \delta_c}\right) \quad (3)$$

The formalisation of $\mu(\dot{\phi})$ expressed as:

$$\mu(\dot{\phi}) = \frac{2}{\pi}\tanh^{-1}(\varepsilon\dot{\phi})\left(\frac{f_1 - f_2}{1 + \delta_c|\dot{\phi}|} + f_0\right) \quad (4)$$

Where

$$\delta_c = 2\pi R_p / \Omega_0 \quad (5)$$

And

$$R_p = c_1 w_0 \sqrt{\Omega_0} + c_2 \quad (6)$$

The following variable alterations are made in eq. (1) to convert the states and inputs to standard notation: $\alpha = x_1$, $\dot{\alpha} = x_2$, $\dot{\phi} = x_3$, $\alpha - \phi = x_4$, $U = x_5$, $\dot{U} = x_6$. Equations provides the whole system's equations of state:

$$\begin{cases} \dot{x}_1 = x_2 \\ \dot{x}_2 = -\frac{k_1}{J_1}x_1 - \frac{c_1 + c_2}{J_1}x_2 + \frac{c_1}{J_1}x_3 - \frac{k_2}{J_1}x_4 + \frac{k_1}{J_1}\Omega_0 t + \frac{c_1}{J_1}\Omega_0 \\ \dot{x}_3 = \frac{c_2}{J_2}x_2 - \frac{c_2}{J_2}x_3 + \frac{k_2}{J_2}x_4 - \frac{T_b}{J_2} \\ \dot{x}_4 = x_2 - x_3 \\ \dot{x}_5 = x_6 \\ \dot{x}_6 = -\frac{k_i}{M}x_5 - \frac{c_i}{M}x_6 - \frac{k_i}{M}V_0 t - \frac{w_0 - w_b}{M} \end{cases} \quad (7)$$

The state space form of this system is:

$$\begin{cases} \dot{x} = Ax + BU \\ y = Cx \end{cases} \quad (8)$$

where

$$C = \begin{bmatrix} 0\,0\,1\,0\,0\,0 \\ 0\,0\,0\,0\,0\,1 \end{bmatrix} \quad (9)$$

3 Coupled Vibrations Analysis

Multiple-input-multiple-output (MIMO) systems interact when transmissions affect one or more outputs (yi(s), i $\neq$ j), or when multiple disturbances or signals (cj(s), i $\neq$ j) affect one or more outputs (yi(s). Bristol developed the Relative Gains Analysis (RGA) technique in 1966 [17] to reduce interference. This method is based on figuring out the open- loop system's statistical gains matrix. Each RGA component is calculated using:

$$\lambda_{i,j} = \frac{\left(\frac{\partial y_i}{\partial u_j}\right)_{u_k=0,k\neq j}}{\left(\frac{\partial y_i}{\partial u_j}\right)_{y_k=0,k\neq i}} \quad (10)$$

The statistical gain from uj to yi in an open loop is represented by the top part of the fraction, and the statistical gain from uj to yi when all other outputs are perfectly regulated is indicated by the bottom part. When all other loops are closed, the relative gain λij indicates whether the gain from uj to yi of the open loop shifts [18].

$$RGA = k_s * \left[k_s^{-1}\right]^T \quad (11)$$

$$k_{sij} = \lim_{s\to 0} G_{ij}(s),\, i, j = 1, \ldots, m \quad (12)$$

*: element by element multiplication.

Let Ks be the statistical gain matrix, and let Ksij be the statistical gain from uj to yi. Algebraically, the sum of the RGA elements in a row (i) and a column (j) equals 1. The corresponding relative gain, λij, is also zero when the Ksij element is zero. If the RGA's diagonal elements (λij: i = j) are close to 1, then the system's interaction level is low. On the other hand, if these variables differ from 1 (that is, fall outside or above 1), a strong interaction occurs [17, 18].

λij, the degree of interaction between input and output points in a system, offers important information for well- designed control configurations. If λij = 0, then there is no relationship at all between the given input and output. When 0 < λij < 1, there is a moderate interaction between the control spheres, indicating that minimizing overall system interactions is the preferred course of action. When λij is equal to 0.5, there is a moderate degree of interaction and the other control factors have equally important effects.

In contrast, no interaction exists between the specific input-output control block and the other control blocks if $\lambda ij = 1$.

On the other hand, a value of λij larger than 1 indicates a strong interaction and the necessity of establishing and monitoring this relationship within the control system. On the other hand, if λij is less than zero, it means that there are important interactions that could cause instability. This highlights the significance of careful selection during the control configuration process.

When choosing the control configuration, consideration is given to the pairs that, according to the model described in the second subsection, have a relative gain λij that is close to 1. This results in:

$$k_s = 10^{-3} * \begin{bmatrix} 0 & 0 & -0.0002 & 0 \\ 0 & 0 & 0 & -0.1838 \\ 0 & 0 & -0.0002 & 0 \\ 0 & 0 & 0 & -0.1838 \end{bmatrix}$$

$$RGA = k_s * \left[k_s^{-1}\right]^T$$

We find

$$RGA = \begin{bmatrix} 0 & 0 & 0.5 & 0 \\ 0 & 0 & 0 & 0.5 \\ 0 & 0 & 0.5 & 0 \\ 0 & 0 & 0 & 0.5 \end{bmatrix}$$

The results of the Relative Gains Analysis (RGA) indicate that [u1-y3], [u2-y4], and [u3-y3], [u2-y4], are the appropriate matchings of inputs and outputs because of their positive values.

4 Results and Discussion

Our goal is to adjust the constant torsional velocity between 1 rad/s and 2.9 rad/s while keeping the input constant at 1 m/s. As the constant torsional velocity increases, Fig. 2 clearly shows an increase in oscillations, indicating the emergence of the stick-slip phenomenon. This observation is consistent with Fig. 3's findings, which show how the torsional velocity constant significantly affects the axial velocity of the system. Interestingly, the oscillations get louder as the constant torsional velocity rises.

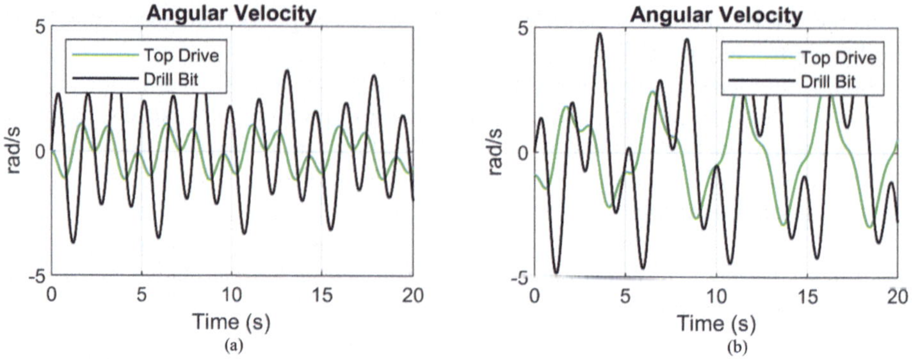

Fig. 2. Variantion of Angular velocity for: (a) $\Omega_0 = 1$ rad/s (b) $\Omega_0 = 2.5$ rad/s

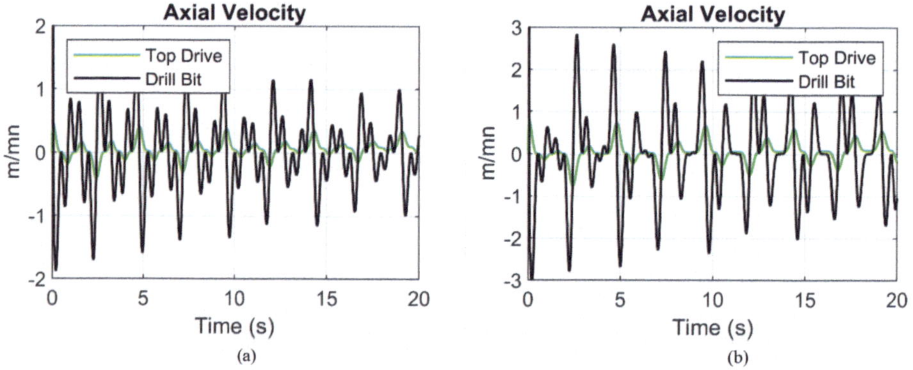

Fig. 3. Axial velocity: (a) $\Omega_0 = 1$ rad/s (b) $\Omega_0 = 2.5$ rad/s

Our goal in this section is to modify the constant axial velocity, which can be adjusted between 5 and 150 m/s, while maintaining the constant torsional velocity at 2 rad/s. Looking at Fig. 4, we can see that changes in the constant axial velocity have no effect on the system's angular velocity because it stays constant. On the other hand, Fig. 5 shows a different pattern, where a decrease in oscillations is correlated with every increase in the constant axial velocity, indicating the existence of the bit bounce phenomenon.

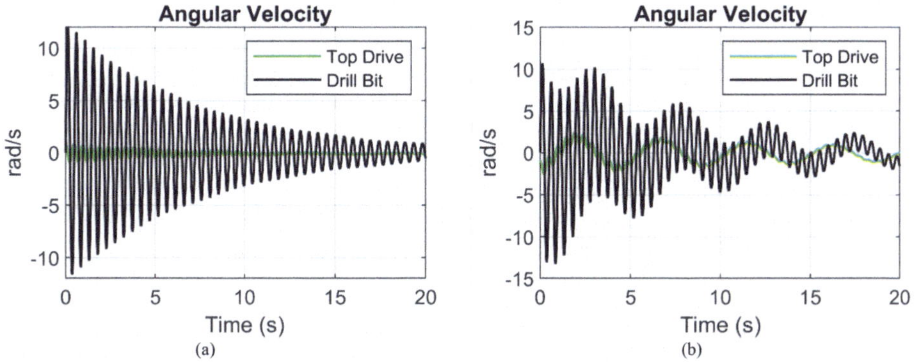

Fig. 4. Variantion of Angular velocity: (a) $V_0 = 150$ m/s (b) $V_0 = 10$ m/s

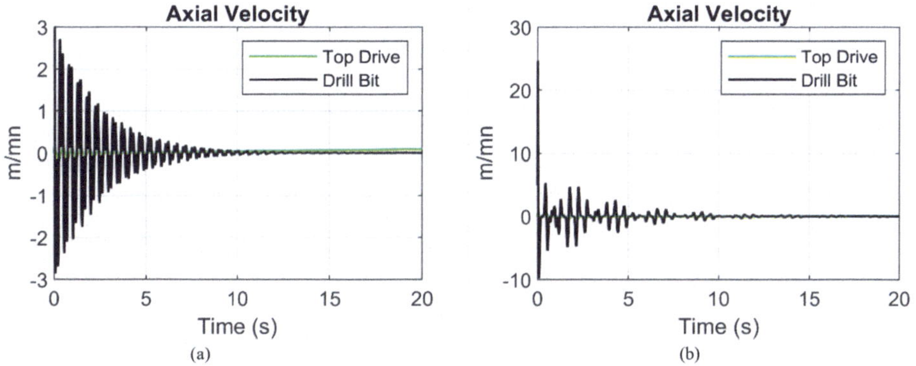

Fig. 5. Variantion of Axial velocity: (a) $V_0 = 150$ m/s (b) $V_0 = 10$ m/s

5 Conclusions

The system exhibits bit-bounce and stick-slip phenomena that are caused by axial and torsional vibrations at the drill bit level in the open loop, as can be seen by looking at the system model and the behavior of these vibrations. These kinds of vibrations are a serious risk to the rotary drilling system, which lowers ROP and increases drilling costs. Thus, it is essential to develop a strong controller to reduce these coupled vibrations and improve drilling efficiency. Furthermore, the RGA-based interaction analysis confirms the significant dependence between torsional and axial vibrations, indicating that the controller's resilience in suppressing these combined vibrations can be strengthened by utilizing the identified control pairs. The suggested controller is expected to mitigate stick-slip and bit-bounce phenomena in an efficient manner, leading to more seamless drilling operations and increased drilling productivity. Further investigation is necessary to develop and apply the suggested controller and evaluate its effectiveness in various operating scenarios.

References

1. Doghmane, M.Z., Bacetti, A., Kidouche, M.: Stick-slip vibrations control strategy design for smart rotary drilling systems. In: Hatti, M. (ed.) Artificial Intelligence and Renewables Towards an Energy Transition, pp. 197–209. Springer International Publishing, Cham (2021). https://doi.org/10.1007/978-3-030-63846-7_20
2. Kessai, I., Benammar, S., Doghmane, M.Z.: Dynamic failure analysis and lifetime estimation of Tool-string in rotary drilling system under Torsional-Axial coupled vibrations. Eng. Failure Anal. **134**, 106037 (2022). https://doi.org/10.1016/j.engfailanal.2022.106037
3. Mendil, C., Kidouche, M., Doghmane, M.Z.: Hybrid sliding PID controller for torsional vibrations mitigation in rotary drilling systems. Indonesian J. Electr. Eng. Comput. Sci. **22**(1), 146 (2021). https://doi.org/10.11591/ijeecs.v22.i1.pp146-158
4. Riane, R., Kidouche, M., Doghmane, M.Z., Illoul, R.: Modeling of torsional vibrations dynamic in drill-string by using PI-observer. In: Bououden, S., Chadli, M., Ziani, S., Zelinka, I. (eds.) Proceedings of the 4th International Conference on Electrical Engineering and Control Applications. ICEECA 2019. Lecture Notes in Electrical Engineering, vol. 682. Springer, Singapore (2021). https://doi.org/10.1007/978-981-15-6403-1_12
5. Liu, K.-C., Friend, J., Yeo, L.: The axial–torsional vibration of pretwisted beams. J. Sound Vibr. **321**(1–2), 115–136 (2009). https://doi.org/10.1016/j.jsv.2008.09.016
6. Mendil, C., Kidouche, M., Doghmane, M.Z.: Hybrid backstepping sliding mode controller for stick–slip vibrations mitigation in rotary drilling systems. IETE J. Res. (2021). https://doi.org/10.1080/03772063.2021.1919217
7. Mendil, C., Kidouche, M., Doghmane, M.Z.: Modeling of hydrocarbons rotary drilling systems under torsional vibrations: a survey. In: Hatti, M. (ed.) Artificial Intelligence and Renewables Towards an Energy Transition, pp. 243–251. Springer International Publishing, Cham (2021). https://doi.org/10.1007/978-3-030-63846-7_24
8. Mendil, C., Kidouche, M., Doghmane, M.Z., Benammar, S., Tee, K.F.: Rock–bit interaction effects on high-frequency stick-slip vibration severity in rotary drilling systems. Multidiscip. Model. Mater. Struct. **17**(5), 1007–1023 (2021). https://doi.org/10.1108/MMMS-10-2020-0256
9. Kessai, I., Doghmane, M.Z., Benammar, S.: Weight-on-bit variations analyses in the drillstring of rotary drilling systems under stick-slip vibrations. In: Hatti, M. (ed.) Artificial Intelligence and Heuristics for Smart Energy Efficiency in Smart Cities: Case Study: Tipasa, Algeria, pp. 717–725. Springer International Publishing, Cham (2022). https://doi.org/10.1007/978-3-030-92038-8_72
10. Kamel, J.M., Ahmet, Y.: Modeling and Analysis of Axial and Torsional Vibrations of Drillstrings with Drag Bits. Paper presented at the International Petroleum Technology Conference, Doha, Qatar, January 2014. https://doi.org/10.2523/IPTC-17258-MS
11. Riane, R., Doghmane, M.Z., Kidouche, M., Djezzar, S.: Observer-based h∞ controller design for high frequency stick-slip vibrations mitigation in drill-string of rotary drilling systems. Vibration **5**, 264–289 (2022). https://doi.org/10.3390/vibration5020016
12. Gupta, S.K., Wahi, P.: Global axial–torsional dynamics during rotary drilling. J. Sound Vibr. **375**, 332–352 (2016). https://doi.org/10.1016/j.jsv.2016.04.021
13. Tashakori, S., Vossoughi, G., Zohoor, H., van de Wouw, N.: Prediction-based control for mitigation of axial–torsional vibrations in a distributed drill-string system. IEEE Trans. Control Syst. Technol. **30**(1), 277–293 (2021). https://doi.org/10.1109/TCST.2021.3065669
14. Kim, B.J., Palazzolo, A., Gharib, M.: Drillstring Simulator: A Novel Software Model for Stick–Slip And Bit-Bounce Vibrations. In: ASME International Mechanical Engineering Congress and Exposition, vol. 85611, p. V07AT07A048. American Society of Mechanical Engineers (2021). https://doi.org/10.1115/IMECE2021-73717

15. Voronov, S.A., Gouskov, A., Kvashnin, A.S., Butcher, E.A., Sinha, S.C.: Influence of Torsional Motion on the Axial Vibrations of a Drilling Tool. J. Comput. Nonlinear Dyn. **2**(1), 58–64 (2007). https://doi.org/10.1115/1.2389212
16. Yu, T., Li, X., Zhang, H., Duan, C., Zeng, H., Agila, W.: Nonlinear analysis of axial-torsional vibration of drill string based on a 3 DOF model. Adv. Mech. Eng. (2022). https://doi.org/10.1177/16878132221107778
17. Fartas, N., Khelassi, A.: The impact of fractional order control on multivariable systems. Algerian J. Sign. Syst. **7**(1), 7–12 (2022). https://doi.org/10.51485/ajss.v7i1.151
18. Bristol, E.: On a new measure of interaction for multivariable process control. IEEE Trans. Autom. Control **11**, 133–134 (1966). https://doi.org/10.1109/TAC.1966.1098266

Numerical Study of an Aircraft Cargo Compartment Fire: The Impact of Fire Retardant Sheets

Houssam Eddine Nadir Hiber(✉) and Hadj Miloua

Department of Mechanical Engineering, Faculty of Technology, Laboratory Mechanics of Structures and Solids (LMSS), University of Djillali Liabes, 22000 Sidi Bel Abbes, Algeria
nadir.hiber@dl.univ-sba.dz

Abstract. This study aims to investigate the impact of fire-retardant materials on delaying the flashover in the case of an airplane cargo compartment fire and highlights the proposal of using fire-retardant sheets in the cargo shipping field. The aircraft cargo compartment fire simulation is a theoretical scenario based on a previously validated model [10]. The models used in this study were based on the new version of the FDS program, which was previously validated [4]. The model employs large eddy simulations (LES) for turbulence and the eddy dissipation concept (EDC) as a combustion model, along with other models for thermal radiation, pyrolysis, and soot formation. The study yelled that the flashover will be delayed by a considerable amount of time compared to when the shipment materials are not protected, making this proposed study promising and valuable when it comes to the protection against fires in airplanes.

Keywords: Aircraft fire · Flashover · Cargo compartment · HRR · Pool fire

1 Introduction

According to the International Air Transport Association's (IATA) 2016 annual evaluation [1], airplanes are one of the safest means of transportation. However, aircraft have been exposed to many accidents over the years, some of which may be considered catastrophic. To uphold this safety reputation, parties like the US Federal Aviation Administration (FAA) implement experiments and guidelines aimed at enhancing aircraft safety [2–4]. Fires remain a leading cause of aircraft accidents. Sarkos [2] categorizes these into two types: post-crash fires, which are more common, and in-flight fires, which often originate in hidden or inaccessible areas of the aircraft due to system malfunctions, broken parts, or inadequate maintenance [5]. Cargo compartment fires can cause extensive damage if they lead to flashover and spread to other areas of the aircraft [6], by definition, a flashover happens when a fire spreads rapidly and intensely, producing substantial heat, smoke, and hazardous gases, quickly filling the cabin [2]. Several factors can influence the occurrence of flashover, where temperatures between 600 and 700 °C and heat flux values between 15 and 33 kW/m^2 have been identified as minimal parameters for flashover initiation [7]. It is also known as the point of the end of survivability, due to the amount of heat and toxic gases it produces.

© The Author(s), under exclusive license to Springer Nature Switzerland AG 2025
M. Hatti (Ed.): IC-AIRES 2024, LNNS 1238, pp. 254–262, 2025.
https://doi.org/10.1007/978-3-031-80301-7_28

Authorities focused on fire prevention in aircraft have sought solutions to combat this issue. However, real- time experiments are costly and time-consuming. As a result, there has been a shift toward Computational Fluid Dynamics (CFD) modeling to simulate fire scenarios instead of conducting actual tests [8]. The Fire Dynamic Simulator (FDS), a CFD tool developed by the Building and Fire Research Laboratory (BFRL) at NIST [9], was used to validate a full-scale FAA experiment [2] by Hiber and Miloua [10]. They investigated parameters such as heat release rate, temperatures, CO2 and O2 concentration, CO density, visibility, and flashover timing within airplane cabins during post-crash fires. This study employs a similar mathematical and equational approach as the referenced article [10], with adjustments including changes in simulation domain geometry to a low-deck cargo compartment instead of an upper-deck passenger cabin and variations in device types and their placement. R. Papa et al. [11] conducted a numerical study using Fluent to model heat and smoke in a cargo compartment, accurately predicting temperature variations over time. Cong et al. [12] examined how pressurization and depressurization affect pool fire characteristics, providing insights on pressure, oxygen levels, and ventilation for future research. K.H. Lu et al. [13] studied the impact of ventilation on fire dynamics in cargo compartments, comparing results with FAA experiments to assess the "one-minute rule" for fire detection. Hiber et al. [14] used FDS to investigate fires in a closed Airbus 330 cargo compartment, analyzing heat release rates, flashover times, and fire spread. They further examined temperature variations within the compartment, using instruments placed at different heights to study flame spread and the movement of hot toxic gases [15].

This study aims to numerically assess the effectiveness of fire-blocking or fire-retardant covers for materials in both large transport and passenger aircraft cargo compartments, using the validated FDS model. The findings will be discussed and compared with previous studies, culminating in a conclusion.

2 Materials and Method

2.1 The Simulation Description

The dimensions and specifications of the cargo compartment can vary depending on the airline's customization and specification. Typically, the cargo compartment is situated in the lower deck or belly of the plane (Fig. 1c). The dimensions chosen for this case are 3.84 m in width, 11.86 m in length, and 2 m in height in the low deck FWD cargo compartment (Fig. 1a) [16]. It is mainly designed to accommodate Unit Load Devices (ULDs) or pallets. In this case, random-sized cargo boxes with different material properties were adopted (Fig. 1b).

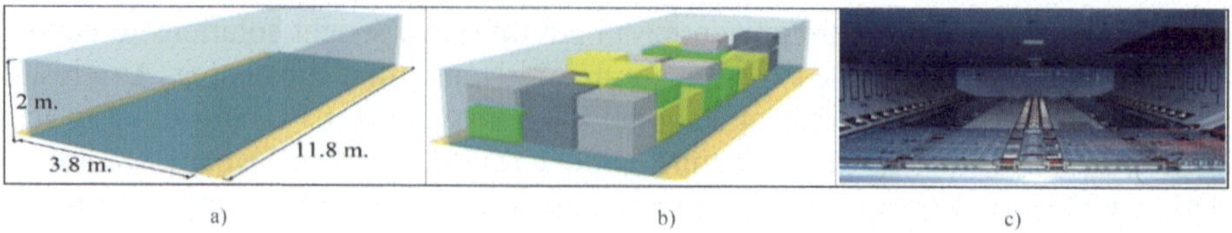

Fig. 1. The FWD lower deck cargo compartment of A 330, a) empty, b) with cargo, and c) real FDW CC [16].

2.2 Simulation Setup, Time-Step, and Mesh Generation

In this simulation, the fire source was a square propane fuel pan ($30 \times 30\,\text{cm}^2$) producing 5,000 kW of heat, positioned at the left rear of the cargo compartment, near the last row of cargo boxes (Fig. 2). The focus was not on the ignition point itself, but on the events following it and the effectiveness of fire-retardant sheets. Wind effects were excluded due to the compartment's closed nature, with buoyancy forces primarily driving the fire.

The reliability of the FDS model was established in reference [10], where results were validated against experimental data. A time step of 0.091 seconds was used, as detailed in reference [20]. A single mesh with a uniform grid system centered on the cargo compartment was employed, comprising 118 cells in the x-direction, 50 in the y-direction, and 38 in the z-direction. The grid cell sizes were 0.1 m in the x and y directions and 0.05 m in the z direction, resulting in approximately 218,500 cells, each with a volume of 0.0005 m^3 (Fig. 3). Adjustments were made to the cargo compartment dimensions to avoid mesh alignment errors (Fig. 1). The cargo was arranged in eight rows of parallelogram-shaped piles (base: 1 m $\times$ 1 m, height: 0.7 m) with varying spacing, creating a significant gap at the end where the fire source was located (Fig. 2).

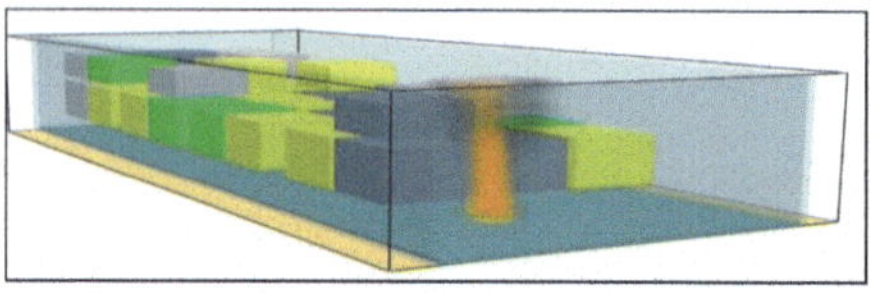

Fig. 2. The fire source within the cargo compartment.

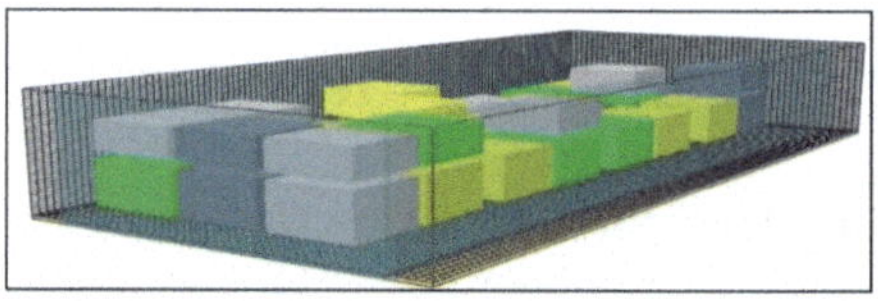

Fig. 3. The grid cell system of the domain.

2.3 Mathematical Models and Governing Equations

The equations used in this article have been validated in previous studies by Hiber, N, H, E, and Miloua, H. [10].

The Fundamental Conservation Equations-Conservation of Mass:

$$\frac{\partial \rho}{\partial t} + \nabla \cdot \rho \mathbf{u} = 0 \tag{1}$$

-Conservation of Momentum (Newton's Second Law):

$$\frac{\partial \rho \mathrm{u}}{\partial \mathrm{t}} + \nabla \cdot \rho \mathrm{uu} + \nabla \mathrm{p} = \rho \mathrm{g} + \mathrm{f}_b + \nabla \cdot \tau_{ij} \tag{2}$$

-Conservation of Energy (First Law of Thermodynamics):

$$\frac{\partial \rho h}{\partial t} + \nabla \cdot \rho \mathrm{hu} = \frac{DP}{Dt} + \dot{q}''' - \nabla \cdot \dot{q}'' + \Phi \tag{3}$$

-Equation of State for a Perfect Gas:

$$P = \frac{\rho RT}{V} \tag{4}$$

The mass conservation equation is written in terms of the mass fractions of the individual gaseous species:

$$\frac{\partial(\rho \gamma_{\mathrm{i}})}{\partial t} + \nabla \cdot \rho \gamma_{\mathrm{i}} \mathrm{u} = \nabla \cdot \rho D_{\mathrm{i}} \nabla \gamma_{\mathrm{i}} + \dot{m}_i''' \tag{5}$$

Summing these equations over all species yields the original mass conservation equation. This implies that the sum of the mass fractions $\sum \gamma i = 1$, the sum of the production/loss rates $\sum m''' i = 0$, and the sum of the diffusion terms $\sum \rho D i \nabla \gamma i = 0$.

Large Eddy Simulation (LES)

The turbulence is treated by employing the Smagorinsky form of Large Eddy Simulation (LES) Ref. [16]. Based on Smagorinsky's analysis:

$$\mu_{LES} = \rho(C_s \Delta)^2 \left(2\overline{S_{ij}} \cdot \overline{S_{ij}} - \frac{2}{3}(\nabla \cdot \overline{\mathrm{u}})^2 \right)^{1/2} \tag{6}$$

where Cs is an empirical constant and Δ is a length on the order of the size of a grid cell. The other diffusive parameters, thermal conductivity, and material diffusivity are related to the turbulent viscosity by

$$k_{LES} = \frac{\mu_{LES.cp}}{Pr}; \; (\rho D)_{l.LES} = \frac{\mu_{LES}}{S_c} \tag{7}$$

The Prandtl number Pr and the Schmidt number Sc are assumed to be constant for a given scenario.

Combustion Model

FDS employs the Eddy Dissipation Concept (EDC) model as explained in Ref. [9, 10, 20]:

$$\tilde{\mathrm{Y}}_\alpha(\mathrm{t}) = \zeta(\mathrm{t})\tilde{\mathrm{Y}}_\alpha^0 + (1 - \zeta(\mathrm{t}))\widehat{\mathrm{Y}_\alpha}(\mathrm{t}). \tag{8}$$

where $\tilde{\mathrm{Y}}_\alpha(\mathrm{t})$ represents the cell mean mass fraction of species α and $\zeta(\mathrm{t})$ is the unmixed fraction of mass within the cell, which evolves based on the following simple ordinary differential equation (ODE):

$$\frac{d\zeta}{dt} = -\frac{\zeta}{\tau_{\mathrm{mix}}}. \tag{9}$$

and the basic EDC model is given by:

$$\dot{\mathrm{m}}_{\mathrm{F}}''' = -\rho \frac{\min\left(\widetilde{\mathrm{Y}_{\mathrm{F}}^0}, \widetilde{\mathrm{Y}_{\mathrm{A}}^0}/s\right)}{\tau_{\mathrm{mix}}}. \tag{10}$$

2.3.1 Pyrolysis Model (Thermally-Thin Solid)

If the surface material is assumed to be thermally thin, that is, its temperature is assumed uniform across its width, its temperature, *Ts*(t), is governed by its density, specific heat, and thickness δ.

$$\frac{dT_S}{dt} = \frac{\dot{q}''c_c + \dot{q}''r}{p_S c_S \delta} \tag{11}$$

2.3.2 Thermal Radiation

The thermal radiation is based on The Radiative Transport Equation (RTE) for an absorbing, emitting, and scattering medium which is given by:

$$s \cdot \nabla I_\lambda(x, s) = -\kappa(x, \lambda) I_\lambda(x, s) - \sigma s(x, \lambda) I_\lambda(x, s) + B(x, \lambda) + \frac{\sigma s(x, \lambda)}{4\pi} \int_{4\pi}^{1} \Phi(s', s) I_\lambda(x, s') ds' \tag{12}$$

where $I\lambda$ (x, s) is the radiation intensity at wavelength, λ; s is the direction vector of the intensity; and κ(x,λ) and σs(x,λ) are the local absorption and scattering coefficients, respectively. B(x,λ) is the emission source term, describing how much heat is emitted by the local mixture of gas, soot, and droplets/particles.

All the equations were reproduced from the FDS technical reference user guides [9, 20] as well as the article paper Ref. [10].

3 Simulation and Results Discussion

3.1 The Heat Release Rate and the Flash Over Time

The heat release rate is a good indicator of the flashover occurrence [10], when the HRR rises, it means there is a considerable amount of heat energy within the domain (cargo compartment) which can be translated to; a number of internal materials (cargo shipments in our case) are burning. The HRR can also help us emphasize the start of the flashover within the domain, by observing (Fig. 4), it can be seen that after 80 s from the start of the simulation, the HRR of the internal materials starts rising exponentially, and that indicates the start of the flashover and when the HRR of the internal materials goes down in value it means the fire start to self-extinguish, most likely due to the oxygen starvation since the domain is in a confined environment. Figure 4 also shows the total along with the pool fire HRR, through observation, it can be understood that the pool fire didn't interfere with the internal materials HRR value rather it served as an ignition source for the cargo boxes. The comparison between the Heat release rate from this simulation and earlier studies Ref [14, 15] depicted in (Fig. 5), shows the role played by the fire retarding sheets and how it helps to minimize the heat release from the burning materials, and how it delays the flashover occurrence. Another crucial observation from (Fig. 5) is the peak value of the HRR from earlier studies is much higher than the peak value of the HRR from this study and that indicates that not all the internal materials that is cargo shipment got ignited and consumed by the flames rather a part of the cargo boxes was ignited then it was self-extinguished later on.

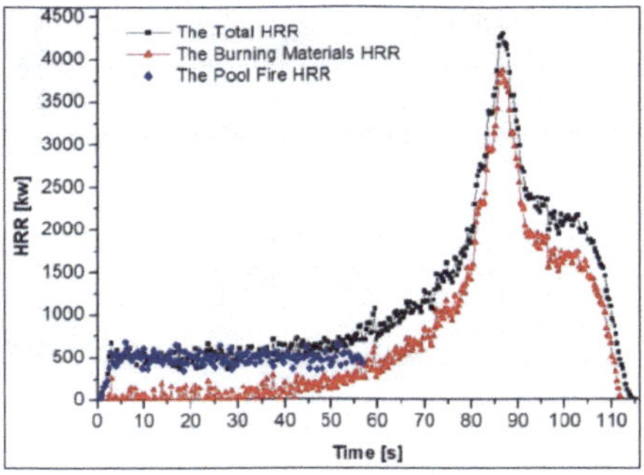

Fig. 4. Total, Materials, and Pool Fire HRR.

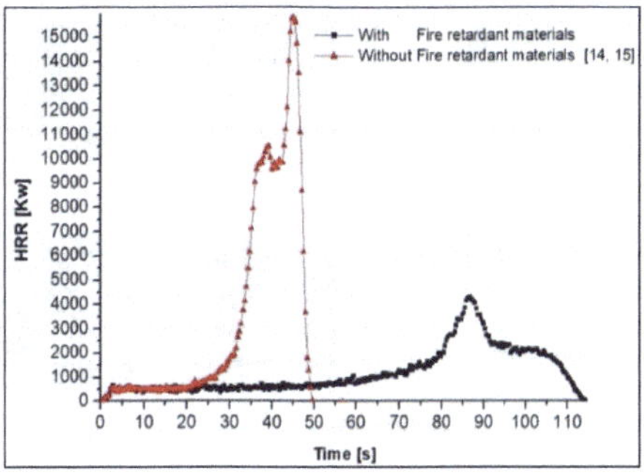

Fig. 5. The HRR from this simulation and Ref [14, 15]

3.2 Vector Slices of the HRR

Figure 6 shows a comparison between vector slices of the heat release rate per unit volume at the symmetry plan of the cargo compartment of this simulation and the one from Ref [14]. After 38 s within the simulation, shown in (Fig. 6a) the flashover had appeared for about 5 s Ref [14], however, that was not the case in this simulation Fig. 6b, that's because the fire retarding sheets (FRS) used to cover the cargo boxes delayed the spread of flames and that one of its benefits, the FRS played the role of a protecting layer, both retarding and suffocating the flames at once. At the time, the fire had caught all the internal materials from the case where the FRS were not used (Fig. 6c), and the adjacent cargo boxes started to catch fire in the case where the FRS were used (Fig. 6b).

It can be observed in (Fig. 6c) and d the amount of heat liberated by the unprotected cargo boxes at the time when the cargo boxes covered by the fire retarded sheets are still resisting fire, which makes it of great importance in the event of fires inside the aircraft's cargo compartment, its use delays and suppresses the fire, which gives more time to deal with this dangerous situation and may help save many lives in the event of a plane fire. However, this technology carries several disadvantages, the most important of which is weight. As is known in the world of aircraft, the biggest enemy of an aircraft is weight. Therefore, carriers seeking to carry Transport companies need to reduce weight in order to compensate for it with more goods. The weight of these heat-retarding materials may be a problem in some cases, even if their thickness is small, as well as their cost. These materials are expensive in general, especially high-quality materials, as well as the time to cover and remove them from the shipping boxes. This time may be considered valuable for some companies that are tied to a tight schedule.

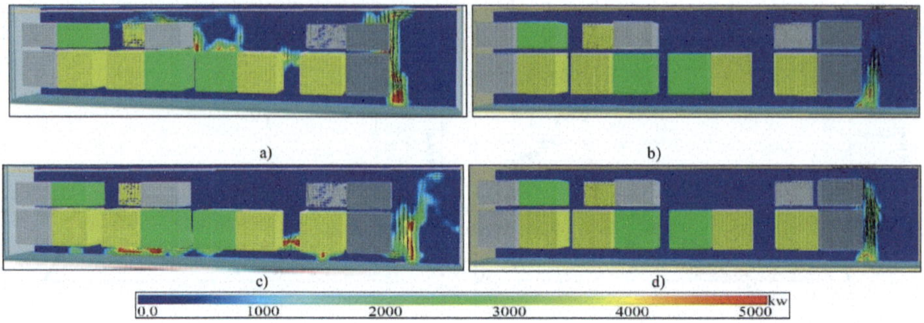

Fig. 6. Vector slices of the heat release rate within the cargo compartment. a and b) t = 38 s, c and d) t = 44 s. (a, c Ref [14]).

This prediction emphasized that the flashover will occur about 80 seconds within the simulation (Fig. 7a), after the hot gases and hot products from the pool fire and the burning adjacent cargo boxes to the pool fire accumulate at the ceilings, the upper layer of the upper cargo boxes start burning from the top to the bottom (Fig. 7b), through radiation the rest of the cargo boxes start burning to unleash a huge amount of heat through the cargo compartment towards the front. However, since the cargo boxes were covered with fire retarding sheets not all the cargo boxes were burned rather parts of the top side of some cargo boxes were burned (Fig. 7c), the first four rows of the cargo boxes were nearly Untouched by the flames and that's because the fire retardant materials of the sheets covering the cargo boxes both suffocate and absorbs the flames and the heat resulting from the flames that spread from the source of ignition to the front part of the cargo compartment. After 110 seconds within the simulation the burning materials as well as the fire ignition source started self-extinguishing due to the starvation of oxygen since the domain was a confined environment there was no exchange of oxygen with the exterior (Fig. 7d).

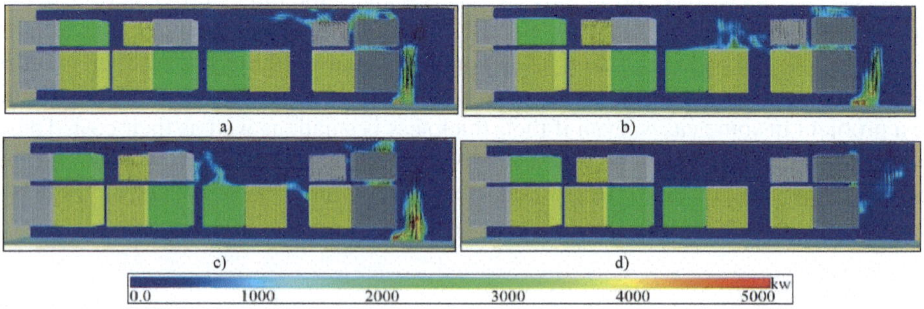

Fig. 7. Vector slices of the HRR within the cargo compartment. a) t = 80 s, b) t = 85 s, c) t = 95 s, d) t = 110 s.

4 Conclusion

This prediction utilized a validated version of the FDS program [10] to analyze a theoretical aircraft cargo compartment fire scenario, where cargo boxes were protected by fire-retardant materials. The prediction relied on large eddy simulation (LES) and the eddy dissipation concept (EDC), among other models. It highlighted that flashover would occur approximately 80 s after ignition about 50 s later than in scenarios without fire protection. Additionally, the heat release rate was significantly lower than predicted in earlier studies [14, 15] without such protection. The fire-retardant sheets absorbed and suffocated the flames, greatly limiting fire spread; only the upper layers of the cargo were affected, while the first four rows remained largely unscathed. In contrast, unprotected cargo saw flames rapidly advance from the back to the front, fueled by substantial heat and toxic gases.

Although fire protection materials are valuable, they can be expensive and increase load weight, leading to potential reductions in cargo. The labor-intensive process of wrapping and removing these covers is often not favored by companies. Ultimately, the data from this study will play a crucial role in improving fire detection systems and facilitating the integration of AI into aircraft fire suppression and general fire protection systems.

Future research will explore this topic further, examining emissions from fire-retardant materials and comparing different types of fire retardants.

References

1. International Air Transport Association, "IATA Annual Review 2016. Accessed 15 Feb 2024
2. Sarkos, C.P.: Application of full-scale fire tests to characterize and improve the aircraft postcrash fire environment. Toxicology **115**(1–3), 79–87 (1996)
3. Sarkos, C.P., Hill, R.G.: Characteristics of aircraft fire measured in full-scale tests. AGARD Conference Proceedings on Aircraft Fire Safety, No. 467, Specialised Printing Services Ltd., Essex, U.K., p. 11.1– 11.19 (1989)
4. Blake, D.R., Hill, R.G.: Fire Containment Characteristics of Aircraft Class D Cargo Compartments. Federal Aviation Administration Technical Center (1983)
5. Tutson, A.L., Ferguson, D.E., Madden, M.: Fire Protection: Passenger cabin. Aeromagazine QTR_04, 2011, vol. 11, p. 18–23
6. Wang, W., Wang, L., Yang, R., Zhang, H., Ren, C., Yang, J.: Investigation of the effect of low pressure on fire hazard in the cargo compartment. Appl. Therm. Eng. **2019**(158), 113775 (2019)
7. Peacock, R.D., Reneke, P.A., Bukowski, R.W., Babrauskas, V.: Defining flashover for fire hazard calculations. Fire Saf. J. **32**(4), 331–345 (1999)
8. Galea, E.: On the field modeling approach to the simulation of enclosure fires. J. Fire. Prot. Eng. **1**(1), 11–22 (1989)
9. McGrattan, K., Hostikka, S., Mcdermott, R., et al.: Fire dynamics simulator user's guide. NIST **1019**(6), 1–339 (2013)
10. Hiber, H.E.N., Miloua, H.: Numerical modeling of aircraft fire: postcrash fire. J. Aircraft **61**, 1–15 (2023)
11. Papa, R., Pustelnik, M., Luis, C.C.S.: "A CFD model for cargo compartment smoke detection. ASME Int. Mech. Eng. Congress Exposition **42193**, 663–671 (2005)

12. Li, C., Yang, R., Yao, Y., Tao, Z., Zhu, P., Zhang, H.: Effects of static pressure, pressurization, and depressurization on n-heptane pool fires in an airplane cargo compartment. Fire Mater. **43**(3), 266–276 (2019)
13. Lu, K.H., Mao, S.H., Wang, J., Lu, S.: Numerical simulation of the ventilation effect on fire characteristics and detections in an aircraft cargo compartment. Appl. Therm. Eng. **124**, 1441–1446 (2017)
14. Houssam Eddine Nadir, H. , Hadj, M.,, Ahmed, S.: Numerical Modeling of Aircraft Fire: Cargo Compartment Fire. Second Conference on Electrical Engineering Applied to Aeronautics CEEAA'02 2024 23 April 2024 at The Air Force Command Higher School of Aviation Technology, Martyr Musa Rahali Algiers/Algeria
15. Houssam Eddine Nadir, H., Hadj, M., Ahmed, S.: Numerical Simulation of an Aircraft Cargo Compartment Fire: Temperature Investigation. 3rd International Workshop on Structural Mechanics and Materials-IWSMM24- on 23–24 April 2024 at The Research Center (University of Batna 2, located in Batna city center/Algeria
16. Airbus: Airbus Commercial Aircraft AC A330. AIRCRAFT CHARACTERISTICS AIRPORT AND MAINTENANCE PLANNING. Retrieved from (2021)
17. Sarkos, C.P., Hill, R.G., Howell, W.D.: The development and application of a full-scale wide-body test article to study the behavior of interior materials during a post-crash fuel fire. J. Fire Flammability **13**, 172–202 (1982)
18. Babrauskas, V.: Ignition handbook principles and applications to fire safety engineering, fire investigation, risk management, and forensic science (2020)
19. Philip, J., DiNenno, P.: E Dougal Drysdale, Craig L. Beyler, W. Douglas Walton, P.E. Richard L. P. SFPE Handbook of Fire Protection Engineering Third Edition (2002)
20. Mcgrattan, K., Randall, M., Marcos, V.: Fire dynamics simulator (version 6.7.6) technical reference guide. NIST **1018**(6) (2021)

An Innovative Ionic Liquid for Use as an Electrolyte in PEM Fuel Cells: 1-(2-Hydroxyethyl)-3-Methylimidazolium Hydroxide

Saad Halloub[1,2](✉), Abdelkader Benabdellah[1], EL-Habib Belarbi[1], Mansour Debdab[1], Yassine Chaker[1], Mustapha Hatti[3], and Abdelhak Guezam[4]

[1] Synthesis and Catalysis Laboratory (LSCT), Faculty of Materials Sciences, University of TIARET, Tiaret, Algeria
saad.halloub@univ-tiaret.dz

[2] Laboratory of Applied and Didactic Sciences, Higher Normal School of Laghouat, BP 4033, 03000 Laghouat, Algeria

[3] UDES, Unité de Développement des Equipements Solaires, EPST/CDER, 11th National Road, Bou Ismail, Tipasa, Algeria

[4] Laboratory for Analysis and Control of Energy Systems and Electrical Systems LACOSERE, Laghouat University, 03000 Laghouat, Algeria

Abstract. Ionic liquids with the imidazolium cation exhibit a high electrochemical stability, besides Hydroxide groups can be incorporated to the cation as side chains acting as carriers in order to facilitate the proton transport. In this work, the impact of 1-(2-Hydroxyethyl)-3-Methylimidazolium Hydroxide $[EtOHMIM^+][OH^-]$ ionic liquid have been tested as electrolyte in PEMFC. The operational duration of the cell was evaluated in an electric motor-simulating load conditions. Systematic investigation of the ionic conductivity showed that the ionic conductivity reached local maxima at the compositions of $[H2O]/[EtOHMIM^+][OH^-]$ = 30 ml/15 ml. Results indicated that a 50% concentration of ionic liquid maximized cell power output, increasing it from 2,7 W to 2,82 W. Although some improvements in terms of thermal and mechanical stability should be achieved, this approach presents a promising electrolyte with excellent characteristics.

Keywords: Ionic liquid · PEMFC · 1-(2-Hydroxyethyl)-3-Methylimidazolium Hydroxide · Electrolyte · Performance

1 Introduction

Worries over environmental changes, particularly global warming, have led to fuel cells' widespread promotion as a clean power generation choice. This is because fuel cells alone create heat, water, and energy, with no byproducts that harm the environment [1, 2]. For example, PEMFCs, or proton exchange membrane have recently exploded in popularity for use in automobiles, but there are many more kinds of fuel cells that are often categorized according to the electrolyte they employ [3–6].

© The Author(s), under exclusive license to Springer Nature Switzerland AG 2025
M. Hatti (Ed.): IC-AIRES 2024, LNNS 1238, pp. 263–268, 2025.
https://doi.org/10.1007/978-3-031-80301-7_29

A lot of work has gone into developing a PEM with high conductivity at low relative humidity as an electrolyte, in the hopes of making the system simpler and cheaper. Furthermore, to enhance the longevity and security of fuel cells, it is essential to create electrolytes that are not volatile and do not catch fire. Thus, in dry circumstances, ionic liquids have been used as electrolytes in PEMFCs in a number of experiments [7–9].

Ionic liquids are organic salts that have melting points that are equal to or lower than room temperature. Among the many desirable characteristics of these compounds for this use are their broad electrochemical window, low vapor pressure, excellent thermal and chemical stability, and good conductivity even when dry [10]. Because of their outstanding ionic conductivities even when dry, ionic liquids are often considered a potential future solution to the issues caused by volatile electrolytes.

This study investigated the use of a novel ionic liquid, namely 1-(-Hydroxyethyl)-3-Methylimidazolium Hydroxide [EtOHMIM$^+$][OH$^+$], as an electrolyte in a Pressurized Electrolyzer Fuel Cell (PEMFC) at different concentrations in water. The current-voltage and maximum load power parameters were established using this electrolyte.

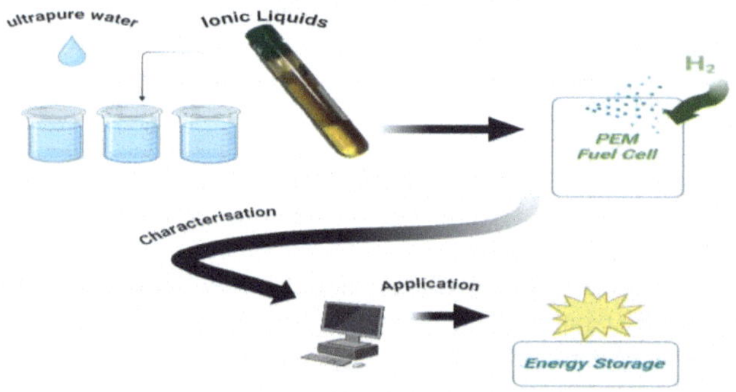

Fig. 1. Graphical abstract

2 Exprimentals

2.1 Reagents and Materials

The [EtOHMIM$^+$][OH$^-$] ionic liquid was synthesized using previously described procedures [11]. The purity of the ionic liquid was checked by 1H- and 13C-NMR spectroscopy and electrochemical methods, i.e., cyclic voltammetry [11].

The experiments were performed in a commercial fuel cell from (Horizon Fuel Cell Europe s.r.o, Czech Republic) (Fig. 2), the characteristics of the cell have been given in (Table 1). The performance of the PEMFC was evaluated by discharge experiments. The measurements were realised by a potentiostatic mode (VSP-300, Bio-Logic SAS). Prior to the discharge experiments, the fuel cell was activated by imposing a convenient current profile.

Table 1. Fuel cell characterization.

Dimensions	54 mm × 54 mm × 17 mm
Input Voltage	1,8 V–3 V DC
Input Current	0,7 A
Hydrogen production rate	7 ml per minute at 1 A
Oxygen production rate	3,5 ml per minute at 1 A

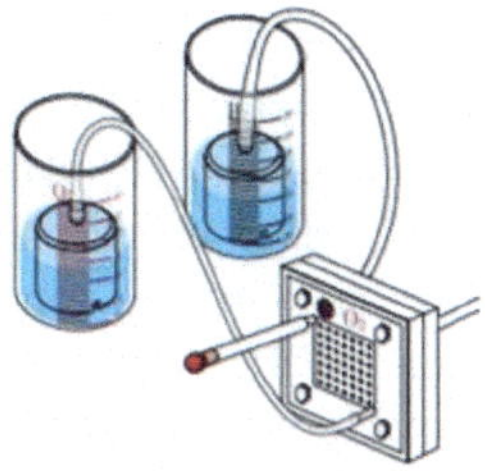

Fig. 2. A schematic Of Horizon PEM Fuel Cell [12]

2.2 Ionic Liquid as Electrolyte for Fuel Cell Application

Firstly, the solutions tested in the PEMFC were prepared by dissolving 1- (2-Hydroxyethyl) -3-Methylimidazolium Hydroxide in ultrapure water at distinct volumes of (5 ml, 10 ml, 15 ml, 20 ml) (see Fig. 1).The electrolytes concentration were calculated by the relation given by Eq. (1).

$$ILL\% = (1 - \frac{V_{Wa} - V_{ILs}}{V_{Wa}}) \times 100 \quad (1)$$

where $V^{Wa}$is the water volume (ml) and V^{ILs} is the volume (ml) of ILs added.

The procedure tests discharge of the cell injected with prepared solutions was performed by electrical current (3VDC Battery) with activation period (30 s) in all tests by using an electric motor as a discharge load (Fig. 3). All experiments were repeated at last three times to certify the reproducibility of the results.

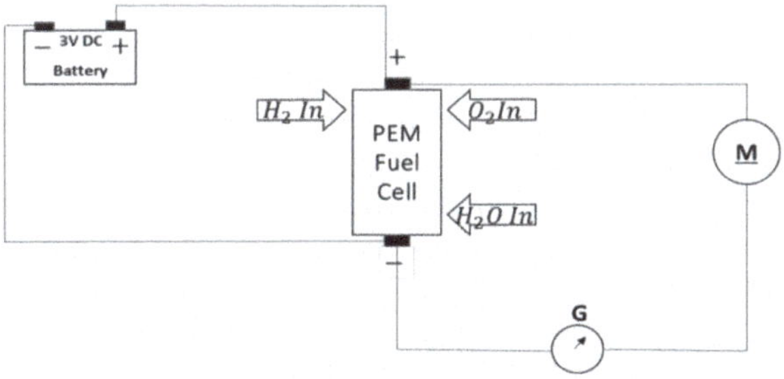

Fig. 3. The electrical circuit used for measurements

2.3 Electrochemical Measurements

A series of electrochemical measurements were conducted on a PEM fuel cell at ambient temperature and atmospheric pressure, while injecting ILs of 3 ml and 5 ml in 10 ml volumes of ultrapure water, respectively, in order to obtain sets of polarization curves for observing the effect of the $[EtOHMIM^+][OH^+]$ ionic liquid on the cell and its performance. The experimental evaluation of the PEM fuel cell's performance after the injection of $[EtOHMIM^+][OH^+]$ ionic liquid was conducted by the measurement of I-V curves using a single-cell (prior cell). Electrochemical impedance spectroscopy (EIS) analysis was conducted using a potentiostatic (VSP-300, Bio-Logic SAS) in the potentiostatic mode, following the identical circumstances as the I-V curve measurement. Upon achieving a stable condition for a minimum of 10 min, measurements were acquired throughout a voltage range spanning from 0 V to 1.8 V (Fig. 4).

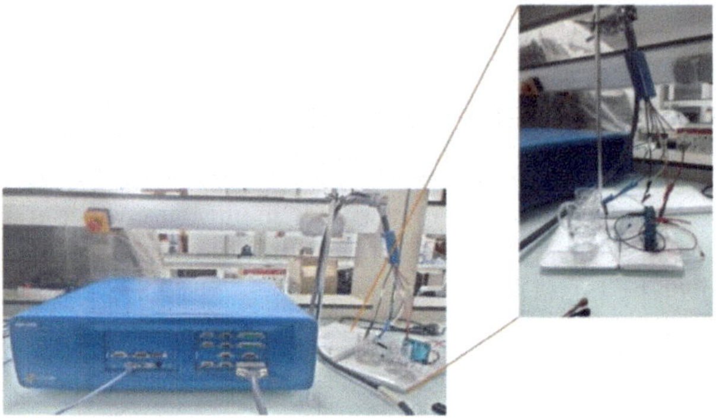

Fig. 4. Picture of electrochemical measurements

3 Results and Discussion

Ionic liquids have been previously demonstrated to be excellent electrolytes in liquid contact fuel cells. Given the unequivocal enhancement of fuel cell performance using ionic liquid electrolytes, this innovation has demonstrated a low technological impact to date because solid systems are preferred for practical applications, mostly in mobile uses, like cars. As a natural evolution of this procedure, electrolyte based different amounts of $[EtOHMIM^+][OH^-]$ ionic liquid dissolved in water were injected in PEMFC and the duration of the cell's operation was then measured. The obtained characteristic parameters of the FC using water and IL solutions are summarized in the (Table 2).

The results in Table 1 showed that the fuel cell performance time increased with increasing ionic liquid percentage (ILL %) added to the water. This procedure due to the improvement of the proton conductivity of the electrolyte and thus improves the performance of the PEMFC. The electrolyte with a percentage of ionic liquid (ILL 10%) has a low FC performance time compared to other percentages; this suggests that there is a response in the performance of the fuel cell with a liquid ionic for higher concentrations. Before performing the electrochemical test, different volumes of $[EtOHMIM^+][OH^-]$ ionic liquid of 3 ml and 5 ml were introduced into the ultrapure water respectively. Figure 5 shows an I-V polarization curve dependence of the cell potential, which indicates the increasing in the PEMFC performance. The performance of PEMFC with the electrolyte contains 5 ml of $[EtOHMIM^+][OH^-]$ ionic liquid injected into the PEMFC had

Table 2. Electrolytes and performance fuel cell times

Electrolytes	Ionic Liquids (ILL %)	Time charge(s)	Performance FC times(s)
01	0	30	17,5
02	10	30	12,13
03	20	30	15,33
04	30	30	15,45
05	40	30	15,06
06	50	30	15,15

maximum voltage, peak current, and power density. All experiments described herein were performed at room temperature.

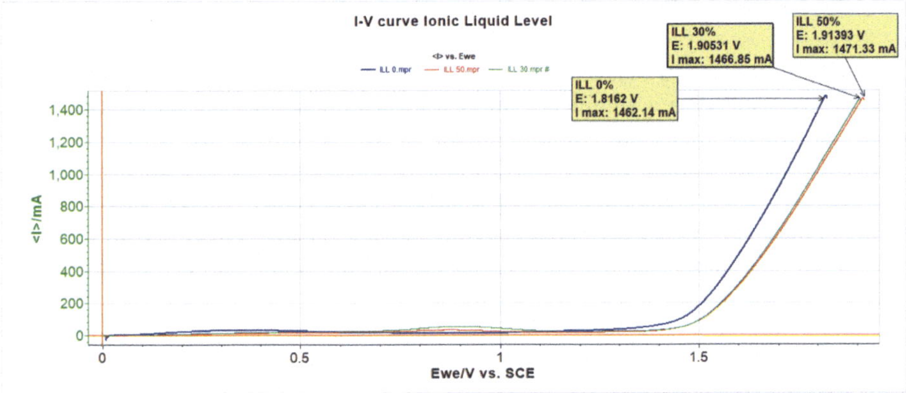

Fig. 5. The I-V polarization curve ionic liquids level (ILL)

The maximum electrical power (W) of the cell is calculated from the relation given in Eq. (2):

$$P_{max} = I_{max} \times E_{max} \tag{2}$$

where:

***Ima**x*: The maximum current (A) measured.

***Ema**x*: The open circuit voltage (OCV).The maximum potential of the fuel cell (V).

Figure 6 show the power curve which indicating the increases in the performance of the PEMFC in the case with the addition of 5 ml volume of ionic liquid to give a maximum power for the cell at 2.82 W. Compared to the minimum performance of the PEMFC without any addition of ionic liquids which give a maximum power for the cell at 2.7 W. This means that the fuel cell performance increases substantially with the amount of addition ILs in the electrolyte.

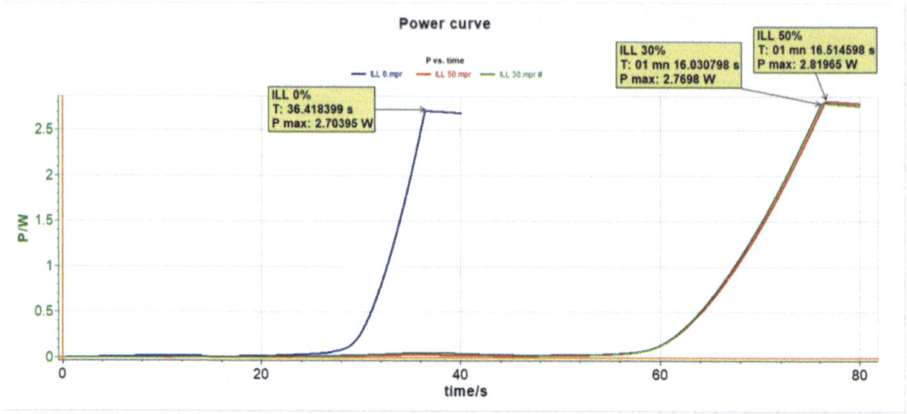

Fig. 6. The power curve

4 Conclusion

This work demonstrated that using [EtOHMIM$^+$][OH$^+$] ionic liquid as an electrolyte in fuel cells resulted in a significant improvement in performance. In contrast to the traditional PEMFC that solely utilizes water as the electrolyte, the addition of 50% [EtOHMIM$^+$][OH$^+$] ionic liquid to the water resulted in an improvement in current density, power density, and efficiency by 15%. The enhancements in the performance of the ionic liquid-PEMFC have a significant impact on its operational efficiency and cycle time, which were somewhat higher in the absence of any additional components.

References

1. R.F. Service: Science **296**, 1222–1224 (2002)
2. Bernaya, C., Marchanda, M., Cassirb, M.: J. Power. Sour. **108**, 139–152 (2002)
3. De Simona, G., Parodib, F., Fermegliaa, M., Taccania, R.: J. Power. Sour. **115**, 210–218 (2003)
4. Hogarth, W.H.J., Benziger, J.B.: J. Power. Sour. **159**, 968–978 (2006)
5. Farooque, M., Maru, H.C.: J. Power. Sour. **160**, 827–834 (2006)
6. Litster, S., McLean, G.: J. Power. Sour. **130**, 61–76 (2004)
7. Sato, T., Morinaga, T., Ishizuka, T.: Preparation, Physicochemical Properties and Battery Applications of a Novel Poly (Ionic Liquid) (2013). https://doi.org/10.5772/51162
8. Macfarlane, D.R., et al.: Energy applications of ionic liquids. Energy Environ. Sci. **7**, 232 (2014)
9. Armand, M., Endres, F., MacFarlane, D.R., Ohno, H., Scrosati, B.: Ionic-liquid materials for the electrochemical challenges of the future. Nat. Mater. **8**, 621 (2009)
10. Neves, L.A., Benavente, J., Coelhoso, I.M., Crespo, J.G.: Design and characterization of Nafion membranes with incorporated ionic liquids cations. J. Membr. Sci. **347**, 42e52 (2010)
11. Zaoui, T., et al.: Synthesis, vibrational and thermal properties of new functionalized 1-(2-hydroxyethyl)-3-methylimidazolium dihydrogenophosphate ionic liquid. J. Mol. Struct.Struct. **1236**, 130–266 (2021)
12. Available from: https://www.horizoneducational.com/solar-hydrogen-science-kit/p1224

Improving Power Quality by Using Active Power Filter Based on DQ Frame of Reference Theory

Benhaoua Larbi Daouli(✉) and Hassan Mana

LAS, Ferhat Abbas University Setif, Setif, Algeria
{Benhaoua.larbidaouli,Hassanmana}@univ-setif.dz

Abstract. In addition to reactive-power management for power factor correction and voltage regulation, load balancing, and voltage flicker reduction, active power filters (APFs) also provide harmonic filtering, damping, isolation, and termination. APFs are now more economically feasible for commercial applications due to recent decreases in the cost of power semiconductor and signal processing devices. This study uses Synchronous Reference Frame Theory to suggest a new approach for obtaining the necessary reference current for the Voltage Source Converter (VSC) of a Shunt Active Power Filter (SAPF). The method achieves optimal control of the SAPF by utilizing the performance of a Proportional-Integral (PID) controller. The reference signals are transformed from the a-b-c stationary frame to the 0-d-q rotating frame by applying Reference Frame Transformation, desired reference signals for Pulse Width Modulation.

Keywords: Shunt Active Power Filter (SAPF) · Proportional Integral Derivative (PID) · Voltage Source Converter (VSC)

1 Introduction

There has been a noticeable rise in harmonic distortion in electric power systems due to the extensive use of power semiconductor switching devices in power supply for computers, televisions, and other microprocessor-based power electronic devices. Serious issues brought on by harmonics can include sensitive power electronic equipment malfunction and excessive heating of electric motors. The key to reducing these problems is harmonic filtering. Active power filters have become a competitive option to passive filters for reactive power compensation and the removal of harmonic currents produced by nonlinear loads. Passive filters have long been utilized for harmonic reduction. Active power filters, sometimes referred to as "active power line conditioners," provide better performance, but in terms of cost and operating losses, they are still marginally less effective than passive filters. However, proactive active power filters are gaining attention as a promising solution for power conditioning applications [1].

© The Author(s), under exclusive license to Springer Nature Switzerland AG 2025
M. Hatti (Ed.): IC-AIRES 2024, LNNS 1238, pp. 269–275, 2025.
https://doi.org/10.1007/978-3-031-80301-7_30

2 Shunt Active Power Filter (SAPF)

One kind of active power filter that is linked in parallel with the load is called a Shunt Active Power Filter (SAPF), as Figure 1 illustrates. The primary goal of the SAPF is to create a pure sinusoidal waveform by injecting a harmonic current into the network that is in opposition to the harmonic current that the load consumes. To do this, a reference signal is produced, which is then utilized to produce gating signals for the Voltage Source Converter (VSC) [1–10].

Compared to passive filters, the SAPF offers a number of benefits, such as enhanced dynamic responsiveness, a broad range of harmonic order compensation, and the capacity to correct for reactive power and unbalanced loads. However, there are also some disadvantages to using SAPFs. These include the potential for the injected current to flow through other passive filters and capacitors connected in the system, the difficulty of creating a high power PWM converter with fast response and low losses, and the higher initial cost when compared to passive filters [1–12].

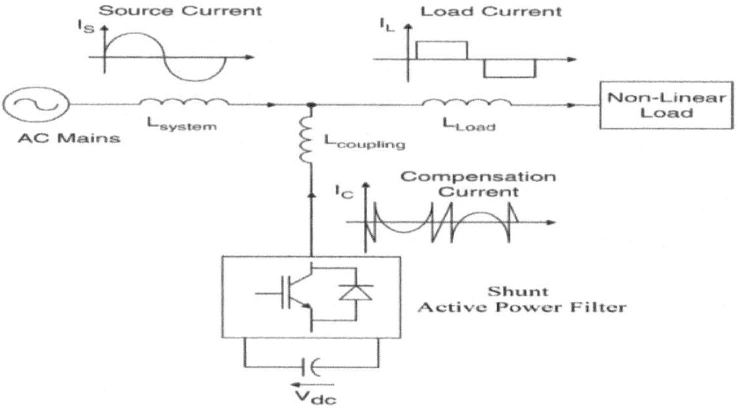

Fig. 1. Shunt Active Power Filter

2.1 Proportional-Integral-Derivative (PID) Controller

In order to achieve optimal control performance, the PID controller is an essential part of the Shunt Active Power Filter (SAPF). The proportional term, the integral term, and the derivative term are the three main components of the PID controller. The SAPF's optimal control performance is achieved using this component The difference (error) between the process variable (PV) and the set-point (SP), that is, the difference between the filter current (If) and the harmonic current reference signal (IH), is the main focus of the PID controller. The PID controller has been implemented in this study Three distinct parameters are used in the PID controller algorithm: the proportional (P), integral (I), and derivative (D). The response to the present error is determined by the Proportional value, the reaction to the sum of the recent errors is determined by the Integral value, and the reaction to the rate of change of the error is determined by the Derivative value. The plant's procedure is modified using the weighted total of these three steps. [7–11]. The PID controller can provide control action tailored to particular process requirements by "tuning" the three constants in its algorithm. This entails modifying the derivative, integral, and proportional gains to get the required response, which could be speed, precision, or stability [2, 4–12] (Fig. 2).

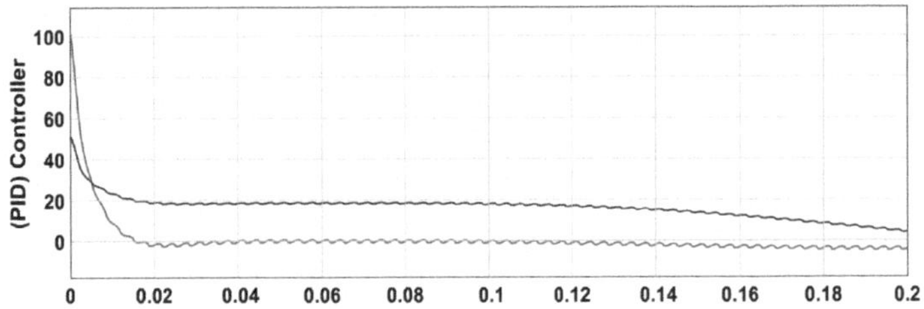

Fig. 2. PID Controller Response

3 Instantaneous Reactive Power Theory

Park put forth the d-q idea in 1929 The d-q hypothesis, sometimes referred to as the park transformation, was first proposed by Robert H. Park in 1929. This theory is a mathematical method that is frequently applied in power systems, motor control, and power electronics to transform three-phase values (ABC) into a rotating reference frame (dq). Power systems, motor drives, power systems, and renewable energy systems are just a few of the numerous applications that have made extensive use of the park transformation, a potent tool for studying and managing three-phase systems [3, 7].

Generated compensating current will be:

$$I_{COMP} = I_{source} - I_{load} \tag{1}$$

where, Icomp=Compensating current, Isource=Source current and, Iload = Load Current (Fig. 3)

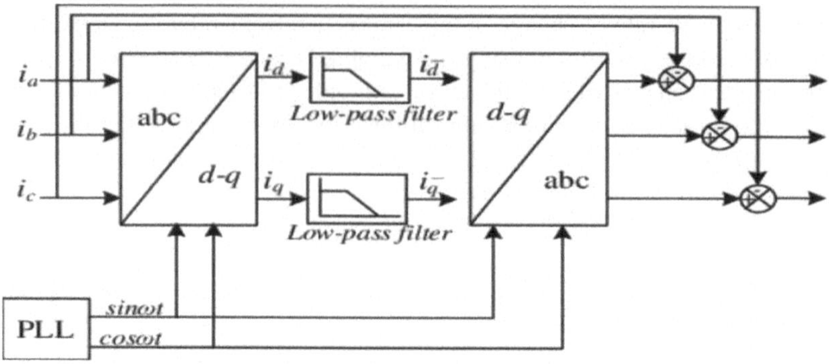

Fig. 3. Basic block diagram of d-q theory

This approach converts the three-phase source voltage and load current into a stationary reference frame with α− β− 0. [3, 7]: Clarke Convolution (from ABC to αβ0) Three-phase quantities (ABC) are transformed into a zero-sequence component (0) and a two-axis stationary reference frame (α − β) using the Clarke transformation The transformation matrix is [3, 7]:

$$\begin{pmatrix} v(\alpha) \\ v(\beta) \\ v(0) \end{pmatrix} = \sqrt{\frac{2}{3}} \begin{pmatrix} 1 & -\frac{1}{2} & -\frac{1}{2} \\ 0 & \frac{\sqrt{3}}{2} & -\frac{\sqrt{3}}{2} \\ \frac{1}{\sqrt{2}} & \frac{1}{\sqrt{2}} & \frac{1}{\sqrt{2}} \end{pmatrix} \begin{pmatrix} v(a) \\ v(b) \\ v(c) \end{pmatrix} \tag{2}$$

$$\begin{pmatrix} i(\alpha) \\ i(\beta) \\ i(0) \end{pmatrix} = \sqrt{\frac{2}{3}} \begin{pmatrix} 1 & -\frac{1}{2} & -\frac{1}{2} \\ 0 & \frac{\sqrt{3}}{2} & -\frac{\sqrt{3}}{2} \\ \frac{1}{\sqrt{2}} & \frac{1}{\sqrt{2}} & \frac{1}{\sqrt{2}} \end{pmatrix} \begin{pmatrix} i(a) \\ i(b) \\ i(c) \end{pmatrix} \tag{3}$$

Transformation of Park (αβ0 to DQ0): The Park transformation aligns the reference frame with the rotor position in machines by transforming the stationary α-β-0 frame into a spinning DQ0 frame. The transformation matrix is [3, 7]:

$$\begin{pmatrix} v(d) \\ v(d) \\ v(0) \end{pmatrix} = \begin{pmatrix} \cos(\theta) & \sin(\theta) & 0 \\ -\sin(\theta) & \cos(\theta) & 0 \\ 0 & 0 & 1 \end{pmatrix} \begin{pmatrix} v(\alpha) \\ v(\beta) \\ v(0) \end{pmatrix} \tag{4}$$

For currents, the transformation is:

$$\begin{pmatrix} i(d) \\ i(d) \\ i(0) \end{pmatrix} = \begin{pmatrix} \cos(\theta) & \sin(\theta) & 0 \\ -\sin(\theta) & \cos(\theta) & 0 \\ 0 & 0 & 1 \end{pmatrix} \begin{pmatrix} i(\alpha) \\ i(\beta) \\ i(0) \end{pmatrix} \tag{5}$$

Composite Conversion (ABC to DQ0) By integrating the transformations of Clarke and Park, the direct transformation from ABC to DQ0 is [3, 7]:

$$\begin{pmatrix} v(d) \\ v(q) \\ v(0) \end{pmatrix} = \begin{pmatrix} \cos(\theta) & \sin(\theta) & 0 \\ -\sin(\theta) & \cos(\theta) & 0 \\ 0 & 0 & 1 \end{pmatrix} \sqrt{\frac{2}{3}} \begin{pmatrix} 1 & -\frac{1}{2} & -\frac{1}{2} \\ 0 & \frac{\sqrt{3}}{2} & -\frac{\sqrt{3}}{2} \\ \frac{1}{\sqrt{2}} & \frac{1}{\sqrt{2}} & \frac{1}{\sqrt{2}} \end{pmatrix} \begin{pmatrix} v(a) \\ v(b) \\ v(c) \end{pmatrix} \tag{6}$$

Likewise, for currents:

$$\begin{pmatrix} i(d) \\ i(q) \\ i(0) \end{pmatrix} = \begin{pmatrix} \cos(\theta) & \sin(\theta) & 0 \\ -\sin(\theta) & \cos(\theta) & 0 \\ 0 & 0 & 1 \end{pmatrix} \sqrt{\frac{2}{3}} \begin{pmatrix} 1 & -\frac{1}{2} & -\frac{1}{2} \\ 0 & \frac{\sqrt{3}}{2} & -\frac{\sqrt{3}}{2} \\ \frac{1}{\sqrt{2}} & \frac{1}{\sqrt{2}} & \frac{1}{\sqrt{2}} \end{pmatrix} \begin{pmatrix} i(a) \\ i(b) \\ i(c) \end{pmatrix} \tag{7}$$

4 Simulation Results and Analysis

See Fig. 4.

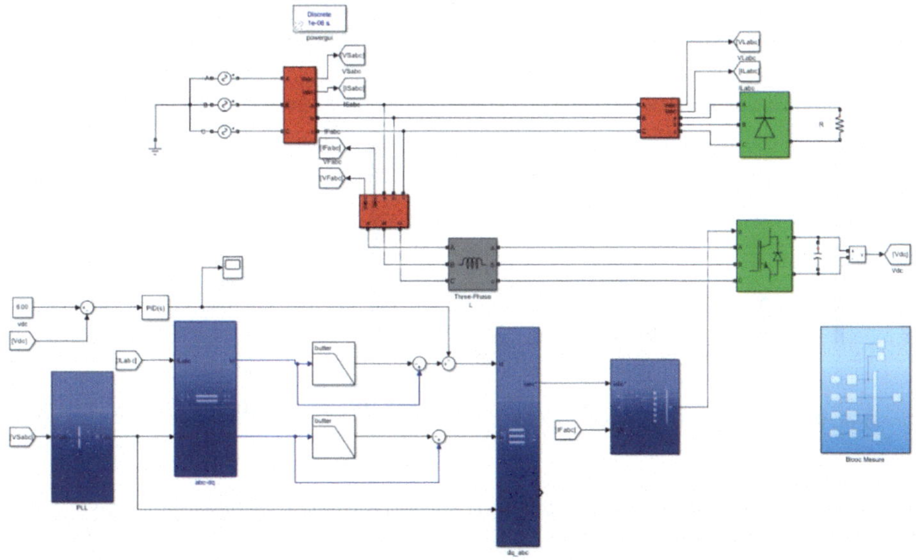

Fig. 4. Shunt APF Simulink model

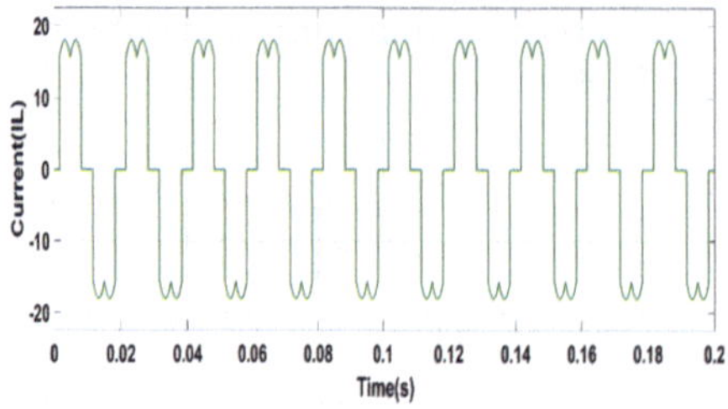

Fig. 5. Load current.

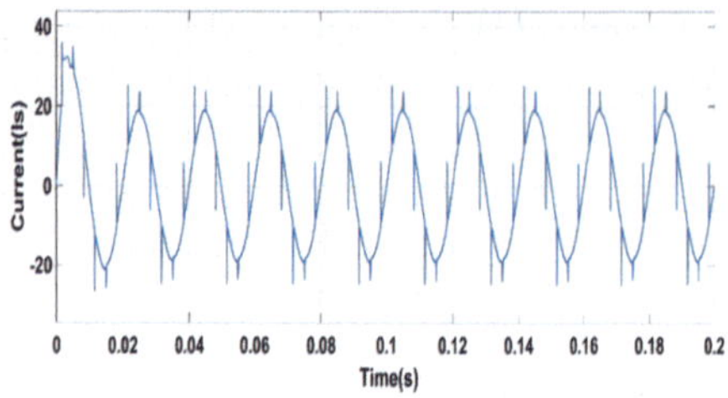

Fig. 6. Source current

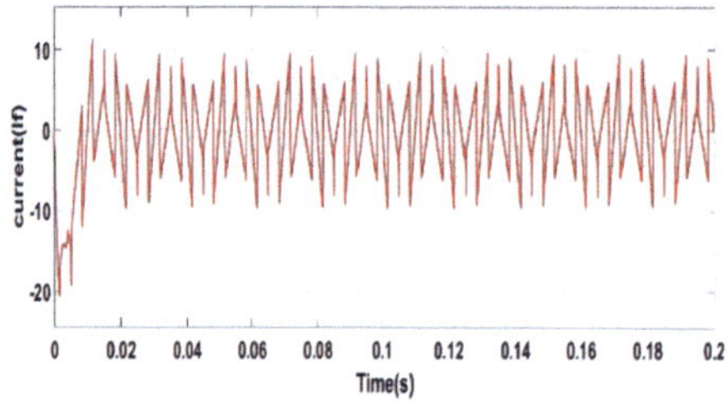

Fig. 7. Compensating current

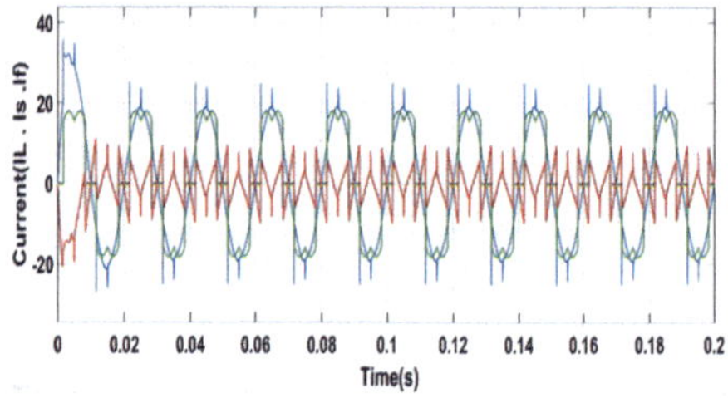

Fig. 8. Load current and Compensating current and Source current

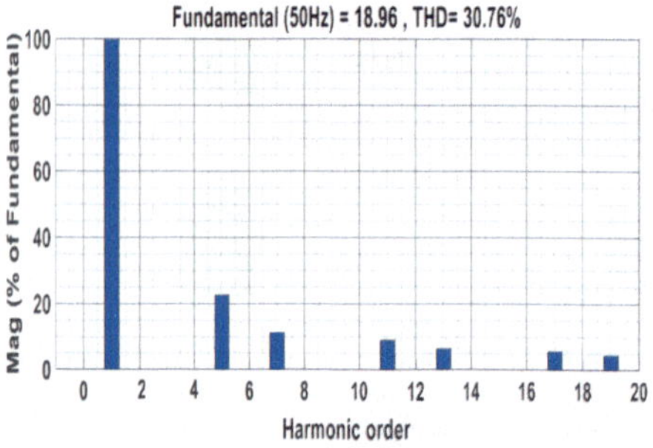

Fig. 9. FFT analysis of Load current

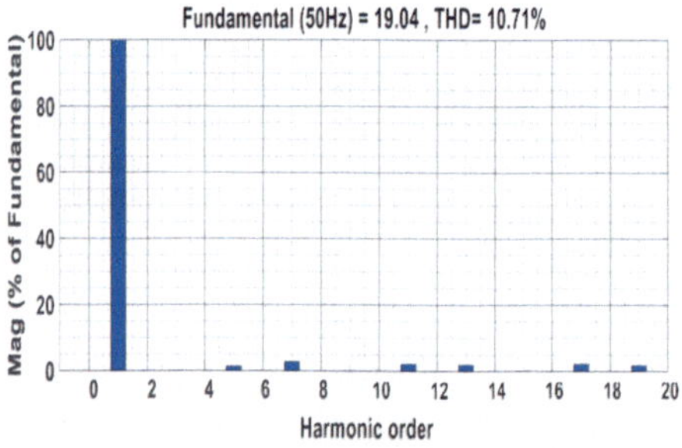

Fig. 10. FFT analysis of source current

The MATLAB software is utilized to verify the performance of the shunt active power filter. The source current waveform is displayed in Figures 5 and 6 both before and after the shunt active power filter is inserted. It shows that the source current is severely distorted until the shunt active power filter is added to the system In this instance, the Total Harmonic Distortion (THD) of the source current was 30.76% prior to the insertion of the shunt active power filter, and it drops to 10.71% following the insertion (Figs. 7 and 8).

The FFT analysis of Rectifier R's source current is displayed in Figures 9 and 10. The comparison of each harmonic before and after harmonic reduction using a shunt active power filter is shown in Figures 9 and 10.

5 Conclusion

The Shunt Active power filter with PID controller was examined in this paper. The performance of the Shunt Active Power Filter with both proposed controller circuits for reference current generation were examined with simulation model and the results were compared. Get good results performance in terms of active harmonic filtering Harmonics. FFT analysis is done through powergui block and results compared with standard.

References

1. Sunitha, M., Kartheek, B.N.: Elimination of harmonics using active power filter based on DQ reference frame theory. Int. J. Eng. Trends Technol. **4**(4), 781 (2013)
2. Eslami, A., Negnevitsky, M., Franklin, E., Lyden, S.: Review of AI applications in harmonic analysis in power systems. Renew. Sustain. Energy Rev. **154**, 111897 (2022)
3. Assiene Mouodo, L.V., Ali, A.-H.M., Tamba, J.G., Olivier, T.S.M.: Comparison and classification of six reference currents extraction algorithms for harmonic compensation on a stochastic power network: case of the TLC hybrid filter. Cogent Eng. **9**(1), 2076322 (2022)
4. Goel, N., Hasan, S., Kalaichelvi, V.: Modelling, simulation and intelligent computing proceedings of MoSICom 2020. In: Proceedings of MoSICom, vol. 1 (2020)
5. Moeini, A., Dabbaghjamanesh, M., Kimball, J.W., Zhang, J.: Artificial neural networks for asymmetric selective harmonic current mitigation-PWM in active power filters to meet power quality standards. IEEE Transactions on Industry Applications (2020)
6. Alhamrouni, I., Hanafi, F.N., Salem, M., Rahman, N.H.A., Jusoh, A., Sutikno, T.: Design of shunt hybrid active power filter for compensating harmonic currents and reactive power. TELKOMNIKA (Telecommun. Comput. Electron. Control) **18**(4), 2148 (2020)
7. Uddin, M.N., Amin, M.T.: Design and simulation of active power filter based on feed forward neural network for harmonic detection and elimination. In: 2020 International. Conference on Computer, Electrical & Communication Engineering (ICCECE), pp. 1–6. IEEE (2020)
8. Aiwa, F.F., Alkhayyat, M.T., Suliman, M.Y.: A review on PQ theory based shunt active power filter. Design Eng. (Toronto) **6**, 6919–6938 (2021)
9. Iqbal, M., et al.: Neural networks based shunt hybrid active power filter for harmonic elimination. IEEE Access **9**, 69913–69925 (2021)
10. Singh, V., Iqbal, S. J., Gupta, S., Yadav, A.: Performance evaluation of a shunt active power filter for current harmonic elimination. In: 2021 IEEE Region 10 Symposium (TENSYMP), pp. 1–6. IEEE (2021)
11. Mahmoud, M.O., Mamdouh, W., Khalil, H.: Source current harmonic mitigation of distorted voltage source by using shunt active power filter. Int. J. Electr. Comput. Eng. **10**(4), 3967–3977 (2020)
12. Li, D., Wang, T., Pan, W., Ding, X., Gong, J.: A comprehensive review of improving power quality using active power filters. Electr. Power Syst. Res. **199**, 107389 (2021). Clerk Maxwell, J.: A Treatise on Electricity and Magnetism, 3rd ed., vol. 2. Clarendon, Oxford, pp. 68–73 (1892)

An Overview of Analytical Methods Utilized for Microbiological and Physiochemical Characterization of Locally Grown Agricultural Crops During Solar Drying

Akil Loumani(✉), Ahmed Mediani, and Meriama Foullanine

Centre de Développement des Energies Renouvelables, Unité de Recherche en Energies Renouvelables en Milieu Saharien, URERMS, CDER 01000 Adrar, Algeria
akil076@live.fr

Abstract. Solar drying is sustainable and cost-effective method for preserving crops, and understanding the changes microbiological and physiochemical properties is crucial for ensuring products quality and safety. In this paper we focus about of the analytical method using for the microbiological and physiochemical characterization effected for locally grown agriculture crops during the solar drying including microbial load, presence of the pathogens, and spoilage organisms. Additionally, the paper we explores the analytical methods used for to evaluate the physiochemical attributes, including chemical composition such as moisture, ash,fat,protein.etc. Nutritional compounds such as minerals and vitamins, physical properties as texture, colour, density, sensory characteristics as test odor, appearance. Hence, achieving the goal of desirable high quality of dried agriculture crops is heavily dependent on effective processing involving drying and physiochemical and microbiological analytic methods.

Keywords: Solar drying · Microbiological · Physiochemical · locally grown agriculture crops · Adrar province

1 Introduction

Solar drying is an important technique in local agriculture for crop preservation .it is an effective and sustainable technique approach to extending their shelf life and reducing the post-harvest losses through the elimination of moisture. This vital technique can helps retain the nutritional content of crops by removing moisture inhibiting the growth of Microorganisms and minimizes enzymatic activity, thus preserving the vitamins, minerals, and other essential nutrients present in the crops. Hence, a solar drying enables the storage of nutritious food for longer duration in areas where obtain to fresh product is so limited due to the hard weather conditions, so it provides a cost effective and efficient solution to mitigate these losses and allowing farmers to store their produce extended periods without significant quality degradation [1].

© The Author(s), under exclusive license to Springer Nature Switzerland AG 2025
M. Hatti (Ed.): IC-AIRES 2024, LNNS 1238, pp. 276–282, 2025.
https://doi.org/10.1007/978-3-031-80301-7_31

A solar drying relies on renewable energy making it an environmentally friendly alternative to traditional drying methods that consume fossil fuels or electricity reducing greenhouse gas emissions and contributes to sustainable agriculture practice [2]. And the capacity the adaptation to the local conditions and available resources using low cost, locally materials and can be implemented in small scale make it accessible to smallholder farmer .in the other hands, solar drying facilities value addition to agriculture products, in addition, these dried products as dried fruits, vegetables, herbs and spice fetch the high price compered toothier counterparts fresh, and they can be sold year-round, even during off-seasons, where the farmer can improve and diversify their incomes and promote food security in their communities. However, the microbiological and physiochemical characterization plays a role significance in ensuring the quality and the safety of products.it helps the quality assurance in evaluating the various parameters that impact the quality products including factors such as pH, moisture, nutrients composition and presence of contaminants and the product safety by assaying the presence of harmful microorganism such as bacteria, molds and yeasts [2].

The microorganisms can pose because spoilage, food bore illnesses, and other health risks. Thus, identification and quantification of the microorganisms allows to void the contamination and ensures safety products. Other key reasons of characterization of microbiological and physiochemical are determining shelf life and stability of products by monitoring the parameters as water activity pH, and oxidative indicators, where can assess the product's susceptibility to spoilage and rancidity and microbial growth. This data give a general idea to establish appropriate conditions and determining the expiration dates of products.at the end, Microbiological and physiochemical characterization is crucial steps to demonstrate a commitment to product quality and safety enhancing the consumers' confidence and purchase the products that have been rigorously evaluated for quality and safety. Finally, our work involved the application of microbiological and physiochemical techniques conducted in our laboratory [2].

These techniques were specifically applied to locally grown agricultural products. The purpose of this work was to provide students with valuable research material to enhance their final studies and as reference to doing their work plans (Fig. 1).

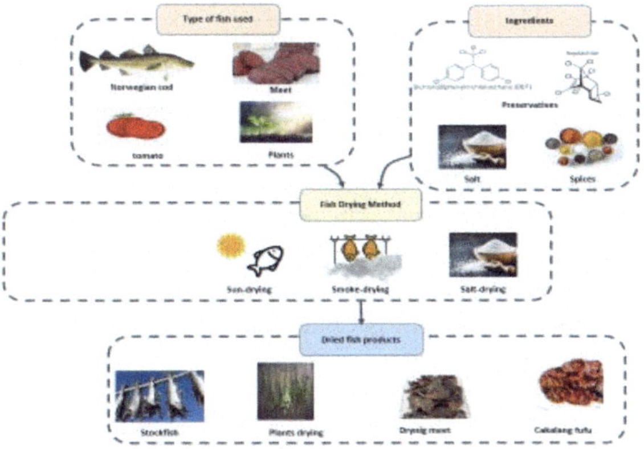

Fig. 1. Different Types and products of drying methods

2 Materials and Method

2.1 Sample Collection Technique

In this technique, the samples were carefully selected (handpicked) at the commercial ripening stage, taking into consideration their homogeneity in color, size, firmness, and skin color, while ensuring they were free from any visible defects. The selection process took place early in the morning. Immediately after selection, the chosen samples were transferred to the laboratory, ensuring cold conditions were maintained throughout transportation. This was done to prepare the figs for the indirect solar drying process [2].

2.2 Sample Preparation and Drying Conditions

For ensuring the success drying process, it must be to wash and clean the selected fresh products to remove any dirt or debris or contaminants [3]. Also to sort and discard any damaged or spoiled samples, than, in second step, cutting the products into uniform sizes and layer in the sample tray. So samples apply pretreatment methods as blanching, sulfuring and salting to enhance drying efficiency and preserve color and quality.in third step, drying conditions, a chamber drying should be well- ventilated and provide protection against insects and birds, dust, then, the parameters of drying as temperature, humidity, drying time and storage should be controlled to avoid loss of color or nutrient degradation and efficient moisture removal and a suitable packaging for cooling and stored to maintain their quality and avoid reabsorption [4].

2.3 Physiochemical Characterization Techniques

2.3.1 Determination of Chemical Composition

Moisture Content The moisture content of the fresh pepper samples was determined using a glass petri dish. A 10g sample of the vegetable matter was sliced and placed on a perforated wooden plate. The sample was then subjected to drying in an oven maintained at a temperature of 135 ± 2 °C for a duration of 2 h [9]. The moisture content was calculated using the following equation [4].

$$\text{Moisture}\% = \frac{\mathbf{ws(w2 - w1)}}{\mathbf{ws}} \times \mathbf{100}$$

where: W1: weight of dish (g), W2: weight of dish after drying (g), Ws: weight of sample (g).

Ash Content Ash is described as the inorganic residue that remain after complete oxidation of organic matter in a food sample. The ash content is a comparatively composition that usually analyzed when evaluating the quality of agriculture food products.to determinate the ash content , the 5 g of sample are calcined in a muffle furnace at 550 °C until a pal ash with consistent weight is obtained, the proportion of organic substance was calculate using the equation [4] :

$$\text{MO}\% = (M\text{i} - M\text{r})/\text{Mr}.100$$

Mi: represents the initial weight of sample (5g)

Mr: represents the weight of the ash residue obtained from calcination process.

2.3.2 FAT Conent

The Soxhlet extraction method is a well-established technique for fat determination, providing reliable results for various samples. Following the AOAC guidelines ensures standardization and accuracy in the analysis [4].

The fat content of the samples was determined using the Soxhlet extraction method according to AOAC guidelines. In this method, the extraction is performed at elevated temperatures, specifically by immersing the sample in boiling solvent (hexane) and allowing it to soak. After the soaking process, the sample is rinsed with hexane to remove any residual fat. The next step involves evaporating the hexane through distillation. By evaporating the solvent, the extracted lipid content is concentrated and obtained. The final step entails weighing the extracted lipid to determine its content in the sample [5].

Protein Dosage The Kjeldahl method involves the conversion of organic nitrogen present in the sample into mineral nitrogen through the use of concentrated boiling NH4S2O4 (ammonium sulfate). This step breaks down the protein molecules and releases the nitrogen in the form of ammonia. The released ammonia is then distilled and captured in an acid solution. The captured ammonia is subsequently titrated with sulfuric acid (H2SO4) in the presence of a colored indicator, typically acid boric, using acidimetry. The indicator changes color when the ammonia is neutralized by the sulfuric acid, allowing for the determination of the amount of ammonia released. To calculate the total protein content, a conversion factor of 6.25 is commonly used. This factor is based on the assumption that proteins contain, on average, 16% nitrogen. By multiplying the nitrogen content obtained from the Kjeldahl analysis by 6.25, the crude protein content of the sample can be determined [5]

Carbohydrate The determined sugars by the colorimetric method using phenol and sulfuric acid developed by DuBois et al, considered the most reliable method of measuring simple sugars and oligosaccharides content. This technique is based on the action of concentrated and hot mineral acids. The hexoses and pentoses in the medium undergo extensive internal dehydration, followed by cyclization resulting in the formation of

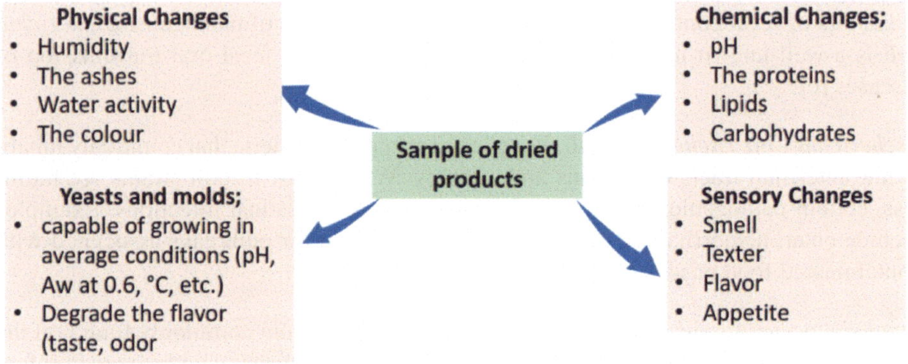

Fig. 2. Diagram of Physicochemical analysis steps of dried products.

derivatives of furfural and 5-hydroxymethylfurfural, reacting with phenol. The formation of a yellow-red complex makes it possible to monitor the concentration of total sugars in the sample by reading the absorbance at 485 nm (Fig. 2) [5].

2.3.3 Microbiological Characterization Techniques

Microbial Load Determination Methods Microbial load determination is a crucial step in assessing the level of microbial contamination or evaluating the efficacy of antimicrobial treatments by quantifying the number of microorganisms present in a given sample. The most commonly used technique for microbial determination is the Plate Count Method, also known as the viable count or colony-forming unit (CFU) method [7]. This method involves spreading a known volume of the sample onto a solid agar medium, followed by incubation of the agar plates under suitable conditions to allow visible microbial colonies to form. Finally, the colonies are counted, and the microbial load is expressed as CFU per unit volume or weight of the original sample. The most researched pathogenic bacteria include Mesophilic bacteria, total coliforms, fecal coliforms, *Escherichia coli*, *Staphylococcus aureus*, and *Salmonella* typhi (Fig. 3). These bacteria are of significant interest in microbiological research and have implications for public health and food safety [5].

Mesophilic Bacteria Mesophilic bacteria are a broad group of bacteria that grow optimally at moderate temperatures (around 20–45 °C).They include various pathogenic bacteria, such as Campylobacter, *Listeria monocytogenes*, and certain strains of Escherichia coli, that can cause foodborne illnesses [7].

Total Coliforms Total coliforms are a group of bacteria that are commonly found in the environment, including soil, water, and the intestinal tracts of warm-blooded animals. They serve as indicators of potential fecal contamination and the presence of other harmful microorganisms. Some coliform species, including Escherichia coli, can cause gastrointestinal infections [7].

Fecal Coliforms Fecal coliforms are a subgroup of total coliforms primarily found in the intestines and feces of warm-blooded animals, including humans. They are used as indicators of fecal contamination and the potential presence of pathogens. *Escherichia coli* is a well-known fecal coliform and is associated with fecal-oral transmission of diseases [6].

Escherichia Coli *Escherichia coli* (E. coli) is a species of bacteria that commonly inhabits the intestinal tracts of humans and animals. While most E. coli strains are harmless, certain pathogenic strains can cause severe gastrointestinal infections. Examples include enterohemorrhagic *E. coli* (EHEC), responsible for outbreaks associated with contaminated food or water [7].

Staphylococcus Aureus *Staphylococcus aureus* is a bacterium commonly found on the skin and mucous membranes of humans and animals. It can cause a wide range of infections, ranging from mild skin infections to more severe conditions such as pneumonia and bloodstream infections [7].

Salmonella *Typhi Salmonella* typhi is the bacterium responsible for causing typhoid fever, a systemic illness characterized by high fever, abdominal pain, and gastrointestinal symptoms. It is primarily transmitted through contaminated food or water. These pathogenic bacteria are extensively studied due to their impact on public health, the understanding of their virulence mechanisms, the development of diagnostic methods, and the implementation of effective prevention and control measures [7].

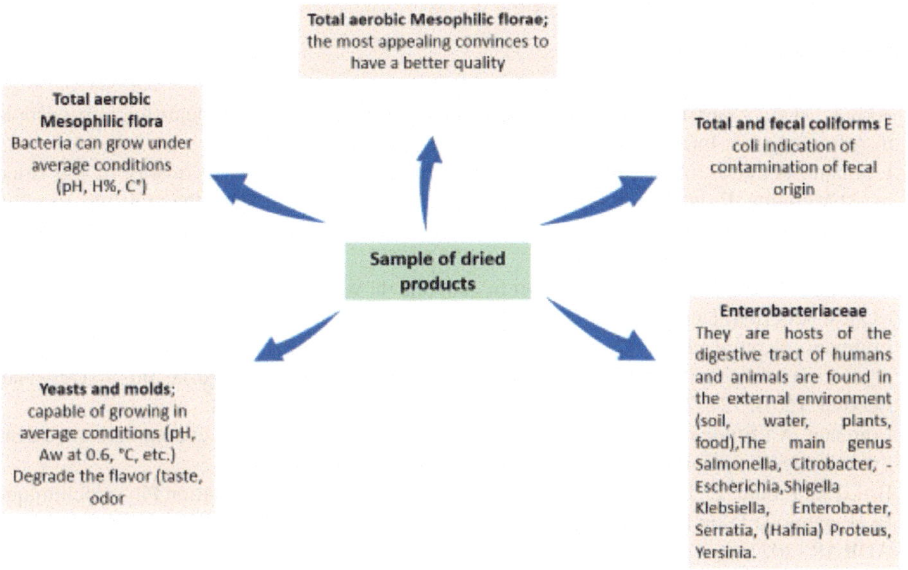

Fig. 3. Diagram of Microbiological analysis steps of dried products

3 Conclusion

In conclusion, the evaluation of microbiological and physicochemical analyses of dried agricultural products is of paramount importance in ensuring the quality, safety, and market viability of these commodities. The synergy between these two analytical approaches provides a comprehensive understanding of the products' characteristics and suitability for consumption and trade.

Microbiological analysis unveils the microbial load and presence of pathogens, safeguarding consumers from potential health hazards. Simultaneously, physicochemical analysis offers insights into moisture content, nutrient composition, and sensory attributes, which collectively determine the product's shelf life, nutritional value, and overall appeal.

This dual evaluation enables informed decision-making throughout the supply chain. Processors can fine-tune drying methods to preserve nutritional integrity and prevent spoilage, contributing to enhanced marketability. Regulatory bodies and quality control institutions benefit from accurate data to enforce safety standards and ensure adherence to international trade regulations.

Ultimately, the harmonized evaluation of microbiological and physicochemical aspects in dried agro products fosters consumer trust, facilitates global trade, supports sustainable agriculture, and bolsters economic growth. By upholding quality and safety through rigorous analyses, we pave the way for a healthier, more resilient food system and contribute to the well-being of both producers and consumers.

References

1. Akil, L., et al.: Assessment biodiversity of medicinal plants used in treatments for native people in Adrar, Algeria. Int. J. Des. Nat. Ecodyn. **17**(3), 469–473 (2022)
2. Loumani, A., et al.: Assessment microbiological and physicochemical quality of figs dried in [3] Meriama, F., et al.: Experimental determination and modeling of the moisture-sorption isotherms and isosteric heat of tobacco leaves. Instrumenta. Mes. Métrologies **20**(5), 269-277 (2021)
3. Loumani, A.: Etude microbiologique et hygiénique du yaourt fabriqué et commercialisé dans l'Ouest Algérien; Université d'Oran1-Ahmed Ben Bella (2011)
4. Loumani, A., et al.: Experimental measurement of isothermal sorption, microbiological and physicochemical analysis of dried tomatoes cultivated in Adrar, Algeria. Int. J. Des. Nat. Ecodyn. **15**(5), 721–728 (2020). https://doi.org/10.18280/ijdne.150514
5. Loumani, A.: Etude microbiologique et hygiénique du yaourt fabriqué et commercialisé dans l'Ouest Algérien. Université d'Oran1-Ahmed Ben Bella (2011)
6. Briki, A., Zidani, Z., Laksaci, H., et al.: Etude De Séchage Et Caractérisation Physicochimique Et Hygiénique D'un Produit Alimentaire Du Site D'Adrar. UNIVERSITE AHMED DRAIA-ADRAR (2021.)
7. Akil, L., Ahmed, B.: Rep PCR characterization of lactic acid bacteria isolated from sheep's milk in Algeria and Assessment of their probiotics potential. Asian J. Microbiol. Biotechnol. Env. Sci. **20**(3), 836–842 (2018)

Emerging Technologies and Applications

OC Fault Detection in Single Phase NPC Inverter

H. Mechri(✉), A. Kheldoun, and A. Ammar

Laboratory of Signals and Systems, Institute of Electrical and Electronic Engineering, University M'hamed Bougara, Boumerdes, Algeria
h.mechri@univ-boumerdes.dz

Abstract. This paper proposes a novel method for rapid and accurate detection of open-circuit faults in the switching devices of single-phase Neutral Point Clamped (NPC) inverters. The approach employs the combination of average current analysis and Model Predictive Control (MPC) to achieve fault identification without requiring additional hardware or computationally intensive calculations. Furthermore, the work investigates the capability of MPC to tolerate open-circuit faults, specifically in clamping diodes. This tolerance has the potential to enhance the overall system resilience and performance during fault scenarios. The effectiveness of the proposed fault detection method is validated through simulation analysis.

Keywords: Fault detection · MPC · Single phase · Open circuit fault · NPC

1 Introduction

The single-phase three-level neutral point clamped (NPC) inverter is a widely adopted topology in power conversion applications, particularly for grid-connected renewable energy systems. This multilevel inverter configuration offers several advantages over traditional two-level inverters, including lower harmonic distortion, higher efficiency, and suitability for high-voltage applications [1–3]. NPC inverters present significant challenges in design and implementation due to the necessity for precise control strategies and advanced fault management [4]. Inverter faults are primarily divided to short-circuit (SC) and open-circuit (OC) faults. Sc faults are highly dangerous as they can rapidly raise the current and harm the system. However, there are various SC faults that can become OC faults using fuses. Apart from switch faults, other issues like driver faults, misfires, and connection failures can produce similar outcomes [5]. While OC faults are less hazardous than SC faults, they still degrade system functionality and can cause a secondary fault in the inverter. While research on open-circuit fault diagnosis has flourished for three-phase converters in both two-level and three-level topologies, single-phase converters haven't received the same level of attention. Existing approaches for single-phase converters include adding extra current sensors for condition monitoring, as suggested in [6]. While this method offers comprehensive monitoring, it can be complex and expensive to implement. In [7], the research has explored employing output current analysis for open-circuit fault diagnosis. This method is valuable for fault detection, but it fails

© The Author(s), under exclusive license to Springer Nature Switzerland AG 2025
M. Hatti (Ed.): IC-AIRES 2024, LNNS 1238, pp. 285–292, 2025.
https://doi.org/10.1007/978-3-031-80301-7_32

to locate the fault. While online fault diagnosis and localization using sampling theory [8] and mixed logical dynamic (MLD) models with grid current residual generation [9] have been proposed, these methods can introduce additional complexity. This research studies the performance analysis of the single phase three-level neutral point clamped (3L NPC) inverter under open circuit fault in switches. In response to these faults, an open-circuit fault detection algorithm is suggested which is based on average current using model-predictive approach.

2 MPC Control Strategy for Single Phase 3L-NPC Inverter

The single phase 3L-NPC inverter under study is illustrated in (Fig. 1). Single-phase bridge arm, with four sets of switching elements, constitute the NPC inverter. Three switching states are defined when clamping diodes are included: $+1, 0$ and -1 [10]. The bridge arm of the single-phase inverter has three switching states defined as follow:

$$s = \begin{cases} +1(\mathrm{P}),\ s_1 \text{ and } s_2 \text{ are turned on} \\ 0(\mathrm{O}),\ s_2 \text{ and } s_3 \text{ are turned on} \\ -1(\mathrm{N}),\ s_3 \text{ and } s_4 \text{are turned on} \end{cases} \tag{1}$$

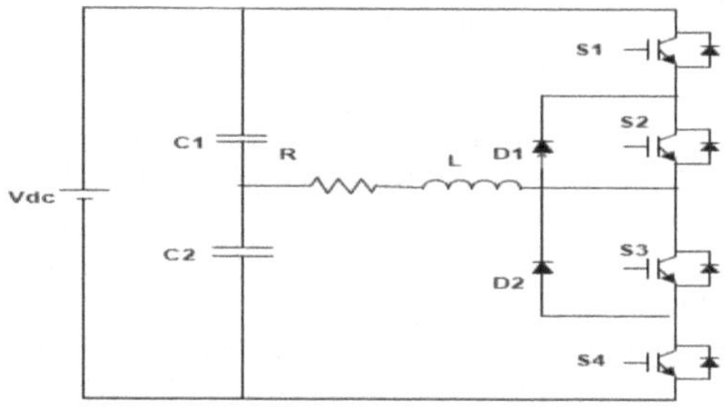

Fig. 1. Single phase 3L-NPC inverter topology

2.1 Model Predictive Control

In [11], the expression for the dynamic vector representing the inverter's output current is formulated as follows:

$$\frac{di_L}{dt} = \frac{1}{L}(v - Ri_L) \tag{2}$$

L, R, are load inductance and resistance respectively.

Using Forward Euler'method at the sampling time Ts, the equation in (2) can be transformed into a discrete predictive model s as:

$$i_L(k+1) = (1 - \frac{RT_S}{L})i_L(k) + \frac{T_S}{L}(v(k)) \tag{3}$$

S.t Eq. (1) where $v \in U = \{V_{dc}/2, 0, -V_{dc}/2\}$.

Then the cost functions is calculated for m = 1…3 as:

$$g_m = \left| i(k) - i^p(k+1) \right| \tag{4}$$

where, $i(k)$ is the reference values, and $i^p(k+1)$ is the predicted value.

The DC link capacitors C1 and C2, which are utilized in this inverter, tend to have equivalent capacities.

3 Fault Detection and Localisation

3.1 Operation of the System Under OC Fault

When an open-circuit failure happens in *S1,* the P state becomes unattainable. If the fault happens while positive current is present, the current flows through *D1* and *S2* until it reaches zero, after which *D1* becomes reverse-biased, stopping the positive current flow. In *S2* Open-Circuit Fault, the current follows the path through *D1* and *S2* when the switching state is O, carrying a positive current. However, if the fault happens in *S2*, the switching states P and O become impossible. This signifies the absence of any current paths, and the positive current in the faulty phase ceases to flow.

In the case of an *S3* failure, current flows through *S3* and *D2*, resulting in a negative current in state O. Conversely, in state *N*, the current is routed through *S3* and *S4.* If an open-circuit fault occurs in *S3* while a negative current is present, it disrupts the negative current flow, thereby making both states O and N unattainable. When *S4* fails, state N becomes unattainable, whereas state O remains feasible. If a fault occurs in *S4* during a negative current flow, the negative phase current in the affected phase is blocked due to the reverse-bias condition of *D2* [12].

3.2 Fault Identification Using MPC

To detect the fault location (upper swiches *S1*or *S2*,lower swiches *S3* or *S4*) the predicted average current i_{avgp} is used. The predicted current is calculated using Eq. (3), then the optimal current is calculated as follows:

$$i_{opt} = \arg\left\{ \begin{matrix} \min, g_m \\ i(k) \end{matrix} \right\} \tag{5}$$

The predicted average current is then calculated as follows:

$$i_{avgp} = 0.63\, i_{opt} \tag{6}$$

For the lower switches, *S3, S4,* i_{avgp} became negative at the moment of the fault, but for the upper switches, *S1, S2* it became negative after half a cycle of the fault occurrence. Based on these differences the detection algorithm thresholds are set to distinguish between the two types of faults. The flowchart in (Fig. 2) explains the proposed algorithm.

In order to simplify the understanding of the results, we set a fault signal indicators Y as explained in (Table 1).

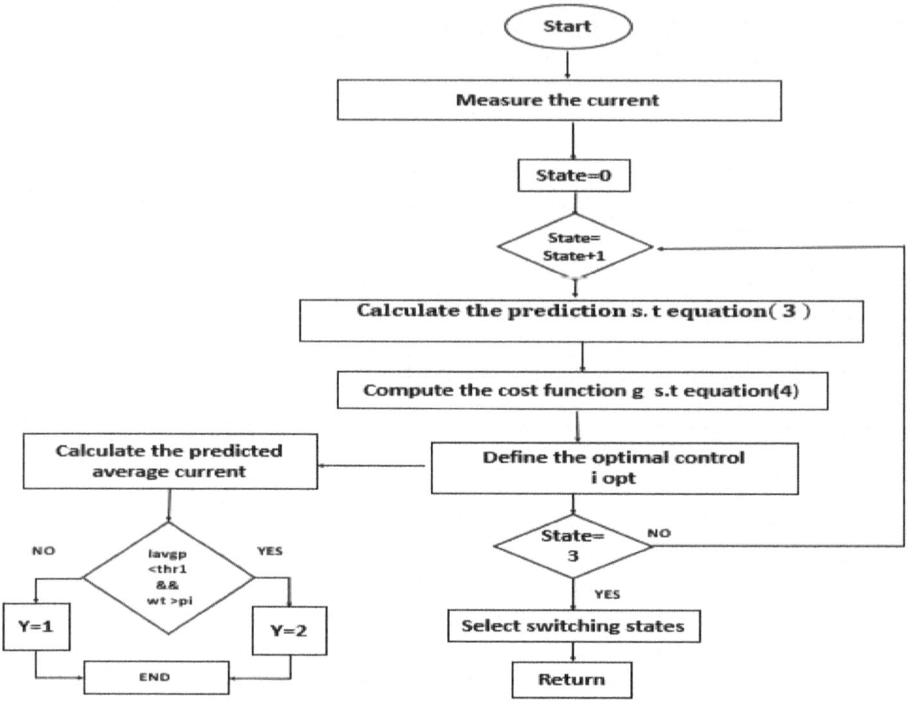

Fig. 2. Flow chart for OC fault detection using MPC

Table 1. Fault signal indicator

Fault signal Y	Faulty switch
0	No fault
1	S1
1	S2
1	D1
2	S3
2	S4
2	D2

4 Similation Results

The system is simulated using MATLAB/Simulink to show the benefits of the proposed strategy. The simulation parameters are listed in (Table 2).

Table 2. Simulation parameters

Parameters	Value	Unit
DC-link voltage	200	V
Capacitor (C1, C2)	300	μF
load Inductance (L)	40	mH
load Resistance (R)	10	Ω
Sampling Frequency	10	khz

4.1 Detection of Faulty Group

Under normal operating conditions, the output current accurately tracks the reference current, with the fault signal indicator parameter Y remaining at zero. This fault-free scenario is illustrated in (Fig. 3).

(Figure 4(b))shows the current alteration under an *S1* fault occurring at t = 0.05 s. The positive half-cycle becomes zero, resulting in distorted current. At the same time, (Fig. 4(a) indicates the value 1 for Y, signifying a fault in the upper switches, with a very short detection time. Similar observations for current and the Y parameter are seen in (Fig. 5) for an *S2* fault. In (Fig. 6(b)), the current change under an *S3* fault at t = 0.05 s is shown, where the negative half-cycle drops to zero, causing current distortion. (Fig. 6(a)) shows a Y parameter value of 2, indicating the fault in the lower switches. Corresponding results for current and Y parameters are shown in (Fig. 7) for an *S4* fault. The detection times for each scenario are detailed in (Table 3).

The results indicate that the detection algorithm identifies faults in less than one fundamental period, which corresponds to a few milliseconds. This rapid detection time is highly advantageous for system protection, as it allows for swift corrective actions to prevent damage. The effectiveness of this detection algorithm enhances the overall reliability and robustness of the system, ensuring that faults are managed promptly and efficiently.

Table 3. Fault detection time

Y	Detection time(ms)	Faulty switch
1	12.2	S1
1	11	S2
2	1.2	S3
2	3	S4

4.2 Effect of MPC on Tolerating the Output Current Under OC Fault in Clamping Diodes

Model Predictive Control (MPC) is designed to adjust output current by minimizing the difference between the reference and actual current. When an open switch fault occurs in one of the clamping diodes, the MPC adjusts the remaining operational switches to correct the error, thus maintaining current regulation. Figure 8 illustrates the output current under open switch faults in clamping diodes *D1* and *D2*. As shown, the system continues to track the reference current even after the switches open at t = 0.05 s. This demonstrates that MPC effectively tolerates clamping diode open-circuit faults, ensuring stable system performance.

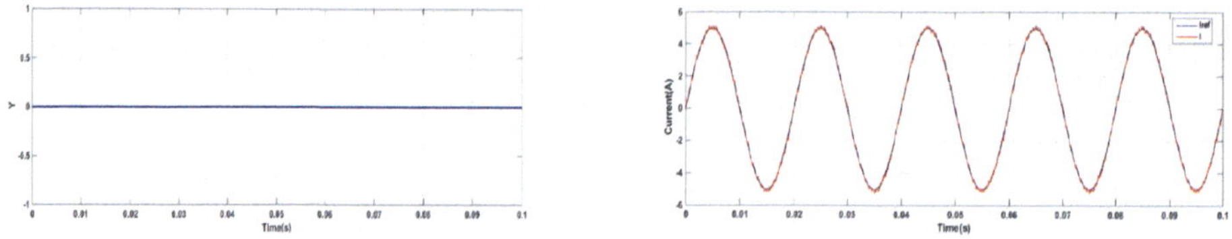

(a) (b)

Fig. 3. Normal condition (no fault) a) Y, b) Output current

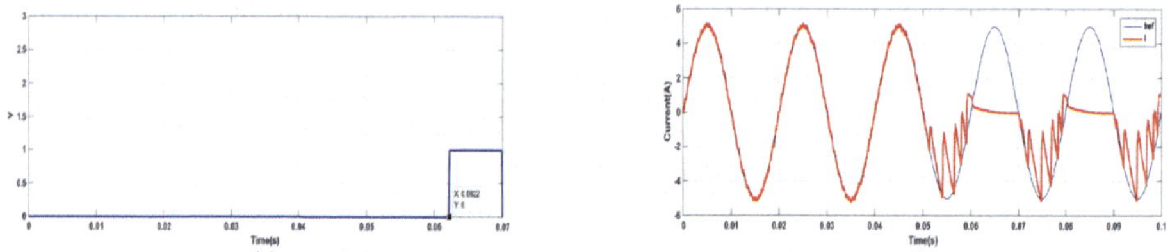

(a) (b)

Fig. 4. *S1* Open circuit faults results a) Y, b) Output current

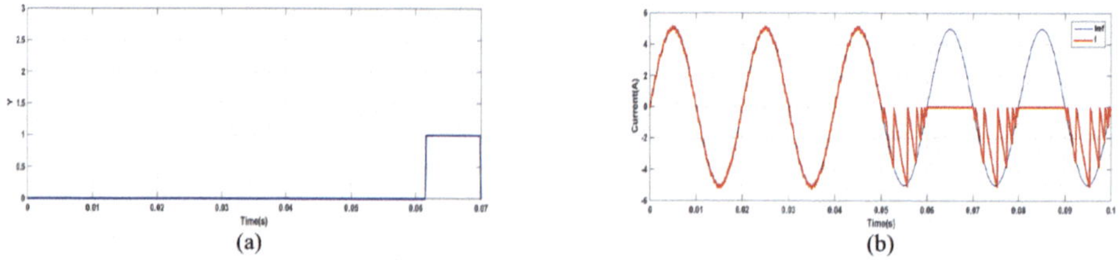

(a) (b)

Fig. 5. *S2* Open circuit faults results a) Y, b) Output current

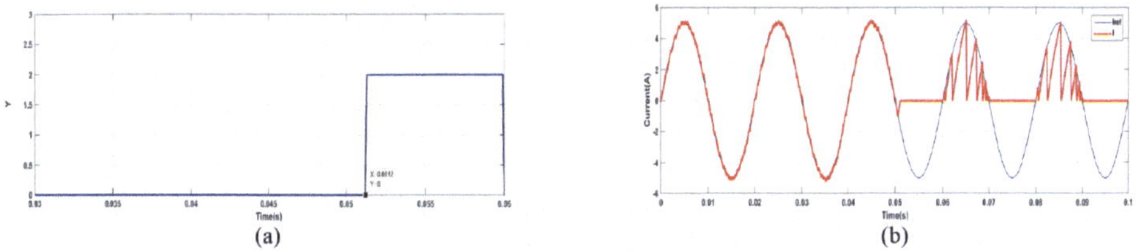

Fig. 6. *S3* Open circuit faults results a) Y, b) Output current

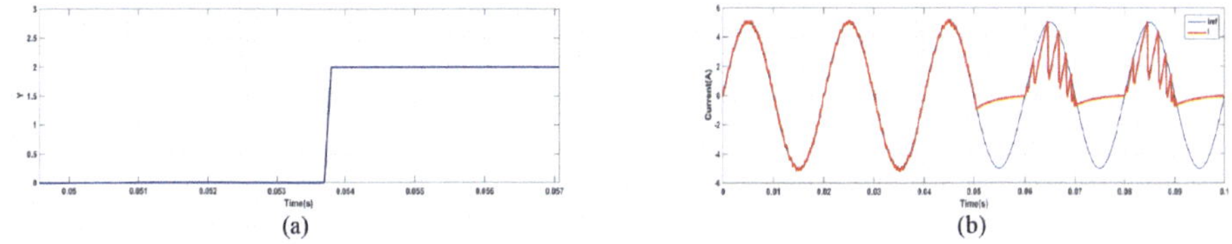

Fig. 7. *S4* Open circuit faults results a) Y, b) Output current

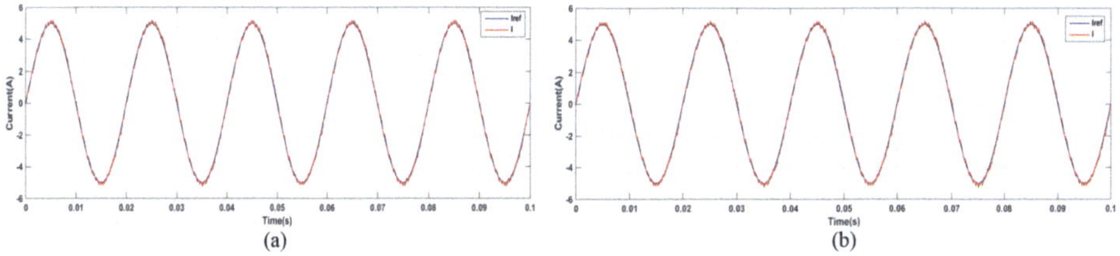

Fig. 8. Resulted current after open circuit fault in a*) D1*, b) *D2*

5 Conclusion

In this study, the primary focus was on detecting open-circuit (OC) faults in single phase three-level Neutral Point Clamped (NPC) inverters. A system combining distorted average current analysis with model-predictive control was introduced for efficient fault identification. This method can detect faulty switches in less than one fundamental period, offering a rapid solution without the need for extra hardware or complicated computations.The functionality of the system is demonstrated by analyzing the predicted average current of the phase, calculated through model-predictive control after introducing various faults at specific times. Additionally, the effectiveness of Model Predictive Control in tolerating OC faults in clamping diodes is confirmed by the output current. The simulation results shows the validity of the presented fault identification approach.

References

1. Zhang, X., Tan, G., Xia, T., Wang, Q., Wu, X.: Optimized switching finite control set model predictive control of NPC single-phase three-level rectifiers. I IEEE Trans. Power Electron. **35**(10), 10097–10108 (2020)

2. Biricik, S., et al.: Operation of three-level single-phase half-bridge npc inverter-based shunt active power filter under non-ideal grid voltage condition with sliding mode controller. In: 2018 IEEE 12th International Conference on Compatibility, Power Electronics and Power Engineering, CPE-POWERENG 2018, IEEE, Doha, Qatar (pp. 1–6) (2018)
3. Wang, S., Song, W., Ma, J., Zhao, J., Feng, X.: Study on comprehensive analysis and compensation for the line current distortion in single-phase three-level NPC converters. IEEE Trans. Ind. Electron. **65**(3), 2199–2211 (2018). https://doi.org/10.1109/TIE.2017.2740849
4. Bughneda, A., Salem, M., Richelli, A., Ishak, D., Alatai, S.: Review of multilevel inverters for PV energy system applications. Energies **14**(6), 1585 (2021)
5. Ma, M., Meng, X., Xiang, N., Wang, H., Zhang, X.: Fault location method of IGBT short-circuit for a grid-tied Neutral-Point-Clamped inverter system. Microelectron. Reliability **126**, 114225 (2021)
6. Craig, E., Mecrow, B.C.. Atkinson, D.J.. Jack, A.G.: A fault detection procedure for single phase bridge converters. In: Proceedings of the 5th European Conference pn Power Electronics Application, 1993, vol. 4, pp. 466–471. IET (1993)
7. Kamel, T., Biletskiy, Y., Chang, L.: Fault diagnoses for industrial grid-connected converters in the power distribution systems. IEEE Trans. Ind. Electron. **62**(10), 6496–6507 (2015)
8. Ge, X., Pu, J., Liu, Y.: Online open-switch fault diagnosis method in single-phase PWM rectifiers. Electron. Lett. **51**, 1920–1922 (2015)
9. Gou, B., Ge, X., Wang, S., Feng, X., Kuo, J.B., Habetler, T.G.: An open-switch fault diagnosis method for single-phase PWM rectifier using a model-based approach in high-speed railway electrical traction drive system. IEEE Trans. Power Electron. **31**, 3816–3826 (2016)
10. Yu, Y., Wang, X.: Multi-step predictive current control for NPC grid-connected inverter. IEEE Access **7**, 157756–157765 (2019)
11. Tamim, T.M., Li, S., Jing, W.: Multi-level model predictive controller with satisfactory optimization for multi-level converters. Simul. Modell. Pract. Theory **92**, 1–16 (2019)
12. Choi, U.-M., Lee, J.-S., Blaabjerg, F., Lee, K.-B.: Open-circuit fault diagnosis and fault-tolerant control for a grid-connected NPC inverter. IEEE Trans. Power Electron. **31**(10), 7234–7247 (2016)

An Improved-Detection System for Diagnosing Photovoltaic Power Plant Efficiency Degradation Using a UAV-Based System

N. Cheggaga[1], S. Tchoketch-Kebir[2(✉)], and A.-H. Kebir[1]

[1] Electrical and Remote Control Systems Laboratory (LabSET), Electronic Department, University Saad Dahled of Blida1, USDB1, Blida, Algeria
[2] Centre de Recherche en Technologies Industrielles, CRTI, Cheraga, P.O.Box 64, 16014 Algiers, Algeria
s.tchoketch-kebir@crti.dz

Abstract. In photovoltaic (PV) power plants, quickly finding faults is crucial for identifying what is causing them and fixing major problems to maintain good efficiency. Many studies have used drones to inspect PV plants, but these drone-based methods usually struggle to address dangerous issues. To solve this latters, this work introduces a smart system for finding damaged PV panels using a drone equipped with two cameras (one thermal, one digital). The drone does regular checks and uses both cameras to take pictures. The digital camera finds problems caused by outside factors, while the thermal camera shows problems inside the panels. A Convolution-Neural-Network method is used at the processing unit to analyses pictures and finds the problems. This study looks at four different states (normal, shadowed, dust, & thermal-faults). The developed method is very accurate (over 0.8), especially when using Euler angle (0.98), showing it can effectively find and locate problems in PV plants.

Keywords: PV Plant · Drones · Diagnosis · Inspection · Thermal-Camera · CNN

1 Introduction

Photovoltaic installations are considered as power plants of optimum production supplied by the sun's natural resource, which is a great candidate for being the ultimate surrogate for fossil fuel resources. It is thus a clean, potent, free, and renewable source of energy with a 95% production of clean energy and a life expectancy of 20–30 years [1], it could provide enough energy for our needs without any risks to the environment. Recently, many photovoltaic (PV) plants are installed all over the world. However, PV panels of these plants generally require maintenance and monitoring to produce healthy, secure, optimal, and safe power under real environmental conditions, according to many degradations that can be caused through long-term operation of PV plant [2]. Generally, the produced energy from PV panels can have a degradation of about 0.5% in a year [3]. The degradation of the PV plant's performance can be caused by some factors that limited the sun's irradiations [4]. These factors are related either to the system itself, such

© The Author(s), under exclusive license to Springer Nature Switzerland AG 2025
M. Hatti (Ed.): IC-AIRES 2024, LNNS 1238, pp. 293–303, 2025.
https://doi.org/10.1007/978-3-031-80301-7_33

as increase of solar panel temperature, composite material degradations, parasitic resistances, decreasing of fill factor (FF), and sub-optimal tilt- angle. Others are related to the location and environmental circumstances such as harmful weather conditions such as shading, soiling, dirt, leaves, and other debris [5]. Besides, some faults are due to the PV plant itself, and can enormously influences the power generated [6] and voltage violation degradation of the PV systems [7]. In that matter, it is so important to regularly check on the PV panels and make sure that the system is in its best condition to avoid any irrelevant disruptions. Moreover, fast diagnosis of photovoltaic panels' occurring damages is so important for increasing their good efficiency and the best performances of the whole PV system. The quick diagnosis of faults in photovoltaic panels is important to increase the efficiency and enhance the performance of the whole PV plants. The operation and maintenance of the latters can be classified into preventive and curative [3]. Preventive maintenance encompasses the regular inspection and replacement of worn-out components within the PV system. This necessitates the identification of visible damages and the implementation of preventive diagnosis to quickly detect potential issues, thereby mitigating economic losses [4]. Curative maintenance consists of repairing damaged parts of PV panels and setting up a breakdown system. The maintenance process of a PV panel system can be high-priced, it can cost about $450 per year [5]. This cost could be reduced by using new systems and technologies, based on the use of Unmanned Aerial Vehicles (UAVs) inspections. UAV-based systems are defined as flying vehicles in the air that are capable of controlling the aircraft [6] and do not carry human-based operators. UAVs are becoming one of the best options for PV plant inspection due to their observation, maintenance, surveillance, monitoring capabilities, and remote sensing ability. Embracing this technology will allow us to easily detect a wider range of defects, thus keeping the performance of the PV panels at their best efficiency. Many previous interesting publications have shown that UAVs have been very useful in PV plant inspection and diagnosis [7–9, and 10]. The work in [7] presented the use of drones equipped with a thermal camera to inspect the health/fault states of PV panels, and in its review work, many different recent studies have been presented, analyzed, and compared. The fruitful of this review work proves the effectiveness of integrating Infra-Red (IR) thermography in PV diagnosis, by applying different approaches to image processing using different tools of machine learning-based method, after recording them by drones. In the work developed in [8], several techniques based on intelligent-based process (machine learning (ML) and deep learning (DL)), have been used for damages inspecting and diagnosing of PV panels, using thermal images processing collected by infrared camera integrated in drones. The identified faults included bypass diode (BPD) faults, shading, short- circuits (SC), and soil. The DL-based techniques have shown best performances in the obtained results of [8] compared to the ML-based techniques. The work of [9] applies a two-stage thermography inspection by using two approaches: aerial-based inspection (fast and reliable detection, localization and diagnosis of faults) and ground-level-based inspection (applied only to relevant incidents). This two-stage thermography inspection strategy has improved a highly performed process of PV diagnosis. Relatively to the above, this work has the objective of using a drone to perform a periodic inspection for PV damages detecting, classifying, and sources identifying. This PV-based inspection allow increasing the PV plant efficiency with reduced cost of monitoring process. The novel idea of this work consists of mounting a Quad-copter drone equipped with two cameras (digital and thermal).

The drone is configured to take photos and send them to the processing unit; the taken images will be processed using intelligent-based by developing Convolution Neural Network (CNN) algorithm through coding using Python language. The given results proved that the developed program is accurate with a high percentage of accuracy (98.47%). The organization of this paper is structured as described in the following points. In the first section, a presentation of an UAV-based system is done, it contains the drones, the thermal and digital cameras that were used for PV power plant inspection. In the following section, a classification is given about the PV panel's degradations, PV damages, and their causes according to the two used cameras. Next, the development of the intelligent CNN-based approach, which is used to identify and tackle the issues for PV panels, have been detailed. In the last section, simulation results of the affected tests are discussed and final thoughts about this work are concluded.

2 UAV-Based System

A UAV-based system or drone is an aircraft without a pilot, can fly autonomously or remotely, controlled from the ground using the on-board control system [10]. The UAV system consists of two segments (Fig. 1):

- Air segment: the UAV, its payload, and its transmission system.
- Ground segment: the control base, the intervening person, and the equipment for launching and recovering drones.

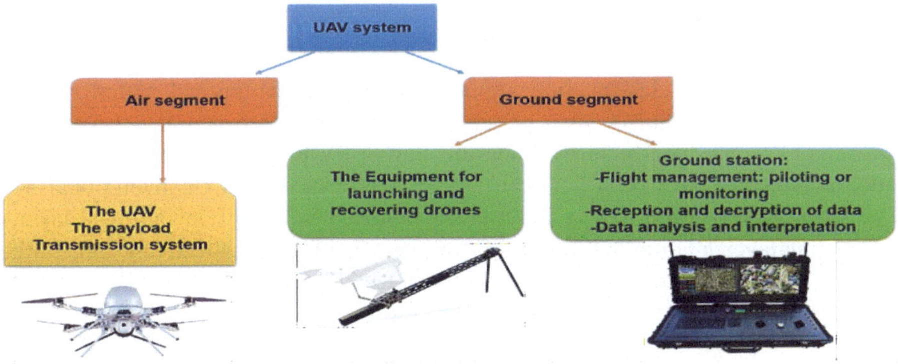

Fig. 1. Constitution of UAV-based system.

3 Classification of PV Degradations and Faults According to the Used Cameras

With the aim of facilitating the detection of degradations and failures in PV panels, we have classified the detected problems according to the type of cameras used to determine the issue; the flow chart in Fig. 2 shows the detected PV degradations & faults by using of digital camera and thermal (Infra-Red (IR) based type) camera.

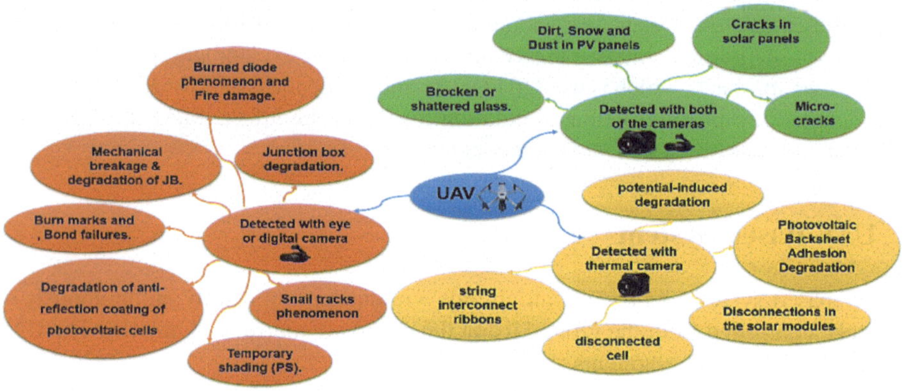

Fig. 2. Classification of PV degradations & faults detected visually or using cameras (digital and thermal).

4 Processing of Collected Images

In image processing (classification applications), the convolutional neural network (CNN) technic is recognized as a popular and powerful deep neural network-based approach [7]. Consequently, for the classification of PV panel damages, a CNN-based model architecture have been developed. The architecture of this convolutional neural network (CNN) comprises an input layer, followed by a 2D convolutional layer, a max-pooling layer, a Dropout layer, and fully connected layers, collectively constituting our 2D CNN-based model. Then, three models, test-1, test-2, and test-3, were trained independently.

4.1 Convolutional Neural Networks (CNN)

Presently, convolutional neural networks are widely acknowledged as the most effective models for image classification. Convolutional neural networks are currently considered the most efficient models for classifying images. Convolutional neural networks (CNNs) are comprised of a series of multi-layer perceptrons (MLPs) designed to preprocess small quantities of data. CNNs find extensive utility in image and video recognition, recommender systems, as well as natural language processing. A CNN is a deep network composed of five types of layers: the convolution layer, the pooling layer, the ReLU correction layer, the dropout layer, and the fully connected layer. Figure 3 presents the most crucial structure of a convolutional layer, which allows the transformation of the input data using a group of connected neurons from the previous layer.

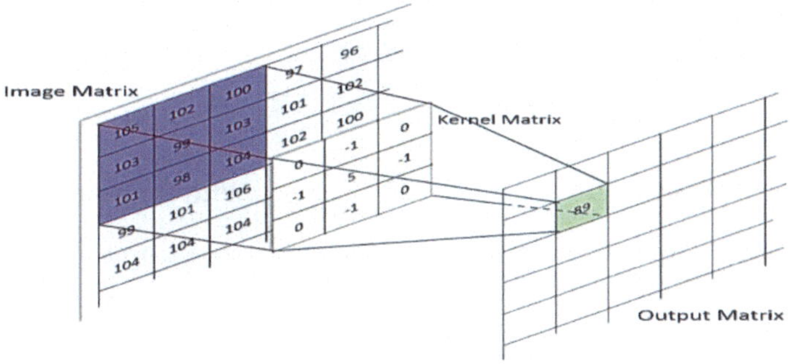

Fig. 3. Structure of convolutional Layer.

Then, the pooling level helps to gradually reduce the local size of the data presentation (Fig. 4) and prevents over- fitting of the learning process for the data.

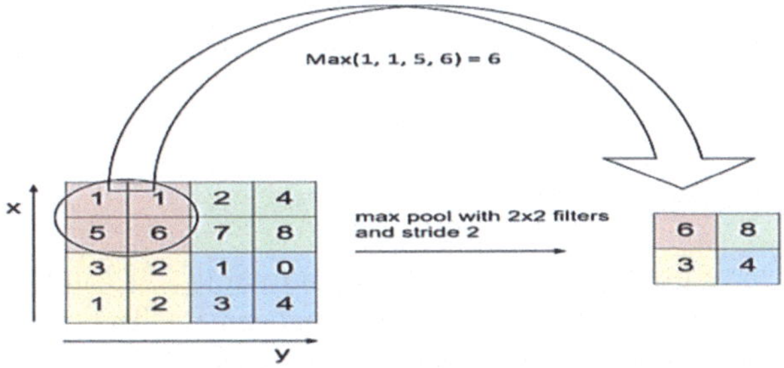

Fig. 4. Principal of pooling Layer.

Dropout layers are important when learning CNNs because they prevent over fitting of the learning of data as shown in Fig. 5.

The performance of the CNN-based approach was assessed by computing the area under the curve (AUC) of the receiver operating characteristic curve [14], along with the calculation of accuracy and precision.

Accuracy (AC): Indicates the percentage of positive predictions that are correct.

$$AC = \frac{TP}{TP + FP} \tag{1}$$

Precision (PR): This is the score of predictions that our model received among all predictions.

$$PR = \frac{TP + TN}{TP + TN + FP + FN} \tag{2}$$

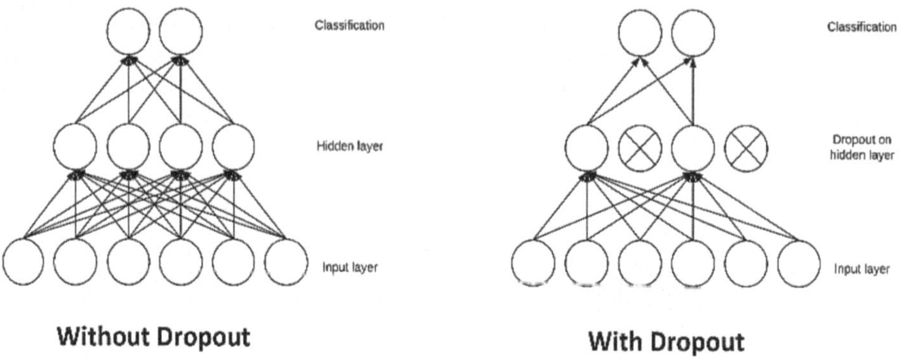

Fig. 5. Artificial neural networks with and without Dropout layer.

where TP, FP, TN, and FN correspond to the following:

- True-Positives (TP): When both the actual class and predicted class of the data point are 1.
- True-Negatives (TN): When both the actual class and predicted class of the data point are 0.
- False-Positives (FP): When the actual class of the data point is 0 and the predicted class is 1.
- False-Negatives (FN): When the actual class of the data point is 1 and the predicted class is 0.

4.2 Preparing of Dataset

In this work, we have used a dataset containing a total of 2387 images captured by a digital camera and 10 images captured by a thermal camera. The original resolution of the image is 3840 × 2160 in jpg format and RGB colored. The original resolution of the thermal image is 240 × 181 in BMP format and RGB colored. We divided this group into two subgroups (two volumes) as presented in Fig. 6. The first contains images used for the learning of our model; the second contains images used for the testing of the same model.

The first subset (folder) called "Train" contains 2157 images in total, these images are divided into 2 classes, the first class contains images of normal PV panels (a folder containing 775 images named "Normal"), the second class contains images of failed PV panels (a folder containing 1382 images named "Failure").

The second subset (folder) called "Test" contains 230 images in total, these images are divided into 2 classes, the first class contains images of Normal PV panels (a folder of 84 images called "Normal", the second class contains images of failed solar panels (a folder containing 146 images called "Failure").

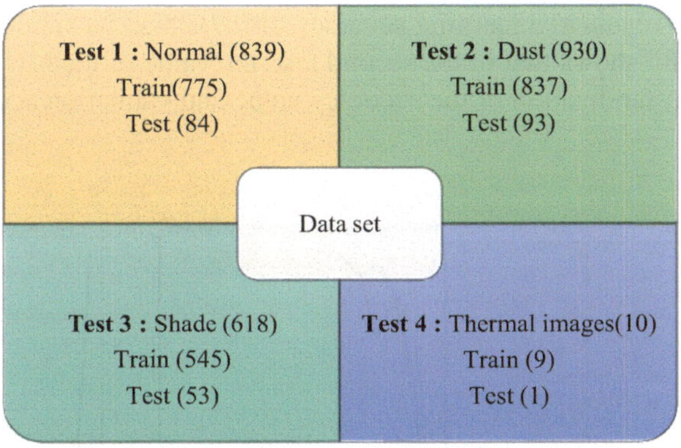

Fig. 6. Breakdown of the collected data.

5 Results and Discussions

5.1 Test-1

In the first test, we took the picture without processing, after using so many models; the selected model was the optimum. In the training process, we attained an accuracy of 0.8667, while in the validation process, the accuracy reached 0.80. Fig. 7a shows the accuracy and the loss of learning process and testing process during the epochs. The Fig. 7b illustrates the results in the first affected test, the accuracy percentage was good (86.67%) but we could not reach more than that and it was not unstable. Thus, the percentage of loss in learning process was considerable.

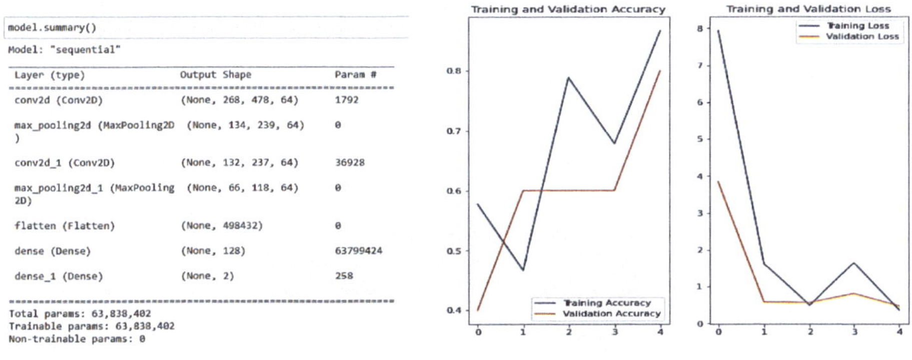

```
model.summary()

Model: "sequential"
_________________________________________________________________
 Layer (type)                Output Shape              Param #
=================================================================
 conv2d (Conv2D)             (None, 268, 478, 64)      1792

 max_pooling2d (MaxPooling2D  (None, 134, 239, 64)     0
 )

 conv2d_1 (Conv2D)           (None, 132, 237, 64)      36928

 max_pooling2d_1 (MaxPooling  (None, 66, 118, 64)      0
 2D)

 flatten (Flatten)           (None, 498432)            0

 dense (Dense)               (None, 128)               63799424

 dense_1 (Dense)             (None, 2)                 258

=================================================================
Total params: 63,838,402
Trainable params: 63,838,402
Non-trainable params: 0
```


Fig. 7. a) Hyper parameters tuning (Test-1); b) Accuracy and loss curves of the sequential model issued from the first conducted test.

5.2 Test-2

In the second test, we have obtained a model that detects the edge of the picture, after using so many models; the selected model was the optimum (Fig. 8a). After training for 15 epochs, we obtained a learning accuracy of 0.9125 and a validation accuracy of 0.8944. The Fig. 8b shows that in the second test, the value of the loss was almost not-existent. With the stabilization of the accuracy value and validation accuracy in the last 5 epochs.

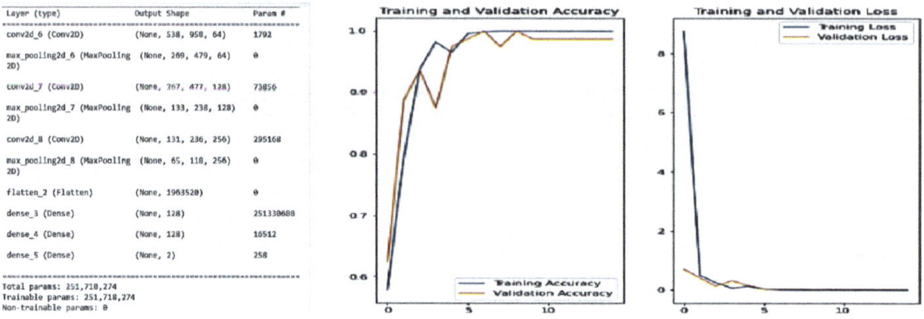

Layer (type)	Output Shape	Param #
conv2d_6 (Conv2D)	(None, 538, 958, 64)	1792
max_pooling2d_6 (MaxPooling2D)	(None, 269, 479, 64)	0
conv2d_7 (Conv2D)	(None, 267, 477, 128)	73856
max_pooling2d_7 (MaxPooling2D)	(None, 133, 238, 128)	0
conv2d_8 (Conv2D)	(None, 131, 236, 256)	295168
max_pooling2d_8 (MaxPooling2D)	(None, 65, 118, 256)	0
flatten_2 (Flatten)	(None, 1963520)	0
dense_3 (Dense)	(None, 128)	251330688
dense_4 (Dense)	(None, 128)	16512
dense_5 (Dense)	(None, 2)	258

Total params: 251,718,274
Trainable params: 251,718,274
Non-trainable params: 0

a) b)

Fig. 8. a) Hyper parameters tuning (Test-2); Accuracy and loss curves of the sequential model (Test-2).

5.3 Test-3

In order to inspect PV panel, the right angle must be chosen (Fig. 9), the token picture must not be so far. To prevent the camera and the operator from being reflected in the glass, they must not face the inspected PV panel. However, the emissivity is maximum when the axis of the camera is perpendicular to the surface. An angle of 5 to 60° with the normal constitutes a good compromise.

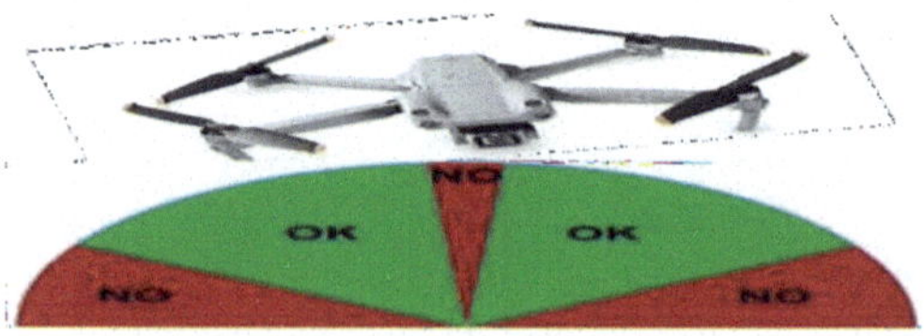

Fig. 9. Monitoring the right angle.

In this third test we created a model (Fig. 10a) that adjusts the angle of the picture because it changes according to the angle of capturing, after using so many models, the selected model was the optimum.

After 15 epochs of training, we achieved a training accuracy of 0.9847 and a validation accuracy of 0.9525. The Fig. 10b shows that in the third test, the accuracy value and validation value were stable at beginning of learning and the loss value was almost not existent.

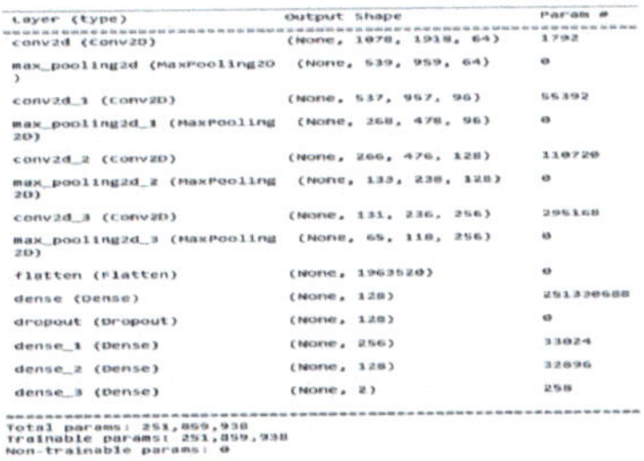

Layer (type)	Output Shape	Param #
conv2d (Conv2D)	(None, 1078, 1918, 64)	1792
max_pooling2d (MaxPooling2D)	(None, 539, 959, 64)	0
conv2d_1 (Conv2D)	(None, 537, 957, 96)	55392
max_pooling2d_1 (MaxPooling2D)	(None, 268, 478, 96)	0
conv2d_2 (Conv2D)	(None, 266, 476, 128)	110720
max_pooling2d_2 (MaxPooling2D)	(None, 133, 238, 128)	0
conv2d_3 (Conv2D)	(None, 131, 236, 256)	295168
max_pooling2d_3 (MaxPooling2D)	(None, 65, 118, 256)	0
flatten (Flatten)	(None, 1963520)	0
dense (Dense)	(None, 128)	251330688
dropout (Dropout)	(None, 128)	0
dense_1 (Dense)	(None, 256)	33024
dense_2 (Dense)	(None, 128)	32896
dense_3 (Dense)	(None, 2)	258

Total params: 251,859,938
Trainable params: 251,859,938
Non-trainable params: 0

a)

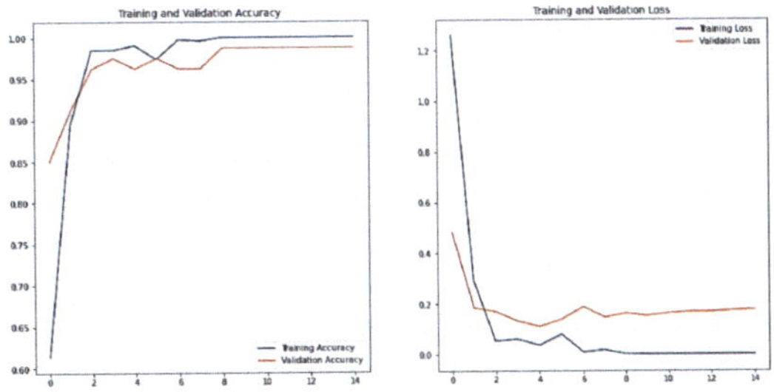

b)

Fig. 10. A) Hyper parameters tuning (Test-3); b) Accuracy and loss curves of the sequential model (Test-3).

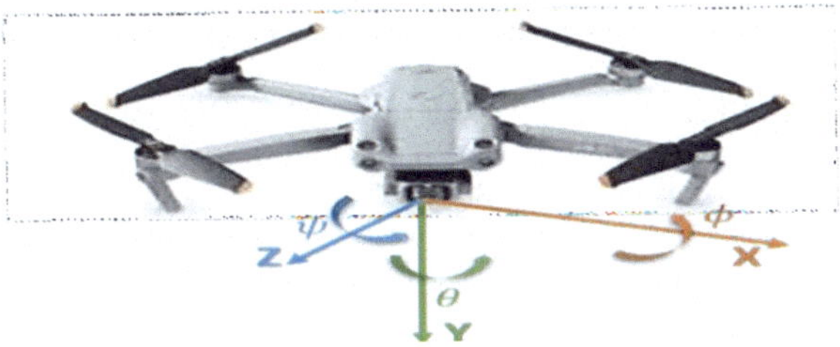

Fig. 11. Camera orientation using *Euler* angles.

Taking into account all of the above results, we have chosen this test-3 for its high accuracy value, to represent the model of image classification for classifying the PV panel damages into faulty or healthy. Subsequently, to further improve accuracy, we utilized the *Euler* angle approach illustrated in Fig. 11, resulting in achieving an accuracy of 98%.

6 Conclusion

The primary aim of this research was to enhance the efficiency of PV power generation by implementing a rapid and automated diagnostic process for identifying faults that may arise in PV plants, specifically within PV panels. To ensure that the PV panel operates reliably throughout its life. The intelligent inspection of PV faults was done by treating images saved from drone. This latter was equipped with two types of cameras (an onboard-integrated module in drone). Then, for the detection and classification, the processing of images was done by developing a CNN-based approach, which has allowed us to achieve the accomplishment of an accurate classification model of PV faults (shading, dust, and IR-based), with various tools and environments. Besides, we were able to achieve a fast diagnosis process since for hundreds of panels; the drone took about 30min to detect faults. Some faults were not even anticipated but were detected by the drone. The intelligent-based diagnosis process developed herein was able to relish breakdowns in PV panels and ensure accurate field responses by the concerned organizations that were proved by reaching a maximum accuracy equal to 98%.

References

1. Satpathy, R., Pamuru, V.: Chapter 9 - Grid-connected solar PV power systems. In: Satpathy, R., Pamuru, V. (eds.) Solar PV Power, pp. 365–433. Academic Press (2021). https://doi.org/10.1016/B978-0-12-817626-9.00009-5
2. Ahadi, A., Miryousefi Aval, S.M., Hayati, H.: Generating capacity adequacy evaluation of large-scale, grid-connected photovoltaic systems. Front. Energy **10**(3), 308–318 (2016)
3. Vidyanandan, K.V.: An overview of factors affecting the performance of solar PV systems. Energy Scan **27**, 2–8 (2017)
4. Singla, P., Duhan, M., Saroha, S.: A comprehensive review and analysis of solar forecasting techniques. Front. Energy **16**(2), 187–223 (2022)
5. Selma T.K., Cheggaga, N., Ilinca, A., Boulouma, S.: A novel neural network-based method for diagnosing faults of PV array. Sustainability (2021)
6. Chen, J.L., Xu, X.Y., Yan, Z., et al.: Data-driven distribution network topology identification considering correlated generation power of distributed energy resource. Front. Energy **16**(1), 121–129 (2022)
7. Zhang, J., Wang, T., Chen, J., Liao, Z., Shu, J.: Cluster voltage control method for "Whole County" distributed photovoltaics based on improved differential evolution algorithm. Front. Energy (2023)
8. Deli, K., Noel, D., Deli, K., Noel, D.: On-field operation and maintenance of photovoltaic systems in cameroon. Maintenance Management. IntechOpen (2019). https://doi.org/10.5772/intechopen.83730
9. Haque, A., Bharath, K.V.S., Khan, M.A., Khan, I., Jaffery, Z.A.: Fault diagnosis of photovoltaic modules. Energy Sci. Eng. **7**, 622–644 (2019). https://doi.org/10.1002/ese3.255
10. Gobler, E.: Solar panel maintenance guide: How to clean and repair solar panels. CNET (2022). https://www.cnet.com/home/energy-and-utilities/solar-panel-maintenance-guide-how-to-clean-and-repair-solar-panels/
11. Eisenbeiss, H.: A mini unmanned aerial vehicle (UAV): system overview and image acquisition
12. Buerhop, C., Bommes, L., Schlipf, J., Pickel, T., Fladung, A., Peters, I.M.: Infrared imaging of photovoltaic modules: a review of the state of the art and future challenges facing gigawatt photovoltaic power stations. Prog. Energy **4**, 042010 (2022)
13. Boubaker, S., Kamel, S., Ghazouani, N., Mellit, A.: Assessment of machine and deep learning approaches for fault diagnosis in photovoltaic systems using infrared thermography. Remote Sens. **15**, 1686 (2023). https://doi.org/10.3390/rs15061686
14. Álvarez-Tey, G., García-López, C.: Strategy based on two stages for ir thermographic inspections of photovoltaic plants. Appl. Sci. **12**, 6331 (2022). https://doi.org/10.3390/app12136331
15. Nassim, S., Ouahiba, B.: Planification de trajectoires pour un véhicule aérien autonome. Thesis, Université Mouloud Mammeri (2018)
16. Branco, P., Gonçalves, F., Costa, A.C.: Tailored algorithms for anomaly detection in photovoltaic systems. Energies **13**, 225 (2020). https://doi.org/10.3390/en13010225
17. Selma, T.K., Cheggaga, N., Cheikh, M.S.A., Haddadi, M., Rahmani, H.: A comprehensive study of diagnosis faults techniques occurring in photovoltaic generators. Engineering Review (2020). https://doi.org/10.30765/er.1714.2021
18. Strobl, C., Meckler, P.: Arc faults in photovoltaic systems. In: 2010 Proceedings of the 56th IEEE Holm Conference on Electrical Contacts, pp. 1–7 (2010). https://doi.org/10.1109/HOLM.2010.5619538

19. Drouot, A.: Control Strategies for Autonomous Navigation Gun Launched Micro Aerial Vehicle. Phdthesis, Université de Lorraine (2013)
20. El-Banby, G.M., Moawad, N.M., Abouzalm, B.A., Abouzaid, W.F., Ramadan, E.A.: Photovoltaic system fault detection techniques: a review. Neural Comput. Appl. (2023)

Anomaly Detection and Classification of Solar Photovoltaic Modules Using Vision Transformers (ViT)

Yacine Boutana, Sofiane Haddad(✉), and Ammar Soukkou

Renewable Energy Laboratory, Faculty of Sciences and Technology, University of Jijel, Jijel, Algeria
s_haddad@univ-jijel.dz

Abstract. In this paper, a deep learning model based on Vision Transformers (ViT) has been introduced to accurately identify a wide range of photovoltaic (PV) module defects from infrared thermal images. The developed model has been evaluated on a comprehensive dataset encompassing 12 distinct anomaly classes, including cell defects, cracking, diode issues, hot spots, soiling, and vegetation interference, among others. This model performed best in detecting diode anomalies (97% accuracy), identifying normal operating conditions (96% accuracy), and identifying cracking defects (67% accuracy). The model also performed well in distinguishing single-cell and multi-cell faults, a traditionally difficult task in photovoltaic anomaly detection. For binary classification (anomaly or non-anomaly), the model achieved an overall accuracy of 88%. In the multiclass classification task, model performance varied across different anomaly types, with an average accuracy of 81% across the 12 anomaly classes.

Keywords: Solar photovoltaic modules · Green energy · Vision Transformers · Fault detection · Deep learning

1 Introduction

Photovoltaic (PV) energy has emerged as a pivotal renewable energy source, harnessing sunlight to generate electricity through semiconductor materials. As the global shift towards sustainable energy intensifies, the efficiency and continuance of PV systems have become crucial factors in their widespread adoption. However, PV modules are susceptible to various anomalies that can significantly impact their performance and lifespan. As of 2022, solar power ranks as the second most deployed RE source, surpassed only by hydropower. With a remarkable installed capacity of 1053 GW [1], representing a 192 GW increase from the previous year, solar energy has solidified its crucial role in the global transition towards cleaner power sources. This rapid expansion of solar power has led to substantial reductions in carbon emissions and environmental pollution as more countries adopt it as an alternative to traditional fossil fuel-based energy sources. However, for solar technology to remain a basis of future energy policies and maintain its widespread adoption, its performance and durability are vital. PV modules are

© The Author(s), under exclusive license to Springer Nature Switzerland AG 2025
M. Hatti (Ed.): IC-AIRES 2024, LNNS 1238, pp. 304–311, 2025.
https://doi.org/10.1007/978-3-031-80301-7_34

inherently sensitive to environmental factors like temperature, humidity, and mechanical stress, which can significantly impact module efficiency and lifespan. Addressing these challenges is essential to assure solar energy's viability and appeal as a sustainable power source. This requires effective and accurate assessment methods to ensure optimal operation. To meet this need, the solar industry uses several imaging techniques to inspect PV farms. At present, three primary methods are employed: RGB (visible light), electroluminescence (EL), and infrared (IR) imaging [2]. Each technique offers unique insights into module health, but IR imaging has emerged as a particularly powerful tool, offering distinct advantages, including the ability to detect thermal anomalies that may indicate performance issues. Damaged PV modules that cause hot spots or heat effects can be easily detected through this non-invasive method [2], which can be performed during operation, providing valuable insights into the module's health that may not be visible through other assessment methods. Recent advancements in artificial intelligence have introduced Vision Transformers as a powerful alternative to Convolutional Neural Networks (CNNs) for image analysis tasks. ViT, which leverage the attention mechanism pioneered in natural language processing, have demonstrated superior performance in various computer vision applications. Compared to CNNs, ViTs offer better handling of long-range dependencies in images and often require less computational resources for training and inference. In this study, we explore the application of ViT for anomaly detection and classification in solar PV modules using IR imaging data. Several studies have explored mainly convolutional neural networks (CNNs) for automatic defect detection in solar PV modules. These approaches leverage image processing techniques to identify cracks, anomalies, and other issues that can impact the efficiency and lifespan of the PV system. The authors in [3], have used a multi-scale CNN with transfer learning to classify various PV module faults from thermo-graphic images. This method achieved good accuracy by leveraging pre-trained knowledge and multi-level feature extraction. The study in [4] proposed a CNN architecture with dual spin max pooling for crack detection in solar cells using electroluminescence (EL) images. Moreover, in [5], the authors has also conducted a research on deep learning for multi-type defect detection in EL images of PV modules on a production line. They have used a CNN model trained on a large dataset (5983 labeled images) to identify 14 different defect types. Furthermore, [6] utilized CNNs and Support Vector Machines (SVMs) to investigate faults in EL images. This method utilizes CNNs for feature extraction and SVMs for classification, achieving good accuracy on two separate datasets. In [7] a CNNs model for anomaly detection in PV modules using infrared (IR) images captured by drones has been explored. The proposed method attained great accuracy while maintaining efficient processing speeds on edge hardware.

2 Methodology

2.1 Infrared Solar Module Dataset

The Infrared Solar Module Dataset contains 20,000 infrared images, each 24×40 pixels, depicting solar modules under various conditions. It is divided into 12 classes: 11 anomaly types commonly found in solar PV systems and one "No-Anomaly" class for normal modules. Images were collected using piloted aircraft and unmanned aerial systems equipped with midwave or longwave infrared (3–13.5 μm) and visible spectrum imaging systems. Anomalies were cropped to isolate individual modules and classified accordingly, more details about the dataset can be found in [8]. Figures 1 and 2 show the image distribution and sample images from each class, respectively. A separate dataset for binary classification will combine all anomaly classes into one.

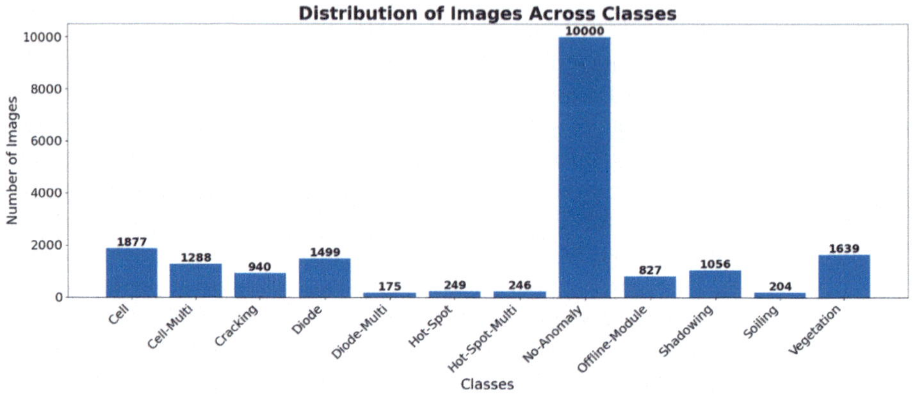

Fig. 1. Number of images of each class in the Infrared Solar Modules dataset

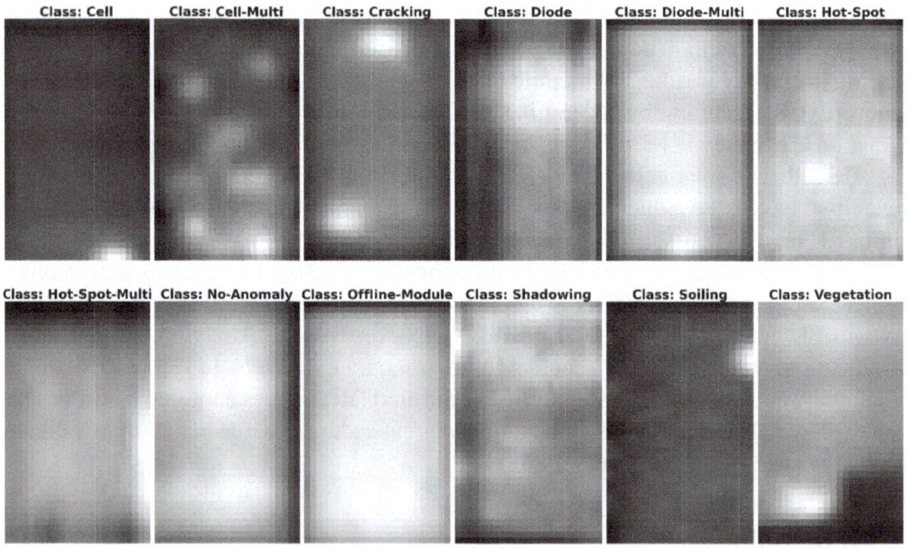

Fig. 2. Few samples of anomaly and no-anomaly classes in the Infrared Solar Modules dataset.

2.2 Vision Transformers

The Vision Transformer (ViT) architecture, introduced in [9] aimed to create a deep neural network algorithm that processes large datasets without using convolution operations. ViT adapts the original transformer architecture, initially developed for NLP tasks, to image processing. The core concept involves dividing an image into patches, flattening them, and projecting them into a D-dimensional embedding space to create patch embeddings. These patches are then treated as tokens (analogous to words in NLP) and used as input. The model adds positional embedding to the appropriate batch in order to retain spatial information and also concatenates a learnable class token. Finally, in the encoder output a classification head is applied to the learnable class token to obtain the model's logits which allows to perform various vision tasks [10] (Fig. 3).

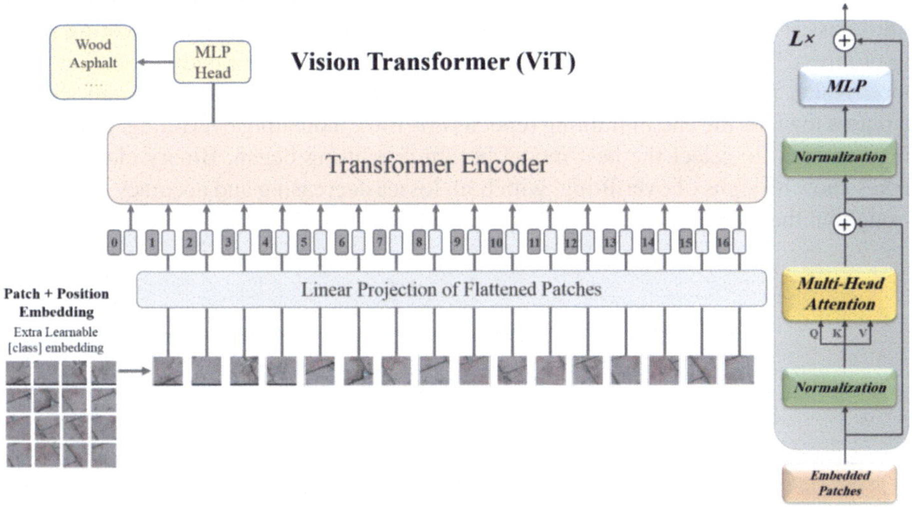

Fig. 3. Visual representation of Vision Transformer model [9]

The main reason behind the popularity of Vision Transformers (ViT) is their use of a multi-headed self-attention mechanism [11]. This mechanism enables the model to capture global contextual information from the input, allowing it to learn long-range relationships between images, with this capability, ViT addresses one of the major limitations of convolutional neural networks (CNNs), which is their operation on fixed-size windows that make it difficult for the model to recognize long-range spatial relationships within or between images. Additionally, CNNs face challenges such as layer pooling, which leads to information loss, and translation invariance, which hinders the models from dynamically adapting to changes in the input [10].

3 Results and Discussions

3.1 Finetuning VIT Model

In our training phase, we used the pre-trained Vision Transformer model "vit-base-patch16-224-in21k", initially trained on the ImageNet-21k dataset. This choice was motivated by findings from [12], highlighting superior transfer learning capabilities. The model processes 16 × 16 pixel patches for feature extraction on 224 × 224 pixel images. We fine tuned the model with Hugging Face's Trainer class over 100 epochs, using a 0.0005 learning rate and a batch size of 32, incorporating early stopping to prevent overfitting. Training was conducted on a Kaggle notebook with a P100 GPU. Minimal augmentation (random rotation and flipping) was applied, resizing all images to 224 × 224 pixels.

Figure 4 shows the training and evaluation loss functions for binary (2-class) and multi-class (12-class) outputs. Both exhibit a general downward trend, indicating learning progress. The training loss decreases smoothly, while the evaluation loss fluctuates more, which is typical. For multi-class outputs, evaluation loss plateaus and slightly increases towards the end of training (epochs 80–100), indicating overfitting. Early stopping was used to select the best model before overfitting began. Binary classification curves show no signs of overfitting, with both losses decreasing and accuracy improving throughout the training.

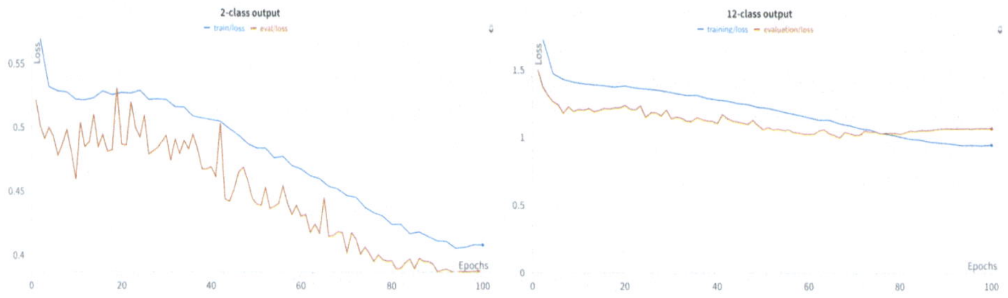

Fig. 4. Loss function of the training and evaluation dataset for 2-class and 12-class output

Figure 5 illustrates the evaluation accuracy for binary (2-class) and multi-class (12-class) classification tasks. For binary classification, accuracy improves from about 0.76 to 0.88 by the end of training. In the multi-class task, accuracy starts around 0.65 and rises to 0.81, with a rapid initial increase followed by gradual improvement. The best-performing model before epoch 80 was selected to avoid overfitting.

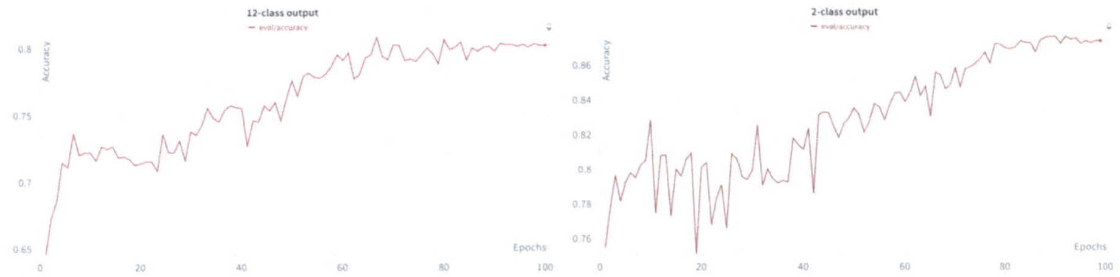

Fig. 5. Evaluation accuracy for 2-class and 12-class output

3.2 Classification Accuracy

Figure 6 shows the confusion matrices for binary and multi-class classification. The model achieved robust performance in the 2-class task, with an overall accuracy of 88%. It excelled at identifying non-anomalous cells with a specificity of 91% and showed a strong capability in detecting anomalous cells with a sensitivity of 86%. These metrics indicate that the model performs well in binary classification, slightly favoring non-anomalous cell identification.

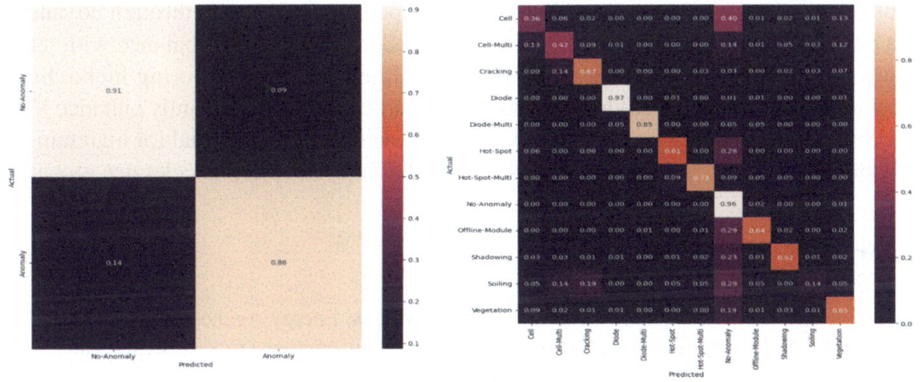

Fig. 6. Confusion matrix of the ViT model for 2-class and 12-class output

The confusion matrix for the 12-class output model shows varied performance across the classes. The model excels in identifying Diode, No-Anomaly, and Diode-Multi, with correct classification rates of 97%, 96%, and 85%, respectively, indicating strong detection of these anomalies and normal conditions. Moderate performance is noted for Hot-Spot-Multi (73%), Cracking (67%), and Vegetation (65%). However, the model struggles with Cell (36%), Cell-Multi (42%), and Soiling (14%) classifications.

The matrix highlights misclassification patterns. Notably, 36% of Cell anomalies are misclassified as No-Anomaly, indicating a bias towards false negatives in this category. Cell-Multi defects show a 42% misclassification rate as Cell, reflecting difficulties in distinguishing these related anomaly types. Inter-class confusions are evident, particularly between Cell and Cell-Multi categories, Hot-Spot misclassified as No-Anomaly (28%), Offline-Module as No-Anomaly (29%), and Shadowing as No-Anomaly (23%). These misclassifications, especially among similar defect types, suggest the model struggles with nuanced features within broader defect categories. This issue could be addressed through enhanced feature engineering or increased input data resolution. The poor performance in classifying Soiling may be due to insufficient representation in the training data, affecting the model's ability to detect this fault effectively.

4 Conclusion

In this paper, the efficiency of Vision Transformer in detecting and classifying anomalies in PV modules using infrared thermal imagery has been demonstrated. The ViT-based model achieved 88% overall accuracy in binary classification, with a specificity of 0.91 and sensitivity of 0.86. For multi-class classification of 12 anomaly types, the model excelled in detecting diode issues (97% accuracy) and normal conditions (96% accuracy), but struggled with cell-level anomalies and soiling. The model's strong performance highlights ViT's potential for automated PV module inspection, though varying accuracies across classes suggest areas for improvement, such as distinguishing similar defects and addressing class imbalance. Misclassification patterns, like confusion between Cell and Cell-Multi anomalies, provide insights for refining the model. Future work should focus on enhancing feature extraction for poorly performing classes through advanced data augmentation or higher resolution imagery, addressing class imbalance with techniques like oversampling or class-weighted loss functions, and exploring hierarchical classification approaches. Addressing these challenges could significantly enhance ViT-based models' efficiency and accuracy in PV module inspection, crucial for maintaining and optimizing solar energy systems as solar technology adoption accelerates globally.

References

1. IRENA. Renewable capacity statistics 2023. Int. Renew. Energy Agency (2023)
2. Afifah, A.N.N., Indrabayu, S., Syafaruddin, A.: A review on image processing techniques for damage detection on photovoltaic panels (2021)
3. Korkmaz, D., Acikgoz, H.: An efficient fault classification method in solar photovoltaic modules using transfer learning and multi-scale convolutional neural network. Eng. Appl. Artif. Intell. **113**, 104959 (2022)
4. Hassan, S., Dhimish, M.: Dual spin max pooling convolutional neural network for solar cell crack detection. Sci. Rep. **13**, 11099 (2023)
5. Zhao, Y., Zhan, K., Wang, Z., Shen, W.: Deep learning-based automatic detection of multitype defects in photovoltaic modules and application in real production line. Prog. Photovolt. Res. Appl. **29**, 471–484 (2021)
6. Et-taleby, A., Chaibi, Y., Allouhi, A., Boussetta, M., Benslimane, M.: A combined convolutional neural network model and support vector machine technique for fault detection and classification based on electroluminescence images of photovoltaic modules. Sustain. Energy Grids Netw. **32**, 100946 (2022)
7. Le, M., Le, D., Ha Thi Vu, H.: Thermal inspection of photovoltaic modules with deep convolutional neural networks on edge devices in AUV. Measurement **218**, 113135 (2023)
8. Millendorf, M., Obropta, E., Vadhavkar, N.: Infrared solar module dataset for anomaly detection (2020)

9. Dosovitskiy, A., et al.: An image is worth 16 × 16 words: transformers for image recognition at scale (2020)
10. Soleymani, M., Bonyani, M., Mahami, H., Nasirzadeh, F.: Construction material classification on imbalanced datasets using Vision Transformer (ViT) architecture (2021)
11. Vaswani, A., et al.: Attention is all you need (2023)
12. Steiner, A., Kolesnikov, A., Zhai, X., Wightman, R., Uszkoreit, J., Beyer, L.: How to train your ViT? Data, augmentation, and regularization in vision transformers (2022)

YOLO-Based Photovoltaic Panel Detection: A Comparative Study

C. Moussaoui[1], M. Bouzerdoum[1], N. Kellil[2], and A. Mellit[1](✉)

[1] University of Jijel, 18000 Jijel, Algeria
adelmellit2013@gmail.com

[2] Unité de Développement des Equipements Solaires UDES/Centre dc Développement des Energies Renouvelables CDER, BP 42415, Tipaza, Algeria

Abstract. This paper aims to evaluate the effectiveness of two object detection models, specifically aiming to identify the superior model for detecting photovoltaic (PV) modules based on aerial images. In this study, we examined the deep learning-based YOLOV5n and YOLOV8 models as two prominent YOLO methodologies for PV panel detection. We began by acquiring a dataset comprising 646 images with corresponding binary masks. These images were subsequently partitioned into 506 training images, 140 validation images, and a test set of novel images. For each image, we have created two distinct annotation types: (i) COCO JSON annotations and (ii) YOLO annotations. The models were then trained on our proprietary database, resulting in box precision values of 93.8% for YOLOV5n and 96.7% for YOLOV8.

Keywords: Object detection · YOLO · photovoltaic panels

1 Introduction

Detecting photovoltaic (PV) panels in the study area is a crucial step in the diagnostic process. Without confirming the presence of panels, analyzing a PV installation is impossible. Consequently, numerous studies have been conducted and published to either detect objects or improve existing machine learning (ML) and traditional methods [1].

In computer vision (CV), object detection is a difficult operation, and deep learning improves performance. There are two types of object detectors: single stage and two stage. Single stage detectors offer faster inference times than two stage detectors in terms of detection accuracy. YOLOs (You Only Look Once) increase the accuracy of detection [2].

Since 2012, object detection has improved as Convolutional Neural Networks (CNN) have advanced. It is critical to increase speed because faster R-CNN has a lower Frame Per Second (FPS) than real-time effects.

For example, the study presented in [3] introduces YOLO, a novel CNN solution that outperforms Faster R-CNN with an unmatched FPS of 155 and a mean average precision of 78.6. YOLOv2 has a great capacity for generalization.

© The Author(s), under exclusive license to Springer Nature Switzerland AG 2025
M. Hatti (Ed.): IC-AIRES 2024, LNNS 1238, pp. 312–320, 2025.
https://doi.org/10.1007/978-3-031-80301-7_35

Thus, since deep convolutional neural networks (DCNNs) are becoming more and more useful for object detection because of their ability to learn both high-level and low-level picture information, the review study in [4] covers backbone networks, training procedures, loss functions, complicated problems, datasets, evaluation metrics, applications, and future possibilities for development in deep convolutional neural network-based object detection algorithms. Moreover, YOLO-LITE is ideal for object detection without the need for a GPU; as a result, YOLO-LITE, a real-time object identification model, was developed in [5] for portable devices devoid of a GPU. On COCO datasets, it achieved a 12.26% mAP, and on PASCAL VOC datasets, 33.81% mAP. It runs at 21 frames per second on computers without a GPU and 10 frames per second on a website.

On the same topic, the writers of [6] compare two deep learning algorithms for object detection and image processing: Mask R-CNN and YOLO. The objective of the study is to understand the complexity of detecting tasks, considering human appearance, behavior, and minute characteristics. It is found that YOLO is able to recognize most human figures more accurately and successfully than Mask R-CNN, which is unable to identify small human figures. The study evaluates the performance of YOLO in terms of detection accuracy and computation time, highlighting the limitations of machine learning algorithms in these types of situations.

Another YOLO approach called Faster-YOLO was proposed in [7] as a feature extractor, it leverages a combined network consisting of the double hidden layer extreme learning machine auto-encoder (DLELM-AE) and the deep random kernel convolutional extreme learning machine (DRKCELM). The technique combines the benefits of ELM-AE and ELM-LRF, enabling quicker training. According to the trial results, Faster-YOLO increases speed by an average of two times and improves detection accuracy by 1.1 percentage points. In addition, A small-scale UAV detection method based on the enhanced YOLOv8 is proposed in [8]. It installs the GAM attention mechanism in the neck, lowers network settings, adds a high-resolution detecting head, and uses SPD-Conv for feature extraction. The method increases precision, recall, and mean average precision by 11.9%, 15.2%, and 9%, respectively, when parameters and model size are decreased. The method outperforms experiments and self-constructed datasets, which makes it suitable for engineering implementation and practical uses of UAV object detection systems.

Object detection approaches are used either to locate solar panels or to determine the defects. In particular, solar panel recognition in remote sensing pictures is examined along with object presence detection and localization. A completely supervised object detector, a minimally supervised anomaly detector, and a weakly supervised image classifier with CAM-based localization are the three models that are assessed. While the object detector provides accurate localization, the classifier excels in binary presence detection. The study also demonstrates the potential accuracy benefits from CAM, with the best results being obtained using GradCAM, GradCAM++, and HiResCAM [9].

In this paper, the main objective is to compare two YOLO models for detecting PV panels in aerial images. Our primary goal is to select the best object detector between the two models studied for real-time testing in upcoming projects. Our main contribution lies in applying this for PV panel diagnosis.

2 YOLO Models and Database Preparation

2.1 YOLO Models

YOLO is an algorithm that segment images into the grid system in order to detect objects [10]. YOLO-Unified, real-time Object Detection is a single-stage object detection model was put forward in 2015 and published in 2016 by Joseph Radmon [11]. Figure 1 shows the release dates timeline.

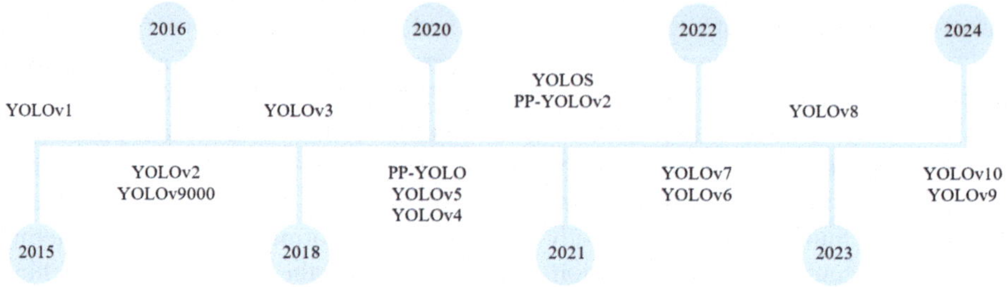

Fig. 1. YOLO: Release dates timeline.

One CNN is used by the YOLO algorithm to split the image into a grid. The grid's cells each forecast a specific number of bounding boxes. Every bounding box has a cell that forecasts a class probability, or the chance that a certain object will be inside the box [12]. The architecture is presented in Fig. 2.

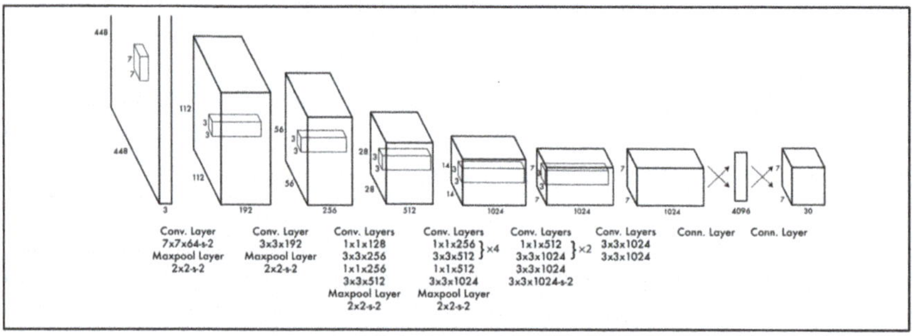

Fig. 2. YOLO's architecture [12].

Too many approaches exist for evaluating object detection models in terms of performance evaluation metrics. First, IoU (Intersection over Union) which can be presented as follows:

$$IoU = \frac{area\ of\ overlap}{area\ of\ union} \tag{1}$$

Figure 3 presents the two areas; the IoU measures how accurate our detection. Taking into account a ground-truth bounding box and a detected bounding box. The IoU can be given any value between 0 and 1. When two boxes do not overlap, the IoU is 0. On the other hand, if the intersection and union regions completely overlap, they are equal. Therefore, in that case, the IoU equals 1 [13].

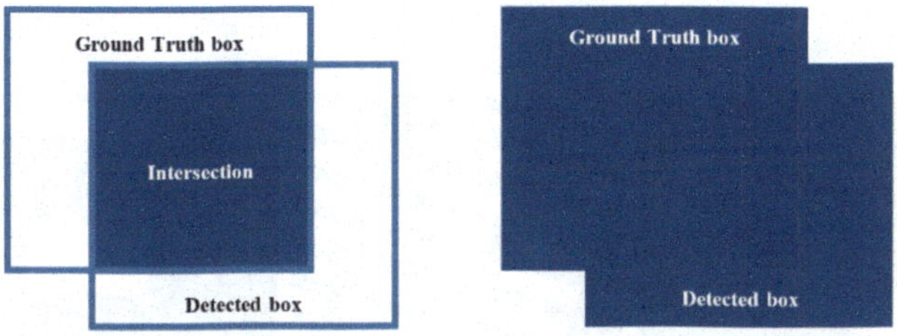

Fig. 3. IoU areas [13].

Second, average precision (AP); a crucial metric in object detection, calculating the precision-recall curve created by varying the detection threshold. It ranges from 0 to 1, with higher values indicating better performance. The mean Average Precision (mAP) is often used to calculate AP for each class. YOLO, a popular algorithm, revolutionizes object detection by completing both identifying and classifying tasks in a single pass [12]. In this study, we have tested both YOLOv5 and YOLOv8, so that we select one object detector between them for PV panel detection in future studies in order to apply the areal inspection. All the used codes are inspired from [15], the owner make them open source containing so many deep learning applications especially in medicine.

2.2 Database Preparation

In our study, we have used a database containing 645 images, downloaded from [15]; a solar (PV) dataset derived from aerial and satellite images. The dataset consists of three groups of PV samples: PV08 from Beijing-2 and Gaofen-2 imaging, PV03 from aerial photography, and PV01 from UAV orthophotos. The samples were gathered at spatial resolutions of 0.8, 0.3, and 0.1 m. The studied one in this work is PV01, which contains rooftop samples that have been split into three groups based on the type of roof that they came from: brick, steel tile, and flat concrete. The database includes binary mask for every image. Figure 4 presents samples of the used images with binary masks for every image.

Data preparation diagram is presented in Fig. 5; and can be summarized as follows:

- **Database collection:** We compiled the previously mentioned database, allocating 80% of the images for training and 20% for validation. Additionally, we reserved a separate set of images for testing to ultimately assess the model's efficiency.
- **Binary mask to coco.json annotation:** The code extracts contours from binary masks using OpenCV, converting them into polygon-like annotations. These annotations include bounding boxes, area, and segmentation information, associated with image IDs and category IDs. The code also creates an images section with metadata, assembles them into a dictionary following the COCO JSON format, and saves the data to

Fig. 4. Samples of the used images with their binary masks.

a file for use with tools. However we ca do the coco.json annotation using Makesense platform.

- **Coco.json to YOLO annotation format:** The code converts a dataset of images and annotations into a YOLO object detection model format and creates a YAML configuration file for training. It reads coco style JSON annotations and inputs as a single JSON file. The code then converts the input JSON file, normalizes polygon segmentation data, and writes them to text files. The *create_yaml* function extracts category names and class numbers, constructs a dictionary, and writes it to a YAML file. The text annotation file contains lines representing individual object annotations. Just like the previous annotation format. we can use Roboflow platform; but in our case we used the code because we don't own the database so we can't see it in a website that makes it open-source.

The diagram of the followed method can be summarized as presented in Fig. 6.

3 Results and Discussion

In the training of the two models, we have detected 1120 instants, which means our object appears on images 1120 times. The results of both two models are showed in Table 1, observing the results we can determine that YOLOv8 outperform v5 in term of precision and recall but in return, it's outnumbered the second one in the recall. That means that YOLOv8 is better in detecting real object but YOLOv5 is better in eliminating the background. In addition, the mAP obtained by YOLOv8 is higher than YOLOv5, which means that in general v8 is better to detect and localize the objects.

In addition, the normalized confusion matrixes show better result with YOLOv8 then YOLOv5, Fig. 7a shows that YOLOv8 predicts 0.88% of the images correctly; meanwhile YOLOv5 predicts 0.85% making v8 w better object detector.

Table 2 presents the summary of obtained curves in the validation set, while both models perform nearly identically, one is marginally superior to the other. According to the f1 score-confidence curve; the optimal threshold for making predictions with yolov5

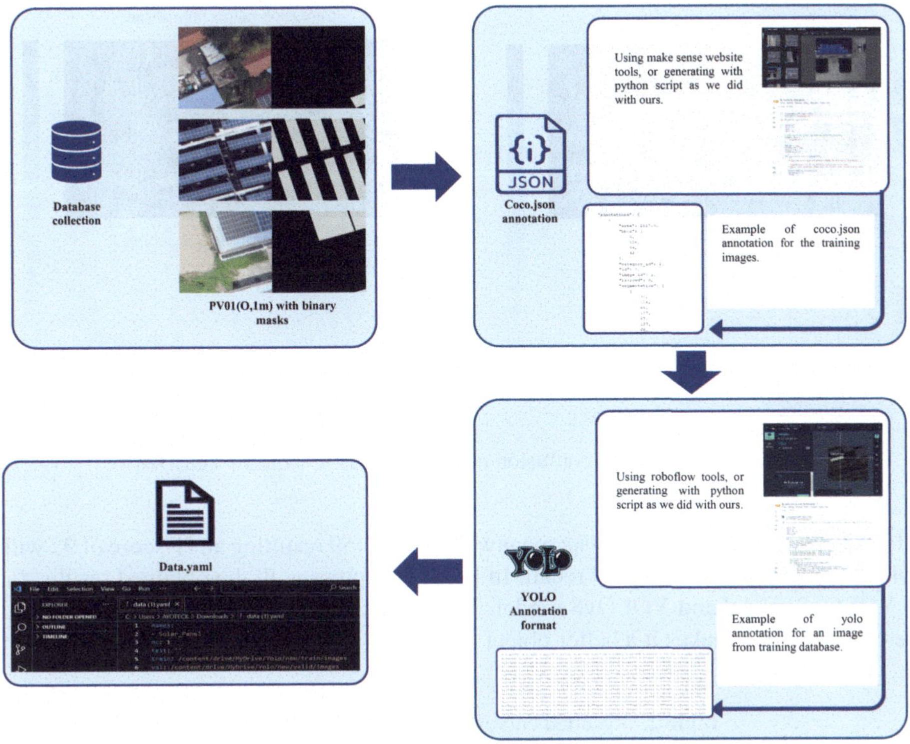

Fig. 5. Data preparation steps.

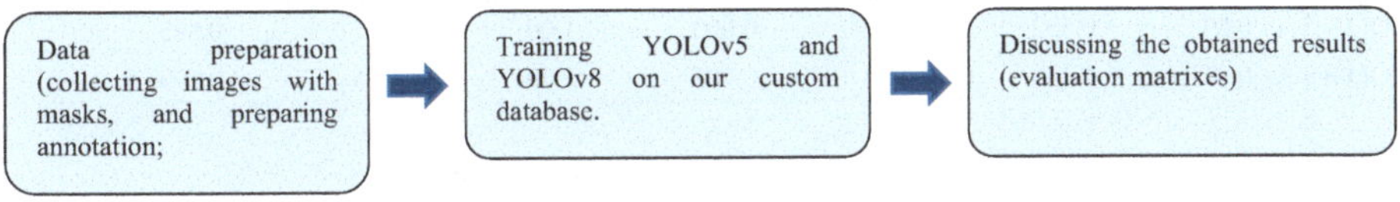

Fig. 6. Diagram of the used method.

Table 1. The results of boxplots and mask of both models.

	YOLOv5	YOLOv8
Precision (%)	93.8	96.7
Recall (%)	84.6	87.1
F1-Score (%)	88.9	91.6
mAP50 (%)	87.5	89.9
mAP50-95 (%)	76.2	76.3

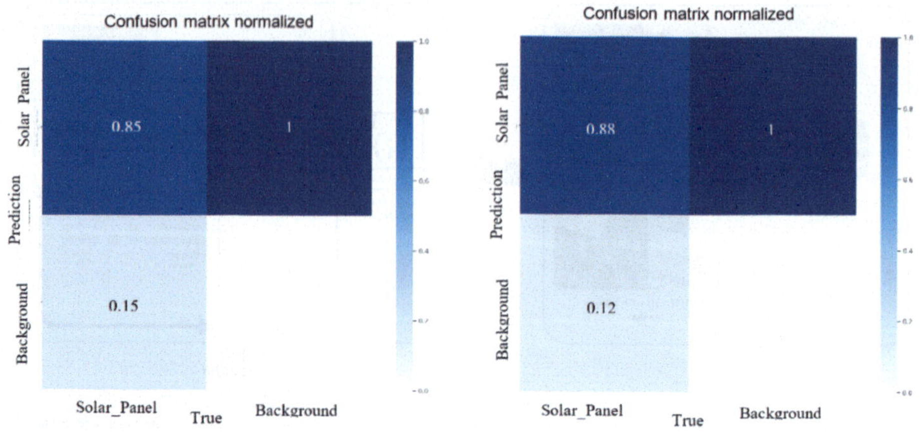

Fig. 7. Normalized confusion matrix of (a) YOLOv5, (b) YOLOv8

is 0.532 giving an f1-score of 89%; meanwhile it is 0.459 resulting an f1-score of 92 with yolov8, and so in precision and recall. In Fig. 8, we present all the obtained results, (a) of YOLOv5 model and YOLOv8 model. The results shows that YOLOv8 outperforms the other one that makes it a better object detector.

Table 2. Resulted curves for YOLOv5 and YOLOv8.

	F1-Confidence		Recall-Confidence		Precision-Confidence		Precision-Recall (mAP 50)
	F1-Score	Thresholds	Recall	Thresholds	Precision	Thresholds	
YOLOv5	0.89	0.532	0.89	0.000	1.00	0.977	0.893
YOLOv8	0.92	0.459	0.90	0.000	1.00	0.954	0.899

The validation results are shown in Fig. 9, (a) the true validation label (1 for YOLOv8 and 3 for YOLOv5) and predicted validation label (2 for YOLOv8 and 4 for YOLOv5), remarking a higher probability provided by YOLOv8 which means a higher possibility for the object to be truly detected because as it is mentioned before, YOLOv8 is better when it comes to true detection (True positive), but YOLOv5 is better when eliminating the background.

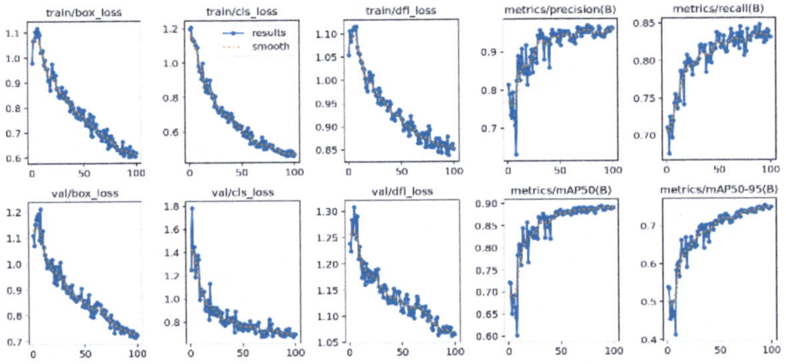

(a)

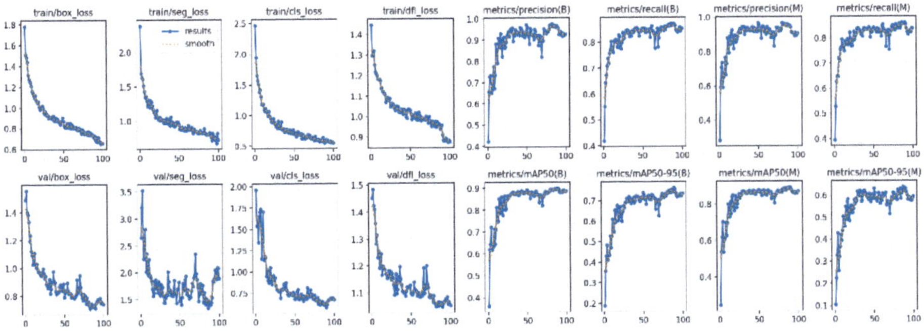

(b)

Fig. 8. (a) Obtained results using yolov5, (b). Obtained results using yolov8.

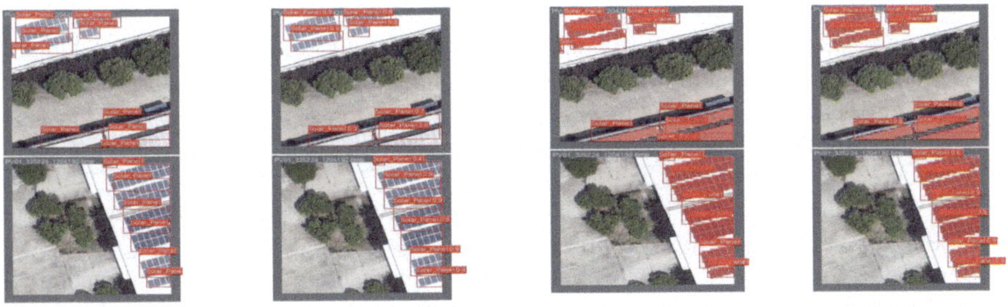

(1) (2) (3) (4)

(a) (b)

Fig. 9. **(a)** batch labels and prediction with YOLOv5, (b) batch label and prediction for YOLOv8 in the validation set. Regarding the percentage, we can determine that YOLOv8 recognize the object better than YOLOv5.

4 Conclusion and Perspectives

In this study, we evaluated two YOLO models to determine the optimal object detector between them. Our findings indicate that YOLOv8 outperforms YOLOv5 across most evaluation metrics, except for recall and loss functions. Therefore, the choice of model depends on the specific detection requirements. With a precision of 96.7%, we selected YOLOv8 for detecting PV panels, prioritizing the ability to detect a maximum number of objects for aerial inspection. As perspective, our goals include improving these results, generalizing the models, and applying them to aerial inspections involving PV plants detection, localization, and classification using drones with high-quality cameras.

Acknowledgments. Dr. A. Mellit would like to acknowledge support from the ICTP, Trieste (Italy) through the Associates Program (2023–2028).

References

1. Mellit, A.: An embedded solution for fault detection and diagnosis of photovoltaic modules using thermographic images and deep convolutional neural networks. Eng. Appl. Artif. Intell. **116**, 105459 (2022)
2. Terven, J., Córdova-Esparza, D.M., Romero-González, J.A.: A comprehensive review of yolo architectures in computer vision: from YOLOv1 to YOLOv8 and YOLO-NAS. Mach. Learn. Knowl. Extract. **5**(4), 1680–1716 (2023)
3. Du, J.: Understanding of object detection based on CNN family and YOLO. In: Journal of Physics: Conference Series, vol. 1004, p. 012029. IOP Publishing (2018)
4. Xiao, Y., et al.: A review of object detection based on deep learning. Multimed. Tools Appl. **79**, 23729–23791 (2020)
5. Huang, R., Pedoeem, J., Chen, C.: YOLO-LITE: a real-time object detection algorithm optimized for non-GPU computers. In: 2018 IEEE International Conference on Big Data (Big Data), pp. 2503–2510. IEEE (2018)
6. Sumit, S.S., Watada, J., Roy, A., Rambli, D.R.A.: In object detection deep learning methods, YOLO shows supremum to Mask R-CNN. In: Journal of Physics: Conference Series, vol. 1529, no. 4, p. 042086. IOP Publishing (2020)
7. Yin, Y., Li, H., Fu, W.: Faster-YOLO: An accurate and faster object detection method. Digit. Signal Process. **102**, 102756 (2020)
8. Zhai, X., Huang, Z., Li, T., Liu, H., Wang, S.: YOLO-drone: an optimized YOLOv8 network for tiny UAV object detection. Electronics **12**(17), 3664 (2023)
9. Burger, M., Wijnhoven, R., You, S.: Exploring different levels of supervision for detecting and localizing solar panels on remote sensing imagery. arXiv preprint arXiv:2309.10421 (2023)
10. https://www.analyticsvidhya.com/blog/2022/06/yolo-algorithm-for-custom-object-detection/. Assessed 20 June 2024
11. https://learnopencv.com/mastering-all-yolo-models/. Assessed 20 June 2024
12. https://encord.com/blog/yolo-object-detection-guide/. Assessed 20 June 2024
13. https://www.baeldung.com/cs/object-detection-intersection-vs-union. Assessed 20 June 2024
14. https://github.com/bnsreenu/python_for_microscopists/tree/master
15. Jiang, H., et al.: Multi-resolution dataset for photovoltaic panel segmentation from satellite and aerial imagery. Earth Syst. Sci. Data Discussions 1–17 (2021)

YOLOv8-Based Photovoltaic Module Detection Using Aerial Imagery

N. Kellil[1](✉), C. Moussaoui[2], A. Mellit[2], and A. Boudiaf[3]

[1] Unité de Développement des Equipements Solaires UDES/Centre de Développement des Energies Renouvelables CDER BP 42415, Tipaza, Algeria
n.kellil@yahoo.com

[2] Faculty of Science and Technology, University of Jijel, Jijel, Algeria

[3] Research Center in Industrial Technologies–CRTI, Annaba, Algeria

Abstract. Nowadays, the use of unmanned aerial vehicles (UAVs) for aerial inspections of photovoltaic (PV) plants has emerged as a crucial alternative solution. This study introduces a method for detecting PV modules and arrays using RGB images processed with YOLOv8. A dataset of RGB images was collected using a drone (DJI mini-3) at the Unit for Solar Equipment Development (UDES), located in northern Algeria, and employed to develop this method. Beyond the detection of PV modules and PV panels, this approach also helps users to detect, classify and locate certain types of PV systems faults and thus, determine the need to launch a maintenance procedure based on visual inspections. Detailed RGB images collected. The results are promising, with a detection accuracy ranging between 70% and 95%. Based on the experience gained, the potential and limitations of the developed method for application to larger-scale PV plants are also discussed.

Keywords: Photovoltaics · anomaly detection · object detection · deep learning · YOLOv8

1 Introduction

The latest report from the International Energy Agency (IEA) on global photovoltaic (PV) capacity reveals significant growth in the sector. As of 2023, the cumulative global PV capacity has reached 1.6 terawatts (TW), up from 1.2 TW in 2022. This increase includes 407 to 446 gigawatts (GW) of new PV installations over the past year [1]. Algeria is making significant strides in increasing its PV capacity as part of its broader renewable energy goals. As of early 2024, the country has initiated several large-scale solar projects aimed at boosting its solar power capacity. However, the total PV installed capacity is about 500 MW at the end of 2023 [2].

A major concern for PV plant proprietors is the need for reliable and efficient maintenance procedures [3, 4]. Currently, aerial inspection of PV plants by means of unmanned aerial vehicles (UAV) become a popular approach and numerous works have been done in the literature [5, 6]. For instance, the authors in [7] proposed an innovative approach

© The Author(s), under exclusive license to Springer Nature Switzerland AG 2025
M. Hatti (Ed.): IC-AIRES 2024, LNNS 1238, pp. 321–329, 2025.
https://doi.org/10.1007/978-3-031-80301-7_36

for monitoring and supervising PV systems using lightweight UAVs. One of the primary goals was to create a cost-effective solution for inspecting PV plants. The findings demonstrated that the UAV-based model could effectively detect most failures in PV plants using infrared and visual imagery. In [8] an unmanned inspection system for multiple defects detection in PV plants was proposed. With reference to the authors, the experimental results clearly demonstrate the effectiveness of our solution in diagnosing photovoltaic modules with various visible defects. An overview on various PV inspection procedure based on UAV was presented [9]. All aspects were meticulously analyzed, and the open issues were highlighted.

The topic of object detection for PV modules or arrays has garnered significant interest from researchers worldwide. Consequently, deep learning-based object detection of PV modules has become a hot topic, leading to the development of various models [10]. In [11] the authors developed a novel approach to PV modules occlusion detection. The main contribution was to design a segment YoLOv5 algorithm. The experimental results showed that the proposed model performs better than other investigated models such as SSD, Faster-RCNN, YoLov4 and U-net. In [12] a deep convolutional neural network was developed to provide some information about small-scale PV plants such as capacity, location and the produced power. Object detection and aerial images were used to detect and positioning key-points of small-scale PV plants [13]. With reference to the authors the designed approach can detect and provide accurately the required information about the PV plants. A novel automatic multi-stage model to detect PV module defects based on aerial images and YoLov3 [14]. The model was evaluated on two large-scall PV plants and the prediction results are about 98%. In [15] the authors proposed a PV-YOLO for PV modules detection, the results showed that the proposed method presents good accuracy of the mean average precision (mAP).

The main objective of this work is the design a cost-effective method for PV modules detection for monitoring, inspection, preventive maintenance and mapping the PV installations. This method is suitable for detecting installation-related faults and for performance optimization. This work was conducted at the unit for solar equipment development (UDES), Tipasa, Algeria.

The YOYOv8 was employed and evaluated based on dataset built at the UDES using an UAV.

2 Methodology and Materials

2.1 Site Description and PV Plants

The UDES is located (Lat, long and alt = 36.6616° N, 2.7203° E) in northern of Algeria, (Mediterranean climate), see Fig. 1a and a photo of real view of the UDES is depicted in Fig. 1b.

Different PV module technologies are installed at the UDES, namely: poly-Si, c-Si and organic. These PV modules are fixed to ground-mounted, rooftop-mounted and front-mounted structures (in front of the conference room), as shown in Fig. 2(a). Almost PV modules mounted face south. For experimental purposes, other experimental PV systems are also installed, such as: 1.2 kWp PV system with sun tracker (Fig. 2. (b)),

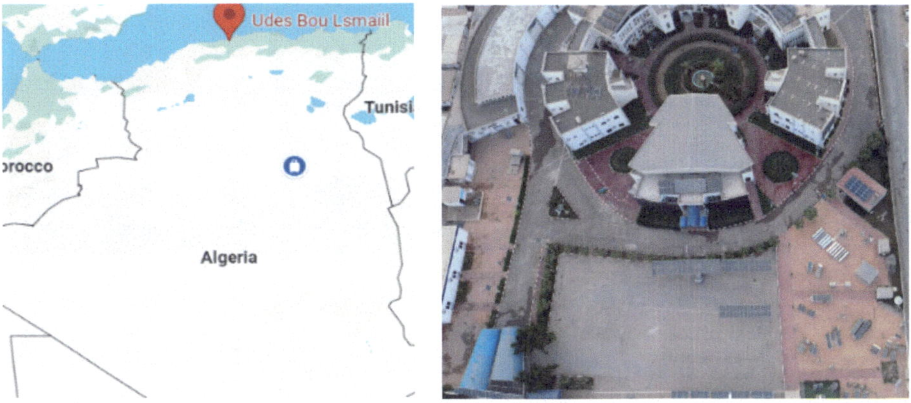

Fig. 1. a) geographical location of the investigate plants, b) real photo to the locations, including some PV arrays.

hybrid systems and an east-west type PV power plant (15 kWp) as illustrated in Fig. 2. (c).

Fig. 2. Different types of PV mounting structures

Meteorological data are collected using a weather station installed on the rooftop of the building, linked to different monitoring systems. The PV generators are permanently affected by extreme weather conditions, such as: windstorms, maritime climate, the origin of corrosion, which reduces the lifespan of PV systems. Added to this, the deposit of grain of salt on the surface of the PV modules which has as effect to reduce the conversion efficiency. The deposition of dust as shown in Fig. 3a due to nearby construction work (Concrete plant Fig. 3b) is another environmental fault affecting the performance of installed PV generators.

Fig. 3. a) Dust deposition on the PV modules, b) Nearby concrete plant causing dust deposition

2.2 The Used UAV Description and Database Preparation

During this work, we used an UAV-DJI mini 3 (See Fig. 4a), which can be remotely controlled or even flying autonomously using auto-controlled flying software, onboard GPS and telemetric sensors. It is equipped with a Li-battery, that allows an autonomous fly of nearly 30 min. The UAV allows us to capture aerial images with high resolution using a 4k camera mounted on a gimbal offering stable views. The UAV is controlled by remote control with integrated screen for real time view. This kind of imaging method can be useful for fast and low-cost fault detecting and diagnosis, using Artificial intelligence (AI) techniques for object recognition, such as Computer vision methods [4].

A database containing 100 images have been built. Then to develop our model each image for the dataset was converted to binary mask. As example, Fig. 4b and Fig. 4c show the real image and the corresponding binary mask.

Fig. 4. a) UAV-DJI mini-3, b) RGB aerial image and c) binary mark

For the preparation of the data base (after capturing and collecting images), we have followed the following steps:

i. **Preparing the binary masks:** for this operation, we have used MATLAB Image Segmenter App, we draw the bounding box for each object in every image, and then we get the binary mask within 1 (black) for the background and 0 (white) for the object as presenting in the following figs.
ii. **Preparing coco.json annotation:** generating the images with a python script; that formed a json file from the binary masks using OpenCV that extracts contours from binary masks, creating polygon-like annotations with bounding boxes, area, and segmentation data. Metadata is added to pictures, compiles into a dictionary using COCO JSON format, and stores information in a file.

iii. **Preparing yolo format annotation:** the python script converts a dataset of images and annotations into a YOLO object detection model format, creates a YAML configuration file for training, normalizes polygon segmentation data, extracts category names and class numbers, constructs a dictionary, and writes it to a text annotation file.

2.3 Object Detection Approach

YOLOv8 (You Only Look Once version 8) is a computer vision model used for object detection, image classification, and instance segmentation tasks. YOLOv8 is a recent version of the YOLO series, which was first launched in 2015 by Joseph Redmond et al., [16]. The structure of the YOLOv8 algorithm can be divided, as shown in Fig. 5, into four parts: a) input, b) backbone network, that extracts multiscale properties from images using convolutional methods, c) neck network, that consolidates the backbone-extracted properties, and d) head, which is an object classification and regression module. YOLOv8 divides the image into different networks and predicts the bounding boxes of the objects contained in them with their classes. The overlapping bounding boxes are eliminated using non-great algorithms [17].

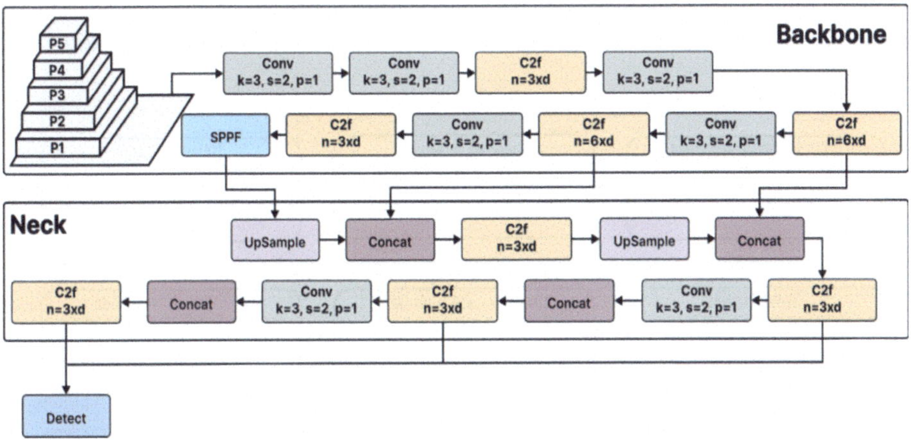

Fig. 5. YOLOv8 principal architecture Source: [18]

The designed approach is widely used in PV fault detection. The proposed method is based on the use of UAVs to acquire large-scale aerial images of PV modules. For this, we used PIX4d software, which is designed for photogrammetry and drone mapping. By delimiting the frontiers of the area to be scanned, and setting flight parameters such as fly altitude and map data resolution (GSD), the software calculates the flight time and defines the trajectory that the drone will follow during its flight (from START point of to the END) as shown in Fig. 6. This trajectory is subsequently transmitted via cloud to the drone remote control, and then the drone will be autonomously piloted according to this trajectory. At the end of the mission, a photogrammetry map of the studied area will be designed and ready for analysis.

Fig. 6. Drone mapping of the UDES area using PIX4d software

This is a non-destructive vision inspection method, which combine deep learning and data argumentation (to solve the problem of a lack of training dataset). The accurate, anomalies of the PV module detection and location can help in making the right decision and planning a maintenance process. The generalized approach on the implementation of the YOLOv8 for PV modules faults detecting can be summarized in the following flowchart.

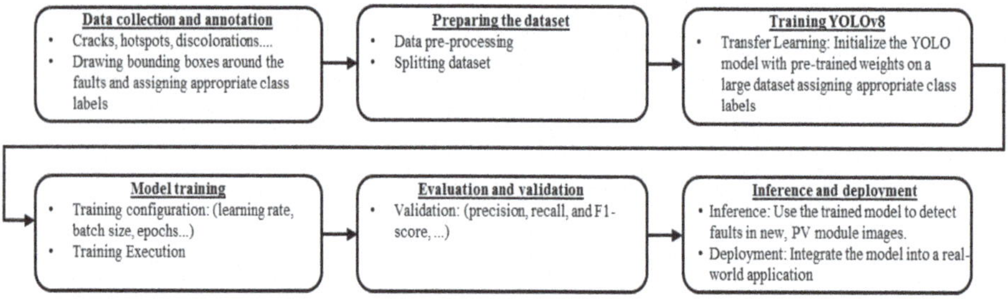

Fig. 7. How to detect or identify anomaly

During the training of the YOLOv8 model, the balanced distribution of different fault types in the dataset has been taken into account to prevent the model from being biased towards more prevalent faults.

3 Results and Discussion

After several experiments including variation of hyperparameters including epoch, batch size, and other. The optimal results are reported in Table 1.

Box loss (box_loss), segmentation loss (Seg_loss), classification loss (Cls_loss) and distribution focal loss (Dfl_loss) are critical metrics in training and evaluation object

Table 1. The results of boxplots and mask of the model.

	YOLOv8
Precision (%)	0.61
Recall (%)	0.53
F1-Score (%)	0.28
mAP50(%)	0.54
mAP50–95(%)	0.32

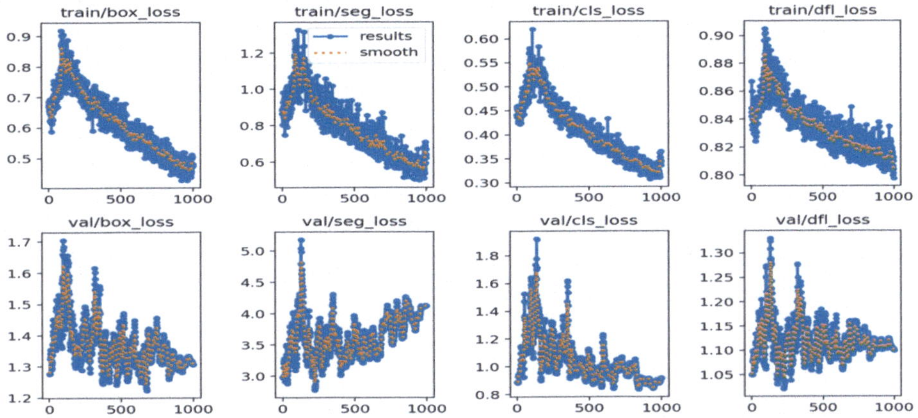

Fig. 8. The obtained results: box_loss, Seg_loss, Cls_loss and Dft_loss

detection models. Figure 7 shows the obtained results for train and validation of box_loss, Seg_loss, Cls_loss and Dft_loss. As can be observed the results are very correlated.

The test results are shown in Fig. 8. From Fig. 8a, it can be clearly observed that the model can identify the PV modules with an accuracy of 90–94%. However, it detects the thermal collector with an accuracy of 21%. One thermal module is not detected at

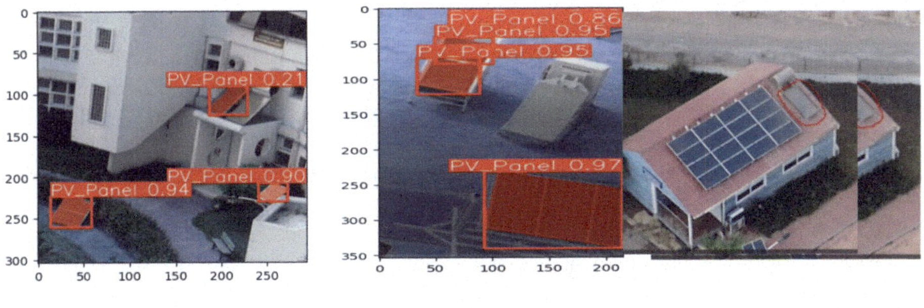

a b c

Fig. 9. The evaluated images

all. While in Fig. 6.b the model can perfectly detect the PV module with an accuracy of 97%, but it has also considered the thermal module as a PV module object with an accuracy of 91%.

As can be seen in Fig. 8c, the captured aerial images contain not only PV modules, but also thermal collectors which have almost the same shape. It is clearly observed that the proposed method has some limitations due mainly to the strong similarity between PV modules and solar thermal collector and low resolution in certain images (Fig. 9).

4 Conclusion and Outlooks

In this paper a method for PV module recognition and dusty PV module identification was designed. The YOLOv8 has been considered and employed. The developed method can detect and identify object, including PV modules and thermal solar collector with an acceptable accuracy. The main limits of the developed method are: 1) the used dataset is relatively small with low resolution, 2) Dust deposit similarity on PV modules, and 3) difficulty in detecting small object (view from afar). To improve the detection accuracy of PV module and dust deposit, we have first to increase the size of the dataset, improve the quality of the collected images and accurate tuning hyperparameters of the YOYOv8. As a continuation of this work, we will address carefully the aforementioned points and use another improved version of YOLO.

Acknowledgements. Dr. A. Mellit would like to acknowledge support from the ICTP, Trieste (Italy) through the Associates Programme (2023–2028).

References

1. Masson, G., Bosch, E., Van Rechem, A., De L'Epine, M.: Snapshot 2024, ISBN: 978-3-907281-55-0 (2024). https://doi.org/10.69766/vhrf4040
2. https://www.sonelgaz.dz/fr/790/energies-renouvelables
3. Bouzerdoum, M., Mellit, A., Djazari, N., Laissaoui, M.: Embedded machine learning for fault detection and diagnosis of photovoltaic arrays using a low-cost device. In: Hatti, M. (ed.) Advanced Computational Techniques for Renewable Energy Systems, pp. 82–90. Springer, Cham (2023). https://doi.org/10.1007/978-3-031-21216-1_9
4. Kellil, N., Aissat, A.E., Boudiaf, A., Mellit, A.: Computer vision-based PV module fault recognition using a transfer learning approach. In: Mellit, A., Belmili, H., Seddik, B. (eds.) Proceedings of the 1st International Conference on Advanced Renewable Energy Systems: ICARES'22, 18–20 Dec, Tipaza, Algeria, pp. 31–37. Springer, Singapore (2024). https://doi.org/10.1007/978-981-99-2777-7_4
5. Addabbo, P., et al.: UAV system for photovoltaic plant inspection. IEEE Aerosp. Electron. Syst. Mag. **33**(8), 58–67 (2018)
6. Li, X., Yang, Q., Chen, Z., Luo, X., Yan, W.: Visible defects detection based on UAV-based inspection in large-scale photovoltaic systems. IET Renew. Power Gener. **11**(10), 1234–1244 (2017)
7. Quater, P.B., Grimaccia, F., Leva, S., Mussetta, M., Aghaei, M.: Light Unmanned Aerial Vehicles (UAVs) for cooperative inspection of PV plants. IEEE J. Photovoltaics **4**(4), 1107–1113 (2014)

8. Li, X., Li, W., Yang, Q., Yan, W., Zomaya, A.Y.: An unmanned inspection system for multiple defects detection in photovoltaic plants. IEEE J. Photovoltaics **10**(2), 568–576 (2019)
9. Niccolai, A., Gandelli, A., Grimaccia, F., Zich, R., Leva, S.: Overview on photovoltaic inspections procedure by means of unmanned aerial vehicles. In: 2019 IEEE Milan PowerTech, pp. 1–6. IEEE (2019)
10. Arnaudo, E., et al.: A comparative evaluation of deep learning techniques for photovoltaic panel detection from aerial images. IEEE Access (2023)
11. Yu, J., Guan, R., Zhang, C., Shao, F.: A novel object recognition method for photovoltaic (PV) panel occlusion based on deep learning. J. Comput. Methods Sci. Eng. **23**(6), 3391–3405 (2023)
12. Malof, J.M., Collins, L.M., Bradbury, K.: A deep convolutional neural network, with pre-training, for solar photovoltaic array detection in aerial imagery. In: 2017 IEEE International Geoscience and Remote Sensing Symposium (IGARSS), pp. 874–877. IEEE (2017)
13. Qi, F., Liang, S., Cao, R., Ding, Y., Yang, Q., Yan, W.: Detection and positioning of keypoints in small-scale photovoltaic system based on object detection network and aerial sequence images. In: 2020 Chinese Control and Decision Conference (CCDC), pp. 4795–4800. IEEE (2020)
14. Di Tommaso, A., Betti, A., Fontanelli, G., Michelozzi, B.: A multi-stage model based on YOLOv3 for defect detection in PV panels based on IR and visible imaging by unmanned aerial vehicle. Renew. Energy **193**, 941–962 (2022)
15. Yin, W., Lingxin, S., Maohuan, L., Qianlai, S., Xiaosong, L.: PV-YOLO: lightweight YOLO for photovoltaic panel fault detection. IEEE Access **11**, 10966–10976 (2023)
16. Redmon, J., Divvala, S., Girshick, R., Farhadi, A.: You only look once: Unified, real-time object detection. In: 2016 IEEE Conference on Computer Vision and Pattern Recognition (CVPR) (2016). https://doi.org/10.1109/cvpr.2016.91
17. Wei, L., Tong, Y.: Enhanced-yolov8: a new small target detection model. Digit. Signal Process. **153**, 104611 (2024). https://doi.org/10.1016/j.dsp.2024.104611
18. Ramos, L., Casas, E., Bendek, E., Romero, C., Rivas-Echeverría, F.: Hyperparameter optimization of YOLOv8 for smoke and wildfire detection: implications for agricultural and environmental safety. Artif. Intell. Agric. (2024)

Deep Learning-Based Segmentation for Oil Pipeline Leak Detection Using Quadcopter Drones

Dalila Cherifi(✉), Sifeddine Mekroud, and Ali Boudaoud

Institute of Electrical and Electronic Engineering, University of Boumerdes, Boumerdes, Algeria
da.cherifi@univ-boumerdes.dz

Abstract. The advancement of machine learning in oil leak detection systems is transformative, with the potential to revolutionize environmental protection, resource management, and industrial safety. However, transitioning from traditional methods to machine learning-based techniques will be a gradual process over several years, involving the incremental introduction of new features and capabilities already utilized in modern leak detection systems. In this work, we explored machine learning-based leak detection, examining various techniques and their applications in the field. We also reviewed several image analysis methods due to their critical role in safely and efficiently developing and testing leak detection systems. The ultimate goal was to create a fully automated leak detection system. To achieve this, we used a U-Net model fortified with an InceptionResNetV2 backbone, a cutting-edge approach in machine learning. Before implementation, we conducted an in-depth study and explanation of this approach. The model was successfully tested for airborne monitoring, yielding an Intersection over Union (IoU) of 75.30%. The results were presented and discussed in detail.

Keywords: Oil and Gas · Leak Detection System (LDS) · Image Segmentation · Machine Learning · Pipeline Monitoring · Environmental monitoring

1 Introduction

The global oil and gas industry is crucial to the world's energy supply and faces significant challenges, particularly petroleum pipeline leakage. Major leakage incidents annually cause severe environmental damage. Companies must manage and clean oil spills, incurring substantial financial burdens from lost oil [1]. The frequency of such incidents complicates attracting partners and investors due to risks of reputation damage, financial losses, and regulatory penalties [2]. The problem is exacerbated in rural areas like the Sahara desert, where harsh conditions increase risks to human and logistical operations. The global oil and gas pipeline leak detection market was valued at 3.5 billion USD in 2021 and is projected to reach 5.7 billion USD by 2028, growing at a CAGR of 7.40% [3]. This project aims to develop a flexible system for rural areas using machine learning, computer vision techniques, and Unmanned Aerial Vehicles (UAVs), specifically quadcopters. UAVs have become integral to the petroleum industry due to

© The Author(s), under exclusive license to Springer Nature Switzerland AG 2025
M. Hatti (Ed.): IC-AIRES 2024, LNNS 1238, pp. 330–341, 2025.
https://doi.org/10.1007/978-3-031-80301-7_37

advancements in sensor technology and navigation, providing a faster, safer, and more cost-effective method for mass data collection. However, there are research gaps and open issues regarding UAV usage in the petroleum sector, presenting opportunities for further development.

The article is structured to provide a clear understanding of the research process, beginning with an overview of oil leak detection techniques, followed by the methodology for object segmentation, a comparison of experimental setups and evaluation metrics, and concluding with a summary of key findings.

2 Oil Leakage Detection Industry Overview

Oil pipeline leaks can cause substantial losses, making an effective leak detection system crucial. Detection systems can be classified based on parameters such as hardware vs. software techniques [4], automation level (automatic, semi-automatic, or manual) [5], and technical nature [6], which includes external methods, visual/airborne inspection, and internal/computational methods.

2.1 External Methods

External pipeline leak detection methods use sensors installed outside the pipeline. Some examples of these sensors include:

- **Acoustic methods:** Acoustic methods leverage the acoustic waves generated by fluid flow through ruptured pipes for leak detection. These waves propagate through the medium inside the pipeline [7] or via the acoustic emission method outside [8]. Various techniques have been developed for acoustic sensing, including smart ball technology introduced in 2008 [9] and its derivative, twin ball technology, from 2016 [10]. The Principal Component Analysis Support Vector Data Description (PCA-SVD) method utilizes statistical features of acoustic signals for anomaly detection in leak detection systems (LDS) [11]. Empirical Mode Decomposition (EMD) is effective for analyzing non-linear, non-stationary acoustic signals, facilitating the separation of these signals into components of varying resolutions, and has been applied to multi-phase flow pipelines [12]. The multi-model sensing approach uses accelerometers to capture acoustic vibration signals from leaks [13], allowing precise localization through Time Difference-of-Arrival (TDOA) analysis. A 2023 method combining Acoustic-Pressure (A-P) signal fusion with a noise reduction algorithm called Dual Pearson thresholds-Ensemble Empirical Mode Decomposition (DP-EEMD) [14] demonstrated high classification accuracy (98.3%) and superior noise reduction, integrating EEMD signal decomposition, data-level fusion, and 1D-CNN feature extraction for leak diagnosis.
- **Fibre optic sensing:** Fiber optics are increasingly utilized for sensing and measurement in oil pipeline monitoring due to their high sensitivity, enabling precise installation on pipes. Changes in vibration or temperature interfere with light pulses in the fiber optic cable, creating reflections that help locate events [15]. The Distributed Temperature Sensor (DTS) is commonly used with gases but has difficulty detecting leaks when crude oil temperature matches ambient conditions. To address this, the

Distributed Vibration Sensor (DVS) converts external vibrations into acoustic signatures for leak detection [16]. The Mach-Zehnder interferometer (MZI) [17] employs light wave interference for accurate measurements, with techniques like Jones matrix modeling [18] and SVM-based MZI [19] focusing on polarization changes and vibration signal features, respectively. Distributed Temperature Gradient Sensing (DTGS) identifies temperature gradients for subtle leaks but requires complex data analysis [20]. The Distributed Differential Temperature Sensor (DdTS), introduced in 2016, offers high sensitivity (0.0005 °C) and accurate leak location within meters [21]-[22]. Fiber Bragg Grating (FBG) technology, proposed in 2018, reduces leak detection costs by employing high-sensitivity sensors for Surface Acoustic Waves, minimizing the need for closely spaced sensors [23]. In 2022, advancements in compressed sensing with FBG arrays improved localization accuracy [24]. Additionally, rubber-based FBG pressure sensors demonstrate greater sensitivity than metallic sensors using the Negative Pressure Wave (NPW) algorithm [25].

2.2 Internal Methods

Interior methods focus on monitoring hydraulic and operational conditions using minimal sensors. These techniques gather data on flow, pressure, and temperature, performing computational analysis for real-time monitoring. Some methods are listed below:

- **Operational data:** Pressure and flow monitoring systems are key techniques in operational data leak detection, focusing on disruptions in gas flow by analyzing changes in frictional losses [26]. A method introduced in 2003 utilized wavelet analysis to examine reflected pressure waves for abrupt shifts, indicating potential issues [27]. A fuzzy clustering algorithm applied to denoised pressure data achieved up to 95% accuracy in field tests [28]. A 2018 study highlighted that upstream pressure is sensitive to both leak size and location, while downstream flow rate is mainly sensitive to leak size, suggesting pressure's versatility as an indicator [29]. To mitigate false alarms from low signal positioning accuracy, an enhanced denoising method was developed, combining cross-spectral analysis with ensemble empirical mode decomposition (EEMD). This approach utilizes infrasound and negative pressure waves for improved denoising compared to traditional techniques [30].
- **Computational Pipeline Monitoring (CPM):** Real-Time Transient Model (RTTM) are mathematical models used to simulate the behavior of fluid flow in pipelines, considered as a very accurate and robust approach for leak detection[31], as it uses the SCADA system to obtain parameters such as pressure, temperature, flow, and they are compared with the model-based parameters in real time, any deviation will be considered a leak. RTTM technology is continually evolving, with two main advancements emerging in recent years: the Extended Real-Time Transient Model (E-RTTM) and FD simulation along with dimensionless variable analysis.
- **Machine Learning and IOT technologies:** Machine Learning (ML) algorithms significantly enhance pipeline leak detection by identifying previously undetectable leak sizes and types. One approach integrates data from distributed acoustic sensing (DAS) and Internal Leak Detection (ILD) using a dynamic Bayesian network (DBN), reducing false alarms and improving sensitivity [32]. In 2019, a method combining accelerometer and acoustic sensor data via wavelet analysis and ML algorithms

like SVM achieved 90% detection accuracy. Other ML models incorporate pressure, temperature, and flow rate measurements, using regressors like gradient boosting alongside a leak detection classifier, matching traditional methods. The Multivariate Adaptive Regression Splines (MARS) algorithm demonstrated superior capabilities in 2021 [33].

Advanced deep learning techniques, such as generative adversarial networks (GANs), estimate leak parameters like location and size, outperforming traditional methods. Convolutional neural networks (CNNs) also analyze acoustic data to extract leak-related features, enhancing classification accuracy.

Challenges persist, including a lack of real leak data for training and complex factors affecting pipeline failures. Bayesian networks were proposed in 2022 to incorporate limited historical data [34]. The Digital Twin (DT) concept is gaining traction; in 2021, a fuzzy digital twin (FDT) combined with SVM improved detection accuracy by 25% over classical methods [35]. In 2022, a system using Markov chain classification and Ensemble Local Mean Decomposition (ELMD) with K-L divergence achieved 91.23% accuracy using pressure data, surpassing SVM and Naive Bayes systems [36].

2.3 Visual-Airborne Inspection Methods

Pipeline monitoring methods are non-continuous and typically involve inspections by trained personnel, helicopters, dogs, and increasingly, UAVs for aerial data collection. UAVs equipped with multispectral, thermal cameras, LiDAR, and Optical Gas Imaging (OGI) technology are used for leak detection. Deep learning techniques like holistically-nested edge detection (HED), extreme inception (Xception) networks, and DexiNed architecture have been applied to detect oil spills in drone-captured images. Another method integrates GIS and thermographic technologies, using infrared and HD recordings alongside GPS for processing with algorithms like RGB channel interpolation [37]. LiDAR creates baseline 3D maps for anomaly detection. A pulsed LiDAR system with a YAG:Nd3 + laser enables precise fluorimetry of crude oil by exciting fluorescence [38]. In [39], a method detects oil spills from drone images, with a Raspberry Pi 4-equipped drone transmitting images and GPS coordinates via MQTT. Deep learning analyzes these images to identify spills, using the DexiNed algorithm for contour detection. Another study trained a VGG16 model with the NAFTA dataset, achieving 92.77% accuracy, while YOLOv3, Mask R-CNN, and PSPNet had mAP and IoU scores of 70.98%, 46%, and 66%, respectively. A specific dataset is not publicly available, requiring a unique solution to be discussed later.

3 Methodology: Designing a Quadcopter Aerial Inspection System with Object Segmentation Architectures

In this section, we outline the methodology applied in our work to detect oil leaks using image processing tools and deep learning methods. Our approach integrates various techniques to perform this analytical task efficiently and accurately.

3.1 Feature Extraction Using GLCM

We used the Gray Level Co-occurrence Matrix (GLCM) to extract statistical parameters as features. GLCM is a powerful tool for obtaining second-order statistical features from images, as it calculates the frequency of pixel pairs with specific values and spatial relationships, creating a matrix that represents the image's texture. The key features we derived from GLCM include Contrast (which measures local gray-level variations), Energy (the sum of squared GLCM elements), Correlation (the joint probability of pixel pairs), and Homogeneity (the closeness of element distribution to the diagonal). These features provide crucial insights into an image's texture and structure, aiding in tasks such as segmentation, classification, and recognition.

3.2 Deep Learning Segmentation Approaches for Oil Leak Detection

Image segmentation can be categorized into instance segmentation and semantic segmentation, both of which involve pixel-level classification. Instance segmentation has the additional capability of distinguishing between different instances of the same class, such as multiple leaks in a single image. For instance segmentation, we developed a model using the YOLOv8 architecture. Trained on an NVIDIA A100 GPU, this model focused on pixel-level labeling to accurately identify leaks. After an initial 30 epochs, we further trained the model for 100 epochs with a batch size of 16. We achieved satisfactory results with semantic segmentation using a U-Net architecture. This model includes a contracting path for downsampling and feature extraction and an expanding path for upsampling and refining the segmentation mask. By incorporating the Inception-ResNetV2 backbone, we enhanced the U- Net's feature extraction capabilities, making it well-suited for complex tasks like oil pipeline leak detection. Our U-Net model with an Inception-ResNetV2 encoder achieved a validation IoU of 75.30% and performed well under various testing conditions. The model reached a maximum depth of 2080 features at its bottleneck and utilized 255 Conv2D layers, 214 batch normalization layers, and 254 activation layers. These components, along with Max Pooling and Average Pooling layers, helped reduce computational complexity while retaining important information. The Inception-ResNetV2 backbone provided rich feature maps, and the concatenate and lambda layers combined features and performed custom operations. The upsampling layers in the expansive path ensured precise localization for segmentation.

4 Experiments and Results

This section outlines experiments to evaluate our machine learning-based leak detection system, focusing on various methodologies for detecting oil leaks using quadcopter drone imagery. We conducted tests to benchmark model performance under different conditions and configurations.

4.1 Synthetic Dataset Generation Using Latent Diffusion Model

A major challenge was obtaining high-quality drone images of oil pipelines with and without leaks in the Algerian Sahara. To build the dataset, the team used Latent Diffusion

Models (LDM) by Stability AI instead of GANs due to their superior performance in recent studies dir [40]. They generated 4000 images and labeled each pixel for oil leaks and pipes using Roboflow. The models were also tested with real images gathered from the web to ensure robustness. Data augmentation techniques like cropping, rotation, and color adjustments were applied to diversify the dataset, expanding it to 7940 images. As seen in the images in Figs. 1 and 2, the images are from an aerial point of view that aligns with the target point of view of a drone and captures the whole field of view needed.

Fig. 1. Generated image example with an oil leak.

Fig. 2. An example image on the images generated with no leak.

4.2 Experiment 1: Feature Extraction Using GLCM

The images in the dataset are marked by darker regions with distinctive textures, as revealed by the histogram of a sample image shown in the Figs. 3, 4 (Fig. 5).

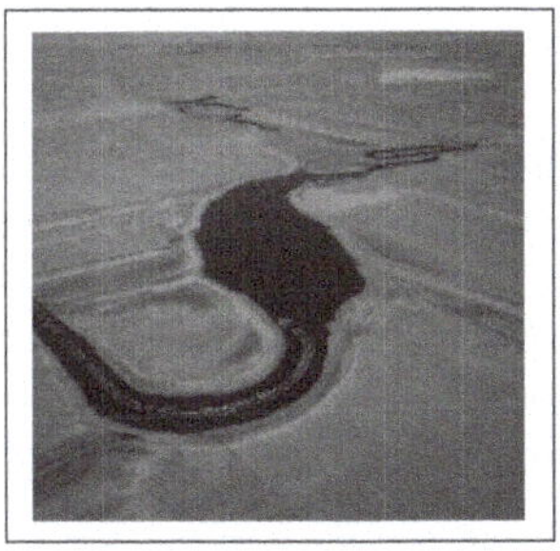

Fig. 3. The input image.

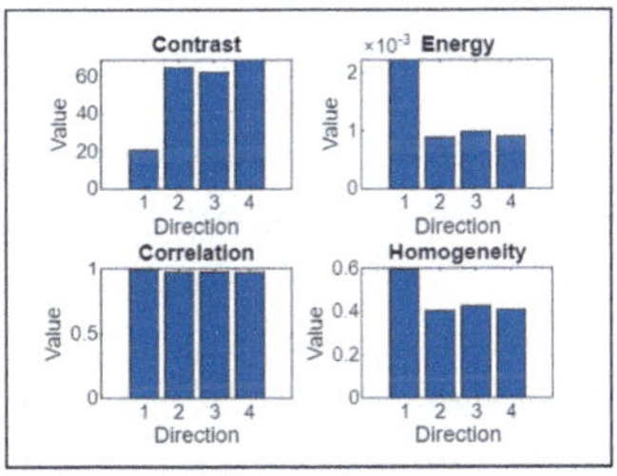

Fig. 4. The results of GLCM features in four directions.

4.3 Experiment 2: Thresholding

Thresholding is a simple yet effective image processing method for segmenting images into distinct regions. It is useful for detecting oil leaks in desert areas, as crude oil typically has medium-dark pixel intensities. In our case, pixel intensities of leaks are mostly below 200 (medium-dark). However, the leak detection industry has strict criteria, as false alarms can be more costly than actual leaks. Thresholding often misclassifies dark shadows as leaks, as shown in Figs. 7 and 8. This limitation led to the exploration of more advanced methods to improve classification accuracy and reliability (Fig. 6).

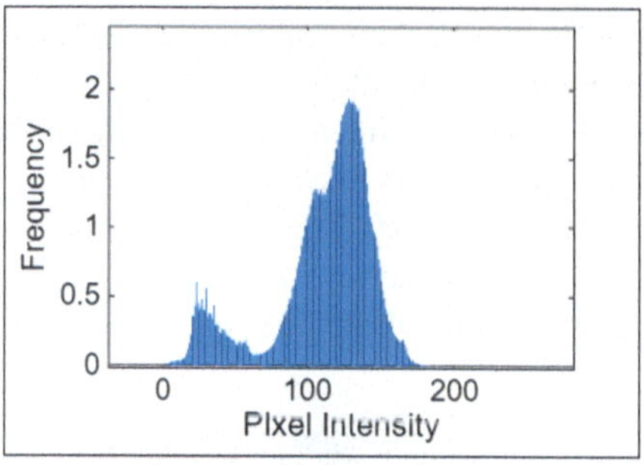

Fig. 5. Histogram of Pixel Intensities in a Grayscale Image of an Oil Pipe with a Leak.

Fig. 6. Results of Thresholding at 0.30.

Fig. 7. Image with only a pipeline and no leak.

Fig. 8. The results of thresholding the image.

4.4 Experiment 3: Instance Segmentation Using YOLOv8

We trained a YOLOv8 model to classify images into two classes: oil pipe (1) and oil leak (0), with other elements as background. Labels include class and (x,y) coordinates for polygon boundaries. Training was performed on Google Colab using an NVIDIA A100 GPU. The obtained results are shown in Fig. 9. We tested our model with publicly available, unseen images from the Web, including both aerial and ground views. Some images were taken from non-optimal drone angles to evaluate the model's robustness (Fig. 10).

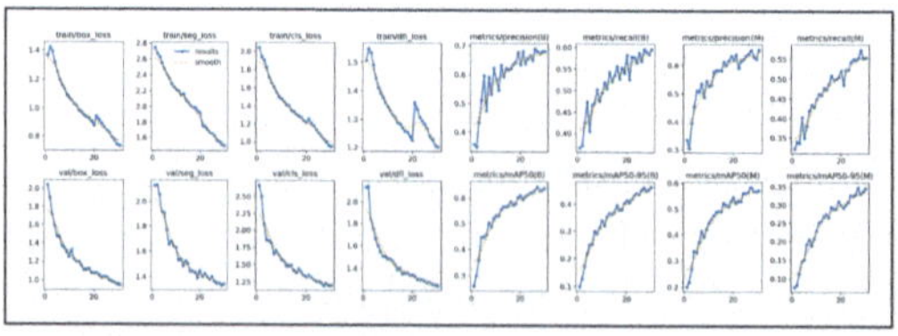

Fig. 9. Results obtained from training the YOLOv8 for 30 epochs.

Fig. 10. Testing result using the first YOLOv8 segmentation model.

The initial results were satisfactory, but testing images taken from low vantage points under clear skies revealed some issues. The darker blue sky was misclassified as an oil leak, and leaks in bright, elevated areas showed varied lighting compared to our training data. These non-optimal conditions led to significant classification challenges. Results from this testing are shown below (Fig. 11):

Fig. 11. Results obtained from training the YOLOv8 for 30 epochs.

The output graphs in Fig. 9 exhibit significant variations. To address this issue, along with the problems related to context variation and misclassification of clear sky, we reconfigured the hyperparameters. Specifically, we increased the number of epochs to 100 and reduced the batch size from 32 to 16 in the previous model to provide the model with more detailed insights.

We implemented and tested the second YOLOv8 model, and the results, shown in Fig. 12, were achieved after 100 epochs and over 17 hours of training. After this training, the F1-score for the mask improved to 0.61 at a confidence level of 0.417, indicating progress from the previous model and resolving the misclassification of clear sky. However, challenges remain with significant changes in context or viewpoint. Given the lengthy 17-hour training for 100 epochs and ongoing issues with detecting oil leaks under varying viewpoints, alternative methods are necessary to address these challenges (Fig. 13).

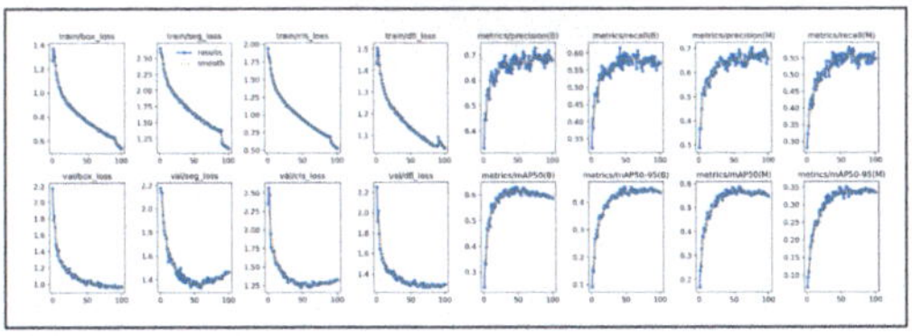

Fig. 12. Results obtained from training the YOLOv8 for 100 epochs.

Fig. 13. Testing output result using the second YOLOv8 segmentation model.

4.5 Experiment 4: Semantic Segmentation Using U-Net

To address the issues from previous experiments, we propose models based on the U-Net architecture with different backbones, specifically ResNet34 and Inception-ResNetv2.

- **U-Net with a ResNet34 Backbone:** ResNet34, a deep convolutional neural network with 34 layers, uses residual learning and skip connections to ease the training of deep networks by mitigating vanishing gradients. Pre-trained on ImageNet, it extracts feature maps in a U-Net architecture. In the first model, trained for 7 epochs with a learning rate of 0.001 and a batch size of 16, results are shown in Fig. 14. In the second model, trained for 35 epochs, IoU remained at 45.2%. An additional class, "sand," was introduced to give the model more context. The updated results are shown in the following figures.

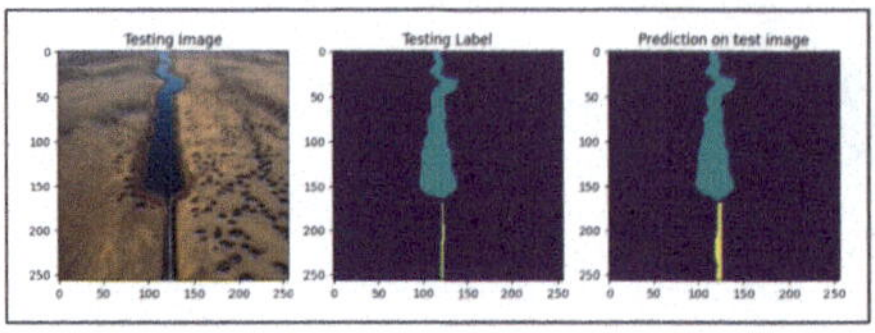

Fig. 14. Output of the first U-Net semantic segmentation model with a resnet34 backbone.

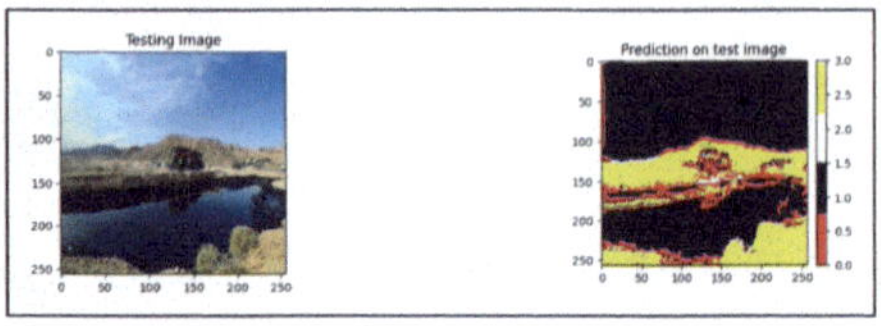

Fig. 15. Output of the second U-Net semantic segmentation model with a resnet34 backbone.

From the Fig. 15, we can observe that the issue of misclassifying the clear sky has reappeared. The additional complexity introduced by the new class did not resolve the problem effectively.
- U-Net with an Inception-Resnetv2 Backbone: InceptionResNetV2 integrates Inception and ResNet, offering high accuracy in visual recognition. Its Inception module extracts features at multiple levels through convolutions of various sizes, resulting in a 3D tensor of 8x8x2048. Though computationally intensive, it delivers superior accuracy for complex tasks. The first model used the following classes: oil leak, oil pipe, sand, and background. The results are illustrated in the Fig. 18.

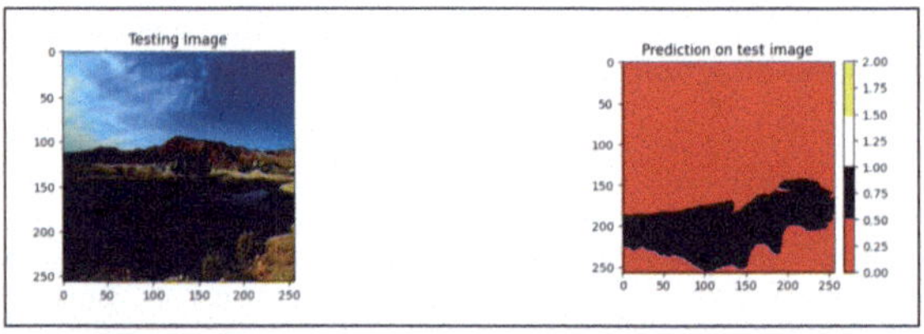

Fig. 16. Output of the second U-Net model with a Inception-Resnetv2 backbone.

In the second model, we used only three classes: oil leak, oil pipe, and background. Training was done for 50 epochs with a learning rate of 0.0001 and a batch size of 16. The validation IoU reached 75.30%, showing good performance on unseen data. Some test images are shown in the Fig. 16 (Fig. 17).

Fig. 17. Testing Image.

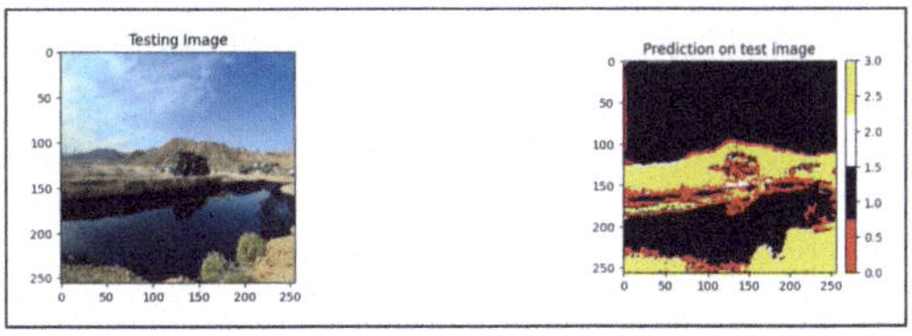

Fig. 18. The segmented image.

5 Conclusion

This work aims to develop an advanced visual and airborne oil leak detection system for pipelines using quadcopter drones. It leverages machine learning to extract crucial features from drone-captured images and make informed decisions about leak presence. We reviewed existing methods, from edge detection to complex thresholding, to guide our deep learning model development. Experiments included instance and semantic segmentation, with the U-Net model using an Inception-ResNetv2 encoder achieving the best performance. This model was trained and tested on our dataset, attaining a validation IoU of 75.30% and demonstrating robustness across various conditions. Our approach integrates deep learning with drone technology, providing an effective solution for detecting oil leaks, thereby helping to prevent hazards.

References

1. Pipeline and gas journal. Sonatrach-oil-leaks. https://pgjonline.com/news/2020/09-september/sonatrach-reports-two-leaks-on-ok1-oil-pipeline. Accessed Feb 2024
2. Qin, G., Zhang, P., Hou, X., Sen, W., Wang, Y.: Risk assessment for oil leakage under the common threat of multiple natural hazards. Environ. Sci. Pollut. Res. **27**(14), 16507–16520 (2020). https://doi.org/10.1007/s11356-020-08184-7. ISSN: 1614-7499
3. Oil and gas pipeline leak detection market - global size, share, trend analysis, opportunity and forecast report, 2018–2028, Segmented By Location (Onshore, Oflshore), By Technology (Acoustic Sensors, Flowmeters, Cable Sensors, Vapor Sensors, Others), By Region (North America, Europe, Asia-Pacific (APAC), Latin America (LATAM), Middle-East and Africa (MEA). https://www.blueweaveconsulting.com/report/oil-and-gas-pipeline-leak-detection-market. Accessed Mar 2024
4. Scott, S.L., et al.: Worldwide assessment of industry leak detection capabilities for single & multiphase pipelines. Offshore technology research center (2003). https://books.google.dz/books?id=J8icHAAACAAJ
5. Murvay, P.-S., Silea, I.: A survey on gas leak detection and localization techniques. J. Loss Prev. Process Ind. **25**(6), 966–973 (2012). https://doi.org/10.1016/j.jlp.2012.05.010. ISSN: 0950-4230
6. Adegboye, M.A., Fung, W.-K., Karnik, A.: Recent advances in pipeline monitoring and oil leakage detection technologies: principles and approaches. Sensors **19**(11), 2548 (2019). https://doi.org/10.3390/s19112548. ISSN: 1424-8220
7. Liu, C.-w, Li, Y.-x, Yan, Y.-k, Jun-tao, F., Zhang, Y.-q: A new leak location method based on leakage acoustic waves for oil and gas pipelines. J. Loss Prev. Process Ind. **35**, 236–246 (2015). https://doi.org/10.1016/j.jlp.2015.05.006. ISSN: 0950-4230

8. Miller, R.K.: Acoustic emission leak detection. In: Final Report No. R96-472 to the New Jersey Institute of Technology (1996)
9. Elliott, J., Fletcher, R., Wrigglesworth, M.: Seeking the hidden threat: applications of a new approach in pipeline leak detection. In: Abu Dhabi International Petroleum Exhibition and Conference, SPE. 2008, SPE–118070 (2008)
10. Chalgham, W., Seibi, A., Lomas, M.: Leak detection and self healing pipelines using twin balls technology. In: SPE Annual Technical Conference and Exhibition? SPE. 2016, D031S052R001 (2016)
11. Wang, F., Lin, W., Liu, Z., Shuochen, W., Qiu, X.: Pipeline leak detection by using time-domain statistical features. IEEE Sensors J. **17**(19), 6431–6442 (2017). https://doi.org/10.1109/JSEN.2017.2740220
12. Ji, J., Li, Y., Liu, C., Wang, D., Jing, H.: Application of EMD technology in leakage acoustic characteristic extraction of gas-liquid, two-phase flow pipelines. Shock Vibr. **2018**(1), 1529849 (2018). https://doi.org/10.1155/2018/1529849. ISSN: 1070-9622
13. Kousiopoulos, G.-P., et al.: Acoustic leak localization method for pipelines in high-noise environment using time-frequency signal segmentation. IEEE Trans. Instrum. Meas. **71**, 1–11 (2022). https://doi.org/10.1109/TIM.2022.3150864
14. Wang, W., Gao, Y.: Pipeline leak detection method based on acoustic-pressure information fusion. Measurement **212**, 112691 (2023). https://doi.org/10.1016/j.measurement.2023.112691. ISSN: 0263-2241
15. Nikles, M., et al.: Leakage detection using fiber optics distributed temperature monitoring. In: Smart Structures and Materials 2004: Smart Sensor Technology and Measurement Systems. SPIE, vol. 5384, pp. 18–25 (2004)
16. Inaudi, D., Glisic, B.: Long-range pipeline monitoring by distributed fiber optic sensing. In: International Pipeline Conference, vol. 42630, pp. 763–772 (2006)
17. Mehra, R., Tripathi, J.: Mach zehnder interferometer and it's applications. Int. J. Comput. Appl. **1**(9), 110–118 (2010)
18. Feng, H., et al.: Modeling of pipeline leakage detection and prewarning system for locating error analysis based on jones matrix. J. Japan Petrol. Inst. **52**(3), 114–119 (2009)
19. Qu, Z., et al.: A SVM-based pipeline leakage detection and pre-warning system. Measurement **43**(4), 513–519 (2010). https://doi.org/10.1016/j.measurement.2009.12.022. ISSN: 0263-2241
20. Bhuiyan, M.A.S., Hossain, M.A., Alam, J.M.: A computational model of thermal monitoring at a leakage in pipelines. Int. J. Heat Mass Transf. **92**, 330–338 (2016). https://doi.org/10.1016/j.ijheatmasstransfer.2015.08.094. ISSN: 0017-9310
21. Wang, C., et al.: A novel DdTS technology based on fiber optics for early leak detection in pipelines. In: 2016 IEEE International Carnahan Conference on Security Technology (ICCST), pp. 1–8. IEEE (2016)
22. Wang, C., et al.: Reliable leak detection in pipelines using integrated DdTS temperature and DAS acoustic fiber-optic sensor. In: 2018 International Carnahan Conference on Security Technology (ICCST), pp. 1–5. IEEE (2018)
23. Zhang, W.: The design of strain sensitising of high-sensitivity SAW sensor based on FBG. Int. J. Inf. Commun. Technol. (Print), **12**(1/2), 175 (2018). https://doi.org/10.1504/ijict.2018.089030
24. Wang, J., et al.: A novel pipeline leak detection and localization method based on the FBG pipe-fixture sensor array and compressed sensing theory. Mech. Syst. Signal Process. **169**, 108669 (2022). https://doi.org/10.1016/j.ymssp.2021.108669. ISSN: 0888-3270
25. Vorathin, E., et al.: Low pressure pipeline leak localization using highly sensitive rubber-based diaphragm FBG pressure sensors and NPW method. Opt. Fiber Technol. **83**, 103687 (2024). https://doi.org/10.1016/j.yofte.2024.103687. ISSN: 1068-5200

26. Detection of critical flow leaks in deepwater gas flowlines. In: Vol. All Days. SPE Annual Technical Conference and Exhibition, SPE-49310, September 1998. https://doi.org/10.2118/49310-MS
27. Al-Shidhani, I., Beck, S.B.M., Staszewski, W.J.: Leak monitoring in pipeline networks using wavelet analysis. Key Eng. Mater. **245**, 51–58 (2003)
28. Wei, L., Laibin, Z., Yingchun, Y.: Data mining technology based leak detection method for crude oil pipeline. In: 2009 WRI World Congress on Computer Science and Information Engineering, vol. 3. pp. 656–660. IEEE (2009)
29. Thiberville, C.J., et al.: Evaluation of software-based early leak-warning system in gulf of Mexico subsea flowlines. SPE Prod. Oper. **33**(04), 802–828 (2018). https://doi.org/10.2118/187417-P. ISSN: 1930-1855
30. Meng, Q., et al.: Leak localization of gas pipeline based on the combination of EEMD and cross-spectrum analysis. IEEE Trans. Instrum. Meas. **71**, 1–9 (2022). https://doi.org/10.1109/TIM.2021.3130680
31. Massinon, R.V.J.: A real time transient hydraulic model for leak detection and batch tracking on a liquid pipeline system. In: PETSOC Annual Technical Meeting, PETSOC, PETSOC–88 (1988)
32. Guerriero, M., et al.: Bayesian data fusion for pipeline leak detection. In: 2016 19th International Conference on Information Fusion (FUSION), pp. 278–285. IEEE (2016)
33. Yilmaz, Y.: New rupture detection data mining algorithms for crude oil pipelines. In: PSIG Annual Meeting. PSIG, PSIG–2102 (2021)
34. Hassan, Shamsu, et al.: An assessment of causes and failure likelihood of cross-country pipelines under uncertainty using Bayesian networks. Reliab. Eng. Syst. Saf. **218**, 108171 (2022). https://doi.org/10.1016/j.ress.2021.108171. ISSN: 0951-8320
35. Wanasinghe, T.R., et al.: Digital twin for the oil and gas industry: overview, research trends, opportunities, and challenges. IEEE Access **8**, 104175–104197 (2020). https://doi.org/10.1109/ACCESS.2020.2998723
36. Priyanka, E.B., Thangavel, S.: Multi-type feature extraction and classification of leakage in oil pipeline network using digital twin technology. J. Ambient Intell. Human. Comput. **13**(12), 5885–5901 (2022). https://doi.org/10.1007/s12652-022-03818-9. ISSN: 1868-5145
37. Sadovnychiy, S., Canul, E.A., López, J.M., Sadovnychyy, A., Hernandez, M.A.: GIS systems on the pipelines thermographic aerial inspection. In: Ragia, L., Laurini, R., Rocha, J. (eds.) Geographical Information Systems Theory, Applications and Management. Communications in Computer and Information Science, vol. 936, pp. 1–18. Springer, Cham (2019). https://doi.org/10.1007/978-3-030-06010-7_1. ISBN: 978-3-030-06010-7
38. Petoukhov, V.M., et al.: Lidar technologies application to leakage detection in oil product pipelines. In: Nondestructive Evaluation of Utilities and Pipelines III. SPIE, vol. 3588, pp. 81–89 (1999)
39. Obaid, M.H., Hamad, A.H.: Deep learning approach for oil pipeline leakage detection using image-based edge detection techniques. J. Environ. Sci. Eng. **56**(4), 663–673 (2023). https://doi.org/10.18280/jesa.560416
40. Dhariwal, P., Nichol, A.: Diffusion models beat GANs on image synthesis. arXiv:2105.05233 [cs.LG] (2021). https://arxiv.org/abs/2105.05233

QuantumSign: Online E-Signature Service Based on Blockchain and Post-Quantum Cryptography

Chaima Boumesaid[1], Yousra Guerfa[1], and Gueltoum Bendiab[1,2(✉)]

[1] Department of Electronics, University of Freres Mentouri, 25000 Constantine, Algeria
kelthoum.bendiab@umc.edu.dz

[2] LIRE Laboratory, University of Constantine 2, 25000 Constantine, Algeria

Abstract. The increasing use of electronic documents and remote online offices in this digital era raises concerns about the security of electronic transactions, especially with the emerging quantum computing that threatens current encryption methods. Actually, quantum computers pose significant risks to current cryptographic systems, particularly those based on public-key cryptography. Quantum algorithms can speed up brute-force attacks on cryptographic algorithms and reduce the time required to search through the keyspace, posing a potential risk to the current encryption schemes and digital signatures. In addition, the existing electronic signature platforms have the problem of data-centred management, which makes them vulnerable to many security issues including data loss, tampering, leakage and other forms of cyberattacks. This paper presents an innovative solution to solve these problems by proposing "QuantumSing", a novel distributed e-signature system based on post-quantum cryptography and blockchain technology. QuantumSing uses the FALCON algorithm to create secure post-quantum signatures that are resistant to quantum-based attacks and blockchain to ensure the integrity of the electronic contract and the non-repudiation of the electronic signatures. Furthermore, QuantumSing has a user-friendly and intuitive web service that enables users to sign and verify their electronic documents in a simple, efficient, and secure manner.

Keywords: Blockchain · Post-Quantum Cryptography · Electronic Signature · Security · Cyber-attacks · Quantum Computing

1 Introduction

The transition to a digital society and the adoption of e-documents have indeed revolutionized various aspects of our lives, offering numerous benefits such as improved communication, efficiency, and accessibility [1]. In particular, the COVID-19 pandemic has notably accelerated the use of electronic signatures across numerous industries, and many organizations and government agencies turned to electronic signature platforms to facilitate the signing and validation of documents [2–4]. These platforms, such

© The Author(s), under exclusive license to Springer Nature Switzerland AG 2025
M. Hatti (Ed.): IC-AIRES 2024, LNNS 1238, pp. 342–353, 2025.
https://doi.org/10.1007/978-3-031-80301-7_38

as DocuSign, Adobe Sign, LegaSign, Xodo Sign, Jotform Sign, HelloSign and others, offer technologies and applications that enable the validity of electronic signatures, making them legally binding and enforceable in various jurisdictions [1, 3, 4]. However, this digital transformation has also introduced new challenges, particularly regarding the security of electronic document transactions [2, 4, 5]. In fact, most existing electronic signing platforms rely on central management that can indeed introduce security vulnerabilities and makes them attractive targets for cyber attackers [2, 5]. In addition, these platforms have different signing processes, and users may need to use multiple platforms [2]. Another important issue for these e-singing platforms is the emergence of quantum computers [6, 7]. This emerging technology presents a major threat to the security of existing encryption methods employed by most signing platforms, as these powerful machines could potentially break traditional cryptographic algorithms that depend on the difficulty of certain mathematical problems [7]. As quantum computing technology progresses, it could make widely used encryption schemes, like RSA (Rivest Shamir Adleman), ECC (Elliptic Curve Cryptography), and even other algorithms such as Diffie-Hellman key exchange and AES (Advanced Encryption Standard) with smaller key sizes, vulnerable to attacks [7, 8]. Actually, with the increasing exchange and storage of confidential information electronically, it is crucial to address these potential vulnerabilities and implement strong security measures to safeguard against various threats. To overcome these challenges, we present QuantumSing, an innovative solution that leverages quantum-resistant cryptography alongside the decentralization and transparency offered by blockchain technology. The proposed platform aims to ensure secure and tamper-resistant digital signing while providing a decentralised and user-friendly environment for electronic singing applications.

Post-quantum cryptography refers to a class of cryptographic algorithms specifically developed to resist the advanced computational power of quantum computers [7, 9]. Unlike traditional cryptographic methods, which could be easily compromised by quantum algorithms, post-quantum algorithms are designed to resist these threats by relying on mathematical problems that remain difficult for quantum systems to solve, such as lattice problems [10], Multivariate Quadratic Equations (MQ) [11] and Supersingular Isogeny Diffie-Hellman (SIDH) [12]. These algorithms aim to ensure long-term data security, even in a future where quantum computing becomes widely available. Currently, researchers, industry professionals, and standards organisations are actively engaged in identifying and standardizing post-quantum cryptographic algorithms to ensure the continued security of sensitive information and communication in a quantum-computing era. In this context, the National Institute of Standards and Technology (NIST) Post-Quantum Cryptography Standardization project aims to select and standardize a suite of cryptographic methods that are secure against attacks from both typical and quantum computers. NIST initiated this process to address the potential risk posed by high-performance quantum computers to traditional cryptographic systems. During the NIST competition, FALCON and other candidate algorithms underwent rigorous evaluation and analysis for their security, efficiency, and suitability for different applications and use cases [13]. On the other hand, blockchain technology is a decentralized and distributed ledger system that enables multiple participants to collaboratively manage and update a synchronized, immutable record of transactions or data without relying on a central authority [14, 15]. It was initially launched as the foundational technology for the digital cryptocurrency Bitcoin [16], but its potential uses have expanded well beyond just digital currencies. At its core, a blockchain consists of a series of blocks, with each block holding a list of transactions or data. These blocks are connected through cryptographic hashes, forming a secure and tamper-proof chain [14, 15]. Figure 1 presents the working parading of a typical blockchain.

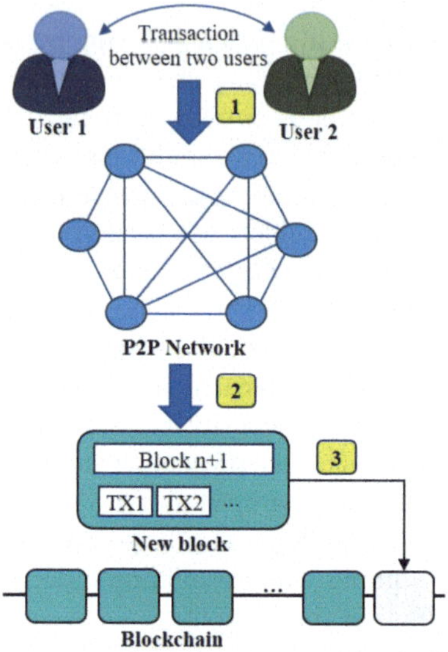

Fig. 1. The working paradigm of blockchain involves a series of steps that enable the broadcasting of transactions to the P2P network (1), validation and creation of new blocks (2), and adding these blocks to the chain (3) [14].

Using both technologies for securing electronic transactions offers several advantages that contribute to ensuring long-term security, transparency and trustworthiness. The remainder of the paper is structured as follows. Section 2 provides a review of relevant prior research that is similar to our study. Section 3 introduces a detailed description of the proposed solution. Section 4 explains the proof-of-concept experiments conducted to test the core functionalities of QuantumSign, along with the results of our evaluation. Section 5 summarizes the paper by discussing conclusions and potential future research directions.

2 Related Work

Currently, there is an increasing interest in using blockchain technology and post-quantum cryptography to tackle the security and privacy issues related to electronic signatures in several domains, ranging from e-government and e-commerce to e-health and supply chain management [17, 18]. In this context, practical studies in [2, 19–21] demonstrate how blockchain technology can offer a reliable and immutable method of document signing and verification. The proposed platform in [19] used a public blockchain to create a decentralized system for signing documents. This guarantees the absence of a central authority who manages the whole process. Instead, network participants work together to reach a consensus and maintain the integrity of the signed documents. Each signed document is linked to a pair comprising the signer's wallet address (which represents their digital identity) and the hash value of the signed document. Storing these pairs on the blockchain ensures both transparency and traceability of document ownership and integrity. The wallet address serves as a digital version of a physical identity card issued by the government. It uniquely identifies the document owner within the blockchain network. Whereas, the hash value of the signed document is used to provide cryptographic proof that the document has not been modified since it was added to the blockchain. As mentioned earlier, any alteration to the document would produce a different hash value, making the change easily detectable. To check the authenticity and validity of a particular document, a graphical user interface (GUI) is utilized to compute the document's hash value and show all applied signatures along with their respective signers. This process allows anyone to independently verify the document's history and the legitimacy of the signatures. In [2], authors used a consortium blockchain as a decentralized electronic contract signing platform. Unlike the previous system, where participants were anonymous, this approach involves a network where all members are known and have been granted permission to join. This enhances privacy and provides better control over access to sensitive data, which is particularly beneficial for industries with stringent regulatory standards. By limiting access to authorized participants, the consortium blockchain ensures that sensitive data remains secure and ensures regulatory compliance.

Post Quantum Cryptography (PQC) has also gained significant attention and development as a response to the potential threat posed by attacks from both classical and quantum computers [22–24]. For instance, authors in [24] proposed a signature service based on the Falcon algorithm [25]. The proposed service is evaluated on different hardware and software platforms with a key length of 1024 bits. In [26], authors explored the feasibility of using post-quantum digital signatures to verify the integrity of electronic travel documents (e.g. passports). For that, they proposed dedicated public key infrastructure built around the qTESLA-III-speed signature scheme, as an example for their proof-of-concept. Authors in \cite{xie2021blockchain} employed both post-PQC and blockchain to ensure data security in cloud storage. In particular, this integrated system offers a robust solution for enhancing trust between cloud service providers and their users by guaranteeing dynamic verification of data integrity stored in the cloud and resilience to potential quantum attacks. Generally, the field of PQC is still largely in the research and development phase, and widespread adoption of PQC-based document-signing tools is not yet commonplace. In the current landscape, a handful of open-source initiatives have emerged to offer services based on post-quantum cryptography, encompassing functions like digital signatures, key exchange, and encryption. Noteworthy instances of these projects include IBM Crypto-services [27], Microsoft QuantumSafe Toolkit [28], OpenQuantumSafe [29], PQCrypto-Sign [30], PQCrypto-KEM [31], and SignServer Community 5.9.1 [32]. SignServer Community 5.9.1 is a new release (June 20, 2022) of the open-source software SignServer that allows users to generate and use post-quantum keys and signatures. As the field of post-quantum cryptography continues to evolve, we can anticipate the availability of even more advanced tools in the future.

3 Presentation of Quantumsign

As mentioned earlier, QuantumSign is a decentralised document-signing platform that aims to improve the current cutting-edge technology and enhance the protection of electronic documents through the use of blockchain and PQC. As illustrated in Fig. 2, the QuantumSign platform comprises various interconnected components collaborating to realize decentralised procedures for signing documents. As discussed earlier, blockchain represents an electronic record-keeping system that facilitates the safe and clear archival of data distributed across multiple computers. This ensures the immutability of all historical data. Within our platform, the blockchain is used to store the hash values associated with the signed documents, accompanied by the identifiers of the signatories. The identifier (Digital identity) is mainly used to uniquely identify the document owner (signer) within the blockchain network. These IDs could be issued by different authorities such as banks, governments, etc. Like other available digital signature methods, signing a document requires first generating a hash of the document and recording it on the blockchain along with the signer's identifier. If the document is modified in any way, the hash will no longer match, making the signature invalid.

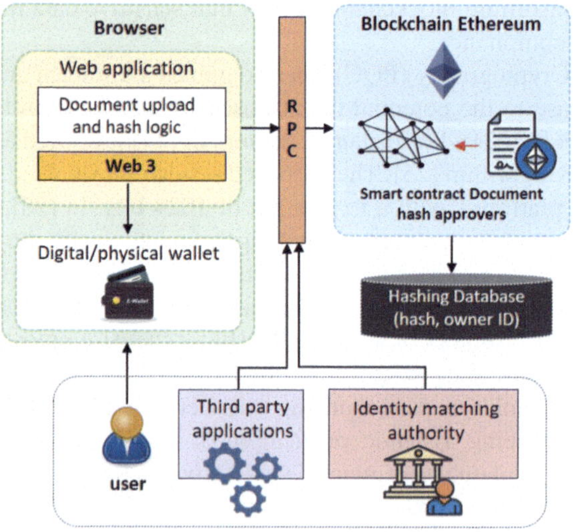

Fig. 2. High-level architecture of QuantumSign

As illustrated in Fig. 2, the blockchain serves as a repository and handles hashes associated with signed e-documents in transactions. These will be used to verify the validity and integrity of the signed documents. These hashes may also be employed in the realm of digital forensics, serving as digital evidence to substantiate claims against malicious entities when requested by a judicial institution. The structure of a transaction in our blockchain (denoted as TX-H) is defined as follows:

$$\boldsymbol{TX - H} = \{\boldsymbol{ID_{RT}}, \boldsymbol{T}, \boldsymbol{ID_{signers}}, \boldsymbol{ID_{e-document}}, \boldsymbol{Hash}\} \tag{1}$$

where $\boldsymbol{ID_{RT}}$ denotes the transaction identifier, $\boldsymbol{T}$ indicates the transaction time, whereas $\boldsymbol{ID_{signers}}$ is the identifier of the document owner (signer) and $\boldsymbol{ID_{e-document}}$ is the identifier of the signed e-document, while $\boldsymbol{Hash}$ represents the corresponding hash generated through the use of a hashing method.

3.1 Signing E-Documents with FALCON

In this work, we utilised the post-quantum cryptography standard FALCON (Fast Fourier Lattice-based Compact Signature over NTRU) [33] to sign e-documents and validate their signatures. FALCON is a lattice-based digital signature algorithm that relies on the NTRU cryptographic framework [34]. It is mainly designed to offer compact public key and signature sizes, easing the transition to post-quantum schemes and enabling smaller digital signatures. It has been selected by NIST to become a digital signature standard capable of withstanding the computational power of future quantum computers. Additionally, FALCON features non-deterministic modules for key and signature generation, while signature verification is deterministic.

1. **FALCON Key Generation**: Algorithm 1 presents the FALCON key pair generation process, which involves the calculation of polynomials $\boldsymbol{f}, \boldsymbol{g}, \boldsymbol{F}, \boldsymbol{G}$, and $\boldsymbol{h}$, defined by the Eqs. 2 and 3:

$$\boldsymbol{fG - gF = q\, mod\, \varnothing} \tag{2}$$

$$\boldsymbol{h = gf^{-1} mod(\varnothing, q)} \tag{3}$$

As shown in algorithm 1, AES is first used to generate a pseudo-random number that will be used to seed SHAKE-256. This initializes the algorithm to generate random polynomials $\boldsymbol{f}$ and $\boldsymbol{g}$ with a Gaussian distribution. These polynomials are discarded and regenerated if their squared norms exceed specified bounds [35]. Similarly, the algorithm rejects polynomials whose orthogonalized vector norms are out of bounds, determined using Fast-Fourier Transform (FFT). Next, using polynomials $\boldsymbol{f}$ and $\boldsymbol{g}$, the algorithm calculates the public key polynomial $\boldsymbol{h}$ to satisfy Eq. 3. Solving Eq. 3 (the NTRU equation) yields the polynomials $\boldsymbol{F}$ and $\boldsymbol{G}$ for the secret key. The algorithm sequentially encodes $\boldsymbol{f}, \boldsymbol{g}, \boldsymbol{F}$ and $\boldsymbol{G}$ for the secret key, and encodes $\boldsymbol{h}$ for the public key. Finally, this process yields both the secret and public keys [35].

Algorithm 1: FALCON Key Generation

No Input Required

Output: Secret key $\boldsymbol{sk} = (\boldsymbol{f}, \boldsymbol{g}, \boldsymbol{F}, \boldsymbol{G})$, Public key $\boldsymbol{pk} = (\boldsymbol{h})$

1- **Procedure** KEY-GENERATION;

2- Generate a seed for the hash function using AES;

3- Initialize the hash function with the seed;

while True **do**

  - Generate polynomials f and g;

  - Verify the norm of the polynomials;

  - Compute the norm of the orthogonalized vector;

  if (vector norm < 16822) **then**

    Continue;

  end if

  if (Failure to generate $\boldsymbol{h}$) **then**

    Continue;

  end if

  if (Failure to generate $\boldsymbol{F}$ and $\boldsymbol{G}$) **then**

    Continue;

  end if

  Break;

end while

4- Encode the secret key $\boldsymbol{sk}$ and public key $\boldsymbol{pk}$;

5- Return $(\boldsymbol{sk} = (\boldsymbol{f}, \boldsymbol{g}, \boldsymbol{F}, \boldsymbol{G}), \boldsymbol{pk} = (\boldsymbol{h}))$;

2. **FALCON Signature Generation**: Algorithm 2 illustrates the process for generating signatures. First, a pseudo-random seed is created using AES to initialize SHAKE-256. The algorithm then calculates a hash digest, denoted as $\boldsymbol{c}$, where $\boldsymbol{c} \in \boldsymbol{Z}_q[x]/(\varnothing)$. The hash value $\boldsymbol{c}$ is computed using the nonce $\boldsymbol{r}$ and the input message $\boldsymbol{m}$. Subsequently, the algorithm decodes the secret key $\boldsymbol{sk}$ to retrieve the polynomials $\boldsymbol{f}, \boldsymbol{g}, \boldsymbol{F}$ and $\boldsymbol{G}$. If $\boldsymbol{G}$ is not already available, the algorithm calculates it. With the polynomials $\boldsymbol{f}, \boldsymbol{g}, \boldsymbol{F}$ and $\boldsymbol{G}$, the algorithm then computes two short vectors s_1 and s_2 such that $s_1 + s_2\boldsymbol{h} = \boldsymbol{cmodq}$. This computation is performed in a way that prevents the exposure of the secret key. The short vector s_2 is then encoded, and the signature length $\boldsymbol{sig}_{len}$, nonce, message $\boldsymbol{msg}$, and encoded s_2 are concatenated and stored in $\boldsymbol{sm}$.

Algorithm 2: FALCON Signature Generation

Input: $\boldsymbol{m}, \boldsymbol{m_{len}}, \boldsymbol{sk} = (\boldsymbol{f}, \boldsymbol{g}, \boldsymbol{F}, \boldsymbol{G})$

Output: $\boldsymbol{sm(sig_{len}, nonce, msg, sig), sm_{len}}$

1- Generate a seed for the hash function using AES;

2- Initialize the hash function with the generated seed;

3- Decode and perform pre-computation on sk;

4- Compute $\boldsymbol{c} \leftarrow \boldsymbol{H(r, m)}$;

5- Determine $\boldsymbol{s_1}$ and $\boldsymbol{s_2}$ such that $\boldsymbol{s_1} + \boldsymbol{s_2 h} = \boldsymbol{c \bmod q}$;

6- Encode $\boldsymbol{s_2}$;

7- Store the nonce $\boldsymbol{r}$ and the encoded $\boldsymbol{s_2}$ into $\boldsymbol{sm}$;

8- Return $\boldsymbol{sm} = (\boldsymbol{sig_{len}, nonce, msg, encoded\ s2})$;

3. **FALCON Signature Verification**: Algorithm 3 outlines the signature verification procedure. This process begins by decoding the public key $\boldsymbol{pk}$ and extracting the polynomial $\boldsymbol{h}$. If the signature length $\boldsymbol{sm_{len}}$ is found to be invalid, the signature is rejected. If it is valid, the algorithm initializes the hash function and computes a hash value $\boldsymbol{c} \in \boldsymbol{Z_q}[x]/(\varnothing)$. Next, it decodes the signature, retrieves the short vector s_2, and calculates vector s_1 using s_2 and $\boldsymbol{h}$. If the L2-norm of the combined vectors (s_1, s_2) exceeds the allowable limit, the signature is rejected; otherwise, it is verified.

Algorithm 3: FALCON Signature Verification

Input: $\boldsymbol{sm}, \boldsymbol{sm_{len}}, \boldsymbol{pk} = (\boldsymbol{h})$

Output: Accept/Reject, $\boldsymbol{m}, \boldsymbol{m_{len}}$

1- Decode the public key $\boldsymbol{pk} = (\boldsymbol{h})$;

if $(\boldsymbol{sm_{len}} + \boldsymbol{nonce})$ $\boldsymbol{or}$ $(\boldsymbol{sig_{len}} > \boldsymbol{2} + \boldsymbol{nonce})$ **then**

  return Reject;

end if

2- Initialize the hash function;

3- Compute $\boldsymbol{c} \leftarrow \boldsymbol{H(r, m)}$;

4- Verify $\boldsymbol{s_1} + \boldsymbol{s_2 h} = \boldsymbol{c \bmod q}$;

if $(||\boldsymbol{s1}, \boldsymbol{s2})|| > \boldsymbol{\beta}$ **then**, where $(\boldsymbol{\beta 2} = \boldsymbol{87067565})$

  return Reject;

end if

5- return $\{\boldsymbol{Accept}, \boldsymbol{m}, \boldsymbol{m_{len}}\}$;

4 Conceptual Experiments and Validation

The implementation and testing of QuantumSign service were carried out on a physical machine featuring an Intel® Core™ i7-9750H processor, operating at 2.60 GHz, equipped with 16 GB of RAM and a 1 TB hard drive. To facilitate the testing, VMware Workstation Pro was employed to set up and manage two virtual machines, providing an environment to thoroughly evaluate the performance and functionality of our service. Given the performance constraints, the experiments were conducted in a controlled private Ethereum blockchain environment consisting of two nodes running on Linux Ubuntu (version 22.04.2). To support the blockchain setup and development, we utilized several packages, including NodeJS, NPM, Geth (Go-Ethereum), and Truffle. This configuration enabled us to effectively test and analyze the service within a secure and manageable blockchain framework. Furthermore, a smart contract named "StorageData.sol" was developed and deployed within the private blockchain network utilizing the Truffle framework with Solidity language. The primary objective of this smart contract is to facilitate interaction with QuantumSign and record hashes of signed e-documents. Interaction with the smart contract is facilitated through a front-end web application developed using the web3.js API (Application Programming Interface). In addition, the QuantumSign service includes several web interfaces to enhance user interaction with the platform. These interfaces were developed using HTML, JavaScript, PHP, and CSS. For the e-documents signing and validation, we used the C source code of FALCON with minor modifications [35]. This implementation has shorter execution times than all other implementations of FALCON. We specifically used FALCON-1024, with the following parameters: $\boldsymbol{n} = 1024$, $\varnothing = \boldsymbol{x}^{\boldsymbol{n}} + 1$, $\boldsymbol{q} = 12289$, $\boldsymbol{\beta}^2 = 87067565$, a signature size of 1273.31 bytes, and a public key size of 1793 bytes.

To evaluate the validity of the platform, different PDF files were digitally signed with a key length of 1024 bits. Figure 3 illustrates the web interfaces corresponding to the three primary modules of the Falcon algorithm: key generation (Algorithm 1), signature generation (Algorithm 2), and signature verification (Algorithm 3). Each figure provides a visual representation of the processes involved in these crucial steps, illustrating how keys are generated, signatures are created, and signatures are verified within the QuantumSign Service. As illustrated in Fig. 3-(b), to sign a file, the user has to select

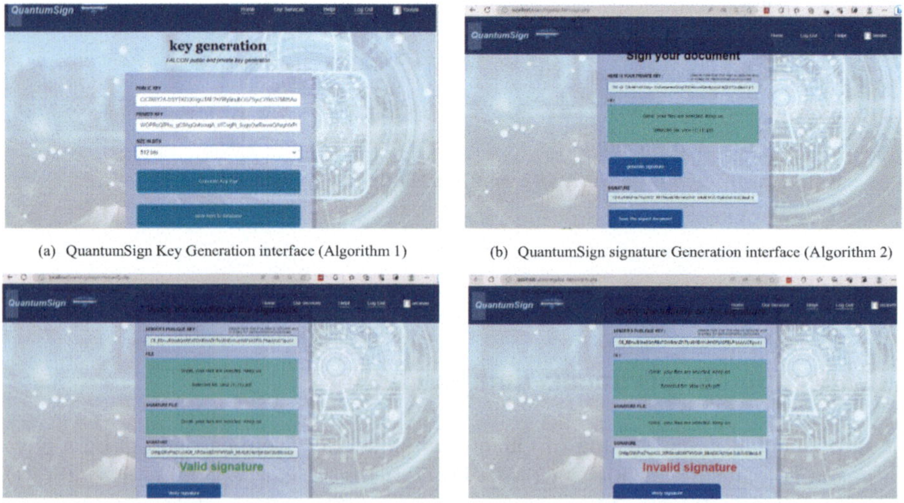

(a) QuantumSign Key Generation interface (Algorithm 1)

(b) QuantumSign signature Generation interface (Algorithm 2)

(c) QuantumSign Signature Verificationinterface (Algorithm 3)

(d) QuantumSign Signature verification interface (Algorithm 3)

Fig. 3. QuantumSing interfaces for (a) Key Generation, (b) Signature Generation and (c), (d) Signature Verification.

it, generate a signature with his private key, and press the "save the signed document" button to save it. If the key exists, the program will immediately return the signing result. If not, it will generate a quantum key as described above (Fig. 3-(a)). If the generation of the digital signature is successful, the generated signature will be displayed in a text file. When authenticating a signed PDF file, the user has to select it along with the public key of the signature and then click the "Verify Signature" button. The program then automatically verifies the signature against the provided public key. If the signature is valid, the program displays a success message indicating that the signature is authentic. If the signature is invalid, an error message is shown to alert the user, as illustrated in Fig. 3-(c) and Fig. 3-(d). It is also crucial to note that upon signing and sending an e-document, the hash and identifiers of the owner and corresponding file are accurately stored in blockchain transactions. These stored hashes can effectively serve as authenticity validation for all participant entities within the blockchain.

5 Conclusion

This study investigated the potential of integrating blockchain technology with post-quantum cryptography for securing electronic documents, especially with the emerging quantum computing that threatens current encryption methods. In this context, we evaluated the efficiency of the FALCON post-quantum cryptographic (PQC) standard in key generation, signature creation, and verification. Additionally, we assessed the performance of our blockchain solution utilizing a private Ethereum blockchain. It is important to emphasize that we chose to integrate Ethereum into our solution primarily due to resource constraints. However, in our upcoming research work, we plan to explore Post-Quantum Distributed Ledger Technology (PQDLT) to further refine our proposed solution and strengthen its resilience against the challenges posed by quantum computing. Our proof-of-concept experiments demonstrated the effectiveness of the QuantumSign platform in signing PDF documents and storing the corresponding hash in blockchain transactions. This ensures the validity of the document can be verified as evidence. In future work, we aim to expand our platform to support various file types, including Word documents, images (such as PNG, GIF, and JPEG), as well as audio and video files. Additionally, we plan to incorporate the most advantageous features of Quantum technology to explore and develop new Quantum signatures for electronic components within our platform.

References

1. Erasmus: How digital transformation is transforming our society. https://www.systemaerasmus.eu/wp/how-digital-transformation-is-transforming-our-society
2. Yang, K., Wu, Y., Chen, Y.: A blockchain-based scalable electronic contract signing system. In: 2022 IEEE International Conferences on Internet of Things (iThings) and IEEE Green Computing & Communications (GreenCom) and IEEE Cyber, Physical & Social Computing (CPSCom) and IEEE Smart Data (SmartData) and IEEE Congress on Cybermatics (Cybermatics), pp. 343–348. IEEE (2022)

3. Gphadmin: The use of electronic signature increases in companies after covid-19. https://www.gphlegal.mx/2021/01/22/the-use-of-electronic-signature-increases-in-companies-after-covid-19/
4. Purbasari, R., Muttaqin, Z., Sari, D.S.: Digital entrepreneurship in pandemic covid 19 era: the digital entrepreneurial ecosystem framework. Rev. Integr. Bus. Econ. Res. **10**, 114–135 (2021)
5. Santosa, A.A., et al.: How the covid-19 pandemic affected the sustainable adoption of digital signature: an integrated factors analysis model. Sustainability **14**(7), 4281 (2022)
6. Mitchell, C.J.: The impact of quantum computing on real-world security: a 5G case study. Comput. Secur. **93**, 101825 (2020)
7. Kirsch, Z., Chow, M.: Quantum computing: the risk to existing encryption methods (2015). http://www.cs.tufts.edu/comp/116/archive/fall2015/zkirsch.pdf
8. Roohparvar, R.: Cybersecurity in a post-quantum computing world. https://www.infoguardsecurity.com/cybersecurity-in-a-post-quantum-computing-world/
9. Buchmann, J.A., Butin, D., Göpfert, F., Petzoldt, A.: Post-quantum cryptography: state of the art. In: Ryan, P., Naccache, D., Quisquater, J.J. (eds.) The New Codebreakers. Lecture Notes in Computer Science, vol. 9100, pp. 88–108. Springer, Heidelberg (2016). https://doi.org/10.1007/978-3-662-49301-4_6
10. Nejatollahi, H., Dutt, N., Ray, S., Regazzoni, F., Banerjee, I., Cammarota, R.: Post-quantum lattice-based cryptography implementations: a survey. ACM Comput. Surv. (CSUR) **51**(6), 1–41 (2019)
11. Wolf, C., Preneel, B.: Taxonomy of public key schemes based on the problem of multivariate quadratic equations. Cryptology ePrint Archive (2005)
12. Thormarker, E.: Post-quantum cryptography: supersingular isogeny diffie-hellman key exchange. Ph.D. dissertation, Thesis, Stockholm University (2017)
13. Lily, C., Dustin, M., Yi-Kai, L.: Post-quantum cryptography PQC. https://csrc.nist.gov/projects/post-quantum-cryptography
14. Michael, J., Cohn, A., Butcher, J.R.: Blockchain technology. The Journal **1**(7), 1–11 (2018)
15. Rajasekaran, A.S., Azees, M., Al-Turjman, F.: A comprehensive survey on blockchain technology. Sustain. Energy Technol. Assess. **52**, 102039 (2022)
16. Bitcoin: Bitcoin is an innovative payment network and a new kind of money. https://bitcoin.org/en/
17. Dutta, P., Choi, T.-M., Somani, S., Butala, R.: Blockchain technology in supply chain operations: applications, challenges and research opportunities. Transp. Research Part E: Logist. Transp. Rev. **142**, 102067 (2020)
18. Ølnes, S., Jansen, A.: Blockchain technology as s support infrastructure in e-government. In: Janssen, M., et al. (eds.) Electronic Government. Lecture Notes in Computer Science, vol. 10428, pp. 215–227. Springer, Cham (2017). https://doi.org/10.1007/978-3-319-64677-0_18
19. Petcu, A., Frunzete, M., Stoichescu, D.A.: A practical implementation of a digital document signature system using blockchain. In: 2023 13th International Symposium on Advanced Topics in Electrical Engineering (ATEE), pp. 1–6. IEEE (2023)
20. Yang, A., et al.: Prevention of business risks of internet information security platforms based on blockchain technology. Comput. Intell. Neurosci. **2022**, 7671810 (2022)
21. Chen, T., Yu, Y., Duan, Z., Gao, J., Lan, K.: Blockchain/ABE-based fusion solution for e-government data sharing and privacy protection. In: Proceedings of the 2020 4th International Conference on Electronic Information Technology and Computer Engineering, pp. 258–264 (2020)
22. Xie, G., Liu, Y., Xin, G., Yang, Q.: Blockchain-based cloud data integrity verification scheme with high efficiency. Secur. Commun. Netw. **2021**, 1–15 (2021)
23. Sjöberg, M.: Post-quantum algorithms for digital signing in public key infrastructures (2017)

24. Luc, N.Q., Nguyen, T.T., Quach, D.H., Dao, T.T., Pham, N.T.: Building applications and developing digital signature devices based on the falcon post-quantum digital signature scheme. Eng. Technol. Appl. Sci. Res. **13**(2), 10401–10406 (2023)
25. Prest, T., et al.: Falcon. In: Post-Quantum Cryptography Project of NIST (2020)
26. Pradel, G., Mitchell, C.J.: Post-quantum certificates for electronic travel documents. In: Boureanu, I., et al. (eds.) Computer Security. Lecture Notes in Computer Science, vol. 12580, pp. 56–73. Springer, Cham (2020). https://doi.org/10.1007/978-3-030-66504-3_4
27. Turner, D.M.: How to cut costs and speed up eidas implementation with digital signature deployment models. https://www.cryptomathic.com/news-events/blog/digital-signature-deployment-models-for-banking-operating-as-an-eidas-compliant-registration-authority-reduces-costs-and-preserves-customer-ownership
28. Bell, C.: Building a quantum-safe future. https://blogs.microsoft.com/blog/2023/05/31/building-a-quantum-safe-future/
29. open-quantum-safe. https://github.com/open-quantum-safe/liboqs
30. kevinmkane Kevin Kane: Pqcrypto-vpn project. https://github.com/microsoft/PQCrypto-VPN
31. Post-quantum cryptography (pqcrypto). https://github.com/kpdemetriou/pqcrypto
32. Try out post-quantum cryptography for code signing with signserver community 5.9.1. https://www.keyfactor.com/blog/try-out-post-quantum-cryptography-for-code-signing-with-signserver-community-5-9-1/
33. Pierre-Alain, F., et al.: Falcon (fast Fourier lattice-based compact signature over NTRU). https://falcon-sign.info/
34. Fouque, P.-A., et al.: Falcon: fast-Fourier lattice-based compact signatures over NTRU. Submission NIST's Post-Quantum Cryptogr. Standard. Process **36**(5), 1–75 (2018)
35. Soni, D., Basu, K., Nabeel, M., Aaraj, N., Manzano, M., Karri, R.: Hardware Architectures for Post-Quantum Digital Signature Schemes. Springer, Heidelberg (2021). https://doi.org/10.1007/978-3-030-57682-0

Review and Comparative Analysis of Meta-Heuristic-MPPT Algorithms of Photovoltaic Panels Under PSC

Fethia Hamidia[1,2](✉), Amel Abbadi[1,2], Ahmed Medjber[2], Abdelkader Morsli[1,2], Redha Skender Mohamed[2], and Madani Safaa[3]

[1] REA Laboratory, Yahia Feres Medea University, Medea, Algeria
fehamidia@gmail.com
[2] REM Laboratory, Yahia Feres Medea University, Medea, Algeria
[3] Yahia feres University, Medea, Algeria

Abstract. Shading presents a significant challenge to the efficiency of photovoltaic (PV) panels, often resulting in suboptimal operation and reduced power generation, thereby leading to economic losses. Traditional Maximum Power Point Tracking (MPPT) algorithms, such as Perturb and Observe (P&O) or Incremental Conductance (Inc-Cond), may struggle to effectively handle Partial Shading Conditions (PSC), where the power-voltage (P-V) curves exhibit multiple local peaks alongside a singular global peak. In contrast, metaheuristic algorithms demonstrate adaptability and efficacy in find the global maximum.

This study investigates and assesses the MPPT capabilities of six optimization techniques: Particle Swarm Optimization (PSO), Moth-flame Algorithm (MFA), Cuckoo Search Algorithm (CSA), Whale Optimization Algorithm (WOA), Flower Pollination Algorithm (FPA), and Grey Wolf Optimization Algorithm (GWO). Metaheuristic algorithms are recognized for their potential to converge more efficiently towards the optimal solution compared to traditional methods. This efficiency is particularly advantageous in real-time applications where rapid adaptation to changing environmental conditions is crucial for maximizing energy harvesting efficiency. The findings from this research contribute to advancing the understanding and application of metaheuristic approaches in enhancing the performance and resilience of PV systems under challenging operational conditions like Partial Shading.

Keywords: Meta-heuristic Algorithm · Partial Shading Condition · Maximum Power Point Technique · Global and Local peak

1 Introduction

Throughout history, humans have relied on the resources of their environment to survive and meet their needs. The predominant reliance on fossil fuels for generating electrical energy has significantly contributed to the grave pollution of the atmosphere. The burning of coal, oil, and natural gas emits greenhouse gases and other pollutants, exacerbating

© The Author(s), under exclusive license to Springer Nature Switzerland AG 2025
M. Hatti (Ed.): IC-AIRES 2024, LNNS 1238, pp. 354–363, 2025.
https://doi.org/10.1007/978-3-031-80301-7_39

climate change and harming air quality. Recognizing these environmental impacts, there has been a growing shift towards renewable energy sources as alternatives. Renewable energies such as solar, wind, hydropower, and geothermal offer clean and sustainable solutions to meet our energy needs without the harmful emissions associated with fossil fuels. Embracing renewable energy not only mitigates environmental degradation but also enhances energy security and promotes economic development through job creation and technological innovation. As global efforts intensify to combat climate change and progress toward a more sustainable future continues, the adoption of renewable energy technologies plays a crucial role in reducing carbon emissions, safeguarding public health, and preserving our planet for future generations. Turning to the development of renewable energy sources then emerged as the ultimate solution to this issue. In the last few years, there has been a notable increase in interest and investment in clean energy technologies [1, 2].

Renewable energies, especially photovoltaic solar energy, offer a host of benefits including cleanliness, silence, availability, and cost-effectiveness as they harness freely available energy from the sun. Photovoltaic (PV) arrays boast a sturdy structure with minimal maintenance requirements, making them particularly suitable for deployment in rural areas. However, solar panels are highly sensitive to shading, whether partial or complete, which significantly diminishes their power generation capacity and results in energy losses. Shading can stem from various environmental factors such as nearby mountains, trees, clouds, dust, or debris like dead leaves. While modern PV modules are equipped with bypass diodes to mitigate these shading effects and protect the solar cells, the impact on overall efficiency remains considerable. As the demand for sustainable energy solutions grows, addressing shading challenges through technological advancements continues to be a focus in maximizing the efficiency and reliability of photovoltaic solar energy systems.

Photovoltaic panels exhibit a distinctive electrical characteristic that is non-linear in nature, distinguishing them from linear electrical devices. This characteristic is illustrated by the current-voltage (I-V) curve of the panel, which depicts how the electrical output varies with the voltage applied across its terminals. One crucial feature of this curve is the Maximum Power Point (MPP), a specific operating point where the panel operates most efficiently to produce its maximum power output under given environmental conditions. The location of the MPP varies with factors such as solar irradiance, temperature, and shading, requiring optimization techniques like Maximum Power Point Tracking (MPPT) to ensure the PV system consistently operates at or near this optimal point. Understanding and managing the non-linear electrical characteristics of photovoltaic panels are essential for maximizing energy production and efficiency in solar energy systems. In this scenario, the characteristic (P-V) curve becomes more intricate, featuring a prominent global peak accompanied by several local peaks. The global maximum represents the highest achievable power output, whereas the local peaks indicate points of significantly lower power. The MPPT approach is designed to continually track the Maximum Power Point (MPP) of photovoltaic panels, ensuring that the system extracts the maximum available power. This optimization process is crucial for enhancing the efficiency and performance of solar energy systems. Typically, MPPT utilizes a power conversion device, such as a Boost converter, as an interface between the PV panel and the load. Commonly used algorithms for MPPT include various sophisticated techniques designed to accurately determine and maintain the optimal operating conditions of the PV system, The most commonly used algorithms [3, 4] include Perturb and Observe (P&O), Constant Voltage (CV), along with artificial intelligence-based methods such as fuzzy logic [5–7], neural networks [8], and neural fuzzy approaches [9].

Metaheuristic algorithms are heuristic methods designed to efficiently explore a solution space to find approximate solutions to optimization problems. Meta-heuristic algorithms represent optimization methodologies renowned for their adeptness in exploration, exploitation, local optima avoidance, and convergence. Studying these algorithms helps in designing systems that can mitigate the impact of shading and maximize energy harvest. For instance, notable examples include Particle Swarm Optimization PSO [10] proposed by James Kennedy and Russell Eberhart [11] in 1995, Ant Colony ACO [12] proposed by Marco Dorigo [13] in 1990 , Artificial Bee Colony ABC [14] proposed Karaboga [15] in 1995, Crow Search Algorithm CSA [16] proposed by Alireza Askarzadeh [17] in 2016, Genetic Algorithm GA [18] proposed by John Holland [19], Grey Wolf Optimization GWO [20, 21], Moth Flame Algorithm MFA [22] and Whale Optimization Algorithm WOA [23] developed by Mirjalili [24–26], Cuckoo Search Algorithm [27, 28] proposed by Yang and Deb [29] in 2009, Flower Pollination Algorithm FPA [30] proposed by Yang [31].

Meta-Heuristic algorithms tend to be more robust against noise or uncertainties in the PV system parameters. They can handle variations in irradiance, temperature, and other environmental factors more effectively than classical algorithms. This paper is structured into five distinct partitions. Partition two provides an overview of the electrical characteristics of photovoltaic panels. Partition three introduces and discusses the use of six meta-heuristic MPPT approaches for photovoltaic systems: CSA, FPA, WOA, MFA, GWO, and PSO. Partition four presents the simulation results of these algorithms under partial shading conditions. Finally, partition five summarizes the findings and draws conclusions based on the study's outcomes.

2 Maximum Power Point Tracking By Meta-Heuristic Algo

To achieve the highest power extraction from the PV generator and efficiently deliver it to the load in real-time, an adaptation stage is employed. This component serves as the intermediary between the PV source and the load, ensuring optimal power transfer using a control mechanism (specifically, adjusting the duty cycle) to harness and deliver the maximum available power from the generator. In photovoltaic systems, a typical adapter is a static converter known as a Boost converter, chosen specifically for its ability to increase the voltage from the photovoltaic generator (PVG). The selection of this converter type depends on the load requirements and the need to elevate the PVG voltage. In this paper, we utilize a configuration comprising five modules connected in series, designed to operate effectively even under partial shading conditions. (200-500-500-1000- 800W/m^2).

Classical methods as P&O are very effective when the characteristic P-V have a single optimal point (absence of shading). Otherwise, the application of these algorithms to locate the maximum power point in the presence of shading conditions (dynamic regime)

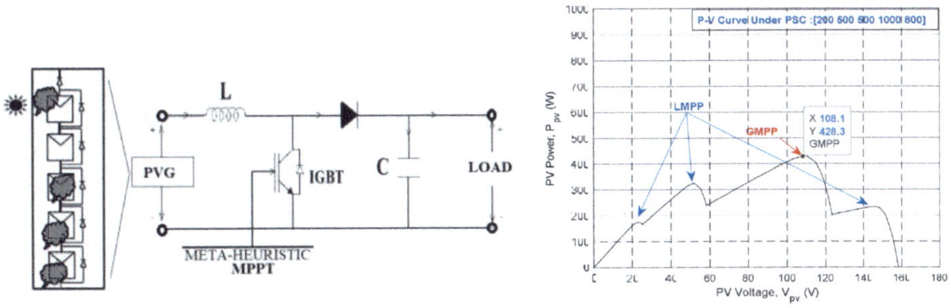

Fig. 1. PV-DC Load and (P-V) curve under partial shading

gives unsatisfactory results, the algorithm has difficulty searching more precisely the MPP, it converges to first LMPP (local maximum). To deal with the problem of existence of several local maxima, we consider here the implementation of an MPPT control based on the intelligent metaheuristic techniques. Metaheuristic algorithms are indeed bio- inspired optimization techniques that emulate natural phenomena observed in animals, insects, plants, and other biological systems. These algorithms harness principles derived from biological processes to solve complex optimization problems efficiently. By mimicking the behavior of natural systems, metaheuristic algorithms can explore and exploit search spaces effectively, often finding near-optimal solutions in a variety of domains where traditional optimization methods may be impractical or inefficient.

In this work, the proposed approachs are evaluated and compared to each other via a dc-dc Boost converter, . Initially, a random population is generated within the solution space (moths, whales, wolfs…). Each particle's fitness value, representing maximum power, is then calculated, and the best position (Gbest) is identified. An iterative update process follows, with iterations continuing until the termination criteria are met (at iteration "j"). The candidate solutions, metaphorically represented as moths, whales, cuckoo birds, or other species according to the proposed algorithm, adjust their positions to solve the problem variables. Figure 2 illustrates the flowcharts of these metaheuristic algorithms applied to achieve Maximum Power Point Tracking (MPPT). You find in [21, 22, 27, 30] more details about these algorithms.

3 MATLAB Simulation Results

All simulation findings were obtained using PV system with five panels under partial shading (200-500-500-1000-800W/m^2), the modules are connected in series as illustrated in Fig. 1.

To fairly compare the results obtained by different optimization algorithms, to conduct a rigorous and unbiased comparison of the six proposed metaheuristic or any other optimization algorithms. This study ensures that the evaluation accurately reflects each algorithm's capabilities and limitations under standardized conditions, facilitating informed decisions for future practical applications (Figs. 3, 4 and 5).

As shown Table 1. Firstly, using the same population size for each algorithm ensures that they are evaluated under comparable computational constraints and resource usage.

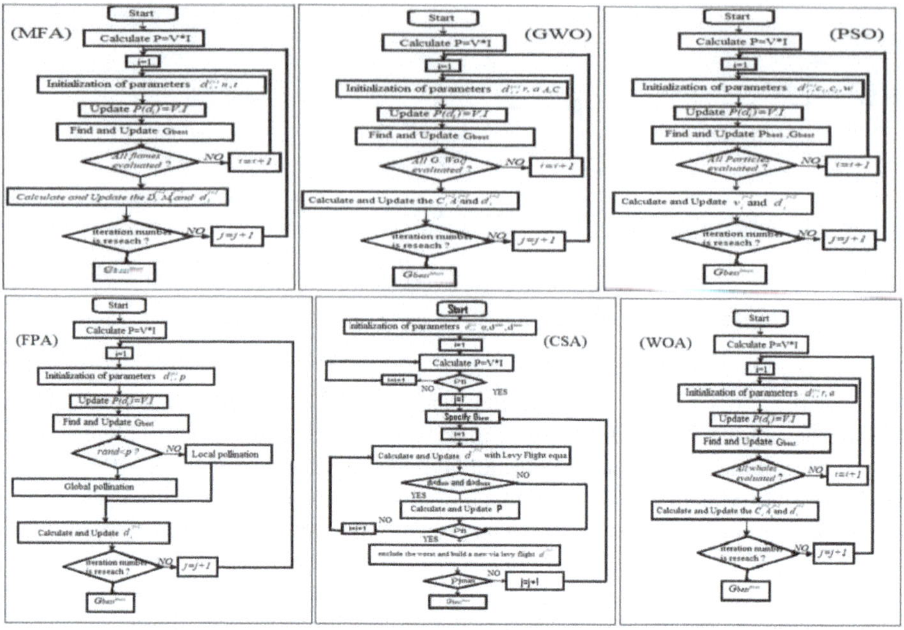

Fig. 2. Flow-diagram of proposed MPPT-Meta-Heuristic methods

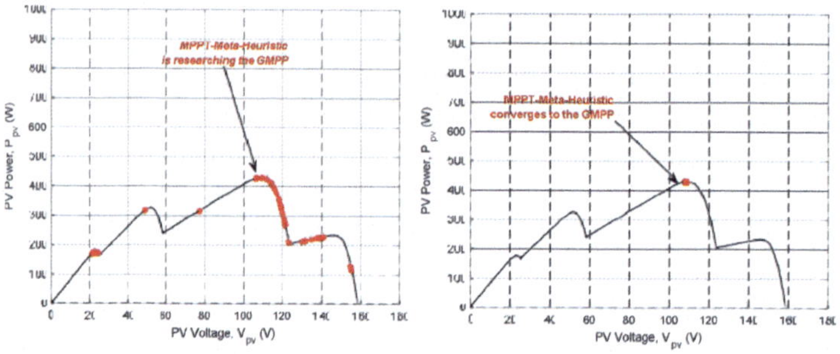

Fig. 3. Power-Voltage curves at the beginning and the end of MPPT search using Metaheuristic Algorithms

Secondly, conducting tests with an identical number of iterations allows for a direct comparison of how quickly and effectively each algorithm reaches the optimal solution within a predefined timeframe. Employing an identical cost function across all algorithms ensures that they are solving the same optimization problem, thereby enabling a fair assessment of their performance. As shown Fig. 2. Standardizing the initialization process, such as using the same initial conditions, minimizes bias and ensures that any observed differences in performance are attributable to algorithmic variations rather than starting conditions. It is also crucial to employ consistent performance metrics to evaluate and compare results, encompassing criteria such as tracking efficiency, convergence rate, and computational.

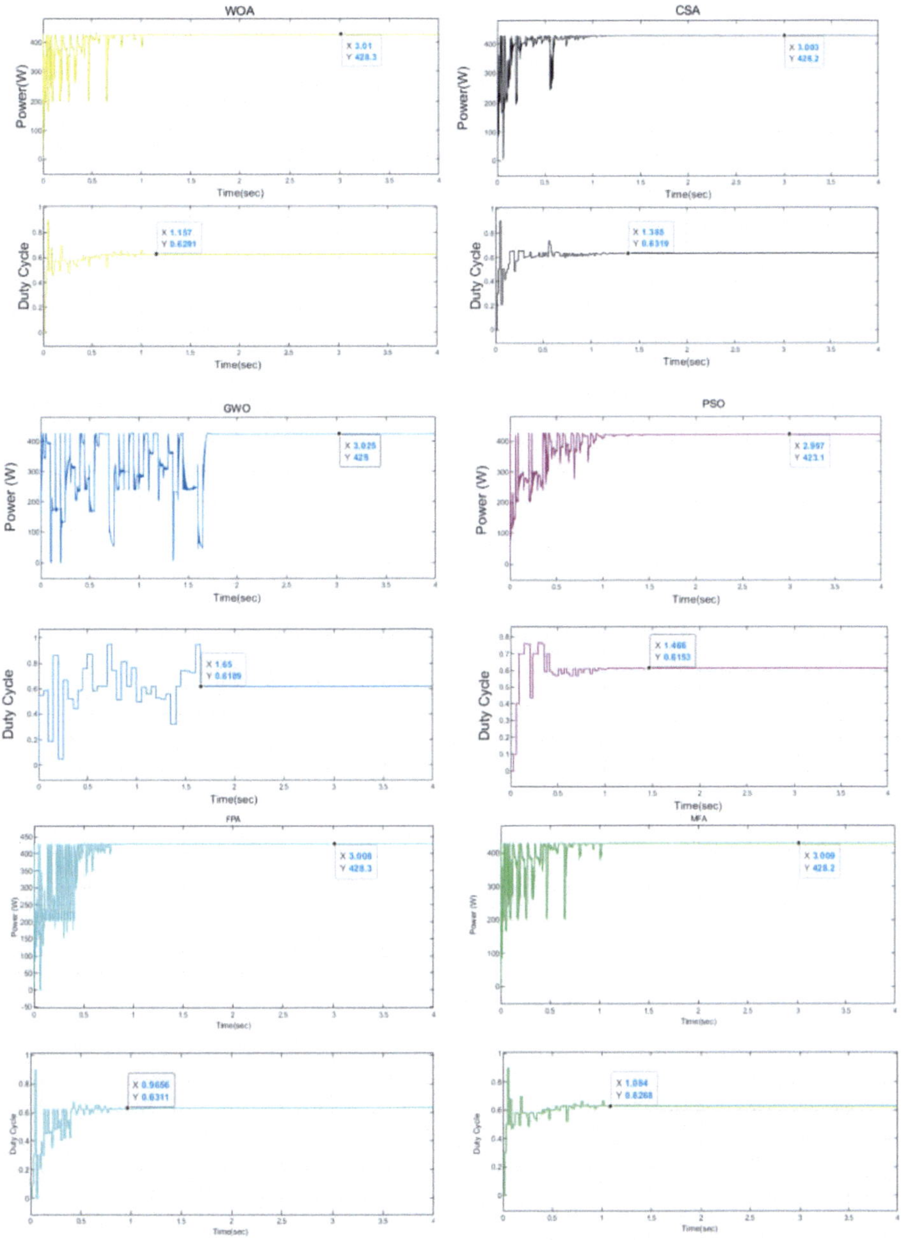

Fig. 4. Power and Duty cycle responses of metaheuristic-MPPT algorithm under PSC (200–500-500–1000-800 W/m^2)

Based on the findings, we present a summary of the performance of Metaheuristic algorithms in Table 1, clearly demonstrating that the six proposed algorithm converge to global MPP, however, the FPA can achieve superior results.

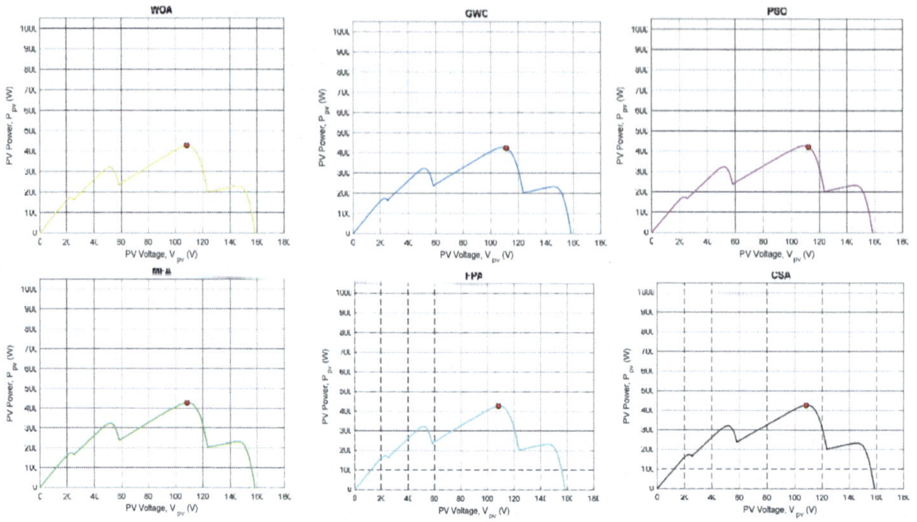

Fig. 5. Power-Voltage curves of each proposed algorithm at the end of MPPT search

Table 1. Proposed Meta-heuristic performances

	PSO	MFA	GWO	CSA	FPA	WOA
Tracking Efficiency	99,87%	99,97%	99,22%	99.97%	100%	100%
Convergence Tracking Time	1.46sec	1.08sec	1.65sec	1.38sec	0.96sec	1.15sec
Convergence to	global	global	global	global	global	global
Iterations	150	150	150	150	150	150
Population number	50	50	50	50	50	50

The table below represents the characteristic of Photovoltaic panel module used in this paper (Table 2).

Table 2. PV panel parameters

Maximum Power	200	W
Open Circuit Voltage Voc	32.9	V
Short Circuit Current Isc	8.21	A
Shunt Resistance Rsh	951.9319	Ohm
Series Resistance	0.2172	Ohm
Number of cells in series per module	54	

4 Conclusion

In this study, we introduced six metaheuristic algorithms under same conditions (number of iteration, population…) for a fair comparison. These algorithms were employed and compared to each other to optimize the maximum power point tracking of photovoltaic panels under Strong shading conditions. Each algorithm successfully converges on the optimal solution, demonstrating its ability to identify the best possible outcome with precision. This convergence ensures that the algorithms are capable of navigating complex problem spaces to reach the most efficient result. Our findings indicate that all six algorithms effectively approached the global MPPT. Notably, the Flower Pollination Algorithm (FPA) exhibited superior performance among them.

References

1. Huang, C., Wang, L., Long, H., Luo, X., Wang, J.-H.: A hybrid global maximum power point tracking method for photovoltaic arrays under partial shading conditions (2020)
2. Mat, Z.B.A., Kar, Y.B., Hassan, S.H.B.A., Talik, N.A.B.: Proton Exchange Membrane (PEM) and Solid Oxide (SOFC) fuel cell based vehicles-a review. In: 2017 2nd IEEE International Conference on Intelligent Transportation Engineering (2017)
3. Rai, N., Abbadi, A., Hamidia, F., Kanouni, B., Kahlessenane, A.: A new modified incremental conductance algorithm used for PV system. In: Hatti, M. (ed.) Artificial Intelligence and Heuristics for Smart Energy Efficiency in Smart Cities. Lecture Notes in Networks and Systems, vol. 361, pp. 256–265. Springer, Cham (2022). https://doi.org/10.1007/978-3-030-92038-8_26
4. Elgendy, M.A., Zahawi, B., Atkinson, D.J.: Assessment of perturb and observe MPPT algorithm implementation techniques for PV pumping applications. IEEE Trans. Sustain. Energy **3**, 21–33 (2012)
5. Chiu, C.S.: T-S fuzzy maximum power point tracking control of solar power generation systems. IEEE Trans. Energy Convers. **25**(4), 1123–1132 (2010)
6. Abbadi, A., Hamidia, F., Morsli, A., Benbouabdellah, O., Chiba, Y.: MPPT based fuzzy-logic controller for grid connected residential photovoltaic power system. In: Hatti, M. (ed.) Smart Energy Empowerment in Smart and Resilient Cities. Lecture Notes in Networks and Systems, vol. 102, pp. 124–131. Springer, Cham (2020). https://doi.org/10.1007/978-3-030-37207-1_13
7. Hamidia, F., Abbadi, A., Tlemçani, A.: Improved pumping system supplied by double photovoltaic panel. Rev. Roum. Sci. Techn. Électrotechn. et Énerg. **64**(1), 87–93 (2019)
8. Lin, W.M., Hong, C.M., Chen, C.H.: Neural-network-based MPPT control of a stand-alone hybrid power generation system. IEEE Trans. Power Electron. **26**(12), 3571–3581 (2011)
9. Al-Majidi, S.D., Abbod, M.F., Al-Raweshidy, H.S.: Maximum power point tracking technique based on a neural-fuzzy approach for stand-alone photovoltaic system. In: 55th International Universities Power Engineering Conference (2020)
10. Hamidia, F., Abbadi, A., Tlemçani, A.: Improved hybrid pumping system with storage battery-based particle swarm algorithm. Rev Roum. Sci. Techn. Électrotechn. et Énerg. **66**(1), 243–248 (2021)
11. Kennedy, J., Eberhart, R.: Particle swarm optimization. In: Proceeding of ICNN'95, International Conference on Neural Network, pp. 1942–1948. IEEE (1995)
12. Titri, S., Larbes, C., Youcef Toumi, K., Benatchba, K.: A new MPPT controller based on the ant colony optimization algorithm for photovoltaic systems under partial shading conditions. Appl. Soft Comput. **58**, 465–479 (2017)

13. Dorigo, M., Stützle, T.: The ant colony optimization metaheuristic: algorithms. In: Applications, and Advances, Chapter 9, Handbook of Metaheuristics, pp. 250–285 (2003)
14. Benyoucef, A.S., Chouder, A., Kara, K., Silvestre, S., Aitsahed, O.: Artificial bee colony based algorithm for maximum power point tracking (MPPT) for PV systems operating under partial shaded conditions. Appl. Soft Comput. **32**, 38–48 (2015)
15. Karaboga, D., Basturk, B.: Artificial bee colony (ABC) optimization algorithm for solving constrained optimization problems. In: Melin, P., Castillo, O., Aguilar, L.T., Kacprzyk, J., Pedrycz, W. (eds.) Foundations of Fuzzy Logic and Soft Computing. Lecture Notes in Computer Science, vol. 4529, pp. 789–798. Springer, Heidelberg (2007). https://doi.org/10.1007/978-3-540-72950-1_77
16. Hamidia, F., Abbadi, A.: Photovoltaic MPPT control based on crow search algorithm. In: Algerian Symposium on Renewable Energy and Materials, ASREM, Medea, Algeria (2022)
17. Askarzadeh, A.: A novel metaheuristic method for solving constrained engineering optimization problems: crow search algorithm. Comput. Struct. **169**, 01–12 (2016)
18. Ramaprabha, R., Gothandaraman, V., Kanimozhi, K., Divya, R., Mathur, B.L.: Maximum power point tracking using GA-optimized artificial neural network for Solar PV system. In: Proceedings of the 1st International Conference on Electrical Energy Systems (ICEES), Newport Beach, CA, USA, pp. 264–268 (2011). pp. 141–149
19. Sastry, K., Goldberg, D., Kendall, G.: Genetic algorithms. In: Chapter 4, Search Methodologies, pp. 97–125 (2005)
20. Hamidia, F., Abbadi, A.: Grey wolf optimization algorithm. In: Artificial Intelligence and Heuristics for Smart Energy Efficiency in Smart Cities, pp.57–65 (2022)
21. Abbadi, A., Hamidia, F., Skender, M.R., Bettache, F.: Grey wolf MPPT Controller for grid connected residential wind system operating under low and high variations in wind speed. In: Hatti, M. (ed.) Advanced Computational Techniques for Renewable Energy Systems. Lecture Notes in Networks and Systems, vol. 591, pp. 261–268. Springer, Cham (2023). https://doi.org/10.1007/978-3-031-21216-1_28
22. Fethia, H., Amel, A., Mohamed Redha, S., Abdelkader, M., Farouk, B., Fatima, S.: MOTH-FLAME-MPPT algorithms of PV panels under PSC. In: Hatti, M. (ed.) IoT-Enabled Energy Efficiency Assessment of Renewable Energy Systems and Micro-grids in Smart Cities. Lecture Notes in Networks and Systems, vol. 984, pp. 130–138. Springer, Cham (2024). https://doi.org/10.1007/978-3-031-60629-8_13
23. Ebrahim, A., Osama, A., Kotb, K.M., Bendary, F.: Whale inspired algorithm based MPPT controllers for grid-connected solar photovoltaic system. Energy Procedia **162**, 77–86 (2019)
24. Mirjalili, S., Mirjalili, S.M., Lewis, A.: Grey wolf optimizer. Adv. Eng. Softw. **69**, 46–61 (2014)
25. Mirjalili, S.: Moth-flame optimization algorithm: novel nature-inspired heuristic paradigm. Knowl.-Based Syst. **89**, 228–249 (2015)
26. Mirjalili, S., Lewis, A.: The whale optimization algorithm. Adv. Eng. Softw. **95**(51–67), 2016 (2016)
27. Hamidia, F., Abbadi, A.: SOFC-PV system with storage battery based on cuckoo search algorithm. In: Nonlinear Dynamics and Systems Theory (2022)
28. Abbadi, A., Hamidia, F.: An MPPT controller based on cuckoo search for residential wind energy conversion system. In: Algerian Symposium on Renewable Energy and Materials, ASREM, Medea, Algeria (2022)
29. Yang, X.S., Deb, S.: Engineering optimisation by Cuckoo search. Int. J. Math. Modell. Neumeric Opt. **1**(4), 330–343 (2010)

30. Hamidia, F., Abbadi, A., Skender, M.R., Morsli, A.: MPPT based FPA control for PV pumping system. In: 2023 2nd International Conference on Electronics, Energy and Measurement (IC2EM) (2023)
31. Yang, X.S.: Flower pollination algorithm for global optimization. In: Durand-Lose, J., Jonoska, N. (eds.) Unconventional Computation and Natural Computation. Lecture Notes in Computer Science, vol. 7445, pp. 240–249. Springer, Heidelberg (2012). https://doi.org/10.1007/978-3-642-32894-7_27

A New Strategy to Mitigate Overheating in Solar PV Systems Using Adaptive Solar Tracking

Khadidja Dahli[1](✉), Karim Selam[2], and Nawal Cheggaga[1]

[1] Electrical and Remote Control Systems Laboratory, LabSET, Faculty of Technology, University of Blida 1, Blida, Algeria
khadi.dja22977@gmail.com, dahlikhadidja@ctu.univ-blida.dz

[2] Process Control Laboratory, Polytechnical National School, Algiers, Algeria

Abstract. High temperatures pose several challenges for solar power generation in desert environments, such as decreased efficiency, increased resistance, thermal-induced degradation, and potential damage to the panels. This study aims to address these challenges by proposing a solar tracking system that manages excessive heat and does not require cooling systems or additional complex mechanisms. Instead, we present an innovative approach to protect solar systems from increased heat by integrating temperature data readings into conventional solar tracking tools. When the temperature exceeds the panel's recommended threshold, the tracker adjusts its position to misalign the panels from the sun and prevent damage. The system will resume regular tracking once the temperature drops below the critical value. Our research indicates that positioning the tracker vertically at 90° helps protect the panels from overheating and reduces dust accumulation. This innovation has the potential to enhance the performance and longevity of solar power systems in desert environments.

Keywords: Desert environments · Solar power · Solar tracking system

1 Introduction

Solar power has immense potential in desert environments due to the abundant sunlight available throughout the year. Deserts receive high solar irradiance, often exceeding 6 kWh/m2 per day [1], which can result in an annual energy output of more than 2,190 kWh/m2 [2]. This high level of solar insolation makes deserts ideal for large-scale solar power generation. For example, the Sahara Desert alone could theoretically produce enough electricity to power the entire world if harnessed efficiently [3]. Additionally, solar farms in desert regions can achieve capacity factors of 25% to 30% [4], which translates to substantial energy production given the vast available land and sunshine. However, excessive heat poses significant challenges for these regions, impacting the efficiency and longevity of photovoltaic (PV) systems.

High temperatures can lead to a marked decrease in the energy output of solar panels, as well as increased resistance and potential thermal damage [5]. Primarily, it decreases the open-circuit voltage (Voc) due to the reduction in the semiconductor's

© The Author(s), under exclusive license to Springer Nature Switzerland AG 2025
M. Hatti (Ed.): IC-AIRES 2024, LNNS 1238, pp. 364–371, 2025.
https://doi.org/10.1007/978-3-031-80301-7_40

energy bandgap, and increases electron-hole recombination rates, reducing the number of charge carriers available for current generation. Additionally, higher temperatures raise the series resistance within the cell, causing greater ohmic losses and reducing the fill factor (FF). These combined effects result in a lower maximum power output. For silicon-based PV cells, efficiency typically drops by approximately 0.4% to 0.5% per degree Celsius above the nominal operating cell temperature [6]. Several studies have reported a drop of efficiency due to high temperature [7–12]. Real- world scenarios have also highlighted the adverse effects of heat on PV systems. In 2011, The one billion dollar solar project in Nevada was to be the biggest solar plant of its kind; however, it failed just few years of its debut due to overheating, leading to substantial financial losses and operational disruptions [13].

Existing techniques in the literature to address these thermal challenges include active cooling systems, such as water cooling technique [14–16], or air cooling [17–19], and passive methods, like heat sinks and phase change materials [20]. Active cooling systems, while effective, require additional infrastructure and maintenance, increasing the overall cost and complexity of the solar installation. Passive methods, on the other hand, have limitations in their effectiveness, particularly in extremely hot and arid environments where heat dissipation is less efficient.

Despite the advancements in cooling technologies, there are still gaps in effectively managing heat in PV systems without incurring high costs or complicating the system design. This study aims to overcome these gaps by proposing a novel approach that integrates real-time temperature monitoring with conventional solar tracking tools. By tilting the panels vertically when temperatures exceed the recommended threshold, the system misaligns from the sun, thereby reducing heat exposure and preventing potential damage. Additionally, this vertical tilting helps reduce dust accumulation, which further contributes to lowering panel temperatures. The system resumes regular tracking once the temperature drops below the critical value. This methodology leverages existing solar tracking technology, avoiding the need for complex or costly cooling mechanisms. It offers a practical and efficient solution to manage excessive heat, ensuring optimal performance and longevity of PV systems in desert environments. This innovative approach serves as a promising solution for future solar power installations in regions facing similar thermal challenges.

2 Methods and Materials

2.1 Temperature Effect on PV Systems

The power output of a photovoltaic (PV) module can be notably influenced by temperature. As temperature increases, the power output decreases, and this correlation is measured using the temperature coefficient of the PV module. This relationship is mathematically represented by the following equation:

$$P_t = P_{STC}[1 + \beta(T_c - T_{STC})] \tag{1}$$

where: Pt is the power output at temperature Tc, P_{STC} is the power output at Standard Test Conditions (STC), β is the temperature coefficient of power, Tc is the actual cell temperature, T_{STC} is the cell temperature at STC, usually 25 °C.This equation shows that for every degree Celsius increase in temperature above the STC, the power output decreases by a percentage equal to the temperature coefficient.

Additionally, the cell temperature *Tc* can be related to the ambient temperature *Ta* using the following equation:

$$T_c = T_a + G\left(\frac{NOCT - 20^{\circ}C}{800\,\mathrm{W/m^2}}\right) \tag{2}$$

where: *NOCT* is the Nominal Operating Cell Temperature (usually around 45–48 °C), G is the irradiance (in W/m^2).

These equations together illustrate how increased temperatures can decrease the efficiency of PV panels, emphasizing the importance of effective thermal management strategies.

2.2 Proposed Strategy

This study introduces a novel approach to address overheating issues in solar photovoltaic (PV) systems by dynamically adjusting the orientation of solar panels based on real-time temperature data. The core of our methodology involves the integration of temperature data for continuous monitoring of PV panel temperature. Adjustments are activated when the temperature surpasses a critical threshold. Upon activation, the solar tracker repositions the panels vertically to a 90° angle, which has been optimized for minimal heat exposure according to our experimental findings. This misalignment from direct sunlight aims to reduce heat accumulation and dust accumulation, which can contribute to further temperature increases. Once the panel temperature falls below the critical threshold, the system seamlessly reverts to tracking the sun's optimal position. Notably, the system does not depend on conventional sun-tracking sensors; instead, it utilizes astronomical algorithms sourced from "Astronomical Algorithms" by Jean Meeus (1998) [21] to accurately determine the sun's position throughout the day. This methodology ensures robust protection and optimal performance of the PV panels without the necessity of cooling systems or additional intricate mechanisms. Consequently, this approach guarantees improved efficiency and prolonged panel lifespan in environments characterized by excessive heat. Figure 1 presents the proposed strategy workflow.

2.3 Prototype

The control unit of the proposed solar tracker, as shown in Fig. 2(a), includes a Bluetooth module (HC-05) (labeled as 2) for temperature data retrieval. An Arduino Uno (labeled as 1) serves as the central processing unit to control the actuators for sun position, angles, and stow position. The system utilizes a NEMA 17 stepper motor (labeled as 4) for elevation rotation and a NEMA 23 (labeled as 5) for azimuth rotation, both controlled by an L298N motor driver (labeled as 3). The prototype (depicted in Fig. 2 (b)) has a sturdy design with V-belt rotation for precise azimuthal rotation and a screw actuator for adjustable elevation angles.

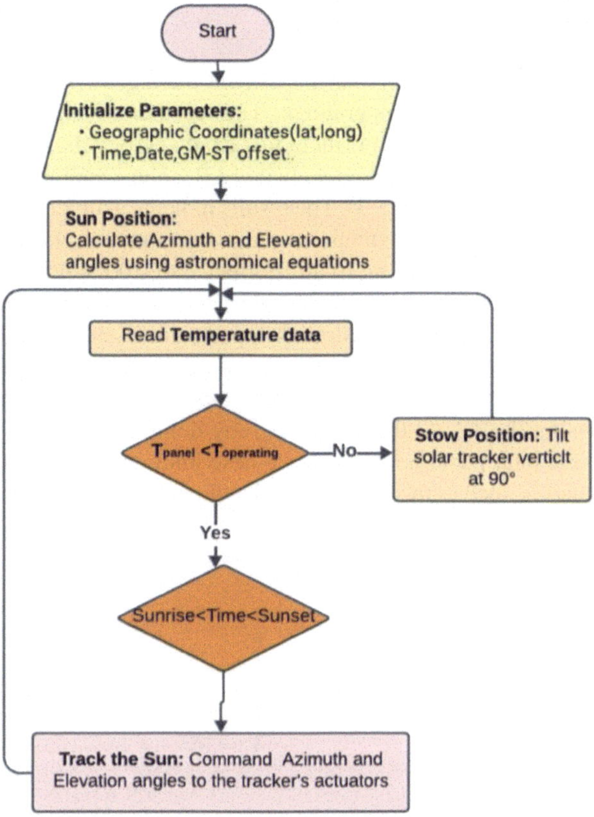

Fig. 1. The proposed strategy.

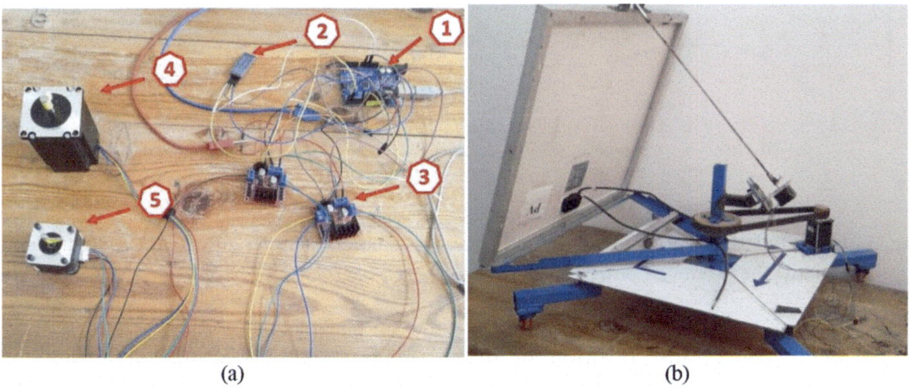

Fig. 2. The proposed solar tracker: (a) Control unit (b) Prototype

3 Results and Discussions

In In this study, a new approach to solar panel heat management was tested in Tiaret, Algeria, an area prone to high temperatures. The tests, conducted during July 2023, involved monitoring various parameters including irradiance levels, ambient temperature sourced from the Weather Underground website [22], solar panel temperature, and the corresponding power output (see Fig. 3). The results indicate a direct correlation between collected irradiance and power production by the solar tracker, with peak production observed on July 31st, corresponding to an irradiance of 200 W/m^2.Conversely, when the ambient temperature exceeded 42 °C, which is the operating limit of the solar panel according to its datasheet, the power output dropped significantly. For instance, on July 12th, the power output decreased to 27 W, representing a 50% reduction in efficiency due to increased temperature. The ambient temperature exhibited dynamic fluctuations, frequently surpassing the 42 °C threshold, thereby demonstrating the adverse impact of excessive heat on solar panel performance. These findings suggest that effective heat management strategies, such as the one proposed in this study, are crucial for maintaining optimal solar panel efficiency in hot climates.

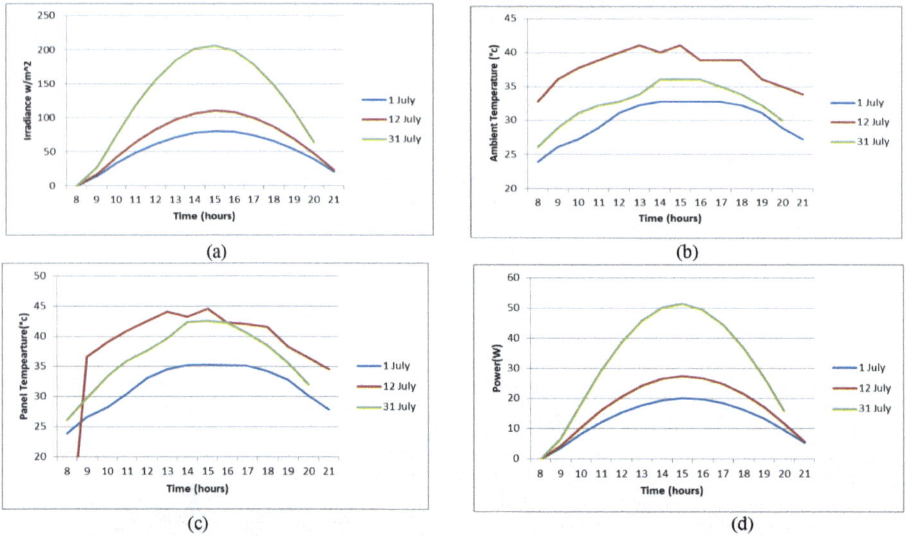

Fig. 3. Collected data: (a) collected irradiance (b) ambient temperature (c) the solar panel temperature (d) the produced power.

To identify the optimal position for minimizing heat on the solar tracker, an analysis of the solar tracker temperature was conducted for various tilt angles (refer to Fig. 4). The findings indicate that the highest heat levels are recorded when the tracker is inclined at 0°, whereas the lowest heat levels are observed at a tilt of 90°. This implies that tilting the panel vertically (90°) when it surpasses the operational temperature limit can effectively decrease heat generation and maintain panel efficiency.

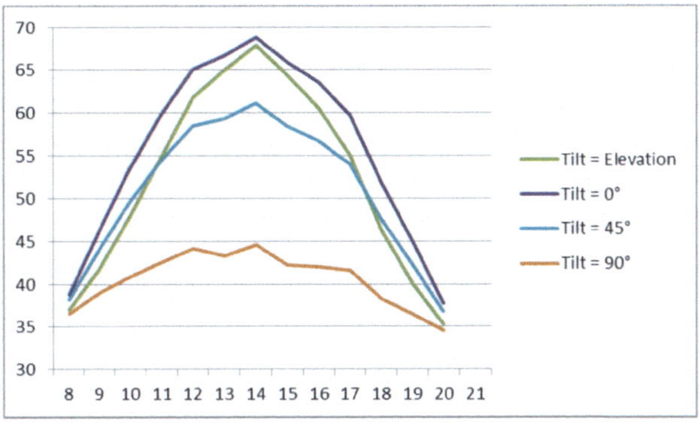

Fig. 4. Solar tracker temperature depending on the tilt angle

The solar tracker's performance was assessed based on its tilt angle and energy production versus consumption during testing (refer to Fig. 5). Results show that the tracker responded to overheating by tilting vertically at 90° to reduce heat. When the temperature decreased, it resumed normal sun-tracking. Surprisingly, despite heat reduction adjustments, energy consumption remained consistent at around 13 Wh, indicating operational efficiency. This demonstrates the effectiveness of the vertical tilt strategy in managing heat without affecting energy use. Additionally, energy production aligned closely with irradiance patterns, peaking on July 31st with maximum irradiance. This consistent energy production indicates the tracker's resilience to temperature changes, maintaining performance under varying conditions.

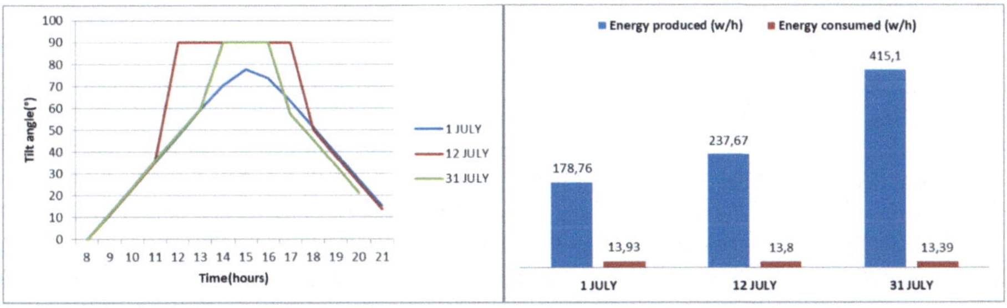

Fig. 5. The solar tracker response to overheating: (a) Tilt angle (b) Produced and consumed energy

The results highlighted the robustness of the tracker's performance strategy, demonstrating its ability to effectively manage heat through vertical tilting while maintaining consistent energy consumption. This approach ensured that the tracker responded dynamically to temperature fluctuations, optimizing energy production based on collected irradiance levels.

4 Conclusion

In In this paper, a new approach is proposed to reduce heat stress in solar panels by dynamically tilting solar trackers based on real-time temperature data. This system effectively manages thermal conditions and optimizes energy production. The analysis of results shows that the proposed solar tracker offers several advantages:

1. Enhanced thermal management by minimizing heat buildup on panels through vertical tilting to 90° during high-temperature conditions, preserving panel efficiency and longevity.
2. Solar panels have a reduced risk of thermal-induced degradation, extending their lifespan.
3. Improved reliability and performance under varying environmental conditions ensure consistent operation and energy output.
4. By using existing infrastructure and advanced sun positioning algorithms, the system becomes more cost- effective, reducing complexity and maintenance needs.
5. Adaptability to desert environments and other high-temperature regions enhances the feasibility of solar power in challenging climates.
6. Simplified system design and installation by utilizing existing solar tracking infrastructure and reducing the need for extra equipment.

Future work will explore the scalability of this approach across different climates and seasons, aiming to further enhance its effectiveness and applicability. This method shows potential for enhancing PV systems and tackling efficiency challenges in solar energy, especially in regions with extreme temperatures.

References

1. Alrwashdeh, S., Al-saraireh, F., Saraireh, M.: Solar radiation map of Jordan governorates. Int. J. Eng. Technol. **7**, 1664–1667 (2018). https://doi.org/10.14419/ijet.v7i3.15557
2. Stökler, S., Schillings, C., Kraas, B.: Solar resource assessment study for Pakistan. Renew. Sustain. Energy Rev. **58**, 1184–1188 (2016). https://doi.org/10.1016/j.rser.2015.12.298
3. Kurokawa, K. (ed.): Energy from the Desert: Practical Proposals for Very Large Scale Photovoltaic Systems. Routledge, London (2003). https://doi.org/10.4324/9781849771139
4. Prăvălie, R., Patriche, C., Bandoc, G.: Spatial assessment of solar energy potential at global scale. A geographical approach. J. Clean. Prod. **209**, 692–721 (2019). https://doi.org/10.1016/j.jclepro.2018.10.239
5. Viswanathan, R., Stringer, J.: Failure mechanisms of high temperature components in power plants. J. Eng. Mater. Technol. **122**(3), 246–255 (2000). https://doi.org/10.1115/1.482794
6. Chandel, S.S., Agarwal, T.: Review of cooling techniques using phase change materials for enhancing efficiency of photovoltaic power systems. Renew. Sustain. Energy Rev. **73**, 1342–1351 (2017). https://doi.org/10.1016/j.rser.2017.02.001
7. Kaldellis, J.K., Kapsali, M., Kavadias, K.A.: Temperature and wind speed impact on the efficiency of PV installations. Experience obtained from outdoor measurements in Greece. Renew. Energy **66**, 612–624 (2014). https://doi.org/10.1016/j.renene.2013.12.041
8. Du, Y., et al.: Evaluation of photovoltaic panel temperature in realistic scenarios. Energy Convers. Manage. **108**, 60–67 (2016). https://doi.org/10.1016/j.enconman.2015.10.065

9. Brinkworth, B.J., Sandberg, M.: Design procedure for cooling ducts to minimise efficiency loss due to temperature rise in PV arrays. Sol. Energy **80**(1), 89–103 (2006). https://doi.org/10.1016/j.solener.2005.05.020
10. Hasan, A., McCormack, S.J., Huang, M.J., Norton, B.: Energy and cost saving of a photovoltaic-phase change materials (PV-PCM) system through temperature regulation and performance enhancement of photovoltaics. Energies, **7**(3), Article no. 3, March 2014. https://doi.org/10.3390/en7031318
11. Effect of temperature on conversion efficiency of single-phase solar PV inverter. https://ieeexplore.ieee.org/abstract/document/9518415. Accessed 15 July 2024
12. Kaplanis, S., Kaplani, E., Kaldellis, J.K.: PV temperature prediction incorporating the effect of humidity and cooling due to seawater flow and evaporation on modules simulating floating PV conditions. Energies, **16**(12), Article no. 12, January 2023. https://doi.org/10.3390/en16124756
13. The one billion dollar solar failure in Nevada. American Experiment. https://www.americanexperiment.org/the-one-billion-dollar-solar-failure-in-nevada/. Accessed 15 July 2024
14. Moharram, K.A., Abd-Elhady, M.S., Kandil, H.A., El-Sherif, H.: Enhancing the performance of photovoltaic panels by water cooling. Ain Shams Eng. J. **4**(4), 869–877 (2013). https://doi.org/10.1016/j.asej.2013.03.005
15. Hussien, H.A., Numan, A.H., Abdulmunem, A.R.: Improving of the photovoltaic/thermal system performance using water cooling technique. IOP Conf. Ser.: Mater. Sci. Eng. **78**(1), 012020 (2015). https://doi.org/10.1088/1757-899X/78/1/012020
16. Salem Ahmed, M., Mohamed, A.S.A., Maghrabie, H.M.: Performance evaluation of combined photovoltaic thermal water cooling system for hot climate regions. J. Solar Energy Eng. **141**(041010) (2019). https://doi.org/10.1115/1.4042723
17. Harahap, H.A., Dewi, T., Rusdianasari: Automatic cooling system for efficiency and output enhancement of a PV system application in Palembang, Indonesia. J. Phys.: Conf. Ser. **1167**(1), 012027 (2019) https://doi.org/10.1088/1742-6596/1167/1/012027
18. Bedair, A.O.M., Shehata, A.S., Hamad, M.S., Tawfik, A.: Energy performance enhancement of PV system through proposed back cooling techniques and natural air draft: an experimental study. Int. J. Green Energy, August 2023. https://www.tandfonline.com/doi/abs/10.1080/15435075.2023.2244057. Accessed 15 July 2024
19. Raina, G., Sinha, S.: Study of PV systems using fin augmentation in real outdoor condition. In: 2019 8th International Conference on Power Systems (ICPS), pp. 1–6, December 2019. https://doi.org/10.1109/ICPS48983.2019.9067710
20. Hasanuzzaman, M., Malek, A.B.M.A., Islam, M.M., Pandey, A.K., Rahim, N.A.: Global advancement of cooling technologies for PV systems: a review. Sol. Energy **137**, 25–45 (2016). https://doi.org/10.1016/j.solener.2016.07.010
21. Meeus, J.: Astronomical Algorithms, 2nd edn. Willmann-Bell, Richmond (1998)
22. Bouchekif, A.: Algeria Weather History—Weather Underground. https://www.wunderground.com/history/daily/dz/a%C3%AFn-bouchekif/DAOB/date/2023-7-12. Accessed 15 July 2024

Intelligent Adaptive Neural Network Controller for Speed Control of PMSM in Electric Vehicles

Slimani Amira[1(✉)], Bourek Amor[1], Ammar Abdelkarim[2], and Kakouche Khoudir[3]

[1] Electrical Engineering Laboratory of Biskra, LGEB University of Biskra, Biskra, Algeria
{amira.slimani,a.bourek}@univ-biskra.dz

[2] Signals and Systems Laboratory of Boumerdes, LSS University of Boumerdes, Boumerdes, Algeria
a.ammar@univ-boumerdes.dz

[3] Faculté de Technologie, Laboratoire de Technologie Industrielle et de l'Information, Université de Bejaia, 06000 Bejaia, Algeria

Abstract. Precise velocity control is crucial for various systems, such as aircraft, power grids, industrial robots, and electric cars. Traditional control approaches sometimes face challenges such as real time limits, nonlinearities, and complex system dynamics. MPCC-NN, a combination of neural networks and model predictive control (MPC), offers a promising solution to these issues by leveraging the flexibility of neural networks and the predictive capabilities of MPC. This approach ensures enhanced accuracy in speed monitoring. Two types of controllers are compared in this study: proportional-integral (PI) controllers and neural network controllers. The FCSMPCC system is used to control the speed of permanent magnet synchronous motors (PMSMs). Variations in elements such as load and speed can compromise the effectiveness of PI controllers, which typically exhibit satisfactory performance during stable operating conditions. We utilize MATLAB/Simulink to execute the simulation and address these constraints. The results indicate that NN controllers have gained popularity due to their ability to make accurate predictions, reduce overshoot, and allow for rapid responses to changes in speed.

Keywords: Electric Vehicles · Finite Set Model Predictive Speed Control · PI controller · Neural Network Controller · PMSM

1 Introduction

In recent times, air pollution has gained prominence as a worldwide concern, exerting a substantial influence on both the environment and human well-being. Technology, particularly in the transportation industry, has exacerbated the problem. Proponents have proposed EVs as a viable alternative to traditional fuel-powered cars to reduce emissions. Electric to their numerous benefits, such as reduced air pollution and comparable performance to traditional vehicles, electric vehicles have garnered significant scientific attention. PMSMs have received significant attention and research in the electric car industry. Dies is largely due to its compact design, high torque-to-current ratio, high power density, and exceptional work efficiency [1, 2].

© The Author(s), under exclusive license to Springer Nature Switzerland AG 2025
M. Hatti (Ed.): IC-AIRES 2024, LNNS 1238, pp. 372–382, 2025.
https://doi.org/10.1007/978-3-031-80301-7_41

The primary control methods used for PMSM-drive systems include field-oriented control (FOC) [3], direct torque control (DTC) [4, 5], and direct flux control (DFC) methodology. The torque response offered by DTC is superior to that of FOC and DFC, making it a more suitable control algorithm for PMSM [7]. Nevertheless, there are certain drawbacks associated with DTC, including inconsistent variations in switching frequencies and torque, as well as changes in flux and current, particularly when operating at lower speeds. Utilizing contemporary control technologies could resolve these problems. In order to enhance the efficiency of PMSM drive systems, researchers are still looking into advanced and effective control mechanisms. These include sliding-mode control (SMC) [8], adaptive control [9], linearization control [10], and predictive control

[11, 12]. The study of power-electronic devices has led to a rapid development in discrete-based digital control technology, as the input signals for these devices are discrete. Simultaneously, the field of digital signal processing has seen significant advancements in computing power, making predictive control a central point of attention.

Currently, people widely acknowledge Model Predictive Control (MPC) as a direct and effective method for managing power converters and electric motors. The benefits of this approach are their ease of integration in systems with several variables, their ability to handle nonlinearities and limitations, and their exceptional dynamic performance. For PMSM, there are two main categories of MPC techniques: finite-control-set-model predictive control (FCS-MPC) [13] and continuous-control-set-model predictive control (CCS-MPC). The main differentiation between these two techniques is whether the control system requires a modification unit. Unlike CCS-MPC, FCS-MPC does not necessitate modulation techniques, making it well-suited for real-time optimization. FCS-MPC can be classified into four unique types, according to the type of prediction: predictive current control (FCS-MPCC), predictive torque control (FCS-MPTC), predictive speed control (FCS-MPSC), and predictive flux control (FCS-MPFC).

This study presents an FCS-MPCC-system for an EV-application that uses a PMSM. We perform a comparative analysis between a PI- controller and a neural network speed controller for the external loop of the PMSM. A neural network controller improves the control method's resilience. The simulation results confirm the efficacy of this strategy. We have organized the paper as follows: Section 2 outlines the mathematical model of the PMSM and the electric vehicle. Section 3 discusses how the FCS-MPCC is used in the inner control loop. In addition, it examines the basics of outer-loop-neural-network-speed control. Section 4 displays the simulation findings, while Section 5 provides the conclusion.

2 Drive Model

2.1 PMSM Model

The mathematical representation of a PMSM in the rotating (d-q) reference frame can be represented as:

$$\begin{bmatrix} U_{sd} \\ U_{sq} \end{bmatrix} = \begin{bmatrix} L_d & 0 \\ 0 & L_q \end{bmatrix} \begin{bmatrix} \dot{I}_{sd} \\ \dot{s}^*_{sq} \end{bmatrix} + \begin{bmatrix} R_s & -L_q\omega_e \\ L_d\omega & R_s \end{bmatrix} \begin{bmatrix} I_{sd} \\ I_{sq} \end{bmatrix} + \begin{bmatrix} 0 \\ \varphi_f\omega_e \end{bmatrix} \tag{1}$$

where *Usd* and *Usq* are the stator input voltage, *Rs* is stator resistance, *Isd and Isq* are the stator currents, φ*f* is flux linkages, *Ld* and *Lq* are the stator inductances, respectively. Additionally, ω*e* is the rotor electrical angular speed, calculated as (ω*e* = p ∗ ω*m*), *p* is the pole pairs, and ω*m* is the rotor mechanical angular speed. The PMSM's dynamical model can be written as follows:

$$\frac{d\omega_m}{dt} = \frac{1}{J}(Te - B\omega_m - T_L) \tag{2}$$

where *B* is the friction factor, *J* is the inertia of PMSM *TL* is the load torque, *Te* is the electromagnetic torque, which can be derived as:

$$Te = \frac{3}{2} I_{sq} p \left[\varphi_f + (L_d - L_q) I_{sd} \right] \tag{3}$$

In particular, for a surface-mounted PMSM where *Ld* equals *Lq*, the electromagnetic torque can be rewritten as:

$$Te = \frac{3p\varphi_f}{2} I_{sq} \tag{4}$$

2.2 Dynamic Model of Electrical Vehicle

Calculating the forces applied to the vehicle's surface during acceleration creates the dynamic model of an EV. Figure 2 illustrates the forces acting on the EV surface.

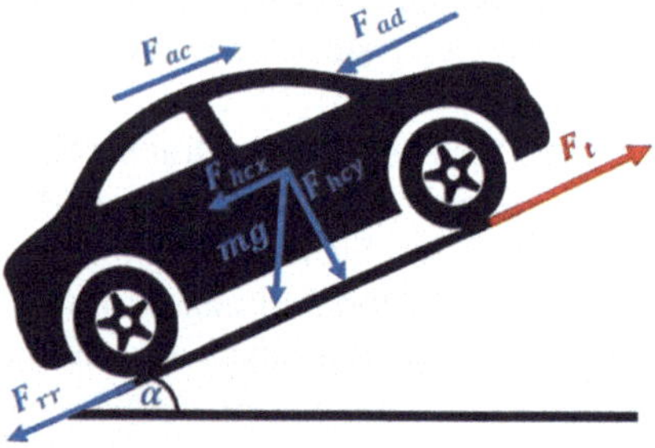

Fig. 1. Forces acting on the vehicle.

The total force *Ft is* written as:

$$F_t = F_{rr} + F_{ad} + F_{hc} + F_{ac} \tag{5}$$

$$F_t = \mu_{rr} mg \cos(\alpha) + \frac{1}{2}\rho C_d A V^2 + mg \sin(\alpha) + Im\frac{dV}{dt} \tag{6}$$

where *Ft* is the total tractive force of the vehicle, *Frr* is the force of rolling resistance due to road and tire interaction, *Fad* is the aerodynamic drag force, *Fhc* is the hill climbing force, and *Fac* is the acceleration force generated from the vehicle mass. Additionally, μrr is the coefficient of rolling resistance, *m is* the vehicle mass, and *g* is the acceleration due to gravity. *Cd*is the drag coefficient, ρ is the air density, *A* stands for the frontal area, *V* is the speed of the vehicle, α denotes the slope of the road, and *I* is the rotational inertia coefficient.

The total tractive torque is directly proportional to the tractive force according to Eq. (7).

$$T_t = F_t r \tag{7}$$

You can calculate the traction motor torque as follows:

$$T_m = \frac{F_t r}{\eta_g G} \tag{8}$$

where *G* is the transmission ratio, ηg is the gear efficiency, and r is the radius of the drive wheels. Table 1 represents the EV parameters [14].

Table 1. Parameters of the electric vehicle model.

Parameters	Values
Mass of the vehicle ***m*** *(kg)*	150
coefficient of rolling resistance μ***rr***	0.015
stands for the frontal area ***A*** (m^2)	1
Tire radius *r* (m^2)	0.23
Drag coefficient ***Cd***	0.25

3 Control Strategy

3.1 Inner Control Loop

The internal control loop is responsible for overseeing the electrical system and regulating the machine's operation. In this scenario, we utilize the FS-MPCC to track the quadrature and direct current references produced for the external loop. Due to the discontinuous nature of the external loop, the control platform necessitates the use of FCS-MPC for prediction. We use a discrete model to represent the control states in the subsequent sample. An apparatus is necessary. The continuous model of the PMSM [13] discretizes equation (1) through the first-order step.

Equations (9) and (10) present the prediction model for the stator d-axis and q-axis currents at the next moment.

$$I_{sd}(k+1) = \left(1 - \frac{R_s T_s}{L_d}\right) I_{sd}(k) + T_s \omega_e(k) i_{sq}(k) + \frac{T_s}{L_d} U_{sd}(k) \tag{9}$$

$$I_{sq}(k+1) = \left(1 - \frac{R_s T_s}{L_d}\right) I_{sq}(k) - \mathrm{T}_s \omega_e(k) I_{sd}(k) - \frac{\mathrm{T}_s \varphi_f}{L_d} \omega_e(k) + \frac{\mathrm{T}_s}{L_d} U_{sq}(k) \quad (10)$$

where Ts is the sampling period of the inner control loop.

You can use the following relation to calculate the cost function $gMPCC$:

$$g_{MPCC} = \left|I^*_{sd} - I_{sd}(k+1)\right| + \left|I^*sq - I_{sq}(k+1)\right| \quad (11)$$

where I^*sd and I^*sq are the reference currents.

3.2 Outer Control Loop

This research presents a neural network-based method as a favorable substitute for the PI speed controller. This method offers benefits such as inherent resilience and improved efficiency in accurately following the speed reference of electric vehicles with a rapid reaction. We optimized the ANN training by using data from the input and output of a conventional PI controller. The graphic below (Figure 2) illustrates the block diagram of a PMSM utilizing an ANN controller.

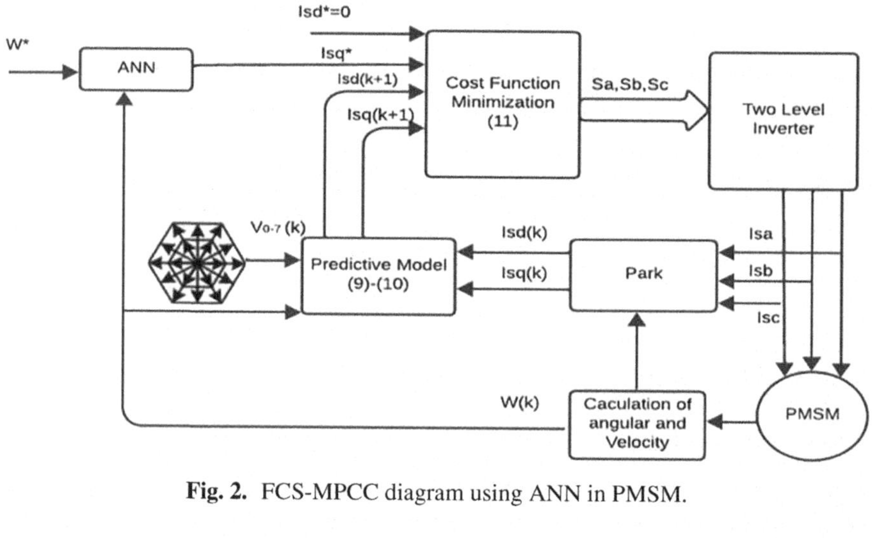

Fig. 2. FCS-MPCC diagram using ANN in PMSM.

Designing ANN models:

The process of designing ANN models entails a sequence of methodical operations, as illustrated in Figure 3. The technique typically involves five fundamental stages: (1) data collection, (2) data preprocessing, (3) network construction, (4) network training, and (5) model performance evaluation.

- ***Data Collection:*** The initial stage in constructing ANN models involves the collection and preparation of sample data. [15]

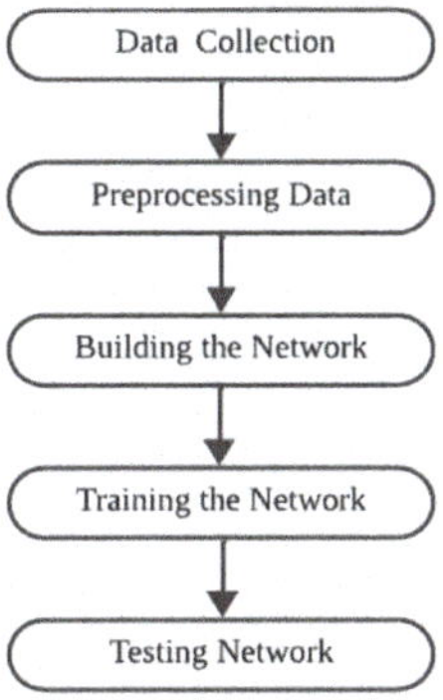

Fig. 3. The fundamental process for creating a model of an artificial neural network.

- ***Data Preprocessing:*** We carry out three preprocessing operations after data collection to improve the training of the artificial neural networks (ANNs). Begin by filling out any missing information. Subsequently, standardize the data. Ultimately, randomize the data.
- ***Building the Network:*** At this point, the designer determines the quantity of concealed layers, as depicted in Fig. 4. The structure consists of a single input, the speed error, and a single output (Te^*). In addition, the designer possesses the power to choose the quantity of neurons in each layer, the transfer function, the training function, the weight/bias learning function, and the performance function.

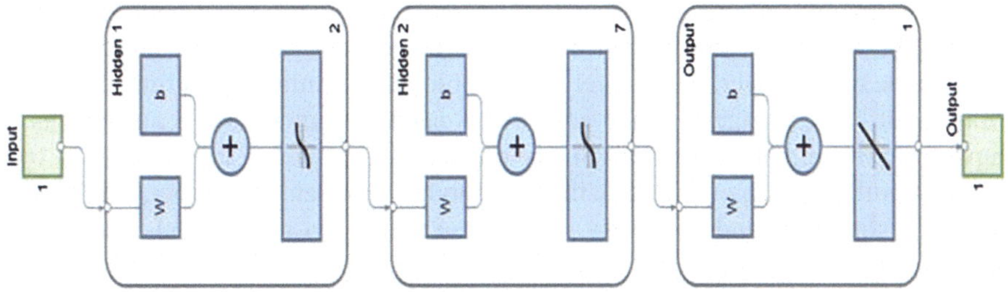

Fig. 4. ANN architecture of speed controller.

- ***Training the Network***: We modify the weights during the training phase to ensure that the predicted outputs closely align with the target (measured) outputs. Figure 5 displays a screenshot depicting the current status of training for the neural network.

- ***Testing the network:*** This stage involves evaluating the overall effectiveness of the developed model using previously unobserved data. We quantitatively evaluate the model's efficacy by using statistical measures like root mean square error (RMSE) and mean bias error (MBE). RMSE quantifies the extent of deviation between projected values and observed data, serving as a measure of short-term performance. A lower

Training Progress

Unit	Initial Value	Stopped Value	Target Value
Epoch	0	1000	1000
Elapsed Time	-	01:00:47	-
Performance	2.41e+03	0.00153	0.0001
Gradient	2.82e+04	0.502	1e-07
Mu	0.001	0.0001	1e+10
Validation Checks	0	0	6

Training Algorithms

Data Division: Random dividerand

Training: Levenberg-Marquardt trainlm

Performance: Mean Squared Error mse

Calculations: MEX

Fig. 5. Neural network training state.

RMSE number indicates a higher level of accuracy. The mean bias error (MBE) offers valuable information about long-term performance by quantifying the average difference between anticipated values and actual measurements. A lower MBE value indicates a higher level of accuracy in long-term predictions.

Programming ANN Models:

- The symbol I denotes data input, the symbol T denotes target value, and the symbol net denotes an artificial neural network. The newff function creates a feed-forward neural network. The minmax(I) function calculates the minimum and maximum values of the input data for training purposes. In this case, the network has 1 node in the input layer, 2 nodes in the first hidden layer, 7 nodes in the second hidden layer, and 1 node in the output layer, respectively. We employ the training algorithm, Levenberg-Marquardt backpropagation, and adjust it as needed to meet task requirements. To represent the number of iterations during training, we set TrainParam.epochs to 1000. We set trainParam.target to reduce the error, and we use net = train (net, I, T) to begin training using input and output data. When the regression value equals one (R = 1), as shown in Fig. 6, it indicates successful training. A measure of the approximation between output and target guarantees that the ANN has been successfully trained. Simulink then incorporates the trained ANN from the workspace.

4 Simulation Results and Discussion

Simulation is carried out on a PMSM in the MATLAB/Simulink environment. To further validate the effectiveness and superiority of the proposed ANN speed controller for FSC-MPCC in PMSM for EV applications, the PMSM nominal parameters are displayed in Table 2.

In Fig. 7(a), when using the FCS-MPCC with a PI speed controller, the recorded rotational speed exhibits variations based on the reference. However, during startup and speed transitions, there is overshooting at the speed change instances. This overshoot

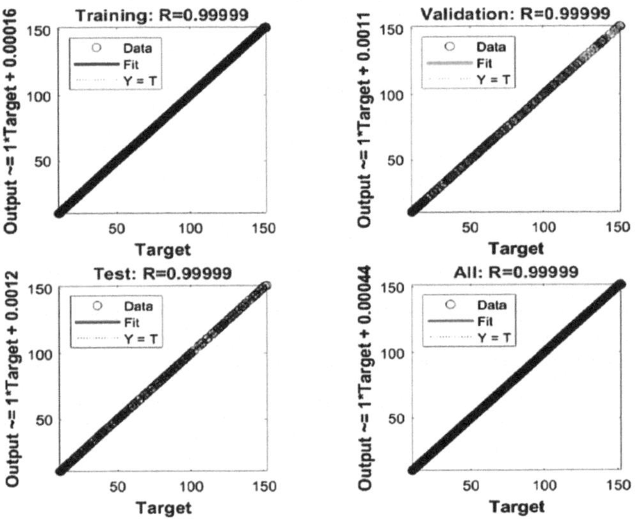

Fig. 6. Graph represents the network validation for ANN.

Table 2. PMSM NOMINAL PARAMETERS

Parameters	Values
Stator inductance *Ld, Lq (H)*	0.0076
Stator resistance *Rs* (Ω)	2.3
Friction coefficient *B (N.m.s)*	0.000169
Moment of inertia *J (kg.* m^2*)*	0.0032
Flux linkages φ***f*** (***Wb***)	0.4
Number of pole pairs *p*	4

Table 3. Performance comparison of various controllers

Controller Speed	PI	ANN
Peak Overshoots Mp (%)	13	0
Settling Time Ts (Sec)	0.05	0
Steady State ess (rpm)	0	0

results in an increase in response time and even a delay in the system. In contrast, with the ANN speed controller, the speed precisely follows its reference without overshooting. Additionally, the speed and reference exhibit strict central symmetry. The ANN speed controller provides a more accurate and stable response, ensuring that the PMSM operates efficiently without the drawbacks seen with the PI speed controller, such as overshooting and delayed response time during critical transitions. Figures 7b, 7c, and 7d illustrate the electromagnetic torque and stator current *I sd* and *Isq* when using the FCS-MPCC with PI and ANN controllers. When employing the PI speed controller, oscillations in stator current are observed. Therefore, it will lead to an increase in ripples about the torque because there is a positive relationship between current and torque in contrast, the proposed MPCC with ANN speed controller exhibits superior performance, showing minimal current ripples, high tracking accuracy, and reduced dynamic overshoot.

These results validate the effectiveness of the ANN speed controller in enhancing overall system performance compared to traditional PI controllers in PMSM applications. By ensuring precise speed tracking and minimizing undesirable oscillations in stator current, the use of the ANN speed controller greatly enhances the dynamic performance and operational efficiency of PMSM drives. This demonstrates the potential of ANN-based control strategies to offer substantial benefits over conventional control methods in electric vehicle.

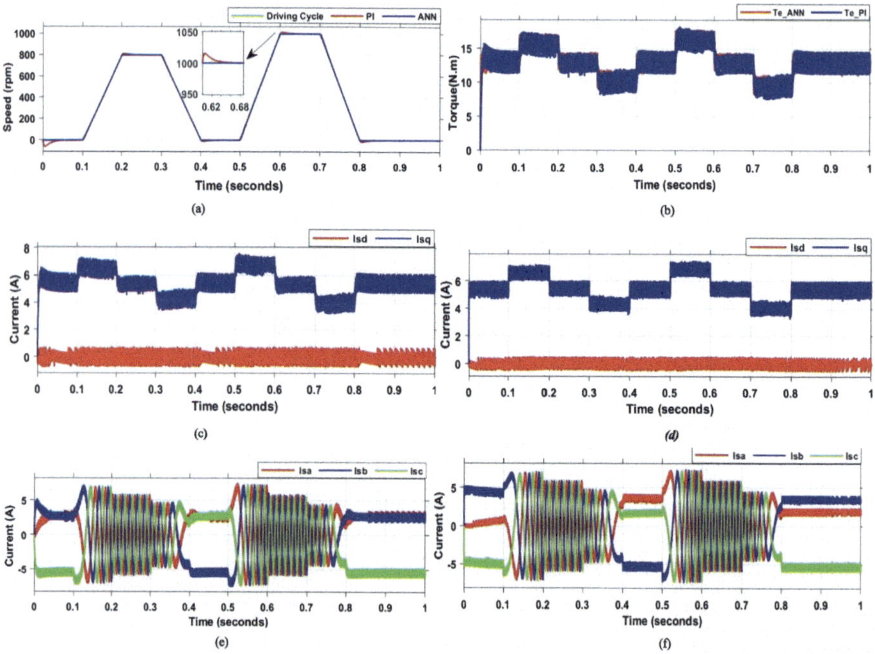

Fig. 7. PMSM performance of the FSC-MPCC using PI and ANN with speed variation: (a) rotor speeds, (b) electromagnetic torque, (c) stator currents with PI controller (*I sd*and *Isq*), (d) stator currents with ANN controller (*I sd*and *Isq*), (e) stator currents with PI controller (*Ia, Ib,Ic*),*and* (*f*) stator currents with ANN controller (*Ia,Ib, Ic*).

Table 3 illustrates a comparison among the two speed controllers at ***TL***= 13 Nm, and Speed equal 1000 rpm.

5 Conclusion

This research presents the control architecture of FCS-MPCC for a PMSM, which combines a PI controller with an ANN controller in the outer rotor speed control loop. The ANN controller exhibits higher performance compared to the typical PI controller in terms of speed tracking and current regulation for PMSM drivers. It efficiently avoids the problem of exceeding the desired speed during the first phase and when changing speeds, guaranteeing precise and consistent speed responses that closely align with the intended path. On the other hand, the PI controller demonstrates a propensity for overshooting. In addition, the ANN controller reduces current fluctuations, resulting in a smoother and more economical operation compared to the oscillatory behavior often observed with the PI controller. These developments are crucial for fulfilling the ever-changing requirements of electric vehicle (EV) applications, where accurate control and rapid response times are necessary. To summarize, the ANN speed controller provides a highly efficient solution for modern electric vehicle applications, improving overall system performance and dependability.

References

1. Chen, H., He, J., Demerdash, N.A.O., Guan, X., Lee, C.H.T.: Diagnosis of open-phase faults for a five-phase PMSM fed by a closed-loop vector-controlled drive based on magnetic field pendulous oscillation technique. IEEE Trans. Ind. Electron. **68**, 5582–5593 (2021)
2. Rassõlkin, A., Kallaste, A., Orlova, S., Gevorkov, L., Vaimann, T., Belahcen, A.: Re-use and recycling of different electrical machines. Latv. J. Phys. Tech. Sci. **55**, 13–23 (2018)
3. Xiaoni, C., et al.: A ripple suppression of sensorless FOC of PMSM electrical drive system based on MRAS. Sci. Direct Res. Eng. **56**(4), 2590–1230 (2023)
4. Vaez-Zadeh, S.: Analysis of a DTC with back EMF oriented voltage for PMS motor drives. IEEE Trans. Energy Convers. **33**(3), 1594–1596 (2018)
5. Sun, X.D., Diao, K.K., Lei, G., Guo, Y.G., Zhu, J.G.: Direct torque control based on a fast modeling method for a segmented-rotor switched reluctance motor in HEV application. IEEE J. Emerg. Sel. Topics Power Electron. **9**(1), 232–241 (2021).
6. Emanuele, G., Marco, P., Fabio, C., Matthias, N., Francesco, C., Francesco, G.: Detection of stator turns short-circuit during sensorless operation by means of the Direct Flux Control technique. In: Proceedings of the 2020 AEIT International Annual Conference (AEIT), Catania, Italy, pp. 1–6 (2020)
7. Kakouche, K., Oubelaid, A., Mezani, S., Rekioua, D., Rekioua, T.: Different control techniques of permanent magnet synchronous motor with fuzzy logic for electric vehicles: analysis, modelling, and comparison. Energies **16**(7), 3116 (2023)
8. Yuan, L., Jiang, Y., Xiong, L., Wang, P.: Sliding mode control approach with integrated disturbance observer for PMSM speed system. CES Trans. Electr. Mach. Syst. **7**(1), 118–127 (2023). https://doi.org/10.30941/CESTEMS.2023.00009
9. Tran, H.N., Jeon, J.W.: Robust speed controller using dual adaptive sliding mode control (DA-SMC) method for PMSM drives
10. Peng, J., Yao, M.: Overview of predictive control technology for permanent magnet synchronous motor systems. Appl. Sci. **13**(10), 6255 (2023)
11. Garcia, C., Silva, C., Rodriguez, J., Zanchetta, P.: Cascaded model predictive speed control of a permanent magnet synchronous machine. IEEE Trans. Power Electr. **PP**(99), 1 (2016)

12. Miranda, H., Co´rtes, P., Yuz, J., Rodrı́guez, J.: Predictive torque control of induction machines based on state-space models. IEEE Trans. Indust. Electr. **56**(6), 1916–1924 (2009)
13. Mohan, H., Agrawal, G., Jately, V., Sharma, A., Azzopardi, B.: Neural network-driven sensorless speed control of EV drive using PMSM. Energies **11**(5), 4029 (2023)
14. Mebarki, N., Rekioua, T., Mokrani, Z., Rekioua, D., Bacha, S.: PEM fuel cell/ battery Storage System Supplying electric vehicle. Sci. Direct 13(1), 0360–3199 (2016)
15. Raveendra Reddy, V., Veera Reddy, V.C., Chandra Jagan Mohan, V.: Speed control of induction motor drive using artificial neural networks- levenberg-marquardt backpropogation algorithm. Int. J. Appl. Eng. Res. ISSN **13**(1), 4562 (2018)

MTPA Flux Weakening Approach for IPMSM Motor Powertrain Used in Vehicle Propulsion System

Aouadj Norediene[1(✉)], Hartani Kada[2], Hammou Meriem[1,2], and Bouziane Yassmine[1]

[1] Second Cycle Department, Higher School in Electrical and Energy Engineering (ESGEE), Oran, Algeria
aouadj.norediene@esgee-oran.dz

[2] Electrotechnical Engineering Laboratory, University Tahar Moulay of Saida, Saida, Algeria

Abstract. The IPMSMs are used in electric vehicles because of their intense power density. The high speed implies a significant electromotive force and requires flux weakening. The usual control algorithms realize flux weakening by adding a negative Id current component when the voltage required by the current regulation exceeds the maximum voltage. If the magnet can be totally defluxed then it is better to use a Maximum Torque Per Ampere (MTPA) strategy. Furthermore, there is no speed regulation in the control and the driver gives a torque reference. This reference value has to be limited by the attainable operating points therefore the battery power limit has to be taken into account in addition to the voltage and current limits. The d-q current references are calculated to minimize the total current magnitude required to reach the reference torque. Also, speed control loop with IP controller is also introduced to the system for speed regulation.

This paper proposes a strategy MTPA-FW to control an IPMSM operating continuously, in order to take into account the different limitations (current, voltage, power) and to expand the over-speed zone. The feasibility of the system is confirmed by Simulink model, the results of which are presented in the end of paper.

Keywords: Maximum torque per ampere (MTPA) · flux weakening (FW) · interior permanent magnet synchronous motor (IPMSM) · electric vehicle (EV) · vector control

1 Introduction

The automobile industry's present and future are defined by electric vehicles, which are recognized for their eco-friendly attributes. These vehicles not only produce lower noise levels but also outperform traditional thermal engines while benefiting from the lower costs associated with electrical energy compared to oil-based fuels. Consequently, electric propulsion necessitates motors that can deliver high torque at low speeds while maintaining a broad speed range with consistent power [1].

© The Author(s), under exclusive license to Springer Nature Switzerland AG 2025
M. Hatti (Ed.): IC-AIRES 2024, LNNS 1238, pp. 383–394, 2025.
https://doi.org/10.1007/978-3-031-80301-7_42

In recent decades, there has been considerable interest in interior permanent magnet synchronous machines (IPMSMs) for a variety of industrial uses, particularly in the realm of electric vehicle (EV) propulsion systems. The advantages of IPMSMs stem from their remarkable efficiency, high power density, extensive speed range, quick torque-speed response, and the decreasing costs associated with permanent-magnet (PM) materials [2, 3]. For applications such as EVs, machine tools, and washing machines, IPMSM-drive systems typically require both a wide speed range and high efficiency, along with a favorable torque to inertia ratio, especially at lower speeds [4].

Recent studies [5, 6] have placed greater emphasis on the investigation of drives that manage permanent-magnet synchronous motors (PMSMs) for the previously discussed applications. Typically, motor control methods can be classified into two categories: scalar control and vector control. These methods, which have been utilized for managing induction machines, are also applicable to PMSMs [7]. Within the realms of scalar control and vector control, there exist three primary control algorithms [6]:

- Voltage/hertz control;
- Field-oriented control (FOC);
- Direct torque control (DTC)

Scalar control, often referred to as volts/hertz control, is typically utilized in straightforward applications such as pumps and fans. This method operates Permanent Magnet Synchronous Motors (PMSMs) by preserving a consistent ratio between volts and hertz, which guarantees stable airgap flux. It is a well-established and economical technique used across various industries. Nevertheless, its effectiveness in dynamic settings is frequently inadequate. This shortcoming arises because the method primarily emphasizes regulating voltage and frequency levels, while overlooking the current's magnitude and phase [7]. In contrast, advanced vector controls like Field-Oriented Control (FOC) and Direct Torque Control (DTC) offer enhanced precision and faster responses in dynamic environments. The core principle of FOC is its capacity to transform the three-phase, time-varying stator currents into two corresponding DC components, which facilitates the easy decoupling and independent regulation of torque and flux [6]. On the other hand, DTC employs the stator voltage space vectors to directly control both stator flux and torque [8], relying on the differences between the reference values and the estimated results for torque and flux.

The literature features numerous studies that outline control strategies designed to achieve satisfactory motor performance under various conditions. One notably effective technique, referred to as maximum torque per ampere (MTPA) control [9], is widely used in industrial settings due to its capacity to provide high torque output, optimize DC bus usage, and reduce copper losses. For the high-speed operation of IPMSM drives, implementing flux-weakening (FW) control is crucial, as it broadens the motor's speed range by decreasing the flux density and flux within the air gap [10]. In traction applications for electric vehicles, IPMSM drives are developed to sustain a constant drive torque up to a designated base speed, after which the torque diminishes in relation to speed until the peak speed is reached [11]. Consequently, at lower speeds, the MTPA method is employed by IPMSM drives, while the FW method is utilized at elevated speeds [6].

The principle of MTPA control aims to maximize the torque of a PM motor for a specified phase current by preserving a particular torque angle. This angle represents the relationship between the current phasor and the positive d-axis direction in the d-q frame. To put it more simply, a precise combination of d-axis and q-axis currents exists that yields the minimum phase current required for any torque level. As a result, MTPA control improves the motor drive's efficiency by decreasing copper losses through a reduction in phase current [6].

The core principle of FW control in PMSMs hinges on the idea that the air-gap flux can solely be diminished by introducing a demagnetizing armature current component along the d-axis of the permanent magnets. As the output torque is directly linked to the product of the air-gap flux and the q-axis current component, FW control effectively lowers this torque [12]. In [13], the author presented a six-step voltage control technique aimed at maximizing the utilization of the dc link voltage for FW operation. Another approach outlined in [14] introduced a voltage compensator paired with a current regulator that features a feedforward decoupling controller for FW operation. The first appearance of the FW algorithm utilized in AC motor drives was in Kim's paper [15]. This method is distinguished by two significant benefits: its straightforward implementation and its robustness against variations in machine parameters.

This paper introduces a control algorithm for FW that is specifically designed for IPMSMs. A key aspect of the proposed algorithm is its ability to overcome the constraints of dc bus voltage during FW operation. To showcase the benefits and effectiveness of this control method, theoretical analysis along with simulations will be carried out. As a result, the potential for improved efficiency remains evident. The effectiveness of the proposed control system is validated through simulations performed in MATLAB/Simulink.

2 IPMSM Mathematical Model

The model of the IPMSM in the reference frame $d - q$ is described by the following equations [16, 17]. The flux linkage of the stator is comprides of:

The equations for voltage are expressed as:

$$v_d = R_s i_d + L_d \frac{di_d}{dt} + \omega_e L_q i_q \tag{1}$$

$$v_q = R_s i_q + L_q \frac{di_q}{dt} + \omega_e \left(\Phi_f + L_d i_d\right) \tag{2}$$

Flux Linkages are given by

$$\Phi_d = \Phi_f + L_d i_d \tag{3}$$

$$\Phi_q = L_q i_q \tag{4}$$

The developed motor torque is

$$T_{em} = \frac{3}{2} p\left(\Phi_d i_q - \Phi_q i_d\right) \tag{5}$$

The mechanical torque can be expressed as

$$T_e = T_L + f\omega_m + J\frac{d\omega_m}{dt} \tag{6}$$

Calculating the mechanical speed of the rotor results in

$$\omega_m = \int \frac{T_e - T_L + f\omega_m}{J} dt \tag{7}$$

where:

$$\omega_m = \omega_e / p \tag{8}$$

In this context, we refer to rotor electrical speed as ω_e and mechanical speed as ω_m.

3 Formulation of MTPA Conditions for FOC

To determine the best torque output corresponding to a particular current, it is crucial to analyze Eq. (5), which illustrates the electromagnetic torque produced by the motor. The currents in the d and q axes can be represented as [6, 18]:

$$i_d = \hat{I}_s \cos\beta \tag{9}$$

$$i_q = \hat{I}_s \sin\beta \tag{10}$$

In Fig. 1 [18], the "torque angle" β ($180^\circ \succ \beta \geq 0^\circ$) is described as the angle formed between the vector of the stator phase current and the positive d-axis direction, while $\hat{I}_s$ denotes the magnitude of the stator phase current in an IPM motor.

The torque generated in IPMSM motors can be understood as the sum of the excitation torque from the magnet, T_{mag}, and the reluctance torque, T_{rel}, as illustrated in Fig. 2. This phenomenon occurs due to the q-axis reactance/inductance being greater than that of the d-axis, leading to the condition $L_q \succ L_d$. The formula for the developed torque is presented as follows:

$$T_e = T_{mag} + T_{rel} \tag{11}$$

where

$$T_{mag} = \frac{3}{2} p \Phi_f \hat{I}_s \sin\beta \tag{12}$$

$$T_{rel} = \frac{3}{2} p (L_d - L_q) \hat{I}_s^2 \frac{\sin 2\beta}{2} \tag{13}$$

In the constant torque domain of an on-board interior permanent magnet synchronous motor (IPMSM), optimal torque output is achieved through a specific combination of d and q-axis currents. This approach, known as maximum torque per ampere (MTPA), is utilized to meet the significant torque demands of traction applications, including

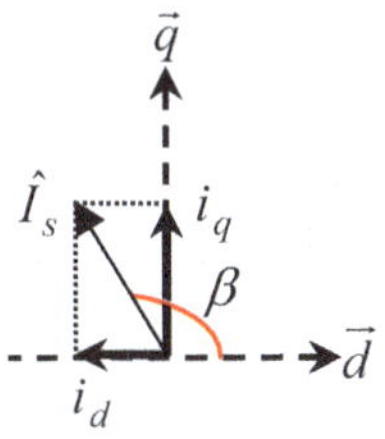

Fig. 1. Torque angle.

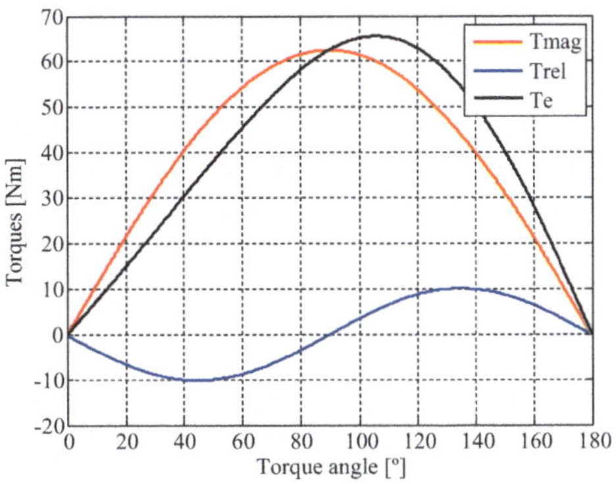

Fig. 2. Developed torque in relation to the torque angle of the IPMSM.

electric vehicles (EVs). When operating below the base speed—the maximum torque achievable within this constant torque region—the motor does not encounter any voltage stress. Consequently, the torque capacity in this region is determined solely by the current limitation. The q-axis current can be mathematically represented as shown in [19]:

$$i_q = \sqrt{i_s^2 - i_d^2} \tag{14}$$

Replace (14) with (5):

$$\frac{\partial C_{em}}{\partial i_d}\left(i_{d,MTPA}\right) = 0 \tag{15}$$

The d -axis current during MTPA operation can be determined in the following manner:

$$i_{d,MTPA} = \frac{\Phi_f}{4\left(L_d - L_q\right)} - \sqrt{\frac{\Phi_f^2}{16(L_d - L_q)^2} + \frac{\hat{I}_s^2}{2}} \tag{16}$$

By replacing Eq. (16) in Eq. (14), one can easily obtain the -axis current for MTPA operation.

$$i_{q,MTPA} = \sqrt{\hat{I}_s^2 - i_{d,MTPA}^2} \tag{17}$$

While the MTPA control does not account for voltage limitations, the torque generated is constrained by the supply current. Each specific supply current corresponds to a unique combination of d-axis and q-axis currents that produces the maximum torque under those circumstances. Consequently, the torque angle during MTPA operation is influenced by both the motor parameters and the value of the supply current. As illustrated in Fig. 3, varying supply currents create distinct intersection points between the current circles and the MTPA trajectory, which results in different torque angles for the MTPA operations [6].

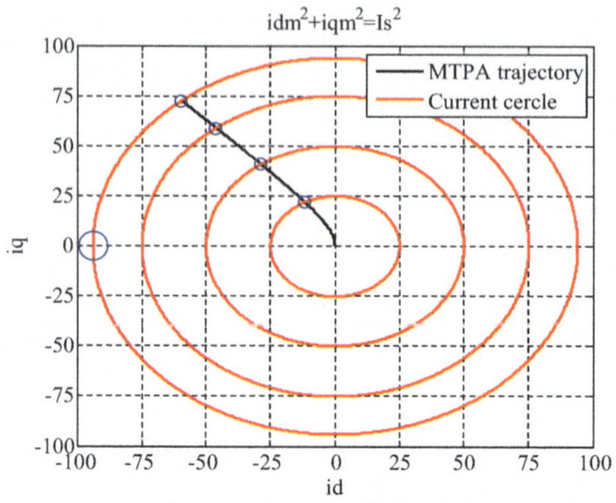

Fig. 3. The MTPA trajectory for IPMSM motors.

As the current falls below I_{sn}, there is a reduction in the corresponding point on the MTPA curve. At the base speed ω_b, the motor attains its peak current and voltage, determined by where the current limit circle intersects with the voltage limit ellipse. This particular intersection occurs at a specified point [20].

$$i_{d\max} = -\frac{\Phi_f}{4(L_d - L_q)} - \sqrt{\frac{\Phi_f^2}{16(L_d - L_q)^2} + \frac{\hat{I}_{\max}^2}{2}} \tag{18}$$

$$i_{q_{\max}} = \sqrt{\hat{I}_{\max}^2 - i_{d\max}^2} \tag{19}$$

We will than give:

$$\omega_b = \frac{V_{sn}}{p\sqrt{\left(\Phi_f + L_d.i_{d\max}\right)^2 + \left(L_q.i_{q_{\max}}\right)^2}} \tag{20}$$

The maximum current $I_{\max}$ serves as the most restrictive current limit, positioned between the motor's nominal current, the inverter's nominal current, and the peak current supplied by the battery, adjusted to reflect an equivalent motor current [6].

$$\sqrt{i_d^2 - i_q^2} \leq I_{\max} \tag{21}$$

4 Flux Weakening Control Strategy

This section outlines the control strategy that derives current references in the $d-q$ reference frame based on the torque reference. These current references are utilized in the motor current regulation through a decoupling approach [21]. Additionally, alternative high-bandwidth control strategies, such as predictive control or optimal control, may be employed for current regulation [1].

The constraints of the machine's operation are determined by its power supply limitations, specifically the battery. In this context, we establish a maximum voltage and a maximum current. To achieve the highest possible speed of the machine within these parameters, it is essential to set specific current references that produce maximum torque, considering that in a machine with salient poles, both the d and q axes contribute to torque generation [1].

The control scheme illustrated in Fig. 4 employs this strategy and is composed of three controllers: the current regulator, the speed regulator, and the FW regulator.

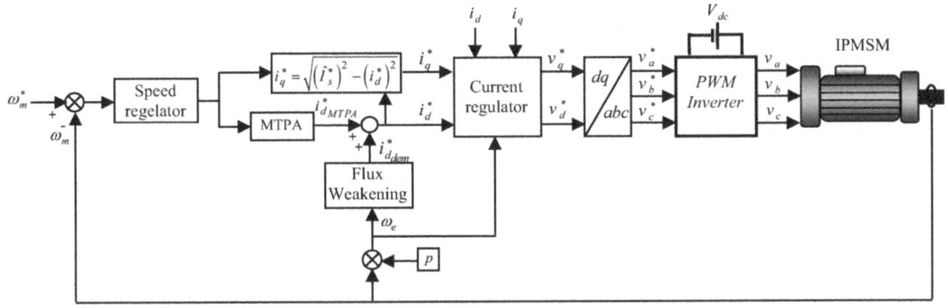

Fig. 4. Control strategy MTPA and FW configuration.

As long as the controlled motor is either in a stopped state or on standby, an initial commanded/reference speed ω_m^* is introduced into the control loop and subsequently compared to the feedback speed ω_m, which indicates the motor's actual speed. The speed controller receives the resulting error, calculated as the difference between the reference speed and the actual speed. Utilizing this speed error, the speed regulator produces a reference value for the amplitude of the phase currents $\hat{I}_s^*$. It is crucial to emphasize that this speed control method does not rely on a reference torque. To ensure maximum torque throughout the entire speed range, the reference phase current is set to its maximum value during both MTPA control and FW control, resulting in $\hat{I}_s^* = \hat{I}_{\max}$. Input for the MTPA block will come from Such $\hat{I}_s^*$, which generates the reference d-axis current component $i_{d_{MTPA}}$ used for MTPA control as detailed in (16). It is crucial to highlight that the FW regulator's output is zero until the base speed ω_{base} (20) is reached. As a result, within the constant torque region, the reference demagnetizing current $i_{d_{dem}}^*$ for FW control remains at zero. Since the reference d-axis current is expressed by the equation $i_d^* = i_{d_{MTPA}} + i_{d_{dem}}$, it follows that in this region, $i_d^* = i_{d_{MTPA}}$. The reference q-axis current can subsequently be derived from (17). In the constant power region, where speed surpasses the base value, all components in MTPA control remain operational. Nevertheless, the FW regulator starts to provide an additional demagnetizing current $i_{d_{dem}}^*$, which can be expressed as follows [6]:

$$i_{dem} = i_{d_{FW}} - i_{d_{MTPA}} \tag{22}$$

where $i_{d_{FW}}$ is the d-axis current component for FW control, which can be written as follows:

$$(L_d - L_q)^2 i_d^4 + 2\Phi_f (L_d - L_q) i_d^3 + \Phi_f^2 - (L_d - L_q)^2 \hat{\imath}_{\max}^2 i_d^2 - 2\Phi_f - (L_d - L_q)^2 \hat{\imath}_{\max}^2 i_d + 4P_b^2 / 9\omega_e^2 - \Phi_f^2 \hat{\imath}_{\max}^2 = 0 \tag{23}$$

Notice that the rated output power P_b, as follows [6] (Fig. 5):

$$P_b = \frac{3}{2}\left(\Phi_f \sqrt{\hat{\imath}_{\max}^2 - i_d^2} + (L_d - L_q) i_d \sqrt{\hat{\imath}_{\max}^2 - i_d^2}\right)\omega_e \tag{24}$$

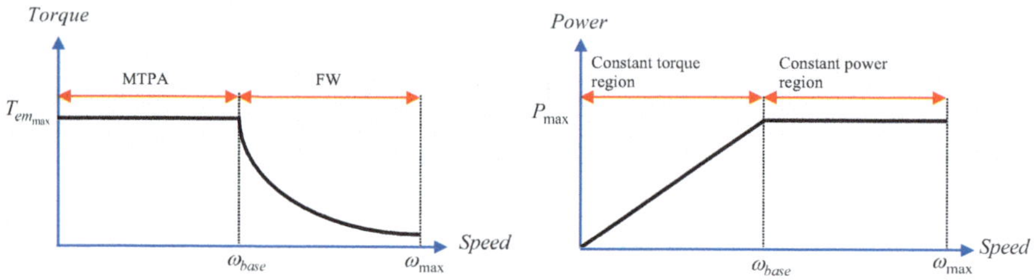

Fig. 5. IPMSM torque/speed and power/speed characteristic.

In this study, the speed regulator functions as an IP anti-windup controller. In contrast, the current regulator often requires the implementation of feed forward compensation in its design. This necessity arises from the inability to control the $d - q$ *axis* current components independently through v_d and v_q due to cross-coupling effects [14], such as $\omega_e L_d i_d$ and $\omega_e L_q i_q$ present in the expressions for v_d and v_q. The influence of cross-coupling effects is particularly significant in IPMSM motors, which feature relatively high inductances. These effects will affect both the current response and the torque response in the high-speed FW region.

Figure 6 illustrates the decoupled current regulator that incorporates feed forward compensation [14], introduced here to address this issue.

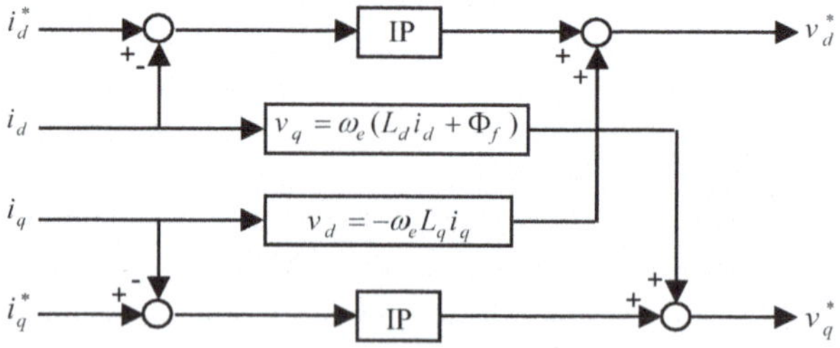

Fig. 6. Current regulator featuring decoupled feedforward control compensation.

5 Simulation Results

To confirm the efficacy of the MTPA-FW method introduced in this paper, a simulation model for flux weakening control of the IPMSM motor has been developed using Matlab/Simulink. The nominal parameters of the IPMSM utilized in the simulation are presented in Table 1.

Table 1. IPMSM parameters

Symbol	Quantity	Value
R_s	Resistance	0.45 Ω
L_d	Inductance direct	0.54 mH
L_q	Inductance quadratic	1.05 mH
Φ_f	Permanent magnet flux	0.148 Wb
p	Pole pairs	3
P_b	Rated power	30 kW

In order to assess the effectiveness of this simulation system at high speeds using the MTPA-FW control, a commanded mechanical speed of $\omega_m^* = 600$ rad/s was utilized as the input signal to the control loop, as illustrated in Fig. 7(a). Illustrated in Fig. 7(a), the motor's mechanical speed over time demonstrates a gradual increase from zero until it reaches stability. The discrepancy between the reference speed and the actual speed is merely 3.2%, indicating dependable controller performance at elevated speeds.

It is crucial to note that the IPMSM motor outlined in Table 1 is designed for high-speed applications. However, by employing flux-weakening methods, this motor type can achieve a wider speed range.

Figure 7(b) illustrates the torque generated in both the constant power region and the constant torque region. It is evident that the torque produced steadily decreases across the constant power region.

Displayed in Fig. 7(d) are the current components along the reference *d*- and *q*- axes. When examining the reference *d*- and *q*- axis current components shown in Figs. 7(e) and (f), it becomes evident that the current trajectories represented in Figs. 7(g) and (h) are in alignment with their reference values. It is apparent that the FW current corresponds with the MTPA current.

The simulation results provided indicate that the control system implemented demonstrates dependable performance in both the low-speed constant torque region and the high-speed constant power region. The discrepancy between the final speed of the motor and the reference speed is minimal. Although there are some torque fluctuations, the torque produced stays relatively constant at low speeds using MTPA control. Meanwhile, during flux-weakening (FW) control, the output power tends to maintain a steady value. Therefore, based on these simulation findings, one can conclude that this FW control algorithm is an effective method for flux-weakening control.

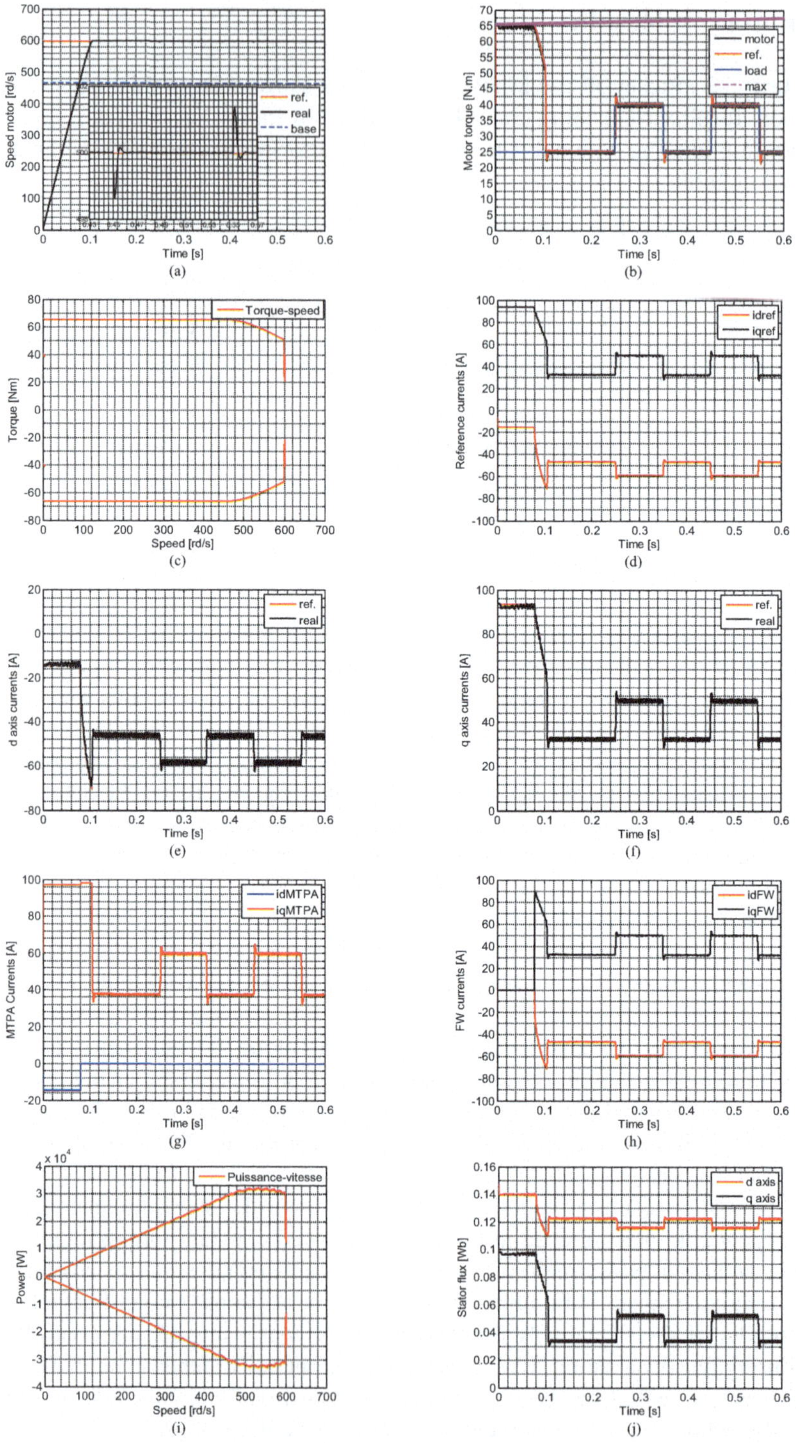

Fig. 7. Simulation results.

6 Conclusion

This paper presents a newly developed flux weakening and MTPA-based FOC scheme specifically designed for IPMSM motors powered by PWM inverters. The simulation's mathematical model of the IPMSM and its parameters are detailed. Additionally, the FW and MTPA methodologies are discussed. A key aspect of operating these machines at elevated speeds is the flux weakening method, which is integrated with a comprehensive control strategy. The authors introduce a relevant control approach defined within the reference frame, ensuring adherence to all constraints set by reference operating points, including voltage and current limits. Flux weakening is achieved through the regulation of the reference voltage as determined by current control. Furthermore, a control scheme encompassing the entire speed range has been developed, merging maximum-torque-per-ampere control (MTPA) for low-speed operation with flux-weakening control for high-speed scenarios.

The validity of the proposed flux-weakening control method was verified through simulation results of a motor-drive system model. As a result, this research demonstrates that the suggested control strategy, which incorporates flux weakening and MTPA, enhances the propulsion operation of the IPMSM on a global scale and significantly boosts energy conversion efficiency. Furthermore, this study affirms that the control strategy has been refined and that MTPA has heightened the effectiveness of the IPMSM within an over speed regime.

References

1. Sepulchre, L., Fadel, M., Pietrzak-David, M., Porte, G.: MTPV flux-weakening strategy for PMSM high speed drive. IEEE Trans. Ind. Appl. **54**, 6081–6089 (2018)
2. Zhang, Y., Güven, M.K., Guven, M.K., Chi, S., Illindala, M.: Experimental verification of deep field weakening operation of a 50-kW IPM machine by using single current regulator. IEEE Trans. Ind. Appl. **47**, 128–133 (2010)
3. Gieras, J.F.: Permanent magnet motor technology: design and applications: CRC press (2002)
4. Chi, S., Xu, L.: A special flux-weakening control scheme of PMSM-incorporating and adaptive to wide-range speed regulation. In: 2006 CES/IEEE 5th International Power Electronics and Motion Control Conference, pp. 1–6 (2006)
5. Pellegrino, G., Vagati, A., Guglielmi, P.: Design tradeoffs between constant power speed range, uncontrolled generator operation, and rated current of IPM motor drives. IEEE Trans. Ind. Appl. **47**, 1995–2003 (2011)
6. Muyang. Li.: Flux-weakening control for permanent-magnet synchronous motors based on Z-source inverters. The Degree of Master of Science, Faculty of the Graduate School, Marquette University (2014)
7. Kohlrusz, G., Fodor, D.: Comparison of scalar and vector control strategies of induction motors. Hungarian J. Indust. Chem. **39**, 265–270 (2011)
8. Le-Huy, H.: Comparison of field-oriented control and direct torque control for induction motor drives. In: Conference record of the 1999 IEEE Industry Applications Conference. Thirty-forth IAS Annual Meeting (Cat. No. 99CH36370), pp. 1245–1252 (1999)
9. Lee, K.-W., Lee, S.B.: MTPA operating point tracking control scheme for vector controlled PMSM drives. SPEEDAM **2010**, 24–28 (2010)
10. Jahns, T.M.: Flux-weakening regime operation of an interior permanent-magnet synchronous motor drive. IEEE Trans. Indus. Appl. 681–689 (198)

11. Qinghua, L.: Analysis, design and control of permanent magnet synchronous motors for wide-speed operation (2005)
12. Giri, F.: AC electric motors control: advanced design techniques and applications: John Wiley & Sons (2013)
13. Bose, B.K.: A high-performance inverter-fed drive system of an interior permanent magnet synchronous machine. IEEE Trans. Ind. Appl. **24**, 987–997 (1988)
14. Morimoto, S., Sanada, M., Takeda, Y.: Wide-speed operation of interior permanent magnet synchronous motors with high-performance current regulator. IEEE Trans. Ind. Appl. **30**, 920–926 (1994)
15. Kim, J.-M., Sul, S.-K.: Speed control of interior permanent magnet synchronous motor drive for the flux weakening operation. IEEE Trans. Ind. Appl. **33**, 43–48 (1997)
16. Dalib, S.: The simulation of the direct torque control of permanent magnet synchronous motor. Universiti Teknologi Malaysia (2007)
17. C. Ed-dahmani, C., Mahmoudi, H., Elazzaoui, M.: Direct torque control of permanent magnet synchronous motors in MATLAB/SIMULINK. In: 2016 International Conference on Electrical and Information Technologies (ICEIT), pp. 452–457 (2016)
18. Direm, C., Hartani, K., Aouadj, N.: New combined maximum torque per ampere-flux weakening control strategy for vehicle propulsion system. In: 2021 SAE International (2021). https://doi.org/10.4271/10-05-02-0009
19. Muyang, L.: Student Member, IEEE, Jiangbiao He, Student Member, IEEE, Nabeel A.O. Demerdash, Life-Fellow, IEEE Department of Electrical and Computer Engineering, Marquette University, WI, 53233, A Flux-Weakening Control Approach for Interior Permanent Synchronous Motors Based on Z-Source Inverters (2014)
20. Itani, K., De Bernardinis, A., Khatir, Z., Jammal, A.: Optimal traction and regenerative braking reference current synthesis for an IPMSM motor using ree combined torque control methods for an electric vehicle. In: 2016 IEEE Transportation Electri cation Conference and Expo (ITEC) , Dearborn, MI, pp. 1–6 (2016)
21. Sepulchre, L., Fadel, M., Pietrzak-David, M.: Improvement of the digital control of a high speed PMSM for vehicle application. In: 2016 Eleventh International Conference on Ecological Vehicles and Renewable Energies (EVER), pp. 1–9 (2016)

Environmental and Sustainable Solutions

Effect of Different Drying Modes on the Essential Oil Extract of the Tobacco Plant *(Nicotiana Tabacum L.)* Before Maturation

M. Foullanine(✉), A. Mediani, W. Braham Chaouch, A. Djaber, A. A. Larbi, A. Loumani, Z. Belkacem, and C. Tigani

Centre de Développement des Energies Renouvelables, Unité de Recherche en Energies Renouvelables en Milieu Saharien, URERMS, CDER, Adrar 01000, Algeria
meriafmf@gmail.com

Abstract. Essential oils and plant aromas represent real potential for the industry to replace synthetic compounds having harmful effects on health and the environment. This work focuses on the effect of different drying methods (in the open air, the oven, indirect solar dryer) on the extraction of essential oils from the tobacco plant (Nicotiana Tabacum L.) from the Timimoune region. We used steam distillation as a method to extract the essential oil from the Tobacco plant before flowering and maturation. Tobacco used during drying could be subjected to changes in their physicochemical properties such as color, water activity (Aw), total soluble solids (TSS), phenol content, antioxidant capacity, and moisture content (MC), and, thus, results were compared with fresh Nicotiana Tabacum L. samples. The extraction of essential oil from the plant gave a yield of 0.6% for indirect drying, a yield of 0.4% for oven drying and a yield of 0.3% for air drying free. They are slightly higher than that of the AFNOR standards (0.15%). The indirect solar dryer allows the preservation of secondary metabolites (essential oils) of all the extracts investigated compared to other drying methods.

Keywords: Tobacco plant · drying methods · indirect dryer · essential oils

1 Introduction

Tobacco plants, scientifically known as *Nicotiana tabacum L.*, are widely cultivated non-food crops with significant economic and research importance. These plants belong to the Nicotiana genus and are grown in over 120 countries worldwide [1]. *Nicotiana tabacum L.* is an intriguing plant due to its economic significance, intricate biology, and potential for contribution to different areas, including agriculture, medicine, and biotechnology [2]. The plant's multifaceted nature and applications are continuously being revealed through ongoing research and studies [2]. The extraction of phenolic compounds from the tobacco plant can be achieved through various techniques, such as liquid-liquid extraction (LLE), ultrasound- assisted extraction (UAE), microwave-assisted extraction (MAE), and supercritical fluid extraction (SFE) [3]. By the used of solvents to optimize the yield and effectiveness of the extracted phenols, Different studies have presented

© The Author(s), under exclusive license to Springer Nature Switzerland AG 2025
M. Hatti (Ed.): IC-AIRES 2024, LNNS 1238, pp. 397–404, 2025.
https://doi.org/10.1007/978-3-031-80301-7_43

the extraction process, focusing on factors such as solvent type, extraction time, and temperature, which significantly influence the phenolic content and antioxidant activity of the extracts [4]. Also the extraction of phenolic compounds from the tobacco plant can provide valuable compounds for use in the food, chemical, and pharmaceutical industries [5]. The specific phenolic compounds present in the tobacco plant and their potential applications would need to be further investigated through analytical techniques such as HPLC-DAD and UPLC-MS/MS [6].

Drying medicinal and aromatic plants is an important process to maintain their quality and enjoy their benefits for a long time. Plants can be dried in several ways: by placing them in a dark, dry place, using special drying ovens, or using solar energy [7]. Drying plants is an essential step before using them and benefiting from their therapeutic and aromatic benefits effectively. The drying process is also an important step in preparing tobacco leaves for processing and uses [7]. Solar drying is a process of using the sun's energy to dry various materials, including agricultural products like tobacco. Based on the context provided [8].

In this work, we studied the effectiveness of different drying methods (air drying, oven drying, and indirect solar drying) at the same temperature on the concentration of phenolic compounds of tobacco plants before maturity (in the first stage of growth).

2 Material and Methods

This work was carried out in the solar drying laboratory of the Renewable Energy Research Unit in the desert center of Adrar URERMS. The experiments carried out are summarized as follows:

1- Determination of some tobacco parameters (humidity, organic, mineral and dry matter, acidity)

2- Extraction of essential oils

A. Plant Material

Tobacco samples were taken at the beginning of growth before flowering in Fevrier 2024 from the province region of Timimoune, geographical coordinates: N: 29° 15′,W0° 15′, Altitude: 700 m,, Charouine Municipality Fig. 1.

The plant was dried using three methods: under direct sunlight (direct solar drying), in an indirect solar dryer and in an oven. The tobacco leaves that were the subject of the study were transformed after drying powder in a blender and used to prepare various extracts.

B. Extraction Method

The essential oils of the three tobacco drying methods are extracted by the hydrodistillation method using a Clevenger type device.

The plant material is crushed to facilitate their introduction into a 1 L glass flask, filled with water up to 2/3 of its capacity. The water is then heated in the heating tank until it boils, which causes the formation of a vapor which will carry away the volatile constituents. These vapors rise and pass into the refrigerant which is constantly cooled to a temperature between 15 °C and 18 °C.

Fig. 1. The Tobacco plant

After separation, the final oil obtained is stored in opaque glass bottles at a temperature of 4 °C. (Merabet, 2018) [9].

C. Determination and Calculation of Yield

The yield is the maximum quantity of essential oil that a given mass of plant produces during a given period. It is the ratio between the mass of the essential oil obtained to the mass of the plant material used [9] (Fig. 2).

The yield was calculated using the following relationship:

$$R(\%) = \text{Mass}(\text{HE}) / \text{Mass}(\text{M V } S) \times 100$$

D.Calculation of Extraction Yield

The following ratio can be used to calculate the plant's dry extract yield:

$$\text{Rdt}(\%) = \text{P1} - \text{P2}/\text{P3} \times 100 \quad (1)$$

P1: weight of the balloon after evaporation.
P2: weight of the balloon before evaporation.
P3: weight of the starting plant material.

E.Determination Of Total Polyphenols (PPT)

The determination of total polyphenols was carried out by a method adapted from Singleton and Ross using the Folin-Ciocalteu reagent [10]. To quantify polyphenols, a calibration curve is produced using Gallic acid as a standard.

Results are expressed in mg Gallic acid equivalent/g of plant extract. The quantity of total phenols is calculated by the following equation:

$$C = c.\, v/m \quad (2)$$

C: total content of polyphenols (mg Gallic acid equivalent/g of plant extract). c: Gallic acid concentration (mg/ml).

v: volume of the extract (ml).

m: mass of the pure plant extract (g).

Fig. 2. The ethanol extraction protocols

3 Results and Discussion

The determination of total phenols was carried out by the spectrophotometric method adapted with the Folin- Ciocalteu reagent using Gallic acid as a standard. The results obtained are expressed in microgram of Gallic acid equivalent per milligram of extract (mg EAG/g) using a calibration curve (Fig. 3 and Table 1).

The results obtained in this study are presented in the table below

From this table we notice that the value of the concentrations of total polyphenols was 72.32 mg/g, for the indirect solar drying sample it was more than for the oven dried sample 49.53 mg/g. and in the open air 65.39 mg/g.

A.Indirect Solar Drying

From the statement shown in the Fig. 4, we notice that the concentrations of phenolic compounds of the tobacco plant in the first stage of growth, before flowering and maturity in both samples after drying in the indirect solar dryer and before drying are similar, which proves that indirect solar drying maintains a large amount of compounds.

B.Drying in the Open Air

Through the statement shown in the Fig. 5, we notice that the concentrations of phenolic compounds of the tobacco plant in the first stage of growth, before flowering and maturity

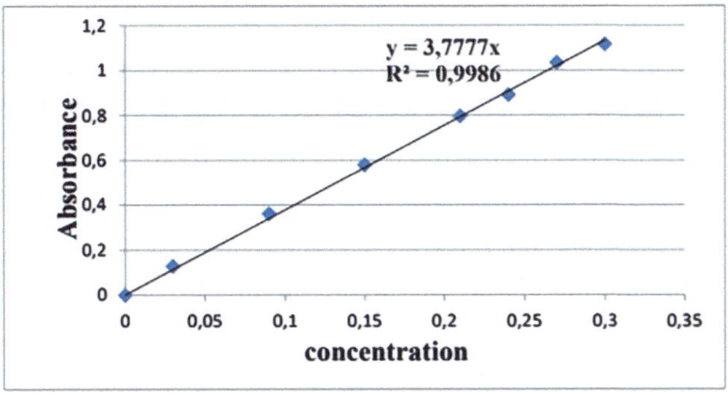

Fig. 3. Calibration curve for total polyphenols

Table 1. POLYPHENOL CONCENTRQTION MG AG/G1.

Vegetal Materiel	Drying methods	Polyphenol concentration
Tabaco	Fresh tobacco	40.72 mg AG/g
	Indirect solar drying	72.38 mg AG/g
	Drying in the open air	49.53 mg AG/g
	Oven drying	65.39 mg AG/g

in both samples after drying it under direct sunlight and before drying are relatively different, as the difference between them reached 20%, which proves that drying under direct sunlight causes damage. A large amount of secondary compounds.

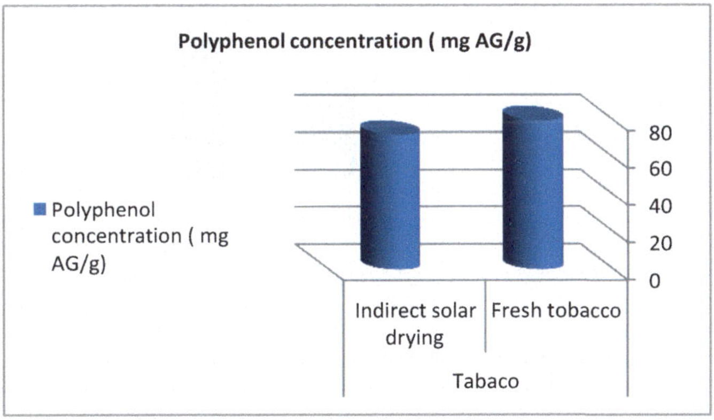

Fig. 4. Polyphenol concentration fresh Tabaco and indirect solar drying

C.Oven Drying

Through the statement shown in the Fig. 6, we notice that the concentrations of phenolic compounds of the tobacco plant in the first stage of growth, before flowering, and maturity in each of the two samples after drying it in the oven and before drying are relatively different, as the difference between them reached 30%, which proves that drying in the oven caused the destruction of a large amount of compounds secondary to more drying under sunlight. This may be due to taking the temperature of the external environment as a reference, which did not exceed 40 °C, which affected the duration of drying and thus affected its effectiveness.

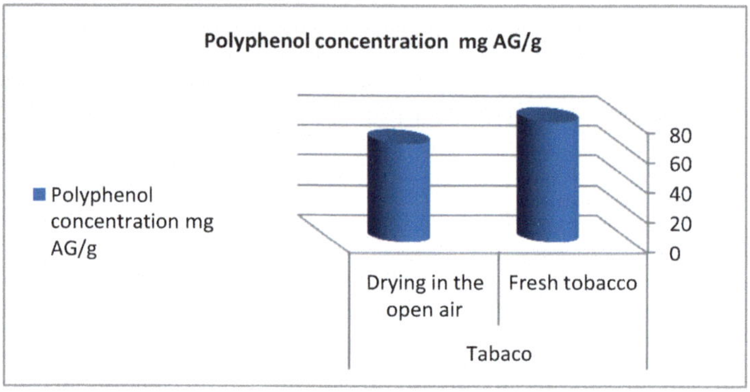

Fig. 5. Polyphenol concentration fresh Tabaco and drying in the open air

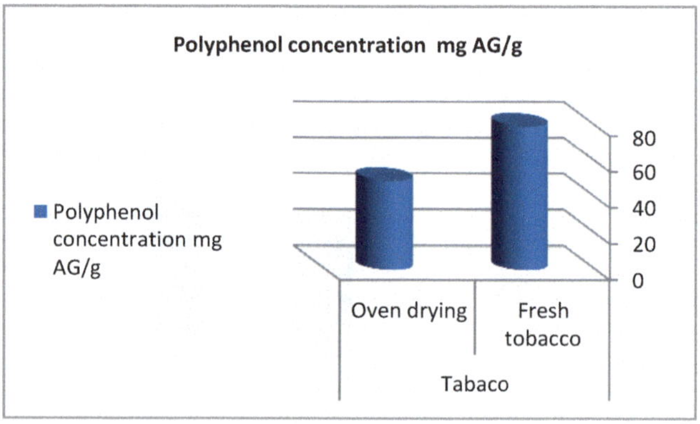

Fig. 6. Polyphenol concentration fresh Tabaco and Oven drying

4 Conclusion

This study is concerned with studying the effect of different types of drying, whether by indirect solar energy, under direct sunlight, or in an oven, on the concentration of phenolic compounds in local tobacco plants before flowering and maturation in the early stages of growth. It was noted that the largest value of phenolic compounds recorded in indirect solar drying was estimated at 32.72 mg EAG/g, while its value in drying under direct sunlight was estimated at 65.39 mg EAG/g, and the lowest value was recorded in oven drying, which was estimated at 49.53 mg EAG/g. Comparing these results with the initial concentration of the sample before drying, which was estimated at 80.00 mg EAG/g, shows that indirect solar drying is the most efficient type of drying studied, as it preserves the secondary phenolic compounds.

Acknowledgment. The authors extend their sincere thanks to the Research Unit in Renewable Energy in Saharan Medium, affiliated with the Renewable Energy Development Center, for conducting this research in its solar drying laboratory.

References

1. Meriama, F., et al.: Experimental Determination and Modeling of the Moisture-Sorption Isotherms and Isosteric Heat of Tobacco Leaves; Instrumentation Mesure Métrologie, vol. 20, no. 5, pp. 269–277 (2021)
2. Omare, M.O., Kibet, J.K., Cherutoi, J.K., Kengara, F.O.: A review of tobacco abuse and its epidemiological consequences. J. Public Health (2021). https://doi.org/10.1007/s10389-020-01443-4
3. Soto-Hernández, M., García-Mateos, R., Palma-Tenango, M.: Plant physiological aspects ofphenolic compounds, extraction techniques of phenolic compounds from plants (2019). https://doi.org/10.5772/intechopen.77494 (Chapter 1). https://doi.org/10.5772/intechopen.84705
4. Abubakar, A.S., et al.: Phytochemical Composition, Antioxidant, Antibacterial, and Enzyme Inhibitory Activities of Various Organic Extracts from Apocynum hendersonii (Hook.f.) Woodson. Plants **2022**, 11 (1964). https://doi.org/10.3390
5. Tsouko, E., Alexandri, M., Fernandes, K.V., Freire, D.M.G., Mallouchos, A., Koutinas, A.A.: Extractionof phenolic compounds from palm oil processing residues andtheirapplicationasantioxidants.FoodTechnol.Biotechnol.**57**(1) (2019). https://doi.org/10.17113/ftb.57.01.19.5784
6. Zoran, Z., et al.: UPLC-MS/MS characterization of phenolic constituents in fennel seed extracts ((Foeniculum vulgare Mill.). Proc. Nutri. Soc. **79**(OCE2), E395 (2020).https://doi.org/10.1017/s0029665120003432
7. Loumani, A., et al.: Experimental measurement of isothermal sorption, microbiological and physicochemical analysis of dried tomatoes cultivated in Adrar, Algeria. Int. J. Des. Nat. Ecodynamics 15(5), pp.721–728 (2020)
8. Mediani, A., Moungar, H., Larbi, A.A., Benatiallah, A.: The isothermal sorption measurement and the isosteric heats determinations for the south algerian date varieties. Instrumentation Mesure Métrologie, **18**(4), 389–396 (2019). http://iieta org/journals/i2m, **18**(4), 389–396
9. Cˇ ulina, P., et al.: Phenolic profile and antioxidant capacity of selected medicinal and aromatic plants: diversity upon plant species and extraction technique. Processes **9**, 2207 (2021). https://doi.org/10.3390/pr9122207

10. MADI A., 2009- "Caractérisation et comparaison du contenu polyphénolique de deux plantes médicinales (thym et sauge) et la mise en évidence de leurs activités biologiques". Mémoire de magister. Université Mentouri Constantine
11. Mahboub, N., Slimani, N., et khelil A.: Effet des differents modes de séchage sur le contenu Phenolique et biologique d'une plante spontanee a Caractere medicinale du sahara septentrional Algerien. Revue des BioRessources, vol 12 N° 1 Juin (2022)

Optimizing Geothermal Heat Exchangers Performance for Energy- Efficient Poultry House Heating

Assia Meziane[1,3](✉), Khaled Mansouri[2], Abdelmalek Atia[1,2], and Abdelbaki Mameri[3]

[1] LEVRES Lab, Technology Faculty, University of El Oued, 39000 El Oued, Algeria
meziane-assia@univ-eloued.dz

[2] Technology Faculty, UDERZA Unit, Universityof El Oued, 39000 El Oued, Algeria

[3] CMASMTF Laboratory, Mechanical Engineering Department, SASF, University Larbi Ben M'Hidi, Oum El Bouaghi, Algeria

Abstract. This paper analyzes the performance of heat exchangers in a geothermal system designed for heating a poultry house in a harsh climate. The study focuses on a traditional poultry house in Algeria, employing a ground source heat pump (GSHP) system with plate heat exchangers. Data was collected over a day, monitoring indoor and soil temperatures, energy consumption, and system efficiency. Results indicate that the geothermal system significantly reduces energy consumption and operating costs compared to conventional heating methods. The system achieved substantial CO2 emissions reduction, demonstrating a positive environmental impact. The paper analyzes the performance of the plate heat exchangers, highlighting their thermal efficiency and heat transfer characteristics. The findings demonstrate the viability of geothermal systems for poultry house heating, contributing to sustainable agriculture and reducing environmental impact.

Keywords: Geothermal System · Poultry House · Heat Exchangers · Energy Efficiency

1 Introduction

As the demand for sustainable and energy-efficient agricultural practices intensifies, the poultry industry faces increasing pressure to optimize operational efficiency while minimizing environmental impact. One promising approach to achieve these goals is the integration of geothermal heat exchangers for heating poultry houses. Geothermal energy, derived from the natural heat of the Earth, offers a renewable and stable source of energy, potentially reducing reliance on fossil fuels and lowering operational costs.

The performance of geothermal heat exchangers is critical in ensuring effective climate control within poultry houses, which is vital for the health and productivity of the birds. Proper heating not only enhances the welfare of poultry but also improves feed conversion rates, ultimately contributing to better economic returns for farmers.

© The Author(s), under exclusive license to Springer Nature Switzerland AG 2025
M. Hatti (Ed.): IC-AIRES 2024, LNNS 1238, pp. 405–415, 2025.
https://doi.org/10.1007/978-3-031-80301-7_44

However, to harness the full potential of geothermal systems, a thorough understanding of their design, operation, and optimization is essential [1].

This study aims to explore the various factors that influence the performance of geothermal heat exchangers in poultry house heating applications. By identifying and implementing optimization strategies, the goal is to enhance energy efficiency, reduce operational costs, and promote more sustainable practices within the poultry industry. Through a combination of theoretical analysis and practical insights, this research seeks to provide a comprehensive framework for effectively utilizing geothermal energy in poultry farming, paving the way for a greener future in agriculture.

2 Studies and Projects in Improving the Performance of Heat Exchangers

Research on optimizing the performance of ground heat exchangers (GHEs) for heating poultry houses specifically is a niche area, but there have been several studies and projects that contribute valuable insights into this topic. Below are some relevant studies, along with their sources, that examine improving the performance of ground heat exchangers for energy-efficient poultry house heating.

	Source	Overview
Thermal Performance of Ground Heat Exchangers	Wang, Y., et al. (2016). "Experimental study on the thermal performance of geothermal heat exchanger." *Energy and Buildings*, 127, 26–35	This study investigates the thermal performance of different designs of geothermal heat exchangers. Although not poultry-specific, the findings related to thermal conductivity and soil properties provide insights that can be applied to poultry heating systems
Optimization Techniques for Geothermal Systems	Spitler, J. D., & Zhang, Y. (2017). "Ground Source Heat Pump Systems: Modeling and Optimization." *Renewable Energy*, 110, 472–479	The authors discuss various optimization techniques for ground source heat pump (GSHP) systems, contributing to efficient heating configurations. Their models can serve as a foundation for optimizing GHEs in poultry houses

(*continued*)

(*continued*)

	Source	Overview
Geothermal Heating Applications in Agriculture	Mermoz, S., et al. (2016). "Geothermal Energy: A Renewable Energy Source for Agriculture." *Renewable and Sustainable Energy Reviews*, 66,	This review highlights various applications of geothermal energy in agriculture, including heating poultry houses. While it does not focus solely on optimization, it provides a background for incorporating geothermal energy efficiently
Multi-Objective Optimization of Ground Heat Exchangers	Kalogirou, S. A. (2007). "Solar Energy Engineering: Processes and Systems." *Elsevier*	Although focusing broadly on solar and geothermal energy systems, this book includes methodologies for the design and optimization of ground heat exchangers, which can be adapted for specific applications like poultry heating
Modeling and Simulation for Energy Simulation	Ameen, M. A., et al. (2020). "The Thermal Modelling Analysis of Ground Heat Exchangers for Heating of Buildings: A Case Study." *Energies*, 13(3), 575	This paper presents a thermal modeling analysis of ground heat exchangers for building heating. The methodologies discussed can be directly applied to optimizing GHE performance in poultry housing scenarios
Poultry House Environmental Management	Collier, S. M., et al. (2010). "The Use of Geothermal Resources for Poultry Thermal Management in the Southeastern United States." *International Journal of Poultry Science*, 9(9), 835–841	This study specifically addresses the use of geothermal resources for managing environmental conditions in poultry houses. It presents a case study highlighting the operational efficiency gains from using ground heat exchangers
Economic Analysis of Geothermal Systems	Yavuz, M., et al. (2018). "Economic Feasibility of Ground Source Heat Pump Systems in the Poultry Industry." *Agricultural Engineering International: CIGR Journal*, 20(1), 92–101	This paper assesses the economic feasibility of implementing ground source heat pump systems within the poultry industry, including performance optimization considerations that can enhance energy efficiency

These studies indicate that while direct research specifically targeting the optimization of geothermal heat exchangers in poultry houses is still developing, relevant findings from geothermal energy applications in agriculture and the poultry sector provide a foundational basis for future research and system improvements. Further tailored studies are needed to bridge the gap directly linking GHE optimization with poultry heating efficiency.

3 System Design and Description

The optimal temperature range for chickens of different classes and ages is debated due to factors like humidity and wind speed affecting temperature responses.

High temperatures can reduce chicken production, especially when combined with high humidity impacting cooling efficiency.

Recommendations include 30–32 °C for day-old chicks, decreasing by 3–4 °C until 4 weeks old, and 18–22 °C for growing broilers to ensure comfort and performance [2, 3]. The table below represents (Table 1) the recommended temperature for poultry farming in the El-Oued area.

Table 1. Recommanded température Schedule.

Age of chicken (week)	Température range (°C)
1	30–32
2	30–26
3	26–23
4	23–20
25	20

Traditional poultry house used in arid regions of **Algeria (El-Oued)**, specifically designed for housing a large number of chickens. Here are the key points from your description:

Location and Type: These poultry houses are located in arid regions of **Algeria (El-Oued)** and are primarily traditional structures. **Shape and Dimensions**: The houses have a semicircular shape, are 44 meters long and 11 meters wide. **Capacity:** Each house can accommodate 4,500 chickens. **Climate Control:** They are equipped with front air conditioning and ventilation. Heating: There are six heaters inside the house. **Floor Material:** The floor is covered with a material like sawdust to prevent chickens from eating dust. Including images would provide a visual context, helping to better understand the layout and design.

A: Poultry house at the level of length
B:the front of the poultry house containing the air conditioning system
C: High level poultry house entrance

AB

CD

Fig. 1. Represents pictures of the poultry house we studied.

D: From inside the house, the flooring material (such as sawdust), where the house contains broiler chickens, cooling, heating and feeding devices.

The El-Oued area is characterized by a cold climate in winter and hot dry summer in the summer and the main source of energy in poultry farms there is fossil fuels and gas, which constitute a large part of the costs spent by farm owners, especially during the cold seasons [4].

Due to heating requirements, poultry farmers incur about **6%** of the total production cost in winter, and **4%** in summer and According to the study, the barns were heated by heat lamps and a diesel boiler, while ventilation relied on the principle of evaporative cooling that used fans [5].

4 Analytical Methods

4.1 Thermal Efficiency Calculation

The thermal efficiency of the heat exchangers was calculated using the formula:

$$\eta = \frac{Q_{out}}{Q_{in}};$$

where $Qout$ is the heat output and Qin is the heat input.

- Heat Transfer Rate:

$$Q = \dot{m}.c_p.\Delta T$$

where: $\dot{m}$. Is the mass flow rate.

cp. Is the specific heat capacity of the fluid.

ΔT. Is the temperature difference across the heat exchanger.

4.2 Energy Consumption Analysis

Energy consumption of the GSHP system was compared to that of conventional heating systems used in similar poultry houses.

- **Energy Savings:**

$$Energy\ Savings(\%) = \frac{Energy_{Conventional} - Energy_{GSHP}}{Energy_{Conventional}} * 100$$

4.3 Cost Analysis

Operating Cost: The operating costs of the GSHP system were calculated, including electricity consumption and maintenance.

Cost Savings: Compared to conventional systems, cost savings were determined by:

$$\text{Cost}\ Savings = \text{Cost}_{Conventional} - \text{Cost}_{GSHP}$$

5 Environmental Impact Assessment

CO2 Emissions Reduction: Emissions were estimated based on the energy consumption data and the local emissions factor for electricity generation.

Environmental Benefits: Quantified in terms of reduced CO2 emissions, using the formula:

$$\text{CO2 Savings(kg CO2)} = \text{Energy (kWh)} \times \text{Emissions Factor (kg CO2/kWh)}$$

5.1 Statistical Analysis

Data Analysis: Statistical methods, such as regression analysis and hypothesis testing, were used to identify significant factors affecting the performance of the heat exchangers.

Software Tools: Data analysis was performed using software tools like Python for efficient processing and visualization.

6 Results

During the month of January 2024, the 2nd and 3rd of January, the maximum outside air temperature was 21.9 °C while the minimum air temperature was 4.1 °C. The maximum soil temperature was 27.5 °C while the minimum soil temperature was 23.5 °C. The maximum pipe temperature was 33.9 °C and the minimum inside the pipe was 22.4 °C (Table 2).

Table 2. Evaluation of the Performance of the Geothermal Exchanger every 10 min for 24 h

Time	T_amb	TTurb1	TTurb2	TTurb3	TTurb4	TTurb5	TTurb6
17:30:20	010.2	027.0	026.4	030.9	024.6	023.9	025.4
17:40:20	010.4	026.3	025.3	033.1	024.6	023.8	025.2
17:50:20	010.1	025.7	025.1	033.4	024.4	023.7	025.1
18:00:20	010.5	025.3	024.7	032.9	024.4	023.7	025.1
18:10:20	008.9	025.1	024.5	033.0	024.3	023.5	024.9
18:20:20	010.0	025.0	024.4	032.7	024.3	023.5	024.9
18:30:20	010.0	024.8	024.2	032.6	024.2	023.4	024.8
18:40:20	009.8	026.5	025.6	033.6	024.5	023.5	025.2
18:50:20	009.2	026.8	026.1	033.1	024.4	023.4	025.2
19:00:20	008.7	025.4	024.8	032.5	024.2	023.3	024.8
19:10:20	009.1	024.8	024.3	032.2	024.2	023.4	024.7
19:20:20	009.3	025.0	024.3	032.1	024.0	023.2	024.6
19:30:20	009.0	026.1	025.2	032.7	024.3	023.3	024.9
19:40:20	007.9	026.2	025.5	032.3	024.3	023.2	024.9
19:50:20	008.2	026.1	025.5	032.5	024.3	023.4	025.0
20:00:20	008.0	025.6	025.0	032.3	024.1	023.2	024.8
20:10:20	007.3	026.1	025.4	032.2	024.2	023.3	024.9
20:20:20	008.9	026.3	025.4	032.3	024.2	023.3	025.0
20:30:20	007.9	025.8	024.8	032.0	024.1	023.1	024.6
20:40:20	008.0	026.3	025.3	032.7	024.2	023.2	024.9
20:50:20	008.3	026.5	025.7	032.5	024.2	023.3	024.9
21:00:20	009.2	025.8	025.1	032.5	024.2	023.2	024.7
21:10:20	008.3	026.1	025.5	032.4	024.2	023.2	024.9
21:20:20	007.2	026.4	025.5	032.6	024.2	023.2	024.9
21:30:20	007.9	024.8	023.9	032.9	023.9	023.1	024.5
21:40:20	007.3	024.5	023.7	033.9	023.9	023.0	024.4
21:50:20	006.4	024.3	023.6	033.6	023.8	023.0	024.3
22:00:20	006.7	024.1	023.2	033.2	023.7	022.9	024.2
22:10:20	006.0	024.2	023.4	034.1	023.7	022.8	024.2
22:20:20	007.4	023.9	023.0	033.5	023.7	022.8	024.1
22:30:20	007.2	023.7	023.0	033.1	023.7	022.8	024.0
22:40:20	007.7	024.2	023.2	033.5	023.7	022.8	024.0

(continued)

Table 2. (*continued*)

Time	T_amb	TTurb1	TTurb2	TTurb3	TTurb4	TTurb5	TTurb6
22:50:20	006.0	024.4	023.4	033.6	023.6	022.7	024.1
23:00:20	007.0	024.0	023.2	033.1	023.6	022.7	023.9
23:10:20	005.7	023.8	023.0	033.1	023.5	022.6	023.9
23:20:20	006.2	023.8	022.9	033.0	023.5	022.6	023.9
23:30:20	006.2	023.8	022.8	032.9	023.5	022.6	023.9
23:40:20	006.6	024.0	022.9	032.7	023.5	022.5	023.9
23:50:20	005.9	023.8	022.8	032.6	023.4	022.5	023.8
00:00:20	006.8	024.0	022.9	033.0	023.4	022.5	023.9
00:10:20	006.1	023.6	022.5	033.0	023.4	022.5	023.8
00:20:20	005.7	023.9	022.8	032.7	023.4	022.4	023.8
00:30:20	005.2	024.0	023.0	032.7	023.4	022.5	023.9

Because of the wide difference between the external temperature and the undisturbed soil temperature at 4 m deep, the possibility of cooling and heating was high the systems. Figure 1 shows the air temperature variation between night and day (Figs. 2 and 3).

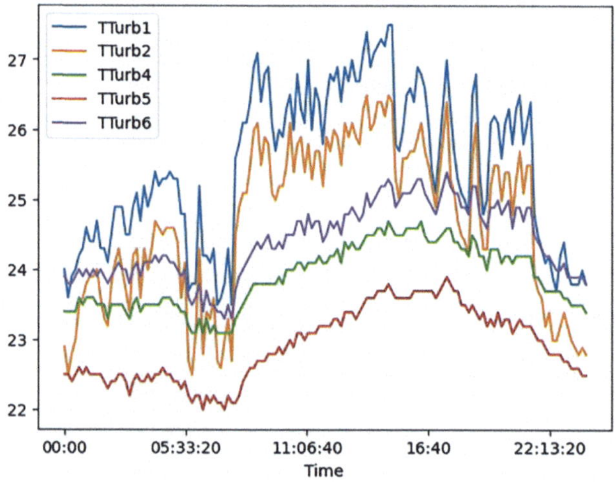

Fig. 2. Air temperatures at the outlets of the systems in relation with outside and ground temperature during the whole day.

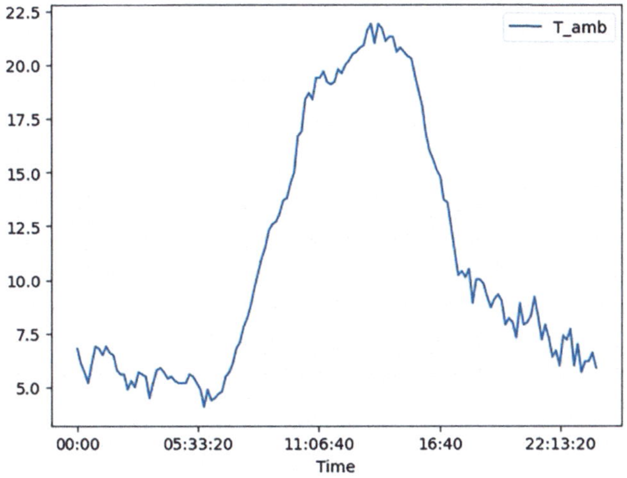

Fig. 3. Outside air temperatures throughout the day.

7 Discussion

The graph 1 shows the temperature changes for six turbines (TTurb1 to TTurb6) throughout the day.

The horizontal axis represents the time, starting from midnight to 10:13 PM. The vertical axis indicates the temperature.

TTurb1 (blue) has the highest temperatures, reaching around 27 degrees. TTurb2 (orange) comes next to TTurb1, with temperatures around 26 degrees.

TTurb4 (green) and TTurb5 (purple) have similar temperatures, ranging between 24 and 25 degrees. TTurb6 (red) records the lowest temperatures, hovering around 22–23 degrees

The graph 2 shows how the ambient temperature (T_amb) changes over the course of a day.

The horizontal axis represents time, starting from midnight all the way to 10:13 PM. The vertical axis indicates the temperature.

The temperature starts at around 5 degrees at midnight.

It gradually goes up, peaking at about 22.5 degrees around 4:40 PM.

After that, the temperature slowly drops back down, hitting about 5 degrees again by 10:13 PM.

The data clearly shows fluctuations in temperature throughout the day, with a noticeable increase in the afternoon. The data shows that the system is more effective at reducing air temperature during the hottest of the houers day.

During the night, the outside temperature is lower than the ground temperature, causing the temperature at the outlets to rise in both Geothermal Heat Exchangers systems.

The performance of Geothermal Heat Exchangers is influenced by various input parameters, with pipe length being a significant factor affecting its effectiveness.

8 Conclusion

In conclusion, optimizing geothermal heat exchanger performance is a crucial step toward enhancing energy efficiency in poultry house heating systems. By integrating advanced design techniques, selecting high- performance materials, and fine-tuning operational parameters, poultry producers can significantly reduce energy consumption while maintaining optimal animal welfare conditions. Implementing these strategies not only lowers operational costs but also contributes to a more sustainable approach to poultry farming. Moreover, the successful application of geothermal energy in this context demonstrates the potential for broader adoption of renewable energy solutions in agriculture, ultimately leading to improved environmental stewardship and economic resilience within the industry. Embracing these innovations will be key to meeting the growing demand for poultry products while minimizing the ecological footprint of production practices.

References

1. Morshed, W, Leso, L., Conti, L., Rossi, G., Simonini, S., Barbari, M.: Cooling performance of earth-to-air heat exchangers applied to a poultry barn in semi-desert areas of south Iraq. In: International Journal of Agricultural and Biological Engineering. - ISSN 1934–6344. - STAMPA. - 11, pp. 47–53 (2018). https://doi.org/10.25165/j.ijabe.20181103.3047
2. Hulzebosch, J.: What affects the climate in poultry houses? World Poultry. **20**, 36–38 (2004)
3. Daghir, N.J.: Nutritional strategies to reduce heat stress in broilers and broiler breeders. Lohmann Inf. **44**(1), 6–15 (2008)
4. Kapica, J., Pawlak, H., ´Scibisz, M.: Carbon dioxide emission reduction by heating poultry houses from renewable energy sources in Central Europe. Agric. Syst. **139**, 238–249 (2015)
5. Kim, M.J., Parvin, R., Mushtaq, M.M.H., Hwangbo, J., Kim, J.H., Na, J.C., et al.: Growth performance and hematological traits of broiler chickens reared under assorted monochromatic light sources. Poult. Sci. **92**, 1461–1466 (2012). https://doi.org/10.3382/ps.2012-02945
6. Gordon RF. Poultry Disease. 2nd ed. London: Baillière Tindall; 1982. p. 370 59. Mason IL. Evolution of Domesticated Animals. London: Longman; 1984. p. 452
7. Amid, S., Gundoshmian, T.M.: Prediction of output energies for broiler production using linear regression, ANN (MLP, RBF), and ANFIS models. Environ. Prog. Sustain. Energy **36**(2), 577–585 (2017)
8. Abu-Hamatteh, Z., Al-Zughoul, K., Al-Jufout, S.: Potential geothermal energy utilization in jordan: possible electrical power generation. Int. J. Thermal Environ. Eng. 9–14 (2011)
9. Al-Zyoud*, S., Rühaak, W., Sass, I.: Dynamic numerical modeling of the usage of groundwater for cooling in north east Jordan -A geothermal case study. Renew. Energy 63–72 (2014)
10. Hosseinzadeh-Bandbafha, H., Safarzadeh, D., Ahmadi, E., Nabavi-Pelesaraei, A.: Optimization of energy consumption of dairy farms using data envelopment analysis–a case study: Qazvin city of Iran. J. Saudi Soc. Agric. Sci. **17**(3), 217–228 (2018)
11. Lee, S.-W., Oh, B.-W., Park, K.-W., Seo, I.-H.: Estimation of adequate capacity of ground source heat pump in energy-saving pig farms using building energy simulation. J. Korean Soc. Agric. Eng. **64**, 1–13 (2022)
12. Banakar, A., Montazeri, M., Ghobadian, B., Pasdarshahri, H., Kamrani, F.: Energy analysis and assessing heating and cooling demands of closed greenhouse in Iran. Therm. Sci. Eng. Prog. **25**, 101042 (2021)

13. Shahzad, K., et al.: Experiments on energy-efficient evaporative cooling systems for poultry farm application in Multan (Pakistan). Sustainability **13**, 2836 (2021)
14. Shakibi, H., Afzal, S., Shokri, A., Sobhani, B.: Utilization of a phase change material with metal foam for the performance improvement of the photovoltaic cells. J. Energy Storage **51**, 104466 (2022)

Organic Waste Biomethanation Controlled by Artificial Intelligence in a Smart City Concept

L. Benhabyles(✉) and M. Saber

Renewable Energy Development Center, CDER, Algiers, Algeria
l.benhabyles@cder.dz

Abstract. Established programs automate specific functionalities by providing instructions to technology based on environmental changes. This not only eliminates repetitive tasks but also minimizes human errors, boosting productivity and product quality.

Managing the large volumes of waste generated by different sectors in urban areas is a major challenge. In concept of smart city, this management would be realized by the introduction of new technologies to improve quality of life in a sustainable way.

The development of technologies is linked to waste variability. Organic wastes are treated by different ways, one of the ancestral and natural process is the anaerobic digestion. A process resulting in two products: a bioenergy product such as biogas and digestate that could be used as bioamendment and biofertilizer.

Conception of intelligent and smart systems to improve biomethanation production is closely related to the choice of the right substrate composition among the high variability of organic wastes. The recovery of biogas is contingent on its methane content, which varies with the methanized substrate.

Every feedstock has its own physico-chemical properties and its impact on the anaerobic digestion process. It can be evaluated by considering their methanogenic potential, by this way, every feedstock can be categorized to emphasize the most productive ones.

Keywords: Anaerobic digestion · Substrate · biogas · smart system · artificial intelligence

1 Introduction

The transition to a smart city leads to the development of strategic projects to introduce new and increasingly intelligent technologies in order to make of these cities good places to live in. Quality of life improvement refers to infrastructure (buildings, roads, bridges, etc.), transport, health system, economic system and environment following the rules of natural systems through biocities concept [1]. A significant challenge in urban areas is managing the vast quantities of waste produced by all its various sectors. Multiple processes have been devised to recover this waste while generating different forms of energy. Smart city concept involves include smart management of resources and smart energy consumption and production including the use of renewable energies [2].

© The Author(s), under exclusive license to Springer Nature Switzerland AG 2025
M. Hatti (Ed.): IC-AIRES 2024, LNNS 1238, pp. 416–422, 2025.
https://doi.org/10.1007/978-3-031-80301-7_45

Anaerobic digestion or biomethanation is more and more seen as an attractive source of renewable energy because it is a natural process that provides both energy and environmental benefits. This technology was initially developed for rural regions due to their remoteness from energy grids and for the consistent availability of large amounts of organic waste, mainly composed of agricultural crop residues and livestock manure. It expanded to the carbon waste from the agri-food industry [3] and has ultimately progressed to manage municipal waste [4].

The anaerobic digestion is a biochemical process in which organic matter is degraded by microorganisms in absence of oxygen. It is a phenomenon occurring spontaneously in nature in herbivore rumen, animal intestines, dumps, lakes… [4, 5]. It is the same process which is responsible for the formation of natural gas in sediments [6].

The objective is to optimize parameters for improving the biomethanation process in order to extract maximum energy from the various substrates by various automatisms governed by artificial intelligence.

The Biology of Anaerobic Digestion Process

Anaerobic digestion is carried out in 4 steps: Hydrolysis and acidogenesis during which the macromolecules that compose organic matter into monomers by hydrolytic and acidogenic acidogene bacteria. During acetogenesis, the monomers are transformed by acetogenic bacteria into organic acids such as volatile fatty acids (VFA) of which acetate and other compouds by acetogenic bacteria. In the final step, at methanogenesis, the methanogenic bacteria transform acetate and other byproducts of the from previous step into methane and carbon dioxide.

Biomethanation Technologies

The size of a biogas plant depends on the substrate and is adapted to the sector in which it will be located. The field of activity also determines the main objective of this installation. The characteristics of the biomass impacting the digester design are origin, physical properties (solid or liquid, water content...), their biochemical constitution. It is imperative to determine the biomass availability, its accessibility and biodegradability in order to determine the quality of the substrate and to choose the suitable methods to be used which is also linked to the daily added substrate to introduce into the digester and the retention time of the substrate in it [7, 8].

The agricultural sector, mainly rich in livestock manure and crop residues, will favor a large size biogas plant unit with one or more infinitely mixed vertical silo-type digesters. The industrial sector, which has many sub-sectors: food, chemicals, dairy... The effluent types are very varied, and units will be installed depending on the size of the plant and the type of effluent. Municipal waste is closely linked to sorting in order to select the bio-waste that can be treated by biomethanation. The processing of this kind of waste is possible on the basis of experience gained in industry. The technologies used in this sector are as varied as bio-waste [9].

Organic Biomass Storage

The installation of organic waste storage structure is related to the various techniques employed for biomass conservation prior to usage. These techniques vary depending on whether the waste is stored at its production site or at the methanation unit's site. The methods are as diverse as organic biomass: Silage is primarily used for plant residues (green waste, vegetable residues...), solids are stored in a specific storage area while manure and dry agricultural residues are kept in covered storage, and for readily biodegradable biomass such as catering waste or food industry waste, it is advisable to directly process them in tight flows methanation. In all cases, to ensure proper storage, it is crucial to protect the waste from air and moisture [10]. Scientific research has also demonstrated the impact of other factors like temperature that must be monitored to prevent the initiation of fermentation [11, 12] the storage of various organic waste for biomethanation must comply with health and environmental regulations. It is vital to consider the nuisances caused by storage facilities, such as odor nuisance and greenhouse gas emissions, which include CH4, H2S and NH3.

Despite the numerous methods used to optimize the storage of organic waste, it is strongly recommended to reduce the storage period to minimize gas emissions from the stored biomass.

Substrate Variability

It is the combination of various wastes that make up a biomethanation substrate (feedstock) and characterize it. Each waste has its proper impact on the biomethanation process. The introduction of organic biomass and their proportions are equally important and they are not done in a random way because some components released during anaerobic digestion instead of stimulating bacterial activity and thereby increase biogas production, they will have the opposite effect and inhibit the process. Research results of [13], demonstrate that adding green garden residues to an anaerobic digestion process of food waste would have a negative impact on the production of biomethane.

Biochemical Methane Potential (BMP)

The BMP is a parameter corresponding to the highest level of CH4 produced during the anaerobic digestion process. It is also used to characterize the substrate by the determination of the biomass biodegradability and by the same time it allows the measurement of pre-treatment efficiency and the determination of the best storage conditions for the substrate before its biomethanation [14–17].

The BMP is mainly evaluated through reactions occurring in batch reactors [18]. It is also the subject of several studies to establish models to forecast factors that are both time and costs to evaluate the efficiency of the system [7]. By this way, this parameter is required to select the most energy efficient biogas plant and appropriate substrate and the optimal mix of organic biomass for the most efficient co-digestion [7, 15].

The biochemical composition of organic substrates varies due to the diversity of their components. The methanogenic potential characterizes the CH4 content in the biogas produced from a substrate, the higher the CH4 content, the higher is the BMP, the better the quality of the biogas (Fig. 1).

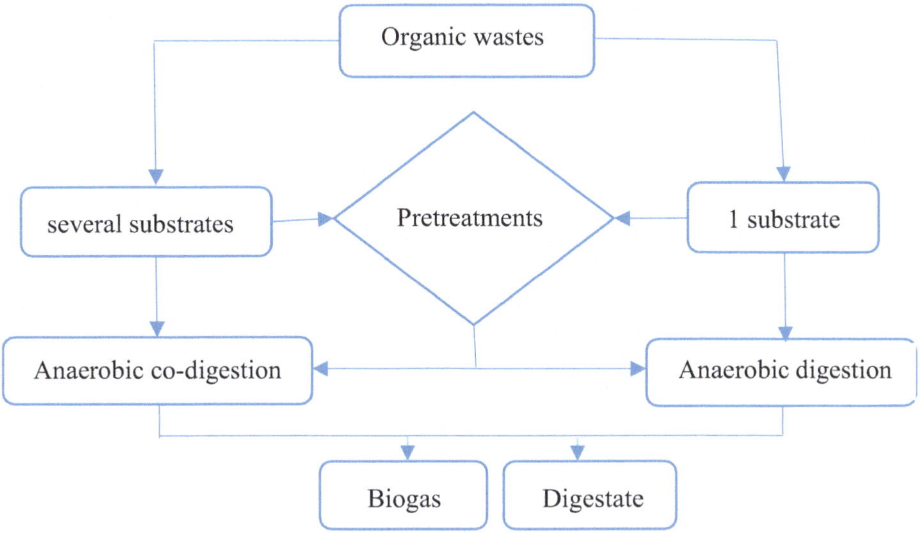

Fig. 1. Biomethanation of organic wastes

2 Artificial Intelligence Regulation

Artificial intelligence can be leveraged at multiple stages, from biogas production to its distribution within the urban gas network. Even if we succeed in generating biogas from organic biomass, numerous challenges remain, one of it is the feedstock variability.

The substrates used in biomethanation consist of organic waste from various economic and municipal sectors, including industrial, agricultural, and household waste. However, assembling substrates based exclusively on their identical characteristics is not recommended i.e. The BMP of a specific organic waste does not guarantee it will have the same effect when mixed with other wastes. Therefore, it is crucial to characterize the waste mixture forming the feedstock to be digested (Fig 2).

Knowing the feedstock properties induce the determination of its biodegradability level and the appropriate pretreatment. It also allows the determination of the right time to introduce substrate into the digester in a continuous mode of anaerobic digestion process by the identification of the hydraulic retention time (HRT) of substrate. In this case, the main tasks that should be automated are the feedstock introduction into the digester and there by the timing of digestate disposal.

Some functionalities should be automated for the regulation of biomethanation parameters. pH, temperature, O2 free environment, stirring speed determine an optimum bacterial activity for good production. The digestate should be at neutral pH and must be monitored mainly at the second step of anaerobic digestion process. The temperature inside the digester could be regulated according to the outside temperature.

Anaerobic digestion relies on a microbial process that is highly sensitive to environmental changes. This biological field is quite complex, which explains the importance of multiplying tests to understand the phenomenon.

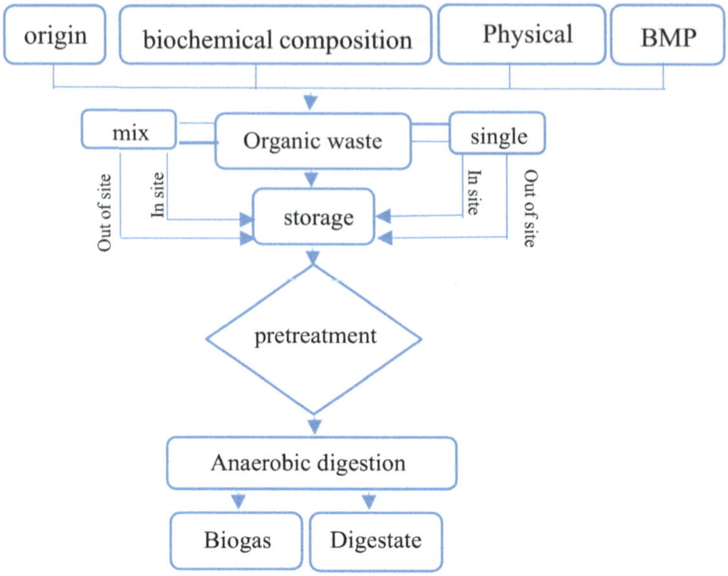

Fig. 2. Flowchart outlining the key features of organic waste for methane production

Implementing an intelligent system to support decision-making which would be able to anticipate events and predict their consequences is based on the combination of AI, ML and DL [19]. Research on AI applications in biomethanation has been ongoing for quite some time, but numerous challenges persist due to issues related to result extrapolation and the technologies employed [20]. One key challenge in AD is the variability of organic biomass.

Before considering all criteria for selecting an appropriate substrate, it is crucial to assess its profitability from energy, socio-economic, and environmental perspectives. AI can play a significant role in selecting substrates for biomethanation (S), choosing the inoculum (I), and determining their proportions, which will define the S/I ratio. Identifying the most profitable substrates and preventing inhibitors of bacterial activity released during biomethanation are essential tasks. Modeling becomes increasingly complex when dealing with numerous variations related to feedstock composition, bacterial activity, and dependent parameters such as the design of biomethanation technologies.

3 Conclusion

Smart city projects aim to apply smart technics gathering different tasks. Managing wastes from anthropogenic activity induce introduction of smart technologies or performing ones already existing. Anaerobic digestion controlled by artificial intelligence meets the established criteria and could respond to the demand. The main advantages of the process are biogas production, a renewable bioenergy and digestate that can be used as biofertilizer.

However, all wastes that can be methanized are organic ones and have various origins, domestic food wastes, food industry, agricultural residues and lignocellulosic wastes and animal manure. The profitability of these various substrates is evaluated by their methanogenic potential. Methanogenic potential express the cumulative methane released by an organic substrate. The most productive substrate is the one that gives a high volume of biogas with high methane content.

Artificial intelligence can give solution and perform anaerobic digestion technologies employed as solution for future smart city establishment.

References

1. Sæbø, A., Hanslin, H.M., Muys, B., Shanafelt, D.W., Sitzia, T., Tognetti, R.: Biodiversity and ecosystem functions as pillars of biocities. In: Scarascia-Mugnozza, G.E., Guallart, V., Salbitano, F., Ottaviani Aalmo, G., Boeri, S. (eds) Transforming Biocities. Future City, vol. 20. Springer, Cham (2023). https://doi.org/10.1007/978-3-031-29466-2_3
2. Moura, F., de Abreu e Silva, J.: Smart cities: definitions, evolution of the concept and examples of initiatives. In: Leal Filho, W., Azul, A., Brandli, L., Özuyar, P., Wall, T. (eds.) Industry, Innovation and Infrastructure. Encyclopedia of the UN Sustainable Development Goals. Springer, Cham (2019). https://doi.org/10.1007/978-3-319-71059-4_6-1
3. Cucui, G., Ionescu, C.A., Goldbach, I.R., Coman, M.D., Marin, E.L.M.: Quantifying the economic effects of biogas installations for organic waste from agro-industrial sector. Sustainability **10**, 2582 (2018). https://doi.org/10.3390/su10072582
4. Saravanakumar, A., Sudha, M.R., Wei-Hsin Chen, W.H., Pradeshwaran, V., Ashokkumar, V., Selvarajoo, A.: Biomethane production as a green energy source from anaerobic digestion of municipal solid waste: a state-of-the-art review. Biocatal. Agricult. Biotechnol. **53**, 102866 (2023). https://doi.org/10.1016/j.bcab.2023.102866
5. Moletta, R.L.: Méthanisation, 2nd edn. Tech & Doc, Paris (2011)
6. Marcon, L., Sotiri, K., Bleninger, T., Lorke, A., Männich, M., Hilgert, S.: Acoustic mapping of gas stored in sediments of shallow aquatic systems linked to methane production and ebullition patterns. Front. Environ. Sci. **10**, 876540 (2022). https://doi.org/10.3389/fenvs.2022.876540
7. Hagos, K., Zong, J., Li, D., Liu, C., Lu, X.: Anaerobic co-digestion process for biogas production: progress, challenges and perspectives. Renew. Sustain. Energy Rev. **76**, 1485–1496 (2017). https://doi.org/10.1016/j.rser.2016.11.184
8. Mosleh, U.M., Wright, M.M.: Anaerobic digestion fundamentals, challenges, and technological advances. Phys. Sci. Rev. **8**(9), 2819–2837 (2023). https://doi.org/10.1515/psr-2021-0068
9. GTZ-ISAT. (1999). Biogas Digest, Volume II, Biogas Application and Product Development
10. Peyrelasse C., Lalanne M., Monlau F., de l'Apesa, et Buffiere P., Bayard R., Teixeira F.: de l'INSA deep. 2017. SAM – Bonnes pratiques pour le stockage de matière. ADEME. 55 p.
11. Hansen, T.L., Sommer, S.G., Gabriel, S., Christensen, T.H.: Methane production during storage of anaerobically digested municipal organic waste. J. Environ. Qual. **35**, 830–836 (2006). Technical Reports: Waste Management. https://doi.org/10.2134/jeq2005.0239
12. Feng, L., Ward, A.J., Moset, V., Møller, H.B.: Methane emission during on-site pre-storage of animal manure prior to anaerobic digestion at biogas plant: Effect of storage temperature and addition of food waste. J. Environ. Manage. **225**, 272–279 (2018). https://doi.org/10.1016/j.jenvman.2018.07.079
13. Borth, P.L.B., Perin, J.K.H., Torrecilhas, A.R., Pan, N.C., Kuroda, E.K., Fernandes, F.: Biochemical methane potential of food and garden waste co-digestion with variation in solid content and inoculum:substrate ratio. J. Mater. Cycles Waste Manage. **23**(5), 1974–1983 (2021). https://doi.org/10.1007/s10163-021-01270-z

14. Filer, J., Ding, H.H., Chang, S.: Biochemical methane potential (BMP) assay method for anaerobic digestion research. Water **11**, 921 (2019). https://doi.org/10.3390/w11050921
15. Rodrigues, R.P., Rodrigues, D.P., Klepacz-Smolka, A., Martins, R.C., Quina, M.J.: Comparative analysis of methods and models for predicting biochemical methane potential of various organic substrates. Sci. Total. Environ. **1**(649), 1599–1608 (2019). https://doi.org/10.1016/j.scitotenv.2018.08.270. Epub 2018 Aug 22 PMID: 30308928
16. Koch, K., Hafner, S.D., Weinrich, S., Astals, S., Holliger, C.: Power and limitations of biochemical methane potential (BMP) tests. Front. Energy Res. **8**, 63 (2020). https://doi.org/10.3389/fenrg.2020.00063
17. Hilgert, J.E., et al.: Assessment of the biochemical methane potential of in-house and outdoor stored pig and dairy cow manure by evaluating chemical composition and storage conditions. Waste Manag. **168**, 14–24 (2023). https://doi.org/10.1016/j.wasman.2023.05.031
18. Liu, Y., He, P., Peng, W., Zhang, H., Lü, F.: Biochemical methane potential database: a public platform. Bioress. Technol. **393** (2024). https://doi.org/10.1016/j.biortech.2023.130111
19. Sarker, H.I.: AI-based modeling: techniques, applications and research issues towards automation, intelligent and smart systems. SN Comput. Sci. **3**, 158 (2022). https://doi.org/10.1007/s42979-022-01043-x
20. Rutland, H., You, J., Liu, H., Bull, L., Reynolds, D.: A systematic review of machine-learning solutions in anaerobic digestion. Bioengineering **10**, 1410 (2023). https://doi.org/10.3390/bioengineering10121410

High-Performance Asymmetric Flexible Supercapacitor Based on Zinc Oxide/Copper Oxide

M. I. Khenissa(✉), A. Benchettara, and S. E. Berrabah

Laboratory of Electrochemistry-Corrosion, Metallurgy and Inorganic Chemistry, Faculty of Chemistry, USTHB, BP 32, 16111 Algiers, Algeria
khenissamounaichrak@hotmail.com

Abstract. Enhancing supercapacitor electrochemical performance through innovative energy storage materials and controlled assembly growth is a popular and promising strategy. This study introduces a novel battery-type electrode material with a core-shell structure based on zinc and copper oxides, synthesized via a straightforward, environmentally friendly two-step electrodeposition process. The electrode materials, composed of ZnO and CuO oxides (CuO@ZnO), were directly deposited onto a three-dimensional conductive nickel foam (NF) substrate to form a binder-free ultra-thin multilayer (CuO@ZnO/NF). At a current density of 2 $mA.cm^{-2}$, the CuO@ZnO/NF nanomaterial exhibits a significant specific capacity of 6123.3 $mF.cm^{-2}$. Impressively, it maintains a high cycling stability of 95% after 8,000 cycles. This research highlights the potential of core-shell structured electrodes to advance supercapacitor technology, offering efficient energy storage solutions suitable for renewable energy applications and beyond.

Keywords: Zinc oxide · copper oxide · flexible asymmetric supercapacitor · energy storage

1 Introduction

Renewable energy sources have driven demand for advanced energy storage technologies capable of efficient power storage and delivery. Consequently, there's a rising focus on developing materials and methods to enhance supercapacitors, crucial for renewable energy systems [1, 3]. This study investigates novel battery- type electrodes with a core-shell structure based on zinc and copper oxides, synthesized using a straightforward, eco-friendly two-step electrodeposition method [4, 7]. The core-shell design is pivotal. Zinc oxide (ZnO) forms the core, enveloped by copper oxide (CuO), both attached to a conductive nickel foam substrate. This configuration maximizes active surface area and promotes efficient charge transfer during electrochemical reactions [8]. The core-shell setup facilitates intercalation pseudo capacitance, allowing ions to easily access and engage with the electrode material, thus boosting specific capacitance and overall performance. Moreover, the synthesis approach is notable for its simplicity and environmental sustainability. Electrodeposition, involving controlled ion deposition

© The Author(s), under exclusive license to Springer Nature Switzerland AG 2025
M. Hatti (Ed.): IC-AIRES 2024, LNNS 1238, pp. 423–427, 2025.
https://doi.org/10.1007/978-3-031-80301-7_46

on a conductive substrate, minimizes environmental impact compared to conventional methods. In summary, this research underscores the potential of core-shell battery-type electrodes to advance supercapacitor technology for renewable energy applications. By refining material design and synthesis techniques, this study aims to contribute to high-performance energy storage solutions that support the integration and utilization of renewable energy sources effectively [9, 10].

2 Experimental Section

2.1 Preparation of ZnO on the Nickel Foam (NF)

First of all, as the substrate, NF was cut into the size 1×2 cm2. Then it was rinsed with ultrasonic to remove impurities on the NF surface. Lastly it was dried in a vacuum oven.

ZnO was electrodeposited on the NF surface by potentiostatic method, using the three-electrode system, with pure NF (the working electrode), platinum wire used as the counter electrode and Hg/HgO electrode conducted as the reference electrode. The electrodeposition of ZnO was proceeded potentiostatically.

2.2 Electrodeposition of CuO on ZnO/NF

CuO was electrodeposited on ZnO/NF using a potentiostatic technique, which is akin to the ZnO production process mentioned above. There was no alteration made to the reference electrode or counter electrode. Rather, the working electrode was ZnO/NF produced in the first phase. Cathodic potential was used to conduct the electrodeposition of CuO.

Following that, the generated material was dried under vacuum and rinsed with CuO@ZnO/NF DI water (Fig. 1).

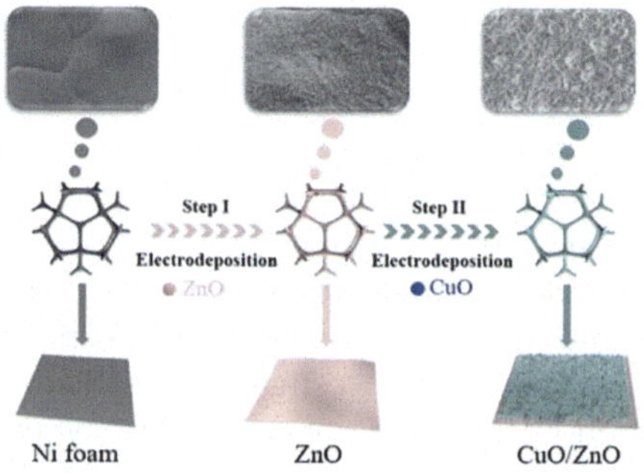

Fig. 1. Schematic representation of the steps associated with preparation for CuO@ZnO/NF.

2.3 Assembly of the CUO@ZNO //AC HSC

Activated carbon (AC) was used as the negative electrode and the previously manufactured CuO@ZnO/NF as the positive electrode to build an aqueous hybrid supercapacitor (HSC) in a 2 M KOH solution.

2.4 Electrochemical Measurements

Tests were conducted on the electrochemical performance of various electrode materials using a typical three-electrode cell with 2 M KOH electrolyte using a Solartron SI 1287 workstation. As working, reference, and counter electrodes, respectively, the as-prepared electrodes, Hg/HgO electrode, and Pt wire were utilized. The techniques of cyclic voltammetry (CV), galvanostatic charge-discharge (GCD), and cycling stability following 8,000 cycles.

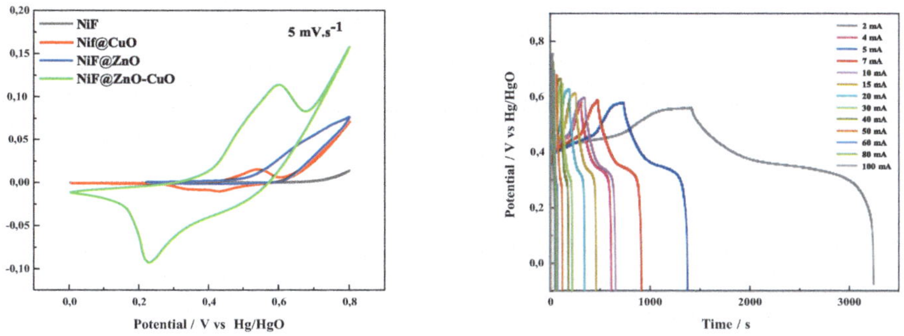

(a) (b)

Fig. 2. (a) CV curves of ZnO@CuO, ZnO, CuO and NF at 5 mV s − 1. (b) GCD curves of ZnO@CuO.

The CV curves of NF, CuO/NF, ZnO/NF and CuO@ZnO/NF are tested at the voltage range from 0 V to 0.8 V at 5 mV s−1 displayed in Fig. 2a. As can be seen that the prepared CuO@ZnO/NF material shows the highest peak response current and the largest CV enclosed area, compared to ZnO/NF and CuO/NF, NF, implying the highest specific capacity of CuO@ZnO/NF.

Figure 2b illustrates the GCD curves of CuO@ZnO/NF for -0.1-0.8 V at different current densities of 2–100 mA cm^{-2}. All GCD profiles have almost symmetrical characteristic and voltage platform, showing excellent reversibility and fast Faraday redox reaction.

The cyclability of the as-fabricated CuO@ZnO/NF//AC HSC device over 8,000 cycles at 100 mA.cm^{-2} is presented in Fig. 3. As expected, it maintains 95.7% of the initial value even after 8,000 GCD cycles, revealing that the HSC device has good stability with a long cycle life.

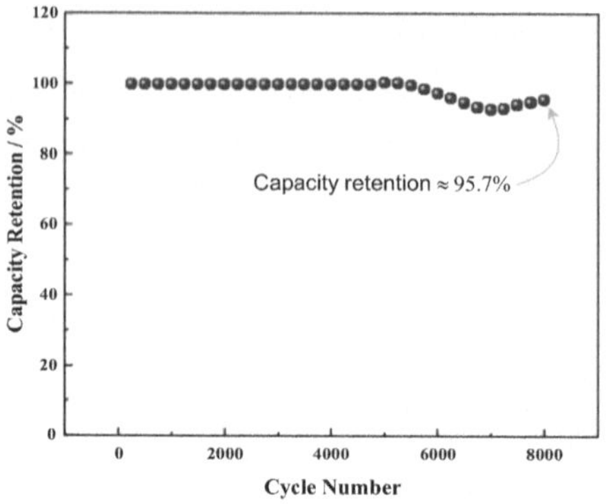

Fig. 3. Cycling performance of CuO@ZnO/NF at 100 mA.cm^{-2}.

3 Conclusion

In conclusion, we were able to effectively create an ultra-thin CuO@ZnO on Ni foam using an easy two-step electrodeposition technique. Notably, CuO@ZnO/NF exhibits excellent electrochemical characteristics when used as the anode of hybrid supercapacitors. In particular, at a current density of 2 mA.cm-2, the synthesized CuO@ZnO/NF shows a high specific capacity of 6123,3 mF.cm-2. CuO@ZnO/NF/AC HSC device constructed with a 95% capacity retention rate.

In general, our study develops an electrode material using an easy and secure process that has significant potential for use in energy storage devices, particularly supercapacitors., this work prepares an electrode material through a simple and safe method, which has great prospects in the application of energy storage equipment especially for supercapacitors.

References

1. Berrabah, E., et al.: Alloys and Compound **942**, 0925–8388 (2023)
2. Jia, H., et al.: Interlaced Ni-Co LDH nanosheets wrapped Co9S8 nanotube with hierarchical structure toward high performance supercapacitors. Chem. Eng. J. **351**, 348–355 (2018)
3. Xie, S., Gou, J., Liu, B., Liu, C.: Synthesis of cobalt-doped nickel sulfide nanomaterials with rich edge sites as high- performance supercapacitor electrode materials. Inorg. Chem. Front. **5**, 1218–1225 (2018)
4. Tang, H., Yao, J., Zhu, Y.: Recent developments and future prospects for zinc-ion hybrid capacitors: a review. Adv. Energy Mater. **11**, 2003994 (2021)
5. Wang, T., Liu, J., Ma, Y., Han, S., Gu, C., Lian, J.: P-N heterojunction NiO/ZnO electrode with high electrochemical performance for supercapacitor applications. Electrochim. Acta **392**, 138976 (2021)
6. Qin, Y., et al.: Nitrogen-doped Ni2P/Ni12P5/Ni3S2 three-phase heterostructure arrays with ultrahigh areal capacitance for high-performance asymmetric supercapacitor. Electrochim. Acta **393**, 139059 (2021)
7. Amiri, M., Moosavifard, S.E., Davarani, S.S.H., Kaverlavani, S.K., Shamsipur, M.: Mn CoP hollow nanocubes as novel electrode material for asymmetric supercapac itors. Chem. Eng. J. **420**, 129910 (2021)

8. Berrabah,S.E., et al.: Alloys Compoun. **942,** 0925–8388 (2023)
9. Ning, F., Shao, M., Zhang, C., Xu, S., Wein, M., Duan, X.: Co3O4@layered double hydroxide core/shell hierarchical nanowire arrays for enhanced supercapacitance performance. Nano Energy **7**, 134–142 (2014)
10. Veeramani, V., Madhu, R., Chen, S., Sivakumar, M.: Flower-like nickel-cobalt ox ide decorated dopamine-derived carbon nanocomposite for high performance supercapacitor applications. ACS Sustain. Chem. Eng. **4**, 5013–5020 (2016)

Optimizing Annual Energy Output of a Residential Bifacial PV System Using Box-Behnken Design: A Case Study in Tamanrasset, Algeria

Khaoula Rahrah[1(✉)], Soufiane Taibi[2], and Hassen Bouzgou[1]

[1] Department of Industrial Engineering, Laboratory of Automation and Manufacturing, University of Batna 2 (Mostefa Ben Boulaïd), 05000 Batna, Algeria
khaoula.rahrah@univ-batna2.dz

[2] Department of Electrical Engineering, Laboratory of Electric Traction Systems, University of Batna 2 (Mostefa Ben Boulaïd), 05000 Batna, Algeria

Abstract. In this paper, Response Surface Methodology (RSM) with optimization is proposed to determine the optimal annual energy output of a small-scale residential bifacial PV system, considering constraints input parameters such as albedo type, module height, tilt angle, and spacing. This could be done by quantifying the relationship between the variable input parameters and the corresponding output parameter where RSM is integrated with PVsyst simulation tool for the analysis. The proposed Box-Behnken design (BBD) optimization required 29 runs for data acquisition and modelling the response surface and Design-Expert software was used to design the experiments and randomize the runs. Regression model was developed and its adequacy was verified to predict the output value at nearly all conditions. The optimization study shows that the system produces the maximum yearly energy when the ground albedo is 0.71, module height is 1.68 m, tilt angle is 19.42°, and module spacing is 7.42 m. These characteristics were found using the meteorological conditions in Tamanrasset, Algeria.

Keywords: Bifacial · Response surface methodology · Box-Behnken Design · PVsyst · Design-Expert · Tamanrasset

1 Introduction

As the global energy landscape evolves, the quest for sustainable and efficient renewable energy solutions becomes increasingly crucial. Solar energy, characterized by its vast potential and rapidly declining costs, has emerged as a frontrunner in this transition. The bifacial photovoltaic system is a particularly promising technology in the solar energy domain, with the potential to significantly increase energy output and system efficiency. In contrast to traditional PV systems, bifacial panels are engineered to capture solar radiation from both the front and rear surfaces of the solar cells, thereby enabling them to harness a greater proportion of the available solar energy [1]. Numerous studies in the

© The Author(s), under exclusive license to Springer Nature Switzerland AG 2025
M. Hatti (Ed.): IC-AIRES 2024, LNNS 1238, pp. 428–436, 2025.
https://doi.org/10.1007/978-3-031-80301-7_47

literature have examined the difference in power output between bifacial and monofacial panels, showing that bifacial modules generally produce 15% to 20% more energy than monofacial ones, depending on various factors [2, 3]. To maximize energy output, it is crucial to determine the optimal installation parameters. A bifacial photovoltaic system's energy output is influenced by several factors, including tilt angle, panel spacing, albedo type, and module height [4].

The performance of a bifacial PV array is influenced by several factors, as illustrated in Fig. 1. To maximize the annual energy gain, RSM with optimisation is used, combining PVsyst simulation with Design-Expert statistical software for analysis and optimization.The current paper builds upon the research conducted in article [5], it attempted use the optimization using the response surface approach, BBD instead of face-centered composite design (CCD), in the case of Tamanrasset site in Algeria.

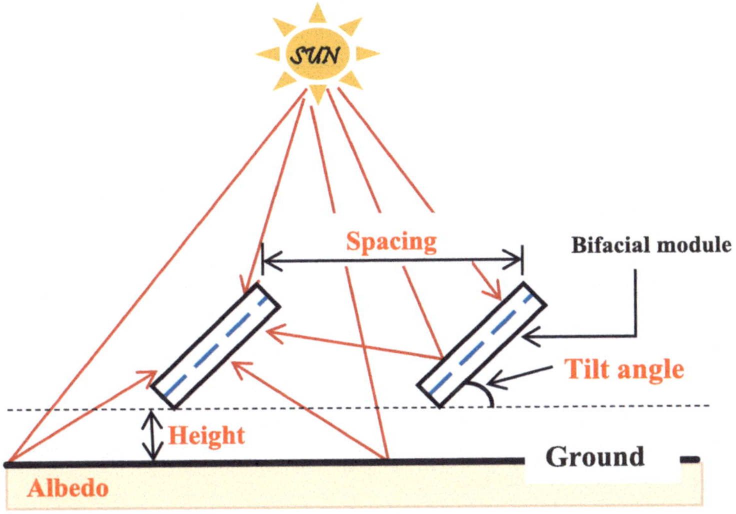

Fig. 1. Factors affecting the performance of bifacial PV module.

2 Methodology and Simulation

A small-scale 7 Kwp photovoltaic system was chosen for this study. Bifacial PV modules (14 modules) enable this system to generate electricity from solar energy.We used the same types of bifacial panels and inverter mentioned in the article [5].

2.1 Response Surface Methodology

The RSM approach was created in the 1950s by Box et al. is a set of mathematical and statistical techniques that are helpful for modeling and assessing engineering issues [6, 7]. The primary goal of this technique is to optimize the response surface, which is affected by several process parameters [7]. This study employed RSM to investigate the impact of four independent variables: Albedo (A), Height (B), Tilt angle (C), and Spacing (D) on the annual energy output of a bifacial photovoltaic system. Figure 2 shows a flowchart of the methodology proposed for optimizing the annual energy output by hybrid integration between PVsyst software (version 7.4) and using BBD (Box-Behnken Design) implemented in Design-Expert statistical software (version 13).

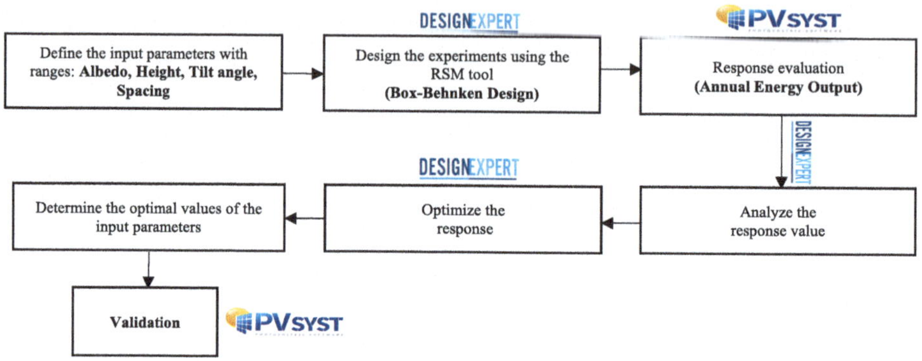

Fig. 2. Flowchart illustrating the methodology used in RSM.

The experiments are designed using parameters derived from the literature [5]. The design variables's upper and lower bounds are defined in Table 2, and a Box-Benhken design (BBD) is tested using Design-Expert. PVsyst is then used to determine the response during testing and Design-Expert software is used to analyze, optimize, and determine the optimal design. Finally, PVsyst is used to validate the optimal inputs and response.The ranges of the variables analyzed, along with their corresponding simulated responses, are outlined in Table 1.

2.2 Box-Behnken Designs

BBD are a class of second-order designs that are either rotatable or nearly rotatable, constructed from three-level incomplete factorials. For three factors, they can be visually represented in two ways [8]:

1. As a cube consisting of the central point and midpoints of each edge, as depicted in Fig. 3a.
2. As a structure of three interconnected 22 factorial designs with a central point, as illustrated in Fig. 3b.

The Box-Behnken design avoids placing points at the vertices of the cubic region defined by the upper and lower bounds of each variable [7].

The adjusted ranges for input parameters (A, B, C, D) at three levels (coded as -1, 0, + 1) to represent lower, intermediate, and higher values, respectively are outlined in Table 2.

Table 1. Variables of BBD matrix and simulated data for annual energy output.

	Input parameters				Response
Run No	Albedo	Height (m)	Tilt angle (°)	Spacing (m)	Annual energy output, (Kwh)
1	0.1	0.2	52.5	6	13414
2	0.475	1.1	90	10	11173
3	0.475	2	90	6	10107
4	0.85	1.1	15	6	16889
5	0.475	1.1	52.5	6	14684
6	0.475	2	15	6	16197
7	0.1	1.1	90	6	7797
8	0.475	2	52.5	10	15605
9	0.475	1.1	52.5	6	14684
10	0.475	0.2	52.5	10	15143
11	0.475	1.1	15	10	16115
12	0.475	1.1	52.5	6	14684
13	0.475	1.1	52.5	6	14684
14	0.475	0.2	90	6	10201
15	0.475	1.1	52.5	6	14684
16	0.475	2	52.5	2	9925
17	0.85	1.1	90	6	12457
18	0.85	1.1	52.5	10	17101
19	0.475	0.2	15	6	15248
20	0.85	2	52.5	6	16129
21	0.475	1.1	15	2	10529
22	0.1	1.1	15	6	14951
23	0.1	1.1	52.5	2	9798
24	0.475	1.1	90	2	6628
25	0.1	2	52.5	6	13479
26	0.85	1.1	52.5	2	10047
27	0.1	1.1	52.5	10	13729
28	0.475	0.2	52.5	2	9916
29	0.85	0.2	52.5	6	15594

The polynomial statistical model is expressed numerically as follows (Eq. 1) [5, 9]:

$$Y(kwh/year) = \alpha_0 + \sum_{i=1}^{4} \alpha_i X_i + \sum_{i=1}^{4} \alpha_{ii} X_i^2 + \sum_{i=1}^{4} \alpha_{ij} X_i X_j \quad (1)$$

where Y(Kwh/year) represents the annual energy output as a response; $\boldsymbol{X_i}$ and $\boldsymbol{X_j}$ are the chosen independent variables; The values of model coefficients are $\boldsymbol{\alpha_0}$ (intercept), $\boldsymbol{\alpha_i}$ (linear effect), $\boldsymbol{\alpha_{ii}}$ (quadratic effect), and $\boldsymbol{\alpha_{ij}}$ (interaction effect). The regression parameters, including the correlation coefficient (R^2), adjusted R^2, F-value, and p-value (probability), were calculated through analysis of variance (ANOVA). This commonly used statistical technique evaluates the effects of input factors and their interactions on the response [5], while also determining the relevance and accuracy of the predicted model,

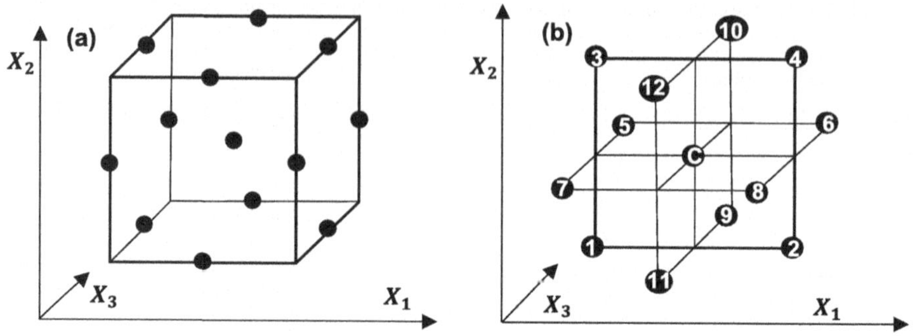

Fig. 3. (a) The cube representation for the Box-Behnken Design, (b) Three interlinked 2^2 factorial designs.

Table 2. Input parameters and their levels identification.

		Levels		
Input Parameter	Units	Low (−1)	Central (0)	High (+1)
Albedo (A)		0.1	0.475	0.85
Height (B)	m	0.2	2.2	2
Tilt Angle (C)	°	15	52.5	90
Spacing (D)	m	2	6	10

with a 95% confidence level in the developed model, an assessment was conducted to determine the impact of independent variables on the annual energy output [8].

3 Results and Discussion

3.1 The Experiment Matrix's Design and the Response Data

Using Box-Behnken RSM, different experiments are designed. Table 3 illustrates the 29 runs that make up this design, which include 24 factorial points and 5 center points. Using Design-Expert software, multiple combinations of the four input factors' levels were represented in the proposed runs. Next, using PVsyst, the system's yearly energy production under Tamanrasset's weather conditions is assessed.

3.2 Analysis of Variance

An analysis of variance (ANOVA) includes estimating the F ratio, mean squares (MS), degrees of freedom (DF), and sum of squares (SS). All the required definitions are important for determining the P-value, which is an indicator of statistical significance for a specific factor. The P-value is considered statistically significant if it is smaller than significance level (5%). ANOVA is employed to assess the statistical significance of the effects by comparing the mean square of each factor to the estimated error. In this analysis, seven effects (A, C, D, AC, AD, C2, and D2) exhibited P-values below 0.05, signifying their significance at a 95% confidence level. These results are detailed in Table 3, where A represents albedo, B is module height, C is tilt angle, and D is module spacing.

Table 3. ANOVA for Quadratic model.

Source	SS	DF	MS	F-Ratio	P-value
Model	2.322E + 08	14	1.659E + 07	98.43	< 0.0001
A	1.887E + 07	1	1.887E + 07	112.00	< 0.0001
B	3.091E + 05	1	3.091E + 05	1.83	0.1971
C	8.303E + 07	1	8.303E + 07	492.77	< 0.0001
D	8.546E + 07	1	8.546E + 07	507.14	< 0.0001
AB	55225.00	1	55225.00	0.3277	0.5761
AC	1.852E + 06	1	1.852E + 06	10.99	0.0051
AD	2.438E + 06	1	2.438E + 06	14.47	0.0019
BC	2.720E + 05	1	2.720E + 05	1.61	0.2246
BD	51302.25	1	51302.25	0.3045	0.5898
CD	2.709E + 05	1	2.709E + 05	1.61	0.2255
A^2	570.10	1	570.10	0.0034	0.9544
B^2	25540.95	1	25540.95	0.1516	0.7029
C^2	1.757E + 07	1	1.757E + 07	104.29	< 0.0001
D^2	2.514E + 07	1	2.514E + 07	149.22	< 0.0001
Residual	2.359E + 06	14	1.685E + 05		
Lack of Fit	2.359E + 06	10	2.359E + 05		
Pure Error	0.0000	4	0.0000		
Cor Total	2.346E + 08	28			

The RSM is formulated based on the pertinent data presented in Table 1, a second-order polynomial model between the output parameter and the input parameters, was obtained and described in the following Equation (Eq. 2):

$$E(Kwh/year) = 6901.25823 - 2638.94074\,A + 400.25617\,B + 48.67741\,C + 1953.04792\,D + 348.14815\,A*B + 48.39111\,A*C + 520.5\,A*D - 7.72593\,B*C + 31.45833\,B*D - 1.735\,C*D - 66.66667\,A^2 - 77.46914B^2 - 1.17049C^2 - 123.05469D^2 \quad (2)$$

where E (Kwh/year) is the annual energy output, A is the albedo, B is the height module, C is the tilt angle, and D is the spacing. AB, AC, AD, BC, BD, and CD is the interaction between input parameters. Equation 2, expressed in terms of actual factors, can be utilized to predict the response at specified levels of each factor (see Table 2). The model's coefficient of determination (R^2) is 98.99%, indicating its excellent accuracy and reliability in predicting the system's annual energy production based on the input parameters within the defined range. The Predicted R^2 of 0.9421 is closely aligned with the Adjusted R^2 of 0.9799, with a difference of less than 0.2. Adequate Precision, which assesses the signal-to-noise ratio, indicates that a value greater than 4 is preferred.With a value of 38.543, this indicates a strong and sufficient signal, as shown in Table 4.

Table 4. The ANOVA's fit statistics.

STD. DEV	410.50	R^2	0.9899
Mean	13158.34	**Adjusted R^2**	0.9799
C.V. %	3.12	**Predicted R^2**	0.9421
		Adeq. Precision	38.5429

Figure 4, presents a comparison between the energy output results from PVsyst and the predicted values from Design-Expert. It also demonstrates that the annual energy generation estimated by the proposed regression model closely matches the simulated results from PVsyst.

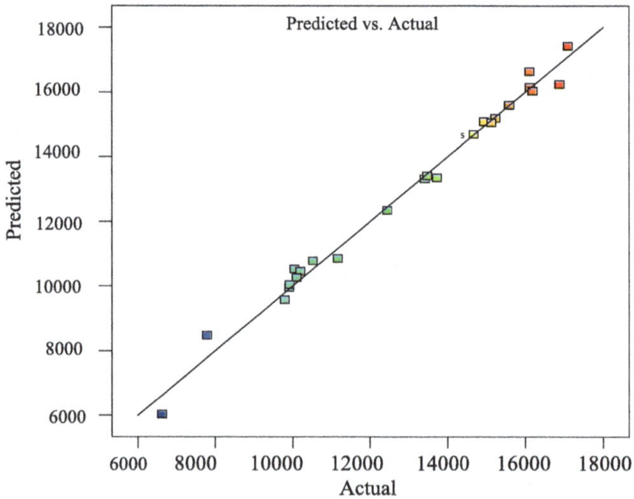

Fig. 4. Comparison of annual energy output: Predicted values from the regression model vs. actual data from the PVsyst.

3.3 Optimisation Research

RSM aids in evaluating the annual energy output of bifacial PV modules and in identifying the ideal conditions to enhance performance. The desirability approach developed by Derringer and Suich [10] is effective for optimizing one or more responses by finding the best combination of input variables to achieve the highest overall desirability.This approach finds the input conditions that produce the best outputs [5], as shown in Table 5, which lists the optimal parameter values and their associated annual energy generation. Figure 5 illustrates these optimal values for Albedo, Height, Tilt angle, and Spacing, along with their corresponding annual energy generation..

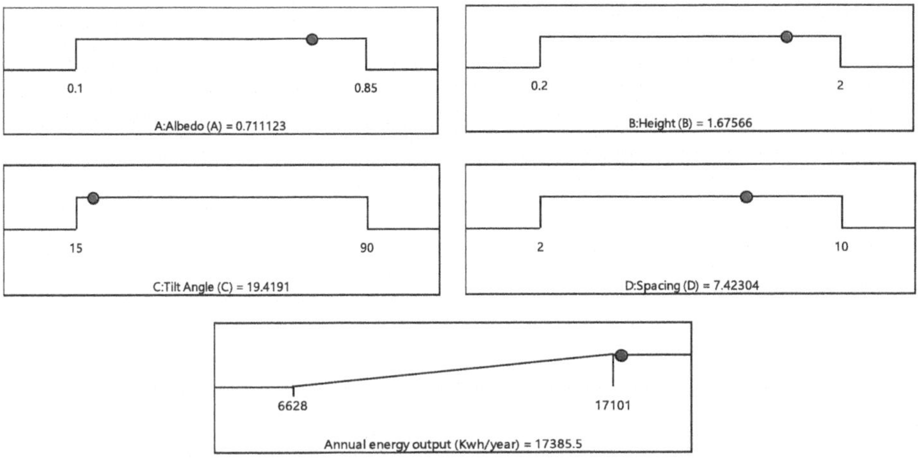

Fig. 5. The system's output energy production was maximized by optimizing its parameters.

It is noted that all optimal values are within the ranges specified for each factor in Table 1. To verify these optimization results, The optimal inputs provided in Table 5 are utilized to compute the annual energy generation using the PVsyst simulation tool. This yields an energy output of 17385.5 kwh, with a percent error of 0.79%.

Table 5. The optimal input parameter values and their associated annual energy generation.

Parameters and response	Goal	Lower limit	Upper limit	The optimal value
Albedo (A)	is in range	0.1	0.85	0.711
Height (B)	is in range	0.2	2	1.676
Tilt Angle (C)	is in range	15	90	19.419
Spacing (D)	is in range	2	10	7.423
Annual energy output (Kwh/year)	maximize	6628	17101	**17385.5**

4 Conclusion

This study aims to identify the optimal parameters for a small-scale residential bifacial photovoltaic (PV) system to maximize energy production. To do this, we utilized PVsyst software and the Design-Expert tool to validate and demonstrate the effects of input parameters and their interactions on the system's annual energy output. Using Response Surface Methodology (RSM) and a second-order polynomial Box-Behnken Design (BBD), we optimized the system's performance, resulting in a predicted annual energy output of 17385.5 kwh/year. This optimal output was achieved with a ground albedo of 0.71, a module height of 1.68 m, a tilt angle of 19.42°, and a panel spacing of 7.42 m, with a percent error of less than 1%. These findings indicate that the optimal parameters identified through desirability optimization significantly improve annual energy generation and should be applied in practice.

References

1. Haegel, N.M., et al.: Terawatt-scale photovoltaics: Transform global energy. Science **364**(6443), 836–838 (2019)
2. Mühleisen, W., et al.: Energy yield measurement of an elevated PV system on a white flat roof and a performance comparison of monofacial and bifacial modules. Renew. Energy **170**, 613–619 (2021)
3. Raina, G., Sinha, S.: A simulation study to evaluate and compare monofacial vs bifacial PERC PV cells and the effect of albedo on bifacial performance. Mater. Today: Proc. **46**, 5242–5247 (2021)
4. Stein, J., et al.: Bifacial photovoltaic modules and systems: Experience and results from international research and pilot applications (No. SAND-2021–4835R; IEA-PVPS T13–14: 2021). Sandia National Lab. (SNL-NM), Albuquerque, NM (United States); Fraunhofer ISE, Freiburg (Germany); Univ. of Applied Sciences and Arts of Southern Switzerland (SUPSI)(Switzerland); TUV Rheinland, Cologne (Germany); Ricerca sul Sistema Energetico (Italy); ATAMOSTEC (Chile); Turku University of Applied Sciences (Finland) (2021)
5. Radwan, A., Mdallal, A., Haridy, S., Abdelkareem, M.A., Alami, A.H., Olabi, A.G.: Optimizing the annual energy yield of a residential bifacial photovoltaic system using response surface methodology. Renew. Energy **222**, 119914 (2024)
6. Yolmeh, M., Jafari, S.M.: Applications of response surface methodology in the food industry processes. Food Bioprocess Technol. **10**(3), 413–433 (2017)
7. Manohar, M., Joseph, J., Selvaraj, T., Sivakumar, D.: Application of Box Behnken design to optimize the parameters for turning Inconel 718 using coated carbide tools. Int. J. Sci. Eng. Res. **4**(4), 620–644 (2013)
8. Ferreira, S.C., et al.: Box-Behnken design: An alternative for the optimization of analytical methods. Anal. Chim. Acta **597**(2), 179–186 (2007)
9. Dbik, A., El Messaoudi, N., Bentahar, S., El Khomri, M., Lacherai, A., Faska, N.: Optimization of methylene blue adsorption on agricultural solid waste using box–behnken design (BBD) combined with response surface methodology (RSM) modeling. Biointerface Res. Appl. Chem **12**(4), 4567–4583 (2022)
10. Derringer, G., Suich, R.: Simultaneous optimization of several response variables. J. Qual. Technol. **12**(4), 214–219 (1980)

Corrosion and Protection of Irrigation Pivots in Adrar Region, Algeria

Boukhlef Djedjiga[1](✉), Belhi Fatima[2], Khaldi Manal[2], Saadi Zine[1], Boutadara Abdelkader[1], and Hadidi Abedelkader[1]

[1] Unité de Recherche en Energies Renouvelables en Milieu Saharien (URERMS), Centre de Développement des Energies Renouvelables (CDER), Adrar, 01000, Algeria
d.boukhlef@cder.dz

[2] Faculty of Science and Technology, Department of Materials Science, University of Ahmed Draia Adrar, 01000 Adrar, Algeria

Abstract. Corrosion is a profoundly serious challenge to most industries around the globe because it can endanger lives, waste resources, and lead to economic loss. Therefore, corrosion of metallic materials is of significant interest to material scientists, chemists, chemical engineers, and agricultural industries that are not immune. Numerous methods exist to prevent corrosion, both theoretically and practically. However, human errors and the overuse of available protection methods have limited their practical application in reducing corrosion rates of materials used in construction. Specifically, a pivot built at the research unit in renewable energies in Saharan medium (URERMS), situated in the Algerian cities of Adrar and Aougrout (Timimoune), is the subject of this study, which attempts to research the problem of corrosion in agricultural machinery and strategies to prevent it, and knowing the type of corrosion that attacks these machines, allows us to find the protection mode and the right choice of material. Due to the use of no longer corrosion-resistant components, an excessively high salinity rate in the irrigation water, and an extra problem with the hot temperature, the URERMS pivot becomes contaminated by galvanic corrosion after six months. Metals like steel, copper, and aluminum that are more naturally resistant to corrosion are frequently used in protection systems, along with water treatment before irrigation or other usage. The findings suggest that irrigation pivots in the Adrar region face a considerable challenge from corrosion, necessitating the use of more resilient materials and aggressive maintenance to guard against harsh climatic conditions. Local researchers and innovators are exploring solutions to address this issue.

Keywords: Saharan mediums · irrigation pivot · material · corrosion · protection

1 Introduction

Corrosion is defined as a physical-chemical interaction between a metal and its environment, resulting in degradation of the properties of the metal and its environment [1, 2]. The term corrosion can be defined in different ways, but the usual interpretation is "an attack on a metallic material by reaction with its environment" [3]. Corrosion is a

© The Author(s), under exclusive license to Springer Nature Switzerland AG 2025
M. Hatti (Ed.): IC-AIRES 2024, LNNS 1238, pp. 437–442, 2025.
https://doi.org/10.1007/978-3-031-80301-7_48

serious challenge for most industries around the globe because it can endanger lives, waste resources, and lead to economic loss. Metal corrosion is something that interests material scientists, chemists, chemical engineers, and agricultural industries. The effect of corrosion on equipment or structures' safe, reliable, and efficient operation is often more severe than simply the loss of a mass of metal [4]. Irrigation pivots installed in the Adrar region of Algeria are susceptible to corrosion, particularly galvanic corrosion, due to the materials used in their construction [5, 6]. The region's water quality, with high salinity and mineral content, contributes to the rapid deterioration of the pivots, which can occur within just six months of installation [6]. Several factors influence the corrosion of irrigation pivots in Adrar, including [6]: Water quality (salinity, mineral content), soil composition, climatic conditions (temperature, humidity) and improper maintenance. To protect against corrosion, the pivots need to be made from materials that are resistant to the harsh environmental conditions of the Adrar region. Proper maintenance and regular inspections are also crucial to identify and address corrosion issues early on [6]. Using renewable energy sources, like solar and wind power, for pumping and irrigation in the region could help reduce corrosion by decreasing the reliance on fossil fuel-powered systems [7]. Smart remote irrigation systems that optimize water usage and distribution could also contribute to the longevity of the pivots by preventing overwatering and reducing the impact of water quality on the system [7]. This research aims to identify the type of corrosion and develop a protective strategy for the agricultural machinery used by farmers in the Sahara region. Equipment can be kept clean and dry after every use, exterior or internal coating systems can be applied, or cathodic protection can be used to regulate and prevent corrosion. This work aims to investigate the issue of corrosion on agricultural machinery and ways to prevent it, with a particular focus on a pivot installed at the research unit in renewable energies in Saharan medium (URERMS), located in Algerian cities of Adrar and Aougrout (Timimoune) [6]. Corrosion control and prevention can be performed by keeping equipment clean and dry after each use, applying external coatings or internal coating systems, or using cathodic protection which can be used to regulate and prevent corrosion. The diverse nature of structures, machinery, and fittings employed for farming purposes makes it hard to pinpoint where corrosion can be most damaging in agriculture. This problem is made worse by environmental factors such as rural, urban, and coastal indices. Also of concern are the climatic conditions, type of farming activities, seasonal use of specific equipment and machinery, and various storage methods employed. The corrosion of a metal is largely influenced by the environment in which it is used [8, 9].

2 Agricultural Industry

2.1 Corrosion Agriculture

Agricultural corrosion is the term used to describe how exposure to chemicals, moisture, and soil can erode metal structures and components used in agricultural applications, such as irrigation systems, storage tanks, and farm machinery [10]. It may result in equipment breakdown and lower output. Corrosion is an important problem in agricultural machinery, which is not considered by most manufacturers [9]. This problem causes the machine's life to be shortened. Corrosion control and prevention can be achieved by keeping equipment clean and dry after each use, applying external coatings (paints) or internal coating systems, or using cathodic protection.

2.2 Corrosion Types

Corrosion will become synonymous with data centers worldwide in the near future with ever-changing ambient air quality. The only solution currently available for achieving clean air in the indoor space for the safety of humans and machines is filtering [8]. Corrosion is said to be localized when the cathode and anodic areas are physically separated, which is not the case for generalized corrosion. For galvanic corrosion, it occurs when two materials with different electrochemical potentials come into contact [10]. Galvanic corrosion, resulting from a metal contacting another conducting material in a corrosive medium, is one of the most common types of corrosion [11, 12]. Corrosion protection can develop according to different processes such as the choice of materials, the suitable shape of the parts, and the choice of the best coating process. A significant amount of research is currently being conducted to advance the field of corrosion protection. This is due to the numerous disadvantages of the corrosion phenomenon, which can cause accidents with serious consequences in the production of energy, civil engineering, transportation, machinery, medical materials, micro-electronic components, etc.

Both irrigation pivots at URERMS exhibit galvanic corrosion at the support and motor (Fig. 1), at the support and central unit of the irrigation pivot of Aougrout, there is a corrosion that is uniform and pitting (Fig. 2).

Fig. 1. Support and the motor of irrigation pivot at URERMS

2.3 Corrosion Causes

The region's water quality, with high salinity and mineral content, contributes to the rapid deterioration of the pivots, which can occur within just six months of installation [6]. The hot temperature is not the only problem with irrigation water's high salinity rate. Metals that are more resistant to corrosion are frequently selected in protection designs, such as steel, copper, and aluminum, and water is sterilized before irrigation or any other use.

Fig. 2. Support and central unit of irrigation pivot at Aougrout

2.4 Corrosion Solutions

In this context, we proposed some solutions for corrosion which are: using more corrosion-resistant materials in the pivot construction, implementing proper maintenance and monitoring of the pivots to detect and address corrosion early [5, 6], and exploring the use of renewable energy sources like solar and wind power to operate the irrigation pivots, as this could help reduce corrosion compared to using fossil fuel-powered systems [7, 13].

3 Results and Discussions

Several elements contribute to corrosion; two prominent ones are high temperatures and salinity, both of which are common in the area of Adrar. Both salinity and high temperatures affect the rate of corrosion and cannot be changed. Additional determining factors are soil resistivity, water pH, conductivity, and pH of the soil. Since low pH speeds up corrosion in anaerobic conditions, the pH values in both sections are above 7 (Fig. 3).

An acidic water supply is not the source of corrosion in either area. These waters are quite salty, according to studies done by the National Institute of Soils of Irrigation and Drainage (INSID), and choosing a different source of water with better chemical quality is advised to maximize crop and irrigation water management. Research conducted in the National Agency of Hydraulic Resources (ANRH) laboratory has shown that groundwater is also characterized by very high salinity and a conductivity value that exceeds the norm, which is why water distillation is being proposed (Fig. 4). The brackish well water, is of poor quality and unfit for consumption [6]. Soil analyses of URERMS, carried out by INSID, showed that its major constraint is its high content of soluble salts, even for those of the Aougrout farm, they are salty and aggressive. Agricultural corrosion is the term used to describe how exposure to chemicals, moisture, and soil can erode metal structures and components used in agricultural applications, such as irrigation systems, storage tanks, and farm machinery. It may result in equipment breakdown and lower output [14]. To reduce the amount of salinity and acidity in the water used for irrigation, it is best to use distilled water. Furthermore, it's critical to reduce the soil's salinity before planting alfalfa because it absorbs a significant number of salts.

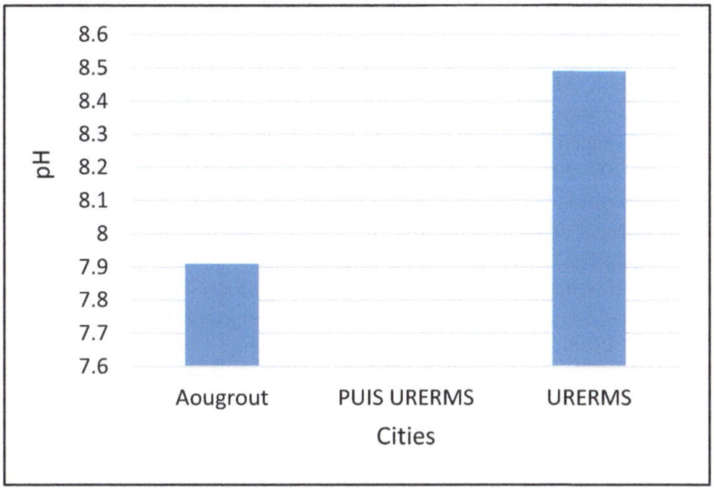

Fig. 3. Water pH values for each region

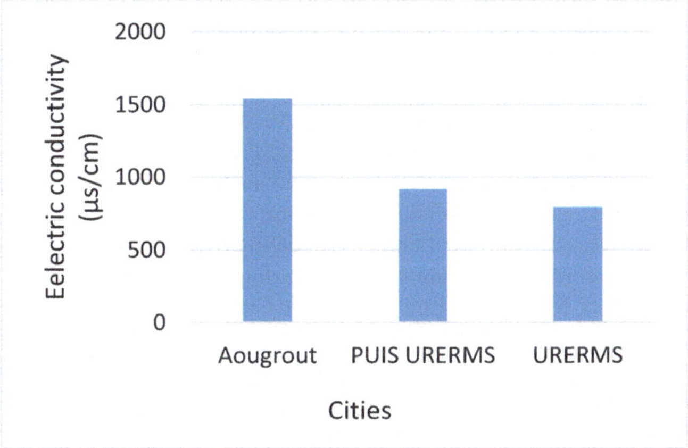

Fig. 4. Water conductivity values for each region

4 Conclusion

By giving basic information about the type of corrosion of the material being used, precautions can be taken to prevent the corrosion problem or extend the life of the material. This is because, in the long run, proper and effective protection against corrosion at the source saves money and resources. Protective designs typically select metals that are naturally resistant to corrosion, like steel, copper, and aluminum, and treat water before irrigation or any other use.

The proposed solutions to reduce salinity, pH of water, and soil salinity are the distillation technique and rearing of fish in tanks used for irrigation. According to the experience of some farmers, they used alfalfa to absorb salinity.

References

1. Boukhlef, D., Allali, M., Yaichi, M.: Corrosion of lead acide battery case study: Melouka photovoltaic solar power station, Adrar, Algeria. In: International Conference on Renewable Energy Advanced Technologies and Applications, URERMS, Adrar, ICREATA'21, Proceedings, pp 207–208. ISBN: 978–9931–9819–0–9 (2022)
2. Bosich, J.F:. Prévention de la corrosion, 2e édition mise à jour, Technique et Documentation, Entreprise moderne d'édition, Paris (1977)
3. Bardal, E.: Corrosion and Protection, Engineering Materials and Processes. Includes bibliographical references and index. Springer (2003). ISBN 1–85233–758–3
4. Eker, B., Yuksel, E.: Solutions to corrosion caused by agricultural chemicals. Trakia J. Sci. **3**(7), 1–6 (2005). ISSN 1312–1723. http://www.uni-sz.bg
5. Boukhlef, D., Allali, M., Hadidi, A., Saadi, Z.: Corrosionprotection of a pivot installed at URERMS, Adrar. In: First International conference on Material Science and Applications, Khenchela, Algeria, 08–09 Février 2023 (2023)
6. Belhi, F., Khaldi, M., Boukhlef, D.: Protection contre la corrosion des machines agricoles dans les zones arides. Master univ Adrar (2023)
7. https://www.iieta.org/download/file/fid/99321
8. Augustin, M.A., Šcerbejova, M.: Anti corrosive efficiency of preservants for farming machinery. Res. Agr. Eng. **49**(1), 2226 (2003)
9. Makanjuola, O., Anawe, P.A.L.: A review of corrosion in agricultural industries. Phys. Sci. Int. J. **5**(4), 216–222 (2015). ISSN: 2348–0130
10. Pealson, N.: Data centers: Electronics are more susceptible to corrosion than ever. https://www.linkedin.com/in/nitish-pearlson-96a17582. Accessed 26 July 2022
11. Lacroix, L.: Mécanismes de corrosion localisée de l'alliage d'aluminium 2024. Apport de la Microscopie à Force Atomique (AFM) couplée au Mode Kelvin (KFM) et des alliages modèles", thèse de Doctorat, Université de Toulouse (2008)
12. Boukhlef, D., Boughrara, D., Mohellebi, H.: Modeling by finite elements method of nonlinear conductivity in corrosive mediums. In: Proceedings of the International Conference on Mechanics, Materials, Mechanical Engineering and Chemical Engineering, Barcelona, Spain, 7–9 April 2015, pp. 203–210 (2015). ISSN: 2227-4596, ISBN: 978-1-61804-295-8
13. https://agritrop.cirad.fr/601443/1/2022%20Ould%20Rebai%20et%20al%20journey%20pivot%20New%20Medit.pdf
14. Saadi, Z.: Étude et réalisation d'un distillateur solaire à ruissellement, thèse de Doctoraten Sciences, Spécialité : Physique Énergétique, Université Oum El-Bouaghi, 24 Juin 2020 (2020)

Stick Slip and Bit Bounce: Survey of Bit-Rock Interaction Modeling

Samira Hadibi(✉), Madjid Kidouche, and Mohamed Zinelabidine Doghmane

Laboratory of Applied Automation, Department of Automation Faculty of Hydrocarbons and Chemistry, University M'hamed Bougara Boumerdes, Boumerdes, Algeria
s.hadibi@univ-boumerdes.dz

ABSTRACT. Drilling systems exhibit complex behaviors characterized by strong oscillations, prompting researchers to study their stability and develop control techniques. The nonlinear interaction at the bit-rock interface is a major problem when trying to model the dynamics of the drill string. This is what causes the vibrations, especially the coupled axial-torsional vibrations. This area has garnered substantial interest, leading to the development of various bit-rock interaction models, ranging from simplified to complex, incorporating both uni-directional and bi-directional couplings. This paper brings together the existing bit-rock interaction models that are used to simulate coupled axial and torsional vibrations in drilling systems, particularly at the bit-rock interface. It includes a detailed examination of the assumptions and simplifications employed in these analytical studies.

Keywords: Torsional vibration · axial vibration · coupled models · bit-rock · vertical drilling

1 Introduction

A drilling system is a set of equipment and techniques used to drill boreholes into the earth, often to extract natural resources such as oil, gas, and minerals. A drilling system is made up of several components, including the drilling rig, drill string, drill bit, and numerous control and monitoring systems. The drill string undergoes a variety of vibrations throughout a drilling operation as a result of the abrasive character of the rocks and the structure's natural flexibility in torsion and bend. Vibrations in drilling systems are significant because they can affect the efficiency and safety of drilling operations. There are several types of vibrations that can arise during drilling, and they are classified based on their axis of occurrence: axial (longitudinal), lateral (transverse), and torsional. Axial vibrations lead the drilling tool to rebound upward and downward, while torsional vibrations create a twisting motion.; and lateral vibrations result in side-to-side movement. Multiple types of vibrations occur simultaneously, including coupled vibrations, which further complicate the system dynamics.

© The Author(s), under exclusive license to Springer Nature Switzerland AG 2025
M. Hatti (Ed.): IC-AIRES 2024, LNNS 1238, pp. 443–452, 2025.
https://doi.org/10.1007/978-3-031-80301-7_49

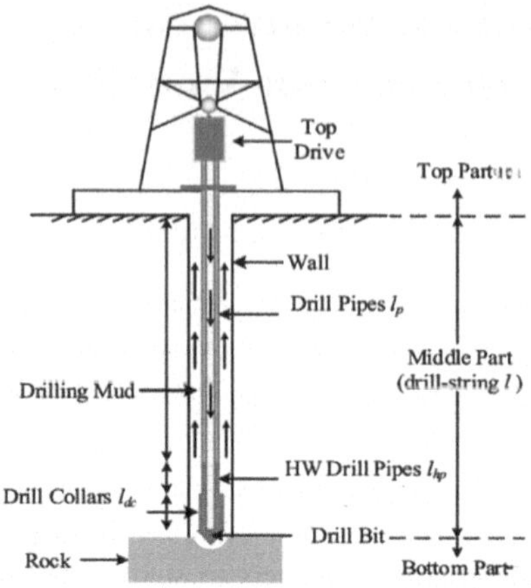

Fig. 1. A diagram of a drilling system

Contact and friction at interfaces such as the borehole-drillstring and bit-formation, as well as imbalance, eccentricity, or initial curvature within drill collar segments, and the presence of different linear or nonlinear resonances that cause self-induced stick-slip oscillations, are some of the main factors that cause these vibrations. During torsional stick-slip motion, the bit moves faster, leading to significant axial vibrations originating at the bottom. These vibrations can rapidly intensify, manifesting as bit-bounce oscillations, which contribute to accelerated bit wear and reduced drilling efficiency. A spectrum of remedies, ranging from basic operational directives to advanced control methodologies, has been proposed to mitigate or eliminate these vibrational phenomena [1].

Over the past decade, the structural dynamic analysis of drill strings has been extensively studied by various researchers employing diverse methodologies. The self-excited axial and torsional vibrations of deep drilling systems, which have been analyzed using discrete models based on a rate-independent bit-rock interaction law, are one significant area of investigation.

The nonlinear dynamics of drill strings can be modeled using several approaches. [2] employed a two-degree approximation for their analysis. Researchers in [3] used a finite element model (FEM) with beam elements and a complex bit-rock interaction model based on bit design to look into how drilling fluid affects the drill string's movement. In [4] and [5], a distributed parameter system, commonly represented through wave equations, is aimed at elucidating the oscillatory characteristics inherent to the drilling system. Within these various models, there is a clear connection between the two types of vibrations: axial and torsional. This shows how complex the dynamic properties of drill strings are.

Empirical observations in the field underscore the critical importance of understanding the intricate vibrational dynamics inherent in a drilling system to enhance operational control and optimize performance outcomes. In order to verify the model's efficacy, drill string vibrations must be replicated on a test rig. In an effort to better understand drill string dynamics and the relationship between the drill bit and the rock, several research have created laboratory test [6]. The primary challenge lies in the highly nonlinear nature of the rock-bit interaction term. Although mathematical models attempt to simulate its dynamics, their reliability is restricted since these systems lack bottom torque detection equipment [7].

A drilling test rig is a specialized setup used to analyze the dynamics and performance of drilling processes, consisting of a rigid frame, a high-speed spindle assembly driven by a variable-speed motor, and a tool holder for various drill bits. The workpiece is held in an adjustable holder, while a feed mechanism controls the drill bit's axial movement. With a number of excitation sources, including stick-slip, well-borehole contact, including the addition of drilling fluid, the rig can replicate the vibrational behavior of the drillstring. To recreate the weight-on-bit effect, a shaker shakes the drillstring along the axis and a magnetic tension brake simulates stick-slip. In [8, 9], researchers detailed frictional interactions and examined numerous friction models, demonstrating that the friction response can be highly sensitive depending on the selected frictional constitutive model. In [10] it was effectively demonstrated that torsional vibrations in two different rock-bit contact models could be minimized. While some publications do not focus specifically on bit-rock interaction, they are highly relevant to the representation of nonlinear coupling. This study explains the rules that govern how the bit and rock interact with each other. It focuses on how axial and torsional dynamics are connected, going into extensive detail about the assumptions and simplifying factors that were used in the analysis.

2 Friction Coefficient

Dry friction arises in mechanical systems where components are in contact and can slide relative to one another [11]. The nonlinear effect of friction within the two surfaces that come into contact is described by frictional properties that come from experiments. This demonstrates that the transition from stick to slide is not seamless. The frictional force is a function of the relative velocity between two interacting objects [12]. Stick-slip is hypothesized to occur due to the frictional interface's built, specifically how the coefficient of friction (μ) depends on the speed of two surfaces in contact.

Stick-slip oscillations are typically approximated using a friction factor that decreases with increasing velocity, as showcased by some experimental studies. This behavior of slowing down the speed makes the frictional force work as a negative damping mechanism. This causes instability to grow, which eventually leads to stick-slip oscillations [2]. Woodhouse et al. [13] have recently provided a comprehensive review of models and measurements pertaining to dynamic friction. Local velocity-weakening and Coulomb friction constitute two prominent friction representations.

3 Depth of Cut

The depth-of-cut function, d(t), refers to the vertical distance that the drill bit penetrates into the formation with each rotation or over a specific time period. It is a measure of how much material is being removed by the bit and is influenced by factors such as weight on the bit, bit type, rotational speed, and the mechanical properties of the formation. In certain drilling scenarios, the depth of cut can exhibit periodic variations that correspond to the bit's rotational motion. These periodic variations have been observed to follow a sinusoidal pattern, represented mathematically as (1), where φ denotes the bit rotation angle, ***s*0** corresponds to the initial depth of cut and ***nb*** is a constant [14]. This sinusoidal assumption aids in modeling and analyzing the dynamic behavior of the drill bit as it interacts with the formation [1].

$$d(t) = s_0 \sin(n_b \varphi) \tag{1}$$

Additionally, the expression (2) is commonly employed in the context of machining dynamics, particularly for analyzing the depth of cut in milling operations. This expression captures the corresponding depth of cut as an expression of the tool's current position ***U(t)*** and its position at a previous time $U(t - tn)$ where ***n*** denotes the number of cutting edges or teeth of the tool. In drilling, the depth of cut is determined by the cutter's axial location relative to the rock surface, which was created by the preceding cutting edge beforehand. This relationship highlights the importance of considering both the tool's current and past positions to be able to appropriately assess the depth of cut and comprehend the cutting process's dynamics [15] (Fig. 2).

$$d(t) = n(U(t) - U(t - t_n)) \tag{2}$$

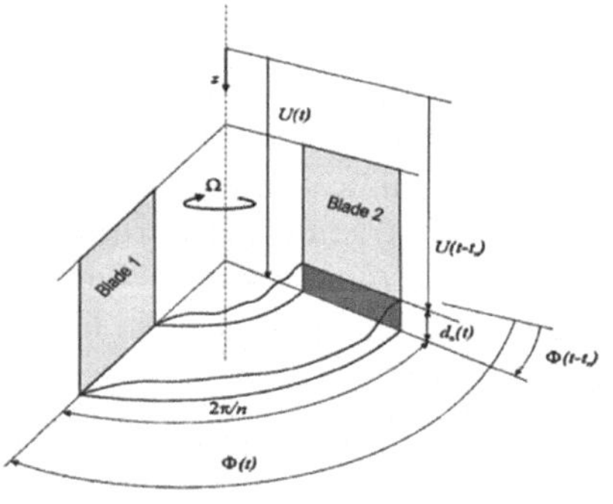

Fig. 2. A schematic of a drag bit illustrates the definition of the instantaneous depth of cut. It shows two successive blades of the bit.

Consider the axial bit position ***U*** and the torsional angle φ. Assuming the parameters $(\dot{U}, \dot{\varphi})$ evolve slowly during one revolution of the bit, the general interaction function may be estimated in a straightforward autonomous form (3) [16]. This analytical expression's accuracy is dependent on the number of bit rotations during a stick-slip cycle (the proportion of ***U*** to the basic pulsation) be adequate. This assumption is particularly valid when the ratio exceeds two; nevertheless, some geometrical properties of the bit may weaken this assumption [17].

$$d(t) = \frac{\dot{U}}{n_b \dot{\varphi}} \tag{3}$$

4 Torque on Bit

The torsional vibration can intensify into a severe condition in which the drilling bit becomes completely lodged and then slides (stick-slip). This extreme condition can be identified based on specific surface torque behavior characteristics which manifest as a drop in surface torque oscillations below the fundamental frequency, indicating a downhole sticking time. The drive surface torque increases to roughly equal the static torque needed to free the BHA and to spin the drill string. When this is combined with a decrease in surface rotational velocity, an inertial force opposing string rotation is created, building on the drive torque. When the bit and BHA break free downhole, surface torque drops below the level needed to rotate the drill string due to inertia [17].

The nonlinear function Tob accounts for the coupled impact of torque on the bit and nonlinear frictional forces throughout the drill string assembly. Although the specifics of these effects are unclear, numerous functions have been put forth for analyzing stick-slip vibrations. There is probably a considerable distinction. Between the static torque and sliding torque at different parts of the drill string, with laboratory measurements primarily focusing on the bit [18].

In laboratory drilling tests under simulated down-hole conditions in shale, the rate of penetration (ROP), torque, and mechanical torque at the drill bit were measured in terms of rotational speed, bit weight, and flow velocity. The data revealed a close relationship between torque and bit weight, except at the highest bit weights. Torque was found to be even more closely related to penetration per bit revolution [19].

Various Tob functions have been suggested to study stick-slip vibrations in models that couple torsional and axial dynamics. A typical Tob function incorporates variables such as depth of cut, friction coefficient, and rock characteristics, as in [2, 20–24]. In this scenario, the governing equation is a delay-differential equation with a state-dependent delay, distinguishing it from conventional models. Nonetheless, the coefficient of friction was supposed to be constant. Different models have demonstrated the opposing torque across the bit-rock interaction, making it easier to understand by showing it as a frictional contact with a local velocity-weakening effect and a two- way coupling [1, 24–26]. In contrast, the Tob model proposed by [28, 29] incorporated solely a unidirectional coupling, devoid of an inherent depth-of-cut function.

5 Weight on Bit

The drilling tool's rotational is governed by the resisting torque from the stiffness of the drillstring as well as the opposing torque on the bit from the bore hole interface. When the former surpasses the latter, the drill bit speeds up and vice versa. However, Tob is influenced by the weight on the bit (WOB), which, in turn, is contingent upon the drill bit's position. As a result, axial and torsional vibrations get interlinked at the point of contact with the formation, necessitating an iterative approach to solve the concurrent equations governing these vibrations [14].

The WOB is the axial force acting on the bit when conditions change [1]. Its changing part is primarily triggered by the bit's interaction with the formation, which causes the WOB to change. Axial wave motion causes axial force variations along the drillstring [30]. To explain the changes in WOB and the lift-off phenomenon that goes with them, two main ideas were put forward: lobes in the formation surface and changes in drilling mud pressure [14]. Field observations indicate that excitation at the bit can generate oscillating forces with amplitudes comparable to those of the bit load. This suggests that, in the specific case under examination, forces at the bottom of the hole can reach levels comparable to normal bit loads [31]. The analysis of the data gathered during field tests highlights two distinct properties of these vibrations: First, the most significant changes in WOB observed at frequencies below the drillstring's inherent frequency. Second, under resonant settings, the range of WOB variation gradually increased to peak levels before abruptly dropping.

As downhole vibration levels increase, the bit bounces off the bottom. Large yet relatively consistent fluctuations in WOB were observed at substantially reduced rotational velocity, but with the same WOB. Increased RPM (rotary speed) led to significant longitudinal vibrations and kelly bouncing [32]. In existing literature, the WOB function has been modeled with an implicit inclusion of the depth of cut variable, commonly represented by means of a continuous function [2, 33] or a discontinuous function. [26, 28].

6 A Case of Study

Reference [34] looked at field data from a five-kilometer-long drill string and showed that a pure torsional model accurately matched the field data, with torsional vibrations being the main thing that was seen. The study also examined the stick-slip factor, which illustrated the torsional stability map. In this research, we will apply the same factor within a coupled axial-torsional model. Using the stick-slip factor (ss%) to measure torsional stability in a coupled axial-torsional system might not fully show how stable the system is, but it's still good enough for comparisons. This is because we are not aiming to evaluate the stability of the coupled system, and the torsional dynamics are known to dominate in coupled drilling models, as previously cited.

Two contrasting case studies are utilized to demonstrate how the bit-rock constitutive model's selection may have a significant impact on expected behavior. The purpose of this study is to compare and contrast the torsional stability in different bit-rock interaction frameworks. It will look at both one-way and two-way coupling scenarios, as well as a wide range of parameters that are necessary for real-world use. Key parameters under scrutiny include the surface rotational speed ($\boldsymbol{\Omega}_0$) and the axial speed (V_0).

For comparison, a 3-DOF model for the coupled axial-torsional drilling system, as referenced in [35], will be analyzed.

$$\begin{aligned}
&I_1\ddot{\varphi}_1 + C_1(\dot{\varphi}_1 - \Omega_0) + K_1(\varphi_1 - \Omega_0 t) + C_2(\dot{\varphi}_1 - \dot{\varphi}_2) + K_2(\varphi_1 - \varphi_2) = 0\\
&I_2\ddot{\varphi}_2 + C_2(\dot{\varphi}_2 - \dot{\varphi}_1) + K_2(\varphi_2 - \varphi_1) = -TOB\\
&M\ddot{U} + C_3\dot{U} + K_3(U - V_0 t) = W_0 - WOB
\end{aligned} \tag{4}$$

With:

$$\begin{cases} WOB = \zeta a\varepsilon d(t)H(\dot{\varphi}_2)H(d(t)) - \sigma al \\ TOB = -\frac{1}{2}(a^2\varepsilon d(t)H(\dot{\varphi}_2)H(d(t)) + a^2\mu\gamma\sigma l\,\mathrm{sgn}(\dot{\varphi}_2)H(\dot{U})) \end{cases} \tag{5}$$

The Heaviside function H(), and the sign function sgn() will both be considered equal to 1.

For the uni-directional coupled system, the depth of cut is given as in (1). In contast, for the bi-directional coupled system, it is defined as in (3).

The various variables are outlined in the table below (Table 1):

Table 1. List of variables and parameters

Variable	Description	Variable	Description
U	Axial displacement	W_0	Nominal weight on bit
φ_1, φ_2	Torsional angle	μ	Friction coefficient
M	Mass of drill string including BHA	ε	Rock specific strength
I	Moment of inertia of the drill string	ξ	Cutter inclination coefficient
C_3	Axial damping	σ	Contact pressure
C_2, C_3	Torsional damping	γ	Geometry parameter of drill bit
K_3	Axial stifness	a	Radius of bit
K_1, K_2	Torsional stifness	l	Wear flat length

Figure 3 shows a comparison of the torsional response of the model in both cases of uni-directional and bi-directional coupled system.

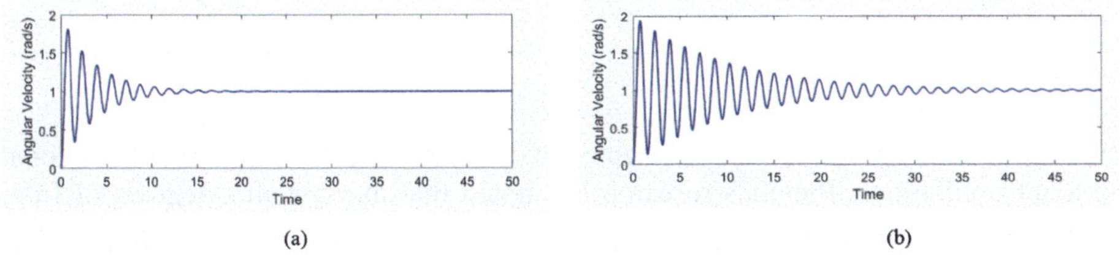

Fig. 3. The torsional dynamic response with V_0=0.15 m/s and Ω_0=1 rad/s: a) Uni-directional coupled system b) Bi-directional coupled system

A. Results interpretation

Both (V_0) and (Ω_0) will be varied from 0 rad/s to 2 rad/s and from 0 m/s to 2 m/s, respectively. The torsional stability map is generated by computing ss% for each combination of (V_0) and (Ω_0). When ss% equals 0, no torsional oscillations are present, and the rotational speed at the bottom matches that at the top. Conversely, when ss% is greater than 0, persistent undesirable torsional oscillations occur at the bit, indicating torsional instability. As illustrated in Figure 1, the dark blue region indicates torsional stability, while other colors represent varying degrees of torsional instability. With higher ss% values, instability increases.

Figure 3 show that the red region, indicating significant instability, is more prominent at low ($\boldsymbol{\Omega}_0$) values in both uni-coupled and bi-coupled systems. At higher ($\boldsymbol{\Omega}_0$) values, both systems display stable torsional behavior. Additionally, the degree of torsional instability, corresponding to ss%, is notably higher (102%) in the bi-directional coupled system, as we can see in (Fig. 3.a), compared to the uni-directional coupled system (95.5%) illustrated in (Fig. 3.b) It is important to note that equivalent parameters were used for both systems (Fig. 4).

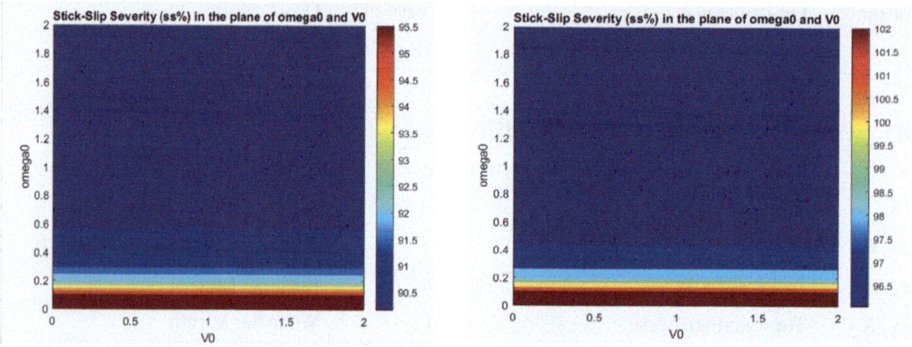

(a) (b)

Fig. 4. The torsional stability map with a) Uni-directional coupled system b) Bi-directional coupled system

7 Conclusion

Several studies have attempted to replicate the dynamics of drill strings under torsional and axial oscillations. Previous research has noted that the system's degrees of freedom can impact its stability [36, 37] and the selection of the friction coefficient [38]. Some studies have looked at how bidirectional versus unidirectional coupling affects the dynamic response of an axial-torsional system that is coupled, but none of them have specifically looked at this. The case study presented in this paper demonstrates that the type of coupling in the system can significantly affect its torsional stability.

References

1. Yigit, A.S., Christoforou, A.P.: Stick-slip and bit-bounce interaction in oil-well drillstrings. ASME J. Energy Resour. Technol. **128**(4), 268–274 (2006)

2. Richard, T., Germay, C., Detournay, E.: A simplified model to explore the root cause of stick–slip vibrations in drilling systems with drag bits. J. Sound Vib. **305**(3), 432–456 (2007)
3. Ritto, T.G., Sampaio1, R., Soize, C.: Drill-string dynamics coupled with the drilling fluid dynamics. In: Proceedings of the XIII International Symposium on Dynamic Problems of Mechanics (2009)
4. Boussaada, I., Mounier, H., Niculescu, S., Cela, A.: Analysis of drilling vibrations: a time-delay system approach. In: 20th Mediteranean Conference on Control and Automation, Barcelona, Spain, pp. 610–614 (2012)
5. Ulf J., Aarsnes, F., Aamo, O.M.: Linear stability analysis of self-excited vibrations in drilling using an infinite dimensional model. J. Sound Vibr. **360**, 239–259 (2016)
6. Aarrestad, T.V., Kyllingstad, A.: An experimental and theoretical study of coupling mechanisms between longitudinal and torsional drillstring vibrations at the bit. In: Proceedings of the SPE Drilling Engineering Conference, pp. 12–18 (1988)
7. Riane, R., Kidouche, M., Illoul, R., Doghmane, M.Z.: Unknown resistive torque estimation of a rotary drilling system based on Kalman filter. IETE J. Res. **68**(4), 2640–2651 (2022)
8. Khulief, Y.A., Al-Sulaiman, F.A.: Laboratory investigation of drillstring vibrations. Proc. Inst. Mech. Eng., Part C: J. Mech. Eng. Sci. 223(10), 2249–2262 (2009)
9. Aarrestad, T.V., Kyllingstad, A.: An experimental and theoretical study of coupled boundary conditions for vibrations in drillstrings. SPE **15563** (1986)
10. Mendil, C., Kidouche, M., Doghmane, M.Z.: Sliding mode controller design for torsional vibrations minimization under rock-bit interaction effects. In: Artificial Intelligence and Heuristics for Smart Energy Efficiency in Smart Cities: Case Study: Tipasa, Algeria, pp. 613–624. Springer International Publishing (2022)
11. Shaw, S.W.: On the dynamic response of a system with dry friction. J. Sound Vib. (1986)
12. Popp, K., Hinrichs, N., Oestreich, M.: Analysis of a self-excited friction oscillator with external excitation. Dyn. Fric.: Model. Anal. Exper.: (Part I), 1–35 (1996)
13. Jim, W., Putelat, T., McKay, A.: Are there reliable constitutive laws for dynamic friction? Philos. Trans. R. Soc. A: Math., Phys. Eng. Sci. **373**, 20140401 (2015)
14. Spanos, P.D., et al.: Modeling of roller cone bit lift-off dynamics in rotary drilling, pp. 197–207 (1995)
15. Stéphan, G.: Delay-differential equation models for machine tool chatter. Nonlinear Dyn. Mater. Proc. Manuf. (1998)
16. Emmanuel, D., Defourny, P.: A phenomenological model for the drilling action of drag bits. Int. J. Rock Mech. Min. Sci. Geomech. Abs. **29**(1) (1992)
17. Fear, M.J., Abbassian, F., Parfitt, S.H.L., McClean, A.: The destruction of PDC bits by severe slip-stick vibration. In: The SPE/IADC Drilling Conference, Amsterdam, Netherlands (1997)
18. Jansen, J.D., Van Den Steen, L.: Active damping of self-excited torsional vibrations in oil well drillstrings. J. Sound Vibr. **179**(4), 647–668 (1995)
19. Black, A.D., Walker, B.H., Tibbitts, G.A., Sandstrom, J.L.: PDC bit performance for rotary, mud motor, and turbine drilling applications. SPE Drill Eng. **1**(06), 409–416
20. Besselink, B., Van De Wouw, N., Nijmeijer, H.: A semi-analytical study of stick-slip oscillations in drilling systems, 021006 (2011)
21. Bart, B., et al.: Analysis and control of stick-slip oscillations in drilling systems. IEEE Trans. Control Syst. Technol. **24**(5), 1582–1593 (2015)
22. Nandakumar, K., Wiercigroch, M.: Stability analysis of a state dependent delayed, coupled two DOF model of drill-stringvibration. J. Sound Vibr. **332**(10), 2575–2592 (2013)
23. Liu, X., et al.: Coupled axial-torsional dynamics in rotary drilling with state-dependent delay: stability and control. Nonlinear Dyn. **78**, 1891–1906 (2014)
24. Yan, Y., Marian, W.: Dynamics of rotary drilling with non-uniformly distributed blades. Int. J. Mech. Sci. **160**, 270–281 (2019)

25. Ali, H., Bakhtiari-Nejad, F.: A new dynamic model of coupled axial–torsional vibration of a drill string for investigation on the length increment effect on stick–slip instability. J. Vibr. Acoust. **139**(6), 061016 (2017)
26. Li, X., et al.: Theoretical and numerical analysis of coupled axial-torsional nonlinear vibration of drill strings. Shock Vibr., 1–15 (2021)
27. Khulief, Y.A., Al-Sulaiman, F.A., Bashmal, S.: Vibration analysis of drillstrings with self-excited stick–slip oscillations
28. Navarro-López, E.M., Suárez, R.: Practical approach to modelling and controlling stick-slip oscillations in oilwell drillstrings. In Proceedings of the 2004 IEEE International Conference on Control Applications, vol. 2, pp. 1454–1460. IEEE (2004)
29. Yigit, A.S., Christoforou, A.P.: Coupled torsional and bending vibrations of actively controlled drillstrings. J. Sound Vibr. **234**(1), 67–83 (2000)
30. Dunayevsky, V.A., Abbassian, F., Judzis, A.: Dynamic stability of drillstrings under fluctuating weight on bit. SPE Drilling Compl. (1993)
31. Paslay, P.R., Bogy, D.B.: Drill string vibrations due to intermittent contact of bit teeth, pp. 187–194 (1963)
32. Wolf, S.F., Zacksenhouse, M., Arian, A.: Field measurements of downhole drillstring vibrations. In: SPE Annual Technical Conference and Exhibition (1985)
33. Khulief, Y.A., Al-Sulaiman, F.A., Bashmal, S.: Vibration analysis of drillstrings with self-excited stick–slip oscillations. J. Sound Vibr. **299**(3), 540–558 (2007)
34. Ritto, T.G., Aguiar, R.R., Hbaieb, S.: Validation of a drill string dynamical model and torsional stability. Meccanica **52**, 2959–2967 (2017)
35. Meddah, S., Idir, A., Tadjer, S.A., Doghmane, M.Z., Kidouche, M.: PID control design of strongly coupled axial-torsional vibrations in rotary drilling systems. PRZEGLĄD ELEKTROTECHNICZNY (2024). (ISSN 0033–2097)
36. Tengesdal, N.K., Hovda, S., Holden, C.: A discussion on the decoupling assumption of axial and torsional dynamics in bit-rock models. J. Petr. Sci. Eng. **202**(2021), 108070 (2021)
37. Liu, X., Vlajic, N., Long, X., Meng, G., Balachandran, B.: State- dependent delay influenced drill-string oscillations and stability analysis. J. Vibr. Acoust. **136**, 051008–1 (2014)
38. Bakhtiari-Nejad, F., Hosseinzadeh, A.: Nonlinear dynamic stability analysis of the coupled axial- torsional motion of the rotary drilling considering the effect of axial rigid-body dynamics. Int. J. Non-Linear Mech. S0020–7462(16)30255–4 (2016)

Optimized GA-Based PID Control Design for Stick-Slip Suppression in Drill String of Rotary Drilling System

Tellaa Mawloud[1,2](✉), Aibeche Abderrezak[1], Doghmane Mohamed Zinelabidine[1,3], and Kidouche Madjid[1]

[1] Automation and Electrification of Process Applied Automation Laboratory, Faculty of Hydrocarbons, Chemistry University M'Hamed Bougara, 35000 Boumerdes, Algeria
{m.tellaa,aibeche_umbb,mkidouche}@univ-boumerdes.dz

[2] Institute of Electrical and Electronics Engineering, University M'hamed Bougara, Boumerdes, Algeria

[3] Bob L. Herd Department of Petroleum Engineering, Texas Tech University, Lubbock, TX 79401, USA

Abstract. Stick-slip phenomena in the oil and gas drilling sector cause drillstring vibrations that provide major issues. These vibrations result in reduced drilling efficiency, equipment malfunctions, and possible dangers. This work introduces a thorough method for modeling and regulating stick-slip behavior. It utilizes a mechanical model with three degrees of freedom (3-DOF) that incorporates the influence of dry friction. The model combines both axial and torsional degrees of freedom to precisely reflect the intricate dynamics of the drillstring. Additionally, a control technique is suggested to alleviate stick-slip vibrations. This strategy involves using a hybrid Proportional-Integral-Derivative (PID) controller that has been tuned using Genetic Algorithms (GA). The PID-GA controller is specifically engineered to manage the rotating speed, mitigate stick-slip oscillations, and improve drilling stability. The efficacy of the suggested 3-DOF model and PID-GA control method in reducing stick-slip vibrations during drilling is proven by numerical simulations conducted under different drilling situations. The findings demonstrate notable advancements in drilling performance, such as decreased vibrations caused by friction, improved efficiency in drilling, and extended lifespan of the drillstring. This research significantly helps to the progress of drilling operations by offering a strong framework for modeling and managing stick-slip behavior. As a result, it improves drilling performance and guarantees operational safety in the oil and gas industry.

Keywords: Proportional–integral–derivative controller (PID) · Genetic Algorithms (GA) · rotary drilling systems · stick-slip vibrations

1 Introduction

The rotary drilling technique is vital in the petroleum sector for the extraction of oil and gas. Three kinds of vibrations can be generated during drilling: axial, lateral, and torsional vibrations. Stick-slip vibrations are the most detrimental, as they can initiate

© The Author(s), under exclusive license to Springer Nature Switzerland AG 2025
M. Hatti (Ed.): IC-AIRES 2024, LNNS 1238, pp. 453–463, 2025.
https://doi.org/10.1007/978-3-031-80301-7_50

other vibrations. The objective of this study is to reduce stick-slip vibrations. In recent years, researchers have studied these vibrations. For instance, Tang et al. [3] discovered that reducing WOB could eliminate stick-slip, albeit with a delay. Increasing rotary table speed also reduces torsional vibrations, but it does not account for drill string stiffness [4]. Sliding mode control, which is applied to four degrees of freedom [5] and extended to n degrees [6], is designed to prevent oscillations, albeit with high stabilization times. Liu [7] experimentally compared rotational velocities and achieved some vibration reduction. Abdulgalil and Siguerdidjane [8] employed back-stepping control to eliminate stick-slip vibrations. Hybrid controllers, which combine various methods, have also been investigated [9, 10]. Although the results are satisfactory, it is necessary to enhance the robustness of the controller. This paper introduces a genetic algorithm (GA) in conjunction with a PID controller to improve vibration suppression. The objective of this innovative methodology is to enhance the performance of controllers that have been implemented in the past. The study contrasts the new method with a PID controller that lacks optimization, noting that PID with GA has not been investigated for this issue. This article presents a PID controller that has been optimized using a genetic algorithm (GA). The manuscript is structured as follows: Section 2 provides a description of the drilling system and the vibrations it produces. Section 3 introduces the mathematical model and the PID controller. Section 4 provides a detailed explanation of the genetic algorithm and how it is applied. Section 5 discusses the results obtained and compares the performance of the PID controller with and without GA optimization. The manuscript concludes by outlining potential areas for future research.

2 Drilling System

The drilling rig consists mostly of the power system, hoisting system, circulating system, rotating system, well control system, and monitoring system. The hoisting system, which includes draw-works, an auxiliary brake, a travelling system, a hook, and a derrick, supports drilling and completion operations. The rotary system drives the bit to break the rock and rotates the drilling strings in the wellbore. The circulation system moves cuttings from the bottom of the hole to the surface, prevents debris settlement, keeps the drill bit cool, and protects the hole walls. The well control system is utilized during emergencies to promptly and effectively halt or terminate the well, hence restoring equilibrium in pressure. The power and transmission system provides energy to all units, while the monitoring and control system ensures the smooth operation of the rig, meeting the drilling process requirements [7].

2.1 Stick-Slip Vibrations

Stick-slip is a detrimental phenomenon characterized by the alternating adhesion and sliding of surfaces, resulting in variable frictional torques. This can produce vibrations in drilling systems, potentially destroying the drill collar and causing the bit to stop totally. Stick-slip occurs owing to the nonlinear interaction between the rock and the bit, and even tiny fluctuations in the upper drive speed might trigger it. Additionally, the elasticity between the force source and the friction point can lead to sliding, which finally causes sticking, especially as the tool string length rises. In order to reduce stick-slip during drilling operations, it is essential for the driller to identify the appropriate downhole vibration by observing specific indicators, such as sudden drops in the rate of penetration, a high torque range in the torque signal, or equipment failure. A common method to reduce stick-slip is to gradually decrease the weight on the bit while increasing the rotating speed of the drill string until stick-slip is minimized and normal drilling can resume.

3 Mathematical Model Under Stick-Slip Vibrations

3.1 Drilling String Model

This study utilizes a mass-spring-damper model to simulate the behavior of a rotary drilling system [3]. The model consists of three degrees of freedom, with the upper disc representing the rotary table, the middle disc representing the tool string, and the lower disc representing the drill bit. The specific parameters of the model are outlined in Table 1. Additionally, Fig 1a provides a schematic diagram of the rotary drilling system, while Fig 1b illustrates the 3-DOF model. This model is mathematically described by Eq. (1) [13].

$$\begin{cases} \ddot{\alpha}_1 = \frac{1}{J_1}(c_1(\dot{\alpha}_{td} - \dot{\alpha}_1) - k_1(\alpha_{td} - \alpha_1) - c_2(\dot{\alpha}_1 - \dot{\alpha}_2) - k_2(\alpha_1 - \alpha_2) - \mu\dot{\alpha}_1) \\ \ddot{\alpha}_2 = \frac{1}{J_2}(c_2(\dot{\alpha}_1 - \dot{\alpha}_2) + k_2(\alpha_1 - \alpha_2) - c_3(\dot{\alpha}_1 - \dot{\alpha}_2) - k_3(\alpha_2 - \alpha_3) - \mu\dot{\alpha}_2) \\ \ddot{\alpha}_3 = \frac{1}{J_3}(c_3(\dot{\alpha}_2 - \dot{\alpha}_3) + k_3(\alpha_2 - \alpha_3) - \mu\dot{\alpha}_3 - Tob) \end{cases} \quad (1)$$

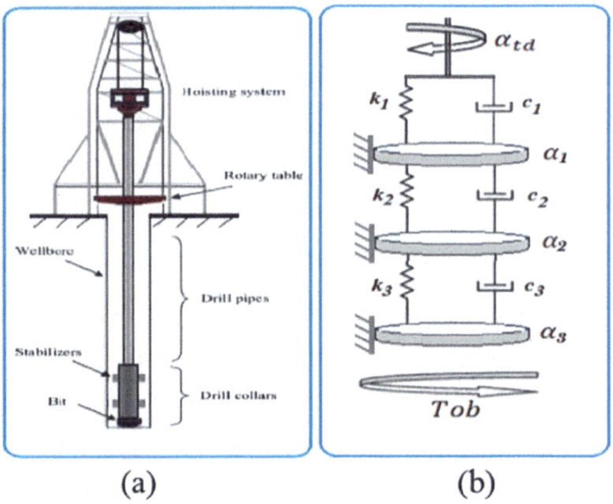

Fig. 1. Rotary drilling systems: (a) Schematic diagram, (b) Equivalent mass-spring-damper three-element model [3, 13].

Where the values of the system parameters are: k1 = k2 = k3 = 481.29 *N. m/rad*, j1 = j2 = 1030.45 *Kg. m2*, j3 = 223.44, c1 = c2 = 51.38 *N. m. s/rad*, c3= 39.79 *N. m. s/rad*, μ = 10 *N. m*

Table 1. Parameters description for the designed rotary drilling model [13, 19].

Parameter	Description	Unit
αtd	The top drive angular displacement	[rad]
$\alpha i = 1,2,3$	The angular displacements of the rope section i	[rad]
$ki = 1,2,3$	Torsional stiffness coefficient of the rope section i	[N.m/rad]
ci = 1,2,3	Internal damping coefficient of the rope section i	[N.m.s/rad]
μ	Wall friction coefficient	[N.m]
$Ji = 1,2,3$	The inertia of the rope section i	[kg.m2]
Tob	The torque on bit	[N.m]

The transition to the state space representation, we implement the following variable changes: The input signal is defined as the angular velocity of the top drive: $u = \dot{\alpha}_{td}$, the states variables are defined as: $x_1 = \dot{\alpha}_1, x_2 = \dot{\alpha}_2, x_3 = \dot{\alpha}_3, x_4 = \alpha_{td} - \alpha_1, x_5 = \alpha_1 - \alpha_2$ and $x_6 = \alpha_2 - \alpha_3$ the Eq. (1) is written as:

$$\begin{cases} \dot{x}_1 = -\frac{c_2+c_1+\mu}{J_1}x_1 + \frac{c_2}{J_1}x_2 + \frac{k_1}{J_1}x_4 - \frac{k_2}{J_1}x_5 + \frac{c_1}{J_1}u \\ \dot{x}_2 = \frac{c_2}{J_2}x_1 - \frac{c_2+c_3+\mu}{J_2}x_2 + \frac{c_3}{J_2}x_3 + \frac{k_2}{J_2}x_5 - \frac{k_3}{J_2}x_6 \\ \dot{x}_3 = \frac{c_3}{J_3}x_2 - \frac{c_2+\mu}{J_3}x_3 + \frac{k_3}{J_3}x_6 - \frac{1}{J_3}Tob \\ \dot{x}_4 = \dot{\alpha}_{td} - \dot{\alpha}_1 = u - x_1 \\ \dot{x}_5 = \dot{\alpha}_1 - \dot{\alpha}_2 = x_1 - x_2 \\ \dot{x}_6 = \dot{\alpha}_2 - \dot{\alpha}_3 = x_2 - x_3 \end{cases} \tag{2}$$

3.2 Rock-Bit Interaction Model

In literature, many functions have been utilized to represent the interaction between the bit and the rock. For the system being studied, we will employ the generalized model described by the equation [13].

$$Tob = \mu_n Nr\left(\frac{x_2}{\sqrt{x_2^2 + \Omega_0^2}} + \frac{p\Omega_0 x_2}{x_2^2 + \Omega_0^2}\right) + Dx_2\left(\frac{x_2}{\Omega_1} - 1\right)$$

where the model coefficients are given in the following table [13] (Table 2).:

3.3 PID Controller

Proportional integral derivative (PID) control is the most common control rule for regulation in engineering. It is used in many fields because it is easy to tune and always works. The PID control equation is usually written as [9].

$$C(t) = K_p e(t) + K_i \int_0^t e(\tau)d\tau + Kd\frac{de(t)}{dt} \tag{4}$$

Table 2. Parameters used in rock-bit general model

parameter	description	value
μ_n	friction coefficient	28 nm
n	the force vector	9.81 * wob
r	the contact radius vector	0.1 m
Ω_0	chain transition speed	1
Ω_1	transition speed for the well	31.4159
p	the initial friction parameter	1.5
d	the linear damping vector	0.28

where K_P is the proportional coefficient; T_i is the integral time constant; T_d is the differential coefficient. The derivative part is filtered in order to avoid singularities; the equation (6) is for conventional PID before calling the PSO algorithm for parameters optimization.

4 Genitic Algorithm

4.1 History and Principle

The Genetic Algorithm (GA) is a metaheuristic algorithm that utilizes a random search technique with parametric coding. It was first introduced by John Holland in 1975 and was initially employed to address control problems in gasoline pipelines by his student Goldberg. The GA is particularly advantageous for solving control problems in situations where an analytical approach is not feasible due to a vast search space. The algorithm operates based on the principle of natural selection, where the fittest candidates survive and reproduce, while the fewer fit ones are eliminated. The Genetic Algorithm (GA) is a metaheuristic algorithm that utilizes a random search technique with parametric coding.

4.2 Fitness Function

To achieve the desired performance, the parameters of a controller are chosen to minimize the dynamic error

$$e(t) = y_{ref}(t) - y_{sys}(t) \tag{5}$$

To do this, a cost function needs to be reduced. There are four basic performance criteria used in PID controller tuning for this purpose [14].

ITAE (Integral of Time-weighted Absolute Error): The ITAE criterion emphasizes errors that persist for a longer time:

$$J_{ITAE} = \int_0^T t|e(t)|dt \tag{6}$$

IAE (Integral of Absolute Error): The IAE criterion is the integral of the absolute value of the error. It is defined as:

$$J_{IAE} = \int_0^T |e(t)|dt \tag{7}$$

ISE (Integral of Squared Error): The ISE criterion emphasizes larger errors more than smaller ones by squaring the error term. It is defined as:

$$J_{ISE} = \int_0^T e^2(t)dt \tag{8}$$

ITSE (Integral of Time-weighted Squared Error): The ITSE criterion is a combination of the ISE and ITAE criteria, taking into account both the amount of the error and the duration. It is defined as follows:

$$J_{ITSE} = \int_0^T te^2(t)dt \tag{9}$$

4.3 PID-GA Based Controller Structure

The architecture of the suggested approach for implementing the PID-GA in rotary drilling systems is depicted in Fig 2. The process of designing an appropriate PID controller hinges on identifying the optimal tuning parameters for the PID controller.

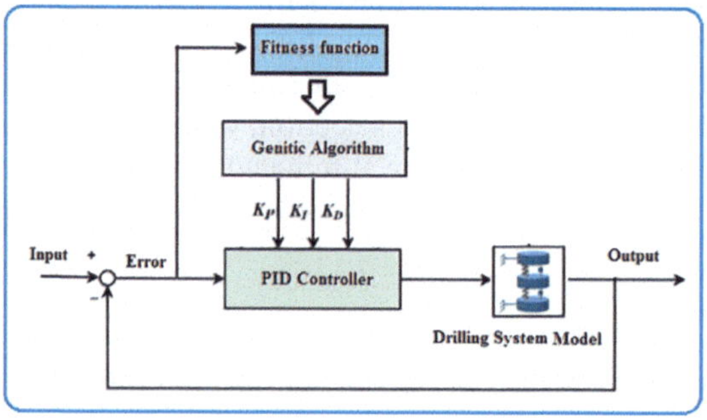

Fig. 2. GA-PID based control system structure [16]

The GA parameters are chosen as follows (Table 3).

5 Simulation Results

The optimum PID controller parameters for the closed-loop drilling system's step response were found after conducting simulations using the GA parameters mentioned in Table 4.

Table 3. Parameters of the GA

Parameter	Value
Generations number	20
Population size	80
Lower bounds = [kp,Ki,Kd]	lb = [0 0 0]
Upper bounds = [kp,Ki,Kd]	ub = [10 10 10]

Table 4. PID tuned Parameters using GA

Parameter	ITAE	IAE	ISE	ITSE
Kp	2.5124	2.9989	3.7906	3.3447
Ki	0.3029	0.2849	0.2383	0.3034
Kd	8.9386	9.0377	9.4359	9.2702
Best fitness	90,3217	72,8152	3036,08	4024,69
Mean fitness	98,9142	75,1851	3046,14	4074,37

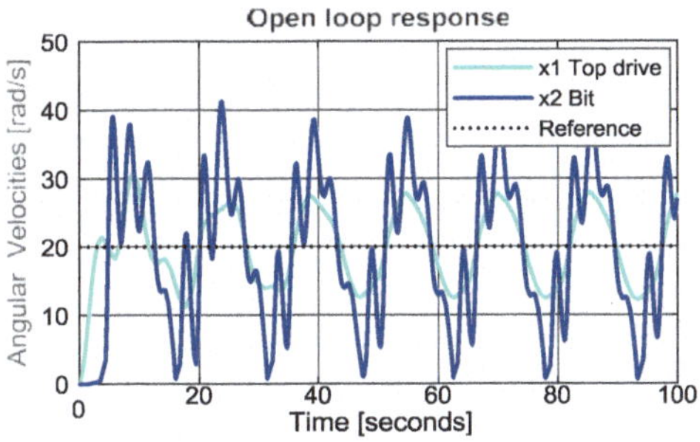

Fig. 3. Drilling string open loop Respone without any control

5.1 Results Without Any Control

(See Fig. 3).

5.2 Results for PID-GA Using ITAE

(See Fig. 4).

5.3 Results for PID-GA Using IAE

(See Fig. 5).

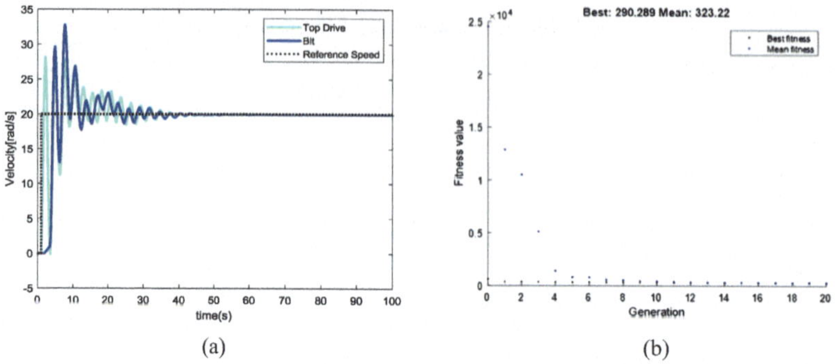

Fig. 4. (a) Drilling String closed loop response. (b) ITAE Fitness function Genitic evolution across genetic algorithm generations.

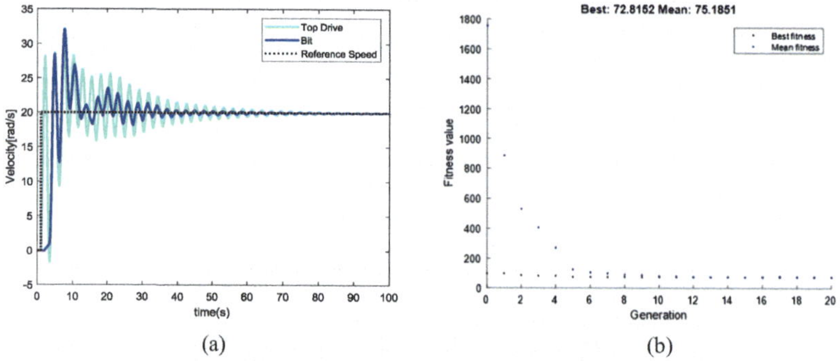

Fig. 5. (a) Drilling String closed loop response. (b) IAE Fitness function Genitic evolution across genetic algorithm generations.

5.4 Results for PID-GA Using ISE

(See Fig. 6).

5.5 Results for PID-GA Using ITSE

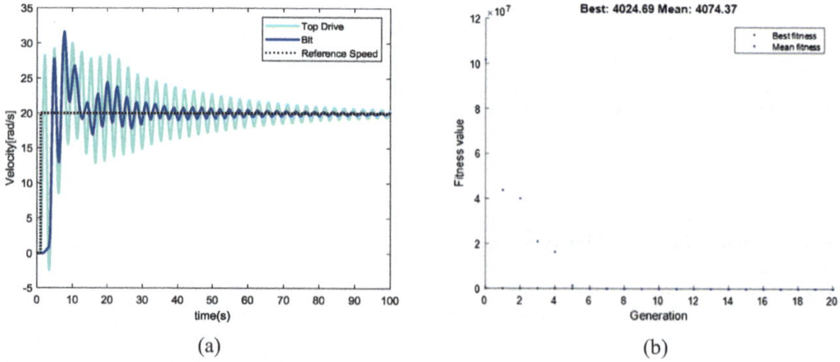

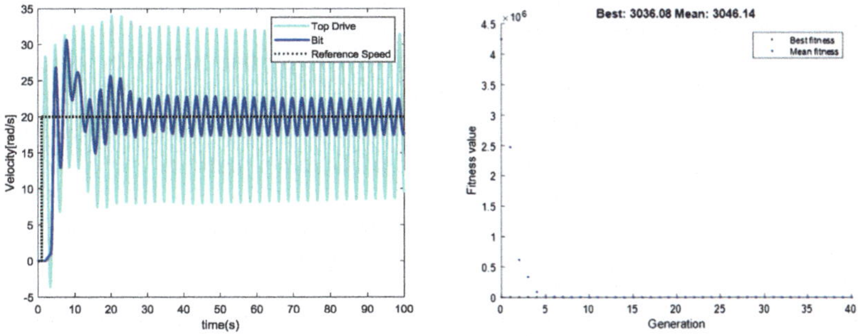

(a) (b)

Fig. 6. (a) Drilling String closed loop response. (b) ISE Fitness function Genitic evolution across genetic algorithm generations.

6 Comparative Study

The selection of a performance criterion while tuning a PID controller can have a substantial effect on the response of the system. Below is a detailed analysis of these criteria:

1. Steady-state error: ITAE and IAE are typically more effective in minimizing steady-state errors as they penalize persistent faults. ISE (Integral Square Error) and ITSE (Integral Time-weighted Square Error) may lead to increased steady-state errors as they prioritize greater faults.
2. The transient reaction is typically faster in ISE due to the greater emphasis on minimizing significant mistakes. ITAE can also offer satisfactory transitory performance but may result in more assertive control actions.
3. Disturbance rejection: ITAE generally offers superior disturbance rejection by prioritizing the reduction of errors that persist over a long period of time.
4. Control effort: The use of Integral of Absolute Error (IAE) and Integral of Time-weighted Squared Error (ITSE) may result in reduced control effort as compared to Integral of Squared Error (ISE) and Integral of Time-weighted Absolute Error (ITAE). This can be advantageous in systems where excessive control actions are undesirable.
5. Complexity: IAE is the most straightforward to calculate and execute, making it an excellent option for systems where computational simplicity is of utmost importance. ITSE and ITAE are more intricate yet offer a more equitable performance.

The selection of a criterion in practical applications is dependent on the specific requirements of the control system, which include the desired balance between response time, stability, and control effort. Often, a combination of these factors is utilized to achieve optimal tuning that satisfies all performance criteria.

7 Conclusion

The purpose of this work was to enhance the reduction of stick-slip vibration in rotary drilling systems. By utilizing a PID controller tuned by a GA algorithm with various fitness functions, the incidence of stick-slip vibration could be rapidly suppressed. The obtained results are pretty acceptable; the power of the GA might enhanced by boosting the population size but that require more time to reach optimal parameters. Future work linked to expand integer order PID controller to non- integer order PID for vibration control in drillstring system the future study could include applying alternative tuning methods as well as more complicated artificial intelligence method.

References

1. Leine, R.I., van Campen, D.H., Keultjes, W.J.G.: Stick-slip Whirl Interaction in Drillstring Dynamics ((2002))
2. Doghmane, M.Z., Kidouche, M., Ahriche, A.: Decentralized overlapping control design with application to rotary drilling system. IETE J. Res. **69**, 2070–2079 (2021). https://doi.org/10.1080/03772063.2021.1886602
3. Tang, L., Zhu, X., Shi, C., Tang, J., Xu, D.: Study of the influences of rotary table speed on stick-slip vibration of the drilling system. Petroleum **1**(4), 382–387 (2015)
4. Tang, L., Zhu, X., Qian, X., Shi, C.: Effects of weight on bit on torsional stick-slip vibration of oilwell drill string. J. Mech. Sci. Technol. **31**(10), 4589–4597 (2017)
5. Navarro-López, E., Licéaga Castro, E.: Non-desired transitions and sliding-mode control of a multi-DOF mechanical system with stick-slip oscillations. Chaos Solitons Fract. **41**(4), 1561–2200 (2009)
6. Vaziri, V., Kapitaniak, M., Wiercigroch, M.: Suppression of drill-string stick–slip vibration by sliding mode control: numerical and experimental studies. Eur. J. App. Math. **29**(5), 805–825 (2018)
7. Liu, Y.: Suppressing stick-slip oscillations in underactuated multibody drill-strings with parametric uncertainties using sliding-mode control. IET Control Theory **9**(1), 91–102 (2015)
8. Abdulgalil, F., Siguerdidjane, H.: Backstepping design for controlling rotary drilling system. In: Proceedings of 2005 IEEE Conference on Control Applications, 2005. CCA 2005, Toronto, Ont., Canada, 28–31 August 2005, pp. 120–124 (2052)
9. Mendil, C., Kidouche, M., Doghmane, M.Z.: Hybrid sliding PID controller for torsional vibrations mitigation in rotary drilling systems. Indon. J. Electr. Eng. Comput. Sci. **22**(1), 146–158 (2021). ISSN: 2502–4752, https://doi.org/10.11591/ijeecs.v22.i1.pp146-158
10. Mendil, C., Kidouche, M., Doghmane, M.Z.: Sliding mode controller design for torsional vibrations minimization under rock-bit interaction effects. In: Hatti, M. (ed.) Artificial Intelligence and Heuristics for Smart Energy Efficiency in Smart Cities, ICAIRES. Lecture Notes in Networks and Systems, vol. 361. Springer, Cham (2022). https://doi.org/10.1007/978-3-030-92038-8_61
11. Guan, Z., et al.: China University of Petroleum Press and Springer Nature Singapore Pte Ltd. Theory and Technology of Drilling Engineering (2021). https://doi.org/10.1007/978-981-15-9327-7_2
12. Ritto, T.G., Ghandchi-Tehrani, M.: Active control of stick-slip torsional vibrations in drill-strings. J. Vibrat. Control **25**(1), 194–202 (2019). https://doi.org/10.1177/1077546318774240
13. Mendil, C., Kidouche, M., Doghmane, M.Z., Benammar, S., Tee, K.F.: Rock–bit interaction effects on high-frequency stick-slip vibration severity in rotary drilling systems. Multidiscip. Model. Mater. Struct. **17**(5), 1007–1023 (2021). https://doi.org/10.1108/MMMS-10-2020-0256
14. Abushawish, A., Hamadeh, M., Nassif, B.: A PID controller gains tuning using metaheuristic optimization methods: a survey. Int. J. Comput. (2020). https://doi.org/10.46300/9108.2020.14.14

15. Israa, A.A., Khamees, M.M.: A review of adaptive tuning of PID-controller: optimization techniques and applications. Int. J. Nonlinear Anal. Appl. **15**(2), 29–37 (2024). ISSN: 2008–6822. https://doi.org/10.22075/ijnaa.2023.21415.4024
16. Kanhua, S., Wei*, J., Li, M., Li, H., Da, W., Zhang, L.: Research on stick-slip vibration suppression method of drill string based on machine learning optimization SV, vol. 57 (2023). https://doi.org/10.32604/sv.2023.043734
17. Hossain, M.E., Islam, M.R.: Drilling engineering problems and solutions a field guide for engineers and students (2018)
18. Riane, R., Doghmane, M.Z., Kidouche, M., Tee, K.F., Djezzar, S.: Stick-slip vibration suppression in drill string using observer-based LQG controller. Sensors **22**(16), 5979 (2022)
19. Kidouche, M., Charef, A.: A constructive methodology of Lyapunov function of composite systems with nonlinear interconnection term. Int. J. Robot. Autom.Autom. **21**(1), 19–24 (2006)

Optimal Control Design for Stick-Slip Vibrations Mitigation in Rotary Drilling Systems

Mawloud Tellaa[1,2](✉), Abderrezak Aibeche[1], Mohamed Zinelabidine Doghmane[1,3], and Madjid Kidouche[1]

[1] Automation and Electrification of Process Applied Automation Laboratory, Faculty of Hydrocarbons and Chemistry University M'Hamed Bougara, 35000 Boumerdes, Algeria
{m.tellaa,aibeche_umbb,mkidouche}@univ-boumerdes.dz

[2] Institute of Electrical and Electronics Engineering, University M'hamed Bougara, Boumerdes, Algeria

[3] Bob L. Herd Department of Petroleum Engineering, Texas Tech University, Lubbock, TX 79401, USA

Abstract. Various forms of vibrations can have a negative impact on the drilling process in rotary drilling equipment. The presence of vibrations can result in equipment damage, reduced borehole quality, decreased drilling efficiency, and increased non-productive time. Currently, the sole existing method to control these vibrations is by manually modifying drilling parameters, which is not a reliable technique and might lead to unfavourable economic consequences. Scientists have devised automated control systems to address these irregularities; however the effectiveness of the controllers is still being refined. Lately, the utilization of a PID controller based on particle swarm optimization has demonstrated noteworthy achievements. Nevertheless, the stick-slip vibration mitigation in rotary drilling equipment has not been subjected to testing. This research seeks to investigate and suggest the utilization of a PID controller, fine-tuned using the particle swarm optimization technique, to enhance the effectiveness of controllers developed for the purpose of reducing vibrations in rotary drilling systems.

Keywords: PID Controller · Particle Swarm Optimization · rotary drilling systems · Vibrations · Nonlinear Control performance

1 Introduction

The petroleum industry relies on rotary drilling technology to extract oil and gas from underground reservoirs. However, the drilling process often generates unwanted vibrations, including axial, lateral, and torsional vibrations, which can occur simultaneously. Among these, stick-slip torsional vibration is particularly harmful, characterized by alternating phases of high angular velocity and abrupt stops in the Bottom Hole Assembly (BHA). Stick-slip is considered the most damaging, as it can trigger other types of vibrations. This study focuses on mitigating stick-slip vibrations, identified as a key contributor to overall vibration issues [1, 2]

© The Author(s), under exclusive license to Springer Nature Switzerland AG 2025
M. Hatti (Ed.): IC-AIRES 2024, LNNS 1238, pp. 464–474, 2025.
https://doi.org/10.1007/978-3-031-80301-7_51

Vibrations in drilling systems can result in detrimental effects such as damage to the drill collar and complete cessation of the drilling bit [2]. Minimizing the effects of torsional vibrations has been a primary area of study in recent times. A study conducted by Tang et al. in [3, 4] found that this kind of vibration can be eradicated when decreasing the weight on bit , although with a minor delay in implementation. However, they did not take into account the stiffness of the drill string. One study [5] used sliding mode control to four degrees of freedom, while another study [6] produced an extended version of this approach for n degrees of freedom. The objective of these methods is to mitigate oscillations and release the bit prior to it becoming immobilized. Nevertheless, the utilization of sliding input surfaces in these approaches has resulted in prolonged stabilizing periods. Researchers, including Liu [7], have conducted experimental studies on the phenomena by comparing the rotational velocity of drill string components with specified values established by drillers. Although Liu's controller effectively minimized vibrations during the testing, its efficacy is still questionable. In their study, Abdulgalil and Siguerdidjane in [8] examined the application of back-stepping control and successfully mitigated stick-slip vibrations, resulting in improved performance. Mendil et al [9] have investigated hybrid controllers, which combine back-stepping with sliding mode and hybrid PID with sliding mode. This approach has also been explored in [10]. Mendil et al. have proposed a sliding mode controller to minimize torsional vibrations in [11].

While past methods have yielded acceptable outcomes, there is still a need for additional advancement in improving the performance of the controller. Therefore, the main aim of this research is to improve the efficiency of vibration suppression by implementing a PID controller in conjunction with the particle swarm optimization (PSO) technique. This work presents a new approach aimed at enhancing vibration suppression performance, as compared to previously existing controllers. In order to assess the suggested method, we conducted a comparison between its outcomes and those of an unoptimized PID controller. It is important to mention that the combination of the PID controller with the PSO algorithm has not been previously investigated for this particular situation.

This article introduces a PID controller that has been enhanced using the Particle Swarm algorithm. The remaining portion of the manuscript is structured in the following manner: Section 2 provides a concise description of the drilling mechanism, highlighting its use of vibration. Section 3 presents the mathematical description of vibrations occurring during stick-slip and also introduces the PID controller. Subsequently, Sect. 4 provides a comprehensive exposition of the mathematical formulation of the particle swarm optimization technique. This formulation is then implemented in the context of rotary drilling equipment experiencing stick-slip vibration. In addition, Sect. 5 provides a detailed analysis of the open loop and closed loop outcomes. The study also thoroughly examines the results of employing PID without PSO. Furthermore, a comparative analysis is undertaken to compare the three scenarios. This manuscript concludes with a summary and provides suggestions for future research endeavors.

2 Rotary Drilling Rig System

The drilling system is composed of two main components: the surface equipment, which includes the hoisting system, rotary table, and the circulation and conditioning system for drilling fluid (mud), and the subsurface drill string, which consists of the Bottom-Hole Assembly (BHA) made up of thick-walled tubulars (drill collars) and specialized downhole instruments. [12, 13]. Stabilizers secure the drill collars, while the Weight on Bit (WOB) measures the force applied to push the bit down, and the Torque on Bit (TOB) refers to the torque transferred from the rotary table to the drill string. Drilling mud plays a crucial role in lubricating, cleansing, and controlling well pressure [14].

2.1 Vibration Types in Drilling

Undesired vibrations during the drilling process might manifest in three forms: axial, lateral, and torsional. The worst scenarios for them include bit-bounce, spinning, and stick-slip. The stick-slip mode of torsional vibration is a highly detrimental phenomena that often triggers other forms of vibration during the drilling process [15]. Bit-bounce, also known as axial vibration, refers to the situation where the drill bit becomes disconnected from the formation it is drilling into. Whirling is defined as a lateral motion that moves from one side to the other in relation to the string, perpendicular to its length. Surface detection of lateral vibrations is challenging, although they can result in significant damage to the Bottom Hole Assembly (BHA). Stick-slip phenomenon arises when the applied rotary torque is not enough to ensure smooth rotation, resulting in a decrease in the speed of the string and the accumulation of energy as it undergoes elastic deformation [16]. Once the resistance is surpassed, the stored energy is discharged, resulting in the unwinding and acceleration of the BHA at rates 3 to 15 times higher than the usual rotary speeds. Severe stick-slip can result in expensive periods of inactivity, which may involve failures of downhole motors, twist offs, or damage to drill bits. It can happen either through the interaction of the rock bit or by the interaction between the drill string and the borehole wall [17].

2.2 Overview of Stick-Slip Vibration Mitigation Techniques

To minimize stick slip during drilling operations, the driller must accurately detect the appropriate downhole vibration by detecting specific symptoms of stick slip, such as abrupt decreases in the pace of penetration, a wide range of torque in the torque signal, or potential equipment malfunction. One effective technique for mitigating stick slip is to systematically reduce the load on the drill bit while simultaneously raising the rotational velocity of the drill string. This approach helps to limit stick slide and allows for the resumption of regular drilling operations. Several studies have investigated methods to regulate the pace of bits in order to reduce the impact of stick slip [18, 19]. This includes the implementation of real-time automated systems such as the DAS created by Pason. This device utilizes surface sensor data and a drill string model to automatically modify the weight on bit and drill string RPM when stick slip exceeds specified thresholds, yielding outcomes that are comparable to pricier downhole instruments. In addition, the bottom hole assembly can include detuners and vibration dampers to decrease stick-slip.

These tools work by modifying the rigidity of the drill string and isolating the excitation frequency from the natural frequency of the component. They can also absorb vibrations within the drill string [20].

3 Mathematical Model Under Stick-Slip Vibrations

The model utilized in this study is a two-element model based on the mass-spring-damper model, which will be further elaborated in the subsequent subsections.

3.1 Drilling String Model

The rotary drilling system under investigation is simulated using the mass-spring-damper model. This model includes a torsional spring model to reflect the behavior of the drill pipes and a rigid body model for the drill collar as seen in Fig. 1a.

The model includes the top drive and two inertial discs locally damped by c_{td} and c_b, respectively. These inertial elements are connected to each other using linear springs with stiffness coefficients $k_{td} = k_b = k$ and torsional friction coefficients τ_l and τ_b.

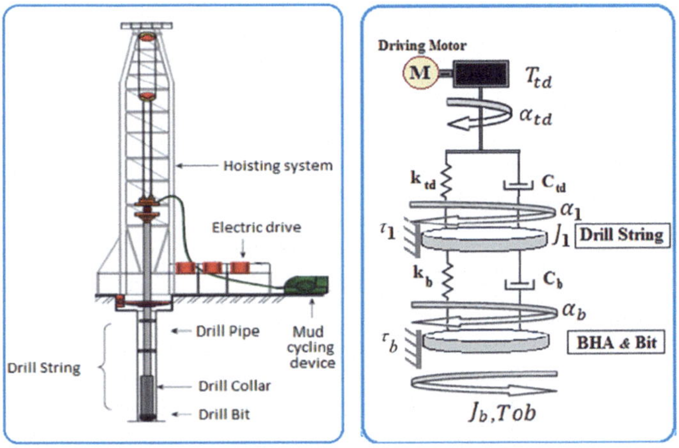

(a) (b)

Fig. 1. a) Schematic diagram of rotary drilling systems, b) Equivalent mass-spring-damper two-element model

The equation of motion is described by the following ordinary differential equation [9]:

$$\begin{cases} J_1\ddot{\alpha}_1 = T_1 - T_2 - \tau_1\dot{\alpha}_1 \\ J_b\ddot{\alpha}_b = T_2 - \tau_b\dot{\alpha}_b - Tob \end{cases} \quad (1)$$

J_1 and J_2 represent the inertia of the string and the BHA with the bit, respectively. T_1 denotes the torque transmitted from the top drive to the string, while T_2 represents the torque transmitted from the string to the BHA and the bit. The torques T_1 and T_2 can be represented by the following equations [9]:

$$\begin{cases} T_1 = c_{td}(\dot{\alpha}_{td} - \dot{\alpha}_1) + k(\alpha_{td} - \alpha_1) \\ T_2 = c_b(\dot{\alpha}_1 - \dot{\alpha}_b) + k(\alpha_1 - \alpha_b) \end{cases} \tag{2}$$

By inserting the expression of $T_{1\ and}$ T_2 from (2) in (1) we get the following equations:

$$\begin{cases} J_1\ddot{\alpha}_1 = c_{td}(\dot{\alpha}_{td} - \dot{\alpha}_1) + k(\alpha_{td} - \alpha_1) - c_b(\dot{\alpha}_1 - \dot{\alpha}_b) - k(\alpha_1 - \alpha_b) - \tau_1\dot{\alpha}_1 \\ J_b\ddot{\alpha}_b = c_b(\dot{\alpha}_1 - \dot{\alpha}_b) + k(\alpha_1 - \alpha_b) - \tau_b\dot{\alpha}_b - Tob(\dot{\alpha}_b) \end{cases} \tag{3}$$

α_{td}, α_1 and α_b denote the angular displacements of the top drive, string and the bit, respectively. The angular velocity of the top drive is considered as the input signal u it is proportional to the torque Ttd. The transmission box usage is to adjust the angular velocity $\dot{\alpha}_{td}$, the frictional torque $Tob(\dot{\alpha}_b)$ represents, the torque on bit combined with the nonlinear frictional forces along the drill collars [2]. By setting $x_1 = \dot{\alpha}_1$, $x_2 = \dot{\alpha}_b$, $x_3 = \alpha_{td} - \alpha_1$ and $x_4 = \alpha_1 - \alpha_b$ the Eq. (3) is written as:

$$\begin{cases} \dot{x}_1 = -\frac{\tau_1 + c_{td} + c_b}{J_1}x_1 + \frac{c_b}{J_1}x_2 + \frac{k}{J_1}x_3 - \frac{k}{J_1}x_4 + \frac{c_{td}}{J_1}u \\ \dot{x}_2 = \frac{c_b}{J_b}x_1 - \frac{\tau_b + c_b}{J_b}x_2 + \frac{k}{J_b}x_4 - \frac{1}{J_b}Tob \\ \dot{x}_3 = \dot{\alpha}_{td} - \dot{\alpha}_1 = u - x_1 \\ \dot{x}_4 = \dot{\alpha}_1 - \dot{\alpha}_b = x_1 - x_2 \end{cases} \tag{4}$$

3.2 Rock-Bit Interaction Model

In literature, several functions have been utilized to represent the interaction between the bit and the rock. For the system being studied, we will utilize the generalized model as expressed in [13] (Table 1).

$$Tob = \mu_n Nr\left(\frac{x_2}{\sqrt{x_2^2 + \Omega_0^2}} + \frac{p\Omega_0 x_2}{x_2^2 + \Omega_0^2}\right) + Dx_2\left(\frac{x_2}{\Omega_1} - 1\right) \tag{5}$$

where the model coefficients are given in the following table [13]:

Table 1. Parameters used in rock-bit general model

Parameter	Description	Value
μ_n	friction coefficient	28 nm
n	the force vector	9.81 * wob

(continued)

Table 1. (*continued*)

Parameter	Description	Value
r	the contact radius vector	0.1 m
Ω_0	chain transition speed	1
$\Omega 1$	transition speed for the well	31.4159
p	the initial friction parameter	1.5
d	the linear damping vector	0.28

3.3 PID Controller

PID control is the most frequently employed control law for regulation in engineering. It is widely employed across several industries due to its consistent performance. The PID control equation is commonly represented as given in Eq. (6) [9].

$$C(t) = K_p e(t) + K_i \int_0^t e(\tau)d\tau + Kd\frac{de(t)}{dt} \tag{6}$$

K_P, T_i and T_d represent the proportional integral, derivative controller coefficients. Equation (6) is the typical PID equation,

4 Particle Swarm Optimization

The initial structure of Particle Swarm Optimization was created by Kennedy and Eberhart in 1995 [15]. The PSO algorithm functions by utilizing a group of n particles that navigate a search space of d dimensions. Each particle is given a random velocity. During each iteration, every particle modifies its flight trajectory by taking into account its own experience as well as the collective experience of its companions. The i_{th} particle is represented as x_i, with its best previous solution (best) denoted as p_iThe current velocity, or position change rate, is described by v_i. Finally, the best solution achieved so far by the entire swarm (*gbest*) is represented as p_g.

During each time step of the process, every particle undergoes movement towards both its personal best (Pbest) and the global best (gbest) positions. The fitness function evaluates the performance of particles to see if the optimal solution has been achieved. The particles are controlled in accordance with the subsequent formulae.

$$v_{id}(t+1) = v_{id}(t) + c_1.rand.(p_{id}(t) - x_{id}(t)) + c_2.rand.\left(p_{gd}(t) - x_{id}(t)\right) \tag{7}$$

$$x_{id}(t+1) = x_{id}(t) + v_{id}(t) \tag{8}$$

The variable c_1 is referred to as the cognitive learning rate, while c_2 is referred to as the social learning rate. Both c_1 and c_2 are positive constants. The rand function generates a random number from the interval [0, 1]. The particles' velocity is limited to the range of *[Vmin, Vmax]*. Nevertheless, because the original PSO formula lacks a velocity control mechanism, the algorithm's capacity to conduct detailed searches is restricted.In order to address this issue, Eberhart and Shi devised an inertia factor that decreases over time, as described in reference [14].

$$v_{id}(t+1) = w.v_{id}(t) + c_1.rand.(p_{id}(t) - x_{id}(t)) + c_2.rand.\left(p_{gd}(t) - x_{id}(t)\right) \quad (9)$$

$$x_{id}(t+1) = x_{id}(t) + v_{id}(t) \quad (10)$$

4.1 Fitness Function ITAE (Integral of Time Multiplied Absolute Error)

The ITAE criterion is determined by integrating the absolute value of the error between the system's response and the desired answer, multiplied by time. The integral is computed over a defined time interval, often spanning from the initial response to a designated final time.

In order to achieve a high level of precision in the dynamic behavior of a system when subjected to a step input, it is necessary for the transient state to exhibit minimal overshoot and an optimal response time. In order to attain these performances, the settings of a regulator are selected in a manner that minimizes the dynamic error.

$$e(t) = y_{ref}(t) - y_{sys}(t) \quad (11)$$

This can be achieved using the ITAE performance criterion. The criterion is determined by Eq. (12) [16].

$$J_{ITAE} = \int_0^T t|e(t)|dt \quad (12)$$

5 Results and Discussion

Developing an effective PID controller necessitates determining the optimal tuning parameters. In this section, we present the open-loop and closed-loop response results from applying the proposed PID-PSO approach to rotary drilling systems using MATLAB with reference of Angular velocity = 20 rad/s for $0 < t < 200$ s.

5.1 Results Without Any Control

The Fig. 2a depicts the step response of a drilling rotary system in the absence of a controller. The bit velocity exhibits significant oscillations, the regulation time is excessively long, the system's dynamic and static characteristics are poor, and the ability to track the rotary table speed is inadequate.

5.2 Results for PID Controller Without PSO

The appropriate selected values of the PID controller parameters are provided: K_p=*0. 1 ;* K_i=*15;* K_d=*1.*

The result was shown in Fig. 2b represents the step response of a drilling rotary system with a PID controller, without the implementation of any tuning procedure. The bit velocity exhibits minor fluctuations, the time response is significantly delayed, and the capacity to accurately track the rotary table speed is achieved only after a prolonged period of time, characterized by large oscillations and eventual stabilization.

5.3 Results for PID Controller with PSO

The PSO algorithm was initiated using the subsequent parameters (Table 2):

Table 2. Particle swarm optimization algorithm parameters

Parameter	Value
Size of the swarm " no of birds "	n = 50;
Maximum number of "birds steps"	bird_setp = 50
Dimension of the problem	dim = 3
Upper limits of (Kp,Ki,Kd)	Bmax = [10 10 10]
Lower limits of (Kp,Ki,Kd)	Bmin = [0 0 0]
PSO parameter C1	c2 = 1.2
PSO parameter C2	c1 = 0.12
PSO momentum or inertia	w = 0.9

After the optimization process by PSO algorithm was finished we get these parameters below

$$K_p = 1.1692;\ K_i = 0.4300;\ K_d = 9.0064$$

*T*he simulation result of the drilling string model controlled with PID tuned by PSO is given in Fig. 2c that depicts the step response of a drilling rotary system that has been modified by a PSO method to optimize the PID controller. It is evident that the stick-slip vibrations were swiftly eradicated, resulting in a significantly shorter time response compared to the original system. Furthermore, there was no overshooting seen after 30 seconds.

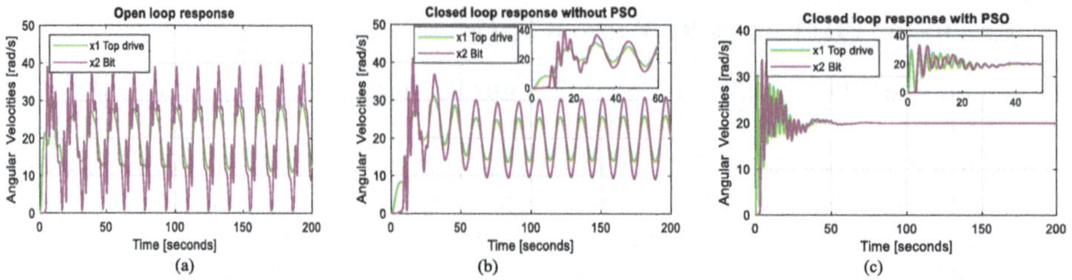

Fig. 2. (a) Open-loop response without control (b)Closed loop using PID Controller without PSO (c) Closed loop using PID Controller with PSO

5.4 Comparative Study

Figure 2a illustrates the outcomes of the open loop simulation indicating the presence of significant stick-slip vibration in the drill string system. All the components of the system exhibit periodic vibrations. Out of all these components, the drill bit undergoes the most intense stick-slip vibration, which is confirmed by the occurrence of both the viscous and sliding phases. These findings validate the efficacy of the methodology provided in this work for mitigating damage to the drilling equipments. This study aims to assess the efficacy of the PID controller in reducing stick-slip vibration. As depicted in Fig. 2b and Fig. 2c, the controller has the ability to decrease stick-slip vibration. However, the PID controller with PSO demonstrates greater effectiveness by completely eliminating the issue in a short period of time, leading to a steady drilling operation. These findings demonstrate the viability of the suggested control techniques for real-world implementation in the drilling sector.

6 Conclusion

The objective of the study was to improve the reduction of stick-slip vibrations in rotary drilling systems by utilizing a PID controller that was fine-tuned using the PSO algorithm, with ITAE serving as the fitness function. The results obtained were encouraging, since the PSO algorithm successfully mitigated stick-slip vibrations within a brief period. Nevertheless, the efficacy of the PSO method could be enhanced by augmenting the population size, albeit at the expense of additional time required to get optimal parameters. It is advisable to investigate the expansion of integer order PID controllers to non-integer order PID controllers for the purpose of controlling vibrations in drill string systems in future studies. In future studies, it would be worthwhile to explore other tuning strategies to compare and assess the efficacy of various controller tuning procedures. Additional inquiries could encompass the examination of interconnected vibrations, such as axial, torsional, and lateral vibrations, when developing a controller and implementing it in a prototype of a rotary drilling system. Ultimately, these investigations would be applied to the operation of a rotary drilling system in an Algerian field.

References

1. Leine, R.I., van Campen, D.H., Keultjes, W.J.G.: Stick-slip Whirl Interaction in Drillstring Dynamics (2002)

2. Doghmane, M.Z., Kidouche, M., Ahriche, A.: Decentralized overlapping control design with application to rotary drilling system. IETE J. Res. (2021). https://doi.org/10.1080/03772063.2021.1886602
3. Tang, L., Zhu, X., Shi, C., Tang, J., Xu, D.: Study of the influences of rotary table speed on stick-slip vibration of the drilling system. Petroleum **1**(4), 382–387 (2015)
4. Tang, L., Zhu, X., Qian, X., Shi, C.: Effects of weight on bit on torsional stick-slip vibration of oilwell drill string. J. Mech. Sci. Technol. **31**(10), 4589–4597 (2017)
5. Navarro-López, E., Licéaga Castro, E.: Non-desired transitions and sliding-mode control of a multi-DOF mechanical system with stick-slip oscillations. Chaos Solitons Fract **41**(4), 1561–2200 (2009)
6. Vaziri, V., Kapitaniak, M., Wiercigroch, M.: Suppression of drill-string stick–slip vibration by sliding mode control: numerical and experimental studies. Eur. J. App. Math **29**(5), 805–825 (2018)
7. Liu, Y.: Suppressing stick-slip oscillations in underactuated multibody drill-strings with parametric uncertainties using sliding-mode control. IET Control Theory **9**(1), 91–102 (2015)
8. Abdulgalil, F., Siguerdidjane, H.: Backstepping design for controlling rotary drilling system. In: Proceedings of 2005 IEEE Conference on Control Applications, 2005. CCA 2005, Aug. 28–31, pp. 120–4. Toronto, Ont., Canada (2005)
9. Mendil, C., Kidouche, M., Doghmane, M.Z.: Hybrid sliding PID controller for torsional vibrations mitigation in rotary drilling systems. Indonesian J. Electr. Eng. Comp. Sci. **22**(1) 146–158 (2021). ISSN: 2502-4752. https://doi.org/10.11591/ijeecs.v22.i1.pp146-158
10. Mendil, C., Kidouche, M., Doghmane, M.Z.: Sliding mode controller design for torsional vibrations minimization under rock-bit interaction effects. In: Hatti, M. (eds.) Artificial Intelligence and Heuristics for Smart Energy Efficiency in Smart Cities. IC-AIRES 2021. Lecture Notes in Networks and Systems, vol 361. Springer, Cham (2022). https://doi.org/10.1007/978-3-030-92038-8_61
11. China University of Petroleum Press and Springer Nature Singapore Pte Ltd. 2021 Z. Guan et al., Theory and Technology of Drilling Engineering, https://doi.org/10.1007/978-981-15-9327-7_2
12. Ritto, T.G., Ghandchi-Tehrani, M.: Active control of stick-slip torsional vibrations in drill-strings. Journal of Vibration and Control **25**(1), 194–202 (2019). The Author(s) 2018 Article reuse guidelines: sagepub.com/journals-permissions https://doi.org/10.1177/1077546318774240
13. Mendil, C., Kidouche, M., Doghmane, M.Z., Benammar, S., Tee, K.F.: Rock–bit interaction effects on high-frequency stick-slip vibration severity in rotary drilling systems. Multidiscip. Model. Mater. Struct. **17**(5), 1007–1023 (2021). https://doi.org/10.1108/MMMS-10-2020-0256
14. Cao, J.-y., Cao, B.-g.: Design of Fractional Order Controllers Based on Particle Swarm Optimization. 2006 1ST IEEE Conference on Industrial Electronics and Applications, pp. 1–6 (2006). https://doi.org/10.1109/ICIEA.2006.257091
15. Ramezanian, H., Balochian, S., Zare, A.: Design of optimal fractional-order PID controllers using particle swarm optimization algorithm for automatic voltage regulator (AVR) system. J. Control Autom. Electr. Sys. **24**, 601–611 (2013). https://doi.org/10.1007/s40313-013-0057-7
16. Oprzędkiewicz, K., Dziedzic, K.: A tuning of a fractional order PID controller with the use of particle swarm optimization method. In: Rutkowski, L., Korytkowski, M., Scherer, R., Tadeusiewicz, R., Zadeh, L., Zurada, J. (eds.) Artificial Intelligence and Soft Computing. ICAISC 2017. Lecture Notes in Computer Science, vol. 10245. Springer, Cham (2017). https://doi.org/10.1007/978-3-319-59063-9_35
17. Problems and Solutions Drilling Engineering Problems and Solutions A Field Guide for Engineers and Students M. E. Hossain and M. R. Islam

18. Riane, R., Doghmane, M.Z., kidouche, M., Tee, K.F.: Stick Slip vibration suppression in drill string using observer based LQG controller. Sensors **22**(16), 5979 (2022)
19. Kidouche, M., Charef, A.: A constructive methodology of Lyapunov function of composite system with nonlinear interconnection term. Int. J. Robot. Autom. **21**(1):19–24 (2006)
20. https://www.pason.com/stories/technology/stick-slip-a-costly-torsional-vibration

Sizing, Design and Simulation of Solar Water Pumping System as Sustainable Solution for El Oued Region, Algeria

M. Benmbarek, M. Daif, and A. Kheldoun(✉)

Department of Power and Control, Institute of Electrical and Electronic Engineering (IGEE), University M'hamed Bougara, Boumerdes (UMBB), Boumerdes, Algeria
aissa.kheldoun@univ-boumerdes.dz

Abstract. The paper presents a suggestion of a sustaible solution for an aride rigion in Algeria, (El oued). In fact, the need to develop the agriculture sector in the region has become evident to meet the population's food requirements. Due to the shortage of water ressources, a correct design of solar water pumping system allows to target both preserving the environement and reducing the total cost of investment. To this, the proposed study intends to select a farm, analyse its need in terms of water, design the required solar water pumping system, design the appropropiate control systgem, and finally simulate the wole systgem. In fact, the selected farm is located in Hassi Khalifa, El Oued, Algeria. The required data of climatology has been obtained using the dedicated software Pvgis while the average water consumtion was evaluated using the CROPWAT simulation tool from the Food and Agriculture Organization (FAO) of the United Nations (UN). The crop water requirements found are translated into electrical energy requirements which have led to the selection of the solar water pumping system's components (pump, motor, inveter, and PV array). Finally, the Matlab/ Simulink is used to carry out the simulation of the whole showcasing the water flow generation as function of the available solar sunlight. This Precise sizing would help in developing a system with no oversizing, which makes it more efficient and cost-effective as less amount of water is wasted.

Keywords: Sustainable Solar energy · Cropwat · PVgis · Crop water requirements · Component Sizing · Solar water pumps · Simulation

1 Introduction

Algeria is a big country with 2,381,741 km2 of total land, making it the world's tenth-largest nation by area, and the largest nation in Africa [1]. 80% of Algeria's land is considered a desert, which means it covers four-fifths of the country's total area, that is what makes Algeria a perfect spot for solar energy applications in frica. Moreover, Algeria receives approximately 3,000 h of sunshine per year, creating ideal conditions for solar energy production used in different applications to meet the domestic, agricultural, and industrial water needs of the region. The government has ambitious plans to harness

© The Author(s), under exclusive license to Springer Nature Switzerland AG 2025
M. Hatti (Ed.): IC-AIRES 2024, LNNS 1238, pp. 475–482, 2025.
https://doi.org/10.1007/978-3-031-80301-7_52

this resource, aiming to achieve 15,000 megawatts of solar energy capacity by 2035 [2]. The Algerian Sahara is also known for its significant contributions to agriculture and the agriculture sector contributes on average about 12% of Algeria's GDP and employs at least 20 percent of the population in rural areas [3].

The region of El Oued specifically, has been transformed into one of Algeria's main agricultural centers, leveraging many of its geographical privileges such as lying on the Albian aquifer (40 – 60 m deep water reservoirs), a mild and sunny winter climate, abundant sunlight, and vast land area. These previously mentioned factors would help in growing a variety of the most needed crops in Algeria namely; potatoes, tomatoes, onions, and peanuts. These crops are largely grown organically due to the high sunlight which reduces the incidence of plant diseases and diminishes the need for pesticides [4]. Irrigation then is highly needed to produce the right amounts of water required by these crops to be harvested at their specific seasons without any delay, therefore most of the irrigation systems in the region of Eloued are based on pumps (surface or submersible) which are supplied generally from the utility grid or by the use of diesel generators, this may lead to some inconveniences, especially for some rural cities where no access to grid system is available, and fuel transportation may seem a bit challenging, this will just lead to more water shortages. As a result, water pumping systems based on solar energy (SWPS) are becoming a trend, especially for irrigation applications in Saharan rural regions [5], where the metrological conditions are suitable for optimal electricity generation through the use of PV arrays, thus fewer water shortages will occur.

Different papers addressing the design of solar water pumping system have been recently published, such as listed in [6–11]. Authors in [6] proposed single stage, grid connected solar wate pumping system. The motor being used is switched relactance motor with high inductance to reduce the volatge ripple. The authors proposed a control system to stabilise the DC link volatge and enhance the operation of the Perturb and Observe MPPT algorithm. A very interesting results have been obtained in terms of synchronization between the PV and the grid where excess of power can be injected into the grid utility. However, design of SWPS components according to the purpose need is not provided. In [7], Rajan K. et al. used a double stage SWPS using a DC motor. This work differs from the previous in the selection of the MPPT algorithm and the elimination of the current sensors. The proposed methodology is interesting as it shows the computation of the most important component of the system. Besides, the simulation and experimental parts have focused on the efficiency of the SWPS under varying irradiation. However, the use of the DC motor is questionable as it requires a periodic maintenance which makes it not suitable for isolated regions such as the region of HASSI KHALIFA in EL OUED province. Intersting design of SWPS network is presented in [8] where an analystical method is proposed to deal with optimal sizing of SWPS within a DC microgrid. The developed method is suitable for the managment of water distribution system as to optimize energy consumption. However, the proposed SWPS within the water ditribution network requires the use of Battery Storage System. The latter is not suitable for remote irrigated areas. Differents techniques of Perturb and Observe MPPT techniques are designed and tested for pumping applications [9]. The purpose is to compare these techniques from different points of view such as stability, performance charcterstic. However, sizing of the SWPS's component is not considered.

A new MPPT algorithm has been applied in [10, 11] to enhance the solar power harvesting of the SWPS where the induction mtor is used to drive the pump. Encouraging results in terms of dynamic effeciency is obtained, however, optimal selection of the SWPS components is not taken into consideration.

The present paper presents a comprehensive methodology of sizing a SWPS. The study starts by the selection of the site where an irrigated land is selected and to which the crop water requirement is determined. After that, an approppriate pump is selected and the size of the remaining components of SWPS is optimaly deduced. Design of the control system is then conducted and simulation of the whole system is carraied out using Matlab/Simulink software. Several tests are condcted cosnidering both irradiance and the temperature variation. The remaining parts of the paper are orgnaized as follows: site selection and sizing of the SWPS are detailled in section II and section III resepectivelly. Design of the control system is preseted insection IV while simulation results and discussion are provided in section V. The paper is ended up with a conclusion.

2 The Site Selection

The site selection process is based on two major inputs: The geographical location of the site and its climatic data. For our case, a site was selected in Hassi-Khalifa, a small commune situated about 20 km away from the region of ELoued, it has the following geographical data shown in Fig. 1 along with the geographical position of Hassi Khalifa in both Eloued and Algeria.

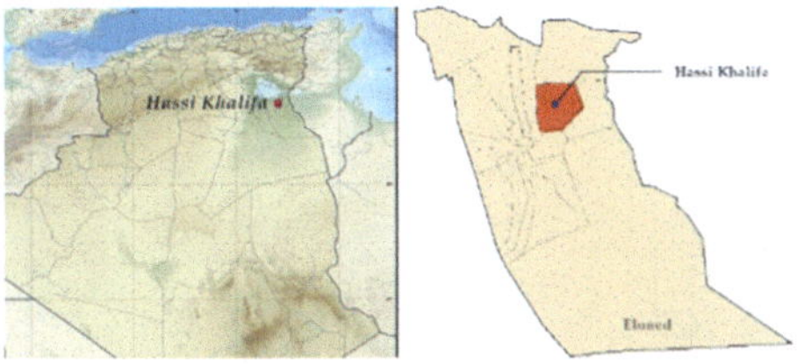

Parameter	Value
Site	Hassi Khelifa
Latitude [Deg]	33°33'44"N
Longitude [Deg]	6°59'25"E
Elevation [m]	48 m
Total Area [Km2]	1 112 km2

Fig. 1. Geographical Information of Hassi Khelifa [12]

The climatic data is considered as our second parameter for the site selection, as for the selected region different metrological data was extracted. First, through the Pvgis platform, we get the monthly irradiation on the horizontal plane in kWh/m2/mo. Additionally, more data was recorded for Eloued's temperature from [13], along with Rainfall in mm, Humidity in %, wind speed in Km/day, and average Sun hours.

3 Sizing of SWPS Components

For motor-pump system sizing, two key parameters are needed: the flow rate (Q) and the total dynamic head (TDH). The flow rate indicates the amount of water the pump can deliver over time. In solar pumping, it's often expressed in cubic meters per day. Flow rate requirements can be determined using reference evapotranspiration (ETo) and crop coefficient (kc), which vary by crop type and growth stage, estimated using the Penman-Monteith equation [14]

$$\boldsymbol{ET}_0 = \left[0.408\Delta(\boldsymbol{R}_n - \boldsymbol{G}) \frac{900}{\boldsymbol{T} + 273} \boldsymbol{u}_2(\boldsymbol{e}_s - \boldsymbol{e}_a) \right] \left[\frac{1}{\Delta + \boldsymbol{\gamma}(1 + 0.34\boldsymbol{u}_2)} \right] \tag{1}$$

Rn is net radiation at the crop surface (MJ · m2 · day^{-1}), G soil heat flux density (MJ · m^{-2} · day^{-1}), T means daily air temperature at 2 m height (°C), u2 wind velocity at 2 m height (m · s^{-1}), e_s and e_a saturation and actual vapor pressure (kPa), slope of vapor–pressure curve (kPa·C^{-1}), psychrometric constant (kPa·C^{-1}). The crop evapotranspiration ETc is determined using the following formula:

$$\boldsymbol{ET}_c = \boldsymbol{K}_c \cdot \boldsymbol{ET}_0 \tag{2}$$

The method shown in Eq [1] and [2] is complex due to the many required parameters, but it provides accurate results, preventing system oversizing. When some parameters are unavailable, the Cropwat software by the FAO is used as an alternative. Cropwat manages crop data from planting to harvest and generates useful curves, aiding farmers, especially those without technical expertise. Now data will be inserted manually from the extracted data discussed above into Cropwat software. Moreover, in our Hassi Khalifa site, potatoes are planted on a 1-hectare plot with 10 cm spacing, yielding 10,000. The crop coefficient (Kc) then ranges from 0.5 to 1.15. Using Kc and reference evapotranspiration, actual evapotranspiration (ETc) is calculated from Eq [2]. Figures 2 and 3 show data from Cropwat software.

Month	Decade	Stage	Kc	ETc	ETc	Eff rain	Irr. Req.
			coeff	mm/day	mm/dec	mm/dec	mm/dec
Sep	2	Init	0.50	4.12	12.4	0.4	12.4
Sep	3	Init	0.50	3.71	37.1	1.4	35.7
Oct	1	Init	0.50	3.29	32.9	1.3	31.6
Oct	2	Deve	0.58	3.35	33.5	1.3	32.1
Oct	3	Deve	0.82	4.14	45.5	1.5	44.0
Nov	1	Deve	1.05	4.62	46.2	1.9	44.3
Nov	2	Mid	1.15	4.36	43.6	2.2	41.4
Nov	3	Mid	1.15	3.87	38.7	1.9	36.8
Dec	1	Mid	1.15	3.30	33.0	1.4	31.6
Dec	2	Mid	1.15	2.77	27.7	1.1	26.6
Dec	3	Late	1.16	2.81	30.9	1.7	29.2
Jan	1	Late	1.04	2.55	25.5	2.6	22.9
Jan	2	Late	0.91	2.23	22.3	3.2	19.1
Jan	3	Late	0.81	2.35	11.7	1.2	10.4
					441.0	23.2	418.1

Fig. 2. Seasonal irrigation data for potatoes crop management

Month	Min Temp	Max Temp	Humidity	Wind	Sun	Rad	ETo
	°C	°C	%	km/day	hours	MJ/m²/day	mm/day
January	5.5	16.5	61	320	8.8	13.4	2.55
February	7.2	18.8	47	336	9.6	16.7	3.78
March	10.5	22.7	39	355	10.5	21.1	5.40
April	15.0	27.2	32	386	11.5	25.3	7.48
May	19.4	32.2	28	401	12.4	28.1	9.46
June	23.0	37.2	25	405	12.9	29.2	11.11
July	26.6	40.0	23	474	12.7	28.5	12.90
August	26.6	39.4	26	340	12.0	26.4	10.48
September	22.7	35.0	36	328	11.1	22.7	8.24
October	17.2	28.8	43	305	10.1	18.2	5.75
November	11.1	22.2	53	305	9.2	14.2	3.70
December	6.6	17.2	63	309	8.6	12.3	2.46
Average	15.9	28.1	40	355	10.8	21.4	6.94

Fig. 3. Different climatic data in the region of hassi khelifa, eloued

Thus, for a 1-hectare area, an irrigation requirement of 44.3 mm per decade equates to approximately 44 cubic meters per day. With peak sun hours (PSH) defined as 10.8, the flow rate calculated using Eq. 3 is about 4 m^3/hr as shown below

$$Q = \frac{Q'}{PSH} = \frac{44}{10.8} = 4.08m^3/hr \approx 4m^3/hr \tag{3}$$

Moving to the second parameter which is the total dynami head (TDH), defined by the author in [3] as the difference in pressure in meters of water column between the suction and discharge ports. This height can be calculated as shown in (4).

$$TDH = H_g + Pl \tag{4}$$

- Hg = is the geometric height between the surface of our submersible pump and the top of our tank (Hd + Ht)
- Pl = is the pipe loss caused by the friction of water against the walls of the pipes. These losses depend on the pipes' length, diameter, and the pump's flow rate (Q) and are expressed in meters of water. Generally, these pressure losses correspond to no more than 10%.

Now, for our selected site, water is found at depths of 60 to 80 m, with 70 m used as a fixed depth. For the pressure losses it is taken as 10% of the geometric height Hg, we get the following result for the TDH shown below:

$$TDH = (70 + 4) \cdot (1 + 0.10) = 81.4m \tag{5}$$

Using the following equation for the hydraulic and electrical power from [6], we get:

$$P_h = \frac{Q \cdot \rho \cdot g \cdot TDH}{3600 \cdot 1000} = \frac{4.32 \cdot 1000 \cdot 9.8 \cdot 81.4}{3600 \cdot 1000} = 0.957\,kW \tag{6}$$

For the motor power and taking a motor-pump efficency of 44.3%, we get:

$$P_m = \frac{0.957}{44.3\%} = 2.16\,kW \approx 2.2\,kW \tag{7}$$

For the inverter, it is sized based on the motor power considering a small margin of about 30% to ensure the inverter's ability to handle the power sparks. The equation below illustrates the inverter sizing:

$$P_{inv} = P_m \cdot (1 + 30\%) = 2.86\,kW \approx 3\,kW \tag{8}$$

For the Pv array, we do the same thing, only this time it would be sized based on the inverter power instead of the motor power. By adding a margin of 5% in this case, we get

$$P_{PV} = P_{inv} \cdot (1 + 5\%) = 3.15\,kW \tag{9}$$

So, choosing a PV module of 320 W, the following number of modules is found:

$$Number\ of\ PV\ Modules = \frac{3.15}{0.319} = 9.9 \approx 10\,modules \tag{10}$$

In our case, we take 10 modules for our PV array, they are all installed in series to meet the voltage and current requirements. For the DC bus voltage we take it as 680 V to meet the input voltage for our inverter.

4 Control System Design

The proposed SWPS with induction motor is shown in Fig. 4. The choosed 2.2 kW induction motor incorporated within our SWPS is controlled using the Direct Torque Control technique (DTC) which controls through a voltage source inverter (VSI) the amount of current needed to generate a certain torque, hence, affecting the motor's speed based on the system's need. Moreover, for a closed loop control system a Motor-side MPPT block is used at the input of our DTC controller. Figure 4 shows the coordination between the PV-side MPPT (P&O) and the Motor-side MPPT to esnure seamless transfer of maximum PV power to the induction motor.

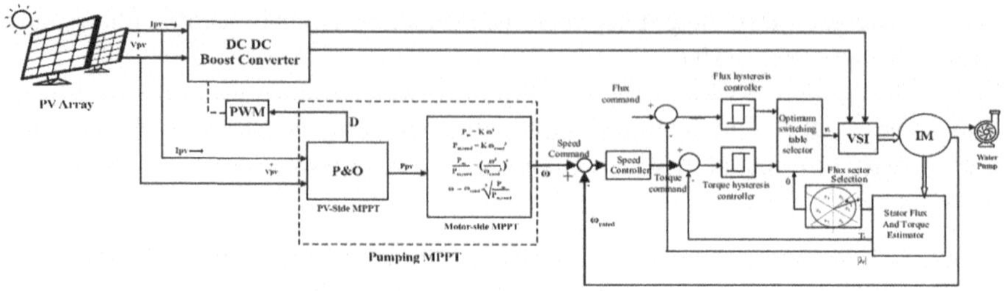

Fig. 4. The ppproposed pv pumping system

For the PV side a classical P&O algorithm [15] is used which ensures maximum energy being harvested from the solar panels even under different metrological conditions.

5 Test Results and Discussion

Different Irradiance and temprature levels were applied on our SWPS as shown in Fig. 5.

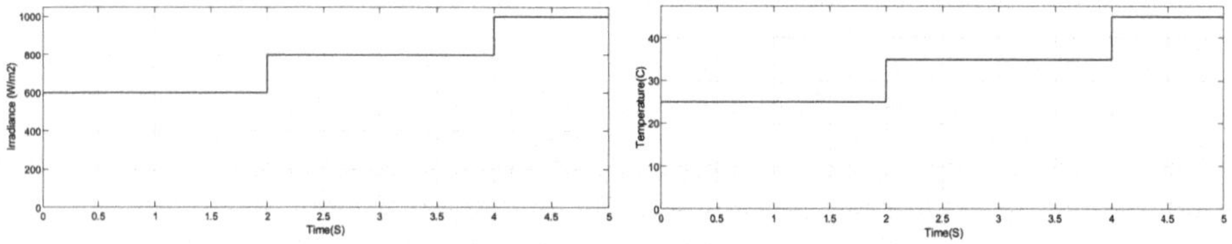

Fig. 5. Different irradiance and temprature levels

5.1 Variable Irradiance Condition

By varying the irradiance applied on our Pv array from 600 W/m^2 to 1000 W/m^2 while maintaining a fixed temprature, we get the following results using Matlab/Simulink

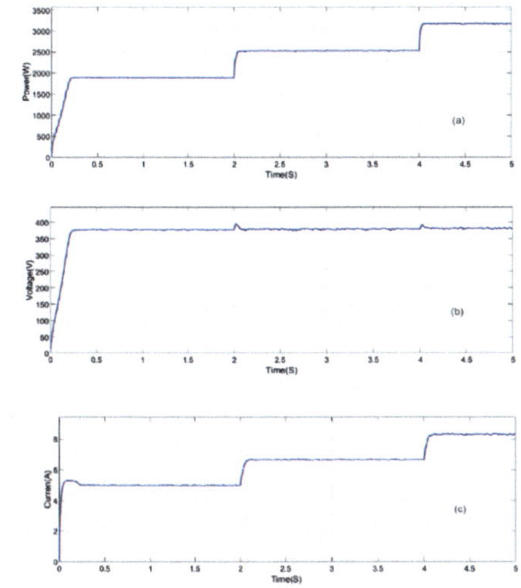

Fig. 6. Starting and steady-state curves for: (a) PPV, (b) vpv and (c) IPV

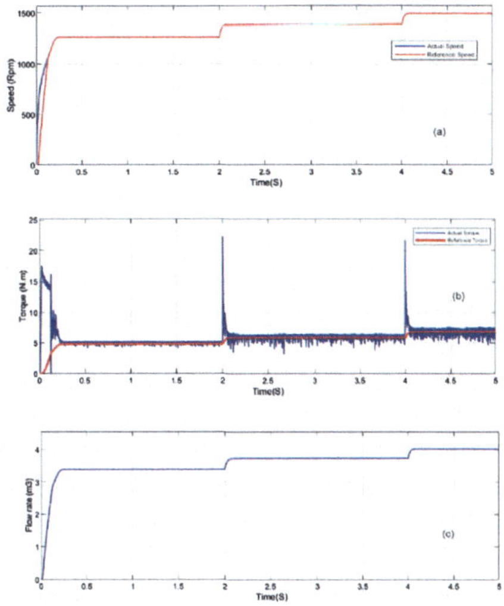

Fig. 7. Starting and steady-state curves for: (a) speed, (b) torque and (c) flow rate

5.2 Variable Temprature Condition:

By varying the temperature applied on our PV array from 25 c° to 45 c° while maintaining a fixed irradiance we get the following results.

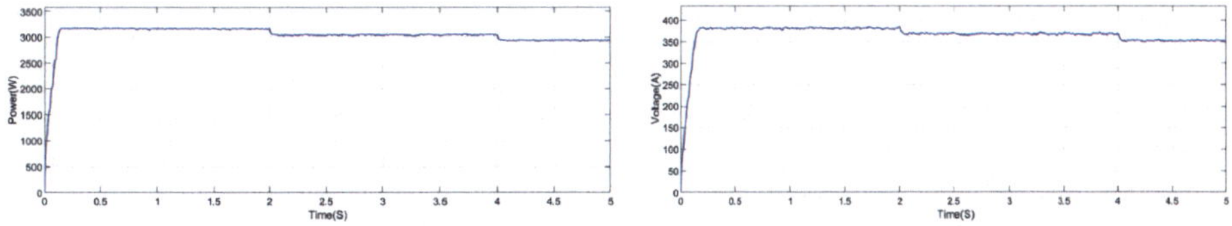

Fig. 8. Starting and steady-state curves of PV array under variable temprature

By analyzing results in Figs. 6, 7 and 8, one can deduce the P&O MPPT has succeeded to converge to the right MPP regardless of the environment conditions (for both irradiance or temperature variation). The obtained maximum power at the output of the PV is not sufficient. To this, another block, that is the motor -side MPPT is added. The later generates the appropriate rotor speed to the DTC so that maximum power is transferred to the motor through the inverter. As such, maximum water flow rate is recorded to fill in the water tank ins very short time.

6 Conclusion

In this paper, we present the sizing, design, and simulation of a SWPS in El Oued, Algeria. The site in Hassi Khelifa was selected due to its favorable climatic conditions. Using Cropwat software, the SWPS was sized. The control system was designed using

DTC to manage the IM's speed, and the P&O algorithm was implemented to optimize PV output power. Simulations in Matlab/Simulink demonstrated the system's effectiveness under varying meteorological conditions. Future work includes developing a test rig to validate the proposed SWPS design.

References

1. Wikipedia. Algeria. Wikipedia, https://en.wikipedia.org/wiki/Algeria. Accessed on: 02 April 2024
2. IEF International Energy Forum. Algeria powers ahead with huge renewable energy plans (2021). https://www.ief.org/news/algeria-powers-ahead-with-huge-renewable-energy-plans. Accessed on 02 April 2024
3. Blom-Zandstra, G., Michielsen, J.-M.: Sustainable water use in potato production in algeria; introduction of a subsurface fertigation system in the desert. Wageningen Research, pp. 9–13 (2020)
4. kawa. Algeria: a flourishing agricultural hub in the desert (2022). https://kawa-news.com/en/algeria-a-flourishing-agricultural-hub-in-the-desert/. Accessed on 03 April 2024
5. Jos´e Luis Dom´ınguez-Garc´ıa Levon Gevorkov and Lluis Trilla Romero. Review on solar photovoltaic-powered pumping systems. Energies **16**(1), 94 (2023). https://doi.org/10.3390/en16010094.2022
6. Mishra, A.K., Singh, B.: Grid Interactive Single-Stage Solar Powered Water Pumping System Utilizing Improved Control Technique. IEEE Transactions on Sustainable Energy **11**(1) (2020)
7. Kumar, R., Singh, B.: Single Stage Solar PV Fed Brushless DC Motor Driven Water Pump. IEEE J. Emerg. Select. Top. Power Electron. **5**(3) (2017)
8. Castillo-Calzadilla, T., et al.: Systematic analysis and design of water networks with solar photovoltaic energy. IEEE Trans. Eng. Manage. **69**(3) (2022)
9. Elgendy, M.A., Zahawi, B., Atkinson, D.J.: Assessment of Perturb and Observe MPPT Algorithm Implementation Techniques for PV Pumping Applications. IEEE Trans. Sustain. Ener. **3**(1) (2012)
10. Djeriou, S., Kheldoun, A., Mellit, A.: Efficiency improvement in induction motor-driven solar water pumping system using golden section search algorithm. Arab. J. Sci. Eng. **43**, 3199–3211 (2018)
11. Kheldoun, A., Djeriou, S., Kouadri, A., Refoufi, L.: A simple and accurate maximum power point tracking algorithm for photovoltaic systems progress in clean energy. Novel Systems and Applications **2**, 721–733. https://doi.org/10.1007/978-3-319-17031-2_50
12. Wikipedia. Hassi Khelifa. Wikipedia, https://fr.wikipedia.org/wiki/Hassi_Khalifa. Accessed on 05 April 2024
13. Weather Spark. Climate and average weather year round in el oued. https://weatherspark.com/y/53031/Average-Weather-in-El-Oued-Algeria-Year-Round. Accessed 05 April 2024
14. Pereira Luis, S., Allen, R.D., Richard, G., Martin, S.: Crop evapotranspirationguidelines for computing crop water requirements. FAO Irrigation and Drainage Paper 56. FAO (1998)
15. El Filali, A., Zazi, M.: Arduino implementation of MPPT with P and O algorithm in photovoltaic systems. Int. J. Eng. Appl. Phys. **1**(1), 9–17 (2021)

Optimized Energy Management of a Hybrid System with Storage Using Genetic Algorithm

Zoulikha Ouchefoun(✉), Mourad Hasni, and Lakhdar Guenfaf

Department of Electrical Engineering, Laboratory of Electrical &Industrial Systems (LSEI), Houari Boumediene University, Bab Ezzouar, Algeria
ouchefoune.electro2@gmail.com

Abstract. In this article, we will optimize energy management for a hybrid system that combines renewable energy sources (solar) with storage systems (batteries), as well as residual loads and electric vehicles. This system is integrated into the traditional electricity network. The main objective is to develop energy management strategies that minimize costs, maximize the use of photovoltaic panels, and ensure a reliable energy supply. This includes the optimal management of energy flows between different sources and storage, utilizing the Genetic Algorithm (GA). The GA employs techniques such as selection, crossover, and mutation to find optimal solutions to management problems. The results demonstrate how the Genetic Algorithm can optimize the energy management of a hybrid system by balancing photovoltaic production, load consumption, electric vehicle usage, and exchanges with the electricity network. This leads to minimized costs and efficient battery charge management, ensuring optimal system performance.

Keywords: Hybrid system · Battery · Photovoltaic panels · Storage · Genetic algorithm

1 Introduction

Faced with the depletion of fossil fuels and the challenges of global warming in recent years, the demand for electrical energy has increased alongside industrialization. Various management and planning tools have been utilized [1, 2], but further research and development are needed to optimize their use in meeting future energy demands. Grid reliability has become a crucial factor, as the conventional grid has transformed into a smarter grid, resulting in zero marginal costs [3, 4]. Therefore, hybrid energy systems with storage, which are well-suited for grid- connected or decentralized production, can help address several energy supply challenges [5]. The system studied consists of photovoltaic panels, which are considered the primary energy source from an electrical standpoint, integrated into traditional electrical networks that supply residual loads. This system also incorporates batteries and electric vehicles. Energy management in this case is based on the application of the Genetic Algorithm [6], a well- known method for solving problems related to energy management and optimization. This approach aims to maximize the use of photovoltaic panels while minimizing the total system cost.

© The Author(s), under exclusive license to Springer Nature Switzerland AG 2025
M. Hatti (Ed.): IC-AIRES 2024, LNNS 1238, pp. 483–490, 2025.
https://doi.org/10.1007/978-3-031-80301-7_53

2 Description and Modeling my Systemtheme Studies

2.1 Description OF THE System Studied

Fig. 1.

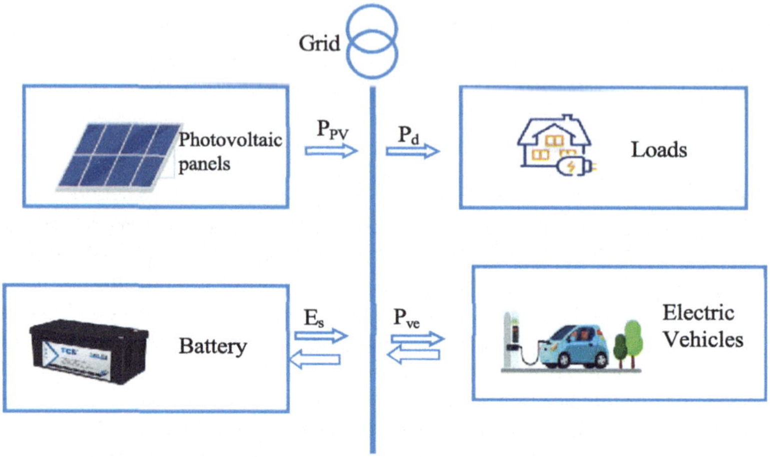

Fig. 1. Description of system studied.

The table below summarizes the system parameters and limits (Table 1).

Table 1. The lower and upper system limits.

Photovoltaic panels	$0 < P_{pv} < 19.5$ kW
Battery	20 kWh $< P_{ch} < 100$ kWh
Electric vehicle	-3 kW $< P_{ve} < 3$ kW
Charge	$0 < P_d < 11$ kW
Network	$0 < P_{grid} < 12$ kW

This equation represents the output power [7]:

$$P_{PV} = P_{ref}.\frac{1}{I_{ref}}[1 + k((T + 0.0256I) - T_{ref})] \quad (1)$$

P_{ref} : The reference power of the module (kw) in this case the values take as low profile, I_{ref}: The reference irradiation rate (1 kW/m2), I: The rate of irradiation exposed to the surface of the module in (W/m2), k: The temperature coefficient (°C-1), T: The temperature coefficient (°C-1) and T_{ref}: The reference temperature in (°C) [7].

The parameters are represented by Table 2:

Table 2. Photovoltaic panel model parameters.

photovoltaic panel	Parameters
Iref	1kW/m^2
K	0.0087 °C-1
Tref	25 °C

2.2 Mathematical Model of Storage Batteries

For proper operation of the hybrid energy system [3], we calculate the excess energy every hour and update the battery charge according to equations (2) and (3):

$$E_s = P_{pv} - (P_d + P_{ve} + P_{g\_import} - P_{g\_export}) \tag{2}$$

$$P_{ch} = P_{ch} + E_s \tag{3}$$

With E_s: The energy stored in kWh, P_{ch} : The power of the battery charge in kWh, P_{ev} : Power used by electric vehicles. P_{g_import}: Power imported from the network and P_{g_export}: Power exported to the network To encourage keeping the battery charge close to the target charge, P: penalty is applied if the final charge of the battery deviates from this target.

2.3 Electrique Vehicles

EVs can play a dual role: consuming energy (charging) or providing energy (discharging) to the system and taken into account in cost calculations and constraints considered as a means of storage.

3 Problematic

In our system sometimes the production is very high and can be exceeded the required loads, so we use storage batteries and electric vehicles to store excess energy and maximize the use of PV to reduce independence from the grid electric and minimize the total cost, for this we propose 03 scenarios allowing you to choose better management of this system.

Scenario 01: The objective of using fully 100% battery charging is to maximize the storage and use of solar energy, to reduce the loss of excess energy.

Scenario 02: Using the battery charge up to 80% allows maximization of the use of solar energy produced and reduces dependence on traditional electricity networks, and increases the energy supplied from EVs.

Scenario 03: Based on the use of battery charging at 50% of nominal load, therefore balance between solar panel production and load demand can be used to store a large part of the excess energy.

3.1 Objective Function and System Constraints

We will choose a cost function as an objective function representing the minimum of the sum of all possible costs shown in equation (4).

$$\text{Minf(x)} = 1\sum\nolimits_{t=1}^{T}(\text{Cg\_import} + \text{Cu} + \text{Ce} + \text{P}) \tag{4}$$

Cu: The total cost of unused energy, Ce: The total excess production and P: the general penalty.

Take into consideration the following costs: cost of energy taken from the network 0.27 (€/kWh), cost of unused energy 0.1 (€/kWh), cost of excess production 0.05 (€/kWh) and the weight of the penalty for deviation from the target battery charge 0.1.

The constraints in our program are used to ensure that the solutions generated by the genetic algorithm respect certain specific rules of the energy system. First inequality constraints to limit the values of the decision variables to certain intervals and the equality constraints ensure that the energy balance is respected in the system as shown in equation (5):

$$C_{eq} = P_{pv} + P_{g\_import} - P_{d} - P_{ve} - P_{g\_export} \tag{5}$$

3.2 The Genetic Algorithm Diagram

The genetic algorithm chosen from approximate methods particularly suited to the type of problem, we presented the bases which made it possible to develop this algorithm [8, 9], it was programmed on the matlab Mfile software without the use of Optimization Tool', Using these GA parameters allows configuring the genetic algorithm to balance the exploration of the solution space and the convergence towards an optimal solution.

Population Size: 200, max generations: 200 and the crossover fraction = 0.8 (Fig. 2).

4 Result and Discusion

The system studied was simulated over a period of 24 hours on 03 scenario explained above, We will choose the temperature and irradiation profiles of Msila in the month of May, represented in Figure 3 (Fig. 4).

4.1 Scenario Result 1 (chcible_battery = 100%)

In this scenario, where the target battery charge is 100%, the system is optimized to maximize the use of photovoltaic energy and manage loads without relying on the grid. Battery and electric vehicles play a crucial role in balancing energy demand and supply. The absence of import and export of energy from the grid indicates energy independence of the system, which can reduce energy costs and improve autonomy (Fig. 5).

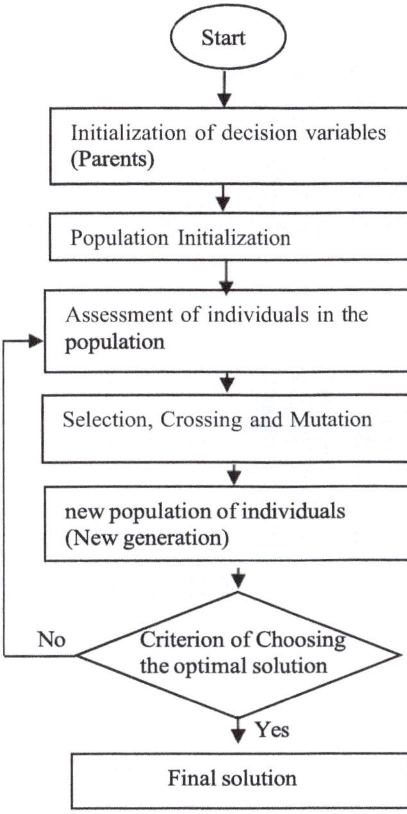

Fig. 2. The diagram of the genetic algorithm used.

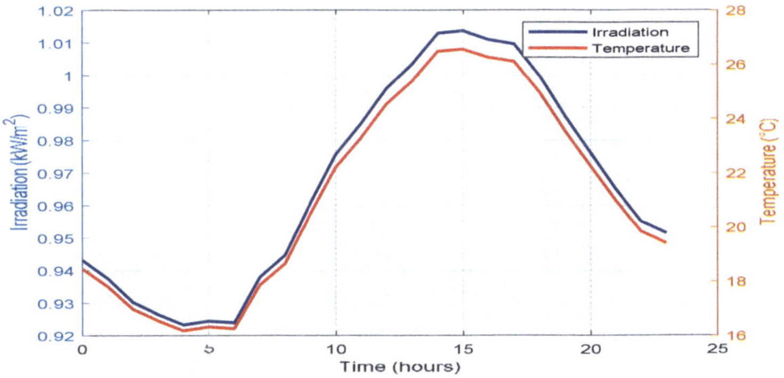

Fig. 3. The Irradiation profile and temperature as a function of time.

4.2 Scenario Result 1 (chcible_battery = 80%)

In this scenario, where the target battery charge is 80%, the system is optimized to maximize the use of photovoltaic energy and manage the loads efficiently. Achieving

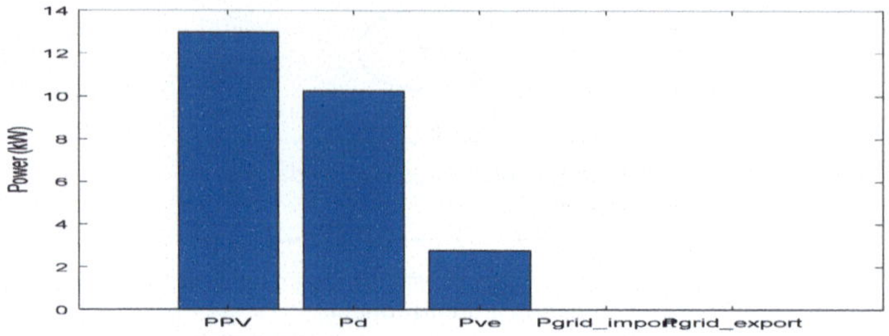

Fig. 4. Optimal power distribution from scenario 1.

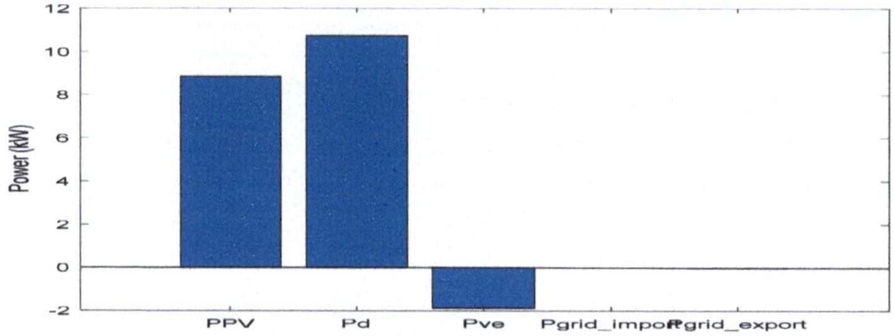

Fig. 5. Optimal power distribution from scenario 2.

the target charge of 80% for the battery ensures a good balance between the use of stored energy and the supply of energy for loads. EVs play a crucial role in providing energy to the system, contributing to overall stability and efficiency (Fig. 6).

4.3 Scenario Result 1(chcible_battery = 50%)

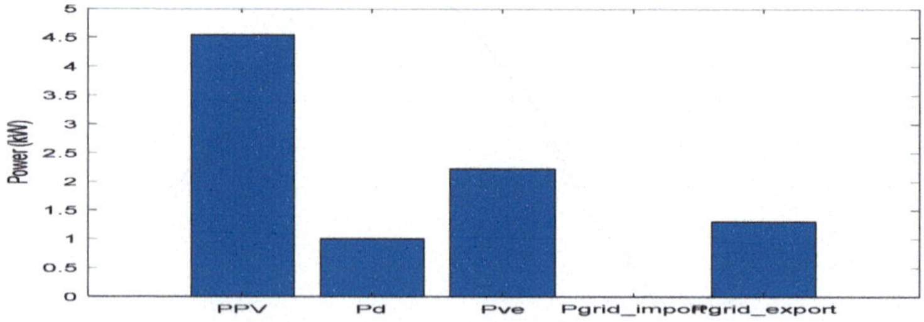

Fig. 6. Optimal power distribution from scenario 3.

In this scenario, Moderate contribution from solar is significant but less compared to higher battery target loads. More than Low consumption by loads and the power exported to the grid is 1.3057 kW, which suggests that excess produced energy is returned to the grid, potentially generating revenue or credits.

4.4 Evaluation and Comparisons Between the Three Scenarios

Table 3 represents the main results of this simulation.

Table 3. The main results of the 03 scenarios.

Poptimal (kW)	Ppv	Pev	Pch	Pg_export	Ctotale (€/kWh)
Scenario1	9.9908	7.4245	2.5665	0	8.8656
Scenario2	8.8558	10.7289	-1.873	0	6.8655
Scenario3	4.5405	1.0064	2.2284	1.3057	8.0258

In order to judge these management scenarios, as well as the several criteria allowing the evaluation of energy and economic costs, even the optimal use of power [10], we can make the comparison.

Scenario 1: Despite in this case maximum use of solar energy and no independence from the grid, but the use of batteries at 100% limits your lifespan more than the cost is high.

Scenario2: Is the best method which provides strict management, and is also less expensive compared to the others.

Scenario3: The lower target battery charge may not provide enough energy reserve for periods of low solar production or high demand.

5 Conclusion

The hybrid energy management system presented uses a genetic algorithm (GA) to optimize power distribution and manage excess energy storage in a system comprising photovoltaic (PV) panels, consumer loads, electric vehicles (EV), and the electrical network. The main objective is to minimize energy costs while adhering to the constraints associated with battery charging. The results obtained demonstrate that optimized energy management can lead to significant savings and a more efficient use of available energy resources. The prospects for improving this work are vast, including the integration of advanced technologies, the exploration of new energy sources, and the development of sophisticated control systems.

References

1. Labbi, Y.: Optimal management and control of electrical energy on production sites. In Mohamed Biskra University, p. 3 (2016)
2. Crosi, L.: Energy management in a multi-source photovoltaic and wind system with hybrid storage batteries/supercapacitors. In: National School of Engineering of Poitiers, p. 6 (2006)
3. Moataz, L., Amrane, O.: Energy management and multisource system optimization based on genetic algorithm. In: Electrical GE Symposium (SGE'14) (2014)
4. Haddi, S.: Contribution to the optimization of the insertion of renewable energies in a smart electricity network (Smart Grid). In: Ferhat Abbas University, pp. 3 (2019)

5. Elhamrani, R.: Study and simulation of wind-photovoltaic hybrid system components with storage (supercapacitor batteries)). In: th International Renewable Energy Conference (CIER) Proceedings of Engineering and Technology – PET, pp. 167–173 (2016)
6. Missaoui, R.: Optimized energy management for a building intelligent multi-source multi-load: different validation principles, p. 16. University of Grenoble INP (2012)
7. Omar, C., Reboul, J., Houivet, D., Gualous, H.: Genetic algorithm applied to energy management of a multi-source power plant installed at the University of Djibout. University of Djibouti (2023)
8. Selma, Amin: Genetic algorithms and asset allocation optimization, p. 2. the Périclès Actuarial team (2015)
9. Vechiu, L.: Modeling and analysis of the integration of renewable energies in a network, p. 125. Harve University (2005)
10. Aurélie, C.: Smart microgrid for energy resource management, p. 43. University of Perpignan Via Domitia (2016)

Advanced Algorithms and Machine Learning

Energy Aware and Stability Multipath Routing Protocol Based on the Cuckoo Search Algorithm for FANETs

Abdelkader Mekkaoui, Omar Smail(✉), and Mohammed Rebbah

Computer Science Department, Faculty of Exact Sciences, University Mustapha Stambouli, Mascara, Algeria
{abdelkader.mekkaoui,o.smail,rebbahmed}@univ-mascara.dz

Abstract. A FANET is a type of very low density ad hoc network where all its nodes move irregularly and quickly. In addition to the environmental variables, several characteristics influence the performance of FANETs, such as speed, mobility, and drone density. In order to solve the problem of energy consumption and data loss for FANETs, we propose a stable and energy-efficient multipath routing protocol based on the cuckoo search algorithm named SE-RPCS. A fitness function is integrated to choose the best path among all the new paths generated after the end of the cuckoo search algorithm. This fitness function relies on two constraints: energy and link stability. The performance of our protocol is compared to AOMDV and AODV protocols. The results show that SE-RPCS improves network performance, provides more efficiency and reliability.

Keywords: FANET · multipath routing · energy constraint · link stability constraint · cuckoo search algorithm

1 Introduction

An ad hoc network is a wireless and infrastructure-less communication network composed of a set of mobile nodes. A FANET is a wireless network that clusters drones in an ad-hoc manner [1]. FANETs have become a more important medium in life considering their advantages, for example: disaster rescue [2], urban patrol [3], critical infrastructure safety inspections [4], precision agriculture [5], etc. Routing is one of the problems in FANET networks. It consists of finding a set of paths between a source drone and a destination drone to transfer data. The routing process is extremely difficult for several reasons including: dynamic network topology, limited characteristics of drone, and bandwidth limitation. Routing protocols can be divided into two classes [6]. The first is the class of topology-based protocols, divided into three subclasses (reactive, proactive, and hybrid) [7]. The second is the location-based protocols, which use real-time location information of drone and their neighbors via GPS to make decisions such as [8].

Researchers in the field of routing in FANETs have tried to optimize a single or a set of objectives [9, 10], others are inspired by nature to extract optimization methods that use natural phenomena. A new era has emerged with the birth of bio-inspired algorithms

© The Author(s), under exclusive license to Springer Nature Switzerland AG 2025
M. Hatti (Ed.): IC-AIRES 2024, LNNS 1238, pp. 493–503, 2025.
https://doi.org/10.1007/978-3-031-80301-7_54

that have become an inexhaustible source of inspiration for FANETs [11, 12]. In the following, we will present some research on the implementation of routing protocols in the field of FANETs to guarantee some requirements such as: packet delivery rate, end-to-end delay, and overhead. The objective is to ensure efficient and reliable communication between drones based on different parameters when choosing the best paths between a source and destination drone.

In [13], the authors suggested a novel congestion-aware routing protocol based on location estimation for sparsely deployed energy-efficient UAVs named LECAR. The objective of this study is to solve two problems: network congestion and energy consumption. This protocol uses the mobility model to find the location of the destination UAV in real time. The source UAV transmits the data to the next UAV that has sufficient buffer capacity and is reached by a minimum distance. LECAR protocol guarantees low energy consumption, high packet delivery ratio, and provides a minimum number of transmissions compared to Spray and Wait, LAROD-LoDiS, LER, GPSR-Q, and GPSR. However, this solution was simulated in two-dimensional space which is unsuitable for FANETs. In [14], a novel energy-efficient routing solution was proposed for FANET networks named ECaD. The authors used the residual energy as well as the drones' motion data to guarantee the communication stability. The path discovery process of ECaD protocol was inspired by the basic AODV protocol based on energy consumption balancing, connectivity degree of discovered paths and link failure prediction between drones. The performance of this solution outperforms LEPR, UVAR-S, MPGR, and CBLADSR routing protocols in terms of PDR, overhead, and end-to-end delay, except that this solution does not take into consideration the stability of the drones and the simulations did not involve a large population of drones. In [15], the authors implemented a routing protocol named MDRMA (Multi Data Rate Mobility Aware) based on the single-path AODV protocol for FANETs. This solution is a major extension of the MA-DP-AODV-AHM (Mobility Aware Dual Phase Ad-Hoc On-Demand Distance Vector with Adaptive Hello Messages) routing protocol. Two algorithms are included in MDRMA. The first is the routing algorithm that ensures data transfer in stable and reliable links by reducing the number of RREQ packets, choosing intermediate drones according to their speed, direction, and data transmission rate. The second is the power-controlled algorithm, which is used to monitor the data power in each route. Simulations show that MDRMA outperforms both MA-DP-AODV-AHM and MDA-AODV protocols in terms of PDR, overhead, and end-to-end delay. This solution was simulated in a two-dimensional space which is not compatible with a FANET network, in addition the energy of the drones is ignored during the discovery process.

Several works in FANETs propose an hybridization between multipath routing protocols and metaheuristic algorithms to design more adaptive, efficient and resilient networks. In [16], the authors tried to minimize the instability problem of FANET networks which is due to the limitation of UAV batteries and their mobility, by implementing an efficient bio-inspired clustering scheme (BICSF). This solution uses an hybridization between two bio-inspired algorithms: the glowworm swarm optimization (GSO) which generates energy-aware clusters and selects cluster heads and the krill herd (KH) algorithm for efficient cluster management. To manage the optimal position of the drones, genetic operations such as mutation and crossover are used. The choice of the best path depends on the value of the fitness function which is expressed as a function of the weighted residual energy of the drone, the distance between drones and the number of neighbors. The results show that BICSF outperforms both ACO and GWO clustering algorithms in terms of energy consumption, delivery success probability, cluster creation time and cluster lifetime. However, no proposal for the mechanism for re-election of a new cluster is addressed. In [17], the authors proposed an energy-aware routing solution for FANETs, based on the OLSR link-state routing protocol. Two parameters are calculated to express the quality of a connection between drones: the ratio between sent and received Hello packets, and the connection time. A bio-inspired Firefly algorithm is injected to select for each drone the multipoint relays (MPRs) from these neighbors. The optimal MPRs are selected according to the four highest parameters (energy, connection quality, neighbor degree, willingness). The paths are chosen according to the energy of the drones and the quality of the connection calculated between them. This protocol outperforms OLSR and G-OLSR in terms of PDR, latency, throughput and energy consumption. However, the suggested technique incurred higher routing overhead. In [18], the authors proposed for FANET networks a routing solution based on a genetic algorithm named GAR. A method from the family of evolutionary algorithms called improved genetic algorithm is implemented, which takes into account several factors during the selection, crossover and mutation operations such as the energy of the drones, the stability and the bandwidth of the links. The choice of the optimal path is made according to the lowest value of these three parameters. GAR provides the network stability, improves throughput and reduces the delay. However, this solution has been compared with basic and older protocols such as AODV and DSR.

In this paper, we implement an energy aware and stability multipath routing solution for FANETs named se-rpcs. Our protocol is based on a bio-inspired algorithm Cuckoo Research that is part of the family of swarm intelligence algorithms [19]. We chose this method considering its simplicity and robustness compared to other methods such as PSO, DE and SA [20, 21]. Our solution selects optimal paths between a source drone and a destination drone based on energy and link stability function. The major contributions of this study are as follows:

- Develop a discovery scheme for multiple link-disjoint paths between source and destination drones.
- After the discovery phase, the bio-inspired Cuckoo Research algorithm is launched to find new paths.
- The selection of optimal paths is done based on a multi-objective function using two parameters: link stability and energy.
- Our protocol has been simulated under NS2 and with a close-to-real FANET network modeling and different configurable simulation parameters.

The remainder of this article is structured as follows: In the second section, we discuss the suggested SE-RPCS protocol. The third section describes the performance evaluation and provides simulation results of our SE-RPCS routing solution with AOMDV and AODV protocols. The fourth section concludes this paper.

2 Implementation of SE-RPCS Protocol

The above algorithm (Algo. 1) illustrates the different steps of our SE-RPCS routing protocol. In our FANET network, each drone is assumed to have a unique ID and is equipped with a variable energy battery.

Algorithm 1 SE-RPCS

01: **Begin**
02: Path discovery process and construction of the initial population of solutions
03: Start the cuckoo search process and calcul the fitness function of each new path
04: **End**

2.1 Path Discovery Process

For our routing solution, path discovery is inspired by the AOMDV multipath routing protocol [22]. During this phase, four types of packets are used by all drones: Route Request RREQ, Route Reply RREP and Hello packet during the discovery phase, and Route Error for maintenance. Each drone has its own routing table (see Table 1).

Table 1. Routing table entry for SE-RPCS.

Destination
Sequence_number
Advertised_hopcount
Path_list
Expiration_Timeout

When a source drone wants to transmit data packets to a destination drone, it searches for the availability of a valid path in its routing table; if it finds it, it sends the packets that are in the queue; otherwise, it launches a route discovery process by broadcasting a RREQ packet to all neighboring drones. A RREQ_Timer is initialized by this source drone to check if the destination drone can be reached. When each intermediate drone receives the RREQ packet, it compares the two sequence numbers recorded in its routing table and in the received packet, ensuring that the hop count of the path recorded in the packet is greater than that in the table. If both conditions are met, it rebroadcasts this packet to all its neighbor drones; otherwise, it saves the new reverse path in its routing table.

The destination drone starts responding to the received RREQ packets by sending a RREP packet. It creates and saves the reverse path to the source drone and sends the RREP packet with new information such as: drones Id, their residual energies, stability values of links between drones and their neighbors.

When the intermediate drones receive the RREP packet, they check the sequence number of the source drone to ensure the freshness of the data and avoid loops. If the sequence number value recorded in the RREP packet is greater than that in the routing table, they delete the saved path and record the new path. When the first RREP packet arrives at the source drone, a RREP_Timer is triggered to interrupt the transfer of data packets and collect more RREP packets.

2.2 Cuckoo Search Process

Cuckoo search is a bio-inspired population-based algorithm that is inspired by the behavior of cuckoo parasitic birds in nature. These birds have an aggressive breeding method where the female lays her eggs in the nest of another species, entrusting the responsibility of incubating, feeding and raising their eggs to the adoptive parents [23]. The principle of Cuckoo search is based on three principles:

- Each female lays a single egg at a time and randomly deposits it in a nest. Each cuckoo egg represents a new solution, each egg in a nest represents a solution.
- The new generation is built from the best nests that have good quality eggs.
- Foreign cuckoos can be found by host birds with a probability between 0 and 1. These host birds have two possibilities: either eliminate the foreign cuckoo, or abandon the nest and build another one in a new position.

In the context of the Cuckoo search algorithm, researchers have integrated a new technique called the Lévy flight for searching and exploring new solutions. This technique is a statistical distribution used to model random movements in various contexts by the following function:

$$x_i(t+1) = x_i(t) + \alpha \oplus L'evy(\lambda) \tag{1}$$

with:

- α: is the step size, it is greater than 0.
- $x_i^{(t+1)}$: the new position of the cuckoo.
- $x_i^{(t)}$: the old position of the cuckoo.
- $\alpha \oplus$ Lévy(λ): value of the new direction measured by the Lévy flight.

The Cuckoo search process CSP (Algo. 2) is used to optimize the search for a new set of paths between source and destination drones. The Lévy flight technique allows the paths to make random movements of the intermediate drones. Each discovered path can be modified by adding a displacement of the current drones based on the Lévy distribution. This displacement can be controlled by the constraint of the neighborhood between the drones, and then we will recalculate the fitness function for the new solutions obtained. The fitness function to be optimized by the cuckoo search algorithm is a multi-objective function with two parameters: energy and link stability. The goal is to find a paths that are more stable and consume less energy.

We consider a FANET network as a graph G(D, L) with:

- A set of mobile drones denoted $D = \{d_1, d_2, d_3, \ldots\ldots, d_i\}$.
- A set of links between each two drones denoted $L = \{l_1, l_2, l_3, \ldots\ldots, l_i\}$, a link between drone d_i and d_j is denoted L_{ij}.

- A path is represented by P_j, where P_j contains at least two drones and at least one link.
- The distance at time t between two drones d_i at coordinates (x_i, y_i, z_i) and d_j at coordinates (x_j, y_j,z_j) is presented by $d_{i,j}$ (t), it is calculated by the following equation:

$$D_{i,j}(t) = \sqrt{(x_i(t) - x_j(t))^2 + \left(y_i(t) - y_j(t)\right)^2 + (z_i(t) - z_j(t))^2} \quad (2)$$

Algorithm 2 CSP algorithm

```
Input: array of paths discovered
       the information matrix
Output: array of New Paths New_P[]
POP: all the set of discovered paths
New_P[]←NULL
Pi Path from POP
di drone from path Pi
for each (all path from the POP) do
  while (Pi←GP(POP))  do
    for each (di in Pi) and (d != src) and (d != dest) do
      apply the equation (01) of Lévy flight and normalize the values
      repair the new path obtained using the information matrix
    end for
  end while
end for
for each (New Pi) do
New_P[]← Pi
Calculate The New Fitness with equation (08)
end for
return New_P[]
```

2.2.1 Link Stability Parameter

Since drones are mobile in our FANET network, they will be considered as the main metric for calculating link stability. The information collected during discovery about drones, such as their movement and coordinates, is used to evaluate the stability of a drone. The HELLO message is used to collect this data to avoid the overhead generated by sending messages. When a drone d_j wants to calculate the distance $D_{i,j}$ that separates it from a drone d_i, it must retrieve the coordinates that are received in the HELLO packet.

The stability value of a link $L_{i,j}$, is defined as:

$$fsl_{i,j}(t) = \frac{SDl_{i,j}(t)}{Ml_{i,j}(t)} \quad (3)$$

The average of the distances collected between two drones d_i and d_j is denoted by $Ml_{i,j}(t)$, calculated by:

$$Ml_{i,j}(t) = \frac{\sum_{t=t1}^{tm} D_{i,j}(t)}{n} \quad (4)$$

The standard deviation of the distances collected between two drones d_i and d_j denoted by $SDl_{i,j}(t)$ is calculated as follows:

$$SDl_{i,j}(t) = \sqrt{\frac{1}{n}\sum_{t=t1}^{tn}\left(D_{i,j}(t) - Ml_{i,j}(t)\right)^2} \tag{5}$$

The coefficient of variation at time t, also defined as the relative standard deviation, is denoted by $fsl_{i,j}(t)$. It is a standardized measure of the dispersion of a probability or frequency distribution. For our solution, it is the measure of the distances between two neighboring drones. If the value of the coefficient of variation tends to 0, then we have a good distribution of distances, which means that the link between the drones is stable. If the value of the coefficient of variation tends to 1, this corresponds to a bad distribution, which means that the link is unstable.

The stability cost function of the path P_j denoted $fsp_j(t)$ at time t, is the maximum stability cost of the links constituting the path P_j.

2.2.2 The Energy Parameter

The calculation of the energy function of a path P_j is based on the residual energies of the drones ($E_{i,j}(t)$) included in this path. The energy function of a drone d_i at time t is calculated as follows:

$$fen_{i,j}(t) = \frac{Elev_{i,j}(t)}{DR_{i,j}(t)} \tag{6}$$

$Elev_{i,j}(t)$ is given by:

$$Elev_{i,j}(t) = \frac{E_{i,j}(t)}{E_{average}(t)} \tag{7}$$

$E_{average,j}(t)$ represents the average residual energy of the drones that participated in the constitution of a path P_j between the source drone and the destination drone. The value $DR_{i,j}(t)$ depends on the rate of use of the drone by several paths.

The energy function of a path P_j at time t denoted $fep_j(t)$ is the minimum of the energy costs $fen_{i,j}$ (t) of the drones included in this path.

2.2.3 The Fitness Function

Our fitness function depends on two parameters, namely: energy and link stability, its formula is mentioned as follows:

$$fp_j(t) = \alpha \times fep_j(t) + \frac{\beta}{fsp_j(t)} \tag{8}$$

Our SE-RPCS routing protocol forwards data packets on the path that has a high fitness value. According to our preferences, this modeling allows us to ensure routing efficiency and reliability. For example, when we want to save more energy, we give a high value to the weight related to the energy criterion ($\alpha >>> (1 - \alpha)$).

3 Performance Evaluation OF SE-RPCS

In this section, we cite the performance parameters and metrics used during the simulation to evaluate the effectiveness and reliability of our SE-RPCS multipath routing protocol. Our solution has been compared with the two proactive routing protocols AOMDV and AODV.

3.1 Performance Parameters

We will use the NS2.35 network simulator to evaluate the performance of SE-RPCS, AOMDV and AODV routing protocols. We will also use the Bonnmotion tool to create and analyze mobility scenarios. The simulation parameters are summarized in Table 2.

3.2 Performance Metrics

We choose the following performance metrics for the evaluation of our solution:

- ***Packet delivery rate***: PDR is the total number of data packets received by the destination drone and the total number of data packets broadcasted by the source drone.
- ***Overhead***: is the ratio between the number of control packets generated by all the drones and the number of data packets received by the destination drone.

Table 2. Simulation parameters.

Parameter	Value
Routing protocols	SE-RPCS, AOMDV, AODV
Mobility model	Random WayPoint (RWP)
Mac type	IEEE 802.11
Transport agent	UDP
Traffic type	CBR
Terrain Range (m)	800 X 800 X 800
simulation time (seconds)	200
Number of drones	50, 100
Speed interval (m/s)	5 - 20
Energy interval (Joules)	30 to 100
Size of packets sent (bytes)	512

3.3 Experimental Results

3.3.1 Packet Delivery Ratio

Figure 1 illustrates the packet delivery rate of the three routing protocols in two scenarios of 50 and 100 drones. In each scenario, the drones move with different speeds. In both scenarios, we observe that the PDR decreases with increasing speed. Our solution outperforms the other two protocols AOMDV and AODV, because it makes a good decision when selecting the best path among the new paths resulting from the Cuckoo search algorithm based on a fitness function with two parameters: energy and link stability.

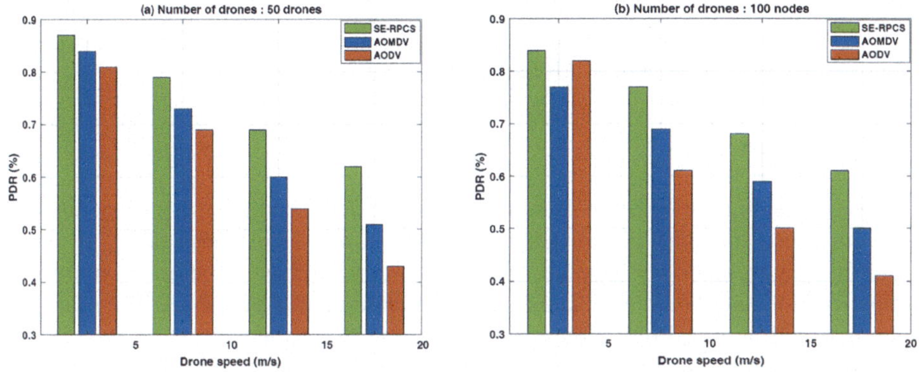

Fig. 1. Packet Delevery Ratio vs node speed.

3.3.2 Overhead

Figure 2 presents the overhead performance measurement in two scenarios with UAV speed variant. Our SE-RPCS routing solution has fewer control packets compared to AOMDV and AODV, because it avoids unstable links by applying the link stability parameter in the fitness function, compared to AOMDV and AODV which are hop-based and ignore stability parameters and suffer from more link failures, especially when the speed increases.

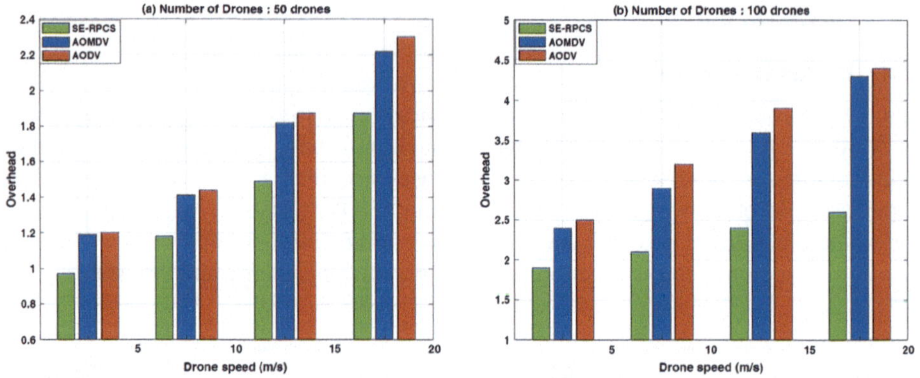

Fig. 2. Overhead rate vs node speed.

4 Conclusion

In FANETs, the high mobility of drones and their limited electric battery capacities are the major causes of link failures that significantly affect their performance. In order to address this problem, we have implemented a stable and energy efficient multipath routing solution based on the cuckoo search algorithm named SE-RPCS. The selection of the best path is based on a fitness function using two parameters: energy and link stability. Simulations show that our solution outperforms other protocols in terms of packet delivery rate and overhead. In our next works, we will contemplate the integration of other parameters such as: quality of service and security, as well as integrating other recent bio-inspired methods to design reliable and efficient routing protocols.

References

1. Kim, T., Lee, S., Kim, K.H., et al.: FANET routing protocol analysis for Multi-UAV-based reconnaissance mobility models. Drones 7(**3**), 161 (2023)
2. Mehmood, A., Iqbal, Z., Shah, A.A., et al.: An intelligent cluster-based communication system for multi-unmanned aerial vehicles for searching and rescuing. Electronics **12**(3), 607 (2023)
3. Xu, B., Zhao, K., Luo, Q., et al.: A GV-drone arc routing approach for urban traffic patrol by coordinating a ground vehicle and multiple drones. Swarm and Evolutionary Computation **77**, 101246 (2023)
4. Jacobsen, R.H., Matlekovic, L., Shi, L., et al.: Design of an autonomous cooperative drone swarm for inspections of safety critical infrastructure. Applied Sciences **13**(3), 1256 (2023)
5. Norhashim, N., Kamal, N.M., Shah, S.A., et al.: A review of unmanned aerial vehicle technology adoption for precision agriculture in Malaysia. Unmanned Systems, pp. 1–19 (2023)
6. Wang, H., Li, Y., Zhang, Y., et al.: Arithmetic optimization AOMDV routing protocol for FANETs. Sensors **23**(17), 7550 (2023)
7. Ahmed, D.E.M., Khalifa, O.O.: A comprehensive classification of MANETs routing protocols. Int. J. Comp. Appl. Technol. Res. **6**(3), 141–158 (2017)
8. Newsome, J., Song, D.: GEM: Graph embedding for routing and data-centric storage in sensor networks without geographic information. In: Proceedings of the 1st international conference on Embedded networked sensor systems, pp. 76–88 (2003)

9. He, Y., Tang, X., Zhang, R., et al.: A course-aware opportunistic routing protocol for FANETs. IEEE Access **7**, 144303–144312 (2019)
10. Mansoor, N., Hossain, M.I., Rozario, A., et al.: A fresh look at routing protocols in unmanned aerial vehicular networks: a survey. IEEE Access (2023)
11. Beegum, T.R., Idris, M.Y.I., Ayub, M.N.B., et al.: Optimized routing of UAVs using bio-inspired algorithm in FANET: A systematic review. IEEE Access (2023)
12. Yadav, A., Verma, S.: A review of nature-inspired routing algorithms for flying ad hoc networks. In: Applications of Artificial Intelligence in Engineering: Proceedings of First Global Conference on Artificial Intelligence and Applications (GCAIA 2020), pp. 197–203. Springer Singapore (2021)
13. Mahmud, I., Cho, Y.-Z.: Location estimation-based congestion-aware routing protocol for sparsely deployed energy-efficient UAVs. Sensors **21**(21), 7192 (2021)
14. Oubbati, O.S., Mozaffari, M., Chaib, N., et al.: ECaD: Energy-efficient routing in flying ad hoc networks. Int. J. Comm. Sys. **32**(18), e4156 (2019)
15. Darabkh, K.A., Alfawares, M.G., Althunibat, S.: MDRMA: Multi-data rate mobility-aware AODV-based protocol for flying ad-hoc networks. Vehicular Communications **18**, 100163 (2019)
16. Khan, A., Aftab, F., Zhang, Z.: BICSF: Bio-inspired clustering scheme for FANETs. IEEE Access **7**, 31446–31456 (2019)
17. Lansky, J., Rahmani, A.M., Malik, M.H., et al.: An energy-aware routing method using firefly algorithm for flying ad hoc networks. Scientific Reports **13**(1), 1323 (2023)
18. Wei, X., Yang, H., Huang, W.: A genetic-algorithm-based optimization routing for FANETs. Frontiers in Neurorobotics **15**, 697624 (2021)
19. Chakraborty, A., Kar, A.K.: Swarm intelligence: A review of algorithms. Nature-inspired computing and optimization: Theory and applications, pp. 475–494 (2017)
20. Solihin, M.I., Zanil, M.F.: Performance comparison of cuckoo search and differential evolution algorithm for constrained optimization. In: IOP conference series: materials science and engineering, p. 012108. IOP publishing (2016)
21. Adnan, M.A., Razzaque, M.A.: A comparative study of particle swarm optimization and cuckoo search techniques through problem-specific distance function. In: 2013 international conference of information and communication technology (ICoICT), pp. 88–92. IEEE (2013)
22. Mathur, B., Jain, A.: AOMDV protocol: A literature review. Int. J. New Technol. Res. **4**(7), 263027 (2018)
23. Mareli, M., Twala, B.: An adaptive Cuckoo search algorithm for optimisation. Applied computing and informatics **14**(2), 107–115 (2018)

The Recent Global MPPT Methods: State-of-the-Art

N. Rouibah(✉), A. Bouttout, S. Oukaci, A. Limam, and L. Derradji

National Center of Studies and Integrated Research of Building, CNERIB, CitéNouvelle El-Mokrani, 16097 Souidania, Algeria
n.rouibah@cnerib.edu.dz

Abstract. The most cost-effective approach to improving the efficiency of photovoltaic (PV) modules is to maintain their maximum power point (MPP), regardless of environmental conditions. This can be done by connecting the power electronic converter, typically a DC-DC converter, to a MPP tracking (MPPT) controller, which will allow the modification of the duty cycle to match the charge. This paper discusses the most recent developments in MPPT methods for PV systems operating under different weather conditions. Much work has gone into creating new MPPT or improving existing ones to increase the efficiency of PV systems. In this paper, an attempt is made to review the most modern MPPT methods for PV systems operating in different environmental conditions. These methods can be categorized into two main groups: local and global MPPT. The LMPPT can only track the MPP under uniform irradiance conditions (UICs). They consist of online, offline, enhanced, and soft computing methods. Although hybrid and metaheuristics MPPT methods are included in GMPPTs: this class of MPPT can extract the GMPP under partial shading conditions (PSCs).

Keywords: Photovoltaic system · MPPT · Global MPP · Local MPP · Partial Shading

1 Introduction

In the last ten years, the energy production sector has relied heavily on renewable energy sources to cut greenhouse gas emissions. Energy from renewable energy production has reached all energy user classes. Owing to its potential, solar energy is now the most effective substitute for other energy-related solutions [1]. Because of its wide irradiance, low operating costs, and accessibility of supply in most places of the world, the solar PV system steals the show.

The MPPT, in the past years, has gained quite a lot of attention and is increasing in demand. This is mainly due to its significant advantages in power system operation at a very reasonable cost. It has been debated that MPPT is only necessary for large-scale solar systems requiring substantial energy conversion [2]. However, as research and development of MPPT have continued, it is now recommended for any solar module of any size due to its continually decreasing costs and promise of more excellent energy conversion.

© The Author(s), under exclusive license to Springer Nature Switzerland AG 2025
M. Hatti (Ed.): IC-AIRES 2024, LNNS 1238, pp. 504–515, 2025.
https://doi.org/10.1007/978-3-031-80301-7_55

An indicative example of MPPT's advantages can be noticed when future solar-powered cars will use MPPT due to the nature of their often-changing environmental operating factors [2].

The importance of MPPT can also be analyzed through the efficiency graph of solar cells under different weather conditions. MPPT can increase the system's efficiency and allow it to operate closer to the MPP [3]. It can also reduce the energy cost per watt because the energy conversion is increased to get more power with the same amount of energy generated. MPPT can reduce the fuel cost for the solar generator set. Fuel saving usually varies from 20–30%. MPPT can also be applied to miniature solar systems, such as consumer electronics, which use batteries as power storage. For the same reason, i.e., to increase the conversion efficiency, the battery lifetime will be longer [3].

The main objective of writing this state of the art is to inform the reader about the methods used in MPPT and its performance. This can be useful for future researchers who are developing the strategies of MPPT so they can look for a suitable method without testing it over and over. By understanding this review, the reader is expected to choose the suitable MPPT method for his PV system, which finally decreases the energy cost. Also, this paper serves as an excellent reference for those who wish to know the effectiveness of each method in different conditions.

This paper's stat of art is organized as follows: Sect. 1 introduces the MPPT methods and a short overview of some works in the literature—Sect. 2. A classification of different MPPT methods has been presented in Sect. 3. Provide a review of the recent GMPPT methods. Finally, the Sect. 4 provides the conclusions.

1.1 Short Overview of MPPT Methods

Over the years, a literature review on various MPPT methods such as the conventional (online/offline), enhanced, soft computing, Metaheuristics and hybrid MPPT methods have been proposed. Firstly, the well-known conventional methods in the literature are Extremum seeking control (ESC), perturb & Observe (P&O) and, Hill Climbing (HC). These methods have particular benefits and drawbacks. For example, the P&O and HC methods suffer from oscillations near MPP, which results in significant power loss [4]. In addition, the P&O algorithm reacts slowly to environmental variables that change quickly. Secondly, the literature has offered a number of enhanced MPPT methods. For example, several modified variants of HC and P&O MPPTs have been presented [5] to improve the functionality of traditional HC and P&O methods.

Thirdly, soft computing methods to control, harvest energy, and enhance the efficiency of various power production systems, particularly photovoltaic systems, are a recent trend in research efforts in renewable energy applications [6]. The most popular SC approaches are neural networks, fuzzy logic, and evolutionary algorithms. They can improve MPPT accuracy, especially in dynamic and nonlinear operating conditions. They may require training data and computational resources for implementation [7].

Fourthly, metaheuristic methods have gained significant attention due to their ability to handle nonlinear and complex optimization problems associated with MPPT. Below are some commonly used metaheuristics-based MPPT methods: Horse herd optimization (HHO), Ali Baba and Forty Thieves (AFT), Flying Squirrel Search Optimization(FSO) and Grey wolf optimization (GWO) [8].

Fifthly, hybrid MPPT methods combine different algorithms to leverage their strengths and mitigate their weaknesses, providing a more efficient and reliable MPPT solution such as GWO combined with the PSO method [9].

2 Classification of MPPT Methods

Recently, many MPPT methods and designs have been proposed in the literature. Every technique has unique requirements, constraints, and uses. Since every process can be appropriate for a particular application but not for another, no specific evaluation study classifies approaches. Thus, the MPPT algorithms cannot be uniquely classified. The present investigation classifies MPPT approaches using many factors, such as tracking strategy and sensing implementation. There are numerous subcategories under each of these classifications, depending on various elements, operational principles, or application methods.

The MPPT methods in this section can be divided into two primary categories: local MPPT and global MPPT. Only under UICs is the MPP trackable by the LMPPT. They include conventional (online/offline), enhanced, and soft computing methods. Although GMPPT includes hybrid and metaheuristic MPPT methods, this class of MPPTs is able to extract the GMPP under PSCs.As seen in Fig. 1, numerous MPPT methods have been created and put to use.

3 Review of MPPT Methods

3.1 Local MPPT Methods

Local MPPT methods are techniques used in PV systems to locate and maintain operation at the MPP within a specific local region of the power-voltage (P-V) curve. These methods work well under UICs but can only work in partial shading scenarios with multiple local maxima. Local MPPT methods are typically more straightforward and less computationally intensive than GMPPT. Numerous LMPPT methods are presented and tested in the literature. These methods can fall within three mean groups: conventional(online/offline), enhanced MPPTs, and soft computing methods.

3.1.1 Conventional (Online/Offline) MPPT Methods

- **ONLINE MPPT**

Real-time tracking and PV system performance optimization are possible using online MPPT methods. Compared to more complex algorithms like ANN or GA, these techniques typically demand less computing resources and are easier to apply. However, they need to improve their ability to respond quickly and accurately when the environment changes quickly. The particular application, required precision, and available computer resources all influence the approach choice. There are many works on online MPPT optimization methods in science. HC, In Cond P&O, ESC, and SF …i.e. [10] are the most used methods.

- **OFFLINE MPPT**

Offline MPPT systems use algorithms based on typical PV characteristics and environmental conditions to estimate the MPP without real-time feedback. Some offline MPPT systems use PV system historical performance data to predict the MPP. Because they do not require continuous real-time adjustments, offline MPPT systems can be more straightforward and less expensive than online MPPT systems. Offline MPPT systems may not be as efficient as online systems because they do not respond dynamically to real-time changes in irradiance, temperature, or load conditions. Some necessary research works on offline MPPT methods are SCC, OCV, and CF [11]. This method has the benefit of simplicity due to its need for more sophisticated calculations. However, it is a drawback because prior familiarity with the GPV's electrical settings is required.

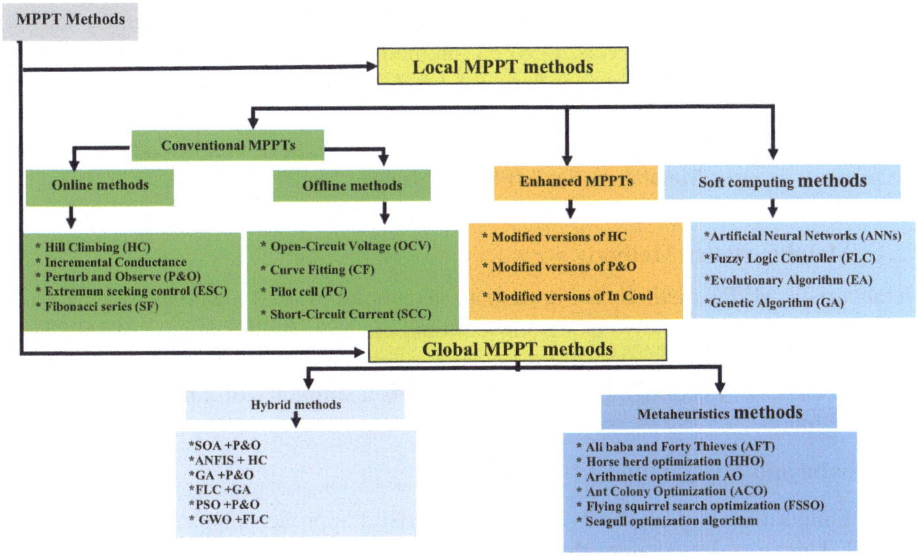

Fig. 1. Classification of MPPT methods.

3.1.2 Enhanced MPPT Methods

Several improved or enhanced MPPT methods have been published in the literature. For example, numerous modified P&O versions have been developed in [12] to modify the P&O technique's step size and discover a balance between low fluctuations and fast responsiveness in steady-state situations. Serval versions of HC MPPT with modifications have also been presented to enhance the performance of traditional HC. Nevertheless, the disadvantages of conventional In Cond MPPT have received little attention in studies [12].

3.1.3 Soft Computing Methods

Soft computing methods in MPPT handle the nonlinear characteristics and uncertainties in Systems. These methods are more flexible and adaptive than traditional MPPT techniques, and they often yield better performance under varying environmental conditions. This method Can achieve higher performance and robustness. The popular SC methods have been presented in the literature: ANNs, FLC, DE, and GA [6, 7].

3.2 Global MPPT Methods

The GMPPT is an advanced control technique used in PV systems to ensure the system operates at its optimal power output under diverse environmental circumstances. Unlike traditional MPPT methods that can be trapped in local maxima, GMPPT algorithms are designed to find and track the true global MPP across the entire power-voltage (P-V) curve of the PV system, even in the presence of partial shading, temperature changes, and fluctuating irradiance levels. GMPPT methods can flow the global peak as they have a mechanism to discriminate between local maxima and global maxima. In the literature, many GMPPT techniques are proposed and implemented. Numerous GMPPT approaches are put up and used in the literature. These methods can be categorized into two primary groups: Metaheuristics MPPT and hybrid MPPT methods.

3.2.1 Metaheuristics Methods

Metaheuristic MPPT methods are optimization techniques used to solve complex problems that are difficult to solve using conventional methods. In the context of MPPT for PV systems, metaheuristic algorithms are employed to find the optimal operating point of a solar panel to maximize power output. This is a summary of a new metaheuristic MPPT technique.

- **Ali baba and Forty Thieves (AFT)**

AFT algorithm is a relatively new metaheuristic approach inspired by the tale of ATF [8]. This story-based algorithm has been applied to various optimization problems, including MPPT for PV systems. The algorithm metaphorically represents the search for the MPP as the quest for the hidden treasure. The power output of the PV system is usually used as the fitness function in the AFT algorithm for MPPT. To maximize this fitness function, thieves (agents) look for the voltage or current corresponding to the MPP. The advantage of AFT methods is that they have a small number of parameters, implying that tuning and implementation are straightforward. Its simplicity may increase accessibility and facilitate deployment in real-world PV systems.

- **Horse herd optimization (HHO)**

HHO algorithm is another metaheuristic approach inspired by the social behavior of horses in a herd. The HHO algorithm is a new metaheuristic algorithm proposed in 2024 by T.P Sujithkumar et al. [13]. The HHO algorithm is an optimization method that simulates the movement patterns of horses in a herd. This algorithm simulates how horse herds interact with one another throughout the course of a lifetime. Horses have a variety of social behaviors depending on their age, which can be divided into six behavioral groups. Roam (R), Grazing (G), Defense (D), Sociability (S), Imitation (I), and Hierarchy (H). For instant PV system applications, the HHO method demonstrated a low steady-state power oscillation rate of only 1.91%, guaranteeing constant and dependable power output [13].

- **Arithmetic optimization AO**

AO is a novel metaheuristic algorithm that has demonstrated exceptional strength and effectiveness in managing and resolving challenging issues. In solar power systems, mainly when there is partial shading. According to S. Chtita et al. [14], the average tracking time and efficiency are 0.85 s and 99.99%, respectively; the AO method performs better than competing methods. To effectively charge the battery by maximizing its charging current, the primary goal of the OA-based MPPT is to maximize the power extracted from the PV array under all climatic situations. The AO method adjusted the duty cycle to extract the PV panel's PPM and transmit it to the battery bank through the chopper converter DC-DC.

- **Seagull optimization algorithm (SOA)**

As reported in [15], the authors have developed a novel MPPT method based on the SOA to run PV systems at GMPP efficiently. Among the bio-inspired algorithms, the SOA is a recent addition. It employs fewer operators and modification parameters than other evolutionary algorithms, which is helpful given the quick design process. Usually, over the sea, seagulls attack migrating birds. The inherent activity structure of the spiral during the attack has an impact on this process. A conceptual representation of these traits is shown in Fig. 2. The mathematical model of the SOA is based on exploration (migration) and exploitation (attacking the prey).

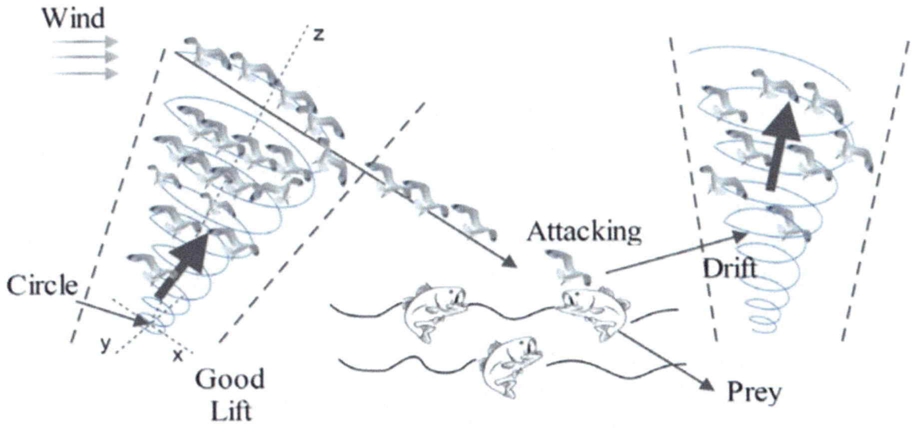

Fig. 2. Behaviors of seagulls [15].

- **Ant Colony Optimization (ACO)**

One metaheuristic method is the ACO. Its underlying theory is how ants find food and identify the best course. As shown in Fig. 3, the ACO technique under PSC can track the GMPP. It determines the ideal duty cycle that will be utilized to attack the DC-DC boost converter after detecting the PV power. In 2021, R. K. Phanden et al. [16], proposed a new modified ACO-based maximum MPPT for PV systems applications. This method offers many advantages, such as easy implementation, capabilities to track GMPP under PSC, and faster response.

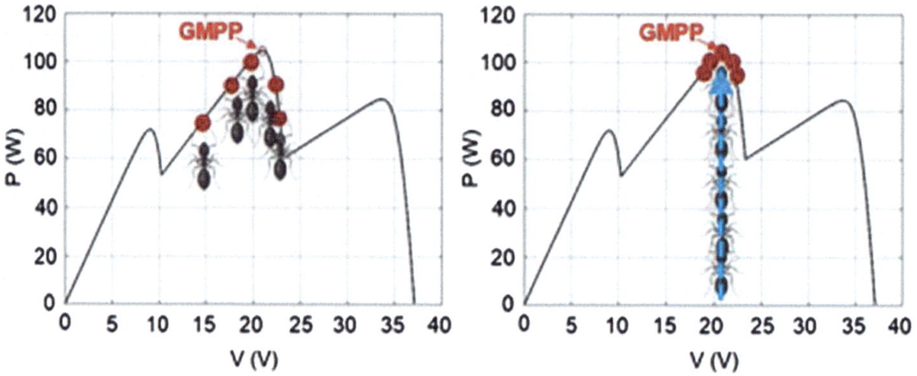

Fig. 3. Ant track the GMPP.

- **Flying Squirrel Search Optimization (FSSO)**

The FSSO algorithm has been used to take advantage of its unique ability to communicate with flying squirrels in three different classes: hickory, acorn, and standard trees, and to update their positions in the direction of the best possible answer. The method has been applied to several PV systems with PV arrays arranged in series or parallel and exposed to PSC. As reported in N. Singh et al.,20220 [17], when using MPPT of the FSSO method, the duty cycle of the qZS converter serves as the decision variable (position), and the output power is considered the target (food source). By removing the predator, the current FSSO algorithm is appropriately adapted to shorten the convergence time.

- **Salp Swarm Optimization (SSO)**

SSO is a bio-inspired algorithm for solving optimization problems modeled after the swarming behavior of slaps. These are marine creatures that move and form chains in deep oceans. The SSO algorithm mimics this behavior to search for optimal solutions in various optimization problems. Combining Salp Swarm Optimization with MPPT can effectively enhance the efficiency of solar power systems. As reported in 2022, I. Dagal et al. [18], the SSO method is a novel algorithm for solving optimization problems in PV systems under UICs and non UICs.

3.2.2 Hybrid MPPT Methods

A hybrid MPPT combines traditional or soft computing MPPTs with metaheuristics methods based on MPPTs or other methods to get around LMPPs and correctly identify the GMPP. The primary goal of hybrid MPPT is to precisely track the right MPP for PV systems, independent of changes in irradiance, PV module characteristics, or PSC features. A summary of recent hybrid MPPT research projects may be found below:

- **SOA - MVSPO**

In 2022, NA. Windarko et al. [15], have developed a new hybrid MPPT technique called SOA-MVSPO. A metaheuristic algorithm under the swarm intelligence category is called the Seagull optimization algorithm (SOA), and the conventional method is modified variable step sizes perturb and observe (MVSPO). As reported in [15], the simulation and experimentation results prove that SOA-MVSPO performed well under complex PSC. Moreover, the SOA-MVSPO method has been compared with many MPPT methods, such as GWO, DE, and P&O.

- **ANN -In Cond**

In this method, ANN-In Cond, the authors have combined an ANN with In Cond MPPT [6]. The ANN method is used to predict the PVG's GMPP automatically. The ANN-In Cond method eliminates the need for extra sensors that would have provided data on the temperature of PV modules and ambient operating conditions by measuring PV voltages and currents. Numerical simulations demonstrate that ANN-based-MPPT offers good resistance to PV system parameter fluctuations. High robustness, however, necessitates an extensive database, which strains memory and computing.

- **GWO-PSO**

In 2022, S.Chtita et al. [9] have hybridized a GWO and PSO method. The GWO-PSO is a new hybrid method based on MPPT for PV systems working under PSC. By evaluating the hybrid GWO–PSO–based MPPT method's performance to P&O, GWO, and PSO, and based MPPT techniques, its viability and efficacy are confirmed by a co- simulation technique that integrates the MATLAB/SIMULINK and PSIM software. Compared to alternative approaches, the results demonstrate the hybrid MPPT GWO–PSO method's satisfactory efficacy in many terms, such as efficiency, high accuracy, and capability to work under PSC.

- **GA -FA**

In [19], The authors have proposed a hybrid method that combines the GA technique and the metaheuristic FA method to converge to the GMPP under PSC accurately:

1. Crossover, mutation, and selection are the main processes of the "survival of the fittest" evolutionary rule, which is the basis for the GA, an optimization method.
2. Each person's fitness is computed for the selection process.
3. The FA method is based on how fireflies behave and flash, with a firefly's brightness being correlated with its attractiveness.

The GA-FA is adjusted to different solar module and PV system configuration technologies using specially created formulas. Its benefits include fast-tracking speed and excellent accuracy.

MSFLA -FLC MPPT

In 2020, L. Yong et al. [20] presented a hybrid MPPT method called MSFLA -FLC. The authors have combined the modified shuffled frog leaping algorithm (MSFLA) and FLC. The proposed MSFLA consists of three phases at its core: initial population, frog distribution, and local search. Moreover, 25 fuzzy rules were used depending on the reference power in FLC methods. As reported in [20], this method is presented with 99% efficiency. On the other hand, the different techniques result in between 95% and 97% efficiency and are slow at reaching the GMPP.

CSA-ABC Method

This MPPT method presented a novel hybrid approach for PV- MPPT termed the Cuckoo Search Algorithm and Artificial Bee Colony algorithm (CSA-ABC). The brood parasitism of some cuckoo species inspires the CSA. Key components include Levy Flights and Host Selection. The foraging behavior of honey bees inspires the ABC method and involves three types: Employed bees, onlooker Bees, and scout Bees. As reported in 2024 by P.Qi et al. [2], The combined CSA-ABC algorithm aims to efficiently find the MPP by leveraging the strengths of both CSA and ABC. According to the obtained result, the proposed CSA-ABC method outperforms the CSA, PSO, ABC, and two other hybrid algorithms in terms of speed, running between 6.2 and 78.6% faster. It also exhibits lower power oscillation during the tracking phase and zero power oscillation at the steady state.

- **ACO-ANN METHOD**

The ACO-ANN method is another hybrid approach combining ACO and ANN for MPPT in PV system applications. The ACO is inspired by the foraging behavior of ants and their ability to find the shortest path between food sources and their nest using pheromone trails. The ANNs are computational models inspired by the human brain, consisting of interconnected layers of neurons. In 2021, B. Babes et al. [16] have been suggested a metaheuristic method known as the ANN-ACO MPPT controller, which helps train the created ANN and evolve the connection weights and biases to obtain the ideal duty cycle converter values that match a PV array's MPP.

3.3 Other Global MPPT Methods

Several studies have been conducted on GMPPT methods to date. Table 1 summarizes some recent GMPPT and important research-based MPPT.

Table 1. Some recent GMPPT methods.

N°	Authors	Year	MPPT methods	Abbreviation
01	K. Aygül et al.	2023	Butterfly optimization algorithm	BOA
02	KH. Chao et al.	2024	Cat swarm optimization	CSO
03	KA. Amalo et al.	2020	Bat Algorithm	BA
04	HB. Percin et al.	2024	Whales Optimization Algorithm	WOA
05	H. Xu. et al.	2023	Mayfly Optimization Algorithm	MOA
06	N. Mohamed et al.	2023	Gorilla Troops Optimizer	GTO
07	M Said et al.	2024	zebra optimization algorithm	ZOA
08	Z. Zhao et al.	2024	Grey wolf optimization- P&O	GWO -P&O
09	KH. Chao et al.	2021	Genetic Algorithm –perturb and observe	GA -P&O
10	MGM. Abdolrasol et al.	2023	Fuzzy logic control -PSO	FLC-PSO
11	I.Dagal et al.	2022	PSO- Salp Swarm Optimization	PSO - SSO
12	G. Ahmed et al.	2024	simulated annealing -P&O	SA-P&O
13	M.Mahmoud et al.	2024	ANFIS- Gorilla Troops Optimizer	ANFIS-GTO
14	S.Sarwar et al.	2021	In Cond Dragonfly Optimization Algorithm	In Cond-DOA
15	M.Mahmoud et al.	2024	ANFIS- Zebra optimization algorithm	ANFIS-ZOA

4 Conclusion

Photovoltaic energy follows the sun's energy trend. Ideal solar cells are insufficient as a resource, and a satisfactory MPPT algorithm is required. Two of the main input characteristics for solar PV panels during installation are changeable temperature and non-uniform irradiation, which provide the researchers with their main challenges. Under variable weather conditions, the efficiency of solar photovoltaic modules is significantly reduced. The MPPT is a popular and effective method researchers use to get around this problem and maximize the power output of PV modules.

In this paper, the objectives of the literature review are to provide a clear review of the most recent popular MPPT methods and their effectiveness in various conditions. Also, a classification of these methods is presented and discussed. These methods can be categorized into two main groups: GMPPT and LMPPT. The main areas for improvement of the GMPPT are the high level of control scheme complexity and the disproportionate amount of computation required to locate and track the GMPP. Sometimes, they fail to reach the GMPP, particularly in the presence of intricate shading patterns and rapidly moving clouds.

References

1. Errebai, F.B., Derradji, L., Limam, A., Amara, M.: Energy efficiency and economic profitability between standard and high energy performance dwellings in a semi-arid climate. Int. J. Ambient Energy **43**(1), 4020–4036 (2021)
2. Qi, P., Xia, H., Cai, X., Yu, M., Jiang, N., Dai, Y.: Novel global MPPT technique based on hybrid cuckoo search and artificial bee colony under partial-shading conditions. Electronics **13**(7), 1337 (2024)
3. Rouibah, N., Barazane, L., Benghanem, M., Mellit, A.: IoT-based low-cost prototype for online monitoring of maximum output power of domestic photovoltaic systems. ETRI J. **43**(3), 459–470 (2021)
4. Rouibah, N., et al.: Experimental Assessment of Perturb & Observe, Incremental Conductance and Hill Climbing MPPTs for Photovoltaic Systems. In: Proceedings of the 2nd International Conference on Electronic Engineering and Renewable Energy Systems: ICEERE 2020, 13–15 April 2020, pp. 461–467. Springer Singapore, Saidia, Morocco (2021)
5. Pandey, A., Dasgupta, N., Mukerjee, A.K.: High-performance algorithms for drift avoidance and fast tracking in solar MPPT system. IEEE Trans. Energy Convers. **23**(2), 681–689 (2008)
6. Mazumdar, D., Sain, C., Biswas, P.K., Sanjeevikumar, P., Khan, B.: Overview of Solar Photovoltaic MPPT Methods: A State of the Art on Conventional and Artificial Intelligence Control Techniques. International Transactions on Electrical Energy Systems (2024)
7. Basha, C.H., Rani, C.: Different conventional and soft computing MPPT techniques for solar PV systems with high step-up boost converters: a comprehensive analysis. Energies **13**(2), 371 (2020)
8. Rehman, K.U., et al.: Fast tracking of maximum power in a shaded photovoltaic system using ali baba and the forty thieves (AFT) algorithm. Processes **11**(10), 2946 (2023)
9. Chtita, S., et al.: A novel hybrid GWO–PSO-based maximum power point tracking for photovoltaic systems operating under partial shading conditions. Sci. Rep. **12**(1), 10637 (2022)
10. Verma, D., Nema, S., Shandilya, A.M., Dash, S.K.: Maximum power point tracking (MPPT) techniques: recapitulation in solar photovoltaic systems. Renew. Sustain. Energy Rev. **54**, 1018–1034 (2016)
11. Jibji-Bukar, F., Anaya-Lara, O.: Frequency support from photovoltaic power plants using offline maximum power point tracking and variable droop control. IET Renew. Power Gener. **13**(13), 2278–2286 (2019)
12. Tang, S., Sun, Y., Chen, Y., Zhao, Y., Yang, Y., Szeto, W.: An enhanced MPPT method combining fractional- order and fuzzy logic control. IEEE journal of photovoltaics **7**(2), 640–650 (2017)
13. Sujithkumar, T.P., Rajendran, V., Mishra, D.: Generating maximum power in photovoltaic systems using HHO- based embedded controllers. In: Advances in AI for Biomedical Instrumentation, Electronics and Computing, pp. 455–460. CRC Press (2024)
14. Chtita, S., Derouich, A., Motahhir, S., Ghzizal, A.E.: A new MPPT design using arithmetic optimization algorithm for PV energy storage systems operating under partial shading conditions. Energy Convers. Manage. **289**, 117197 (2023)
15. Windarko, N.A., et al.: Hybrid photovoltaic maximum power point tracking of Seagull optimizer and modified perturb and observe for complex partial shading. Int. J. Electr. Comp. Eng. **12**(5), 4571–4585 (2022)
16. Babes, B., Boutaghane, A., Hamouda, N.: A novel nature-inspired maximum power point tracking (MPPT) controller based on ACO-ANN algorithm for photovoltaic (PV) system fed arc welding machines. Neural Comput. Appl. **34**(1), 299–317 (2022)

17. Singh, N., Gupta, K.K., Jain, S.K., Dewangan, N.K., Bhatnagar, P.: A flying squirrel search optimization for MPPT under partial shaded photovoltaic system. IEEE J. Emerg. Select. Topics in Power Electron. **9**(4), 4963–4978 (2020)
18. Dagal, I., Akın, B., Akboy, E.: MPPT mechanism based on novel hybrid particle swarm optimization and salp swarm optimization algorithm for battery charging through simulink. Sci. Rep. **12**(1), 2664 (2022)
19. Huang, Y.P., Chen, X., Ye, C.E.: A hybrid maximum power point tracking approach for photovoltaic systems under partial shading conditions using a modified genetic algorithm and the firefly algorithm. Int. J. Photoenergy **2018**(1), 7598653 (2018)
20. Li, Y., Samad, S., Ahmed, F.W., Abdulkareem, S.S., Hao, S., Rezvani, A.: Analysis and enhancement of PV efficiency with hybrid MSFLA–FLC MPPT method under different environmental conditions. J. Clean. Prod. **271**, 122195 (2020)

Machine Learning Techniques for GDP Per Capita Prediction in Developing Countries: A Case Study of Sub-Saharan Africa

Hanane Zermane(✉)

Laboratory of Automation and Manufacturing, Industrial Engineering Department, Faculty of Technology, Batna 2 University, Batna, Algeria
h.zermane@univ-batna2.dz

Abstract. Solid biofuels are a crucial energy source for Sub-Saharan Africa, with a significant portion of the population relying on them for cooking and heating. The prediction of Gross Domestic Product (GDP) is a critical task for economic planning and policy making. Traditional statistical methods have been widely used for this purpose; however, the advent of machine learning techniques offers new opportunities for improving predictions. Several machine learning models are used to determine their effectiveness in predicting GDP per Capita. The models' performance was assessed based on many metrics. Support Vector Regression achieved the lowest MAE of 0.81 and the highest R^2 of 0.98. Ridge and Lasso Regression also performed well with an MAE of 0.89 and 0.88, respectively, and a high R^2 of 0.98. In contrast, Decision Tree and Random Forest exhibited higher MAE values and lower R^2, while XGBoost and TabNet provided intermediate performance metrics. Accurate GDP forecasts enable governments and organizations to make informed decisions.

Keywords: Renewable energy · Solid biofuels · Economic development · Sub-Saharan Africa · GDP prediction · Machine Learning · Transfer Learning

1 Introduction

Solid biofuels constitute a vital energy source for an estimated 900 million people across Sub-Saharan Africa. Despite their critical importance in the region, obtaining accurate and up-to-date data on biofuel consumption remains a significant challenge. The African continent accounts for approximately 50% of global solid biofuel consumption, primarily used by 82% of its population for heating and cooking purposes (IEA 2022). Within Sub-Saharan Africa specifically, biomass in the forms of fuelwood, charcoal, and agricultural residues serves as the primary energy source for about 95% of the population. The availability of precise data on solid biofuels' contribution to the region's energy mix is essential for the formulation of effective policies and the assessment of sustainability objectives. However, Sub-Saharan Africa faces systemic challenges in data collection, resulting in reported figures that are often outdated and inconsistently gathered across countries. Existing surveys and statistical frameworks struggle to provide comprehensive

© The Author(s), under exclusive license to Springer Nature Switzerland AG 2025
M. Hatti (Ed.): IC-AIRES 2024, LNNS 1238, pp. 516–524, 2025.
https://doi.org/10.1007/978-3-031-80301-7_56

and harmonized data, hindering efforts to accurately measure biofuel consumption trends and their socio-economic implications. In recent years, the application of machine learning techniques has emerged as a promising approach to address these data challenges. Machine learning offers powerful tools to analyze complex datasets, predict trends, and fill gaps in existing data sources (Zermane et al., 2024). By leveraging machine learning algorithms, researchers can enhance the accuracy of biofuel consumption estimates, identify patterns in energy usage, and support evidence-based policy decisions. This study explores the use of machine learning techniques to improve our understanding of solid biofuel consumption patterns in Sub-Saharan Africa, aiming to provide insights that contribute to sustainable energy development and policy formulation in the region.

The primary data for this study was obtained from the International Energy Agency (IEA). The dataset includes a variety of features relevant to biofuel consumption and socio-economic indicators for countries in Sub-Saharan Africa. The features include Country_Name, Year, Access to clean fuels and technologies for cooking (% of population), Life expectancy at birth (total and female), Population growth (annual %), Rural and Urban Average Household characteristics, GDP growth (annual %), Population in largest city, Renewable energy consumption (% of total final energy consumption), Energy intensity level of primary energy, Age dependency ratio (% of working-age population), Arable land (% of land area), and GDP per capita growth (annual %). Based on the provided data, this study employs several techniques to analyze and predict GDP per capita growth (annual %) in Sub-Saharan Africa, Regression Analysis (Ridge and Lasso), Decision Trees (DT) and Random Forests (RF), Support Vector Regression (SVR), and Transfer Learning Techniques selecting TabNet emphasizing how each method contributes to addressing the challenges of data accuracy and policy relevance in the context of solid biofuels in Sub-Saharan Africa. These techniques enable the study to not only predict biofuel consumption more accurately but also to uncover underlying patterns and trends that traditional statistical methods may overlook. By integrating these advanced analytical approaches, this research aims to provide robust insights into solid biofuel usage dynamics in Sub-Saharan Africa, supporting evidence-based policy formulation and sustainable energy development initiatives.

2 Related Work

The relationship between energy consumption, particularly solid biofuels, and economic growth has been the subject of extensive research. This section reviews relevant studies that have informed the current understanding of these dynamics, with a particular focus on Sub-Saharan Africa. These studies collectively underscore the importance of energy consumption, particularly solid biofuels, in driving economic growth in Sub-Saharan Africa. The integration of machine learning techniques offers new opportunities for enhancing the accuracy and reliability of GDP predictions, which can inform more effective economic planning and policy-making.

Adam et al. (2022) provide a comprehensive analysis of solid biofuel residential consumption, highlighting the significant reliance on these fuels in Sub-Saharan Africa (Adam et al., 2022). Their work underscores the importance of accurate data for policy development and the challenges in measuring biofuel consumption due to outdated and inconsistent data collection methods. Garba and Bellingham (2021) examine the impact of solid cooking fuels on GDP per capita in developing countries, focusing on Sub-Saharan Africa (Garba and Bellingham, 2021). Their findings indicate a negative correlation between the use of solid cooking fuels and economic growth, suggesting that transitioning to cleaner energy sources could have substantial economic benefits. Kebede et al. (2010) explore the broader relationship between energy consumption and economic development in Sub-Saharan Africa (Kebede et al., 2010). They conclude that energy consumption is a significant driver of economic growth, and policies promoting energy access and efficiency are crucial for sustainable development. Best and Burke (2018) investigate the role of electricity availability as a precondition for faster economic growth (Best and Burke, 2018). Their study reveals that increased electricity access is strongly associated with higher economic growth rates, emphasizing the importance of electrification in economic development strategies. Hirsh and Koomey (2015) discuss the evolving relationship between electricity consumption and economic growth (Hirsh and Koomey, 2015). They argue that while electricity consumption has historically driven economic growth, this relationship is becoming more complex due to changes in energy efficiency and the shift toward digital economies. Hamilton and Kelly, (2017) use input-output analysis to explore low-carbon energy scenarios for Sub-Saharan Africa (Hamilton and Kelly, 2017). They demonstrate that achieving universal energy access could significantly boost economic growth while promoting sustainable development.

3 Materials

The International Energy Agency (IEA) is implementing a 4-year program co-funded by the European Commission titled "An Affordable and Sustainable Energy System for Sub-Saharan Africa" (IEA, 2022). The program aims to improve energy data management and long-term energy planning in 10 countries: Benin, DRC, Ethiopia, Ghana, Kenya, Nigeria, Rwanda, Senegal, Uganda, and Zambia. The activities under the program include capacity development to reinforce in-country capabilities in energy information management and energy modeling, as well as measuring progress towards SDG-7, NDCs, and other energy policy goals. This study utilizes a dataset sourced from the IEA. The dataset includes 15 key features and 1230 instances. These features collectively provide a comprehensive view of socioeconomic, demographic, and environmental factors relevant to analyzing solid biofuel consumption patterns and their socio-economic implications in Sub-Saharan Africa. Figure 1 displays all features and the correlation between them. Table 1 illustrates all features and their description.

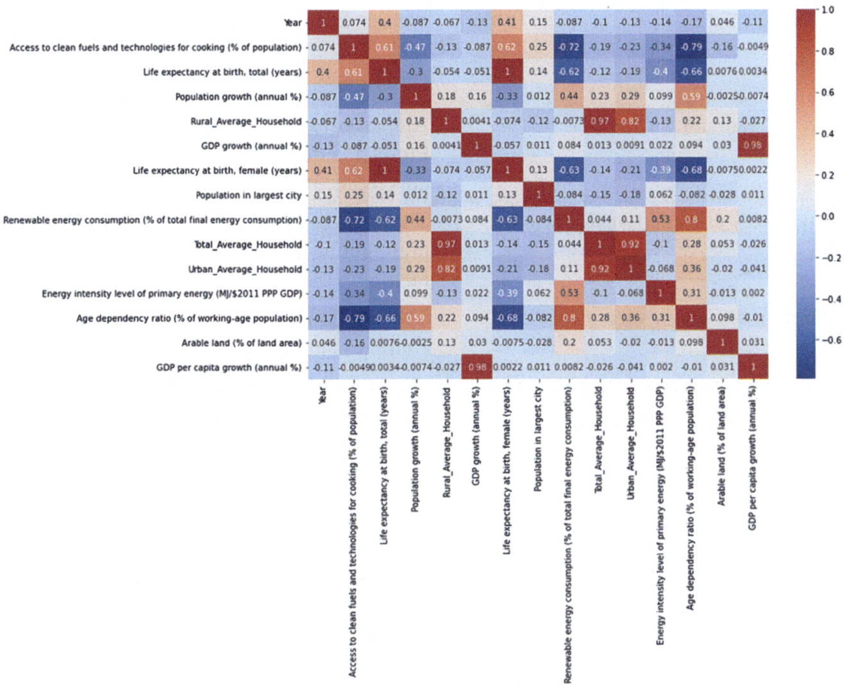

Fig. 1. Features' heatmap.

Table 1. Features and their description.

Feature	Description
Country_Name	African Countries
Year	Indicates the year of GDP per capita growth (annual %)
Access to clean fuels and technologies for cooking (% of population)	Percentage of the population with access to modern cooking fuels and technologies, a critical indicator of energy access levels
Life expectancy at birth, total (years)	Average lifespan at birth, reflecting overall health and socio-economic conditions
Population growth (annual %)	Annual percentage change in population size, influencing energy demand dynamics
Rural_Average_Household	Characteristics of rural households, essential for understanding consumption patterns in non-urban settings

(continued)

Table 1. (*continued*)

Feature	Description
GDP growth (annual %)	Annual percentage change in Gross Domestic Product, a measure of economic activity and development
Life expectancy at birth, female (years)	The average lifespan for females at birth, highlighting gender-specific health and social factors
Population in the largest city	The population size of the largest city in each country is indicative of urbanization trends
Renewable energy consumption (% of total final energy consumption)	The proportion of renewable energy in the total energy mix is crucial for assessing sustainability efforts
Total_Average_Household	Aggregate characteristics of households across all regions, providing a comprehensive view of national energy consumption patterns
Urban_Average_Household	Characteristics specific to urban households, including size, income, and energy usage behaviors
The energy intensity level of primary energy (MJ/$2011 PPP GDP)	Energy efficiency relative to economic output, is essential for evaluating resource utilization efficiency
Age dependency ratio (% of working-age population)	The ratio of dependent age groups to the working-age population influences social and economic policies
Arable land (% of land area)	Percentage of land suitable for agriculture, impacting food security and economic activities
GDP per capita growth (annual %)	Annual percentage change in GDP per capita, indicating economic progress and living standards

Information in the Database is updated. Including the following reports (Adam et al., 2022):

- **2000–2050:** [GDP per capita growth (annual %)]- IMF World Economic Outlook Forecast of April 2022.
- **2000–2050:** [GDP growth (annual %)] - IMF World Economic Outlook Forecast of April 2022.
- **2021–2050:** Population growth (annual %) is computed on UN World Urbanization Prospects 2018.

4 Methods

This study employs a comprehensive approach to analyze and predict solid biofuel consumption in Sub-Saharan Africa using advanced statistical and machine learning techniques. The methodology involves several key steps, including data collection, preprocessing, model selection, and evaluation.

4.1 Data Preprocessing

The first step was the data preprocessing, which is a crucial step to ensure the quality and reliability of the analysis. It includes the following steps:

- *Data cleaning*: Missing values are handled using imputation techniques, while outliers are identified and treated to prevent skewed results.
- *Normalization*: Features are normalized to ensure that they are on a similar scale, which is essential for the performance of certain machine learning algorithms.
- *Data splitting*: The dataset is divided into training and testing sets, with 80% of the data used for training the models and 20% reserved for testing their performance.

4.2 Model Selection

A variety of machine learning models are performed to predict the GDP per capita including, Ridge Regression is a linear regression model that includes L2 regularization to prevent overfitting. Lasso Regression is a linear regression model with L1 regularization, which can perform feature selection by shrinking some coefficients to zero. Decision Tree is a non-linear model that splits the data into subsets based on feature values, capturing complex interactions between features. Random Forests is an ensemble of decision trees that improves predictive accuracy and controls overfitting by averaging multiple trees (Zermane and Drardja, 2022). Support Vector Regressor (SVR) is a model that uses support vector machines (SVM) to capture high-dimensional relationships and provide robust predictions (Zemouri and Bouzgou, 2018). Transfer Learning Technique TabNet is a deep learning architecture designed to handle tabular data, leveraging attention mechanisms to capture intricate patterns in the data (Marei et al., 2021).

Each model is trained on the training dataset and evaluated using the testing dataset. The performance metrics used to assess the models include the Mean Squared Error (MSE) which measures the average squared difference between predicted and actual values. The R-squared (R^2) indicates the proportion of variance in the dependent variable that is predictable from the independent variables. Finally, the adjusted R^2 is a statistical measure used to evaluate the goodness-of-fit of a regression model while accounting for the number of predictors in the model. It adjusts the R^2 value based on the number of explanatory variables, preventing overestimation of the model's predictive power. Higher Adjusted R^2 indicates a better fit of the model, considering the number of predictors. The evaluation process involves comparing the performance metrics across different models to identify the most effective techniques for predicting the GDP per capita.

To ensure the robustness of the results, cross-validation techniques are employed. The models' predictions are further interpreted to understand the significance of different features and their impact on biofuels consumption patterns. This interpretation provides valuable insights for policy-making and sustainable energy development.

5 Results

The results indicate that Support Vector Regression (SVR) performed the best among the tested models, achieving the lowest MAE of 0.817 and the highest R^2 of 0.988, with an Adjusted R^2 of 0.987. This suggests that SVR is highly effective in capturing the patterns in the data and providing accurate GDP per capita predictions for Sub-Saharan Africa. Both Ridge and Lasso Regression also performed well, with very similar metrics (MAE of 0.896 and 0.886, respectively, and both having an R^2 and Adjusted R^2 of 0.987). This demonstrates that regularized linear regression models can also be very effective in this context, offering a balance between model simplicity and predictive power. In contrast, Decision Tree and Random Forest models showed higher MAE values (21.310 and 19.202, respectively) and lower R^2 and Adjusted R^2 values (0.699 and 0.681 for Decision Tree, 0.729 and 0.713 for Random Forest). This suggests that tree-based models may overfit the training data and fail to generalize well to unseen data in this specific application. XGBoost provided intermediate performance with an MAE of 8.148 and an R^2 of 0.885 (Adjusted R^2 of 0.878). While better than the tree-based models, it did not match the performance of the SVR or the regularized linear models. However, it is still a strong contender due to its robustness and ability to handle complex interactions in the data. TabNet, a deep learning model designed for tabular data, also performed well with an MAE of 1.489 and an R^2 of 0.979 (Adjusted R^2 of 0.978). This indicates that deep learning models can be effective for this task, particularly when more complex patterns

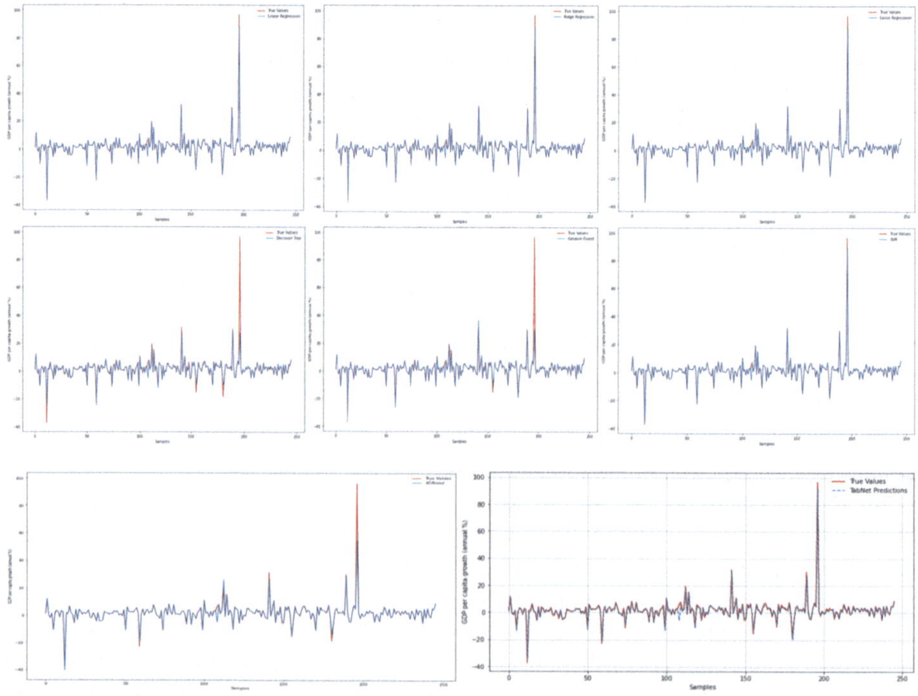

Fig. 2. Models' evaluations.

need to be captured. The results highlight the effectiveness of various machine learning techniques in predicting GDP per capita. SVR, Ridge, and Lasso Regression models stand out as top performers, providing highly accurate predictions. Obtained results are illustrated in Fig. 2 and the comparison of metrics of each model is collected in Table 2.

Table 2. Comparison of models' metrics.

Model	MSE	R^2	Adjusted R^2
Linear Regression	0.888	0.987	0.987
Ridge Regression	0.896	0.987	0.987
Lasso Regression	0.886	0.987	0.987
Decision Tree	21.310	0.699	0.681
Random Forest	19.202	0.729	0.713
SVR	**0.817**	**0.988**	**0.988**
XGBoost	8.148	0.885	0.878
TabNet	1.489	0.979	0.978

6 Conclusion

Solid biofuels play a critical role as an energy source in Sub-Saharan Africa, particularly for cooking and heating among a significant portion of the population. Predicting Gross Domestic Product (GDP) per capita is essential for economic planning and policymaking in this region. While traditional statistical methods have been historically used, the emergence of machine learning offers new avenues for enhancing prediction accuracy. This study evaluates several machine learning models to assess their effectiveness in predicting GDP per capita. Performance metrics such as Mean Absolute Error (MAE) and Coefficient of Determination (R^2) were used to compare the models. The key findings from our analysis include, that Support Vector Regression (SVR) emerged as the most effective model, achieving the lowest Mean Absolute Error (MAE) of 0.817 and the highest R^2 and Adjusted R^2 of 0.988 and 0.987, respectively. This indicates that SVR is highly capable of capturing complex patterns in the data, resulting in accurate GDP per capita predictions. Ridge and Lasso Regression also performed exceptionally well, with MAE values of 0.896 and 0.886, and both models achieving an R^2 and Adjusted R^2 of 0.987. These regularized linear models provide a robust and straightforward approach to GDP prediction, balancing model simplicity and predictive accuracy.

Decision Tree and Random Forest models showed higher MAE values (21.310 and 19.202) and lower R^2 and Adjusted R^2 values (0.699 and 0.681 for Decision Tree, 0.729 and 0.713 for Random Forest), suggesting that these models may be prone to overfitting and less effective in generalizing to new data in this context. XGBoost, with an MAE of 8.148 and an R^2 of 0.885 (Adjusted R^2 of 0.878), performed better than the tree-based models but did not match the top-performing SVR or regularized linear models. Nonetheless, it remains a strong contender due to its robustness and ability to handle complex data interactions. TabNet, a deep learning model, also showed strong performance with an MAE of 1.489 and an R^2 of 0.979 (Adjusted R^2 of 0.978). This demonstrates the potential of deep learning models for capturing intricate patterns in the data, though they may require more computational resources and tuning.

In conclusion, machine learning techniques, particularly Support Vector Regression, Ridge Regression, and Lasso Regression, have demonstrated significant potential in accurately predicting GDP per capita in Sub-Saharan Africa. These models can provide valuable insights for policymakers and stakeholders, enabling more informed decisions to drive economic growth and sustainable development in the region. Future research could explore the integration of additional socioeconomic and environmental variables, as well as the application of these models to other regions and economic contexts to further validate and enhance their predictive capabilities.

References

Adam, Z., Maria, B., Thierry, O.: Solid Biofuel Residential Consumption. International Energy Agency (2022)

Best, R., Burke, P.J.: Electricity availability: a precondition for faster economic growth? Energy Economics **74**, 321–329 (2018). https://doi.org/10.1016/j.eneco.2018.06.018

Garba, I., Bellingham, R.: Energy poverty: estimating the impact of solid cooking fuels on GDP per capita in developing countries - case of Sub-Saharan Africa. Energy **221**, 119770 (2021). https://doi.org/10.1016/j.energy.2021.119770

Hamilton, T.G.A., Kelly, S.: Low Carbon Energy Scenarios for Sub-Saharan Africa: An Input-Output Analysis on the Effects of Universal Energy Access and Economic Growth. Energy Policy **105**(July 2016), 303–19 (2017). https://doi.org/10.1016/j.enpol.2017.02.012

Hirsh, R.F., Koomey, J.G.: Electricity consumption and economic growth: a new relationship with significant consequences? Electricity Journal **28**(9), 72–84 (2015). https://doi.org/10.1016/j.tej.2015.10.002

IEA: Solid Biofuels Consumption Estimation Model - Data Product (2022). Retrieved 25 June 2024. https://www.iea.org/data-and-statistics/data-product/solid-biofuels-consumption-estimation-model#data-sets

Kebede, E., Kagochi, J., Jolly, C.M.: Energy consumption and economic development in Sub-Sahara Africa. Energy Economics **32**(3), 532–537 (2010). https://doi.org/10.1016/j.eneco.2010.02.003

Marei, M., El Zaatari, S., Li, W.: Transfer learning enabled convolutional neural networks for estimating health state of cutting tools. Robotics and Computer-Integrated Manufacturing **71**(January 2020), 102145 (2021). https://doi.org/10.1016/j.rcim.2021.102145

Zemouri, N., Bouzgou, H.: Ensemble of support vector methods to estimate global solar radiation in Algeria. Lecture Notes in Networks and Systems **35**, 155–163 (2018). https://doi.org/10.1007/978-3-319-73192-6_16

Zermane, H., Drardja, A.: Development of an efficient cement production monitoring system based on the improved random forest algorithm. Int. J. Adv. Manuf. Technol. **120**(3–4), 1853–1866 (2022). https://doi.org/10.1007/s00170-022-08884-z

Zermane, H., Madjour, H., Ziar, A., Zermane, A.: Forecasting material quantity using machine learning and time series techniques. J. Electr. Eng. **75**(3), 237–248 (2024). https://doi.org/10.2478/jee-2024-0029

Carbon Emission Forecasting Using Advanced Machine Learning Algorithms

Abdelghani Bouziane[1](✉), Mohammed Bouziane[2], and Khatir Naima[1]

[1] University Centre of Naama, Naama, Algeria
{bouziane,knaima}@cuniv-naama.dz
[2] Mechanical Engineer, Horn Glass Industries AG, Unterrehberg Georgenberg, Germany
bouziane@cuniv-naama.dz

Abstract. CO_2 emissions are the pressing issue of our time, burdening humanity. To reduce CO_2 emissions and global warming, there is only one solution: sustainability and the expansion of renewable energies. However, it does not help if one part of the world reduces emissions while another increases CO_2 emissions. This task must be solved together. This paper addresses modern machine learning algorithms that utilize the available data from all countries from 2000 to 2020. The extensive data ranges from access to electricity to geographical and economic figures to make country-specific CO_2 predictions.

In this paper, we use various machine learning approaches to predict CO_2 emissions and evaluate these approaches in terms of accuracy and predictive suitability. The focus will be on the algorithms and applied techniques that lead to the best predictions.

Keywords: CO_2 emissions · renewable energies · Artificial intelligence · Machine Learning · Deep Neural Networks · Decision Tree Regressor

1 Introduction

In today's rapidly evolving economic landscape, the significant challenge is to assess the extent of this development, evaluate the dangers of industrial change, and counteract them. These dangers include CO_2 emissions and global warming. The impacts and the relationship between CO_2 emissions and economic variables have been examined in various papers [1–4]. Visualization allows for deeper insights into the extensive dataset and simplifies the analysis, enabling a statistical examination in the data exploration phase to draw conclusions about the differences between countries and their impact on emissions [5–8] During data exploration, the grouping of countries based on their economic situation is also observed. Clustering methods allow the data to be divided into clusters [9–12]. After the data exploration phase, various models are created, evaluated, and compared.

The prediction of CO_2 emissions has been addressed in various papers. Paper [13] focuses on the application of machine learning algorithms such as Deep Learning (DL), Support Vector Machine (SVM), and Artificial Neural Networks (ANN). The results

© The Author(s), under exclusive license to Springer Nature Switzerland AG 2025
M. Hatti (Ed.): IC-AIRES 2024, LNNS 1238, pp. 525–535, 2025.
https://doi.org/10.1007/978-3-031-80301-7_57

regarding forecasting performance show coefficient of determination R^2 values ranging from 0.8639 to 0.9235, which are considered insufficient. Paper [14] used linear regression models with LASSO, ARIMA, and a Softmax-weighted average combination. For CO_2 prediction (6-hour horizon), they achieved an Root Mean Square Error RMSE of 38.45 for average emissions and 9.63 for marginal emissions. An coefficient of determination R^2 value is not explicitly provided in the document. Paper [15] utilized various algorithms, including Artificial Neural Networks (ANN), Random Forest (RF), and Long Short-Term Memory (LSTM), for predicting CO_2 emissions. The models showed different coefficient of determination R^2 ANN (0.75-0.90), RF (0.70-0.85), and LSTM (above 0.80).

[16] This paper examines the prediction of CO_2 emissions using time series models such as Autoregressive Integrated Moving Average (ARIMA), Seasonal Autoregressive Integrated Moving Average with Exogenous Variables (SARIMAX), Holt-Winters, linear regression, Random Forest and Long Short-Term Memory (LSTM). The results show that LSTM is the most accurate model followed by SARIMAX and Holt-Winters. Long Short-Term Memory achieved the lowest Mean Absolute Percentage Error (MAPE) of 3.101%, making it the best model for CO_2 predictions.

[17] This paper explores the development of prediction models for CO_2 emissions by leveraging data collected from countries worldwide. This data includes geographic, demographic, energy, economic, and emissions information, crucial for creating computational models to accurately forecast CO_2 emissions. As CO_2 emissions present a global threat, early detection and countermeasures are essential.

Following a thorough data analysis, model building with various algorithms is conducted, and the results are compared. The focus is on predicting CO_2 emissions using the aforementioned indicators for sustainable energy and other relevant factors in all countries from 2000 to 2020, as sourced from the World Bank and the International Energy Agency. The following algorithms were used and validated: Decision Tree Regressor, Gradient Boosting Regressor, and Neural Network. All three models produced good results, with the best performance from a Neural Network with 3 hidden layers containing 62/32/16 neurons, achieving a coefficient of determination R^2 score of 0.998. This model provides a highly accurate prediction of CO_2 emissions based on geographic, demographic, energy, and economic data.

2 Data Exploration

The data are collected from websites like the World Bank and the International Energy Agency. The dataset includes various features that contain factors and indicators describing sustainable energy. The goal is to reveal hidden patterns in the dataset that represent the dynamics of global climate and emission processes. Additionally, the dataset includes important aspects in all countries from 2000 to 2020, such as access to electricity, renewable energy, carbon emissions, electricity from fossil fuels, and other crucial geographical and economic information.

These factors enable the assessment of CO_2 emissions, predictions, and evaluations regarding sustainability. Overall, the dataset consists of 19 features and contains 2085 entries (Table 1).

Table 1. Dataset Global Data on Sustainable Energy 2000–2020

Entity	Year	Access to electricity (% of population)	Access to clean fuels for cooking	Renewable energy share in the total final energy consumption (%)	Electricity from fossil fuels (TWh)	Electricity from nuclear (TWh)	Electricity from renewables (TWh)	Low-carbon electricity (% electricity)	Value_ CO_2_ emissions kt_by country
Algeria	2000	98,9731	97,1	0,43	23,84	0	0,05	0,20929259	80050
Algeria	2001	98,96687	97,3	0,43	24,96	0	0,07	0,2796644	78650
Angola	2001	20	41,7	72,34	0,58	0	1,01	63,52201	15960
Angola	2002	26,352118	41,7	70,42	0,58	0	1,13	66,08187	15689,99
Antigua and Barbuda	2000	97,68926	100	0	0,14	0	0	0	330
Antigua and Barbuda	2001	97,785255	100	0	0,16	0	0	0	350
Argentina	2002	96,22887	96,1	11,62	39,7	5,39	36,06	51,07825	117470,0012
Argentina	2003	96,442635	96,45	10,82	46,73	7,03	34,38	46,982075	127660,0037

2.1 Data Distribution

The descriptive statistics in Table 2 show important values such as significant variations across multiple areas. Access to electricity and clean fuels, renewable energy capacity, and energy consumption exhibit wide disparities. Similarly, primary energy consumption, CO_2 emissions, GDP growth, population density, and land area all demonstrate large variances, reflecting diverse conditions and usage patterns among countries (Fig. 1).

Table 2. Energy and Emissions Statistics Overview

	Access to electricity (% of population)	Access to clean fuels for cooking	Electricity from fossil fuels (TWh)	Electricity from renewables (TWh)	Value_ CO_2_emissions_kt_ by_country
count	2082	2082	2082	2082	2082
mean	67,75140568	49,38475024	52,01798271	18,25957253	122477,7662
std	32,91014789	38,3964949	341,137915	109,3953669	800246,2712
min	1,2522693	0	0	0	30
25%	37,61357125	8,3	0,15	0,02	1269,999986
50%	81,857773	43,55	1,23	0,71	5670,000076
75%	98,6263525	89,675	12,0325	5,16	26914,99948
max	100	100	5098,22	2014,57	10707219,73

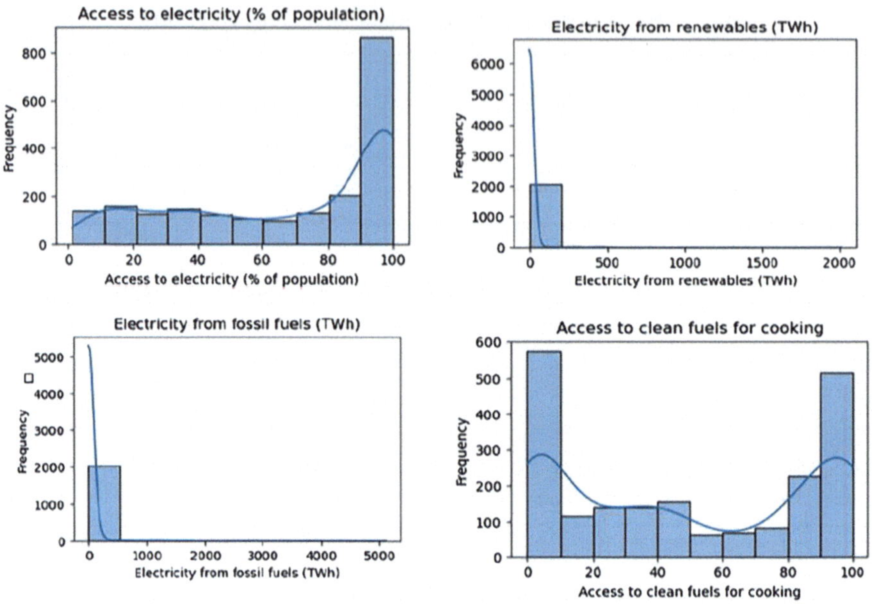

Fig. 1. Data Distribution

2.2 Data Clustering

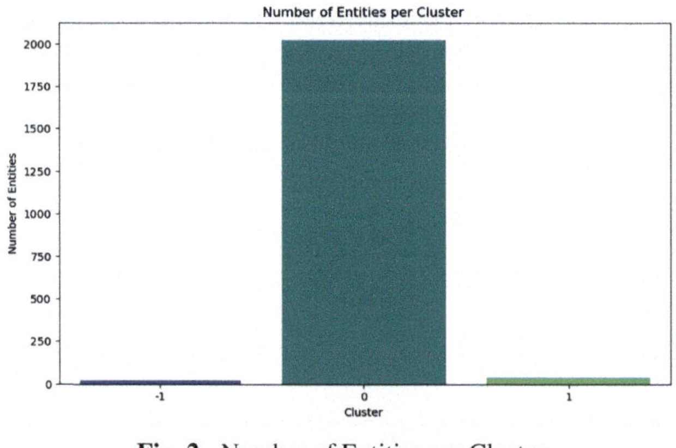

Fig. 2. Number of Entities per Cluster

The Fig. 2 above displays the distribution of entities across different clusters as determined by a clustering algorithm. The majority of entities fall into Cluster 0, with smaller numbers in Clusters -1 and 1. This visualization helps to understand how entities are grouped based on their similarities in the dataset. Table 3 shows the clusters and the listed countries:

Table 3. Country Cluster

Cluster -1	Cluster 0	Cluster 1
China, India	Afghanistan, Algeria, Angola, Antigua and Barbuda Argentina, Armenia, Azerbaijan, Bahrain, Bangladesh, Barbados, Belize, Benin, Bhutan, Botswana, Burkina Faso Burundi, Cambodia, Cameroon, Central African Republic, Chad, Colombia, Comoros, Costa Rica, Cuba, Djibouti, Dominica, Dominican Republic, Ecuador, El Salvador, Equatorial Guinea, Eritrea, Eswatini, Ethiopia, Fiji, Gabon georgia, Ghana, Grenada, Guatemala, Guinea, Guinea- Bissau, Guyana, Haiti, Honduras, Iraq, Jamaica, Jordan, Kenya, Kiribati, Kuwait, Lesotho, Liberia, Madagascar, Malawi, Maldives, Mali, Mauritania, Mauritius, Mexico, Mongolia, Morocco, Mozambique, Myanmar, Namibia, Nauru, Nepal, Nicaragua, Niger, Nigeria, Oman, Pakistan Panama, Papua New Guinea, Paraguay, Peru, Philippines, Qatar, Rwanda, Samoa, Sao Tome and Principe, Senegal, Seychelles, Sierra Leone, Singapore, Solomon Islands, Somalia, South Africa, South Sudan, Sri Lanka, Sudan Suriname, Tajikistan, Thailand, Togo, Tonga, Trinidad and Tobago, Tunisia, Turkmenistan, Uganda, United Arab Emirates, Uruguay, Uzbekistan, Vanuatu, Zambia Zimbabwe	Brazil, India

2.3 Data Correlation

The heatmap Fig. 3 shows that countries with high access to electricity and clean cooking fuels tend to rely less on renewable energy and more on fossil fuels and nuclear power. This leads to higher CO_2 emissions, highlighting the negative impact of fossil fuels on the climate. While some countries are attempting to diversify their energy sources, the transition to renewable energy remains a challenge to reduce emissions in the long term and secure a sustainable energy future.

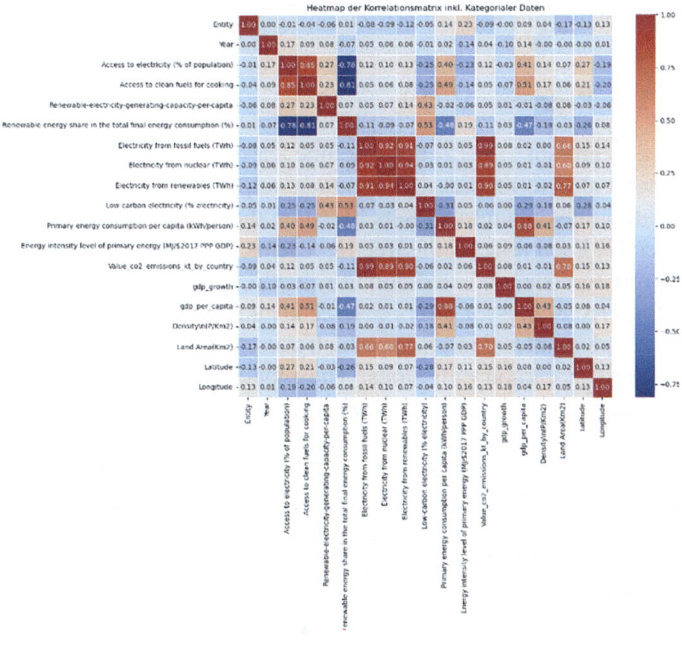

Fig. 3. Data Correlation

3 Machine Learning for CO_2 Emissions Prediction

This section aims to build and evaluate machine learning models using three different algorithms: Decision Tree Regressor, Gradient Boosting Regressor, and neural networks.

The data is prepared for training and evaluation of the machine learning models. As described in the data analysis section, there are 19 features. First, the data is cleaned by removing the 'Entity' and 'Year' columns, as these are not relevant for predicting CO_2 emissions. After the removal of 'Entity' and 'Year', the dataset consists of 17 columns. Next, divide the data into input and output sets, with CO_2 emissions as the target output.

The features and target variable are then converted into NumPy arrays, and the target variable is reshaped to ensure it has the correct dimensions. To prepare the data for machine learning, the features and target variables are scaled to a range between 0 and 1 using a Min Max Scaler.

The Min-Max Scaler is used to ensure all features are on the same scale. The Min-Max Scaler transforms the data to a range between 0 and 1, which helps improve the performance and training stability of machine learning models. It is defined as:

$$X = \frac{x - xmin}{xmax - xmin} \tag{1}$$

Finally, the scaled data is split into training and testing sets, ensuring that a portion of the data is reserved for evaluating the performance of the machine learning models. The dataset is typically split with 70% for training and 30% for testing.

3.1 Decision Tree Regressor Model

A Decision Tree Regressor [18, 19] is a machine learning algorithm used for regression tasks, where the goal is to predict a continuous output variable. The model works by recursively splitting the data into subsets based on the value of input features, creating a tree-like structure. Each node in the tree represents a decision point on a feature, and each leaf node represents a predicted output value.

The Decision Tree Regressor offers several advantages in this context. Its simplicity makes it easy to implement and interpret, which is useful for understanding the influence of different factors on carbon emissions. The model's flexibility allows it to handle both numerical and categorical data, although, in this dataset, all features appear to be numerical. Moreover, the Decision Tree Regressor provides valuable insights into the most influential factors contributing to CO_2 emissions, which can benefit policymakers and researchers aiming to develop effective strategies for reducing the carbon footprint.

In the context of forecasting carbon emissions, the Decision Tree Regressor is used to model the relationship between various features (such as access to electricity, renewable energy share, GDP growth, etc.) and the target variable (CO_2 emissions).

The model achieves very good results for the target Y.

Accuracy evaluation: coefficient of determination R2 score.

$$R^2 = 1 - \frac{\sum\left(yi - \widehat{yi}\right)^2}{\sum(yi - \overline{y})^2} \tag{2}$$

where:

- y = observed values
- $\widehat{yi}$ = predicted values
- $\overline{y}$ = mean of observed values

In this case, the model achieved an excellent coefficient of determination R^2 score of 0.996, indicating a very high level of accuracy in its predictions.

By using the Decision Tree Regressor, the program aims to capture the complex interactions between various socio-economic and environmental factors to accurately predict CO_2 emissions, aiding in the development of strategies to reduce carbon footprint.

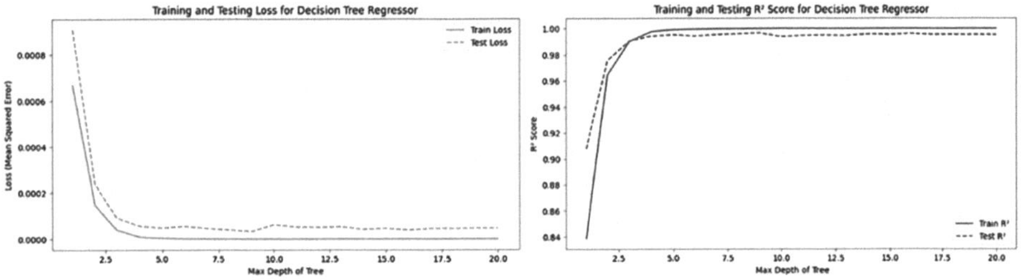

Fig. 4. Training and testing loss and R2 score for decision Tree Regressor

Figure 4 illustrates the relationship between the model's complexity, controlled by the maximum depth of the tree, and its performance. The training loss decreases consistently as the tree depth increases, reflecting the model's ability to precisely fit the training data. However, the testing loss shows a more complex pattern it initially decreases, indicating improved generalization. The training R^2 score rises towards 1 with increasing depth, showing an excellent fit to the training data,

3.2 Gradient Boosting Regressor

The Gradient Boosting Regressor [19] is used in the context of CO_2 emission prediction to enhance forecasting accuracy by combining the strengths of multiple weak models. This ensemble learning technique builds a series of decision trees, each correcting its predecessors' errors. For CO_2 emission prediction, the Gradient Boosting Regressor is trained with 100 trees, a learning rate of 0.1, and a maximum depth of 1. Iteratively adjusting and refining the model captures complex patterns and interactions between various socio-economic and environmental factors. The model's performance is evaluated using metrics such as mean squared error and the coefficient of determination

R^2 score, providing a robust and precise prediction of CO_2 emissions. This advanced approach aids in identifying key drivers of emissions and supports the development of targeted strategies to mitigate the carbon footprint.

The model's performance is evaluated using metrics such as mean squared error and the coefficient of determination R^2 score, providing a robust and precise prediction of CO_2 emissions. In this context, the Gradient Boosting Regressor achieved an impressive R^2 score of 0.997, indicating a very high level of accuracy in its predictions. This advanced approach aids in identifying key emissions drivers and supports the development of targeted strategies to mitigate the carbon footprint.

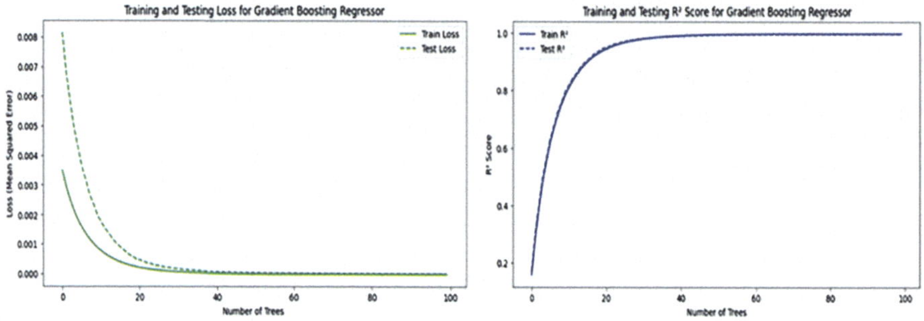

Fig. 5. Training and testing loss and R2 score for Gradient Boosting Regressor

As can be seen from Fig. 5, The Gradient Boosting Regressor's plots demonstrate how the model's performance evolves with the number of trees. Both training and testing loss decrease as more trees are added, indicating improved learning and generalization. The testing loss provides a crucial measure of the model's ability to predict unseen data accurately. The R^2 score plots reveal a similar trend: training and testing R^2 scores increase with the number of trees, showing enhanced predictive accuracy.

3.3 Neural Networks

For CO_2 emission prediction, the neural network model [20, 21] Provides a sophisticated approach to capture complex, non-linear relationships between various socio-economic and environmental factors. The model is built using the Keras library and consists of multiple layers: an input layer with 16 neurons, an first hidden layer with 64 neurons and ReLU activation, a second hidden layer with 32 neurons and ReLU activation, another hidden layer with 16 neurons and ReLU activation, and an output layer with a single neuron for regression (Fig. 6).

This architecture enables the neural network to learn complex patterns in the data and provide detailed and accurate predictions of CO_2 emissions. The model is trained with scaled features to ensure it effectively generalizes from the training data to new, unseen data. The neural network achieved an outstanding coefficient of determination R^2 score of 0.998, indicating an exceptionally high level of accuracy in its predictions. By leveraging the power of deep learning, this neural network model aids in understanding

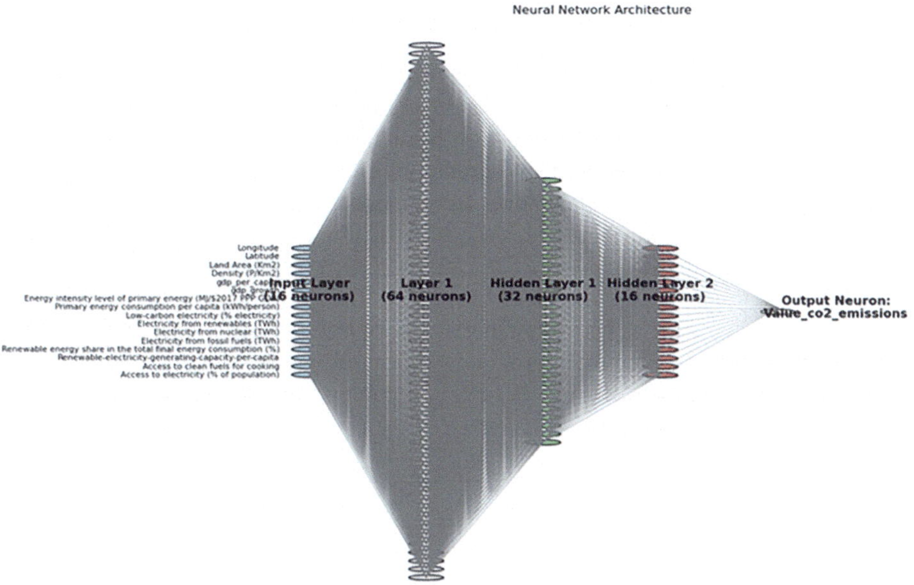

Fig. 6. Neural Network Architecture

the complex drivers of CO_2 emissions, offering valuable insights for developing effective mitigation strategies.

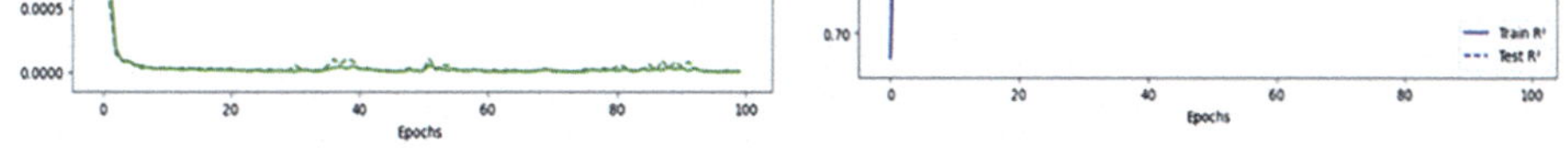

Fig. 7. Training and testing loss and R2 score for Neural Network

Figure 7 shows that the neural network's performance over epochs is captured in its loss and R^2 score plots. The training and testing loss typically decreases with more epochs, reflecting the model's learning process. A plateau in testing loss suggests that the model has converged. The training R^2 score approaches 1 as the model continues to learn, showing an excellent fit to the training data. Meanwhile, the testing R^2 score indicates the model's generalization ability, with a high score demonstrating effective prediction on new data.

3.4 Performance Comparison of Predictive Models

In summary, the deployment of three distinct machine learning models—Decision Tree Regressor, Gradient Boosting Regressor, and Neural Network offers a robust approach to forecasting CO_2 emissions as can be seen from Table X. The Decision Tree Regressor, with its simplicity and ease of interpretation, effectively elucidates the impact of various factors on emissions and achieves a commendable R^2 score of 0.996. The Gradient Boosting Regressor further refines accuracy by integrating multiple weak models, resulting in a noteworthy R^2 score of 0.997, adeptly capturing complex interactions. The Neural Network model, leveraging deep learning to discern intricate patterns, delivered the highest accuracy with an exceptional R^2 score of 0.998. Each model brings unique strengths to the table, collectively providing profound insights for policymakers and researchers to formulate effective strategies for reducing the carbon footprint (Table 4).

Table 4. Comparison of Model Accuracy for Predicting CO_2 Emissions Based on R^2 Score.

Model	R2 score
Decision Tree Regressor	0,996
Gradient boosting regressor	0,0997
Neural network	0,998

4 CONCLUSION

The causes of CO_2 emissions are diverse and include various factors present in the dataset collected from websites such as the World Bank and the International Energy Agency. Each of these factors—whether they are economic indicators, energy production and consumption data, access and usage, or demographic and geographic information affects CO_2 emissions to varying degrees. The goal of this work is to develop a model for global CO_2 emissions that captures the complex relationships between all involved features and provides accurate predictions. The models developed, including Decision Tree Regressor, Gradient Boosting Regressor, and Neural Network, delivered excellent results, with the Neural Network achieving an R^2 score of 0.998. This model will facilitate forecasting and help develop strategies in advance to reduce emissions.

References

1. Abdullah, L.: Linear relationship between CO_2 emissions and economic variables: evidence from a developed country and a developing country. Journal of Sustainable Development **8**(2), 66 (2015)
2. De Freitas, L.C., Kaneko, S.: Decomposing the decoupling of CO_2 emissions and economic growth in Brazil. Ecol. Econ. **70**(8), 1459–1469 (2011)

3. Yamaji, K., Matsuhashi, R., Nagata, Y., Kaya, Y.: A study on economic measures for CO_2 reduction in Japan. Energy Policy **21**(2), 123–132 (1993)
4. Du, Q., Zhou, J., Pan, T., Sun, Q., Wu, M.: Relationship of carbon emissions and economic growth in China's construction industry. J. Clean. Prod. **220**, 99–109 (2019)
5. Wu, R., Geng, Y., Cui, X., Gao, Z., Liu, Z.: Reasons for recent stagnancy of carbon emissions in China's industrial sectors. Energy **172**, 457–466 (2019)
6. Hamit-Haggar, M.: Greenhouse gas emissions, energy consumption and economic growth: A panel cointegration analysis from Canadian industrial sector perspective. Energy Economics **34**(1), 358–364 (2012)
7. Akbostancı, E., Tunç, G.İ, Türüt-Aşık, S.: CO_2 emissions of Turkish manufacturing industry: a decomposition analysis. Appl. Energy **88**(6), 2273–2278 (2011)
8. Zhang, C., Liu, C.: The impact of ICT industry on CO_2 emissions: a regional analysis in China. Renew. Sustain. Energy Rev. **44**, 12–19 (2015)
9. Ali, M., Saidur, R., Hossain, M.: A review on emission analysis in cement industries. Renew. Sustain. Energy Rev. **15**(5), 2252–2261 (2011)
10. Yan, D., Lei, Y., Li, L., Song, W.: Carbon emission efficiency and spatial clustering analyses in China's thermal power industry: Evidence from the provincial level. J. Clean. Prod. **156**, 518–527 (2017)
11. He, Y., et al.: Factors influencing carbon emissions from China's electricity industry: Analysis using the combination of LMDI and K-means clustering. Environ. Impact Assess. Rev. **93**, 106724 (2022)
12. Burnett, J.W.: Club convergence and clustering of US energy-related CO_2 emissions. Resource and Energy Economics **46**, 62–84 (2016)
13. Ağbulut, Ü.: Forecasting of transportation-related energy demand and CO_2 emissions in Turkey with different machine learning algorithms. Sustainable Production and Consumption **29**, 141–157 (2022)
14. Leerbeck, K., et al.: Short-term forecasting of CO_2 emission intensity in power grids by machine learning. Appl. Energy **277**, 115527 (2020)
15. Khalil, M., McGough, A.S., Pourmirza, Z., Pazhoohesh, M., Walker, S.: Machine learning, deep learning and statistical analysis for forecasting building energy consumption—a systematic review. Eng. Appl. Artif. Intell. **115**, 105287 (2022)
16. Kumari, S., Singh, S.K.: Machine learning-based time series models for effective CO_2 emission prediction in India. Environ. Sci. Pollut. Res. **30**(55), 116601–116616 (2023)
17. Han, Y., et al.: Novel economy and carbon emissions prediction model of different countries or regions in the world for energy optimization using improved residual neural network. Sci. Total Environ. **860**, 160410 (2023)
18. Xu, M., Watanachaturaporn, P., Varshney, P.K., Arora, M.K.: Decision tree regression for soft classification of remote sensing data. Remote Sens. Environ. **97**(3), 322–336 (2005)
19. Natekin, A., Knoll, A.: Gradient boosting machines, a tutorial. Front. Neurorobot. **7**, 21 (2013)
20. Hecht-Nielsen, R.: Theory of the backpropagation neural network. In: Neural networks for perception, pp. 65–93. Elsevier (1992)
21. Canziani, A., Paszke, A., Culurciello, E.: An analysis of deep neural network models for practical applications. arXiv preprint arXiv:1605.07678 (2016)

Analysis and Prediction of Solar Power Plant Energy Production: A Comparative Study of Machine Learning Algorithms

Lakhdari Lahcen(✉) and Habbab Mohamed

Laboratory of Control Analysis and Optimization of the Electro-Energetic Systems (CAOSEE), Tahri Mohammed Bechar University, Bechar, Algeria
{lakhdari.lahcen,habbab.mohamed}@univ-bechar.dz

Abstract. This project aims to analyze and predict the energy production of a solar power plant using multiple machine learning algorithms. The main objective is to compare the performance of various models to identify the one that offers the most accurate and reliable predictions. The data used for this study includes meteorological variables such as distance-to-solar-noon, temperature, wind direction and speed, sky cover, visibility, humidity, average wind speed over a period, and average pressure over a period, as well as the energy produced by the solar power plant. Each model was evaluated using regression metrics such as Mean Squared Error (MSE) and the coefficient of determination (R^2), as well as classification metrics like accuracy, precision, recall, and F1 score. The results of these evaluations allow for a comprehensive comparison of the performance of different algorithms. The performance of the models was visualized using bar charts, facilitating a clear and concise comparison of the various models based on the mentioned metrics.

Keywords: Solar Energy · Machine Learning · Energy Prediction · Renewable Energy · Data Analysis

1 Introduction

The optimization of renewable energy production is a major issue in the current context of energy transition and the fight against climate change. Among renewable energy sources, solar energy occupies a prominent place due to its abundance and potential to reduce greenhouse gas emissions. However, solar energy production is intrinsically linked to weather conditions, making its prediction complex and subject to numerous variables. This project addresses this issue by aiming to improve the prediction of solar power plant energy production using machine learning algorithms. The objective is to identify the most effective machine learning models to provide accurate predictions, thereby enabling better management and optimization of solar energy production. To achieve this, a database comprising meteorological variables and energy production data was analyzed. The variables include distance-to-solar-noon, temperature, wind direction and speed, sky cover, visibility, humidity, average wind speed over a period, average pressure over a period, and the energy produced by the solar power plant [1].

© The Author(s), under exclusive license to Springer Nature Switzerland AG 2025
M. Hatti (Ed.): IC-AIRES 2024, LNNS 1238, pp. 536–543, 2025.
https://doi.org/10.1007/978-3-031-80301-7_58

The following algorithms were selected for this comparative study:

Linear Regression, Random Forest, Gradient Boosting, k-Nearest Neighbors (k-NN), Support Vector Machines (SVM) [2].

Each model was evaluated using both regression and classification metrics to determine their accuracy and reliability. The regression metrics include Mean Squared Error (MSE) and the coefficient of determination (R2), while the classification metrics include accuracy, precision, recall, and F1 score. This document presents the methods used to train and evaluate the different models, the results obtained, and comparative visualizations in the form of bar charts. The analysis of the results highlights the most effective models and offers recommendations for the optimization of solar energy production [2].

2 Literature Review

Solar energy is a crucial renewable energy source for meeting growing energy demands while reducing greenhouse gas emissions. Solar energy production is highly dependent on weather conditions, making its prediction complex. Improving the accuracy of forecasts is vital for optimizing the integration of solar energy into power grids and for effectively planning solar power plant operations [3].

2.1 Methods for Predicting Solar Production

Several methods have been explored to predict solar energy production, ranging from physical models to machine learning techniques [3].

1. **Physical Models:** Physical models rely on mathematical equations to simulate atmospheric processes and solar irradiance. Although they can provide accurate forecasts, they require detailed meteorological data and significant computational resources. Examples of physical models include radiative transfer models and high- resolution weather models.
2. **Statistical Models:** Statistical models use regression techniques to establish relationships between meteorological variables and solar energy production. Traditional methods include linear regression, ARIMA models, and time series methods. These models are simple to implement but may lack accuracy when dealing with complex nonlinearities in the data.

2.2 Machine Learning in Solar Energy Prediction

With the rise of machine learning techniques, they have become powerful tools for predicting solar production due to their ability to handle large amounts of data and capture complex nonlinear relationships [2, 3].

3. **Linear Regression:** Linear regression is one of the simplest and most interpretable methods for prediction. However, it is limited by its inability to model complex nonlinear relationships between variables.
4. **Random Forest:** Random Forest is an ensemble algorithm that combines multiple decision trees to improve accuracy and reduce overfitting. It is particularly effective for data with complex interactions and nonlinearities.

5. **Gradient Boosting:** Gradient boosting is an ensemble technique that builds weak models sequentially, with each model correcting the errors of its predecessors. It is known for its high accuracy but can be prone to overfitting if hyperparameters are not properly tuned.
6. **k-Nearest Neighbors (k-NN):** k-NN is a similarity-based algorithm that predicts values based on the k nearest observations in the feature space. While it is simple to understand and implement, it can be sensitive to noise and high-dimensional spaces.
7. **Support Vector Machines (SVM):** SVMs are powerful algorithms for classification and regression, capable of handling high-dimensional feature spaces and capturing nonlinear relationships using kernel functions.

2.3 Comparison and Limitations of Machine Learning Algorithms

The performance of machine learning algorithms heavily depends on the quality and quantity of available data, as well as the chosen parameters and hyperparameters. For instance, although Random Forest and Gradient Boosting perform well, they require careful tuning of hyperparameters to avoid overfitting. On the other hand, methods like linear regression are simpler but less flexible for capturing complex relationships in the data [4, 5].

3 Methodology

4 Data Description

The database used for this study contains meteorological measurements and solar energy production data from a solar power plant. The available variables include [5, 6]:

- Distance-to-Solar-Noon: The time distance to solar noon.
- Temperature: The air temperature in degrees Celsius.
- Wind Direction: The wind direction in degrees.
- Wind Speed: The wind speed in meters per second.
- Sky Cover: The fraction of the sky covered by clouds.
- Visibility: The visibility distance in kilometers.
- Humidity: The relative humidity percentage.
- Average Wind Speed (Period): The average wind speed over a given period.
- Average Pressure (Period): The average atmospheric pressure over a given period.
- Energy Produced: The solar energy produced in kilowatt-hours (kWh).

4.1 Data Preparation

The following steps were taken to prepare the data [6, 7].:

- Data Cleaning: Removal of missing values and handling of outliers.
- Normalization: Scaling of numerical variables using Min-Max normalization to ensure that all features are on a comparable scale.
- Data Splitting: Splitting the data into training (80%) and testing (20%) sets.

4.2 Machine Learning Algorithms

The following algorithms were selected for this comparative study [6, 7]:

- **Linear Regression**: Used as a baseline reference for predictions.
- **Random Forest**: Uses ensembles of decision trees to improve accuracy and reduce overfitting.
- **Gradient Boosting**: An ensemble technique that builds models sequentially, with each model correcting the errors of its predecessors.
- **k-Nearest Neighbors (k-NN)**: Predicts values based on the k nearest observations.
- **Support Vector Machines (SVM)**: A powerful algorithm for regression, capable of handling nonlinear relationships.

4.3 Model Evaluation

To evaluate the performance of the different models, the following metrics were used [9–11] :

- Mean Squared Error (MSE): Measures the average squared error between predicted and actual values.
- Coefficient of Determination (R^2): Indicates the proportion of the variance in the data explained by the model.
- Accuracy: For classification tasks, measures the proportion of correct predictions.
- Precision: For classification tasks, measures the proportion of correctly predicted positive instances.
- Recall: For classification tasks, measures the proportion of true positives correctly identified.
- F1 Score: A combination of precision and recall, useful for evaluating models in situations with class imbalance.

5 Data Analysis

To perform a fundamental analysis of the dataset, we will follow several steps: data exploration, preprocessing, visualization, and descriptive statistics.

The following figures illustrate this analysis:

- Figure 1 shows the correlation matrix, which helps identify linear relationships between the different variables. This visualization is crucial for understanding how factors like temperature, wind speed, and humidity may be interconnected and influence energy production.
- Figure 2 depicts the distribution of the temperature variable, revealing the frequency of different temperature values in our dataset.
- Figure 3 explores the bivariate relationship between temperature and energy, helping us observe potential trends and correlations.
- Figure 4 illustrates the quartiles and outliers, providing a clear view of the variability and potential anomalies in the energy production data.

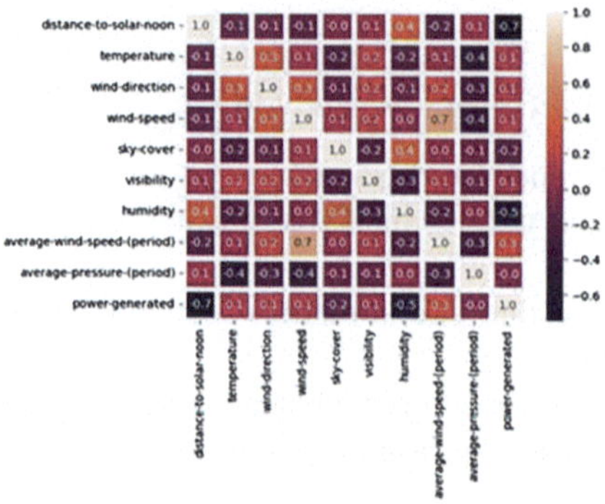

Fig. 1. Correlation Matrix

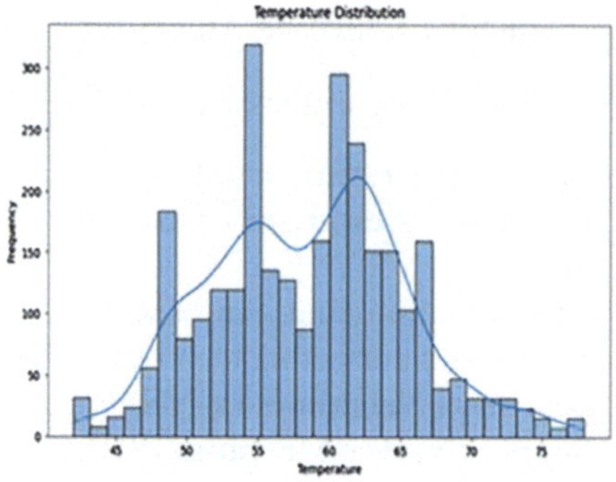

Fig. 2. The distribution of temperature values

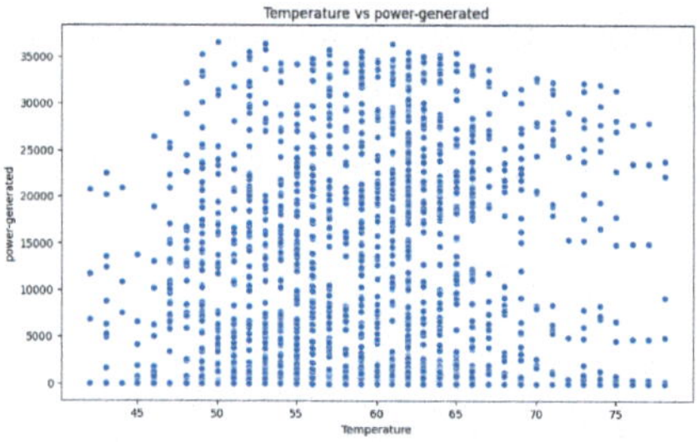

Fig. 3. The relationship between Temperature and Energy Produced

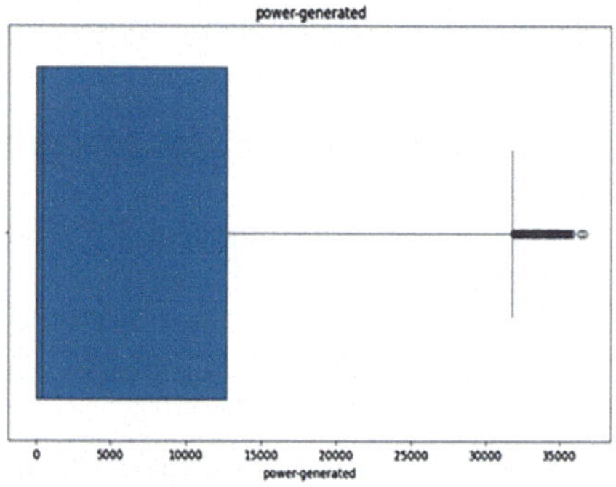

Fig. 4. The distribution of data in quartiles of Energy Produced

6 Results and Discussion

In this section, we present and analyze the performance of the classification models tested on the solar power generation dataset. The evaluated models include logistic regression, random forest, gradient boosting, k-nearest neighbors (K- NN), and support vector machines (SVM). We compare the performance of these models using various metrics, including Mean Squared Error (MSE), the coefficient of determination (R2), accuracy, precision, recall, and F1 score. The following figures illustrate the comparison of the performance of different machine learning models used to predict the energy generation of a solar power plant. These comparisons include MSE, R2, accuracy, precision, recall, and F1 score.

- Figure 5 This figure shows the mean squared errors of the tested models. A lower MSE indicates better model performance in terms of prediction accuracy.
- Figure 6 The coefficient of determination R^2 is a measure of the proportion of variance in the dependent variable that is predictable from the independent variables. A higher R^2, closer to 1, indicates better predictive ability of the model.

- Figure 7 The measured accuracy. A higher accuracy indicates better overall performance of the model in terms of correct classification.
- Figure 8 Precision is the ratio of correct positive predictions to the total positive predictions made. Higher precision means the model has fewer false positives.
- Figure 9 Recall, or sensitivity, is the ratio of correct positive predictions to the total actual positive instances.
- Figure 10 The F1 score is the harmonic mean of precision and recall, providing a single measure that balances both. A higher F1 score reflects better overall model performance in terms of combined precision and recall.

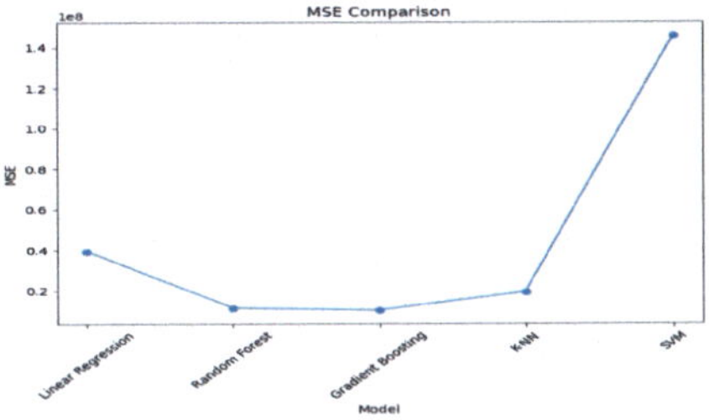

Fig. 5. MSE Comparaison

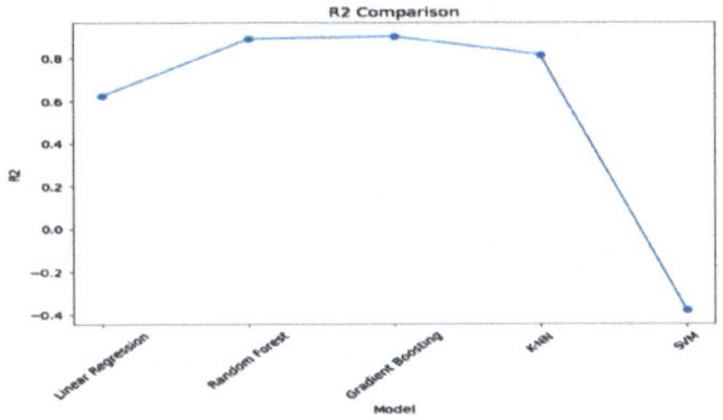

Fig. 6. R^2 Comparaison

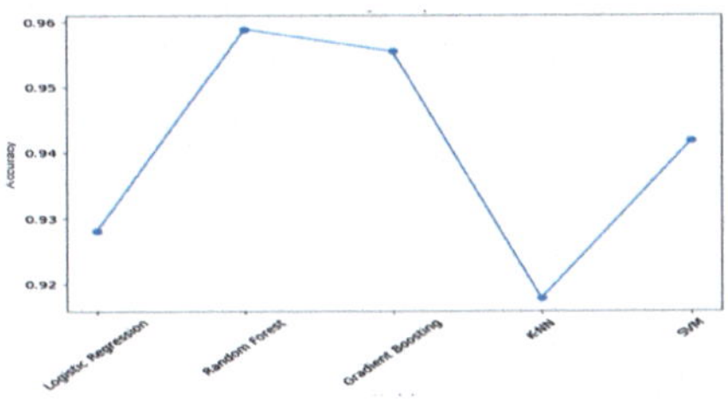

Fig. 7. Accuracy Comparison

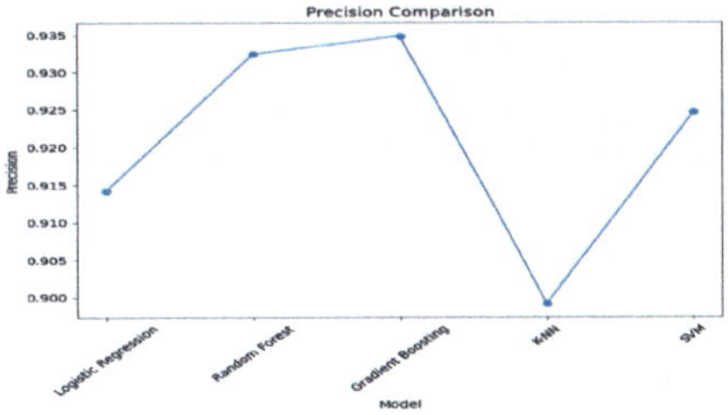

Fig. 8. Precision Comparison

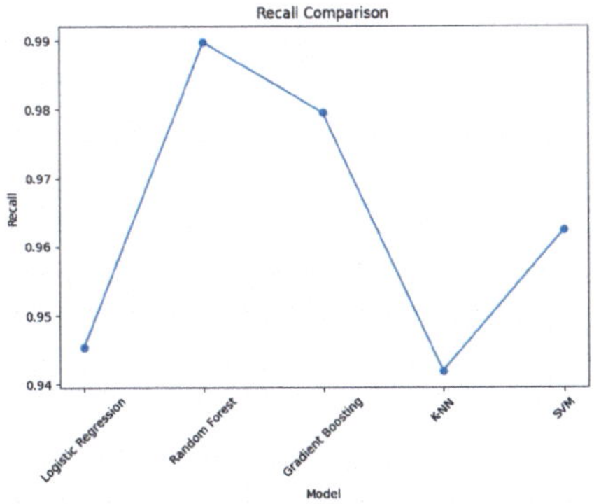

Fig. 9. Recall Comparison

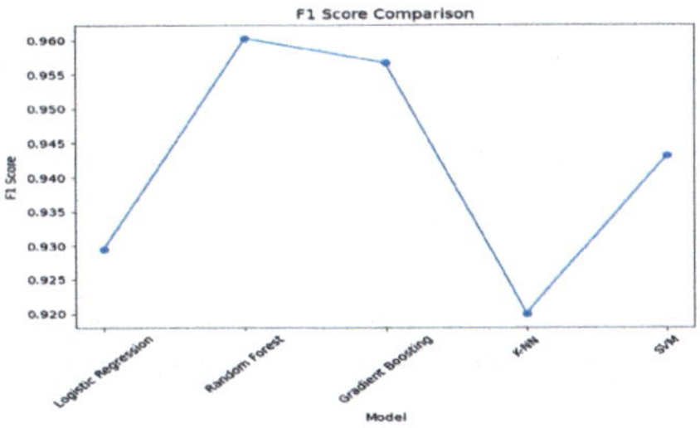

Fig. 10. F1 Score Comparison

Table 1. Performance of the models

Modèle	MSE	R^2	Accuracy	Precision	Recall	F1 Score
Logistic Regression	39495180	0.625059	0.928082	0.914191	0.945392	0.929530
Random Forest	11610740	0.889775	0.958904	0.932476	0.989761	0.960265
Gradient Boosting	10426190	0.901021	0.955479	0.934853	0.979522	0.956667
K-NN	55736140	0.470878	0.917808	0.899023	0.941980	0.920000
SVM	146402600	-0.389848	0.941781	0.924590	0.962457	0.943144

The performance of the models is summarized in the following table (Table 1):

6.1 Results Analysis

- **Logistic Regression:** Logistic regression showed acceptable performance with an accuracy of 0.928082 and an F1 score of 0.929530. However, its MSE is relatively high, indicating less precise predictions compared to more complex models.
- **Random Forest:** The random forest achieved excellent results, with an R^2 of 0.889775 and an MSE of 11610740. Its precision, recall, and F1 scores are also the highest among all tested models, indicating robust performance.
- **Gradient Boosting:** The gradient boosting model also performed well with an R^2 of 0.901021, slightly higher than that of the random forest. Its precision, recall, and F1 metrics also show excellent performance, making this model one of the best choices for this dataset.
- **K-NN:** The K-NN model showed weaker performance, with an R^2 of 0.470878 and a high MSE of 55736140. Although its precision and recall scores are respectable, it does not perform as well as the random forest and gradient boosting models.
- **SVM:** The SVM model demonstrated poor performance, with a negative R^2 (−0.389848) and the highest MSE among all models. Its precision, recall, and F1 scores are also lower, indicating that this model is not well- suited for this dataset.

7 Conclusion

In this study, we explored various machine learning algorithms to predict the energy generation of a solar power plant using a dataset containing meteorological and temporal parameters. The models tested include logistic regression, random forest, gradient boosting, K-NN, and SVM.

The results show that Gradient Boosting and Random Forest provided the best overall performance, with lower MSEs and higher R2 values, indicating better accuracy and predictive capability. In terms of classification metrics, these models also showed high values of Accuracy, Precision, Recall, and F1 Score, demonstrating balanced performance between precision and sensitivity.

The importance of this study lies in demonstrating the effectiveness of various machine learning algorithms for predicting solar energy production, providing crucial insights for optimizing and managing renewable energy resources. By further optimizing model parameters and employing advanced feature selection and data cleaning techniques, performance could be improved even more.

This research paves the way for practical applications in the management of solar energy production, enabling more efficient integration of renewable energies into the power grid and thus contributing to a more sustainable energy future.

References

1. Tang, N., Mao, Sh, et al.: Solar Power Generation Forecasting With a LASSO-Based Approach. IEEE Internet Of Things Journal **5**(2) (2018)
2. Voyant, C., et al.: Machine Learning methods for solar radiation forecasting (2017). https://www.researchgate.net/ publication/312024341
3. Voyant, C., et al.: Machine learning methods for solar radiation forecasting: A review. Renewable Energy **105**, 569–582 (2017)
4. Gensler, A., Henze, J., Sick, B., Raabe, N.: Deep Learning for solar power forecasting An approach using AutoEncoder and LSTM Neural Networks. IEEE International Conference on Systems, Man, and Cybernetics (SMC), pp. 2858–2865 (2016)
5. Anuradha, K., et al.: Analysis of Solar Power Generation Forecasting Using Machine Learning Techniques (2024). https://www.researchgate.net/publication/355120881
6. Antonanzas, J., et al.: Review of photovoltaic power forecasting. ELSEVER **136**, 78–111 (2016)
7. Kelachukwu, J., Iheanetu: Solar Photovoltaic Power Forecasting. MDPI Journal (2022). https://doi.org/10.3390/su142417005.
8. Mellit, A., Pavan, A.M.: A 24-h forecast of solar irradiance using artificial neural network: Application for performance prediction of a grid-connected PV plant at Trieste, Italy. ELSEVER, Solar Energy **84**(5), 807–821 (2010)
9. Géron, A.: Machine Learning with Scikit-Learn and TensorFlow: Concepts, Tools, and Techniques to Build Intelligent Systems. O'Reilly Media (2022)
10. Raschka, S., Mirjalili, V.: Python Machine Learning: Machine Learning and Deep Learning with Python, scikit- learn, and TensorFlow 2. Packt Publishing (2019)
11. Zhou, Z.-H.: Machine Learning. Springer, University, China (2021)

Spatiotemporal Prediction of Solar Irradiance in Algeria: A CNN-Driven Approach for Solar Energy Mapping

Salah Eddine Megaache[1](✉), Hassen Bouzgou[1], Amor Gama[2], and Naima Zerari[1]

[1] Department of Industrial Engineering. Laboratory of Automation and Manufacturing, University of Batna, 2 (Mostefa Ben Boulaïd), 05000 Batna, Algeria
{s.megaache,h.bouzgou,n.zerari}@univ-batna2.dz

[2] Centre de Développement des Energies Renouvelables, CDER, Unité de Recherche Appliquée en Energies Renouvelables, URAER, Ghardaïa, Algeria

Abstract. The first objective of this study is to estimate the potential of solar irradiance in Algeria using an artificial intelligence approach, namely, convolutional neural networks (CNNs). The second objective is to visualize the different solar components DNI, DHI, and GHI as spatial maps. Data from 267 locations for 2005–2022, obtained from the NREL database, subdivided into training and testing data are used to build the different forecasting models. Testing data are not used when training CNNs to provide an indication of the performance at unknown locations. CNN models with 9 input variables average temperature, relative humidity, sunshine duration, wind speed, precipitation, longitude, latitude, altitude, and month were used to estimate the monthly values of GHI, DNI, and DHI. Statistical error analysis is conducted using mean absolute error (MAE), Mean Absolute Percentage Error (MAPE), Mean Bias Error (MBE) and coefficient of determination R^2. This study shows that CNNs can be a better solution to estimate solar irradiance data. Seasonal solar mapping was developed in a GIS environment, representing locations and values of solar irradiance. These maps offer valuable insights into solar energy resources, aiding in the implementation of solar energy systems.

Keywords: Convolutional neural networks · Direct Normal Irradiance · Diffuse Horizontal Irradiance · Global Horizontal Irradiance · Solar map · Algeria

1 Introduction

Solar irradiance is the amount of solar energy received per unit area, expressed in watts per square meter (W/m2), at a given moment. This energy, originating from solar radiation, plays a crucial role in many natural and technological processes [1]. For example, it directly influences climatic and weather cycles. Understanding and measuring this irradiation is essential for various scientific and industrial purposes [2]. In addition, solar irradiation can be a key resource for the development of renewable energies, particularly in photovoltaic and thermal systems. Photovoltaic panels directly convert sunlight into

© The Author(s), under exclusive license to Springer Nature Switzerland AG 2025
M. Hatti (Ed.): IC-AIRES 2024, LNNS 1238, pp. 544–553, 2025.
https://doi.org/10.1007/978-3-031-80301-7_59

electricity, while solar thermal systems use the sun's heat to produce thermal energy, often subsequently transformed into electricity. These technologies are crucial for reducing dependence on fossil fuels, decreasing thus greenhouse gas emissions and promoting sustainable development [3]. Algeria is a country with one of the highest levels of sunshine in the world, where solar irradiation represents a major opportunity for economic and energy development. Effectively harnessing this resource can transform the country's energy landscape. Solar energy can also play a crucial role in isolated regions where the access to the electrical grid is limited. However, to maximize the utilization of solar irradiation, it is essential to have an accurate estimate of its different components: direct, diffuse, and global in different locations all around the country. Traditional estimation methods often have limitations in terms of precision and spatial resolution [4]. Convolutional neural networks (CNNs), an advanced technology in the field of artificial intelligence, offer a promising new approach to overcoming these challenges and providing more precise and detailed estimate [5]. Accurate estimation of solar irradiation allows also a better planning and optimization of solar installations, thereby improving investment profitability, energy efficiency, and reducing maintenance costs. The creation of a solar maps is crucial as it provides detailed mapping of solar resources, helping to identify the best installation sites, plan strategically, attract investments, reduce risks, and raise public awareness [6]. In the case of Algeria, a solar map would maximize the exploitation of this abundant resource, promoting economic development and accelerate the transition to the development of sustainable energies. Numerous studies and initiatives in Algeria have focused on creating solar maps. Researchers have proposed various models for estimating solar irradiation, using methods such as satellite imagery, regression models, fuzzy genetic algorithms, and artificial neural networks. These models provide a useful overview but are limited in precision and spatial resolution [7].Several studies have been conducted in Algeria to generate solar irradiance data for the purpose of creating a solar map using various approaches. Among these studies, [8] developed global solar irradiance maps for Algeria, covering all types of sky conditions. They used numerical models to determine the solar irradiance on different surfaces and orientations. In [9], they used the r.sun clear-sky model to create solar irradiation maps for Algeria, also assessing the regional potential for concentrating solar power designs, photovoltaic technologies, and solar water heater systems, improving accuracy using a one-arc-minute digital elevation model. In [10], the authors have created an updated solar resource maps for Algeria using data from the Modern-Era Retrospective analysis for Research and Applications (MERRA-2) dataset. Similarly, [11] used Artificial Neural Networks (ANN) to predict solar radiation in Fiji, utilizing 36 years of meteorological and geographical data. The results showed strong correlation between predicted and measured values, aiding in the optimal design and placement of photovoltaic systems across the region. In China, the authors in [12] evaluated the hybrid PSO-ELM machine learning model for predicting, quantifying, and mapping daily global solar radiation and photovoltaic power on the Loess Plateau of China. When compared with five other models, the PSO- ELM model obtained the most accurate estimations, confirming its effectiveness for solar energy modeling, especially in regions without in-situ measurements. Furthermore, this study explores the use of available meteorological data in Algeria to predict the monthly components of solar radiation using convolutional neural networks (CNN).

Hence, employing CNNs for this task represents a significant advancement, enabling thus a more efficient and sustainable use of this abundant natural resource. Additionally, this study will generate solar maps for Algeria, aiding in the assessment of the solar energy potential of each region with their associated reliability. The remainder of the paper is organized as follows: Section 2 presents the materials and methods used in the study. Section 3 discusses the predicted maps results. Based on these results, Section 4 offers a conclusion.

2 Materials and Methods

2.1 Study Area and Data Collection

Algeria, the largest country in Africa, stretches along the northern coast between latitudes 17.96° and 37.5° north and longitudes 9.5° west and 12.36° east. High plateaus and mountains reaching up to 2,000 meters dominate the southern region. The remaining 85% of the country consists of various forms of Sahara desert. In this study, we use Geographic Information System (GIS) to create a virtual grid of stations encompassing the entire study area, resulting in 267 data points. Meteorological data from the National Solar Radiation Database (NSRDB) [13] for the period 2005-2022 are used to build the different predicting models. The data include average temperature, relative humidity, precipitation, monthly mean hourly solar radiation, latitude, longitude, and altitude. The solar radiation parameters—GHI, DNI, and DHI—were predicted using CNN models trained on data from 2005 to 2016. The data were split into training and testing subsets. After model training and performance optimization, QGIS and the Inverse Distance Weighting (IDW) interpolation technique were used to create a solar map with the predicted GHI, DNI, and DHI values for 276 locations.

2.2 Design of the CNN Model

The study focused on designing Convolutional Neural Network (CNN) models to predict different components of solar irradiation Global Horizontal Irradiance (GHI), Direct Normal Irradiance (DNI), and Diffuse Horizontal Irradiance (DHI) using meteorological and geographical data. Each model was specifically designed for its respective irradiation component, leveraging CNNs to capture complex patterns in the input data. The architecture included convolutional layers followed by dense layers. The input features, such as temperature, relative humidity, precipitation, latitude, longitude, and altitude, were critical for achieving accurate predictions. The CNN models can be mathematically represented in Eq. 1

$$\text{Output} = \mathbf{f}(\text{Conv}_n(\ldots(\text{Conv}_2(\mathbf{f}(\text{Conv}_1(\text{Input}))))\ldots)) \quad (1)$$

2.3 Training and Testing of the Model

The dataset are split into two parts: data from 2005 to 2016 for training the different models and data from 2017 to 2022 for testing. For each solar radiation component (GHI,

DNI, DHI), a distinct CNN model was developed using the respective portion of the data. The models are optimized using the Adam optimizer, with mean absolute error (MAE) as a loss function. The training and testing data are reshaped to fit the input requirements of CNNs. After training, the models are evaluated and assessed on the test set to predict the solar irradiation values, which are then compared to the actual measured values.

2.4 Evaluation Criteria

The model's performance is evaluated using several metrics, including Root Mean Squared Error (RMSE), Coefficient of Determination (R2), Mean Absolute Percentage Error (MAPE) and Mean Bias Error (MBE). These metrics provide a comprehensive assessment of the models' accuracy and bias. The RMSE is calculated using Eq. 2

$$\mathrm{RMSE} = \sqrt{\frac{1}{n}\sum\nolimits_{i=1}^{n}(y_i - \hat{y}_l)^2} \tag{2}$$

The R2 score is computed using Eq. 3 to determine the proportion of variance in the dependent variable predictable from the independent variables:

$$R^2 = 1 - \frac{\sum_{i=1}^{n}(y_i - \widehat{y_i})^2}{\sum_{i=1}^{n}(y_i - \overline{y})^2} \tag{3}$$

The Mean Absolute Percentage Error (MAPE) is calculated using Eq. 4

$$\mathrm{MAPE} = \frac{100}{n}\sum_{i=1}^{n}\left|\frac{y_i - \widehat{y_i}}{y_i}\right| \tag{4}$$

MBE is used to measure the average bias in the predictions, as shown in Eq. 5

$$\mathrm{MBE} = \frac{1}{n}\sum\nolimits_{i=1}^{n}(\hat{y}_i - y_i) \tag{5}$$

3 Results and Discussion

3.1 Solar Radiance Estimation Using CNN in Algeria

According to Table 1 below, the performance metrics for predicting different components of solar irradiance GHI, DHI, and DNI reveal distinct strengths and weaknesses.

GHI predictions show a high level of accuracy, with an R2 value of 0.9088 indicating a strong correlation between predicted and actual values. The RMSE of 26.31 suggests a moderate level of absolute error, while a MAPE of 21.26% shows a relatively moderate error in percentage terms.

The MBE of 2.84 indicates the good performance of prediction. As illustrated in the corresponding Fig. 1, the data points are tightly clustered around the regression line, confirming the accuracy of the GHI predictions.

Table 1. Performance Metrics for Predicting Solar Radiation Components: GHI, DHI and DNI.

Component	RMSE	R^2	MAPE	MBE
GHI	26.31	0.9088	21.26	2.84
DHI	22.20	0.8306	38.49	2.53
DNI	59.98	0.6668	23.47	−1.24

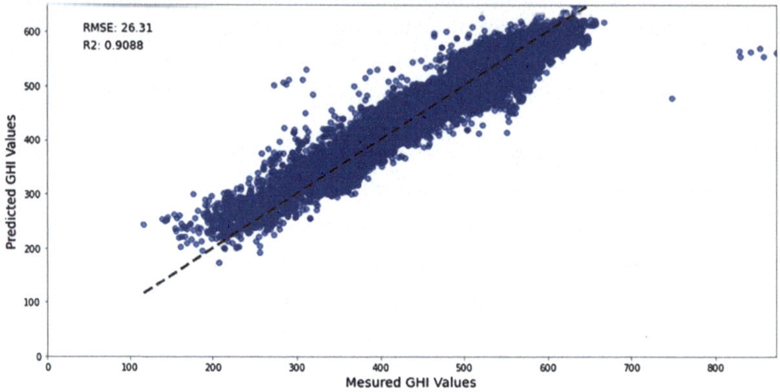

Fig. 1. Measured vs Predicted GHI Values

In contrast, DHI predictions, with an R2 of 0.8306, illustrate a good correlation but are less accurate than those of GHI. The RMSE for DHI is 22.20, indicating moderate absolute errors, while the MAPE of 38.49% suggests a higher percentage error compared to GHI. The MBE of 2.53 indicates a slight positive bias. The DHI prediction score (see Fig. 2), shows a wider dispersion of points around the regression line, suggesting greater variability in the predictions.

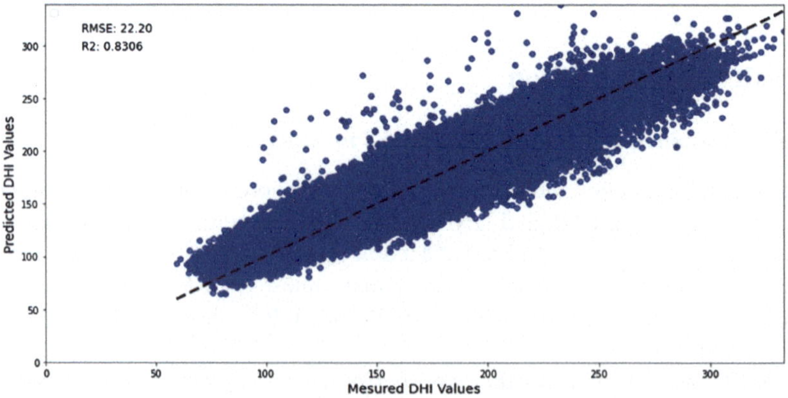

Fig. 2. Measured vs Predicted DHI Values.

DNI predictions are the least accurate among the three components, with an R2 value of 0.6668 reflecting a weaker correlation. This lower R2 value indicates that the model does not explain a significant portion of the variability in DNI predictions. The RMSE of 59.98 highlights substantial errors in the predictions, suggesting that the deviations from the actual values are quite pronounced. Additionally, the MAPE of 23.47% suggests a moderate error in percentage terms, reflecting the average magnitude of prediction errors relative to the actual values. However, the MBE of -1.24 indicates a slight underestimation in predictions, pointing to a consistent bias towards lower values. The DNI scatter plot in Fig. 3 shows a relatively greater dispersion of points around the regression line, confirming the larger errors and the weaker correlation of predictions. This dispersion further underscores the challenges in achieving accurate predictions for DNI.

The comparative analysis reveals that GHI models perform the best in terms of correlation and error metrics, while DHI models offer reasonably good predictions with some limitations. However, DNI models require further refinement to improve accuracy and correlation.

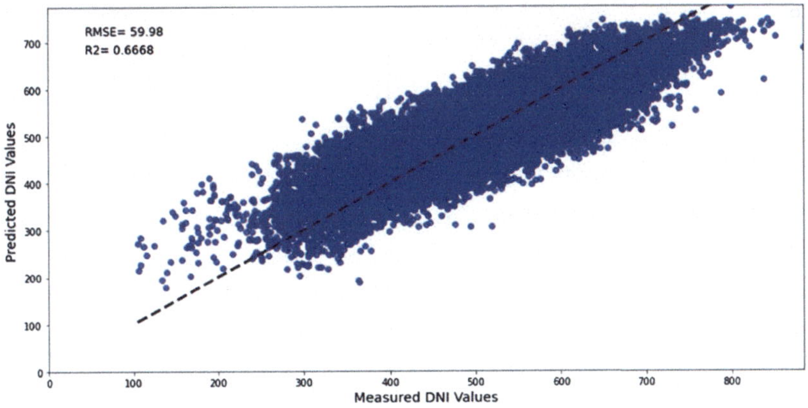

Fig. 3. Measured vs Predicted DNI Values

3.2 Mapping of solar radiation

The CNN estimated values of GHI, DNI, and DHI were presented as seasonal maps to build a spatial database for solar irradiation in Algeria. Monthly solar irradiance data over six years were used to calculate seasonal averages, with the seasons defined as winter (December–February), spring (March–May), summer (June–August), and autumn (September–November). The darkest color on the map represents the highest solar energy potential. These maps are valuable for decision-making in solar energy projects and identifying solar radiation potential in Algeria. The DNI distribution varies seasonally. In winter, DNI values range from $\leq$390 W/m2 in the northern regions to over 641 W/m2 in the southern desert areas. In spring, values increase, ranging from 0–382 W/m2 in the north to over 564 W/m2 in the south. During summer, DNI values are high, from 0–381 W/m2 in the north to over 492 W/m2 in the central and southern regions, marking peak solar potential. In autumn, DNI ranges from 0–397 W/m2 in the north to over 514 W/m2 in the south, resembling spring patterns but slightly lower than in summe (Fig. 4)

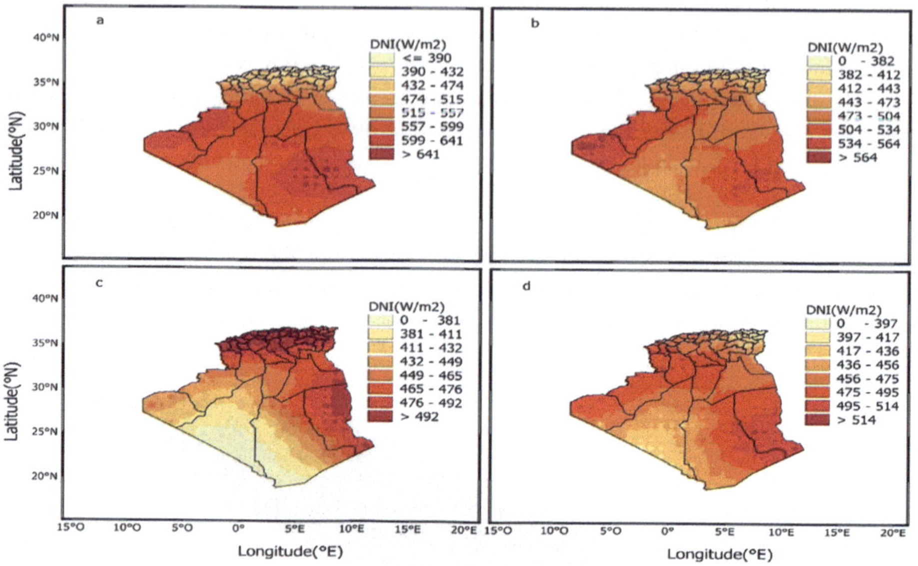

Fig. 4. Seasonal spatiotemporal maps of DNI in Algeria: (a) Multiyear average of DNI in winter; (b) spring; (c) summer; (d) autumn

The DHI distribution across the seasons is detailed as follows (see Fig. 5). In winter (see Fig. 5a), DHI values range from 0–21 W/m2 in the northern regions to over 35 W/m2 in the southern regions. The highest DHI values are concentrated in the central and southern parts of Algeria, indicating moderate solar potential during winter. In spring Fig. 5b, DHI values range from ≤166 W/m2 in the north to over 259 W/m2 in the south, showing a gradual increase with the highest values in the southern regions. During summer (see Fig. 5c), DHI values range from 0–21 W/m2 in the north to over 35 W/m2 in the south, with widespread high values especially in the southern and central regions, indicating strong diffuse solar radiation. In autumn (Fig. 5d), DHI values range from 0–133 W/m2 in the north to over 223 W/m2 in the south, showing a decrease compared to summer, but the southern regions still maintain higher values.

The GHI distribution across the seasons is detailed in Fig. 6. In winter (Fig. 6a), where GHI values range from 0–342 W/m2 in the northern regions to over 486 W/m2 in the southern regions, with the highest values concentrated in the southern desert areas, indicating significant solar energy potential. In spring (Fig. 6b), GHI values increase noticeably, ranging from ≤402 W/m2 in the north to over 561 W/m2 in the south, particularly in the southern and central regions. During summer (Fig. 6c), GHI values range from 0–475 W/m2 in the north to over 590 W/m2 in the south, exhibiting widespread high values across most regions, especially in the central and southern parts, indicating

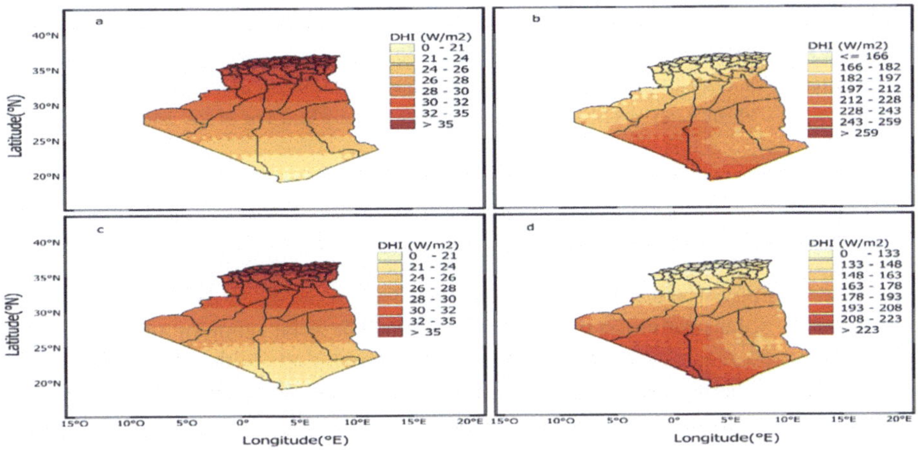

Fig. 5. Seasonal spatiotemporal maps of Direct Normal Irradiance (DHI) in Algeria: (a) Multiyear average of DHI in winter; (b) spring; (c) summer; (d) autumn.

peak solar potential. In autumn (Fig. 6d), GHI values range from 0–272 W/m2 in the north to over 449 W/m2 in the south, with a distribution similar to spring, showing high values in the southern regions, though slightly lower compared to summer.

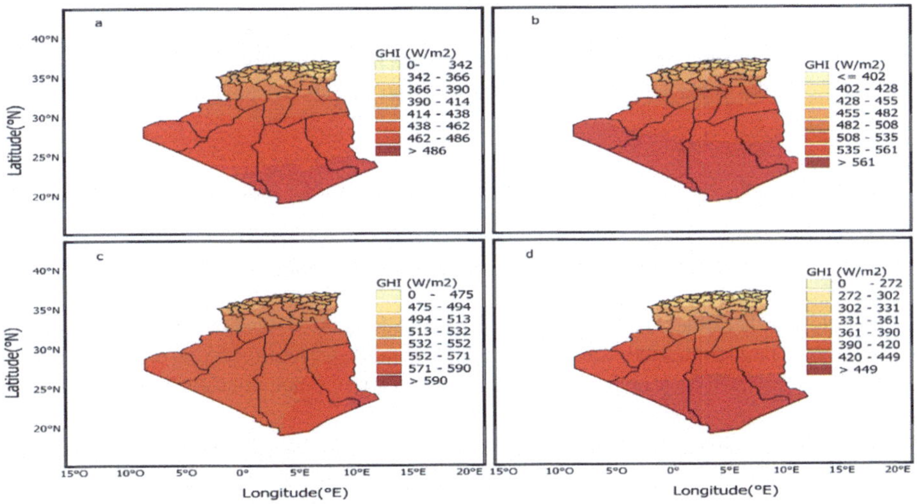

Fig. 6. Seasonal spatiotemporal maps of GHI in Algeria: (a) Multiyear average of GHI in winter; (b) spring; (c) summer; (d) autumn

The different spatiotemporal maps illustrate the seasonal variations and regional differences in DNI, DHI, and GHI across Algeria. Winter shows the lowest values across all metrics, with the highest values concentrated in the southern desert areas. Spring and autumn display a gradual increase in irradiance values, with higher values moving from north to south. Summer exhibits the highest values across all metrics, particularly in the central and southern regions, indicating peak solar potential. These maps highlight the potential for solar energy exploitation in Algeria, with the southern regions consistently showing higher values across all seasons, making them ideal for solar energy projects.

4 Conclusion

This paper presents a study on the potential estimation of solar irradiation using CNN and GIS technology. The results indicate that the CNN model developed here can be used to estimate the solar irradiation potential in Algeria. The solar mapping of 58 provinces in Algeria is presented as seasonal maps using GIS technology with values predicted by the CNN. The generated seasonal maps show significant variations in solar irradiance across different seasons and regions of Algeria. In winter, GHI values are the lowest, ranging from 100 to 300 W/m2, with the southern regions receiving moderate levels of irradiance. In spring, the levels increase to between 200 and 500 W/m2, promoting better solar production throughout the country. Summer records the highest GHI values, ranging from 500 to 800 W/m2, with peaks in the desert areas of the Sahara, indicating optimal potential for solar projects. In autumn, GHI values decrease again, ranging from 300 to 600 W/m2, but remain high enough to support efficient solar energy production. These results show that Algeria benefits from a strong solar energy potential, especially in the southern desert regions, throughout the year. The use of CNN technology for predicting solar irradiance, coupled with GIS mapping, allows for precise and detailed evaluation of solar potential, thus facilitating the planning and optimization of renewable energy projects in the country. Seasonal variations must be considered to maximize the efficiency of solar installations and adapt energy strategies to specific local conditions.

References

1. Hofierka, J., Suri, M.: The solar radiation model for Open-source GIS: implementation and applications. In: Proceedings of the Open-source GIS-GRASS users conference, Vol. 2002, pp. 51–70 (200)
2. Pazikadin, A.R., Rifai, D., Ali, K., Malik, M.Z., Abdalla, A.N., Faraj, M.A.: Solar irradiance measurement instrumentation and power solar generation forecasting based on Artificial Neural Networks (ANN): A review of five years research trend. Sci. Total Environ. **715**, 136848 (2020)
3. Sharaf, M., Yousef, M.S., Huzayyin, A.S.: Review of cooling techniques used to enhance the efficiency of photovoltaic power systems. Environ. Sci. Pollut. Res. **29**(18), 26131–26159 (2022)
4. Krishnan, N., Kumar, K.R., Inda, C.S.: How solar radiation forecasting impacts the utilization of solar energy: A critical review. J. Clean. Prod. **388**, 135860 (2023)
5. Shoaei, M., Noorollahi, Y., Hajinezhad, A., Moosavian, S.F.: A review of the applications of artificial intelligence in renewable energy systems: An approach-based study. Energy Convers. Manage. **306**, 118207 (2024)
6. Song, Z., Cao, S., Yang, H.: Assessment of solar radiation resource and photovoltaic power potential across China based on optimized interpretable machine learning model and GIS-based approaches. Appl. Energy **339**, 121005 (2023)

7. Gürel, A.E., Ağbulut, Ü., Bakır, H., Ergün, A., Yıldız, G.: A state of art review on estimation of solar radiation with various models. Heliyon **9**(2) (2023)
8. Yaiche, M.R., Bouhanik, A., Bekkouche, S.M.A., Malek, A., Benouaz, T.: Revised solar maps of Algeria based on sunshine duration. Energy Convers. Manage. **82**, 114–123 (2014)
9. Gama, A., Yettou, F., Malek, A., Panwar, N.L.: Generation of solar irradiation maps for various applications under Algerian clear-sky conditions. Int. J. Ambient Energy **39**(3), 243–256 (2018)
10. Abdeladim, K., Razagui, A., Semaoui, S., Arab, A.H.: Updating Algerian solar atlas using MEERA-2 data source. Energy Rep. **6**, 281–287 (2020)
11. Oyewola, O.M., Ismail, O.S., Olasinde, M.O., Ajide, O.O.: Mapping of solar energy potential in Fiji using an artificial neural network approach. Heliyon **8**(7) (2022)
12. Feng, Y., Hao, W., Li, H., Cui, N., Gong, D., Gao, L.: Machine learning models to quantify and map daily global solar radiation and photovoltaic power. Renew. Sustain. Energy Rev. **118**, 109393 (2020)
13. National Renewable Energy Laboratory. (n.d.). National Solar Radiation Database (NSRDB) https://nsrdb.nrel.gov/data-viewer

A Deep Reinforcement Learning Approach for Tracking the Maximum Power Point

S. Belarbi(✉), N. Drir, and L. Barazane

Laboratory of Electrical and Industrial Systems, USTHB, Algiers, Algeria
Sbelarbi1@usthb.dz

Abstract. Many recent research studies have focused on the use of renewable energies, which are clean, inexhaustible, and help to safeguard the environment by lowering carbon emissions. The output power of the solar panel is low and maximum power point approaches are used to increase it. This study describes a deep reinforcement learning-based technique (DRL) and demonstrates aimed at optimizing the operation of PV panels by tracking their maximum power point (MPP). The core focus of this research lies in investigating how the training process of the DRL agent named deep Q_Learning (DQN), measured in terms of training episodes, impacts the output power of PV systems. The output power is effectively dependent on the number of training episodes, as demonstrated by the simulation results.

Keywords: Photovoltaic System · Deep Q_learning · Maximum power point tracking · Deep Reinforcement Learning

1 Introduction

Renewable energy sources have received a lot of attention when it comes to environmental conservation on an international scale [1]. By converting solar energy into electricity, a photovoltaic (PV) system helps to decrease environmental pollution by consuming less fossil fuels. Nevertheless, because photovoltaic systems' output characteristics are nonlinear and vary with temperature and irradiance, we require a controller known as a maximum power point tracker, or MPPT, to extract the maximum power at the PV terminals [2]. This is done to guarantee that PV systems operate efficiently at the maximum power point (MPP) under various weather conditions [1, 6].

In addition, classical MPPT algorithms are inefficient for large-scale solar energy systems and show oscillations close to the MPP [8]. To overcome these drawbacks, research sought to develop innovative approaches to control the maximum power point.

In this research, we discuss about how an intelligent technique based on deep reinforcement is employed to monitor a solar system's maximum power [3]. By utilizing DRL, which integrates principles from reinforcement learning with deep neural networks, the purpose of the study is to highlight how effective this strategy is by dynamically adjusting the operational parameters of PV systems to maximize their power output under varying environmental conditions [5, 12].

© The Author(s), under exclusive license to Springer Nature Switzerland AG 2025
M. Hatti (Ed.): IC-AIRES 2024, LNNS 1238, pp. 554–562, 2025.
https://doi.org/10.1007/978-3-031-80301-7_60

This paper contributes to the understanding of how machine learning methods can be applied to raise the effectiveness and performance of renewable energy systems. Insights gained from this study not only underscore the potential of DRL in enhancing PV system operations but also pave the way for further refinements and applications in optimizing energy harvesting from solar sources [8, 9].

This is how the document is set up. In Sect. 2, the photovoltaic system's mathematical model is presented; in Sect. 3, the DRL and MPPT idea is introduced; and in Sect. 4, simulation results for two scenarios of the episode number are given. The paper's conclusion, found in Sect. 5, addresses further attempts.

2 Photovoltaic System Model

As shown in Fig. 1 A current source and a diode connected in parallel is the equivalent circuit of a solar cell. The photocurrent Iph of the current source is directly proportional to the amount of light that strikes the cell [1, 6].

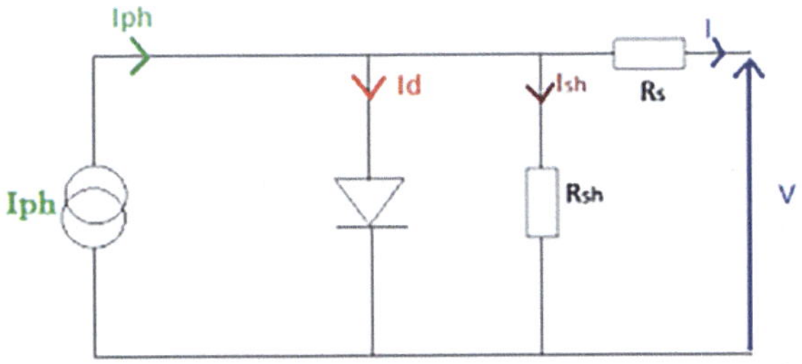

Fig. 1. Single diode photovoltaic cell model.

A current source, a diode, and a series of resistors connected in parallel constitute the PV cell components illustrated in Fig. 1. The following formulas can be used to determine the current produced by a solar cell [9, 12].

$$\mathrm{I} = \mathrm{Iph} - \mathrm{Id} - \mathrm{Ish} \tag{1}$$

In this case, I and V stand for the output voltage and current of the PV array. The parallel resistance (Rsh), PV, and diode current are presented by Ish, Iph, and Id. The resistance in series is Rs.

$$\mathrm{Id} = \mathrm{I0}.((\mathrm{q}(\mathrm{V} + \mathrm{IRs})/\mathrm{n.KB.T}) - 1) \tag{2}$$

The reverse saturation current is shown by I01. The symbols for temperature, electron charge, Boltzmann constant, and diode factor are T, q, KB, and n, respectively.

Equation (3) yields the current produced by the photovoltaic cell.

$$I_{ph} = \frac{G}{G0}(Isc + \lambda(T - T0)) \tag{3}$$

Irradiance and temperature reference are indicated by G0 and T0. Isc stands short circuit current. The temperature coefficient and irradiance are represented by λ and G.

$$I = I_{ph} - I_{0}.\left((q(V + IRs)/(nKT) - 1) - \left(\frac{V + Rs.I}{Rsh} \right) \right. \tag{4}$$

I_0 is defined by:

$$\mathrm{I}_0 = \mathrm{I}_{01\mathrm{ref}} \left(\frac{T}{T0} \right)^3 e^{\left[\frac{qEG}{nK} \left(\frac{1}{T} - \frac{1}{T0} \right) \right]} \tag{5}$$

where EG is the energy bandgap.

A direct voltage is given by the photovoltaic module. To provide the increased power required for the connected charge, a DC-DC boost converter is added to the chain [7, 9]. To maintain the power point at its maximum we should join an MPPT to the photovoltaic system as shown in Fig. 2 [2, 5].

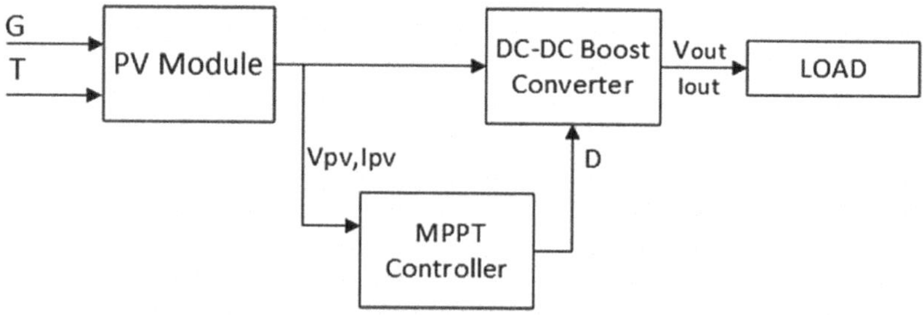

Fig. 2. MPPT controller in a photovoltaic system

3 Deep Reinforcement Learning and MPPT

The main elements of the deep reinforcement learning diagram are displayed in Fig. 3. In this case, the environment provides the agent with observations and rewards. The DQN agent acts in response to this information [10, 13]. A policy that maximizes positive rewards is learned by the DQN agent throughout the episodes, and thus drives the photovoltaic system to its maximum power point [11, 14].

The Markov Decision Process is used to implement a DRL MPPT with a DQN agent. The following components are thus created: [4, 16].

The main elements of the states tuple are the output voltage (V), output current (I), duty cycle (D), and the duty cycle change [3, 15].

$$\mathrm{S} = \{\mathrm{V}, \mathrm{I}, \mathrm{DU}, \Delta\mathrm{D}\} \tag{8}$$

The action space is defined as:

$$\mathrm{A} = \{-0.28, 0, +0.28\} \tag{9}$$

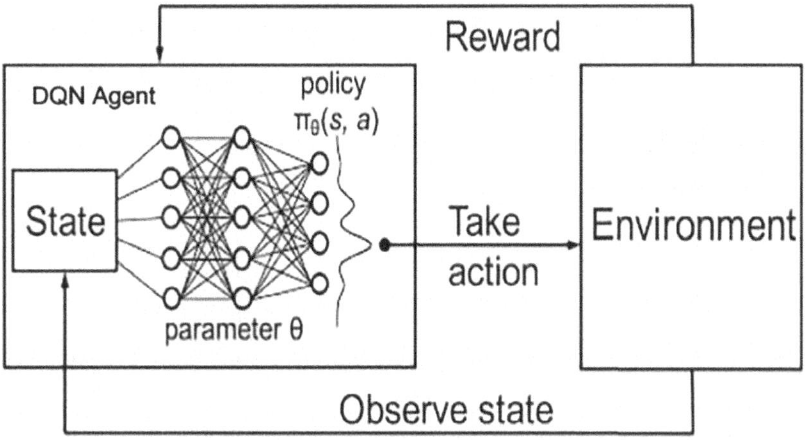

Fig. 3. DRL diagram with DQN agent

The following equations provide the reward:

$$RW = R_1 + R_2 + R_3 \tag{10}$$

$$R_1 = (Pt + 1)/Pmppstc \tag{12}$$

$$\mathrm{R}_2 = ((Pt + 1)/\ \mathrm{Pmppstc}\ )^2 \quad \text{if } \Delta\mathrm{P} >= -\alpha \tag{13}$$

$$\mathrm{R}_3 = 0 \quad \text{if } \Delta\mathrm{P} < -\alpha \tag{14}$$

α is a small number that allows for ripples over the MPPT obtained by the DQN agent.

4 Simulation and Results

MATLAB/SIMULINK was used to implement the system model and the Deep Q-Network agent was found to be suitable for the model. The different simulations are presented below.

4.1 Simulation of 1000 Episodes for DQN Agent Training

Case 01: Irradiation = 1000 w/m^2, Temperature = 25 °C
See Fig. 4.

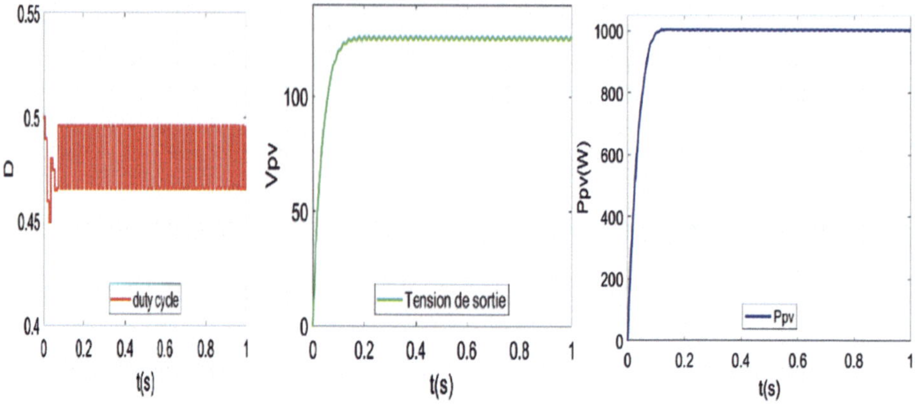

Fig. 4. The Duty Cycle, output voltage and output power using DQN agent at STC conditions

Case 02: Irradiation = 800 w/m², Temperature = 25 °C

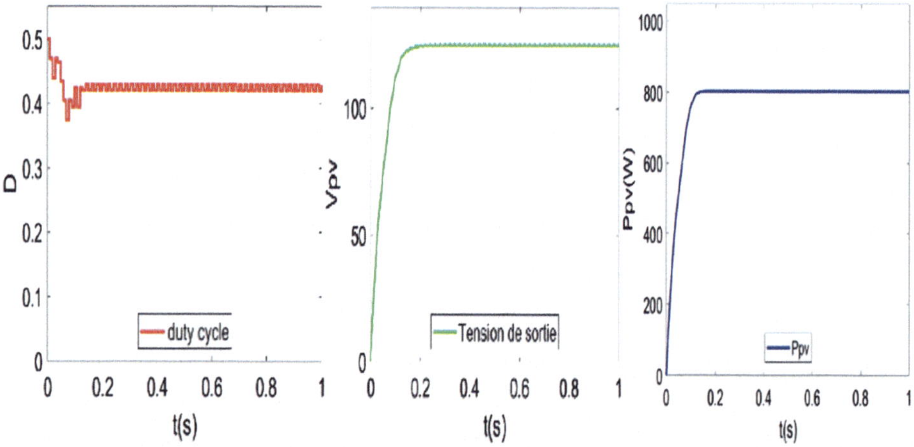

Fig. 5. The Duty Cycle, output voltage and output power using DQN agent

In the second scenario, we see in Fig. 5 duty cycle, output voltage, and output power.

Case 03: Irradiation = 1000 w/m^2, Temperature = 35 °C

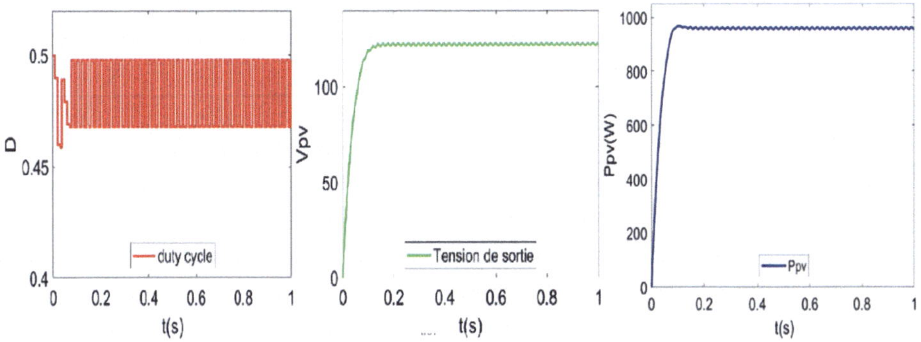

Fig. 6. The Duty Cycle, output voltage and output power using DQN agent

Figure 6 show the duty cycle, output voltage, and output power. We should remark that in this scenario, the output evaluate power graph has some ripples.

The analysis of these results highlights the DQN agent's capability to adapt to varying environmental parameters, maintaining, or optimizing system performance across different scenarios. These findings underscore the effectiveness and robustness of the DQN approach in enhancing the operational efficiency of photovoltaic systems.

4.2 Simulation of 6000 Episodes for DQN Agent Training

The same three simulation cases were extended to 6000 episodes each to further evaluate the performance of the DQN agent under varying environmental conditions:

Case 01: Irradiation = 1000 w/m^2, Temperature = 25 °C

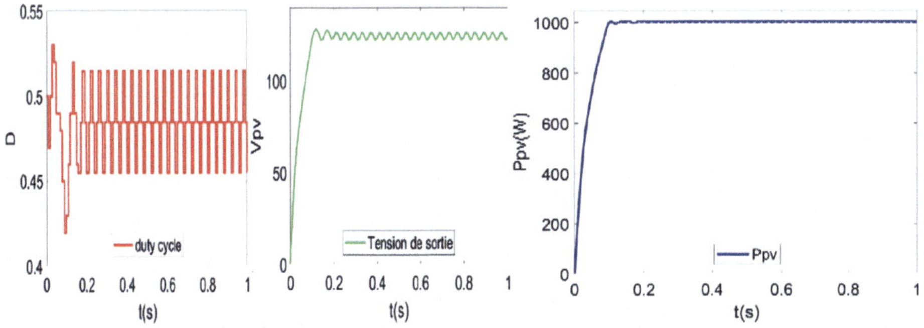

Fig. 7. The Duty Cycle, output voltage and output power using DQN agent at STC conditions

Figure 7 illustrate the duty cycle, output voltage, and output power under standard conditions and as the number of episodes increase from 1000 to 6000.

Case 02: Irradiation = 800 w/m^2, Temperature = 25 °C

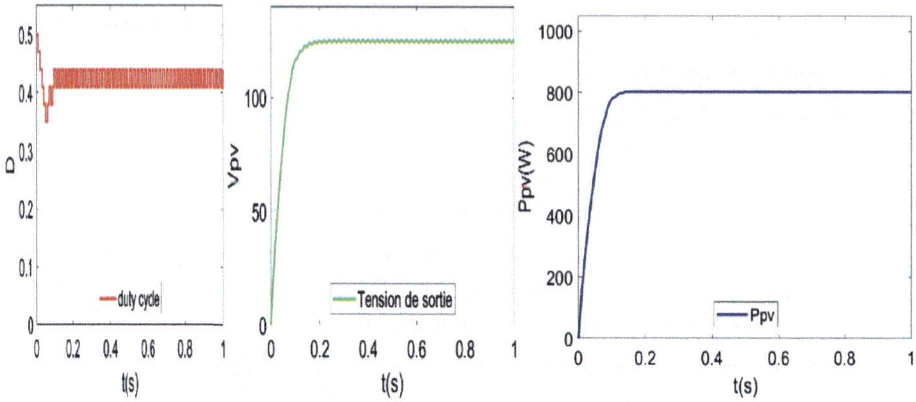

Fig. 8. The Duty Cycle, output voltage and output power using DQN agent

Figure 8 illustrate the duty cycle, output voltage, and output power. We observe that, in this scenario, the output power is less than 1000 watts.

Case 03: Irradiation = 1000 w/m^2, Temperature = 35 °C

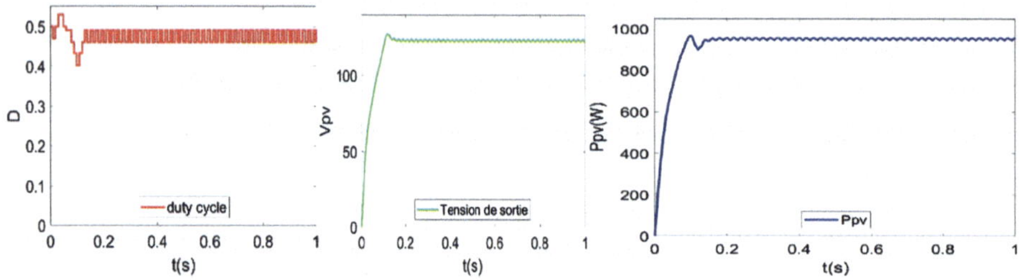

Fig. 9. The Duty Cycle, output voltage and output power using DQN agent

Figure 9 show the duty cycle, voltage output, and output power, which present small ripples at reaching 1000 watts. In Case 01, the DQN agent continued to maintain stable performance with consistent duty cycle and output power throughout the extended training period. Case 02 provided additional insights into agent's adaptability under reduced irradiation, showcasing its ability to optimize performance despite lower light conditions over an extended training period. Case 03, under elevated temperature conditions, revealed the agent's capability to adjust strategies and improve efficiency over time, indicating robust performance under challenging environmental factors.

The following table displays the output power levels at t = 0. For both 1000 and 6000 training DQN agent episodes (Fig. 10):

Temperature T(°C°)	Irradiation G(W/m2)	Number of episodes N	Output power Ppv (W)
25	1000	1000	1003
		6000	958
25	800	1000	802.9
		6000	802.6
35	1000	1000	955.5
		6000	958

Fig. 10. Output power levels for two distinct training episode numbers

The table above compares the output power levels achieved with different numbers of training episodes for the DQN agent under various environmental conditions. It is evident that, in nearly all cases, increasing the training episodes from 1000 to 6000 results in higher output power. This demonstrates the DQN agent's ability to learn and adapt over prolonged training periods, effectively optimizing the performance of photovoltaic systems across diverse environmental scenarios.

5 Conclusion

In this paper, a DRL-based algorithm for tracking the maximum power point of a solar system is described. It can be plainly observed from the system response that the strategy based on the Deep Q-learning Network agent delivers the most efficiency in terms of MPPT tracking.

The simulation results underscore the significance of the number of training episodes for optimizing the DQN agent's performance. Future efforts will focus on refining this technique to reduce computation time during agent training and simulation duration. Furthermore, ongoing research will compare this approach with other deep reinforcement learning methodologies.

References

1. Bavarinos, K., Dounis, A., Kofinas, P.: Maximum power point tracking based on reinforcement learning using evolutionary optimization algorithms. Energies **14**(2), 335 (2021)
2. Benchikh, S., Jarou, T., Boutahir, M.K., Nasri, E., Elamrani, R.: Improving photovoltaic system performance with artificial neural network control. Data Metadata **2**, 144 (2023)
3. Djerourou, A., Dekhane, A., Bouraiou, A., Atoui, I.: Evaluating MPPT techniques: optimizing photovoltaic systems under partial shading conditions. In: 2023 second international conference on energy transition and security (ICETS), pp. 1–9. IEEE, Adrar, Algeria (2023)

4. Guntupalli, R., Sudhakaran, M., Ajay-D-Vimal Raj, P.: Modeling & implementation of DRLA based partially shaded solar system integration with 3-ϕ conventional grid using constant current controller. Heliyon **8**(6), e09669 (2022)
5. Hassan, A., Bass, O., Masoum, M.A.S.: An improved genetic algorithm based fractional open circuit voltage MPPT for solar PV systems. Energy Rep. **9**, 1535–1548 (2023)
6. Hybrid, Optimization, Intelligent and Classical PV MPPT Techniques: Review. CSEE J. Power Energy Syst. (2020)
7. Kalogerakis, C., Koutroulis, E., Lagoudakis, M.G.: Global MPPT based on machine-learning for PV arrays operating under partial shading conditions. Appl. Sci. **10**(2), 700 (2020)
8. Kumar, M., Panda, K.P., Rosas-Caro, J.C., Valderrabano-Gonzalez, A., Panda, G.: Comprehensive review of conventional and emerging maximum power point tracking algorithms for uniformly and partially shaded solar photovoltaic systems. IEEE Access **11**, 31778–31812 (2023)
9. Manna, S., Singh, D.K., Akella, A.K., Kotb, H., AboRas, K.M., Zawbaa, H.M., Kamel, S.: Design and implementation of a new adaptive MPPT controller for solar PV systems. Energy Rep. **9**, 1818–1829 (2023)
10. Mnih, V., et al.: Human-level control through deep reinforcement learning. Nature **518**(7540), 529–533 (2015)
11. Phan, B.C., Lai, Y.-C., Lin, C.E.: A deep reinforcement learning-based MPPT control for PV systems under partial shading condition. Sensors **20**(11), 3039 (2020)
12. Roh, C.: Deep-learning algorithmic-based improved maximum power point-tracking algorithms using irradiance forecast. Processes **10**(11), 2201 (2022)
13. Sharmin, R., Chowdhury, S.S., Abedin, F., Rahman, K.M.: Implementation of an MPPT technique of a solar module with supervised machine learning. Front. Energy Res. **10**, 932653 (2022)
14. Tercha, W., Tadjer, S.A., Chekired, F., Canale, L.: Machine learning-based forecasting of temperature and solar irradiance for photovoltaic systems. Energies **17**(5), 1124 (2024)
15. Yung Yap, K., Sarimuthu, C.R., Lim, J.M.-Y.: Artificial intelligence based MPPT techniques for solar power system: a review. J. Mod. Power Syst. Clean Energy **8**(6), 1043–1059 (2020)
16. Zhu, J., Wu, F., Zhao, J.: An overview of the action space for deep reinforcement learning. In: 2021 4th international conference on algorithms, computing and artificial intelligence, pp. 1–10. ACM, Sanya China (2021)

Autoencoder, LSTM, RNN, and SVM for Anomaly Detection, Prediction, and Localization in Industrial Systems

Dalila Cherifi(✉), Mohammed Amine Bedri, and Said Boutaghane

Institute of Electrical and Electronic Engineering, University of Boumerdes, Boumerdes, Algeria
da.cherifi@univ-boumerdes.dz

Abstract. This work focuses on developing an anomaly detection, prediction, and localization system using deep learning methods applied to the data collected from sensors built in IoT-enabled machines. The primary objective is to detect faults in real-time, estimate the machine's behavior and health for a future time period, and precisely localize errors within the machine. In this paper we use Auto-encoder Neural Networks and LSTM Recurrent Networks and One Class SVM's to actualise the previously mentioned objectives.

Keywords: Anomaly detection · IoT-enabled machine · Deep Learning · Predictive Maintenance · Condition Monitoring

1 Introduction

Data-driven predictive maintenance is a key component of Industry 4.0, using the utility of data analytics and advanced technologies to develop maintenance strategies in industrial settings. Traditionally, maintenance activities were totally reactive or scheduled based on fixed intervals. However, with the up-rise of Industry 4.0 and the availability of sensory data generated by IoT-enabled machines, the focus was shifted towards predictive maintenance. The maintenance works by using real-time sensors and AI to predict and prevent equipment failures. By monitoring the performance and condition of machines over time, potential faults can be identified early on, allowing maintenance teams to take actions before costly breakdowns occur. This approach minimizes downtime, reduces maintenance costs, and maximizes the lifespan of industrial goods.

The approach that we went with includes training machine learning models on the healthy state machine data, meaning when the machine is still "new". In this paper we will use an Auto-encoder Deep Neural Network and a One Class SVM to learn the patterns in the healthy data. The outputs given by the models will be used to monitor the behavior of the system during it's lifetime. A degradation signal will be the fundamental component of the monitoring process. Furthermore, we use the generated degradation signal as a measure of health and train another LSTM recurrent neural network to predict future values of the degradation signal, this serves the purpose of predicting future behavior thus possible failures before they occur. Finally, we look at the degradation contribution

© The Author(s), under exclusive license to Springer Nature Switzerland AG 2025
M. Hatti (Ed.): IC-AIRES 2024, LNNS 1238, pp. 563–572, 2025.
https://doi.org/10.1007/978-3-031-80301-7_61

of each column/sensor in our data, and from it we can localize the errors in the system to better assist the maintenance teams and experts in accurately pinpointing failures.

The contents of this paper are structured as follows: first, the Sect. 2 will go over a review of relevant literature to this article, and similar use cases of ML for maintenance. Secondly, Sect. 3 will go over the multi-model approach that consists of an Auto-encoder and a One Class SVM for fault detection and localization of faults in the system, and a LSTM RNN to predict future faults. Thirdly, we explain the previously mentioned approaches on a well known data set and go over the results of all the models as well. Finally the article is concluded in Sect. 5.

2 Related Work

Recent research has shown that deep learning techniques like Auto-encoders can be highly effective for fault detection in industrial systems. By training an Auto-encoder on normal operating data, it can learn to reconstruct normal patterns and flag deviations as potential faults. Lachekhab et al. [1] have actualised a comparison between LSTM layered Auto-encoder and normal Auto-encoder, the research resulted in the effectiveness of both models in flagging anomalies, although the LSTM had lower errors considering their superior complexity. Birgelen et al. [2] have used a Self-Organizing Maps for anomaly detection and localization as a predictive maintenance strategy. By producing quantization errors using the SOM, analysis of the machine bearings was very clear, and also the localization of faults inside the bearings was an applicable task using the errors. Predictions using the LSTM recurrent networks of future degradation signals in our case was inspired by the work in this study. In addition to that, LSTM's ability to capture long-term dependencies in time series data makes them well-suited for such applications. In the study published by Bibhuti et al. [3], the LSTM-RNN model outperformed a traditional recurrent neural network (RNN) model as well as a naive method for low-flow forecasting. The LSTM-RNN achieved a correlation coefficient (R) of 0.943, a Nash-Sutcliffe efficiency (ENS) of 0.878, and a Root-Mean-Square Error (RMSE) of 0.487. As for One Class SVM's, Seliya et al. discussed the advantages of One-Class SVMs for anomaly detection, including their effectiveness in high-dimensional spaces, flexibility with non-linear data, robustness to outliers, and ability to operate in an unsupervised manner without requiring labeled anomaly data. However, the authors also note some limitations, such as parameter sensitivity, computational complexity, and challenges with very large datasets.

3 Methodology

In this section, we go through the methodology that we have applied in the work. We understood how Machine and Deep Learning methods can be used to perform analytical tasks such as fault detection, prediction and localization in the industrial field.

3.1 Fault Detection

Firstly in fault detection, we employ an Auto-encoder Neural Network. We train the Auto-encoder using the healthy state of our complete dataset, which encompasses the lifetime operation of our machine. The healthy state corresponds to the initial 20% of the dataset. By reconstructing the entire dataset using the trained Auto- encoder, we can quantify the error or deviation from the healthy state. This error serves as a wear component that eases the the interpretation of the machine's behavior. Using statistical methods we can identify a point threshold, where it works as identification point for faults. If ever the degradation bypasses the calculated threshold it will be flagged as a fault in the machine by the model.

- **Auto-encoder Neural Network**
  An Auto-encoder is a type of artificial neural networks that uses the method of unsupervised learning [4]. It works by compressing the representation of the input data through an encoder network, and then reconstructing the original input data from this compressed representation using a decoder network. The encoder's role is to transform the input data into a lower-dimensional representation, while the decoder aims to reconstruct an identical output from the original input of the encoded representation [5]. Here are the mathematical formulas for the encoder and decoder.

  - **Encoder:** Let x denote the input data and h represent the encoded representation.

$$h = f_{enc}(x) = \sigma(W_{enc}x + b_{enc}) \tag{1}$$

  - **Decoder:** Let $x\hat{}$ denote the reconstructed input and h represent the encoded representation.

$$\mathrm{h = fenc(x) = \sigma(Wencx + benc)} \tag{2}$$

  In the context of data reconstruction which is the strategy that we have used to generate a degradation signal, the encoder network is used to compress the input data into a lower-dimensional representation. To reconstruct the data, the decoder network takes the compressed representation and generates an output that aims to resemble the original input data as closely as possible. The difference between the reconstructed and actual dataset is the error we use to identify faults when we pass in the entire data set through the model. That error is the degradation signal. Analyzing the degradation signal allows for monitoring of the system and eventually fault detection. The architecture of an Auto-Encoder usually is represented by the following figure (Fig. 1):

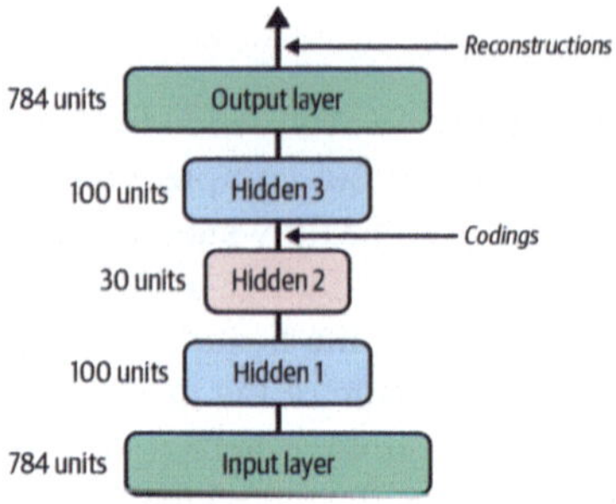

Fig. 1. Basic Auto-encoder Architecture [6].

Assuming that the input vector is x and the output vector is $x^{'}$. Our degradation signal is basically $(x - x^{'})$. The difference will be the deviation from what we consider the norm, allowing for analysis of anomalous behaviour easily.

- **One-class Support Vector Machine**
  The One-Class SVM is an algorithm used for unsupervised outlier detection that was introduced first by Scholkopf and Muller in 2001 [7]. It aims to identify anomalies or outliers in a dataset by learning a boundary that encloses the region of normal data points. The OCSVM maps the data into a higher dimensional feature space using a kernel function and then develops a decision boundary that separates the training positive observations from the origin. Gaussian and polynomial kernels are the most well known kernel functions that have are used to actualise this [8]. When properly applied, in principle a kernel-based method is able to model any non-linear pattern of normal behaviour [9]. The process of choosing which kernel to work with is crucial as it effects the performance of the One Class SVM heavily [7]. The Gaussian or the RBF non-linear kernel is the focus for generating the separating hyperplane in this article due to the dense and non-linear nature of our data. The RBF kernel is defined by the following equation [10]:

$$K(x, y) = \exp - \gamma \cdot ||x - y||^2 \tag{3}$$

Through its kernel it captures complex non-linear relationships between data points [9]. It assigns higher similarity (or kernel value) to data points that are closer together and lower similarity to those that are farther apart. The choice of the γ parameter is crucial when using the RBF kernel in One-Class SVM as shown in [11]. A smaller γ value results in a wider Gaussian distribution, yielding a smoother decision boundary and potentially capturing more outliers. Conversely, a larger γ value leads to a narrower distribution, resulting in a more focused decision boundary and potentially capturing fewer outliers. Determining the optimal γ value for a specific dataset often requires experimentation and parameter tuning.

3.2 Fault Prediction

Here we develop a fault prediction system using a LSTM Recurrent neural network. By utilizing the Degradation Signal produced by the Auto-encoder and One-Class SVM as the training data for the LSTM model. After pre-processing it for the model's suitability, we aim to forecast future values of the previous signal.

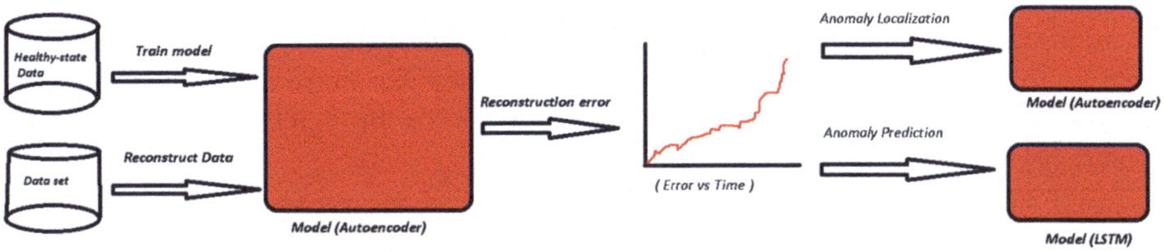

Fig. 2. Condition Monitoring Work Flow.

The primary goal is to flag future values of error that exceed a predefined threshold of degradation. The corresponding date time of these flagged errors will serve as a predictive date of the machine failure. The Fig. 2 shows the work flow we're taking.

Recurrent Neural Networks (RNN's) are a type of artificial neural network designed to process sequences of data. It works especially well for jobs requiring sequences, such as time series data. One of the challenges that a RNN faces is the vanishing/exploding gradient problem, due to that it only takes into account short dependencies in the data. Thus, RNN's is not considerably applicable when there are long term dependencies in the data. Here, another special type of RNN's, the Long-Short term Memory (LSTM) networks were introduced to tackle the issues that RNNs had. Its creators were Hochreiter and Schmidhuber [12], The new improved cell architecture facilitate storing information for longer periods. LSTM cells can be described by the by following equations [13] and figure [14]:

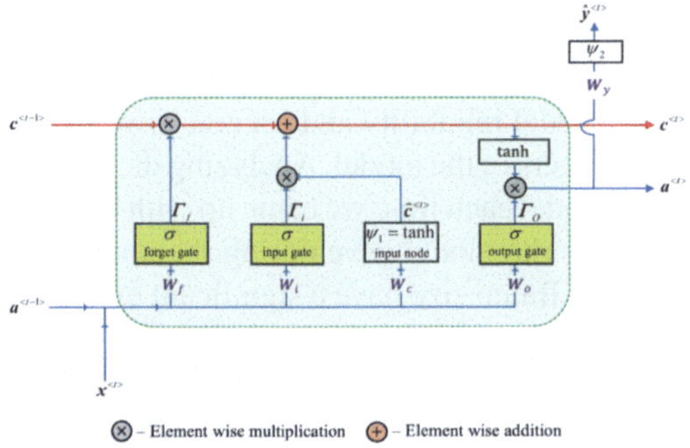

$$f_t = \sigma(W_f \cdot [h_{t-1}, x_t] + b_f),$$

$$i_t = \sigma(W_i \cdot [h_{t-1}, x_t] + b_i),$$

$$\tilde{C}_t = \tanh(W_C \cdot [h_{t-1}, x_t] + b_C),$$

$$C_t = f_t * C_{t-1} + i_t * \tilde{C}_t,$$

$$o_t = \sigma(W_o \cdot [h_{t-1}, x_t] + b_o),$$

$$h_t = o_t * \tanh(C_t),$$

Fig. 3. LSTM unit cell.

An LSTM network is a specialized type of recurrent neural network designed for time-series data (see Fig. 3). It uses memory cells and gating mechanisms to retain important information over long sequences, addressing the challenges of traditional RNNs. By incorporating memory cells and gates, LSTMs can capture long-term dependencies and minimize the vanishing and exploding gradient problems [12]. They are particularly effective for time series forecasting tasks [15], as they can make accurate predictions based on historical data. During training, back propagation through time is used to update the network's weights and biases [16].

3.3 Fault Localization

By visualizing the error that each component/sensor of the machine outputs from the model, we can identify the sensor in the machine that most likely captured the error when it occurred if it contributes the most to the overall degradation signal.

4 Experiments and Results

4.1 Dataset

This dataset contains the machine data recorded over the duration of 12 month total. It was initiated in the European research and innovation project IMPROVE. The subject of experiment is data recorded from the VegaHT wrapper machine [17]. The data consists of 1062912 instances, a little over a million cases. It includes 7 features/sensors that map the machine's coordinates during its production process.

4.2 Experiments

- **Experiment 1: Degradation Signal Generation using Auto-Encoder**
  Using the functions provided by both Keras and Tensor Flow, we have built an Auto-Encoder neural Network, and to ensure model reliability and accuracy, we have performed a number of tries to tune the parameters of the model. Analyzing the Validation Loss data from the training phase of the model each time we came up with the following data. Table **??** and **??** show the results the model gave during the tuning phase. We can see that the number of Epochs and Batch size have a significant effect on the model's performance. We find that the smaller those parameters are the worse the model performs, although there's degree to which the epochs have much more of an effect. We can see that on the last two columns in both cases. However the changing of the parameters has a less significant effect on the simpler model. Meaning that models with complicated architectures need more Epochs and larger Batch sizes to perform optimally. Making the Auto-encoder as complex as it possibly could be before it had any over fitting within the training phase was an objective, and that was achieved by monitoring the validation loss each time it was trained.

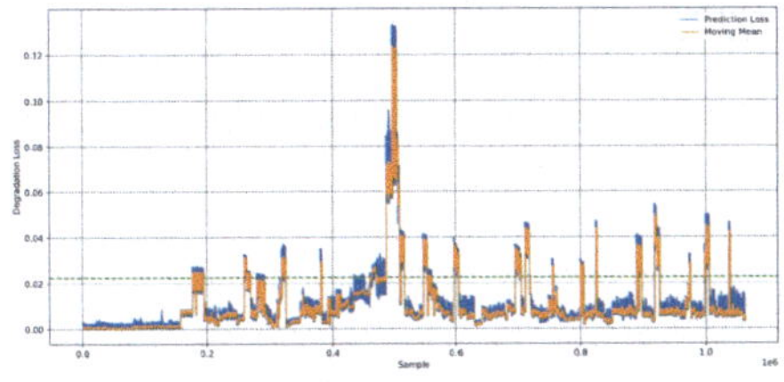

Fig. 4. Auto-Encoder Machine Degradation Signal.

128 unit Architecture		
Epochs = 100, Batch = 200	Epochs = 10, Batch = 100	Epochs = 10, Batch = 32
1.74 x 10¯5	5.37 x 10¯5	6.09 x 10¯5
64 unit architecture		
Epochs = 100, Batch = 200	Epochs = 10, Batch = 100	Epochs = 10, Batch = 32
1.35 x 10¯5	3.36 x 10¯5	3.347 x 10¯5

Fig. 5. Performance comparison for different architectures and parameters.

The next step was to pass the entire dataset through the Auto-Encoder than subtract the input from the output to get a degradation signal. This will allow for clear analysis for the machine's behavior as represented on the Fig. 4. Analyzing this graph, initially the spike of error occurred on the 18th of April meaning the machine failed completely at that point. We also explain the return to the norm after that peak with the blades of the machine being changed in a reactive maintenance manner (Fig. 5).

The machine kept working normally after that, considering this second phase its second lifetime. The machine had trouble working optimally again as from the data when the last spike of error occurred on 14:36 in the afternoon on the 20th of October, signifying the blades wearing out again. The average of both lifetimes can be used as a maintenance interval set time. Now utilizing the degradation signal for localization. Plotting the degradation signals of each feature (sensors) individually can help identify which feature contributes the most to the failures of the machine. Figure 6 illustrates the degradation contribution of each feature individually.

According to the documentation, it was mentioned that the errors that occurred with this specific machine were primarily associated with the cutting blade of the machine unit. The sensor plots confirm that information.

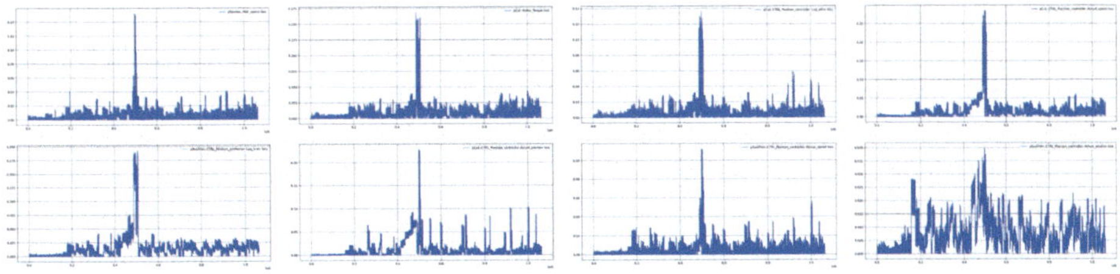

Fig. 6. Degradation Signal of Individual Sensors.

Features in the data like blade speed loss and blade position loss scored the highest among all features in terms of it's deviation from the "healthy" state. The Auto-encoder completed its objective and was able to detect faults in the machine and localize the exact sensors that caught these faults. Producing a reliable result for use even in real industrial cases.

- **Experiment 2: Component Degradation Error using One Class SVM's**
  While using the OCSVM model, we try to train the model on the "healthy" state data. For each prediction of our data using this model, an anomaly score is recorded. It is the distance of that particular instance from the hyperplane surrounding the healthy-state data points. The bigger that score the bigger the error. Let's see if it behaves the same as the Auto-Encoder model. When we plot the anomaly scores we get.

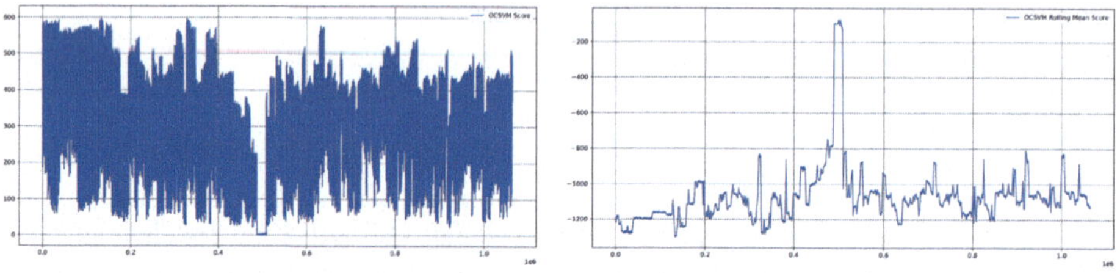

Fig. 7. OSCVM generated Degradation signals.

The OCSVM model produced much noise, incomparably to that of the Auto-Encoder's. In an attempt to reduce the noise in the output we use mean the values of the scores in windows of 2050 instances. This can smooth out some of the edges/noise that disturbs the analysis of the plot. Figure 7 is the result of rolling mean values.

At first sight the data is pretty clear when employing the rolling mean values. It also produced nearly the same results of the Auto-Encoder experiment. The same conclusions can be drawn from both outputs on the same date times. Although the OCSVM required some further manipulation of the output was inferior to the Auto-Encoder. The similarity in both models outputs considering their very different architectures signifies the accuracy of both. We will be considering the Auto-encoder generated data set due to its low noise nature and high quality compared to that of the OCSVM. The degradation signals produced will now be used to learn a LSTM RNN model to predict future values, and in turn future analysis and forecasting of errors.

- **Experiment 3: Fault prediction LSTM, RNN**
  Using the Auto-Encoder generated degradation data, The training split for this model was a different than the other models. We trained the model on the first half of the entire data, which represents the first life-cycle. It ends at the previously shown degradation peak. Testing the model using the test set. Basically the test set is the rest of the data that we have, representing another life-cycle of the machine. After predicting the values we plot them along the actual values of the dataset. Plotting the predicted values and actual values is shown in the Fig. 8.

The Mean Squared Error between the actual and predicted values: $MSE = 1.09 \times 10^{-7}$. We can see that the MSE value is substantially small. The results corresponds to the prior analysis we have done on the plots. Figure 8 shows the distinction between actual values and predicted ones using the LSTM model. The predicted red portion appears to be very similar to the actual normal dataset from before. Signifying the model's ability to simulate future behavior of the machine. Taking the date time of spikes in predicted time vs actual time gives us:

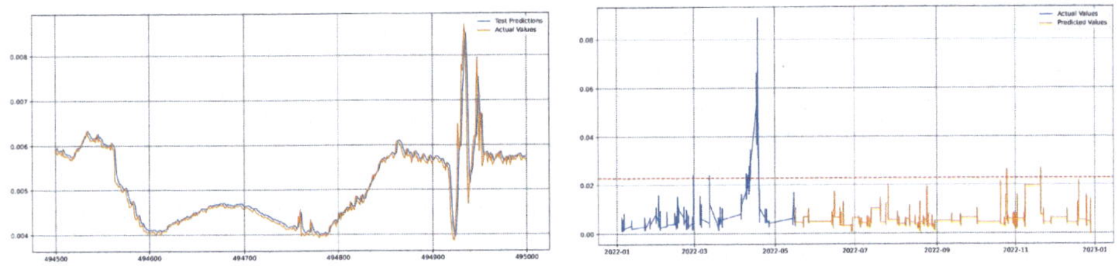

Fig. 8. LSTM test Results - Predicted vs Actual Values Produced By LSTM based Model.

- PREDICTED TIME OF FAILURE: 2022–10-25 14:36:00
- ACTUAL TIME OF FAILURE: 2022–10-20 15:39:37

The model managed to predict machine failure five days after the day of actual machine failure.. The results that the model gave are very optimistic, it was accurate when identifying trends in the data, and was able to draw another life cycle from scratch, which makes it reliable for future use with other kinds of sensory data.

5 Conclusion

The implementation of machine learning (ML) and deep learning methods in the field of predictive maintenance holds immense usefulness and potential. By leveraging these techniques, organizations can analyze large volumes of historical data to detect patterns and anomalies, enabling them to predict and prevent equipment failures before they occur. In this paper we used Deep Learning and Machine Learning methods by training an Auto-Encoder and One Claas SVM for component degradation data set generation, and an LSTM recurrent neural network for future value prediction. Component degradation is a normal part of any mechanical machine's life cycle. Down time costs can be countered with data driven process monitoring processes and predictive maintenance strategies. Models such as the Auto-Encoder and One Class SVM can generate machine degradation signals after being trained on the "healthy/new" state of machine data through multiple sensors/features. Where Deep Learning methods are powerful and can learn the patterns and relationships between the working sensors data. The degradation signal provides information on how any machine behaves during its work time. Helping in overall analysis of machine health and degree of degradation or out of norm behavior. The localization through reverse engineering each feature data helps in pinpointing where exactly the degradation is happening. The fault localization provides information for maintenance teams on where faults are and gives a better idea on how to deal with the machine when a particular failure happens. LSTM RNN like models can be used to predict the remaining life cycle of the machine. As we saw from previous experiments the results were relatively accurate and made the model reliable for real use case in industrial settings. All the algorithms were trained and tested on the Vegaht OCME cutting machine dataset.

References

1. Lachekhab, F., et al.: LSTM-autoencoder deep learning model for anomaly detection in electric motor. Energies **17**(10), 1996–1073 (2024). https://doi.org/10.3390/en17102340
2. Von Birgelen, A., et al.: Self-organizing maps for anomaly localization and predictive maintenance in cyber-physical production systems. Procedia CIRP **72**, 480–485 (2018)
3. Sahoo, B.B., et al.: Long short-term memory (LSTM) recurrent neural network for low-flow hydrological time series forecasting. Acta Geophys. **67**(5), 1471–1481 (2019)
4. An, J., Cho, S.: Variational autoencoder based anomaly detection using reconstruction probability (2015). https://api.semanticscholar.org/CorpusID:36663713
5. Kim, D., Lee, S., Kim, D.: An applicable predictive maintenance framework for the absence of run-to-failure data. Appl. Sci. **11**(11), 2076–3417 (2021). https://doi.org/10.3390/app11115180
6. Géron, A.: Hands-On machine learning with scikit-learn, keras & tensorflow farnham". O'Reilly, Canada (2019)
7. Müller, K.-R., et al.: An introduction to kernel-based learning algorithms". In: Hu, Y.H., Hwang, J.-N. (eds.) Handbook of neural network signal processing, pp. 4–1. CRC Press (2018)
8. Anaissi, A., et al.: Adaptive one-class support vector machine for damage detection in structural health monitoring. In: Kim, J., et al. (eds.) Pacific-Asia conference on knowledge discovery and data mining, pp. 42–57. Springer (2017)
9. Erfani, S.M., et al.: High-dimensional and large-scale anomaly detection using a linear one-class SVM with deep learning". Pattern Recogn. **58**, 121–134 (2016)
10. Hejazi, M., Singh, Y.P.: One-class support vector machines approach to anomaly detection. Appl. Artif. Intell. **27**(5), 351–366 (2013)
11. Schölkopf, B., et al.: Estimating the support of a high-dimensional distribution. Neural Comput. **7**, 1443–1471 (2001)
12. Hochreiter, S., Schmidhuber, J.: Long short-term memory. Neural Comput. **9**(8), 1735 (1997)
13. Shewalkar, A., Nyavanandi, D., Ludwig, S.A.: Performance evaluation of deep neural networks applied to speech recognition: RNN, LSTM and GRU. J. Artif. Intell. Soft Comput. Res. **9**(4), 235–245 (2019)
14. Zargar, S.: Introduction to sequence learning models: RNN, LSTM, GRU. Department of Mechanical and Aerospace Engineering, North Carolina State University (2021)
15. Siami-Namini, S., Tavakoli, N., Namin, A.S.: A comparison of ARIMA and LSTM in forecasting time series. In: 2018 17th IEEE international conference on machine learning and applications (ICMLA), p. 1394
16. Graves, A., Mohamed, A.-R., Hinton, G.: Speech recognition with deep recurrent neural networks. In: 2013 IEEE international conference on acoustics, speech and signal processing, p. 6645. IEEE (2013)
17. Artificial Intelligence in IT. OCME VegaHT Shrink Wrapper Machine One Year Industrial Component Degradation. Data set (2018). https://www.kaggle.com/datasets/inIT- OWL/one-year-industrial-component-degradation

Multi-classification and Multi-labeling of UML Diagrams Using Transfer Learning and CNN

Abderrahim Lakehal[1], Adel Alti[1,2](✉), Boubakeur Annane[1], Bensalah Imene[1], Smaani Anissa Afaf[1], and mami meriem[1]

[1] LRSD Laboratory, Computer Sciences Department, Sciences Faculty, Setif 1 University - Ferhat ABBAS, Setif, Algeria
alti.adel@univ-setif.dz

[2] Department of Management Information Systems, College of Business and Economics, Qassim University, Buraidah, Saudi Arabia

Abstract. Unified Modeling Language (UML) diagrams are standard graphical tools to specify, visualize and understand the design and behavior of software systems. Accurate classification and identification of UML components are crucial for the quality and efficiency of software development processes. This work deals with the application of Convolutional Neural Networks (CNNs) in the classification of UML diagrams. The proposed approach focuses on deep learning techniques to accurately classify UML diagrams into different categories, including UML and non-UML classes, as well as identifying the components of UML diagrams, addressing the challenges of multi-labeling and multi-classification. Furthermore, transfer learning techniques are investigated to utilize pre-trained models and test their contribution to improving classification performance given the limited size of the dataset. The proposed approach is applied to a dataset of 4705 images of UML and non-UML diagrams. The results demonstrate the effectiveness of CNNs for multi-classification and multi-labeling of UML diagrams with an accuracy of 97.54%, outperforming the traditional models.

Keywords: UML · Class Diagram · CNN · Deep Learning · Transfer Learning

1 Introduction

Unified Modeling Language (UML) offers a wide range of diagrams that are useful in diverse industrial and renewable energy systems. These graphical representations are used in the software development lifecycle to visualize the design and structure of software systems [1]. They serve purposes such as analysis, documentation, software design, and development. Consequently, UML diagrams provide many benefits to researchers, software developers, and academics.

Deep learning is a powerful technique of Artificial Intelligence (AI) utilized in computer vision, allowing machines to accurately understand and analyze visual data [2]. Convolutional Neural Networks (CNNs) present a highly promising approach for image classification, showcasing exceptional performance in diverse domains. Classification of UML diagrams is particularly challenging due to their complex and diverse

© The Author(s), under exclusive license to Springer Nature Switzerland AG 2025
M. Hatti (Ed.): IC-AIRES 2024, LNNS 1238, pp. 573–583, 2025.
https://doi.org/10.1007/978-3-031-80301-7_62

nature, requiring sophisticated deep-learning techniques to achieve accurate results. This research study leverages the CNN model to classify UML diagrams into different categories, including both UML and non-UML classes, while addressing the complexities of multi-labeling and multi-classification. CNN models are trained on labeled datasets to classify images as UML diagrams or not. Once identified as a UML diagram, further categorization into different types such as class diagrams, activity diagrams, etc. is performed. Furthermore, a multi-labeling task is conducted to determine the components of each UML diagram such as aggregation, notes, class, relation…etc., all depending on the UML diagram under study. Given the complexity and variability of UML diagrams, data augmentation techniques and pre-trained models can increase accuracy. Pre-processing the data, extracting features, and fine-tuning the model remain crucial steps to achieve accurate classification results in this task. We summarize the proposed contributions:

- Exploring CNNs and transfer learning for UML diagrams for multi-classification and multi-labeling
- Based on CNN models, the proposed system involves training CNN models for predicting UML class (i.e., multi-classification) and predicting components in class and activity diagrams (i.e., multi-labeling).
- We conduct a comparative study to illustrate the usefulness and efficiency of the suggested approach. A dataset of 4705 images of UML and non-UML diagrams is used to evaluate transfer learning and CNN models. It defines 11 classes including UML class diagrams, UML activity diagrams, etc. Our results led to higher accuracy and outperformed the traditional models.
- Another comparative study including a class diagram and activity diagram is conducted to test the multilabeling performance of CNN models. Two datasets of 307 and 375 images were used for identifying the components of the diagrams within each data set.

This paper is arranged as given. Section 2 contains a discussion of the relevant and current literature. Section 3 presents the suggested transfer-learning CNN-based classification system. The research results and analysis are presented in Sect. 4. Section 5 draws and discusses the conclusions from this work.

2 Literature Review

Several works have emerged in the last few years focusing on classifying UML diagrams using machine learning and deep learning techniques [3–10]. They include groundbreaking research aimed at improving systems analysis and design processes through the integration of UML and deep learning techniques. Many studies have explored the benefits of integrating UML, a unified modeling language in software development, with deep learning algorithms to automate tasks, enhance accuracy, and boost system performance. Colanzi et al. [4] utilized data augmentation and transfer learning to classify UML diagrams. For this purpose, they used a specially designed dataset including six different kinds of UML models. Experiments were carried out using three popular CNN architectures: VGG16, ResNet50, and InceptionV3. The results showed that transfer learning plays a significant role in attaining favorable outcomes even when working with small datasets. However, the UML diagram classification could still be improved in terms of accuracy.

Shcherban et al. [5] collected UML diagrams from publicly available datasets. Some UML diagrams are classified into ten classes and the rest are not classified as UML diagrams. The performance of seven well-known neural network architectures was assessed by applying transfer learning to a dataset containing 4706 images, which comprised 3986 UML images and 720 non-UML images. However, the multi-labeling of UML diagrams is not considered in this work. Gosala et al. [6] have implemented a tool that automates the classification of UML class diagrams with CNNs, removing the necessity for manual labeling. Their tool can handle various image formats and identify if an image represents a UML class diagram.

Wang et al. [7] transformed the problem of detecting design patterns into an image classification task. Positive and negative patterns are recognized by the system and identified as UML color models. The features were extracted from the dataset using VGGNet and then used to train SVM for binary classification to detect pattern instances. Experiments utilized three open-source projects to identify design patterns using five traditional machine learning methods. The precision and recall of their technique were higher, consistently above 85% for various programs and patterns. Their approach demonstrated effective recognition of design patterns in colored UML models, as indicated by the experimental results.

Rigou et al. [8] discussed tools developed in software engineering to address the problem of creating PIM from natural language. The authors proposed an architecture based on deep learning techniques using entity detection, relation classification, and coreference resolution to construct these diagrams. The suggested system consists of an encoding layer based on BERT and three decoding layers of feed-forward neural networks, generating clear and uniform UML class diagrams and demonstrating encouraging early outcomes.

Koenig et al. [9] have developed a platform based on a machine-learning model to classify and detect semantic components within a class diagram depicted in an image. A comprehensive collection of labeled design models has been suggested, accessible through Internet. The qualitative and quantitative evaluations were performed on two distinct subsets, yielding accuracy rates of 92.59% and 94.11% respectively. This work demonstrates the ability of the suggested learning model to apply broadly to various examples.

Rhim et al. [10] introduced CS-Mobilenet-VGG16 hybrid model, addressing two important issues related to classification and ensemble learning. The authors established bidirectional information transfer between CS-Mobilenet and VGG16, allowing information to flow from Mobilenet to VGG16, while also addressing variability in the UML diagrams variability. The CS-Mobilenet-VGG16 model demonstrated superior performance in comparison to both Mobilenet and VGG16 models ($p < 0.05$).

The main limitations found from the above-investigated methods are to ensure multi-classification and multi-labeling of UML diagrams with higher accuracy. The proposed approach based on CNNs with a good preprocessing phase and a pre-trained model with transfer learning is expected to achieve more accurate results.

3 Proposed Approach

This section describes the five steps of our research contribution. Initially, the dataset is collected and structured. Following that, the dataset underwent preprocessing. Then, the dataset is divided into three subsets: training, testing, and validation. Afterward, transfer learning was performed using both AlexNet and CNN models on this particular dataset. Lastly, we evaluated and compared the performance of our models. These steps are depicted in Fig. 1.

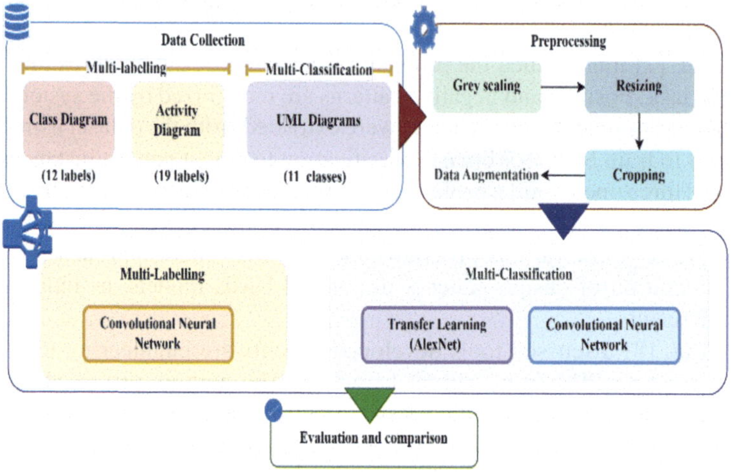

Fig. 1. Workflow Model.

A. **Dataset**

Our dataset consists of three main types of diagrams: all diagrams, class diagrams, and activity diagrams. The subset labeled all diagrams is divided into 11 classes and further segmented into training, test, and validation sets. The class diagrams and activity diagrams subsets are already split and have a CSV format. These subsets contain detailed descriptions of the individual diagrams, including associations and relationships. These diagrams provide a structured approach to data management that facilitates the application of deep learning models or analysis techniques to extract meaningful insights from the data.

B. **Data Preprocessing**

This section presents the preprocessing steps that were applied to the UML images for multi-classification problems in our work. The purpose of preprocessing is to enhance the raw data's quality and make it compatible with analytical or modeling activities. It converts the data into a suitable format for further analysis, leading to more accurate and robust results. The following steps were performed:

- **Grayscale conversion (RGB2Gray):** Data was simplified and computational complexity was reduced by converting each image from RGB to grayscale.

- **Resizing:** The dimensions of all images are resized to (100x100) pixels.
- **Cropping:** Images were cropped to concentrate on the important regions of the diagrams, removing unnecessary borders and whitespace.
- **Data Augmentation:** Various augmentation techniques were applied to enhance the dataset and improve the model's generalization capabilities. This included transformations such as rotation, shifting, shearing, flipping, and scaling.

As far as the class diagram dataset and activity diagram dataset, the retrieved data are already preprocessed. The dataset is then split into distinct subsets, 80% training, 8% testing, and 12% validation. The training set contains 263 images, the testing set contains 38 images, and the validation set contains 74 images in the class diagrams. Similarly, the training set contains 255 images, the testing set contains 20 images, and the validation set contains 32 images in the activity diagrams [11, 12]. This maximizes model performance and obtains the best possible results.

C. **Transfer Learning and CNN Model**

This section presents the methods we applied for the multi-classification and multi-labeling of UML diagrams, utilizing transfer learning with the AlexNet model and the CNN model.

- ***AlexNet Model Of Transfer Learning***

  After splitting the dataset, we used transfer learning models which included the AlexNet architecture. AlexNet, a deep convolutional neural network, has played an important role in advancing fields of deep learning and computer vision [13–15]. The AlexNet architecture consists of sequence data flow starting from the image input of size $227 \times 227 \times 3$ for RGB images, passing through convolutional and max pooling layers with Local Response Normalization (LRN), and finally leading to the final classification output.

  The AlexNet model is configured to classify 11 different classes. Training is conducted over 20 epochs. The batch size is set to 64 to balance memory efficiency and training speed. An instance of the AlexNet model is created with 11 output classes. The AlexNet model is moved to the Configured Computational Device (GPU) to leverage hardware acceleration. The CrossEntropyLoss loss function is utilized, making it suitable for multi-classification. The learning rate used with the SGD optimizer is 0.005, along with a weight of 0.005 and a momentum of 0.9.
- ***CNN Model for UML diagrams multi-classification***

  The architecture of CNN for multi-classification of UML diagrams consists of three convolutional layers, succeeded by max-pooling layers to capture features from the input images. The flattened output is then passed through dense layers to make predictions. The model is trained with data augmentation to improve generalization and achieves high accuracy on both training and test datasets.

  During training, ImageDataGenerator is used for data augmentation to increase the diversity of training examples. The training images are rescaled to a range of [0, 1] and augmented with random shearing, zooming, and horizontal flipping. The standard learning rate is used together with the Adam optimizer. The use of categorical cross-entropy as the loss function is appropriate for handling multi-classification. It trained for 40 epochs, with a batch size of 32.

- ***CNN Model for UML diagrams multi-labeling***
  The CNN model architecture for UML diagrams multi-labeling consists of several layers. It begins with an input layer expecting RGB images of size 400 × 400 × 3, passing through convolutional, ReLU functions and max pooling layers, and finally leading to the final classification output. The number of neurons in the model corresponds to the number of labels in each dataset and uses the sigmoid activation function for multi-labeling. The model undergoes training for 50 epochs with a batch size of 4. During every epoch, the model's performance is assessed on both the training and validation sets. The training loss gradually decreases over the epochs, which means that the model minimizes the error between the predicted and actual labels.

4 Results and Discussions

In our work, we effectively used various tools and technologies. Google Colab is a cloud-based environment utilized for our analysis. We used Python programming language and leveraged essential libraries such as TensorFlow, Keras, and Kaggle. This section presents our study's results and discussion on UML diagrams classification using transfer learning and the CNN model.

A. **Evaluation Metrics**

The performance of CNN and AlexNet models was compared and evaluated using three key metrics.

- **Accuracy**: The accuracy metric calculates the percentage of correctly classified UML diagrams by the model compared to all UML diagrams, computed by (1):

$$\text{Accuracy} = (\text{TP} + \text{TN})/(\text{TP} + \text{FP} + \text{FN} + \text{TN}) \tag{1}$$

  where TP: true positives. TN: true negatives. FP: false positives. FN: false negatives.
- **Recall:** is a crucial metric in classification tasks, especially when focusing on minimizing false negative predictions. The model accurately identifies the proportion of true positive instances. The calculation of Recall is done by (2).

$$\text{Recall} = \text{TP}/(\text{TP} + \text{FP}) \tag{2}$$

- **Precision:** is the ratio between accurately identified positive instances by the model and all instances classified as positive. This metric highlights the precision of positive predictions and proves particularly valuable when the goal is to minimize false positives. It is calculated by (3).

$$\text{Precision} = \text{TP}/(\text{TP} + \text{FN}) \tag{3}$$

B. **Transfer Learning**

Figure 2.a and 2.b show the accuracy and loss of AlexNet for a dataset of 11 classes of UML diagrams. As the number of epochs increases, the training accuracy also increases. When we look at the validation accuracy, we find that the AlexNet model has a good validation accuracy of 82.69%with a low loss rate of 0.3228 at 20 epochs.

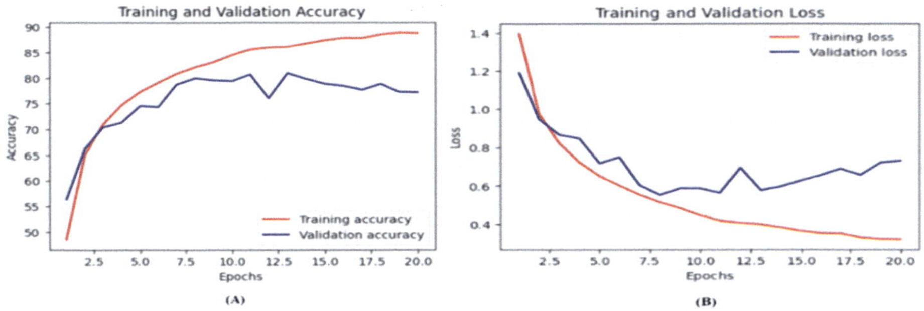

Fig. 2. Alex Net's model performance results under different epochs (A) Training and validation accuracy (B) Training and validation loss.

A. **CNN Model**

The effectiveness of the CNN model was assessed by examining 11 sets of UML diagrams at various training and validation accuracy levels, along with training and validation loss. In Fig. 5, it can be observed that the CNN model's accuracy score improves as the number of epochs increases. Notably, the CNN model demonstrated an impressive accuracy of 97.54% with a corresponding loss of 0.0839.

B. **Models Comparison**

We performed model comparisons of AlexNet and CNN in terms of accuracy, loss, and other training details. As shown in Table 1, CNN outperforms AlexNet in both accuracy and loss metrics. AlexNet achieved an accuracy of 82.69% with a corresponding loss of 0.3228, while CNN achieved a significantly higher accuracy of 97.54% with a lower loss of 0.0839. The training sets indicate that AlexNet is learned for 20 iterations with a batch of 64, while CNN is learned for 40 iterations with a reduced batch of 32. Nevertheless, when observing Figs. 2 and 3 on training performance, we find that the CNN model becomes more pronounced around epoch 10, and the AlexNet model overfits around epoch 12. We see that the CNN training model reaches a training value of 0.77 and a testing value of 0.87 around epoch 20. If the underfitting issue was eliminated, the CNN model's performance could converge to that of the AlexNet model. Similarly, the AlexNet model achieves a training performance of 0.92 around epoch 20, suggesting the potential for exceptional performance under better conditions.

Both models provided predictions for 11 classes specific to UML diagrams. The current results highlight the superior performance of CNN on this classification task, regardless of whether it is a model developed from scratch or one that has already been trained. This is likely due to the CNN's ability to learn more complicated features

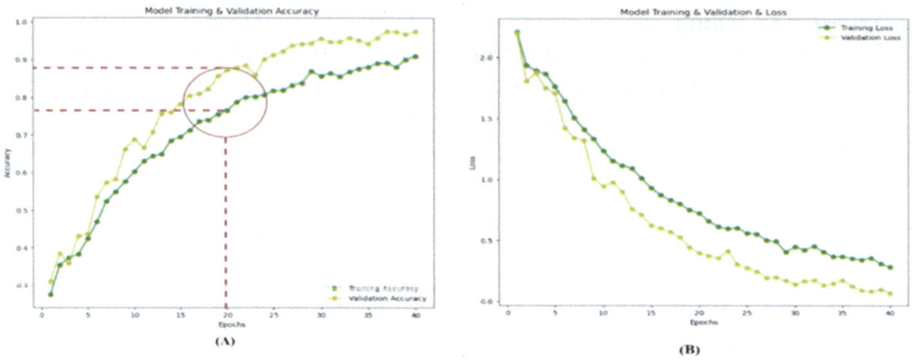

Fig. 3. CNN's performance results under different epochs (A) Accuracy of training and validation (B) Loss of training and validation.

from the data, as well as the refined training process. Based on the present results, CNN would therefore be considered the best model for the classification of UML diagrams.

Table 1. Accuracy and Loss of AlexNet and CNN Models best performance

Model	*Accuracy*	*Loss*	Training Details
AlexNet	82.69%	0.3228	20 epochs - Batch size 64
CNN	97.54%	0.0839	40 epochs - Batch size 32

The comparison also included traditional machine learning models like K-Nearest Neighbours (KNN), Decision Trees (DT), Random Forest (RF), and Support Vector Machine (SVM) with Radial Basis Function (RBF) kernel. Figure 4 and Table 2 illustrate the comparison between traditional machine learning models and CNN models. CNN model shows superior performance compared to other models, achieving an accuracy of 0.85 and a recall of 0.77 after being trained for 40 epochs. This is attributed to its ability to effectively learn from data over a long training period. Even the RF model 1, which achieves an accuracy of 0.82, has fairly low precision and recall (0.440,.30). The model's performance results using CNN and AlexNet for 20 epochs of training are all identical (accuracy is 0.82, precision is 0.77 and recall is 0.69), because of their good learning skills under the auspiciously insufficient coaching structure. On the other hand, if we use less complicated models such as KNN or DT, they also bring lower prediction accuracy, which is not good enough for this task. Support Vector Machine (SVM) with RBF kernel also exhibit mediocre performance (accuracy: 42%, precision: 44% and recall 30%), better than KNN and DT but worse than CNN/RF. These results illustrate the importance of both the model's complexity and how long to train it in order to achieve higher performance.

C. **Multi-labeling Results and discussion**

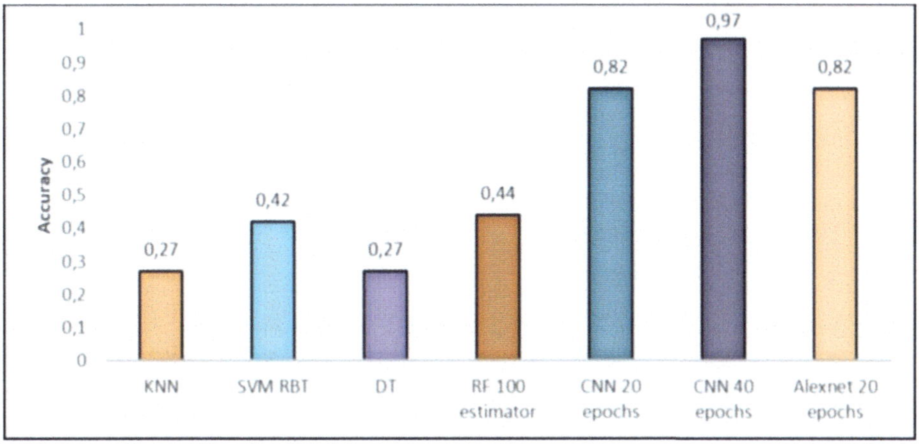

Fig. 4. Comparison of the performance of traditional machine learning models and CNN models.

Table 2. Precision and recall performance of traditional models and CNN models

Metric	Precision	Recall
KNN	0.21	0.21
SVM(RBF)	0.44	0.3
DT	0.21	0.21
RF	0.44	0.3
CNN 20 epochs	0.77	0.69
CNN 40 epochs	**0.85**	**0.77**
AlexNet 20 epochs	0.77	0.69

Table 3 shows the result of CNN performance on multilabel classification tasks for activity or class Diagrams. We observe that the pre-trained CNN has an accuracy of 0.65, a loss of 0.37 for activity diagrams, and an accuracy of 0.55, while a higher loss is observed from the class diagram indicating early-stage underfitting. After increasing the training duration to 50 epochs the accuracy improves to 0.75. The loss decreases to 0.34 for activity diagrams. Similarly, the accuracy increases to 0.61 with a loss of 0.47 for class diagrams indicating better feature learning. Upon reaching 100 epochs the model further boosts its accuracy to 0.81 for activity diagrams and 0.71 for class diagrams, corresponding to a loss of 0.33 and 0.43, respectively. This shows that prolonged training significantly enhances the model performance by capturing the intricacies of multi-label classification tasks even though the accuracy, of class diagrams still leaves much to be desired compared to activity diagrams.

Table 3. Performance of CNN model in class diagram and activity diagram multi-labeling problem.

Epochs	Diagram type	Precision	Loss
20	Activity Diagram	0.65	0.37
	Class Diagram	0.55	0.68
50	Activity Diagram	0.75	0.34
	Class Diagram	0.61	0.47
100	Activity Diagram	0.81	0.33
	Class Diagram	0.71	0.43

5 Conclusion

This study investigated Convolutional Neural Networks (CNNs) and transfer learning in UML diagram classification, with a focus on mitigating the challenges associated with multi-labeling and multi-classification tasks. The methodology strategically included careful data pre-processing techniques such as image normalization and augmentation, employed transfer learning mechanisms to leverage existing knowledge encoded in pre-trained models, classified UML and non-UML diagrams, and developed robust strategies to effectively generalize the suggested model to new data. The CNN-based models achieved exceptional accuracy, outperforming the traditional machine learning models, especially in the context classification scenarios. In future works, we aim to enhance model explainability to render their decision-making processes more transparent and interpretable. Besides we extend the application of CNN-based classification approaches to renewable energetic fields within the realm of software engineering.

References

1. Alhir, S.S.: Understanding the unified modeling language (UML). Methods Tools (1999)
2. Sharma, N., Sharma, R., Jindal, N.: Machine learning and deep learning applications-a vision. Glob. Trans. Proc. **2**(1), 24–28 (2021)
3. Karasneh, B., Chaudron, M.R.: Extracting UML models from images. In: 2013 5th international conference on computer science and information technology. IEEE (2013)
4. Tavares, J.F., Costa, Y.M., Colanzi, T.E.: Classification of UML diagrams to support software engineering education. In: 2021 36th IEEE/ACM international conference on automated software engineering workshops (ASEW). IEEE (2021)
5. Shcherban, S., Liang, P.L.Z., Yang, C.: Multiclass classification of four types of UML diagrams from images using deep learning. SEKE (2021)
6. Gosala, B., Chowdhuri, S.R., Singh, J., Gupta, M., Mishra, A.: Automatic classification of UML class diagrams using deep learning technique: convolutional neural network. Appl. Sci. **11**(9), 4267 (2021)
7. Wang, L., Song, T., Song, H.N., Zhang, S.: Research on design pattern detection method based on UML model with extended image information and deep learning. Appl. Sci. **12**(17), 8718 (2022)

8. Rigou, Y., Khriss, I.: A deep learning approach to UML class diagrams discovery from textual specifications of software systems. In: Proceedings of SAI intelligent systems conference. Springer International Publishing, Cham (2022)
9. Koenig, A., Benjamin, A., Emmanuel, R.: NEURAL-UML: intelligent recognition system of structural elements in UML class diagram. In: 2023 ACM/IEEE international conference on model driven engineering languages and systems companion (MODELS-C). IEEE (2023)
10. Bnouni Rhim, N., Cheballah, S., Ben Mabrouk, M.: Cross synergetic mobilenet-vgg16 for uml multiclass diagrams classification. In: International conference on innovations in bio-inspired computing and applications, pp. 24–30. Springer Nature Switzerland, Cham (2022)
11. Activity Diagrams Dataset: Open Source Dataset. Robflow Universe (2022)
12. UML-Digital Dataset: Open Source Dataset. Roboflow Universe (2023)
13. Krizhevsky, A., Sutskever, I., Hinton, G.E.: Imagenet classification with deep convolutional neural networks. Adv. Neural Inf. Process. Syst. **25** (2012)
14. Alti, A., Djoudi, M., Smeda, A.: An automatic transformation from COSA software architecture to CORBA platform. In: Proceedings of the 8th international conference on New technologies in distributed systems (2008)
15. Kamel, M., Alti, A., Roose, P.: Dynamic semantic-based green bio-inspired approach for optimizing energy and cloud services qualities. Trans. Emerg. Telecommun. Technol. **29**(5), e3305 (2018)

Determination of the Optimal Volume and Distribution of Learning Data for Solar Irradiance Forecasting Applications

Houssem Eddine Louchene[1], Hassen Bouzgou[1](✉), and Christian Gueymard[2](✉)

[1] Department of Industrial Engineering, Laboratory of Automation and Manufacturing, University of Batna 2 (Mostefa Ben Boulaïd), Batna 05000, Algeria
h.louchene@univ-batna2.dz, h.bouzgou@unv-batna2.dz

[2] Solar Consulting Services, Colebrook, NH 03576, USA
Chris@SolarConsultingServices.com

Abstract. Given the significant importance of renewable or alternative energies today, extensive research is being conducted to enhance the efficiency and reduce the costs of utilizing these energy sources. Among these studies, solar energy forecasting plays a crucial role in achieving these objectives. Accurate forecasting can optimize energy yield, improve grid management, and facilitate the integration of solar power into existing energy systems, ultimately contributing to more reliable and cost-effective renewable energy solutions. This contribution investigates how data volume influences forecasting accuracy. In particular, the impacts on forecasting accuracy of varying forecast horizons and optimal data splitting for training and testing phases are examined. Additionally, the effects on forecasting Global Horizontal Irradiance (GHI) of clustering the data into 3, 5, or 7 groups using the K-means algorithm are investigated. Five different predictive models are employed— multi layer perceptron (MLP), support vector regression (SVR), random forest (RF), convolutional neural network (CNN), and long short-term memory (LSTM)— alongside the newly proposed kResCLSTM hybrid method. Using GHI observations at an arid site in southern Algeria, it is found that a 10-year time series is optimal, along with a 60%-40% split in it for the training vs. testing periods.

Keywords: Solar irradiance · CNN · clustering · residual network · LSTM · forecasting

1 Introduction

Global energy demand has increased significantly in recent years, while concerns for the environmental and climate impacts of carbon-based fuels have increased. Consequently, various alternative energy sources now need to be intergated into power grids for large-scale energy production. Among these alternatives, solar energy stands out as one of the most reliable, evenly distributed, and sustainable options. As an inexhaustible renewable energy source, solar energy offers a consistent and abundant supply of power, making it

© The Author(s), under exclusive license to Springer Nature Switzerland AG 2025
M. Hatti (Ed.): IC-AIRES 2024, LNNS 1238, pp. 584–593, 2025.
https://doi.org/10.1007/978-3-031-80301-7_63

an essential component to the transition toward a more sustainable and stronger energy infrastructure [1]. The integration of solar energy also enhances energy security by varying the energy mix. Consequently, the adoption and advancement of solar technology are critical to meeting the growing global energy demand and achieving long-term sustainability goals. However, the effectiveness of these systems is highly dependent on the meteorological conditions (most importantly the cloud regime) that impact the local solar resource, which can fluctuate significantly over short periods. These rapid fluctuations pose challenges for the smooth integration of solar power into the electricity grid system, which can only be resolved by appropriate forecasts. To obtain such forecasts, a variety of methods can be used, including machine-learning and feature-selection techniques [2].

Solar energy forecasting commonly relies on mesoscale numerical weather prediction (NWP) models, which are constructed based on global forecasts of many atmospheric variables, most importantly those related to cloudiness. These models have undergone a gradual evolution spanning several decades. Their complicated nature poses a challenge to advancing forecasting capabilities, and constitutes a formidable barrier to further improvements. Overcoming these challenges requires innovative approaches that strike a balance between computational efficiency and predictive accuracy [4]. Conversely, models employing artificial intelligence techniques such as machine learning (ML) or deep learning (DL) techniques are gaining significant traction and frequently demonstrate high levels of accuracy. This growing popularity, at least for short-term forecasts, can be attributed to their ability to leverage large datasets, allowing these models to learn and make predictions with remarkable precision. As a result, these models are increasingly being adopted across various industries.

Solar power production based on photovoltaic systems essentially depends on the local resource in global horizontal irradiance (GHI). Many GHI forecasting methods using ML and DL models exist and have been reviewed [5]. In [1], the authors proposed a deep-learning-based hybrid forecasting framework that integrates the sine-cosine algorithm (SCA), bidirectional long short-term memory (BiLSTM) networks, and complete ensemble empirical mode decomposition with adaptive noise (CEEMDAN). In [6], an extreme-learning machine (ELM) was presented as an alternative to traditional neural networks (NNs). In [2], the authors introduced a machine-learning model designed for day-ahead forecasting, utilizing support vector regression to establish the input-output mapping. They defined the input feature vector through an appropriate feature extraction technique. Meanwhile, in [7], the authors proposed a novel convolutional neural network (CNN) algorithm for short-term forecasting.

In the present approach, the main goal is to explore the impact of dataset size on the precision of predictive modeling. This topic has apparently not been thoroughly addressed so far, despite the significance of data volume in shaping model outcomes. Here, various experiments are conducted across various forecasting horizons to determine an optimal size of the database necessary for efficient learning. The subsequent sections of the paper are organized as follows: Sect. 2 presents an in-depth introduction to the hybrid model by detailing the methodology employed. Section 3 elaborates on the data sources and describes the experimental setup, providing insights into the rationale behind these choices. Section 4 discusses the obtained results, analyzing the findings in the context of the research goal. Finally, Sect. 5 offers concluding remarks, summarizing the key insights and suggesting directions for future research.

2 Methodolgy

2.1 Hybrid kResCLSTM Model

In this study, a hybrid deep-learning model (Fig. 1) is proposed, integrating several ML components: K-means clustering, CNN, residual blocks, and LSTM models. This model is meticulously designed to analyze a training set of data and enhance both the efficiency and accuracy of the forecasting process. Initially, the training set is clustered using the K-means algorithm to group similar data points together and discover the underlying structure and patterns in the data. This helps reduce the computational complexity by focusing only on relevant data points during the training phase. Subsequently, the input data is processed through the CNN, thus extracting the pertinent features and reducing the dimensionality of the data. To address the vanishing gradient problem typical in deep neural networks (DNNs), the residual block architecture is employed here. This architecture facilitates easier training of DNNs by providing shortcuts that allow the gradient to flow more effectively through the network. Finally, the LSTM model is employed to analyze the sequential data time series. Its strength lies in its ability to retain information from previous states, making it particularly well-suited for time-series analysis. The LSTM can thus evaluate the current state in relation to past states, providing a more comprehensive analysis of the data.

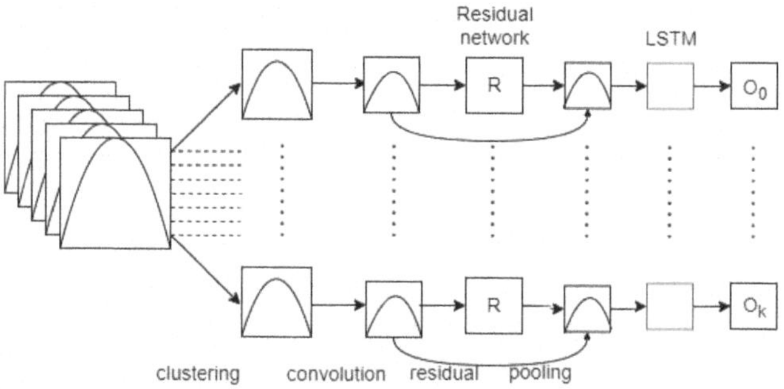

Fig. 1. The proposed KRESCLSTM hybrid model.

2.2 K-Means Clustering

The implementation of the K-means algorithm follows the steps below:

- Determine the number of clusters (K): Decide on the number of clusters to be formed from the data.
- Initialize centroids: Randomly select K data points to serve as the initial centroids for each of the K clusters.

- Assign data points to clusters: For each data point, calculate its distance to all centroids using a distance metric, typically the squared Euclidean distance, and then assign the data point to the cluster whose centroid is the closest.
- Update centroids: After all data points have been assigned to the appropriate cluster, recalculate the centroids of the new clusters; the centroid is usually the mean of all the data points assigned to that cluster.

These steps are repeated until either the centroids no longer change significantly or a maximum number of iterations is reached, indicating that the algorithm has converged to a solution.

2.3 One Dimensional Convolutional Neural Network (1DCNN) Model

CNNs constitute a groundbreaking class of NN architecture that are specifically designed to process data that features a grid-like topology, such as images. A subset of CNN, known as one-dimensional CNN (1DCNN), is particularly adept at handling data that exhibits temporal or sequential characteristics. This includes time-series data, audio waveforms, and textual information, where the *order* of elements is crucial to understanding the underlying patterns [8]. 1DCNNs operate by applying convolutional filters along one dimension of the input data, effectively capturing local dependencies and temporal dynamics. This makes them an excellent choice for tasks that require the analysis of sequences, as they can efficiently learn from the temporal structure of the data without the need for manual feature extraction.

Subsequently, the generated feature maps often undergo a down sampling procedure known as pooling. This operation serves to diminish the spatial dimensions of the data while simultaneously highlighting the most significant features. Following the convolutional and pooling stages, the extracted features are directed through one or several fully connected, or "dense", layers. These layers provide a weighted summation of the input features, which is then channeled through an activation function. This iterative process continues until the data reaches the output layer, where the network's final output is produced. The strength of 1DCNNs lies in their ability to automatically learn spatial hierarchies of features directly from the data, making them a powerful tool for a wide array of applications that involve sequential data [8].

2.4 Bidirectional Long Short Term Memory (BiLSTM) Model

The BiLSTM model was originally introduced to address the limitations of standard LSTM networks, which process inputs in a strict chronological order. This sequential processing can overlook the relevance of future inputs when considering the impact of past inputs [9]. To overcome this, BiLSTM incorporates two LSTM layers: a forward LSTM that processes the input sequence from the beginning to the end, and a backward LSTM that processes the same sequence in reverse order, from the end to the beginning [10]. As a result, BiLSTM captures information from both past and future contexts, enhancing the model's ability to understand and utilize the full temporal context of the data. This makes BiLSTM particularly useful for tasks where the context from both directions is crucial for accurate predictions or understanding of the data.

2.5 Residual Network

The core concept behind Residual Networks (ResNets) is the incorporation of skip connections, also known as shortcut connections, to create residual blocks. These blocks are defined by:

$$y_t = I(x_t) + f(x_t) \tag{1}$$

$$x_{t+1} = f(y_t) \tag{2}$$

The skip connections allow the network to bypass certain layers, directly connecting earlier layers to later ones. This architectural design significantly aids in the training of very deep Convolutional Neural Networks (CNNs) by mitigating the vanishing gradient problem, which is a common issue that arises when training deep networks.

To evaluate the accuracy of the models used in this paper, mean absolute percentage deviation (MAPD), root mean square difference (RMSD), forecasting skill (FS), and coefficient of determination (R^2) are commonly used metrics. MAPD and RMSD can help assess the accuracy of the models by measuring the difference between predicted and actual solar radiation values. The coefficient of correlation can provide insight into the strength and direction of the relationship between forecasted and measured values. The forecasting skill score is a metric that assesses the effectiveness of a forecasting model in comparison to a reference or benchmark model. This metric quantifies the increase in accuracy of the forecasting model relative to the benchmark model. The calculation of the forecasting skill score can involve various statistical measures, such as mean absolute error (MAE), mean square error (MSE), correlation coefficient, and others. A higher forecasting skill score indicates better performance of the forecasting model as compared to the reference or benchmark model. These metrics are crucial for ensuring the effectiveness of the models for GHI forecasting and improving the solar energy contribution with the whole energy system.

3 Results and Discussion

3.1 Data

The database used here is sourced from the Baseline Surface Region Network (BSRN) Tamanrasset radiometric station in southern Algeria, representative of an arid climate with very high solar energy potential. The database encompasses 11 years of GHI records, compiled at hourly intervals for the present study. To guarantee significant results, the time series has first been carefully quality-controlled for quality and consistency, following the same method as in [11]. This includes testing the solar irradiance components for their physical limits and consistency, removing outliers and nighttime values, handling missing values, and normalizing the data to improve their representation for the model. Additionally, the 11-year dataset is split into two complementary parts of three possible relative dimension: a training (TR; 80, 70, 60% of the whole data) and a testing subset (TST; 20, 30, 40% of the whole data), with the training set used to train the model and the testing set used to evaluate its performance. This variable partition is designed to determine if it is advantageous to use a longer training set than what is normally used in the literature (2/3 for training and 1/3 for testing), or conversely, if a shorter training set is sufficient. The latter case would have considerable practical significance because many stations have only collected a few years of GHI data. In addition, different forecast horizons are considered here (1–24 h ahead), as well as different lengths of the database (1–11 years). The accuracy of the forecasting process is thus analyzed for a wide range of sub-daily horizons and training/testing durations.

3.2 Forecast Performance

The forecast performance of five models including five independent models MLP, LSTM, CNN, and SVR for comparison is analyzed first, using three performance metrics: Root Mean Squared Deviation (RMSD), R-squared (R^2), and Fraction Skill Score (FS). The results of the simple average of all models are presented in Table 1.

Table 1. Performance evaluation of the machine-learning models on databases of varying lengths. The second row indicates the number of years used and the relative fraction of the training set; e.g. 11–80 means a 11-year dataset, with 80% of it used for TR. The best results in each category appear in boldface. The gray shade indicates the best results overall.

Split		RMSD	R2	FS
Split: 80% TR, 20% TST	11-80	19,87	0,90	0,83
	10-80	18,37	0,92	0,85
	9-80	18,13	0,92	0,85
	8-80	17,67	**0,93**	0,85
	7-80	19,14	0,91	0,84
	6-80	**17,65**	0,92	**0,85**
	5-80	17,99	0,92	0,85
	4-80	20,30	0,90	0,83
	3-80	24,20	0,86	0,80
	2-80	26,64	0,82	0,78
	1-80	31,42	0,64	0,73
Split: 70% TR, 30% TST	11-70	**17,49**	**0,93**	**0,85**
	10-70	18,22	0,92	0,85
	9-70	17,74	0,93	0,85
	8-70	18,54	0,92	0,84
	7-70	19,00	0,92	0,84
	6-70	19,87	0,90	0,83
	5-70	19,48	0,91	0,84
	4-70	20,77	0,89	0,83
	3-70	22,42	0,87	0,81
	2-70	27,12	0,81	0,77
	1-70	34,04	0,67	0,71
Split: 60% TR, 40% TST	11-60	17,47	0,93	0,85
	10-60	**16,35**	**0,94**	**0,87**
	9-60	19,08	0,91	0,84
	8-60	17,21	0,93	0,86
	7-60	19,00	0,91	0,84
	6-60	19,30	0,91	0,84
	5-60	19,95	0,90	0,83
	4-60	22,13	0,88	0,81
	3-60	22,74	0,86	0,81
	2-60	27,64	0,78	0,77
	1-60	35,67	0,67	0,70

The results in Table 1 show that, as the database spans a period of 1–11 years, an improvement can be observed in the performance of all models. This improvement is evident through the decrease in RMSD along with an increase in R^2 and FS scores. An obvious explanation is that a longer time series provides any model with more data to learn from, resulting in improved predictive abilities. From Table 1, the best results are obtained for a 10-year time series and a 60/40 partition (60% for training and 40% for testing). At sites where only a very short database exists (such as one or two years' worth of data), the results also indicate that the optimal splitting is now 80% for training and 20% for testing. Once the issues of data size and partitioning is resolved, we evaluate the performance of the developed model across various forecasting horizons, ranging from 1 h ahead to 24 h ahead.

Table 2 indicates that RMSD increases for longer horizons, while the R^2 and FS metrics decrease. The shortest forecast horizon is thus most accurate, as could be expected. Interestingly, however, the strong decrease in performance that occurs between 1 h and 3 h ahead is followed by a long plateau over which there is not much loss of performance for increasing horizons between 4 and 12 h ahead.

Table 2. Effect of the forecast horizon on the accuracy of the MLP model, using an optimally split 10-year dataset. The best results appear in boldface with a gray shade.

Dataset: 10-year time series with 60% TR, 40% TST	1	2	3	4	5	6	7	8	9	10	11	12	13	14	15	16	17	18	19	20	21	22	23	24
RMSD	**16,38**	20,69	23,34	24,76	24,42	24,79	24,89	25,07	25,03	25,35	25,73	25,88	27,06	27,25	30,27	29,14	29,92	30,11	29,83	29,56	30,09	30,06	30,55	32,50
R2	**0,94**	0,90	0,84	0,86	0,86	0,85	0,85	0,85	0,85	0,85	0,84	0,84	0,73	0,82	0,78	0,80	0,79	0,79	0,79	0,79	0,79	0,79	0,78	0,75
FS	**0,86**	0,83	0,80	0,79	0,80	0,79	0,79	0,79	0,79	0,79	0,78	0,78	0,77	0,77	0,75	0,76	0,75	0,75	0,75	0,75	0,75	0,75	0,74	0,73

Still using the same optimal time series, the performance of different models is also evaluated for various numbers of clusters, specifically 3, 5, and 7 clusters. Tables 3, 4 and 5 show the results obtained in the case of 3, 5, and 7 clusters, respectively.

Table 3. Performance of different models for 3 clusters. The best results appear in boldface.

# Clusters		1	2	3
MLP	RMSD	26,46	**13,59**	11,75
	R2	0,89	0,9	0,95
	FS	0,58	**0,41**	0,66
SVR	RMSD	27,91	14,34	15,87
	R2	0,88	0,91	0,92
	FS	0,56	0,37	0,50
1D-CNN	RMSD	42,95	22,94	22,66
	R2	0,70	0,77	0,84
	FS	0,32	0,00	0,29
biLSTM	RMSD	28,47	48,27	56,21
	R2	0,87	0,8	0,7
	FS	0,55	0,11	0,5
kResCLSTM	RMSD	**26,41**	14,81	**11,26**
	R2	**0,89**	**0,91**	**0,96**
	FS	**0,58**	0,35	**0,67**

Table 4. Performance of different models for 5 clusters. The best results appear in boldface.

# Clusters		1	2	3	4	5
MLP	RMSD	11,22	**26,54**	15,21	**16,69**	18,17
	R2	0,83	**0,87**	0,92	**0,91**	0,96
	FS	0,89	**0,79**	0,87	**0,85**	0,86
SVR	RMSD	**10,86**	27,44	15,07	17,56	18,71
	R2	**0,84**	0,86	0,93	0,90	0,96
	FS	**0,90**	0,78	0,87	0,85	0,86
1D-CNN	RMSD	18,30	39,90	25,64	21,87	33,13
	R2	0,54	0,70	0,79	0,84	0,86
	FS	0,82	0,68	0,78	0,81	0,75
biLSTM	RMSD	46,84	27,36	25,79	55,11	37,43
	R2	0,59	0,50	0,78	0,60	0,83
	FS	0,62	0,74	0,77	0,52	0,72
kResCLSTM	RMSD	11,21	29,01	**14,56**	17,42	**17,40**
	R2	0,83	0,84	**0,93**	0,90	**0,96**
	FS	0,89	0,77	**0,87**	0,85	**0,87**

Table 5. Performances of different models for 7 clusters. The best results appear in boldface.

# Clusters		1	2	3	4	5	6	7
MLP	RMSD	59,62	**10,42**	10,84	**11,91**	**11,23**	**10,19**	**10,54**
	R2	0,42	0,67	0,29	0,78	0,75	**0,94**	**0,76**
	FS	0,20	0,00	0,30	0,34	0,15	**0,68**	0,30
SVR	RMSD	**21,05**	35,08	14,43	27,01	61,87	14,04	22,30
	R2	0,50	0,25	0,79	0,34	0,31	0,90	0,07
	FS	**1,44**	0,49	0,49	**3,24**	0,42	0,56	0,48
1D-CNN	RMSD	45,28	20,37	19,81	36,41	36,69	22,87	18,06
	R2	0,75	0,75	0,60	0,84	0,76	0,74	0,30
	FS	0,48	0,13	0,30	0,44	0,16	0,29	0,20
biLSTM	RMSD	41,64	47,89	39,90	73,77	43,52	48,53	55,04
	R2	0,59	0,46	0,62	0,29	0,55	0,44	0,28
	FS	0,65	**0,60**	**0,67**	0,38	**0,63**	0,59	**0,54**
kResCLSTM	RMSD	34,55	12,73	**9,68**	20,45	24,88	11,63	12,64
	R2	**0,86**	**0,90**	**0,91**	**0,95**	**0,89**	0,93	0,66
	FS	0,60	0,46	0,66	0,69	0,43	0,64	0,16

When considering 3 clusters (Table 3), the best results are obtained for the hybrid kResCLSTM model with RMSD = 11.26, $R^2 = 0.96$ and FS = 0.67 for cluster #3. Cluster #1 is also best described by kResCLSTM, but MLP is somewhat better for cluster #2. In all other cases, kResCLSTM and MLP generally perform best. Although clustering can improve things by combining similar data configurations, increasing the number of clusters might be counter effective because the data volume decreases within each cluster.

4 Conclusion

This study has investigated the impact of data volume, data splitting, data clustering, and forecasting horizon on the accuracy of solar irradiance forecasting. The effects of these factors on the forecasting accuracy of various data-driven models were evaluated. The K-means method was also employed to cluster the data into 3, 5, and 7 groups. Global horizontal irradiance (GHI) was predicted at horizons from 1 to 24 h ahead using five models: MLP, SVR, CNN, LSTM, and the novel kResCLSTM. The latter is a hybrid model combining K-means clustering, CNN, residual blocks, and LSTM models.

The present findings indicate that the accuracy of solar irradiance forecasting is largely influenced by the data volume. For the test station used here, located in an arid climate with relatively low cloudiness, a time series of 10 years was found optimal. It can thus be expected that even longer time series would be needed at sites experiencing very variable and frequent cloudiness. The optimal split of that time series was found to be 60% for training and 40% for testing, which differs somewhat from the usual assumption found in the literature. In the frequent case when only short time series of observational data are available, with a non-optimal duration of one or two years, it is then best to do a 80/60 split, so that the training is done on more data. Using a time series of optimal duration and training/testing ratio, a rapid decrease in forecasting performance is found to occur when the forecast horizon increases from 1 to 3 h, but then the performance plateaus, up to 12-h ahead.

Overall, it is important to consider the optimal characteristics of the necessary databases when designing or validating forecasting models. Additionally, this study highlights the importance of selecting the most suitable forecasting method for different data clusters. More studies are needed to determine whether the optimal database characteristics are also dependent on the cloud regime of the location under scrutiny.

References

1. Ajith, M., Martínez-Ramón, M.: Deep learning algorithms for very short term solar irradiance forecasting: a survey. Renew. Sustain. Energy Rev. **182**, 113362 (2023). https://doi.org/10.1016/j.rser.2023.113362
2. Cocchi, G., Galli, L., Galvan, G., Sciandrone, M., Cantù, M., Tomaselli, G.: Machine learning methods for short-term bid forecasting in the renewable energy market: a case study in Italy. Wind Energy **21**(5), 357–371 (2018). https://doi.org/10.1002/we.2166
3. Elsinga, B., van Sark, W.G.J.H.M.: Short-term peer-to-peer solar forecasting in a network of photovoltaic systems. Appl. Energy **206**(September), 1464–1483 (2017). https://doi.org/10.1016/j.apenergy.2017.09.115
4. Srivastava, S., Lessmann, S.: A comparative study of LSTM neural networks in forecasting day-ahead global horizontal irradiance with satellite data. Sol. Energy **162**(January), 232–247 (2018). https://doi.org/10.1016/j.solener.2018.01.005
5. Peng, T., Zhang, C., Zhou, J., Nazir, M.S.: An integrated framework of Bi-directional long-short term memory (BiLSTM) based on sine cosine algorithm for hourly solar radiation forecasting. Energy **221**, 119887 (2021). https://doi.org/10.1016/j.en-ergy.2021.119887
6. Wan, C., Xu, Z., Pinson, P., Dong, Z.Y., Wong, K.P.: Probabilistic forecasting of wind power generation using extreme learning machine. IEEE Trans. Power Syst. **29**(3), 1033–1044 (2014). https://doi.org/10.1109/TPWRS.2013.2287871
7. Huang, C.J., Kuo, P.H.: A short-term wind speed forecasting model by using artificial neural networks with stochastic optimization for renewable energy systems. Energies **11**(10), 2777 (2018). https://doi.org/10.3390/en11102777
8. LeCun, Y.: Generalization and Network Design Strategies.pdf. (1989)
9. Bou-Rabee, M.A., Naz, M.Y., Albalaa, I.E., Sulaiman, S.A.: BiLSTM network-based approach for solar irradiance forecasting in continental climate zones. Energies **15**(6), 2226 (2022). https://doi.org/10.3390/en15062226
10. Singla, P., Duhan, M., Saroha, S.: An ensemble method to forecast 24-h ahead solar irradiance using wavelet decomposition and BiLSTM deep learning network. Earth Sci. Inform. **15**(1), 291–306 (2022). https://doi.org/10.1007/s12145-021-00723-1
11. Ruiz-Arias, J.A., Gueymard, C.A.: CAELUS: Classification of sky conditions from 1-min time series of global solar irradiance using variability indices and dynamic thresholds. Sol. Energy **263**, 111895 (2023). https://doi.org/10.1016/j.solener.2023.111895

Empowering MPPT Efficiency in Partial Shading with Machine Learning Driven Metaheuristic Strategies: Recent Developments

F. Belhachat(✉), C. Larbes, and R. Bennia

Laboratory of Communication and Photovoltaic Conversion Devices, National Polytechnic School, Algiers, Algeria
{faiza.belhachat,cherif.larbes,rachid.bennia}@g.enp.edu.dz

Abstract. The use of MPPT control technology has proven to be the most effective way to improve the performance of photovoltaic systems (PVS), especially in partial shading conditions (PS-C). Therefore, accurately determining the "maximum power point (MPP)" is crucial in PVS design. Thus, many MPPT methods have been proposed in the literature, ranging from simple to complex algorithms. However, choosing the best strategy for PVS implementation can be challenging due to the wide range of choices. Recently, metaheuristic optimization algorithms, machine-learning models, and their hybridization have attracted much attention from researchers due to their multiple advantages in solving optimization challenges. In view of these advantages, this paper focuses on the examination of the recent advancements in hybrid intelligent MPPT techniques operating under PS-C published over the last five years. The paper evaluates and compares all selected and classified categories based on diverse criteria such as computational complexity level, percentage of efficiency, convergence speed, converter, application and configuration type, and validation, discussing the strengths and weaknesses of each approach and suggesting directions for future research. Results from evaluation and comparison highlight the promise of hybrid intelligent solutions in improving MPPT technique performance.

Keywords: MPPT · metaheuristics · machine learning · optimization · hybrid

1 Introduction

Solar photovoltaic (PV) energy is gaining popularity as a clean, sustainable source of electricity. Environmental conditions such as partial shading (PS) have a significant impact on the efficiency of photovoltaic systems (PVS). Partial shading (PS) occurs when parts of a photovoltaic array (PVA) do not receive sunlight, resulting in uneven illumination of the solar panel. This phenomenon produces multiple local maxima (LM) on the power-voltage curve, making it difficult for typical maximum power point tracking (MPPT) methods to accurately identify and monitor the global maximum power point. Efficient MPPT under partial shading conditions (PS-C) is essential to improve the energy yield and operating efficiency of PVS. In recent years, research efforts have

© The Author(s), under exclusive license to Springer Nature Switzerland AG 2025
M. Hatti (Ed.): IC-AIRES 2024, LNNS 1238, pp. 594–612, 2025.
https://doi.org/10.1007/978-3-031-80301-7_64

focused on developing more sophisticated and advanced MPPT algorithms that can successfully minimize the negative effects of PS. These efforts have led to the development of novel MPPT strategies that utilize various algorithms, control methods, and system architectures to improve the performance and resilience of MPPT in challenging environments.

Metaheuristic algorithms are optimization methods inspired by natural or biological phenomena. They are characterized by their ability to efficiently traverse complex search spaces and find near-optimal solutions. Metaheuristic MPPT techniques based on the concepts of adaptation, exploration, and exploitation offer suitable solutions to the problems caused by PS in PVS. Additionally, machine-learning (ML) models can adapt and learn from new information over time. This flexibility is especially important in dynamic scenarios where shading patterns change frequently. ML models have the capacity to predict the optimal operating point more accurately than conventional approaches, enhancing the energy production of PV panels even in PS-C. In recent years, there has been an increase in interest in employing machine-learning (ML) for developing effective and robust metaheuristics techniques. Some of these metaheuristic algorithms driven by machine learning have produced outstanding results. To further improve system performance, these algorithms are hybridized to overcome the problems associated with single intelligent algorithms. Therefore, hybrid intelligent algorithms (HIAs) integrate multiple computational intelligence techniques such as machine learning (ML), fuzzy logic (FL), and metaheuristic optimization algorithms (MeHeOA). These hybrid methods leverage the strengths of each component approach to improve the effectiveness, robustness and adaptability of MPPT techniques. This review article comprehensively explores and categorizes the recent advances in hybrid intelligent MPPT methods into four main categories, which are described in detail in this article. Furthermore, this article evaluates and compares these approaches based on a number of criteria, including complexity, efficiency, convergence speed, applications, converter configuration, validation, and the characteristics, advantages, and disadvantages of each technique along with prospects for future research. The main contributions of this document are:

- Exploring the recent developments in hybrid intelligent MPPT techniques published in the last five with a special focus on the improvements achieved by the latest techniques that have not been reviewed previously.
- The selected hybrid intelligent MPPT methods from the literature are classified and evaluated.
- The selected methods are evaluated based on various criteria, including level of complexity, efficiency, convergence speed, converter configuration and implementation application, and validation.
- Each method's key benefits and drawbacks are presented, while suggestions and recommendations for future research are provided.
- The results of the evaluation and comparison highlight the potential of hybrid intelligent solutions in improving the performance of MPPT techniques, especially through meta-heuristic machine learning (MET-ML) and fuzzy logic machine learning (ML-FL) approaches.

The paper is structured as follows: Sect. 2 describes PS effects on PV modules. Section 3 proposes classification for the hybrid intelligent MPPT techniques for PVS.

Section 4 provides an exhaustive review of these techniques. Section 5 compares and evaluates the different MPPT techniques based on several criteria while illustrating features as well as challenges and potential prospects. Finally, Sect. 6 provides a conclusion.

2 Partial Shading Effect on Performance of PV Array

The inevitable occurrence of shading has detrimental effects on energy savings and solar panel performance [1]. It may be caused by dust, falling tree leaves, bird droppings, or clouds. Furthermore, solar panels may receive shadows from surrounding obstacles, including light fixtures, chimneys, and building structures. The amount of energy produced by solar panels can be greatly impacted by PS. Since PV cells are connected in series, the panel's total output is lowered if one of the cells is shaded and provides less electricity. Solar panels with bypass diodes installed can help lessen this problem. Even in cases where certain cells in a panel are shaded, these diodes enable the panel to continue producing power. In PV modules, bypass diodes are often employed to reduce hot spot damage. However, they can also change the IV and PV characteristic curves, resulting in more complex characteristics with multiple peaks. Figure 1 illustrates the simulated P-V characteristic for the 4S configuration for different scenarios.

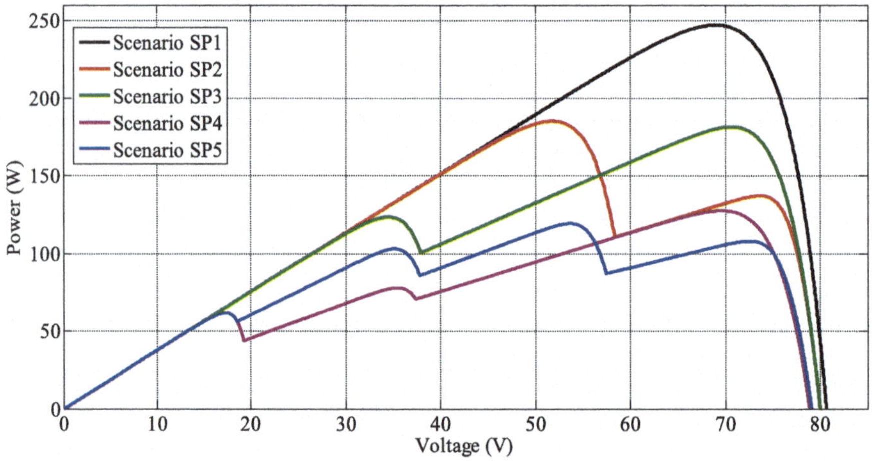

Fig. 1. Simulated P–V characteristic for 4S configuration for different scenarios.

3 Classification of MPPT Techniques for a Partially Shaded PV System

Classifying the MPPT methods operating under PS-C offers a systematic framework and organized structure for identifying and comprehending numerous methodologies utilized for optimizing PV system performance under PS-C. These approaches can be categorized based on a number of criteria, including mathematical principles or algorithms nature, etc. The classification aids in a better understanding of global MPPT approaches and their applications in improving energy harvesting efficiency in PV systems subjected to partial shadowing.

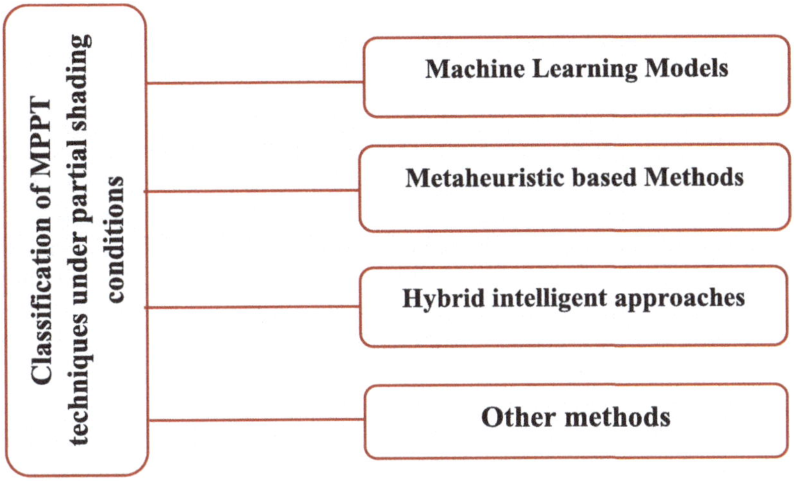

Fig. 2. Proposed classification for MPPT techniques.

Different classifications for MPPT techniques exist in the literature. In this work, we have classified them as metaheuristic-based methods, machine learning models, hybrid intelligent approaches, and other methods (Fig. 2). Machine learning models use data-driven methodologies to forecast and alter system, settings for best performance under varying irradiance circumstances. Metaheuristic-based methods use optimization algorithms (OA) inspired by nature, swarm, physics, social and human behaviors in order to optimize the MPPT process. Hybrid approaches integrate different techniques to make use of each technique's capabilities. In this work, we will only focus on intelligent hybrid strategies.

4 Recent Developments in Hybrid Intelligent MPPT

Recently, hybrid based MPPT technology has attracted greater interest in PVS applications due to its high efficiency in tracking the MPPT. Hybrid methods generally include two or more techniques for the improvement of the control algorithm. Hybrid algorithms for global MPPT provide various benefits over standard single-method techniques. Hybrid algorithms play a crucial role in global MPPT because they combine the strengths of numerous optimization techniques to improve efficiency, accuracy, and robustness over single-method approaches. They provide a robust and adaptive solution for maximizing power output in solar systems under a wide range of operating conditions. Hybrid algorithms combine two or more optimization approaches, such as metaheuristic algorithms, machine learning techniques, fuzzy logic, or control strategies, resulting in a unified strategy for addressing optimization issues. These algorithms combine the benefits of different approaches while minimizing their drawbacks, yielding enhanced performance, robustness, and flexibility.

4.1 Hybrid Metaheuristic-Metaheuristic Algorithms (MET-MET)

This class refers to an optimization technique in which different metaheuristic algorithms are integrated to improve the MPPT process.

1. *"A Fusion Firefly OA with Simplified Propagation Process" (FuFiFOA-SPP)*
   The authors in [2] presented an enhanced MPPT algorithm using the "Fusion Firefly" OA with "Simplified Propagation Process (FuFiFOA-SPP)" for PVS under PS-C. This method uses the "fusion firefly algorithm (FFA)" and incorporates a revolutionary simplified propagation method. The suggested FFA, which combines the "neighborhood attraction firefly" algorithm (NaFA) with the "simplified firefly algorithm (SFA)", can track the GMPP with excellent accuracy. Furthermore, the suggested SPP technique decreases sampling events by removing duplicate propagations, which increases tracking speed while decreasing energy loss and oscillations during the sampling process. The performance of the suggested FFA and the speed enhancement achieved via the SPP process were simulated and confirmed using a hardware assessment device. Experimental findings indicated that the FFA provides great efficiency and accuracy while tracking quickly.
2. *"Fireworks and Particle Swarm OA" (FWOA-PSOA)*
   In [3], the Fireworks (FWOA) and Particle Swarm (PSOA) algorithms are associated for tracking the single global MPP under PS-C. By using the velocity operator of PSOA and the "mutation" and "explosion sparks operator" of FWOA, the difficulty to balance between "exploration" and "exploitation" is mitigated within FWOA-PSOA.
3. *"Levy Flight and Particle Swarm Optimization" (LFO-PSOA)*
   LFO-PSOA is an association of the "Levy flight" and "PSOA" algorithms. The PSOA is utilized because of its easy implementation and simple structure. "Levy flight optimization" (LFO) is a "Random Walk Distribution: RWD" that is fast and simple. The two approaches are combined to get better performance when searching for G-MPP under PS-C. In this method, PSOA is employed as a checking algorithm for identifying G-MPP. After detecting the G-M, the LFO improves precision by functioning at the previously found global maximum. LFO improves G-MPP for rapid and more precise tracking [4].
4. *"Genetic Algorithm and Grey Wolf OA" (GA-GWOA)*
   The GA-GWOA merges the capabilities of the "Genetic Algorithm" (GA) and "Grey Wolf" OA (GWOA) algorithms in tracking the G-MPP [5]. The algorithm effectively combines the advantages of both the GA and GWOA optimization algorithms. In GWOA, all major search agents tend to reach the same point. This can cause "multi-modal" problems to become stagnant. Another disadvantage of GWOA is that its performance degrades linearly as the number of wolves declines. By incorporating the GA algorithm into GWOA, wolf diversity is increased through "crossover" and "mutation processes". Therefore, the GA-GWOA technique provides a better optimum than GWOA and converges faster to G-MPP.

5. *"Grey Wolf OA and Particle Swarm OA" (GWOA-PSOA)*

   [6] provides a "Grey Wolf OA and Particle Swarm OA" (GWOA-PSOA) strategy that integrates the GWOA with the PSOA. Because it only requires two control parameters and its convergence to G-MPP is independent of the search process's beginning circumstances, this technique is a simple and resilient MPPT method for effectively managing PS in PVS. In this technique, GWOA search agents extensively explore the search space in order to avoid L-MPP and hence converge towards G-MPP. This exploration is controlled by the PSOA algorithm, which steadily improves the solutions acquired during the process in order to expedite convergence to G-MPP in the "exploitation" phase. The combination of GWOA and PSOA results in the algorithm achieving G-MPP in a few stages, making it more efficient. Furthermore, the proposed MPPT uses a direct control method, which means that the adjustment of "duty cycle" is directly performed by the GWOA-PSOA algorithm without the use of a "linear compensator"; this simplifies the process and eliminates the computational burden associated with adjusting "compensator gains".
6. *"Improved Artificial Bee Colony Algorithm and the Simultaneous Heat Transfer Search Algorithm" (IABCOA-SHTSA)*

   An "Improved Artificial Bee Colony Algorithm and the Simultaneous Heat Transfer Search Algorithm" (IABCOA-SHTSA) was proposed in [7]. For improving the tracking process's accuracy and speed, [7] proposed a bionic 2-stage MPPT technique consisting of a fast positioning stage and an accurate determination stage to optimize the "duty cycle" of the chopper. First, an IABC algorithm using "simplified probability selection mechanism" (SPSM) and "bee phase" is used to balance "exploration" and "exploitation" for high rapidity in placing a rough search region around the global peak. The "SHTS" algorithm is then used to precisely determine the G-MPP in the obtained search area. Additionally, the "SHTS" can decrease the subjectivity of artificial parameter setting in order to ensure the universality and accuracy of the search process. On the other hand, parallel search in "SHTS" can significantly reduce optimization time.
7. *"Cuckoo Search and Artificial Bee Colony"(CSA-ABCOA)*

   The Cuckoo Search (CSA) algorithm works on the concept of removing the worst individuals in the population based on probability, allowing the whole population to swiftly converge near the global optimal individual. Artificial Bee Colony (ABCOA) improves search accuracy through grouping and roulette-following procedures. As a result, by combining these two approaches, CSA-ABCOA maximizes the benefits of both algorithms, considers the algorithm's search speed and accuracy, and enhances MPPT performance. The novel CSA-ABCOA method addresses the issues of the ABCOA algorithm's slow convergence speed and the CSA's low convergence accuracy in the latter stages; it strikes a balance between global and local searches and lowers oscillation throughout the search process [8]. The findings show that the CSA-ABCOA algorithm outperforms the other algorithms in terms of striking a balance between tracking speed and accuracy, a lower power oscillation during the tracking phase, and no power oscillation during the steady process.
8. *"Teaching Learning (TLOA) and Artificial Bee Colony (ABCOA)" (TLOA-ABCOA)*

   The study in [9] proposed a hybridization of teaching-learning (TL) and an artificial bee colony (ABCOA) named TLABC to reduce oscillations around the GMPP.

This work uses the hybrid teaching and learning-based artificial bee colony (TLOA-ABCOA) strategy to balance exploration and exploitation. This strategy typically involves three parts: the teaching-based employ bee, the learning-based observer (onlooker) bee, and the generalized oppositional scout bee phases. This hybrid technique is more efficient in tracking the MPP than traditional algorithms. The benefit of employing the PSO-trained NN with FSSOA is that it can rapidly and precisely track the PV system's MPP with less computing effort than traditional techniques.

9. *"Cauchy Preferential Crossover and Global Pollination O-A" (CPC-GPOA)*

   [10] developed an algorithm combined with a PO method to enhance MPPT for a PVS operating under PS-C. The PO algorithm is used in the first step to obtain the local MPP, and the hybrid algorithm is used in the second part to reach the global MPP. The hybrid method used here combines "Cauchy Preferential Crossover" (CPC) and "Global Pollination" OA (FPA). Moreover, the C-C improves the FPA's exploitation ability, and the combined hybrid algorithm can generate the optimal duty cycle of the boost for MPPT.

4.2 Hybrid Metaheuristic- Machine Learning (MET-ML)

This kind of hybrid MPPT combines metaheuristic optimization algorithms with machine learning models to achieve optimal performance.

1. *"Extreme Learning Machine and Variable Steepest Gradient Ascent" (ELM-VSGA)*

   [11] designed an Extreme Learning Machine (ELM-trained SLFN) in conjunction with the Variable Steepest Gradient Ascent (VSGA) approach (ELM-VSGA). The presented method combines a "non-iterative artificial neural network's" accuracy with a flexible "steepest gradient ascent's" tracking capability. In the MPPT, a "proportional integral fractional order integral (PI-FOI)" cascade controller was used. SSA is employed to maximize controller parameters.

2. *"Whale OA and Pattern Search Algorithm Based on INC and ANFIS" (WOA-PSA-ANFIS-INC)*

   In [12], an enhanced MPPT approach uses ANFIS and INC. Based on the training set of data, the suggested method is used to identify the MPP. For problem optimization, the WOA-PS method is used. The temperature (T) and irradiance (E) are selected as the inputs, and the optimal voltages are achieved by the WOA-PS, which is used to tune the INC, in order to track the maximum power correctly. Due to the fact that INC can cover ANFIS when it can't precisely identify a point, this method has the advantage of requiring fewer training samples because it combines INC and ANFIS. The results indicate that the suggested scheme offers greater power, a quick response, and less oscillation.

3. *"Shuffled Frog Leaping and Pattern Search Algorithm Based ANN with PO" (SFL–PSA–ANN–PO)*

   In a GC-PVS, the "Shuffled Frog Leaping and Pattern Search (HSFL-PSA)" algorithms are combined to optimize the ANN MPPT controller. After training the ANN and specifying neuron weights, the PO technique is used in the tracking cycle process, which begins an accurate tracking scheme [13]. Therefore, this technique locates and tracks the MPP in two phases. Firstly, the HSFL-PS technique is used for determining the best voltages under various temperature and irradiance circumstances. After training, the ANN technique offers a value as the optimal voltage (V_{mpp}) based on the radiation conditions. The PO algorithm completes the tracking cycle procedure and launches an exact tracking approach from that point in the subsequent stage. The combined ANN and PO approach requires fewer training data. The PO can handle a situation where the ANN is unable to make a precise point. In order to create a stable and dependable hybrid system, a BESS is also used in the system to deliver load alongside the PV system. When the PVS is unable to deliver loads, the storage has the capacity to respond to those loads for a brief period of time. Additionally, the output current is regulated using PV/battery systems with a unity power factor to maintain the same frequency and phase as the grid voltage. Such a regulation can enhance the power quality of the power system and lower the harmonics that are introduced into the main grid.

4. *"Boosted Salp Swarm Optimization Based Deep Radial Basis Function Network" (BSSOA-DL-RBFN)*

   [14] proposed "Boosted Salp Swarm Optimization based Deep Radial Basis Function Network" (BSSOA-DL-RBFN). The DL-RBFN approach is introduced because it does not fluctuate near the MPP region and performs well in nonlinear and quickly changing conditions [14]. Furthermore, RBFN is improved by utilizing the "Boosted SSO (BSSOA)" strategy to achieve higher precision and convergence speed by removing the arbitrary weights. Furthermore, the BSSOA-DL-RBFN technology incorporates a boost converter for quick transient response.

5. *"Particle-Swarm-Optimization (PSOA)-Trained Machine Learning (ML) and Flying Squirrel Search Optimization (FSSOA)" (PSOA-ML-FSSOA)*

   To achieve the highest level of effectiveness in solar PV systems, a hybrid MPPT technique based on particle swarm optimization-trained machine learning and Flying Squirrel Search Optimization (PSOA-ML-FSSOA) has been presented [15]. The hybrid PSOA-FSSOA neural network improves the tracking performance of the MPPT algorithm by combining the benefits of PSOA and FSSOA techniques. The PSOA method optimizes the parameters of a neural network (NN) model, which is then employed to forecast the maximum power of a PVS. The FSSOA method uses this prediction to change the PV system's operating point to match the MPP.

6. *"Differential Evolution (DE) with a Feed-Forward Neural Network" (FFNN) (DE-FFNN)*

   In [16], an intelligent hybrid algorithm is developed that combines a "Differential Evolution" (DE) algorithm with a "Feed-Forward Neural Network" (FFNN) to build a more efficient MPPT method (DE-FFNN). Unlike classic gradient-based optimization approaches, DE does not depend on the objective function's gradients or derivatives, making it more resilient to noise and local optima (LO). This is particularly relevant in the situation of MPPT, when the system's operating circumstances may be highly variable and the objective function may be non-smooth or even discontinuous. Furthermore, the DE can handle parameter restrictions and boundaries, which may be important in real-world applications, and it is effortlessly parallelized, which can accelerate the optimization process even further [16]. The method establishes a balance between response time, stability, and efficiency of energy. DE's ability to explore various solutions for adapting the NN, as well as FFNN's adaptive learning, ensures that energy conversion is optimized while maintaining stability. This equilibrium is required for reliable and consistent MPPT performance.

7. *"Growth Optimizer Algorithm Trained Deep Neural Network" (GO-DNN)*

   Technologies based on ML, particularly deep neural networks (DNN), offer promise for overcoming the constraints of traditional MPPT methodologies. DNNs successfully manage the non-linear behavior and uncertainties of PV-TEG modules, precisely tracking the GMPP. However, training DNN parameters is a complicated optimization problem. The work presented in [17] suggests using a metaheuristic Growth Optimization (GO) algorithm to improve DNN weights and biases, therefore improving prediction accuracy and reducing the downsides of standard training approaches [17].The suggested technique aims to enhance the energy efficiency of a hybrid PV-TEG system, which generates electricity by combining PV panels and thermoelectric generators (TEG) and employs an evolutionary DNN trained with the GO algorithm. The DNN model is trained for predicting reference voltage (V) and current (I), which are then utilized by the Fractional Order PID (FOPID) controller to calculate the boost converter's duty cycle. Furthermore, the GO algorithm is used to modify FOPID controller gains to reduce overshoots and static errors, hence improving the control system's resilience (GO-FOPID). This technique enables precise and accurate MPPT control under variable environmental and load circumstances by integrating the GO-DNN and GO-FOPID controllers.

8. *"Zebra Optimization Algorithm (ZOA)-ANFIS / Gorilla Troop Optimizer (GTOA)-ANFIS*

   The study in [18] describes an efficient hybrid renewable energy system. The system makes use of two principal energy sources: PV and wind energy conversion systems. In addition, the battery energy storage system (BESS), hydrogen energy storage system (HESS), and super capacitor energy storage system (SESS) are being integrated as backup sources to provide stable simultaneous power supply and loading [18]. Artificial intelligence algorithms are used to find the optimal energy flow when the system is instantly operational. Furthermore, two hybrid MPPT algorithms have been presented, including the zebra optimization algorithm-ANFIS (ZOA- ANFIS) and the gorilla troop optimizer-ANFIS (GTOA-ANFIS) to obtain the MPP of PV panels and wind turbines, which leads to improving the performance of these energy sources. The hybrid ZOA-ANFIS approach uses the artificial ZOA Zone to continually adjust the PV system's operating voltage for MPPT. Furthermore, the ANFIS is enhanced using a logically based control mechanism to improve tracking precision. The hybrid GTOA-ANFIS incorporates the "GTOA" while the ANFIS improves the wind energy system's control settings, allowing it to properly track its MPP. The "ZOA" and "GTOA" algorithms were evaluated under diverse operating situations, including fluctuations in temperature and irradiance, to evaluate their performance in tracking the MPP. According to the findings, the "ZOA" and "GTOA" systems performed roughly the same in terms of their capacity to monitor solar energy. The "ZOA" algorithm surpassed the "GTOA" method in terms of computation time. This suggests that the "ZOA" algorithm processed and calculated the necessary modifications for solar tracking faster than the "GTOA" algorithm. This can be viewed as a benefit for the "ZOA" algorithm since it may adapt faster to changes in irradiance conditions, ensuring optimum solar energy.

4.3 Hybrid Metaheuristic -Fuzzy Logic (MET-FL)

This class of hybrid MPPT combines metaheuristic optimization algorithms with fuzzy logic models to achieve optimal performance.

1. *"BAT Algorithm and Fuzzy Logic Controller" (BATOA-FLC)*

   For determining the best values for the controller's parameter for the MPPT, [19] uses the FLC and BAT algorithms. In the BAT-Fuzzy technique, the "triangular membership functions" of the input and output are independently set. The results from modeling the suggested model on a test system demonstrate the advantages of the BAT-Fuzzy technique in terms of precision and convergence rate, while taking into account changes in irradiance and temperature values. When taking into account the system's effectiveness in real-world scenarios, it also performs better. Additionally, distinct PSC states and their effects on the environment have been discussed.
2. *"Modified Shuffled Frog Leaping Algorithm and Fuzzy Logic Controller" (M-SFLA-FLC)*

   [20] investigated the performance of an improved control strategy for the use of "H-ES" that included a PV and a "B-ESS" connected to the grid. In this regard, the proposed system makes use of the FLC in conjunction with the Modified Shuffled Frog Leaping Algorithm (M-SFLOA) to optimally adjust the control parameters and track the "optimal point" (O-MPP).
3. *"Fuzzy Adaptive and PSOA" (FuA-PSOA)*

   In this study, a "fuzzy adaptive" (FuA) was combined with the PSOA algorithm for reaching the "global optimal point" (GOP) under uniform and PS-C. FLC and PSOA advantages are integrated into the controller [21]. For improving GS capabilities and convergence speed, the PSOA parameters are adjusted dynamically by the FLC. Because the PSOA parameters were designed to be shared by all particles, thus reducing computational complexity. The rule base of FLC aims to produce a rapid transient response as well as a stable steady-state response.
4. *"Fuzzy Logic Controller and Modified Krill Herd" (FLC-MKH)*

   In [22], this strategy combines FLC and Modified Krill Herd (MKH). The fuzzy and MKH algorithms are used in tandem to fine-tune the control parameters of the proposed MPPT method. The rule base and MFs are essential parameters to implement FLC, It is impossible to fine-tune these parameters correctly without an accurate system. To address this issue, the MKH algorithm employs the MF scaling factors. To enhance system stability, the PVS is employed in conjunction with the storage system. The hybrid (FLC-MKH) system can cope with the PHV system's stochastic character and offers excellent stability under all environmental conditions.
5. *"Fuzzy Logic Controller and Improved Farmland Fertility OA" (FLC-IFarFerOA)*

   In [23], a hybrid FLC-IFarFerOA approach that combines the Fuzzy Logic Controller (FLC) and the Improved Farmland Fertility (I-FarFerOA) algorithms optimally adjusts the controller parameters for increasing the efficacy of the PV system. In the case of homogeneous irradiance and PS, this method would provide the system with very high performance. The FLC serves as the foundation for the controller, and its parameters are optimally adjusted to the I-FarFerOA used to achieve the G-MPP. FarFer-OA is a MeHeOA algorithm that addresses optimization problems in a variety of fields. The FLC's main advantages in properly tracking the G-MPP are its simplicity and robustness. The FLC consists of three parts: "fuzzification," "fuzzy inference," and "defuzzification".

6. *"Ant Colony Optimization and Fuzzy Logic Ant-Fuzzy Optimization (AFO) Algorithms"*

   [24] introduces the Ant-Fuzzy Optimization (AFO) method, a hybrid algorithm that combines the Ant Colony Optimization (ACO) and Fuzzy Logic (FL) algorithms, for tracking a PV system's maximum power point (MPP). The Ant-Fuzzy Optimization (AFO) method makes use of the ant colony optimization (ACO) algorithm's global search capabilities as well as the fuzzy logic (FL) algorithm's high precision performance, reducing the fuzzy algorithm's tendency to fall into local optima under shadow conditions. Internally, the AFO algorithm consists of two concurrent logics that choose alternative tracking techniques based on changing ambient conditions, establishing an equilibrium between tracking precision and computational efficiency [24]. This intelligent logic selection method enables the algorithm to dynamically adapt to the varied operating conditions of photovoltaic (PV) arrays, improving the system's robustness and adaptability. AFO also has quick dynamic response capabilities, reaching steady state in 0.9 s, making it a viable option for maximizing solar array output power.

4.4 Hybrid Machine Learning -Fuzzy Logic (ML-FL)

This class of hybrid MPPT integrates machine learning (ML) models with fuzzy logic (FL) for optimizing performance.

"Reinforcement Learning with Fuzzified Reward Approach" (RL-FRA)

[25] uses Deep Reinforcement Learning (DRL) with a "fuzzy reward" approach (FRA) for improved translation of continuous space into various abstraction levels. The development of RL has a benefit because using a "Deep-Q learning" algorithm renders the structure model-free and eliminates the numerous modeling parameters and their impacts on the design.

5 Evaluation and Comparison

Table 1 illustrates performance comparisons of hybrid intelligent MPPT techniques (HIA). This table contrasts the efficiency percentage, tracking speeds, complexity level, converter, configuration, application type and implementation.

HIAs are used in a variety of applications, including micro grids, hybrid PV-BES systems, and other configurations. The complexity level ranges from medium to high indicating the integration of numerous optimization techniques. The procedures are very efficient, with percentages frequently exceeding 95% and nearing 100%. Convergence speeds are often very short, which means they are adaptable to severe PS and rapid changes. Most developed approaches have not been validated through experimentation, with only a few exceptions. Across all categories, efficiency stays consistently high, demonstrating the usefulness of these strategies in increasing solar panel power production. All techniques show promise in terms of implementation, with validation through simulations or experiments giving assurance of their real-world suitability.

Table 1. Performance comparison of hybrid intelligent algorithms (HIA) based MPPT techniques.

Refs	Technique	Comp	Eff (%)	TS (s)	Conv/App	Conf	Val	Algorithm Classification	Remarks
[2]	FuFifOA-SPP	H	85.5	–	Bo/SA	4S	S/E	MET-MET	-Reducing sampling events during the MPPT search process of the algorithm by eliminating redundant propagation -Experimentally validated -The addition of the fusion process and SPP slightly increases the complexity of the proposed method, avoids oscillations or traps in the LMPP, and provides a fast tracking speed with high accuracy and efficiency
[3]	FWOA-PSOA	M	97.1	5.52	Bo/SA	3S	S/E	MET-MET	-Adaptive spark control is proposed to accelerate convergence in the later stage of the search process
[4]	LFO-PSOA	M	99.50	0.5	Bo/SA	PV module	S/E	MET-MET	-Reduce steady-state oscillation and is particularly efficient in tracking global optimal power under severe PS-C
[5]	GA-GWOA	H	>99	0.09	Bo/SA	3x3 TCT	S	MET-MET	-The technique correctly detects changes in the PSC -The implementation of the algorithm is more complex than the parent and traditional approaches -The suggested approach uses two meta-heuristic algorithms with several cycles of population generation and mutations, resulting in a resource-intensive program

(*continued*)

Table 1. (*continued*)

Refs	Technique	Comp	Eff (%)	TS (s)	Conv/App	Conf	Val	Algorithm Classification	Remarks
[6]	GWOA–PSOA	M	99. 66	2.04	Bo/SA	2S	S	MET-MET	-The algorithm's convergence is independent of starting search conditions and requires just two control parameters, making it simpler and more adaptable. Furthermore, it does not require prior knowledge of PV array characteristics, making it simple for implementation in larger PV systems, both off-grid and on-grid -An experimental setup is required to evaluate the GWO-PSO method in a real PVS environment
[7]	IABCOA- SHTSA	H	99.5	0.15	Bo/SA	4x2 SP	S	MET-MET	-The two-stage technique is effective in avoiding local optimums and achieving GMPP tracking performance -The suggested technique is relatively complex due to its combination of two optimization techniques - For best optimization results, the transition procedure between two stages in the offered MPPT should be given enough consideration -Practical experiments are not implemented

(*continued*)

Table 1. (*continued*)

Refs	Technique	Comp	Eff (%)	TS (s)	Conv/App	Conf	Val	Algorithm Classification	Remarks
[8]	CSA-ABCOA	M	99.77	0.14	Bo/SA	4S	S	MET-MET	-The algorithm provides the fastest tracking speed while maintaining the maximum tracking accuracy, with lower power oscillations throughout the tracking process and no power oscillations during the steady state. The CSA-ABC algorithm can restart and re-track to maximum output power once shading circumstances vary -The algorithm is not implemented in practical systems, and it is not tested under more complex environmental conditions for further validation of its effectiveness
[9]	TLOA-ABCOA	M	99.89	0.52	Bo/SA load + battery storage	4S3P	S/E	MET-MET	- Experimental validation -The MPPT controller's architectural complexity results in a long implementation time
[10]	CPC-GPOA	M	99.96	0.08	Bo/GC	2S	S	MET-MET	-Highly effective and straightforward to execute in a short period of time -CCGPA parameters need more optimization
[11]	ELM-VSGA	M	>98.5	–	Bo/SA	–	S	MET-ML	-The structure is simple and produces more power with less power fluctuation under different atmospheric conditions. The implementation of the proposed technique is simple and can be implemented on low-cost processors -Experiments are needed to verify the results. It requires the realization of experiments to validate the results

(*continued*)

Table 1. *(continued)*

Refs	Technique	Comp	Eff (%)	TS (s)	Conv/App	Conf	Val	Algorithm Classification	Remarks
[12]	WOA-PSA-ANFIS-INC	M	99.3	0.16	Bo/GC	81 PV module (SP)	S	MET-ML	-Can successfully track the MPP in a variety of situations, with rapid response and low oscillations around it. Increased production power and high tracking efficiency
[13]	SFL–PAS–ANN–PO	M	99.23	0.11	Bo/ GC	–	S	MET-ML	-Improved performance. Delivers the highest efficiency and fastest response in a steady state when the irradiation varies
[14]	BSSOA-DL-RBFN	H	98.5	0.02	Bo/SA	4S	S/E	MET-ML	-It does not require re-initialization. Faster convergence, more stable operation, reduced oscillation, and optimal power harvesting -It has a complex structure; a huge training dataset is needed
[15]	PSO-ML-FSSO	H	–	–	Bo/SA	24 × 3S cells	S	MET-ML	-Improved efficiency and reduced settling time
[16]	DE-FFNN	H	99.99	0.02	Bo/ SA	1S5P	S	MET-ML	-High accuracy, robustness, fast convergence, high stability, fast search for the global optimal, -The need for training data, complex parameter tuning, computationally expensive -The method establishes an equilibrium between response time, energy efficiency, and stability. DE's ability to investigate various solutions for adapting the NN, as well as FFNN's adaptive learning, ensures that energy conversion is optimized while maintaining stability
[17]	GO-DNN	M	99.92	0.005	Bo/hybridPV-TEG	4	S/E	MET-ML	-Improving the power generation efficiency of hybrid PV-TEG systems
[18]	ZOA-ANFIS/AGTOA-ANFIS	–	–	–	Hybrid microgrid (wind-solar)	–	S	MET-ML	-This technique was employed on a small scale

(continued)

Table 1. (*continued*)

Refs	Technique	Comp	Eff (%)	TS (s)	Conv/App	Conf	Val	Algorithm Classification	Remarks
[19]	BATOA-FLC	M	99.16	0.12	Bo/GC	–	S	MET-ML	-The temperature variation effect is taken into account -Not experimentally validated
[20]	MSFLA-FLC	M	99.12	0.13	Bo/GC	–	S	MET-FL	-Not experimentally validated
[21]	FuA-PSOA	M	98.32	–	Bo/SA	–	S	MET-FL	-The PSO parameters are dynamically adjusted by the FLC for improving the global search process and convergence speed
[22]	FLC-MKH	M	99.32	0.13	Bo/Hybrid BES-PV	–	S	MET-FL	-The MKH method is used to tune FLC MFs to account for irradiance and temperature uncertainties
[23]	FLC-IFarFerOA	M	>99	0.12	Bo/ Hybrid PV-BES	–	S	MET-FL	-The algorithm optimally tunes the controller parameters
[24]	ACO-FL	H	98.7	0.9	Bo/ SA	–	S/E	MET-FL	-Uses the global search capacity of the ACO algorithm with the high precision performance of the FL algorithm to boost the system's robustness and flexibility, striking a balance between tracking accuracy and computing efficiency -It has fast-speed convergence, steady state zero oscillation, excellent accuracy, and resilience, even when the weather changes -Complexity and hardware implementation costs should be reduced
[25]	RL-FRA	M	73	–	-/SA	12S	S	ML-FL	-The rise time is shorter, and since it is a robust technique with minimal complexity, it is easier to implement in real time

Comp: Complexity, H: High, M: Medium, Eff: Efficiency, TS: Tacking Speed, Conv: Converter, App: Application, Conf: Configuration, Val: Validation, Bo: Boost, SA: Standalone, GC: Grid Connected, S: Simulation, E: Experimentally validated.

Intelligent hybrid methods provide optimized performance by eliminating the drawbacks of the algorithm alone. The hybrid intelligent MPPT system designed to operate under PS-C typically demonstrates superior performance in terms of efficiency, complexity, response time, and less parameter adjustment which is achieved through the integration of multiple techniques. Nevertheless, they are more complex.

Each of these MPPT algorithms offers unique features and performance metrics suitable for the optimization of the power output from PVS. They generally exhibit high efficiencies and fast tracking speeds, which are essential to maximizing energy harvest from PV panels. However, experimental validation is crucial to confirm their effectiveness in real-world conditions, particularly for scenarios involving dynamic changes in irradiance and environmental conditions.

- All hybrid intelligent MPPT techniques achieve high efficiency in optimization tasks. Nevertheless, specific efficiency varies based on algorithm selection, implementation details and problem complexity.
- MET-MET and MET-ML tend to be more complex due to integrating multiple strategies
- ML-FL and MET-FL focus on integrating fuzzy logic with either machine learning or metaheuristics, balancing complexity with adaptive capabilities.
- MET-MET and MET-ML may require more expertise and computational resources due to integration challenges, while MET-FL, ML-FL, balance complexity with adaptive capabilities.
- Challenges include increased computational complexity, parameter-tuning difficulties and potential slower convergence in multi-method approaches (MET-MET)
- The choice of the most appropriate MPPT method depends on specific application requirements:
- MET-ML and ML-FL might be suitable for robust and adaptive optimization due to their ability to adapt to dynamic environments and achieve efficient convergence.
- MET-FL could be advantageous with its capability to manage imprecise data and make effective decisions for handling uncertainty:
- The use of fuzzy logic-driven metaheuristic approaches, as well as machine learning-driven techniques, has the potential to revolutionize MPPT technology by ensuring optimal energy harvest from PV systems in a variety of complicated operational scenarios.
- Future research should focus on:
- Extending the suggested approach by evaluating it under a range of changing conditions, such as moving cloud scenarios, rainy conditions, and the effects of other constraints, to ensure its resilience.
- The recommended MPPT algorithms require being assessed in real systems under challenging and more complex environmental circumstances to ensure effectiveness alongside the enhancement of these approaches.
- The hardware implementation of these approaches on a target board, such as an FPGA is extremely beneficial for better practical validation and testing of the recommended method's efficiency.
- Cost implications need to be thoroughly considered to ensure the feasibility and economic feasibility of implementing such strategies in large-scale solar installations.

- Improve hardware compatibility while minimizing implementation costs.
- Enhancing the current results by developing other hybrid intelligent algorithms
- Integration of several renewable energy sources for increased efficiency.
- Further optimization is needed to minimize complexity and enhance adaptability to changing environmental circumstances.

The researchers can choose appropriate MPPT methodologies based on their specific application requirements, achieving a balance between efficiency, speed, complexity, and implementation feasibility.

6 Conclusion

This study explores and analyzes the latest developments in hybrid intelligent MPPT methodology published in the last five years, classifying them into four classes. The selected methods are assessed and contrasted using a variety of criteria. The evaluation examines the benefits and disadvantages of each technique and suggests future study areas. The evaluation results highlight the potential for hybrid solutions, particularly metaheuristic machine learning and fuzzy logic machine learning techniques, to greatly improve MPPT performance in renewable energy systems. This study is a valuable source of information for researchers searching for an appropriate and advantageous MPPT for their application.

References

1. Belhachat, F., Larbes, C.: Survey and classification of hybrid GMPPT techniques for photovoltaic system under partial shading conditions. ENP Eng. Sci. J. **2**(2), 31 (2022)
2. Huang, Y.-P., Huang, M.-Y., Ye, C.-E.: A fusion firefly algorithm with simplified propagation for photovoltaic MPPT under partial shading conditions. IEEE Trans. Sustain. Energy **11**(4), 2641–2652 (2020)
3. Chai, L.G.K., et al.: A novel global MPPT technique using improved PS-FW algorithm for PV system under partial shading conditions. Energy Conver. Manag. **246**, 114639 (2021)
4. Charin, C., et al.: A hybrid of bio-inspired algorithm based on Levy flight and particle swarm optimizations for photovoltaic system under partial shading conditions. Sol. Energy **217**, 1–14 (2021)
5. Yadav, K., Kumar, B., Guerrero, J.M., Lashab, A.: A hybrid genetic algorithm and grey wolf optimizer technique for faster global peak detection in PV system under partial shading. Sustain. Comput. Inform. Syst. **35**, 100770 (2022)
6. Chtita, S., et al.: A novel hybrid GWO–PSO-based maximum power point tracking for photovoltaic systems operating under partial shading conditions. Sci. Rep. **12**, 1063 (2022)
7. Gong, L., Hou, G., Huang, C.: A two-stage MPPT controller for PV system based on the improved artificial bee colony and simultaneous heat transfer search algorithm. ISA Trans. **132**, 428–443 (2023)
8. Qi, P., Xia, H., Cai, X., Yu, M., Dai, N.J.Y.: Novel global MPPT technique based on hybrid cuckoo search and artificial bee colony under partial-shading conditions. Electronics **13**, 1337 (2024). https://doi.org/10.3390/electronics13071337
9. Krishna Kishore, D.J., Mohamed, M.R., Sudhakar, K., Peddakapu, K.: Swarm intelligence-based MPPT design for PV systems under diverse partial shading conditions. Energy **265**, 126366 (2023). https://doi.org/10.1016/j.energy.2022.126366

10. Sundararaj, V., Anoop, V., Dixit, P., Arjaria, A., Chourasia, U., Bhambri, P., Rejeesh, M.R., Sundararaj, R.: CCGPA-MPPT: Cauchy preferential crossover-based global pollination algorithm for MPPT in photovoltaic system. Prog. Photovolt. Res. Appl. **28**, 1128–1145 (2020)
11. Behera, M.K., Saikia, L.C.: A new combined extreme learning machine variable steepest gradient ascent MPPT for PV system based on optimized PI-FOI cascade controller under uniform and partial shading conditions. Sustain. Energy Technol. Assess. **42**, 100859 (2020)
12. Tao, H., Ghahremani, M., Ahmed, F.W., Jing, W., Nazir, M.S., Ohshim, K.: A novel MPPT controller in PV systems with hybrid whale optimization-PS algorithm based ANFIS under different conditions. Control. Eng. Pract. **112**, 104809 (2021)
13. Jiang, M., Ghahremani, M., Dadfar, S., Chi, H., Abdallah, Y.N., Furukawa, N.: A novel combinatorial hybrid SFL–PS algorithm based neural network with perturb and observe for the MPPT controller of a hybrid PV-storage system. Control. Eng. Pract. **114**, 104880 (2021)
14. Antony, R.S., Giftson Samuel, G.: BOSS-D-RBFN: BOosted Salp swarm optimization based deep RBFN for MPPT under partial shading condition in photovoltaic systems. Optik – Int. J. Light Electron Opt. **259**, 168876 (2022)
15. Kumar, D., et al.: A novel hybrid MPPT approach for solar PV systems using particle-swarm-optimization-trained machine learning and flying squirrel search optimization. Sustainability **15**, 5575 (2023). https://doi.org/10.3390/su15065575
16. Ncir, N., El Akchioui, N., El Fathi, A.: Enhancing photovoltaic system modeling and control under partial and complex shading conditions using a robust hybrid DE-FFNN MPPT strategy. Renew. Energy Focus **47**, 100504 (2023). https://doi.org/10.1016/j.ref.2023.100504
17. Khan, N.M., Khan, U.A., Asif, M., Zafar, M.H.: Analysis of deep learning models for estimation of MPP and extraction of maximum power from hybrid PV-TEG: a step towards cleaner energy production. Energy Rep. **11**, 4759–4775 (2024). https://doi.org/10.1016/j.egyr.2024.04.035
18. Elymany, M.M., Enany, M.A., Elsonbaty, N.A.: Hybrid optimized-ANFIS based MPPT for hybrid microgrid using zebra optimization algorithm and artificial gorilla troops optimizer. Energy Convers. Manag. **299**, 117809 (2024). https://doi.org/10.1016/j.enconman.2023.117809
19. Ge, X., Ahmed, F.W., Rezvani, A., Aljojo, N., Samad, S., Foong, L.K.: Implementation of a novel hybrid BAT-fuzzy controller based MPPT for grid-connected PV-battery system. Control. Eng. Pract. **98**, 104380 (2020)
20. Li, Y., Samad, S., Ahmed, F.W., Abdulkareem, S.S., Hao, S., Rezvani, A.: Analysis and enhancement of PV efficiency with hybrid MSFLA-FLC MPPT method under different environmental conditions. J. Clean. Prod. **271**, 122195 (2020)
21. Guo, L., Abdul, N.M.M.: Design and evaluation of fuzzy adaptive particle swarm optimization based maximum power point tracking on photovoltaic system under partial shading conditions. Front. Energy Res. **9**, 712175 (2021)
22. Hu, Z., Norouzi, H., Jiang, M., Dadfar, S., Kashiwagi, T.: Novel hybrid modified krill herd algorithm and fuzzy controller based MPPT to optimally tune the member functions for PV system in the three-phase grid-connected mode. ISA Trans. **129**(Part B), 214–229 (2022)
23. Hai, T., Zhou, J., Muranaka, K.: An efficient fuzzy-logic based MPPT controller for grid-connected PV systems by farmland fertility optimization algorithm. Optik – Int. J. Light Electron Opt. **267**, 169636 (2022)
24. Xia, K., Li, Y., Zhu, B.: Improved photovoltaic MPPT algorithm based on ant colony optimization and fuzzy logic under conditions of partial shading. IEEE Access **12**, 44817–44825 (2024)
25. Singh, Y., Pal, N.: Reinforcement learning with fuzzified reward approach for MPPT control of PV systems. Sustain. Energy Technol. Assess. **48**, 101665 (2021)

Communication Systems and Optimization

IoT-Enhanced Photovoltaic Systems: Achieving Low LCOE and Improved Sustainability Through Advanced Energy Management

Mustapha Hatti(✉)

UDES, Unité de Développement des Équipements Solaires, EPST-CDER, BP 386, 42415 Bou Ismail, Tipaza, Algeria
m.hatti@udes.dz

Abstract. The global energy landscape is rapidly evolving, with a growing emphasis on sustainability and renewable energy sources. Solar energy, in particular, plays a crucial role in reducing carbon emissions and mitigating climate change. However, the intermittent nature of solar power poses challenges for grid integration and efficient energy management. This paper explores the integration of Internet of Things (IoT) technologies in photovoltaic (PV) systems to enhance performance, reduce Levelized Cost of Energy (LCOE), and improve overall sustainability. We present a comprehensive analysis of IoT applications in PV systems, focusing on real-time monitoring, predictive maintenance, and adaptive energy management strategies. Our findings demonstrate that IoT-enabled PV systems can significantly improve energy yield, reduce operational costs, and extend system lifespan, ultimately contributing to a more sustainable and efficient energy infrastructure.

Keywords: Internet of Things · LCOE · Photovoltaic Array · Management · landscape · sustainability · Management

1 Introduction

The urgent need to address climate change has accelerated the global transition towards renewable energy sources, with solar power emerging as a key player in this shift. Photovoltaic (PV) systems offer a clean, sustainable alternative to fossil fuels, but their integration into existing power grids presents several challenges. The intermittent nature of solar energy, influenced by factors such as weather conditions and time of day, can lead to imbalances between supply and demand, potentially compromising grid stability and efficiency (Alam et al. 2022). Traditional energy management systems often lack the flexibility and responsiveness required to effectively handle the variable output of renewable energy sources. This limitation has sparked interest in more advanced, adaptive solutions capable of optimizing energy production, distribution, and consumption in real-time). (Kumar et al. 2020). The Internet of Things (IoT) has emerged as a promising technology Fig. 1 to address these challenges, offering unprecedented capabilities

© The Author(s), under exclusive license to Springer Nature Switzerland AG 2025
M. Hatti (Ed.): IC-AIRES 2024, LNNS 1238, pp. 615–622, 2025.
https://doi.org/10.1007/978-3-031-80301-7_65

in data acquisition, processing, and communication within energy systems. This paper explores the potential of IoT-enhanced photovoltaic systems to achieve lower Levelized Cost of Energy (LCOE) and improved sustainability (Mellit et al. 2018). We examine how IoT technologies can be leveraged to optimize PV system performance, enhance grid integration, and extend system lifespan through advanced monitoring and predictive maintenance strategies.

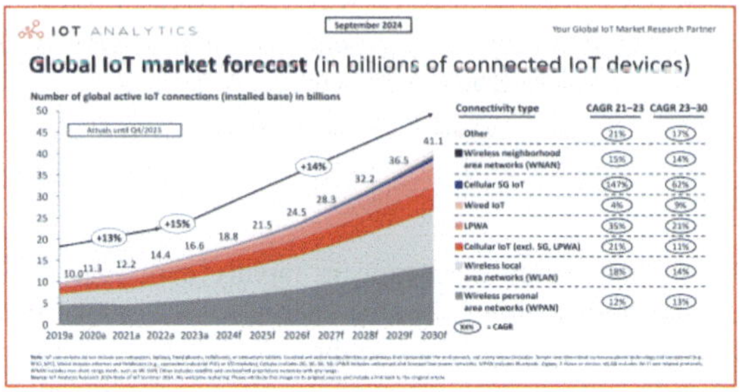

Fig. 1. IoT Market forecast (2024)

2 IoT Integration in Photovoltaic Systems

A. IoT Architecture for PV Systems

The integration of IoT in PV systems involves a multi-layered architecture that enables seamless data collection, processing, and control. This architecture typically consists of (Ren et al. 2020):

- Sensor Layer: Comprises various sensors deployed across the PV system to collect real-time data on environmental conditions, energy production, and equipment status.
- Network Layer: Facilitates data transmission between sensors, control devices, and central management systems using protocols such as MQTT, LoRaWAN, or cellular networks.
- Data Processing Layer: Employs edge computing and cloud-based systems to analyze collected data, generate insights, and trigger automated responses (Schmitt et al. 2021).
- Application Layer: Provides user interfaces and APIs for system monitoring, control, and integration with other energy management platforms.

B. Key IoT Applications in PV Systems

IoT technologies enable several critical functionalities in PV systems:

Real-time Monitoring: Continuous tracking of system performance, environmental conditions, and energy production.
Predictive Maintenance: Utilization of machine learning algorithms to forecast potential equipment failures and optimize maintenance schedules.

Adaptive Energy Management: Dynamic adjustment of energy production and distribution based on real-time demand and grid conditions (Shafique et al. 2021).
Enhanced Grid Integration: Improved coordination between PV systems and the broader power grid to maintain stability and efficiency, reducing curtailment losses by up to 20% (Zsiborács et al., 2019).

3 Lowering LCOE Through IoT-Enhanced PV Systems

The Levelized Cost of Energy (LCOE) is a crucial metric for assessing the economic viability of energy production systems. IoT integration in PV systems can significantly contribute to reducing LCOE through various mechanisms:

3.1 Optimizing Energy Production

IoT-enabled systems can optimize energy production by:

Implementing advanced Maximum Power Point Tracking (MPPT) algorithms that adapt to real-time conditions.
Adjusting panel orientation in smart tracking systems based on precise solar irradiance data.
Identifying and mitigating partial shading effects through granular monitoring and control.

3.2 Reducing Operational and Maintenance Costs

Predictive maintenance strategies enabled by IoT can substantially reduce operational costs:

Early detection of potential faults or performance degradation.

Optimization of maintenance schedules to minimize downtime and maximize system availability. Remote diagnostics and troubleshooting, reducing the need for on-site interventions.

3.3 Extending System Lifespan

IoT technologies contribute to prolonging the operational life of PV systems:

Continuous monitoring of component health to prevent catastrophic failures.
Optimized operating conditions to reduce stress on system components.
Data-driven decisions for timely replacements and upgrades.

4 Enhancing Sustainability Through IoT Integration

Beyond economic benefits, IoT integration in PV systems also contributes to improved sustainability:

4.1 Increased Energy Efficiency

Real-time load balancing and energy routing to minimize losses.

Adaptive inverter control strategies to optimize power quality and reduce conversion losses.

5 Improved Grid Stability

Enhanced forecasting of PV output to better manage grid integration.

Demand response capabilities to align energy production with consumption patterns (Shaikh et al. 2014).

5.1 Reduced Environmental Impact

Optimization of energy storage systems to maximize the utilization of solar energy (Vinuesa et al. 2020). Minimization of equipment replacements through predictive maintenance, reducing electronic waste.

6 Case Studies and Empirical Evidence

This section presents several case studies demonstrating the practical implementation and benefits of IoT-enhanced PV systems:

6.1 Large-Scale Solar Farm in California, USA

A 100 MW solar farm implemented IoT-based monitoring and predictive maintenance, resulting in:

5% increase in annual energy yield
15% reduction in operational costs
2-year extension of expected system lifespan

6.2 Residential PV System Network in Germany

A network of 10,000 residential PV systems equipped with IoT sensors and smart inverters achieved:

8% improvement in grid integration efficiency
20% reduction in maintenance-related system downtime
3% decrease in overall LCOE

Aggregate System Performance:

Before IoT: 52.5 GWh/year
After IoT: 56.7 GWh/year
Annual CO2 emissions avoided: increased from 21,000 to 22,680 metric tons

6.3 Commercial Rooftop PV Installation in Singapore

A 500 kW rooftop PV system on a commercial building integrated with building energy management systems (BEMS) via IoT:

12% increase in self-consumption of generated energy
18% reduction in peak demand charges
7% improvement in overall system ROI

6.4 UDES Rooftop PV Installation in BOU Ismail Tipaza, Algeria

6.4.1 Lowering LCOE Through IOT-Enhanced PV Systems

The Levelized Cost of Energy (LCOE) is a crucial metric for assessing the economic viability of energy production systems. To understand the impact of IoT on LCOE, we first present the standard LCOE equation (Figs. 2, 3, and 4):

$$\mathrm{LCOE} = \left(\sum (\mathrm{It} + \mathrm{Mt} + \mathrm{Ft})/(1 + \mathrm{r})^{\mathrm{t}}\right)/\left(\sum \mathrm{Et}/(1 + \mathrm{r})^{\mathrm{t}}\right) \quad (1)$$

Where:

- It = Investment expenditures in year t
- Mt = Operations and maintenance expenditures in year t
- Ft = Fuel expenditures in year t (zero for PV systems)
- t = Life of the system
- Et = Electricity generation in year t
- r = Discount rate

Fig. 2. IoT unit

7 Challenges and Future Directions

While IoT integration in PV systems offers significant benefits, several challenges remain:

- Data Security and Privacy: Ensuring the protection of sensitive operational data and user information.
- Standardization: Developing common protocols and interfaces for seamless integration across different systems and manufacturers.
- Scalability: Designing solutions that can efficiently handle the massive data volumes generated by large-scale PV deployments.

Cost-Benefit Analysis: Quantifying the long-term economic and environmental benefits of IoT integration to justify initial investment costs (Mutlag et al. 2019; Zsiborács et al. 2019).

Fig. 3. Telemonitoring in real time [Humidity - Temperature].

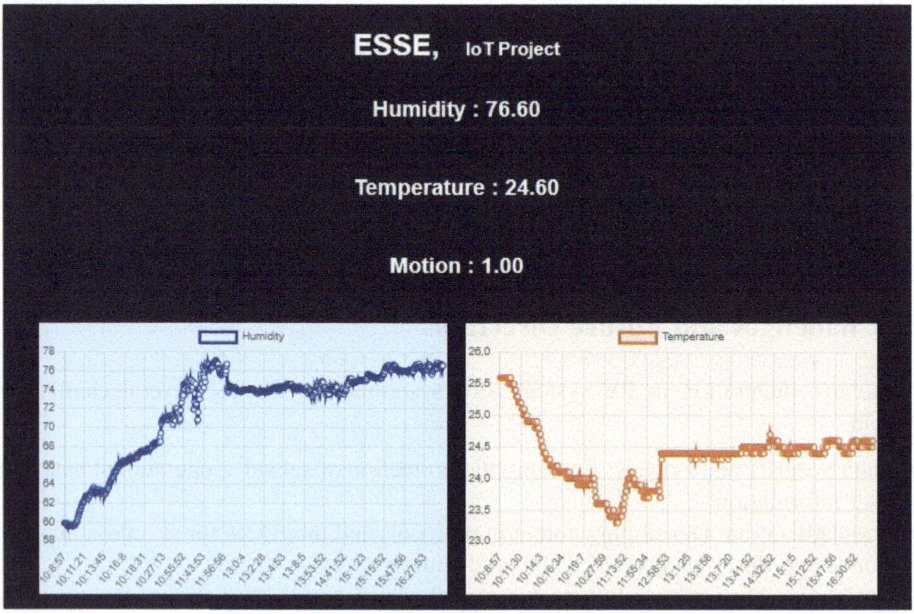

Fig. 4. UDES, ESSE-project Installation (2024)

Future research directions should focus on:

Advanced AI and machine learning algorithms for more accurate predictive maintenance and energy forecasting. Integration of blockchain technology for secure and transparent energy trading in decentralized PV networks. Development of edge computing solutions to reduce latency and enhance real-time control capabilities. Exploration of synergies between IoT-enhanced PV systems and other smart grid technologies.

8 Conclusion

The integration of IoT technologies in photovoltaic systems represents a significant step towards achieving lower LCOE and improved sustainability in the renewable energy sector. By enabling real-time monitoring, predictive maintenance, and adaptive energy management, IoT-enhanced PV systems offer substantial improvements in energy yield, operational efficiency, and system longevity. The case studies presented in this paper provide empirical evidence of the tangible benefits of such integration, demonstrating reductions in LCOE and enhancements in overall sustainability. Our analysis, supported by real-world case studies and quantitative data, demonstrates that IoT-enhanced PV systems can achieve:

Energy yield improvements of 3–10%; Operational cost reductions of 15–30%; System lifespan extensions of 2–5 years LCOE reductions of 9–13%.

As the global energy landscape continues to evolve, the role of IoT in optimizing renewable energy systems will become increasingly crucial. Addressing the challenges of data security, standardization, and scalability will be essential to fully realizing the potential of IoT-enhanced PV systems. Future research and development in this field promise to further improve the performance, reliability, and sustainability of solar energy, contributing significantly to the global transition towards a cleaner, more efficient energy future.

References

Alam, M.J.E., Saha, S., Hossain, M.J.: IoT-based smart energy management for renewable energy sources: a comprehensive review. Renew. Sustain. Energy Rev. **161**, 112308 (2022)

Kumar, N.M., et al.: Distributed energy resources and the application of AI, IoT, and blockchain in smart grids. Energies **13**(21), 5739 (2020)

Mellit, A., Tina, G.M., Kalogirou, S.A.: Fault detection and diagnosis methods for photovoltaic systems: a review. Renew. Sustain. Energy Rev. **91**, 1–17 (2018)

Mutlag, A.A., Ghani, M.K.A., Arunkumar, N., Mohamed, M.A., Mohd, O.: Enabling technologies for fog computing in healthcare IoT systems. Futur. Gener. Comput. Syst. **90**, 62–78 (2019)

Ren, J., Ren, X., Lee, J.: A comprehensive review of IoT applications in the energy sector: opportunities and challenges. IEEE Access **8**, 230380–230409 (2020)

Schmitt, P., Faix, P., Schlund, S.: IoT-based predictive maintenance: a systematic literature review. Comput. Ind. **132**, 103502 (2021)

Shafique, M., Thao, L.T., Jin, Y.: Intelligent monitoring and performance optimization for photovoltaic systems using internet of things. IEEE Internet Things J. **8**(17), 13161–13173 (2021)

Shaikh, P.H., Nor, N.B.M., Nallagownden, P., Elamvazuthi, I., Ibrahim, T.: A review on optimized control systems for building energy and comfort management of smart sustainable buildings. Renew. Sustain. Energy Rev. **34**, 409–429 (2014)

Vinuesa, R., et al.: The role of artificial intelligence in achieving the sustainable development goals. Nat. Commun. **11**(1), 1–10 (2020)

Zsiborács, H., et al.: Intermittent renewable energy sources: the role of energy storage in the European power system of 2040. Electronics **8**(7), 729 (2019)

Up-Link OFDMA Hybrid Access for Enabling IoT-Based Remote Video Monitoring Over Wi-Fi 7

Said Ould Amara(✉), Mohand Yazid, and Souhila Mammeri

Research Unit LaMOS (Modeling and Optimization of Systems), Faculty of Exact Sciences, University of Bejaia, 06000 Bejaia, Algeria
{said.ouldamara,mohand.yazid,souhila.mammeri}@univ-bejaia.dz

Abstract. With the emergence of Internet of Things (IoT) in recent years, a multitude of IoT applications have seen an exponential growth. Among these, remote video monitoring has become one of the most important IoT-based applications, facilitating the continuous monitoring of areas, assets or individuals from remote locations. For such applications, high throughput, wide bandwidth and efficient channel utilization are highly desirable. Consequently, the underlying network must have the capability to fulfil these requirements. The next generation of wireless networks IEEE 802.11be, also known as Wi-Fi 7, is announced to be key facilitator of IoT applications, thanks to its innovative features and techniques. A promising way to provide high throughput and efficient channel utilization is to introduce some modifications in its OFDMA technique originally introduced in IEEE 802.11ax amendment. More specifically, in the up-link OFDMA random access (UORA) procedure that handles the OFDMA transmissions in UL direction, some issues should be addressed. To this end, we propose the Up-link OFDMA Hybrid Access protocol which combines both scheduled and random modes to access the channel radio resources. The obtained simulation results demonstrate the effectiveness of the proposed protocol.

Keywords: Remote Video Monitoring · Wi-Fi 7 · OFDMA technique · UL Communications · UORA protocol

1 Introduction

Internet of Things (IoT) has rapidly positioned itself as a pioneering technology, linking billions of devices to the internet and facilitating a diverse range of applications across various sectors. Among these applications, remote video monitoring has emerged as one of the most critical, enabling continuous surveillance of areas, assets, or individuals from remote locations. Such applications are extensively used in various domains including healthcare for patient monitoring, security for surveillance systems, industrial automation for monitoring machinery and processes, and smart homes for enhanced safety and security [1]. These applications demand high throughput, wide bandwidth, and efficient

© The Author(s), under exclusive license to Springer Nature Switzerland AG 2025
M. Hatti (Ed.): IC-AIRES 2024, LNNS 1238, pp. 623–633, 2025.
https://doi.org/10.1007/978-3-031-80301-7_66

channel utilization to function effectively [2]. Therefore, the underlying network infrastructure must be able to meet these stringent requirements. The up-coming wireless networks standard IEEE 802.11be (Wi-Fi 7) is set to play a pivotal role in facilitating IoT applications due to its innovative features and advanced techniques [3]. One of the key technological advancements expected in Wi-Fi 7 is the enhanced Orthogonal Frequency Division Multiple Access (OFDMA) technique, which was originally introduced in the IEEE 802.11ax amendment. OFDMA allows multiple users to be served simultaneously in both Up-link (UL) and Down-link (DL) directions. It consists of dividing the transmission channel into smaller sub-channels called Resource Units (RUs), thereby improving spectral efficiency and overall network performance.

Regarding the UL direction, two access modes are used to access the RUs, namely: scheduled access (SA) and random access (RA) modes. However, the current uplink OFDMA random access (UORA) procedure, which manages OFDMA transmissions in the uplink direction at MAC layer, faces several issues (Back-off decrementing procedure, RUs allocation strategy, etc.) that need to be addressed to fully exploit the benefits of this technique [4]. To tackle these issues, we propose a new uplink OFDMA Hybrid Access (UOHA) protocol for IoT-based video monitoring applications operating over IEEE 802.11be networks. The proposed protocol combines both scheduled and random access modes to optimize the allocation of channel radio resources. The rest of this paper is structured as follows: the standard UORA protocol and its related existing enhancements are outlined in Sect. 2. In Sect. 3, we describe the operation functioning of the proposed UOHA protocol. Simulation results are discussed in Sect. 4. Section 5 concludes the paper.

2 Related Work

2.1 Background

Orthogonal Frequency Division Multiple Access (OFDMA) is a multi-user (MU) version of the Orthogonal Frequency Division Multiplexing (OFDM) modulation scheme. It has become a crucial technique in Wireless Local Area Networks (WLANs). It was introduced for the first time in the IEEE 802.11ax standard (Wi-Fi 6) [5]. OFDMA divides the available channel bandwidth into multiple independent sub-channels called Resource Units (RUs), which consist of a group of sub-carriers or tones. For instance, a 20 MHz transmission channel may be composed of 4 types of RUs, namely: 26-RU, 52-RU, 106-RU and 242-RU. The number of RUs that can be created from a 20 MHz channel is directly affected by the chosen types of RUs [6]. These RUs can be allocated to different stations (STAs) for simultaneous DL and UL transmissions.

To access radio resource units in UL direction, two OFDMA communication modes are defined in the IEEE 802.11ax amendment, specifically: Scheduled Access (SA) and Random Access (RA). The latter mode is recommended to access RUs in UL direction, since the queue state information is not always known by the Access Point (AP) [7]. The Up-link OFDMA Random Access (UORA) procedure is based on two main parameters, namely: OFDMA Contention Window (OCW) and OFDMA Back-Off (OBO). After receiving a Trigger Frame (TF) from the AP, each STA sets its OCW to the OCWmin value, then it selects randomly an OBO within the range 0 to OCW and decrements it by the number of RUs available for RA (RA-RUs). Following that, each STA with a non-positive OBO selects a random RA-RU for UL data transmission. The other STAs wait for the next TF. The AP uses a Multi-User Block Acknowledgment (MU-BA) to inform each STA about the success or failure of its transmission. In the case of a successful transmission, the OCW is reset to OCWmin. Otherwise, it is doubled until reaching OCWmax.

2.2 Existing Studies

In this section, we briefly describe the most existing studies dealing with UL OFDMA transmissions. Studies in [8–11] focused on scheduling approaches for UL transmissions. In [8], the authors examine the performance of several scheduling approaches. They formulated UL OFDMA scheduling as an optimization problem, in which, a utility function associated with the RU allocation is maximized under some constraints. Additionally, the study examines multiple classical schedulers to assess their impact on UL data transmission rates. The authors in [9] focused on the combined use of both random and scheduled access modes for up-link transmission in IEEE 802.11ax WLANs. They analysed the balance between improving network throughput and supporting new users (STAs) by defining a new performance metric named BSR delivery rate. In [10], the authors proposed a history-based method for scheduling OFDMA transmissions. This method determines priorities by considering the previous allocation decisions and then schedules transmissions using a round-robin algorithm. In [11], the authors proposed a delay sensitive OFDMA scheduling algorithm for the forthcoming IEEE 802.11be to deal with healthcare IoT applications. The proposed scheduler takes into account the traffic class (access category) as well as the former scheduling decisions, in order to allocate a RU for UL OFDMA communication. Works in [12, 13] and [14] had the common goal which is the improvement of the standard up-link OFDMA random access (UORA). In [12], to mitigate increased delays for re-transmissions caused by the doubled OCW value, the authors proposed the probability complementary transmission scheme (PCTS), in which the STAs perform complementary transmission without back-off. They showed through simulations that the proposed protocol outperforms UORA in throughput and packet delay. In [13], the authors designed the Cyclic Resource Assignment (CRA) algorithm to meet the latency and reliability requirements of real-time applications (RTA). CRA leverages UORA functionality to detect situations where multiple STAs require RUs and assigns RUs sequentially to each RTA STA, ensuring that, when needed, each STA receives a RU. The same authors later proposed the Group Resource Allocation (GRA) algorithm in [14], which further optimizes resource allocation for the upcoming IEEE 802.11be amendment. Its basic principle is as follows: when the AP detects that a given STA requires a RU, it divides the STAs into equal groups and assigns a RU to each group.

3 Proposed Protocol

To ensure the fulfilment of the requirements of IoT-based remote video monitoring applications, we propose an Up-link OFDMA Hybrid Access protocol called UOHA. This latter aims to offer a balanced solution for managing up-link OFDMA transmissions by combining scheduled and random access modes. To this end, we make the following assumptions: There are N STAs associated with a single AP. The UL traffic is saturated (the STAs always have pending data to be sent to the AP). The UL frames vary in size, following a uniform distribution. The transmission channel is ideal, with data loss occurring only due to collisions when multiple STAs transmit in the same RU.

UORA protocol, as it is defined in the IEEE 802.11ax standard, suffers from spectral efficiency issue, especially with a lower number of STAs. This is attributed to its random allocation strategy. In UOHA, we address this issue by the combined use of random access mode for BSRs delivery and scheduled access mode for UL data transmission. The UOHA operations are organized in two main phases as follows:

3.1 Collecting BSRs Form STAs

STAs send Buffer Status Reports (BSRs) in response to a Buffer Status Report Poll (BSRP) issued by the AP, to indicate their resource requirements. This is done by running the standard UORA operations rules. The AP utilizes the information relayed in BSRs as well as the number of successfully transmitted BSRs to schedule RUs for UL data transmission. The detailed operation functioning of this phase are shown in Fig. 1

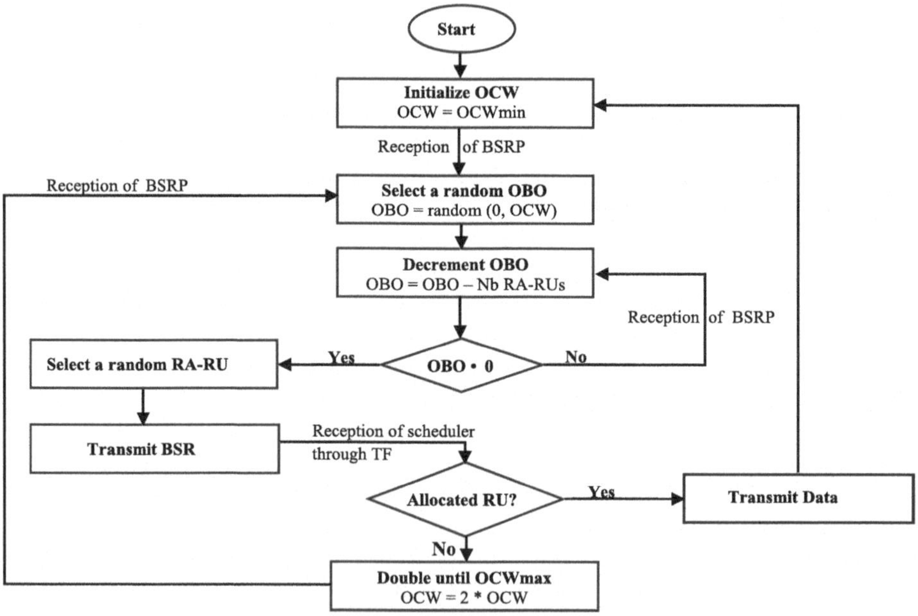

Fig. 1. A STA's operations in UOHA.

Each STA initializes its OFDMA Contention Window (OCW) to OCW_{min} value. Upon receiving a BSRP frame from the AP, each STA chooses a random OFDMA Back-off (OBO). After that, the OBO is decremented by the number of RUs dedicated for random access (RA-RUs). If a STA's OBO is less than or equals to 0, it randomly selects a RA-RU to transmit its BSR frame. Else, it will wait for the next BSRP frame. Upon receiving the scheduler through a Trigger Frame (TF) from the AP, each STA verifies if an RU is allocated to it. If it is the case, it will transmit its data and then reset its OCW. Otherwise, it doubles its OCW (until OCW_{max} is reached) and waits for the next cycle.

3.2 Scheduling RUs for UL Data Transmission

After receiving Buffer Status Reports (BSRs) from STAs, the Access Point (AP) utilizes two main information to schedule Resource Units (RUs) for STAs. These include the buffer status of STAs and the number of successfully transmitted BSRs. Based on this information, Table 1 enumerates the possible combinations of RA-RUs within a 20 MHz channel. These combinations are set to maximize bandwidth utilization and the number of served users.

Table 1. Possible combinations of available RA-RUs in a 20 MHz channel.

Number of available RA-RUs	Number of successfully received BSRs	Allocated RUs	Loss in bandwidth
1	1	1 × 26-RU	0
2	1	1 × 52-RU	0
	2	2 × 26-RU	0
3	1	1 × 52-RU	1 × 26-RU
	2	1 × 52-RU + 1 × 26-RU	0
	3	3 × 26-RU	0
4	1	1 × 106-RU	0
	2	2 × 52-RU	0
	3	1 × 52-RU + 2 × 26-RU	0
	4	4 × 26-RU	0
5	1	1 × 106-RU	1 × 26-RU
	2	2 × 52-RU	1 × 26-RU
	3	2 × 52-RU + 1 × 26-RU	0
	4	1 × 52-RU + 3 × 26-RU	0
	5	5 × 26-RU	0
6	1	1 × 106-RU	2 × 26-RU
	2	1 × 106-RU + 1 × 26-RU	1 × 26-RU
	3	1 × 106-RU + 2 × 26-RU	0

(continued)

Table 1. (*continued*)

Number of available RA-RUs	Number of successfully received BSRs	Allocated RUs	Loss in bandwidth
	4	2 × 52-RU + 2 × 26-RU	0
	5	1 × 52-RU + 4 × 26-RU	0
	6	6 × 26-RU	0
7	1	1 × 106-RU	3 × 26-RU
	2	1 × 106-RU + 1 × 52-RU	1 × 26-RU
	3	1 × 106-RU + 1 × 52-RU + 1 × 26-RU	0
	4	3 × 52-RU + 1 × 26-RU	0
	5	2 × 52-RU + 3 × 26-RU	0
	6	1 × 52-RU + 5 × 26-RU	0
	7	7 × 26-RU	0
8	1	1 × 106-RU	4 × 26-RU
	2	1 × 106-RU + 1 × 52-RU	2 × 26-RU
	3	1 × 106-RU + 1 × 52-RU + 1 × 26-RU	1 × 26-RU
	4	1 × 106-RU + 1 × 52-RU + 2 × 26-RU	0
	5	1 × 106-RU + 4 × 26-RU	0
	6	2 × 52-RU + 4 × 26-RU	0
	7	1 × 52-RU + 6 × 26-RU	0
	8	8 × 26-RU	0
9	1	1 × 242-RU	0
	2	2 × 106-RU	1 × 26-RU
	3	2 × 106-RU + 1 × 26-RU	0
	4	1 × 106-RU + 2 × 52-RU + 1 × 26-RU	0
	5	4 × 52-RU + 1 × 26-RU	0
	6	3 × 52-RU + 3 × 26-RU	0
	7	2 × 52-RU + 5 × 26-RU	0
	8	1 × 52-RU + 7 × 26-RU	0
	9	9 × 26-RU	0

The detailed scheduling operations (the case of 9 × 26-RUs available for RA is considered) are shown in Fig. 2:

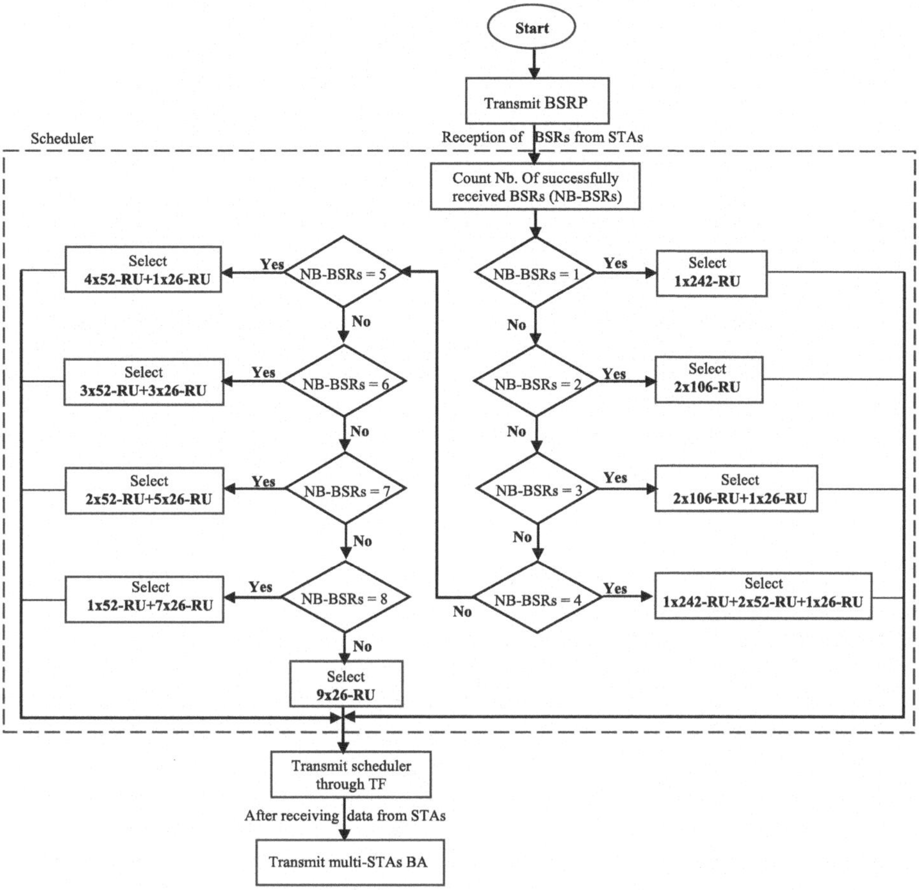

Fig. 2. Scheduling operation (case where the number of RA-RUs equals to 9).

Upon receiving BSRs from STAs, the AP counts the number of successfully received BSRs. According to this latter and the number of available RA-RUs, the AP selects the combination of RUs to be adopted as an allocation scheme (see Table 1, case where the number of RA-RUs equals to 9). After that, the scheduling information is transmitted by the AP through a Trigger Frame (TF).

4 Simulation Results

In this section, we present the simulation results obtained with the proposed UOHA protocol. To demonstrate its effectiveness, a comparison is made with the standard UORA protocol. Here, we particularly focus on the following metrics: throughput, Allocation ratio, loss in bandwidth and fairness. Simulations are conducted using MATLAB, and the most important parameters are listed and described in Table 2.

Table 2. Simulation parameters.

Parameter	Description	Value
CBW	Channel bandwidth	20 MHz
N	Number of STAs	from 10 to 50
RA-RUs	Number of RUs dedicated to RA	from 1 to 9
OCW_{min}	OFDMA minimal contention window	7
OCW_{max}	OFDMA maximal contention window	31
TF	Trigger Frame length	28 Bytes at least
BSR	BSR Frame length	30 Bytes
MPDU	MPDU length	11454 Bytes at most
C_{max}	Number of communication cycles simulated	1000 cycles
N_{sim}	Number of times a simulation is repeated	100 times

In Fig. 3, we evaluate the UOHA protocol and compare its performance with the standard UORA protocol by varying the network size from 10 to 50 STAs and considering the maximum number of RA-RUs, which is 9.

In Fig. 3A, we compare the throughput performance of the UOHA and UORA protocols. The UOHA protocol significantly achieves higher throughput compared to the UORA protocol, regardless the size of the network. Unlike UORA protocol that uses only narrowed RUs (26-RUs), the UOHA protocol also uses RUs of a larger size (52-RUs, 106-RUs and even 242-RU) when the number of successfully transmitted BSRs (the number of UL streams) is strictly smaller than 9. As the number of STAs increases, the throughput of both protocols decreases due to increased contention and collisions. The allocation ratio results are depicted in Fig. 3B. The proposed UOHA protocol consistently achieves a higher allocation ratio compared to UORA across all numbers of STAs. For instance, with 10 STAs, UOHA achieves a maximum allocation ratio of 96%, whereas UORA achieves 37%. This is attributed to the fact that, UOHA has a better utilization of available RUs due to its hybrid access approach. In both protocols, the allocation ratio decreases when the size of the network increases. This is due to the increased contention and collisions. Figure 3C shows the bandwidth loss for both protocols. The UOHA protocol consistently exhibits lower bandwidth loss. In fact, unlike UORA, UOHA protocol uses larger RUs in addition to narrower ones, which reduces waste and improves efficiency. As the number of STAs increases, bandwidth loss increases for both protocols. This statement is evident because, the greater the size of the network, the higher the number of collisions, resulting in a higher bandwidth loss ratio. In Fig. 3D, we compare the fairness index (in term of number of times the STAs in the network are successfully served) of the UOHA and UORA protocols. To this end, we carried out the following steps as described in [15]: First, we count during the simulation, the number of times $X[i]$ that each STA i has been successfully served. Then, we use statistical formulas to determine: the average number of times ($\overline{X}$) that any STA would be successfully served, the dispersion (σ_X) of the $X[i]$ values around

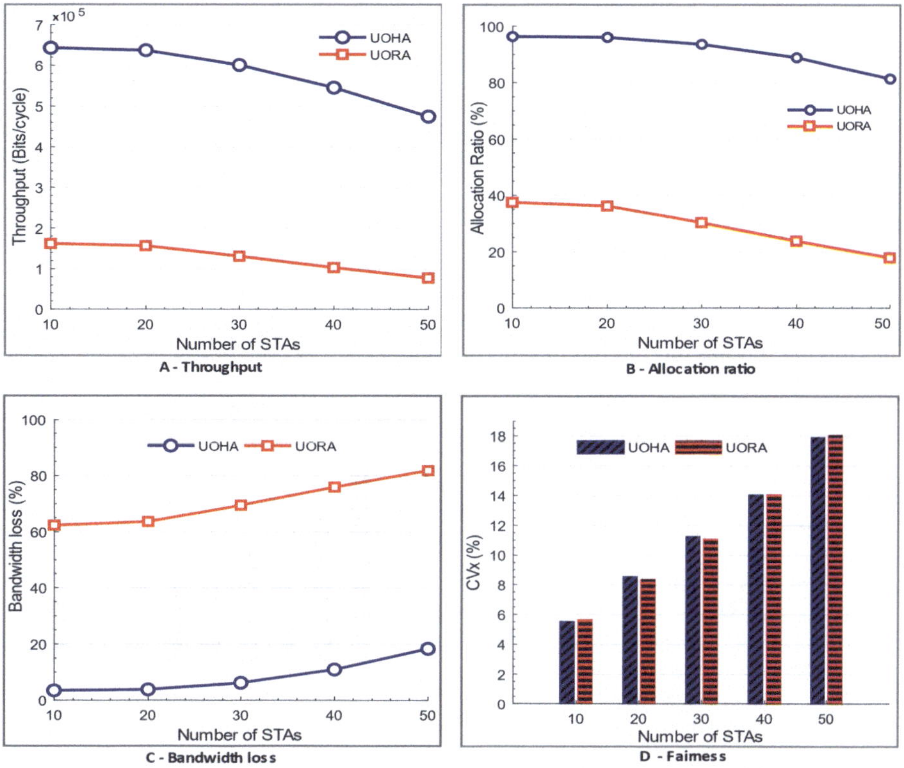

Fig. 3. Evaluating and comparing UOHA to UORA according to the number of STAs.

their mean, and the level of dispersion (CV_X) of the sets of X values relative to their corresponding averages. We show in Fig. 3D that both UOHA and UORA protocols have almost the same level of dispersion of their $X[i]$ values around the mean. In other words, UOHA protocol is as fair as UORA regardless the size of the network.

In Fig. 4, we evaluate the performance of the UOHA protocol and compare it with the standard UORA protocol by varying the number of RUs available for random access (RA-RUs), and considering a little size of the network (10 STAs).

In Fig. 4A, we compare the achieved throughput with both protocols. We clearly remark the ability of the proposed UOHA protocol to reach a higher throughput for any given number of RUs. This is attributed to the hybrid access method and the flexibility in RU sizes (utilization of narrow and large RUs). Moreover, the achievable throughput in both protocols increases in proportion with the number of available RUs. This due to the fact that, the greater the number of RUs, the higher the amount of UL data transmitted. In Fig. 4B, we compare the achieved allocation ratio of both UOHA and UORA protocols. UOHA clearly exhibits a significantly higher allocation ratio compared to the standard UORA protocol. This is due to the ability to utilize both narrow and large RUs by UOHA to effectively match resource allocation with UL streams, resulting in higher allocation ratio. The allocation ratio of UORA protocol remains almost the same

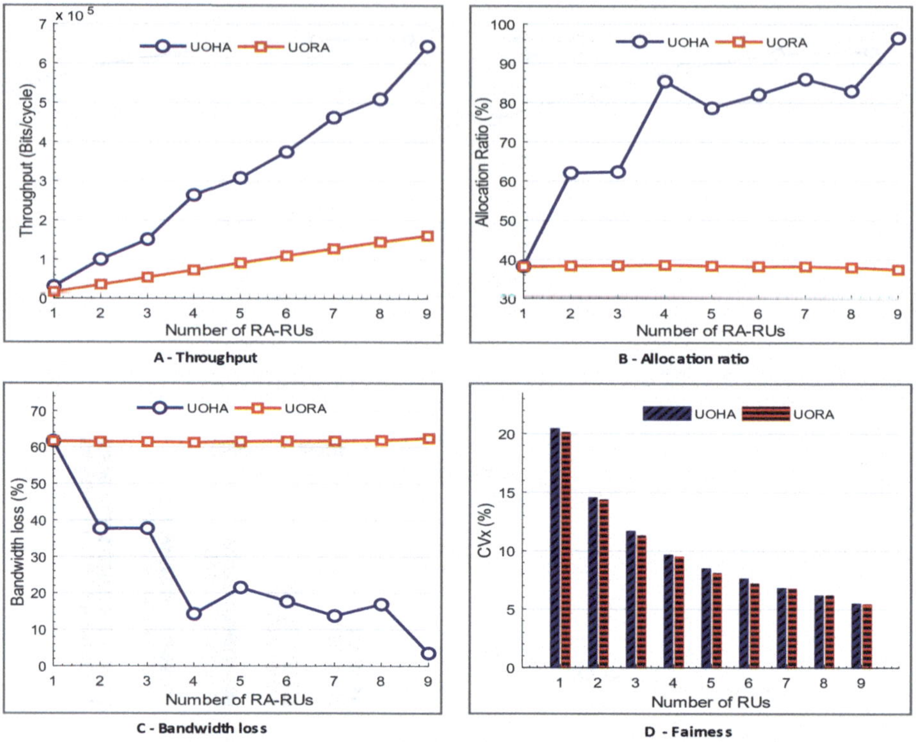

Fig. 4. Evaluating and comparing UOHA to UORA according to the number of RA-RUs.

regardless the number of RA-RUs. This stability is justified by the random allocation strategy of UORA protocol and its exclusive use of only narrower RUs. In Fig. 4C, we compare the bandwidth loss (in terms of the proportion of available bandwidth that is not effectively used for data transmission) between UOHA and UORA protocols. UOHA shows a lower bandwidth loss compared to UORA. As for the allocation ratio, the bandwidth loss in UOHA is highly dependent on the allocation strategy and the combined use of narrower and larger RUs. Consequently, a better allocation ratio means a lower bandwidth loss. We also remark a stability bandwidth loss of UORA, this is due to its random allocation strategy. In Fig. 4D, we evaluate the fairness index of UOHA to that of UORA. Both protocols have almost the same level of dispersion of their $X[i]$ values around the mean. In Other words, UOHA protocol is as fair as UORA regardless the number of RUs dedicated to random access.

5 Conclusion

In this research work, we addressed the problem of spectral efficiency of UORA and improved the overall throughput by proposing a new up-link OFDMA hybrid access (UOHA) protocol for the up-coming WLANs IEEE 802.11be standard. The combination use of both scheduled and random access modes significantly enhances throughput and ensures efficient utilization of the channel radio resources, which are the key requirements in IoT-based video monitoring applications. Simulation results have clearly demonstrated the superiority of the proposed protocol in terms of throughput, allocation ratio, bandwidth loss compared to the standard UORA protocol, while maintaining the same level of fairness.

References

1. Vishwakarma, A.K., Chaurasia, S., Kumar, K., Singh, Y.N., Chaurasia, R.: Internet of things technology, research, and challenges: a survey. Multimed. Tools Appl. (2024). https://doi.org/10.1007/s11042-024-19278-6
2. Uddin, R., Koo, I.: Real-time remote patient monitoring: a review of biosensors integrated with multi-hop IoT systems via cloud connectivity. Appl. Sci. **14**(15), 1876 (2024)
3. Khorov, E., Levitsky, I., Akyildiz, I.F.: Current status and directions of IEEE 802.11be, the future Wi-Fi 7. IEEE Access **8**, 88664–88688 (2020)
4. Brahmi, S., Yazid, M.: Towards a fair allocation and effective utilization of resource Units in multi-user WLANs-based OFDMA technology. Comput. Netw. **224**, 109639 (2023)
5. IEEE Standard for Information Technology–telecommunications and information exchange between systems local and metropolitan area networks–specific requirements part 11: wireless LAN medium access control (MAC) and physical layer (PHY) specifications Amendment1: enhancements for high-efficiency WLAN, pp. 1–767. In: IEEE Std 802.11ax-2021(Amendment to IEEE Std 802.11-2020) (2021)
6. Coleman, D.D., Westcott, D.A.: Chapter 19: 802.11ax: high efficiency (HE). In: Certified wireless network administrator study guide, 5 th ed., pp. 784–814. John Wiley & Sons (2018)
7. Brahmi, S., Yazid, M.: Towards an efficient allocation of resource units in multi-user of DMA communications. In: Artificial intelligence and heuristics for smart energy efficiency in smart cities, pp. 536–544. Springer International Publishing (2022)
8. Bankov, D., Didenko, A., Khorov, E. and Lyakhov, A.: OFDMA uplink scheduling in IEEE 802.11ax networks. In: 2018 IEEE international conference on communications (ICC), Kansas City, MO, USA, pp. 1–6 (2018)
9. Naik, G., Bhattarai, S., Park, J.M.: Performance analysis of uplink multi-user OFDMA in IEEE 802.11ax. In: 2018 IEEE international conference on communications (ICC), Kansas City, MO, USA, pp. 1–6 (2018)
10. Avallone, S., Imputato, P., Redieteab, G., Ghosh, C., Roy, S.: Will OFDMA improve the performance of 802.11 Wifi networks? IEEE Wirel. Commun. **28**(3), 100–107 (2021)
11. Qadri, Y.A., Zulqarnain, Nauman, A., Musaddiq, A., Garcia-Villegas, E., Kim, S.W.: Preparing wi-fi 7 for healthcare internet-of-things. Sensors **22**(16), 6209 (2022)
12. Wang, J., Wu, M., Chen, Q., Zheng, Y. and Zhu, Y.H.: Probability complementary transmission scheme for uplink OFDMA-based random access in 802.11ax WLAN. In: 2019 IEEE wireless communications and networking conference (WCNC), Marrakesh, Morocco, pp. 1–7 (2019)
13. Avdotin, E., Bankov, D., Khorov, E., Lyakhov, A.: OFDMA resource allocation for real-time applications in IEEE 802.11ax Networks. In: IEEE international black sea conference on communications and networking (BlackSeaCom), Sochi, Russia, pp. 1–3 (2019)
14. Avdotin, E., Bankov, D., Khorov, E., Lyakhov, A.: Enabling massive real-time applications in IEEE 802.11be Networks. In: IEEE 30th annual international symposium on personal, indoor and mobile radio communications (PIMRC), Istanbul, Turkey, pp. 1–6 (2019)
15. Hocini, K., Yazid, M., Ksentini, A.: Toward an efficient and coordinated control of full-duplex radio communications in high-efficiency wireless local area networks. Int. J. Commun. Syst.Syst **35**(10), e5165 (2022)

Double Threshold Energy Detection Algorithm in Cognitive Radio/mmWave Mobile Communications at 60 GHz

Amel Haouzi[1(✉)], Samir Kameche[1], and Haroun Errachid Adardour[2]

[1] Department of Telecommunications, Faculty of Technology, UABT University, Tlemcen, Algeria
{amel.haouzi,samir.kameche}@univ-tlemcen.dz
[2] Department of Electronics, Faculty of Technology, UHBC University, Chlef, Algeria
h.adardour@univ-chlef.dz

Abstract. Millimeter waves (mmWaves) in the V band have recently become widely used in cognitive/5G radio networks to improve spectral efficiency. This paper presents an algorithm based on three techniques: Alpha Beta (α-β) Filter, Simple Recursive (S-R) estimator, and double threshold energy detection, which have been used to analyze the detection of the primary signal performance by the cognitive user 5G (CU_{5G}) in a cluttered indoor environment at 60 GHz. The CU_{5G} is considered to be in movement using a Gauss Markov mobility model at low speed. This study evaluates the received signal power at the CU_{5G} under the AWGN channel. To achieve this, we have implemented two propagation models: Close In and Floating-Intercept. These models simulate different line-of-sight (LoS) and non-line-of-sight (NLoS) scenarios, ensuring a comprehensive evaluation of the algorithm's performance. Finally, the probability of total detection error has been examined in the proposed algorithm, further validating its effectiveness.

Keywords: MmWaves · cognitive/5G radio · α-β Filter · S-R estimator · primary signal · double threshold · propagation models

1 Introduction

Nowadays, 5G wireless communication networks have been widely used due to their high-speed data rates and increased dependability and resistance to interference and noise. However, the rapid growth of daily wireless users has led to a shortage of more wireless spectrum resources, addressed using cognitive radio technology in millimeter waves (mmWaves) spectrum WLAN (Wireless Local Area Networks) band. The cognitive radio 5G technology has effectively solved spectrum inefficiency by providing dynamic spectrum access to cognitive users 5G (CUs_{5G}) to use the primary base station (PBS) band (Ajani et al. 2021; Li et al. 2018). To achieve this, CUs_{5G} or unlicensed users require sensing the spectrum to decide the presence/absence of the PBS, and then CUs_{5G} can use that spectrum hole (vacant particular band) to transmit information data. At the same time, the PBS is absent (Haouzi et al. 2022; Zhu et al. 2008). In recent years,

© The Author(s), under exclusive license to Springer Nature Switzerland AG 2025
M. Hatti (Ed.): IC-AIRES 2024, LNNS 1238, pp. 634–645, 2025.
https://doi.org/10.1007/978-3-031-80301-7_67

mmWaves have become a promising candidate for 5G small cells in urban areas. The Federal Communications Commission (FCC) allotted 7 GHz of spectrum for unlicensed usage in the 60 GHz band, ranging from 57 GHz to 64 GHz (Li et al. 2018).

Establishing a constant and effective communication link between the PBS (transmitter T_x) and the CUs_{5G} (receiver R_x) is the goal of any dependable transmission. In contrast, the path loss is influenced by various variables, including the signal's frequency, the distance between the CUs_{5G} and the PBS, and the state of the surrounding cluttered environment. Thus, the mmWaves propagation path-loss models such as Floating-Intercept (FI) and Close-In (CI) for the indoor environment in LoS (Line-of-Sight) and NLoS (Non-Line-of-Sight) scenarios have been studied. Furthermore, the reduction of the received signal power for both propagation models must be investigated. A key tool in this investigation is the Simple Recursive (S-R) estimator, which plays a crucial role in stabilizing the signal strength measurements at the CUs_{5G} node (Adardour and Kameche 2017, 2022).

To assess spectrum detection performance, we compared traditional (single threshold) and Double Threshold Energy Detection (DTED) methods in Additive White Gaussian Noise (AWGN) channels. We used the probability of total detection error to evaluate the primary signal sensing level. However, the proposed algorithm is organized into three main parts. The first part uses the Gauss Markov Mobility Model (GMMM) to predict the user's (CUs_{5G}) trajectory and implements the Alpha Beta Filter (α-β-Filter) to reduce the impact of user mobility (CUs_{5G}) on the received primary signals of PBS. The second part focuses on estimating the path loss of the communication and interference propagation models in indoor environments using a S-R estimator to ensure stable signal power measurement. This allows for calculating the signal-to-noise-ratio (SNR) uncertainty for each link between the CUs_{5G} node and the PBS node. Finally, the simulation results of the cognitive mmWaves Spectrum Sensing (SS) performances at 60 GHz are evaluated and analyzed using traditional/ DTED under AWGN channels.

The rest of the paper is organized into four sections, with a general introduction to the 5G mmWaves communication technology at 60 GHz. Section 2 covers the proposed 5G SS algorithm and the DTED theory. Section 3 shows simulation results and analysis. Finally, we draw our conclusions in Sect. 4

2 System Description

2.1 CUs_{5G} Trajectory Tracking

During the initial phase, the GMMM model was used to estimate the position and speed of the CUs_{5G}. Subsequently, the α-β-Filter was employed to minimize the impact of CUs_{5G} mobility on the PBS's sensing performance.

2.1.1 Gauss Markov Mobility Model

The versatile Gauss-Markov Mobility Model (GMMM), introduced by Liang and Haas, was initially proposed to simulate Personal Communication Service (PCS). In this model, the speed and direction of each CUs_{5G} node are updated based on their past values at earlier time intervals. The computation of these two values can be adjusted to suit the characteristics of the simulated wireless network, providing a reassuring level of adaptability (Alenazi et al. 2020).

The speed S_{CUs5G}(t) and direction D_{CUs5G}(t) of each node in a two-dimensional simulation field are determined using the Gauss-Markov stochastic process (Alenazi et al. 2020) in the following manner:

$$S_{CUs5G}(t) = \eta \cdot S_{CUs5G}(t-1) + (1-\eta) \cdot \overline{S}_{CUs5G} + \left(\sqrt{1-\eta^2}\right).W_{S_{CUs5G}}(t-1) \quad (1)$$

$$D_{CUs5G}(t) = \eta \cdot D_{CUs5G}(t-1) + (1-\eta) \cdot \overline{D}_{CUs5G} + \left(\sqrt{1-\eta^2}\right) \cdot W_{D_{CUs5G}}(t-1) \quad (2)$$

Here, S_{CUs5G}(t) and D_{CUs5G}(t) represent the speed and direction of the CUs_{5G} node's path at time t, respectively. The constants $\overline{S}_{CUs5G}$ and $\overline{D}_{CUs5G}$ indicate the mean values of the speed and direction, while $W_{S_{CUs5G}}$(t − 1) and $W_{D_{CUs5G}}$(t − 1) are random variables from a Gaussian distribution with zero mean and standard deviation σ at time t − 1. The η parameter, which ranges between 0 and 1, is crucial in determining the degree of randomness in the GMMM.

The future position (X_{CUs5G}(t), Y_{CUs5G}(t)) of the CUs_{5G} is determined by the present location (X_{CUs5G}(t − 1), Y_{CUs5G}(t − 1)), the speed S_{CUs5G}(t − 1), and the direction D_{CUs5G}(t − 1) of the previous movement (Alenazi et al. 2020). This relationship is expressed in Eqs. (3) and (4).

$$X_{CUs5G}(t) = X_{CUs5G}(t-1) + S_{CUs5G}(t-1) \cdot cos(D_{CUs5G}(t-1)) \quad (3)$$

$$Y_{CUs5G}(t) = Y_{CUs5G}(t-1) + S_{CUs5G}(t-1) \cdot sin(D_{CUs5G}(t-1)) \quad (4)$$

where, (X_{CUs5G}(t), Y_{CUs5G}(t)) and (X_{CUs5G}(t − 1), Y_{CUs5G}(t − 1)) are the respective 2nd dimensional coordinates of the CUs_{5G} position node at times t and (t − 1); S_{CUs5G}(t − 1) and D_{CUs5G}(t − 1) are the speed and direction of the CUs_{5G} node in the two previous Eqs. (1) and (2), respectively, at time (t − 1). The α-β-Filter was used to estimate the position and velocity of the CUs_{5G} in the following section.

2.1.2 Alpha Beta Filter (α-β-Filter)

This adaptive predictive filter, using two gains, α and β, estimates the position and speed of CUs_{5G}. It is based on a constant acceleration model, ensuring it can handle various scenarios. The filter employs prediction and smoothing as its two steps (Sighencea et al. 2022). The prediction step is expressed mathematically (Adardour and Kameche 2017) by the following state Eqs. (5) and (6).

$$Axis\,x \begin{cases} x_p(i) = x_s(i-1) + T.v_{xs}(i-1) + \left(\frac{T^2}{2}\right).acc_{xs}(i-1) \\ v_{xp}(i) = v_{xs}(i-1) + T.acc_{xs}(i-1) \end{cases} \quad (5)$$

$$Axis\,y \begin{cases} y_p(j) = y_s(j-1) + T.v_{ys}(j-1) + \left(\frac{T^2}{2}\right).acc_{ys}(j-1) \\ v_{yp}(j) = v_{ys}(j-1) + T.acc_{ys}(j-1) \end{cases} \quad (6)$$

In Eqs. (5) and (6), $\left(x_p(i), y_p(j)\right)$ and $\left(x_s(i-1), y_s(j-1)\right)$ represent the predicted and smoothed coordinates of the position in the subsequent and current iterations. Similarly, $\left(v_{xp}(i), v_{yp}(j)\right)$ and $\left(v_{xs}(i-1), v_{ys}(j-1)\right)$ indicate the predicted and smoothed values of the speed. Lastly, $\left(acc_{xs}(i-1), acc_{ys}(j-1)\right)$ designate the smoothed values of the acceleration. T, the duration between each sample event, is a key variable that significantly influences our calculations.

The corrective (smoothing) step (Adardour and Kameche 2017) is determined based on the calculations in (7) and (8), considering the previously anticipated values:

$$X_s(k) = X_p(k) + \alpha.\left(X_0(k) - X_p(k)\right) \tag{7}$$

$$V_s(k) = V_p(k) + \left(\frac{\beta}{T}\right).\left(X_0(k) - X_p(k)\right) \tag{8}$$

X_0 represents the accurately measured position node of the CUs_{5G}.

To determine the signal-to-noise ratio (SNR) between the CUs_{5G} and the PBS, the amount of signal energy received at the CUs_{5G} level must be forecasted. This prediction is based on the expected distance between the PBS and the CUs_{5G}, which has been adjusted using the Alpha Beta Filter.

2.2 SNR Link Estimate

The α-β-Filter enables the calculation of the estimated distance between the PBS and the CUs_{5G} nodes using the Euclidean distance in meters (Adardour and Kameche 2017, 2022). This can be represented as follows:

$$\hat{d} = \sqrt{\left(x_{i(PBS)} - x'_{i(CUs_{5G})}\right)^2 + \left(y_{j(PBS)} - y'_{j(CUs_{5G})}\right)^2} \tag{9}$$

The coordinates $\left(x'_{i\,(CUs_{5G})}, y'_{j\,(CUs_{5G})}\right)$ of CUs_{5G} are estimated using the Alpha Beta Filter, while the coordinates $\left(x_{i\,(PBS)}, y_{j\,(PBS)}\right)$ of PBS are fixed at (0, 0). This estimated distance, denoted as $\hat{d}$, will be included in the propagation models required to determine the received signal power at the CUs_{5G} node and the SNR of each connection linking the PBS and the CUs_{5G} node.

2.2.1 The Received Signal Power Estimate

The received signal power at the CUs_{5G} node is evaluated and computed in decibels, starting with the PBS and ending at the receiver node CUs_{5G} (Adardour and Kameche 2017, 2022). The Eq. (10) is given.

$$P_{Rx(CUs_{5G})}\left(\hat{d}\right)[dB] = P_{Tx(PBS)}[dB] - PL\left(f, \hat{d}\right)[dB] \tag{10}$$

$P_{Rx\,(CUs_{5G})}$ represents the estimated received signal strength at the CUs_{5G} node in decibels (dB). $P_{Tx\,(PBS)}$ refers to the transmit power originating from the primary base station PBS, also in dB. $PL(f, \hat{d})$, a significant metric, represents the path loss at a distance $\hat{d}$, measured in dB.

Our proposed method involves using two propagation models for 5G mmWave technology.

a) CI model

The Close-In (CI) model is considered the most straightforward 5G propagation model currently available (Qamar et al. 2021; Sun et al. 2016). It relies heavily on the $FSPL(f, d_0)$ formula, a crucial element primarily influenced by the carrier signal frequency (f in GHz) and the reference physical distance separation d_0. The equation representing the CI model is provided in Eq. (11):

$$\begin{aligned} PL^{CI}(f, \hat{d})[dB] &= FSPL(f, d_0) + 10 \cdot n \cdot log_{10}\left(\frac{\hat{d}}{d_0}\right) + X_\sigma^{CI} \\ &= 20 \cdot log_{10}\left(\frac{4\pi \cdot d_0 \cdot f}{c}\right) + 10 \cdot n \cdot log_{10}\left(\frac{\hat{d}}{d_0}\right) + X_\sigma^{CI} \end{aligned} \tag{11}$$

The $FSPL(f, d_0)$ formula, a crucial expression in our understanding of free space path-loss, represents the path-loss in decibels at a Close-In (CI) range, with the reference distance set at d_0=1 m. The variable n denotes the path-loss exponent, c represents the speed of light, and X_σ^{CI} is a zero mean Gaussian distributed random variable with a standard deviation of σ.

b) FI model

The Floating Intercept (FI) propagation model is used to calculate the optimal lowest error fit value (Qamar et al. 2021) using the FI α in dB and the β line values. The following equation is utilized to compute the FI model:

$$PL^{FI}\left(\hat{d}\right)[dB] = \alpha + 10 \cdot \beta \cdot log_{10}\left(\hat{d}\right) + X_\sigma^{FI} \tag{12}$$

Here, X_σ^{FI} is a zero mean Gaussian random variable in dB.

To estimate the $P_{Rx\ (CUs_{5G})}$ at the CUs_{5G} node in a crowded area, a straightforward SR estimator is used. This method is believed to be more stable and less affected by noise (Adardour and Kameche 2017, 2022). The following equation demonstrates the estimation:

$$\overline{P_{Rx(CUs_{5G})}}\left(\hat{d}, t\right) = \Delta * P_{Rx(CUs_{5G})}\left(\hat{d}, t\right) + (1 - \Delta) * \overline{P_{Rx(CUs_{5G})}}\left(\hat{d}, t - 1\right) \tag{13}$$

where, Δ is the weighting factor, in the condition $(0 < \Delta < 1)$.

2.2.2 SNR Sensitivity Estimate

Equation (14) depicts the estimated sensitivity of the $\overline{SNR_{(PBS, CUs_{5G})}}$ for each link distance between the CUs_{5G} and the PBS (Adardour and Kameche 2017, 2022).

$$\overline{SNR_{(PBS, CUs_{5G})}}\left(\hat{d}, t\right)[dB] = \overline{P_{Rx(CUs_{5G})}}\left(\hat{d}, t\right)[dB] - P_{Noise}[dB] \tag{14}$$

where, P_{Noise} is a noise power.

The objective of the next section is to predict the significant primary signal level and evaluate it using information from previous steps.

2.3 Performance of 5G Spectrum Sensing Technique

The signal received at the CUs_{5G} can be represented in the literature as a binary hypothesis testing issue. The detector's task is to distinguish between two hypotheses, as described in references (Adardour and Kameche 2022, Lertsinsrubtavee et al. 2013):

$$\begin{cases} \overline{S}_{Rx(CUs_{5G})}(t) = N_{AWGN}(t)\{H_0\} \\ \overline{S}_{Rx(CUs_{5G})}(t) = \overline{S}_{Tx(CUs_{5G})}(t) + N_{AWGN}(t)\{H_1\} \end{cases} \tag{15}$$

where, $\overline{S}_{Rx(CUs_{5G})}(\mathrm{t})$ is the received signal estimate at the CUs_{5G} from the PBS with zero mean and variance $\sigma_s{}^2$;$\overline{S}_{Tx(CUs_{5G})}(\mathrm{t})$ is the transmitted signal by the primary base station and $N_{AWGN}(t)$ is the additive white Gaussian noise (AWGN) $\sim N(0, \sigma_n^2)$ with zero mean.H_0 is the null hypothesis, which indicates that PBS is absent and H_1 is the alternative hypothesis, which indicates that PBS is present. The test statistic (Adardour and Kameche 2017) is given as follows, providing a clear and concise overview of our research process:

$$\begin{cases} \{H_0\} \Rightarrow \overline{E} = \left(\frac{1}{L}\right) \cdot \sum_{n=1}^{L} \left|\overline{S}_{Rx(CUs_{5G})}(t)\right|^2 < \lambda \\ \{H_1\} \Rightarrow \overline{E} = \left(\frac{1}{L}\right) \cdot \sum_{n=1}^{L} \left|\overline{S}_{Rx(CUs_{5G})}(t)\right|^2 > \lambda \end{cases} \tag{16}$$

Let's explore the energy detection process.$\overline{E}$ is the energy level of the received signal estimate at the CUs_{5G};λ is the energy detection threshold, and L is the number of samples taken for sensing. The received signal energy, a critical factor in our decision-making process, is then compared with a single threshold λ. This comparison is a crucial step in our process. If the energy level is higher than the threshold λ, it is decided H_1; if not, it is decided H_0.

2.3.1 Simple Energy Detector (Single Threshold-Based Detection)

The conventional energy detection method, with its use of a single threshold λ to verify the presence of the PBS, is a reliable approach. Moreover, we can approximate the energy received signal on a statistic test T_s for each CUs_{5G} node at the end of the sensing time. This can be modeled as in (Shen and Xie 2012), as the standard Gaussian distribution in two hypotheses, as follows:

$$T_s = \begin{cases} N\left(\sigma_n{}^2, \sigma_n{}^2/L\right)\{H_0\} \\ N\left(\sigma_n{}^2 + \overline{E}, \left(\sigma_n{}^2 + \overline{E}\right)^2/L\right)\{H_1\} \end{cases} \tag{17}$$

where, $\sigma_n{}^2$ is the noise variance;$\overline{E}$ is the estimated signal power and L is the sampling number.

For a practical assessment of spectrum detection in the 5G system under an AWGN channel (Samarah et al. 2015), the Probability of Total Detection Error (PTDE) is a valuable tool. This probability, determined by the Probability of Detection (PD) and the Probability of False Alarm (PFA), as defined by Eqs. (18) and (19), has direct applications in our work.

$$PD = Prob\{T_s > \lambda | H_1\} = Q\left(\frac{\lambda - L\left(\sigma_s{}^2 + \sigma_n{}^2\right)}{\sqrt{2L\left(\sigma_s{}^2 + \sigma_n{}^2\right)^2}}\right) \tag{18}$$

$$PFA = Prob\{Ts > \lambda|H_0\} = Q\left(\frac{\lambda - L\sigma_n{}^2}{\sqrt{2L\sigma_n{}^4}}\right) \tag{19}$$

$$PTDE = 1 - PD + PFA \tag{20}$$

where, Q(•) represents the Gaussian Q-function and $\sigma_s{}^2$ is the variance of the received signal.

Since the probability of false alarms is predetermined (Alijani and Osman 2020), the threshold value can be determined from Eq. (21)

$$\lambda = \left(\sqrt{\frac{2}{L}}Q^{-1}(PFA) + 1\right) \cdot \sigma_n{}^2 \tag{21}$$

2.3.2 Double Threshold Energy Detection Method

In cognitive energy detection (CED), each unlicensed CUs_{5G} can utilize the licensed frequencies while the PBS is absent, which makes its local decision based on the single threshold. To achieve a good SS performance, we introduce an innovative double-threshold detector algorithm. This algorithm, with its two thresholds,λ_H(upper threshold) and λ_L(lower threshold), is designed to make a decision that depends on the noise uncertainty parameter ρ in (Alijani and Osman 2020; Lertsinsrubtavee et al. 2013). They are represented as follows:

$$\lambda_L = \left(\sqrt{\frac{2}{L}}Q^{-1}(PFA) + 1\right) \cdot \frac{1}{\rho} \cdot \sigma_n{}^2 \tag{22}$$

$$\lambda_H = \left(\sqrt{\frac{2}{L}}Q^{-1}(PFA) + 1\right) \cdot \rho \cdot \sigma_n{}^2 \tag{23}$$

where, ρ represents the received SNR by the CUs_{5G}, which is defined as $\rho = \frac{\sigma_s{}^2}{\sigma_n{}^2}$.

The two thresholds are considered to be higher and lower than the single threshold level λ. In this context,λ_H is regarded as the maximum threshold, and λ_L is the minimum threshold (Lertsinsrubtavee et al. 2013), as illustrated in Fig. 1.

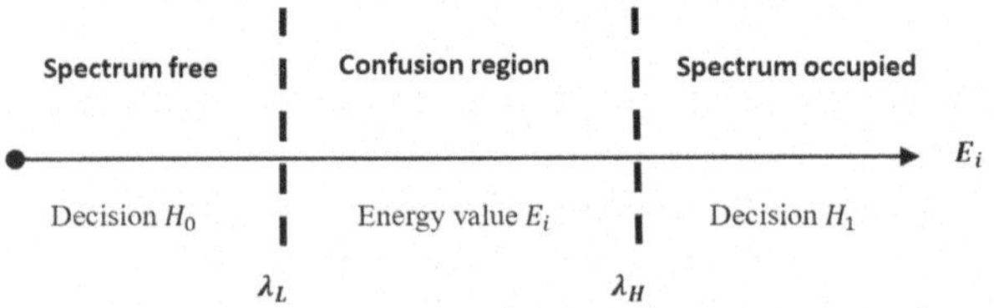

Fig. 1. Double threshold energy detector.

Each CUs_{5G} (i = 1...L) determines a local decision (LD) on the presence of the PBS based on the observed test statistic T_s. The decision is made based on the comparison of

T_s to the high threshold λ_H and the low threshold λ_L, as described in (Lertsinsrubtavee et al. 2013). These thresholds are crucial in our method.

$$LD = \begin{cases} 0, T_s < \lambda_L \\ 1, T_s > \lambda_H \end{cases} \tag{24}$$

If the value of T_s falls within the range of the lower threshold and the higher threshold ($\lambda_L < T_s < \lambda_H$), the CUs_{5G} node will conduct another spectrum sensing.

This strategy enhances the likelihood of detecting signals by reducing the uncertainty caused by noise (Shen and Xie 2012). Furthermore, the effectiveness of the DTED technique is contingent upon the possibility of detection, as denoted in Eq. (25).

$$\begin{aligned} PD_{DTED} &= \frac{Prob\{T_s > \lambda_H | H_1\}}{Prob\{T_s > \lambda_H | H_1\} + Prob\{T_s < \lambda_L | H_1\}} \\ &= \frac{Prob\{T_s > \lambda_H | H_1\}}{Prob\{T_s > \lambda_H | H_1\} + 1 - Prob\{T_s > \lambda_L | H_1\}} \end{aligned} \tag{25}$$

3 Results and Discussion

In this section, we have developed the simulation proposed model in MATLAB 2022b software to verify and study their results. The objective of the present work is not just to improve the detection performance of the PBS at 60 GHz under AWGN channels in real-time, but to potentially revolutionize the field. The impact of the CUs_{5G} mobility and the congested environment have been taken into account for the two propagation models proposed. All parameters used in the simulation are represented in Table 1, providing a roadmap for future research and development. In this work, the first stage of the proposed algorithm aims to study the motion tracking of CUs_{5G} modeled by the GMMM model (see Fig. 2a), then predict the position and speed of the CUs_{5G} compared to the original position of PBS fixed at (0,0), through the use of α-β Filter.

Table 1. Simulation parameters (Adardour and Kameche 2022).

Parameter	Value
The coverage area of the CUs_{5G}	70×70 [m^2]
Frequency **f**	60 [GHz]
Transmission power $\mathbf{P_{Tx(PBS)}}$	20 [dBm]
Noise power $\mathbf{P_{Noise}}$ at the CUs_{5G}	−115 [dBm]
Reference distance $\mathbf{d_0}$	1.0 [m]

The actual and estimated paths of the CUs_{5G} are depicted in Fig. 2a. The α-β-Filter-based position tracking yielded favorable outcomes with a reduced estimation error. The average relative error of the position is around 6.731e−5 [m]. Figure 2b illustrates the actual and estimated speed tracking of the CUs_{5G} in the coverage area of the PBS. The CUs_{5G} starts its movement at the point $\left(\sqrt{2}/2, \sqrt{2}/2\right)$ at t $=$ 250 s. The speed values show both increments and decrements, as shown in Fig. 2b. The α-β-Filter's estimation of the CUs_{5G} speed closely matches the actual case, with an average error of approximately 2.692e−7 [m/s]. This similarity between the actual and estimated speed of CUs_{5G} provides reassurance about the accuracy of our research. We can observe that the CUs_{5G} node movement equally affects the primary signal sensing performances, as we will see in the next.

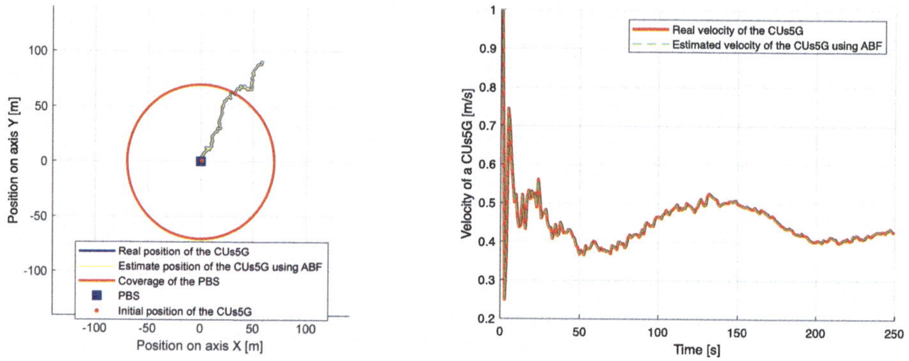

(a) (b)

Fig. 2. Real and estimated: **(a)** Trajectory of CUs_{5G}, **(b)** Speed of the CUs_{5G}.

Equations (10) to (12) allow us to thoroughly assess the potential changes in the actual $P_{Rx\ (CUs_{5G})}$ and estimated $P_{Rx\ (CUs_{5G})}$ for each CUs_{5G} position (as shown in Fig. 3a). This comprehensive evaluation is done using the CI and FI propagation models in both the LoS and NLoS situations (Qamar et al. 2021), with the parameters specified in Table 2.

Table 2. CI and FI path-loss model parameters at 60 GHz.

			n	α	β	σ
60 GHz	LoS	CI	2.6	–	–	4.46
		FI	–	55.06	3.84	4.46
	NLoS	CI	3.2	–	–	5.47
		FI	–	54.64	4.18	5.47

Figure 3a clearly illustrates that the application of the α-β-Filter shows a certain similarity between the actual and estimated cases of $P_{Rx\ (CUs_{5G})}$ at the CUs_{5G} level for the four propagation scenarios. In the NLoS environment, the signal quality is apparently inferior to that in the LoS environment. The results in Fig. 3a show that the CI propagation model, with its simplicity and accuracy, is more efficient than the FI model for 5G mmWave signal propagation, reassuring the audience about its ease of use. The obtained results indicate that the attenuation of $P_{Rx\ (CUs_{5G})}$ at the CUs_{5G} is influenced by two factors: the mobility model of CUs_{5G} and its proximity to the end of the radiation coverage zone and distance from PBS, as shown in Figs. 3a and 2a, respectively.

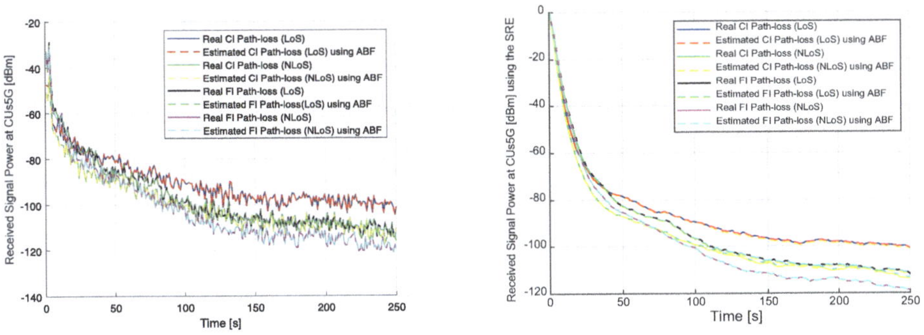

(a)

Fig. 3. Received signal power at the CUs_{5G} [dBm] vs. Time [s]: **(a)** without the S-R estimator and **(b)** using the SR estimator.

To overcome this problem, the proposed algorithm uses a S-R estimator, as shown in Fig. 3b. To achieve greater stability of the $P_{Rx\ (CUs_{5G})}$ at the CUs_{5G}, we can see that when we apply the S-R estimator with a Δ value equal to 0.1, the estimate of the $\overline{P_{Rx\ (CUs_{5G})}}$ at the CUs_{5G} is not just acceptable, but perfectly so (see Fig. 3b). This should instill confidence in the algorithm's performance.

Figure 4 presents the sensitivity of the $\overline{SNR_{(PBS,CUs_{5G})}}$, the link between the PBS and the CUs_{5G} after using the S-R estimator at 60 GHz in an AWGN channel. The results clearly show a stable estimate of $\overline{SNR_{(PBS,CUs_{5G})}}$, which significantly improves the reliability of detection of the primary signal under additive white Gaussian noise at 60 GHz.

Figure 5 illustrates the adaptability of the DTED algorithm to low SNR environments. The probability of total detection error over time is shown for single and DTED with 1500 samples for the CI and FI propagation models. The upper and lower threshold values are determined for different $\overline{SNR_{(PBS,CUs_{5G})}}$ values.

The DTED method excels in environments where noise instability is more significant. Both graphics clearly show the superiority of the double-threshold energy detection method over the one-threshold energy detection. Figure 5 demonstrates that, for both propagation models, DTED offers a lower probability of detection error than single-threshold detection. As CUs_{5G} approaches the end of the coverage zone, the DTED algorithm proves to be better suited to low SNR environments, further highlighting its effectiveness.

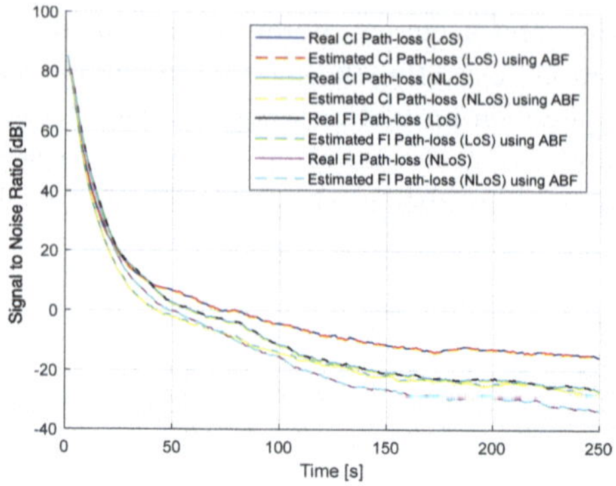

Fig. 4. Signal-to-noise ratio sensitivity at the CUs_{5G} [dBm] using the S-R estimator vs. Time [s].

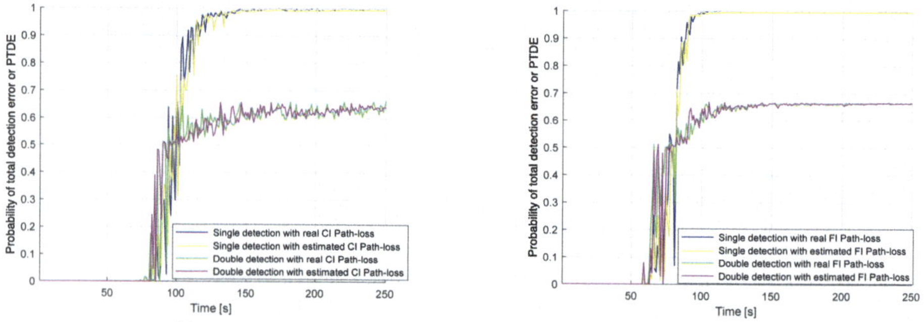

(a) **(b)**

Fig. 5. Probability of total detection error vs. Time [s]: **(a)** CI model and **(b)** FI model.

4 Conclusion

The paper proposed an efficient algorithm for improving primary signal detection performance for cognitive/5G radio networks using a hybrid system combining the α-β-Filter/S-R estimator technique and DTED. The results obtained revealed a significant improvement in detection performance compared with single-threshold detection. In addition, the implementation of the CI model is very straightforward. It gives a very stable performance with accuracy compared to the FI model at the 60 GHz frequency for mmWave 5G signal diffusion. In conclusion, the proposed algorithm for CUS_{5G} tracking in cognitive/5G radio networks not only shows promising results but also instills hope for future mobile applications.

References

Adardour, H.E., Kameche, S.: Predicting the primary signal sensing for cognitive radio users using an alpha-beta filter. In: 5th internaional conference on electrical engineering - boumerdes (ICEE-B), pp. 1–6 (2017)

Adardour, H.E., Meliani, M., Hachemi, H.: Improved local spectrum sensing in cluttred environment using a simple recursive estimator. Comput. Electr. Eng. **61**, 208–222 (2017)

Adardour, H.E., Kameche, S.: Wireless LAN for IOT-5G applications in a congested indoor environment : improved primary signal sensing at millimeter-wave V-band [40–75 GHz]. Int. Arch. Photogramm. Remote Sens. Spatial Inf. Sci. **XLVIII-4/W3-2022**, 5–11 (2022)

Ajani, A., Kayode, A., Awasume, E.: Comparative analysis of V-band and E-band mmWaves for green backhaul solutions for 5G ultra-dense networks. Int. J. Electr. Electron. Eng. Telecommun. **10**, 6 (2021)

Alenazi, M.J.F., Abbas, S.O., Almowuena, S., Alsabaan, M.: RSSGM: recurrent self-similar Gauss-Markov mobility model. Electrinics **9**, 2089 (2020)

Alijani, M., Osman, A.: Performance improvement of energy detection in cognitive radio under noise uncertainty. In: 2020 14th international conference on innovations in information technology (IIT), pp. 148–153 (2020)

Haouzi, A., Kameche, S., Adardour, H.E.: Overiew: application of cognitive radio technology in fifth-generation wireless communications. In: 2008 3rd international conference on cognitive radio oriented wireless networks and communications (CrownCom 2008), pp.1–6. IEEE, Singapore (2022)

Lertsinsrubtavee, A., Malouch, N., Fdida, S.: Efficient dynamic spectrum sharing through rate compensation and spectrum handoff, pp. 85–86. IEEE (2013)

Li, S., Liu, Y., Lin, L., Sun, D., Yang, S., Sun, X.: Simulation and modeling of millimeter-wave channel at 60 GHz in indoor environment for 5G wireless communication system. In: 2018 IEEE international conference on computational electromagnetics (ICCEM), pp. 1–3. IEEE, Chengdu (2018)

Qamar, F., Nour Hindia, M., Abd Rahman, T., Hassan, R., Dimyati, K., Ngoc Nguyen, Q.: Propagation characterization and analysis for 5G mmWave through field experiments. Comput. Mater. Contin **68**, 2249–2264 (2021)

Samarah, A., Tomar, G.S., Bagwari, A., Kanti, J.: Double stage energy detectors for sensing spectrum in cognitive radio networks. In: 2015 fifth international conference on communication systems and network technologies, pp. 181–184 (2015)

Shen, L., Xie, S.: Double-threshold energy detection of spectrum sensing for cognitive radio under noise uncertainty environment. In: 2012 international conference on wireless communications and signal processing (WCSP), pp. 1–5 (2012)

Sighencea, B.I., Stanciu, R.I., Sorandaru, C., Caleanu, C.D.: The alpha-beta family of filters to solve the threshold problem: a comparison. Mathematics **10**, 880 (2022)

Sun, S., et al.: Propagation path loss models for 5G urban micro and marco cellular scenarios. In: IEEE 83rd vehicular technology conference (VTC Spring), pp. 1–6 (2016)

Zhu, J., Xu, Z., Wang, F., Huang, B., Zhang, B.: Double threshold energy detection of cooperative spectrum sensing in cognitive radio. In: 2008 3rd international conference on cognitive radio oriented wireless networks and communications (CrownCom 2008), pp. 1–5. IEEE, Singapore (2008)

TTC Antenna at S-band for the Drone to Geo Satellite Application

Z. Kourdi[1](✉), I. Four[2], R. Mansour Bekhtache[1,2], and N. Addaoui[2]

[1] Center Development of the Satellites, Algerian Space Agency, Oran, Algeria
zkourdi@cds.asal.dz

[2] Department Electronics of Technology Faculty, Hassiba BenBouali University, Chlef, Algeria

Abstract. The approach of designing an antenna and a photovoltaic cell on the same substrate is useful for reducing costs. Many recent studies have explored the integration of photovoltaic cells with antennas due to the potential advantages of wireless communication. The researchers designed state-of-the-art photovoltaic antennas integrated into large cells with complex geometries. In this work, compact antenna arrays on a single device integrated into a photovoltaic cell are proposed for satellite applications and ensure the reception of band channels. The proposed model can accept to increase more than exposure to the sun; the photovoltaic cell is part of the antenna and thus improves the performance of the antenna. The antenna integrated with the designed photovoltaic cell accepts 100% exposure to sunlight and we extract all the characteristics of the antenna (impedance bandwidth, gain and radiation pattern).

Keywords: Satellite chain · patch antenna networks · S band · Photovoltaic cell · Simulation Software

1 Introduction

In recent years, the use of Unmanned Aerial Vehicles (UAVs), commonly known as drones, has expanded significantly across various industries, including surveillance, environmental monitoring, agriculture, and communications [1–3]. One critical aspect of drone operations, especially for long-range missions, is maintaining robust and reliable communication links. This is where Telemetry, Tracking, and Command (TTC) systems play a pivotal role. TTC antennas operating at the S-band frequency range (2–4 GHz) have emerged as a vital component in establishing and maintaining communication between drones and Geostationary (GEO) satellites [4–6]. The S-band is particularly suitable for TTC applications due to its favorable propagation characteristics, including lower susceptibility to atmospheric attenuation and better penetration capabilities compared to higher frequency bands [7–10]. These attributes make the S-band an optimal choice for ensuring consistent and reliable communication links over long distances, which is essential for effective drone operation and control. A vital component of the TTC system is the antenna, which plays a crucial role in establishing and maintaining the communication link. Among the various frequency bands available, the S-band (2

© The Author(s), under exclusive license to Springer Nature Switzerland AG 2025
M. Hatti (Ed.): IC-AIRES 2024, LNNS 1238, pp. 646–655, 2025.
https://doi.org/10.1007/978-3-031-80301-7_68

to 4 GHz) is particularly favored for TTC applications due to its balance of bandwidth availability and atmospheric penetration capabilities [11, 12]. S-band frequencies offer a compromise between the high data rates of higher frequency bands and the long-range propagation characteristics of lower frequencies, making them ideal for satellite communications. The importance of reliable TTC links cannot be overstated, as they ensure the seamless transmission of telemetry data, precise tracking of the drone's position, and effective command and control operations. By leveraging S-band frequencies, we can achieve a reliable communication link that supports high data throughput and robust signal integrity, thereby enhancing the operational capabilities of drone missions involving GEO satellites [13–15].

This article explores the design, development, and application of TTC antennas at S-band for drone-to-GEO satellite communication. We will delve into the technical considerations, recent advancements, and challenges associated with these systems. Understanding the nuances of TTC antenna technology is crucial for enhancing the performance and reliability of UAV operations, especially in scenarios where real-time data transmission and precise navigation are paramount. Through a detailed examination of the latest research and innovations, this article aims to provide a comprehensive overview of the current state and future prospects of TTC antennas at S-band for drone-to-GEO satellite applications.

2 Modeling and Design

The design of single element patch, 1 × 2, 1 × 4 and 2 × 4 elements array antennas as well as the 1 × 4 elements beam steering antenna have been presented and analyzed one by one in this section. All the antennas have been operated with in S band.

In this work, rectangular microstrip patch antennas were adopted due to the convenience of fabrication and impedance matching. The design parameters of rectangular microstrip patch antenna include the length of rectangular patch L = 62.23, the width of rectangular patch W = 80.85, the thickness of dielectric substrate h = 1.6 mm, the relative permittivity of the substrate Er = 4.3. The performance parameters of microstrip arrays include the directivity, gain, input impedance, center frequency, bandwidth and mechanical performance.

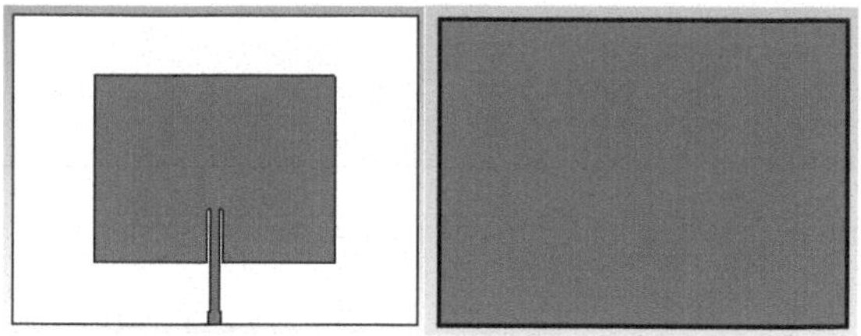

Fig. 1. Structure of the Single-element patch antenna.

Single-Element Patch Antenna The internally fed single-element patch antenna is designed with a center operating frequency of 2.1 GHz as shown in Fig. 1. Consists of a squared patch with two slits cutout in the lower portion of the patch to extend the feed point connection to the patch in order to enhance the matching. In addition, the 50 Ω microstrip feed line is stepped to further improve the impedance match. The figure (b) of the dielectric substrate is grounds plane. The antenna is designed on FR4 substrate with Er = 4.3 and h = 1.6 mm. Length and width of the slits and the size of stepped feed line have a critical effect in the antenna's impedance bandwidth.

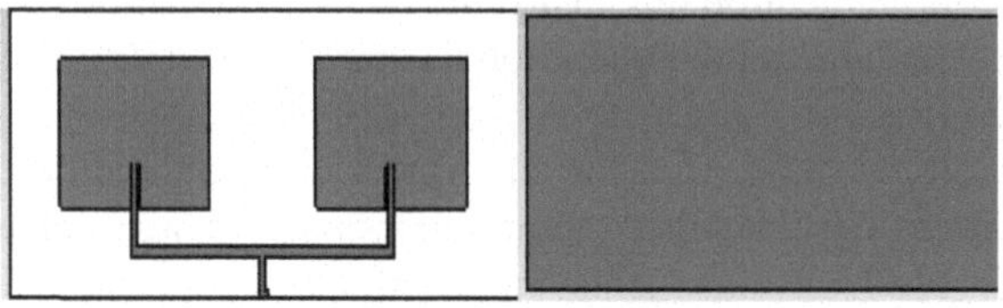

Fig. 2. Structure of the 1*2 element patch antenna array.

A 1 × 2 elements microstrip array antenna which shown in Fig. 2 has been designed by using two single elements microstrip patch, a transmission line of 50 Ω, substrate. The figure (b) is a ground plane. In this design, the patch elements are connected using quarter wavelength microstrip line and T- junction power divider.

(a)

(b)

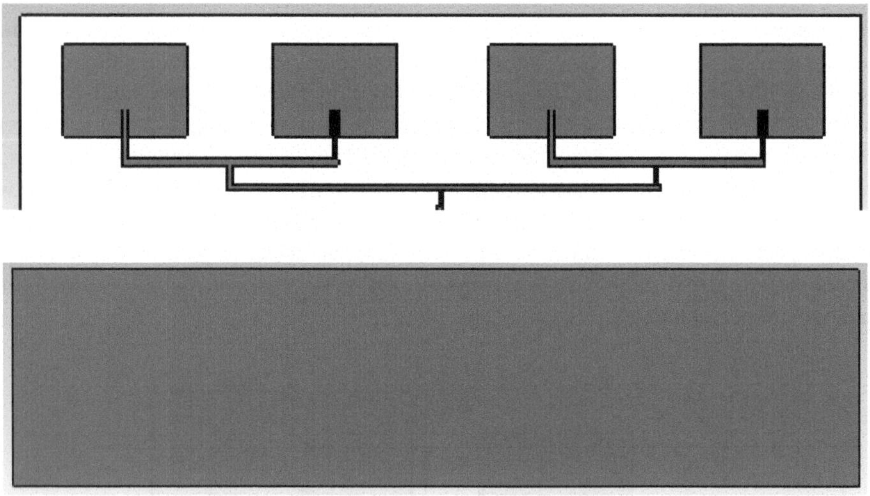

Fig. 3. Structure of the 2*2-element patch antenna array a) patch side, b) grand side.

This subsection represents design structure shown in Fig. 3 and the simulation results of 1 × 4 elements microstrip array antenna which. The dimension of the patches and spacing between two patches are kept same as the previously designed 1 × 2 array antenna. Therefore, only inset distance and feed line width have been readjusted to get better impedance 50 Ω matching. The corporate feed method is used to excite each array elements. In this design, the patch elements are connected using quarter wavelength microstrip line and T- junction power divider.

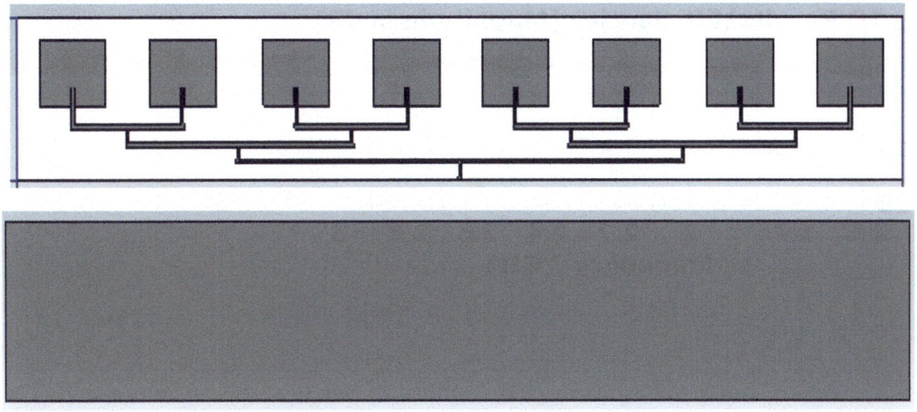

Fig. 4. Structure of the 2*4 element patch antenna array

To enhance the directivity further, additional four elements have been added and made a 2 × 4 elements array. Similar to the 2 × 2 elements array antenna, all the elements of 2 × 4 array antenna are excited by using feed lines as shown in Fig. 4.

3 Resultats and Discussion

3.1 One Element Antenna Patch

The figure illustrates the S_{11} parameters, plotted against frequency. Its purpose is to identify the fundamental resonant frequency of the system (Fig. 5).

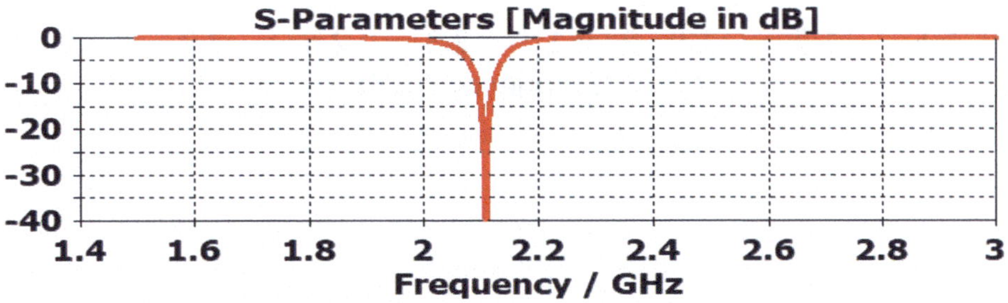

Fig. 5. S-Parameters of single patch antenna.

The analysis shows that the antenna has a maximum bandwidth of 0.42 GHz centered around its resonant frequency of 2.1 GHz, within the interval of 2.079 GHz to 2.121 GHz. These results indicate effective matching, as evidenced by a lower reflection coefficient of −39 dB at 2.1 GHz.

During the simulation, peak gains fluctuated between 3 and 8.24 dB across the frequency range of 2.1 GHz (Fig. 6).

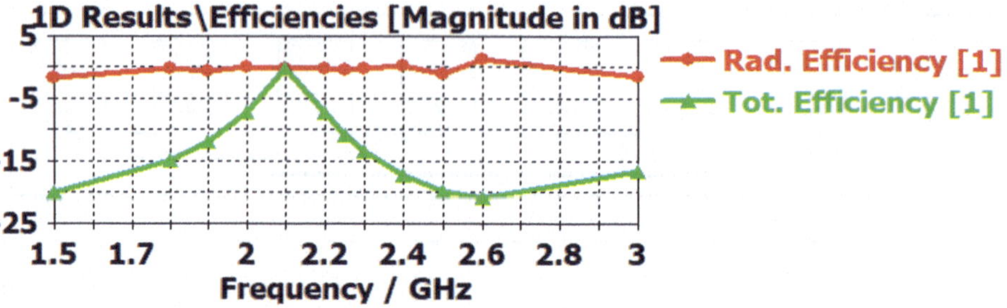

Fig. 6. Efficiencies of single patch antenna.

The figure shows the radiation and overall efficiency of the antenna patch. The patch has a compact omnidirectional design that ensures high radiation efficiency. The antenna's structure has been modified to improve its radiation efficiency by achieving an optimal overall impedance dimension. This optimization enables the antenna to maintain high radiation efficiency while occupying smaller dimensions relative to the operating wavelength in free space (Fig. 7).

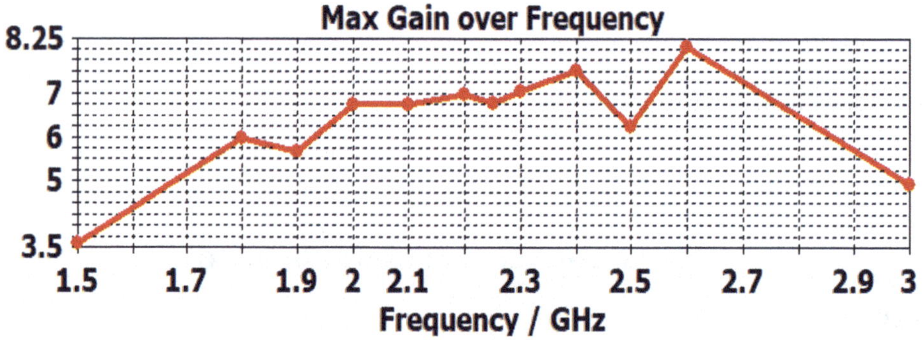

Fig. 7. Gain of single patch antenna.

During the simulation, the peak gains ranged 3.5 dB to 8.25 dB. This makes it suitable for satellite communications in the S-band frequency range. The gain pattern displays an increase from 1.8 GHz to 2.6 GHz, which is a common behavior in patch antennas where gain typically increases with higher frequencies until around 2.6 GHz, after which it begins to decrease (Fig. 8).

- This diagram illustrates how the power radiated by an antenna varies with direction away from the antenna, providing insight into the distribution of energy radiated into
- Angular is between 82.0° to 92.6°, antennas are radiating omnidirectional.

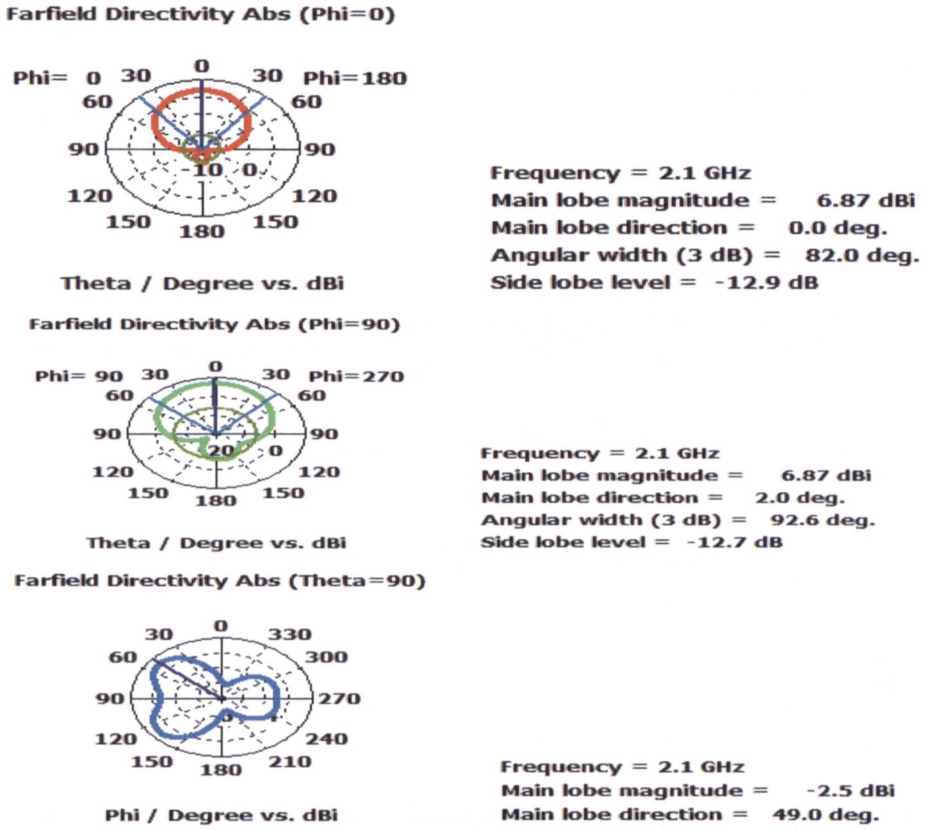

Fig. 8. Radiation pattern of single patch antenna.

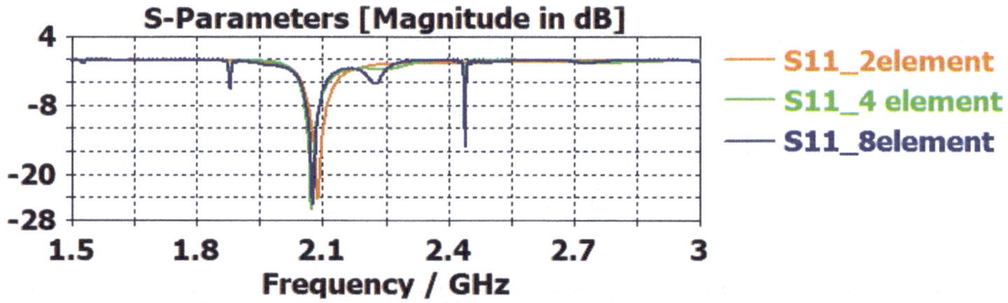

Fig. 9. S-Parameters of Array antenna patch.

3.2 Array Antenna Patch

The graph shows the reflection coefficient S11 measured across antenna networks with 2, 4, and 8 elements (Fig. 9).

The orange curve represents the S11 parameters and demonstrates the antenna network's impressive adaptability within the frequency range of 1.5 GHz to 3 GHz. Notably, it exhibits exceptional performance between 2.05 GHz and 2.15 GHz, achieving a reflection coefficient of −24 dB precisely at the resonant frequency of 2.1 GHz.

When simulated with four elements, the S parameter stabilizes at -26 dB at the frequency of 2.09 GHz. Within a bandwidth a bandwidth from 2.07 GHz to 2.1 GHz.

The blue curve shows the reflection coefficient observed in an eight-element configuration. At the resonant frequency of 2.08 GHz, the antenna demonstrates effective impedance matching, reflected by a −25 dB reflection coefficient. Furthermore, a bandwidth ranging from 2.06 GHz to 2.1 GHz is obtained (Fig. 10).

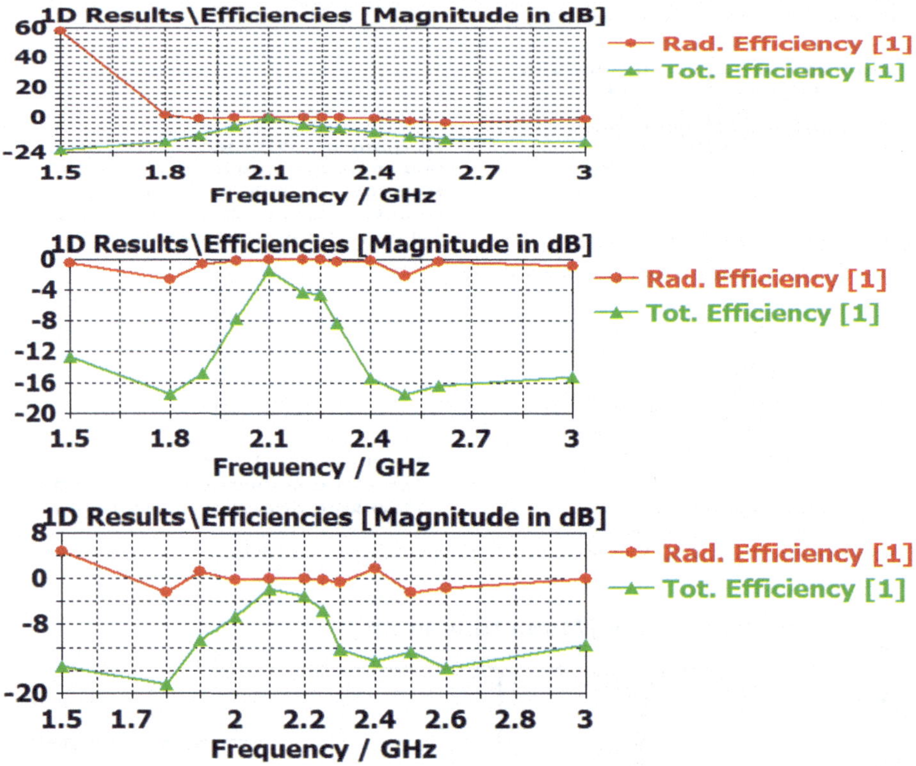

Fig. 10. Efficiencies of Array antenna patch.

Graphs show the radiation efficiency and overall efficiency of a radio receiver versus frequency. The x-axis is labeled "frequency/GHz" and the y-axis is labeled "Efficiency [volume in decibels]". With 2, 4 and 8 elements. be Efficiency 0 dc at frequency 2.1 GHz (Fig. 11).

At the frequency of 2.1 GHz, the gain observed in the antenna network varies across configurations. In the 2-element setup, the gain ranges from 2.1 to 9.5 dB, indicating an increase compared to the gain of the patch antenna, even when adjusted with a notch at the S band.

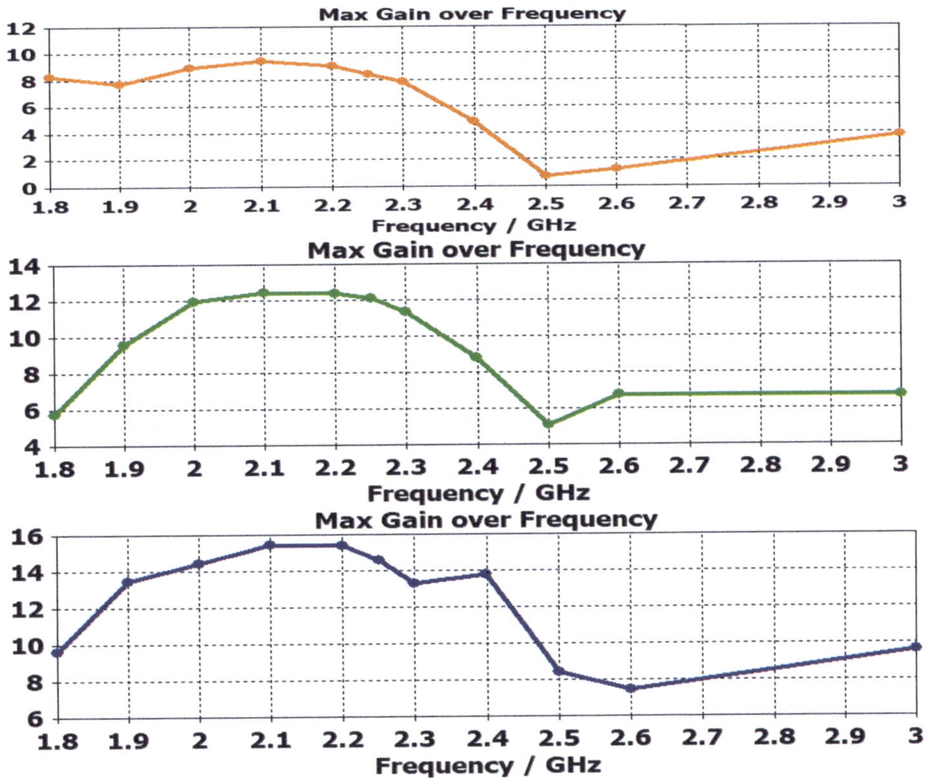

Fig. 11. Gain of Array antenna patch.

In the second simulation involving a 4-element antenna network, the maximum gain peaks at 12.2 dB at the resonance frequency of 2.1 GHz. This indicates significantly improved radiation compared to the previous configuration.

Moving to the third simulation, the antenna achieves a maximum gain of 15.45 dB at 2.1 GHz. The minimum gain is nearly equal to the maximum gain of the first simulation. Consequently, we observe a higher level of radiation overall (Fig. 12).

The antenna radiation pattern serves to assess the antenna's directivity. Typically represented as a 3D shape, it's segmented into three columns: the first with 2 elements, the second with 4, and the last with 8. These elements are showcased across three cutting planes: the horizontal plane at pH = 0°, pH = 90°, and at theta = 0°.

- for ph = 0° and a resonance frequency of 2.1 GHz:
    - for 2 elements, it obtained opening angles of 44.2°.
    - for 4 elements, it obtained opening angles of 113.0°.
    - for 8 elements, it obtained opening angles of 10.9°.
- for ph = 90° and a resonance frequency of 2.1 GHz:

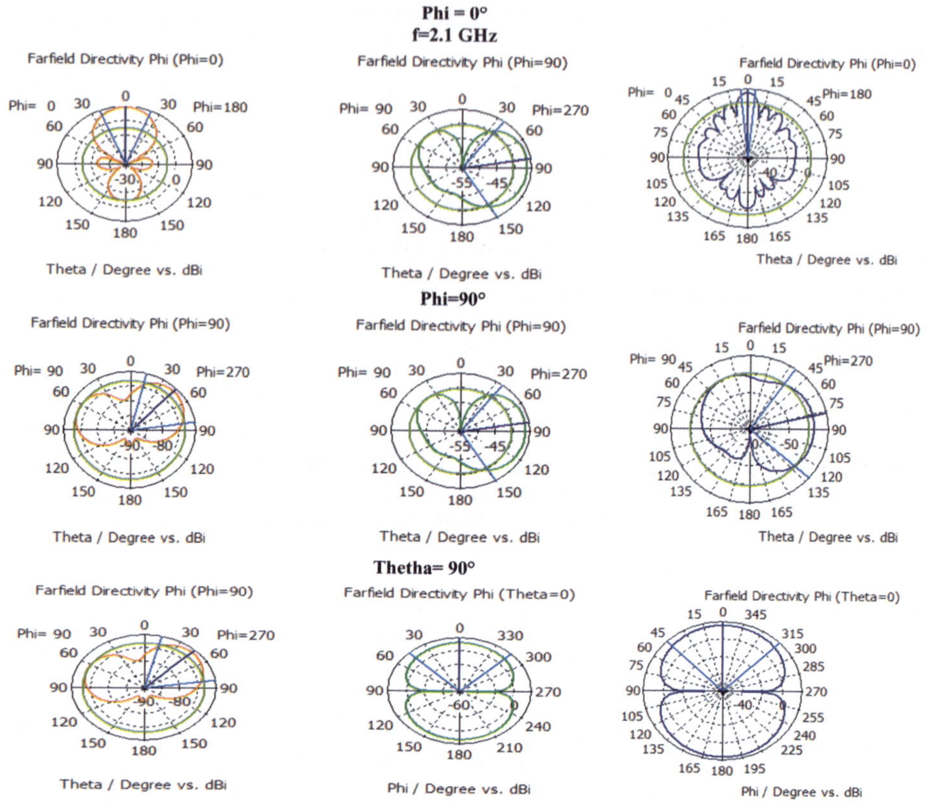

Fig. 12. Radiation pattern of Array antenna patch

- for 2 elements, it obtained opening angles of 67.4°.
- for 4 elements, it obtained opening angles of 113.0°.
- for 8 elements, it obtained opening angles of 95.6°.

- for theta = 90° and a resonance frequency of 2.1 GHz:

  - The antenna radiation is omnidirectional with an opening angle of 44°at 67.4°
  - According to the simulation models reading in 2 The antenna remains perfect and radiates in all directions, so it is That is, it provides radiation addressing directions.

4 Conclusion

It has provided a comprehensive overview of the design and simulation of antennas operating in the S band. Throughout this discussion, we have highlighted the importance of antennas in enabling efficient and reliable communication, radar, and satellite applications within this critical frequency range.

The design and simulation of antennas in the S band represent a dynamic and evolving field that continues to push the boundaries of modern communication and radar technologies.

By combining theoretical insights with practical examples and simulation demonstrations, this paper has provided valuable insights into the principles and practices that underpin the development of antennas for the S band, empowering engineers and researchers to innovate and excel in this exciting field.

References

1. 2023 Study on Phased Array Antennas: A comprehensive analysis of phased array technology for UAV communication with GEO satellites highlights significant improvements in beam steering capabilities and link reliability
2. Innovations in Metamaterials for S-band Antennas: This study explores the application of metamaterials to achieve higher gain and better efficiency in compact antenna designs (2022)
3. Performance Analysis of S-band TTC Systems for UAVs: An in-depth performance evaluation of various S-band TTC systems for drone applications, considering factors like link budget, environmental impact, and real-world deployment scenarios (2021)
4. Lamotte, Q., at al.: Multi-permittivity 3D-printed ceramic dual-band circularly polarized dielectric resonator antenna for space applications. In: 15th European conference on antennas and propagation (EuCAP) (2021)
5. Lofaldli, A., Ernst, H., Pelorossi, F., Evans, D., Flentge, M., Smolinski, G.: ESOC ground station architecture for OPS-SAT. In: 8th international workshop on tracking, telemetry and command systems for space applications (TTC) (2019)
6. Gupta, N., Goyal, V., Sanadhya, V.: System identification of the IDSN32M antenna control system model. In: IEEE Indian conference on antennas and propogation (InCAP) (2018)
7. Zhang, Y., Li, J., Gao, S., Wei, G., Xu, J.: Broadband circularly polarized patch antenna for small satellites applications. In: 8th international symposium on communication systems, networks & digital signal processing (CSNDSP) (2012)
8. Sreeja, T.K., Arun, A., Jaya Kumari, J.: An S-band micro-strip patch array antenna for nano-satellite applications. In: International conference on green technologies (ICGT), institute of electrical and electronics engineers (2012)
9. I. Gonzalez, J. Gómez, A. Tayebi, F. Cátedra, "Optimization of a Dual-Band Helical Antenna for TTC Applications at S Band", IEEE Antennas and Propagation Magazine, Vol: 54, N: 4, 2012
10. Gómez, J., et al.: Dual K-band/S-band tracking ground station antenna for future earth observation applications. In: 8th International workshop on tracking, telemetry and command systems for space applications (TTC) (2019)
11. Samsuzzaman, M., Islam, M.T., Faruque, M.R.I.: Circularly polarized dual frequency patch antenna for TTC applications. In: Third world congress on information and communication technologies (WICT 2013) (2013)
12. Nayna, T.F.A., Baki, A.K.M., Ahmed, F.: Comparative study of rectangular and circular microstrip patch antennas in X band. In: Electrical engineering and information communication technology (ICEEICT), international conference on IEEE, vol. 14 (2014)
13. Garg, R., Barthia, P., Bahl, I., Ittipiboon, A.: Microstrip antenna design book. Artech House, London (2001)
14. Antenna Theory: Analysis and design. John Wiley & Sons (2016)
15. Kabacik, P., Opalka, P., Hornik, P.: TTC patch antennas made in a conformal form with small radius. First European Conference on Antennas and Propagation (2006)

Efficient Predictive Dynamic Routing of IoT Communication Networks in Smart Farming: An Effective Fuzzy Q-Learning Solution

Abdesselem Beghriche[1,2](✉), Boubekeur Annane[1], and Amina Abed[3]

[1] Computer Science Department, Ferhat Abbas University, Sétif -1, Sétif, Algeria
{abdesselem_beghriche,boubakeur.annane}@univ-setif.dz
[2] Misc Laboratory, Abdelhamid Mehri University, Constantine-2, Constantine, Algeria
[3] Letters and English Language Department, Mentouri Brothers University, Constantine-1, Constantine, Algeria

Abstract. The Internet of Things (IoT) plays a significant role in smart agriculture, which is now on the agenda of intelligent farming systems, so better communication networks are required. Employing dynamic routing in these networks lets the data be transmitted with the right timing and reliability. This paper proposes a new paradigm that employs Fuzzy Q-Learning to improve the routing aspect of IoT networks for smart farming. The proposition uses fuzzy logic to deal with fuzzy information and uncertainties associated with agro-scenarios, like pest detection, irrigation scheduling, environmental monitoring, etc. Combining Q-Learning with Fuzzy logic would allow the system to make specific predictions of the state of the network to make the right proactive routing decisions, leading to increased overall network performance by reducing latency, packet loss ratio, and energy consumption efficiency. The performances are simulated using the introduced method in an imaginative farming scenario and compared against conventional routing algorithms. The simulation results show that the proposed solution could achieve superior network efficiency and reliability performance compared to other current solutions.

Keywords: Internet of Things (IoT) · Reliable Communication · Dynamic Routing · Fuzzy Q-Learning · Smart Farming · Precision Agriculture

1 Introduction

In recent years, the use of Internet of Things (IoT) technologies in smart farming has witnessed a tremendous change that has improved the agricultural sector in all aspects, including efficiency and productivity [1]. The IoT communication networks support most applications such as precision agriculture, livestock monitoring, environmental sensing, etc. [2]. The network contains many sensor and actuator nodes deployed over the farm to sense and send data. These sensed data facilitate farmers to make quick and accurate decisions on their farms, keeping a real-time view of demand that helps increase crop production, improve resource utilization, and maintain the health condition of livestock

© The Author(s), under exclusive license to Springer Nature Switzerland AG 2025
M. Hatti (Ed.): IC-AIRES 2024, LNNS 1238, pp. 656–671, 2025.
https://doi.org/10.1007/978-3-031-80301-7_69

[3]. Nevertheless, in this context, where all devices interact to collect and send data, the issue of efficient routing needs to be addressed.

To make these IoT networks work better in several ways, they must have good, efficient routing. Firstly, it helps reduce latency (a key objective for real-time-based applications used in smart farming environments, for example). By ensuring that data packets are routed through the most optimal path, latency can be minimized, enabling farmers to receive timely information about their farm's status [4]. Secondly, efficient routing also helps increase the network's reliability by avoiding congested or unreliable routes that could lead to packet loss or dropping and ensuring all required critical information has been received at the receiver's end. In smart farming, this is important as any delay or loss of such information being sent will highly impact crop yields or livestock health. In addition, routing efficiently also contributes to minimizing energy consumption, especially on energy-constrained IoT devices, by utilizing the most energy-efficient route, thus slowly consuming less battery power, resulting in a longer operational lifetime, i.e., not necessary for replacing batteries frequently, especially for devices deployed remotely within the farm [5].

Most existing IoT network routing protocols [6–9] still need to improve their ability to cope with innovative farming environments' dynamic and heterogeneous characteristics. Fuzzy Q-Learning [10] provides a new mechanism combining fuzzy logic and the reinforcement learning technique. This technique allows each node in the network to decide both crisp and fuzzy parameters. As a result, these nodes can change their strategies based on current and future conditions.

In Fuzzy Q-Learning, each node keeps a Q-table, which stores the expected utility of initiating an action (i.e., selecting a route) in a particular state. This table is updated along iterations, considering the rewards obtained by performing this action and the rewards expected from similar actions. Fuzzy logic is applied here for handling uncertainty and vagueness in the environment, i.e., nodes can make decisions based on fuzzy rules that describe relations among network parameters and routing behavior.

One of the more exciting features of Fuzzy Q-learning is that it can learn and adapt to changes in network conditions. It does not need central control, nor are routes predetermined. Since nodes continually update their paths based on their interactions, they will continually improve their decisions over time. Fuzzy Q-learning is particularly well-suited for the smart farming application domain, where conditions may evolve rapidly [11]. By considering multiple objectives such as latency, reliability, or energy consumption, nodes can make better-facilitated decisions when deciding the best hop to forward packets. For this reason, it has been shown that Fuzzy learning can perform better than traditional routing protocol due to more globally oriented decisions [12].

The primary purpose of this paper is to design and implement a new predictive dynamic routing in IoT communication networks for smart farming to enhance the efficiency and reliability of smart farming applications. This approach was designed to minimize latency, improve reliability, and reduce energy consumption by predicting future network conditions and making adaptive decisions on the path choice in real-time. This will lead to increased productivity and greater sustainability in agriculture.

The key technical contributions of this work include:

1) Development of a novel routing algorithm that combines fuzzy logic with Q-Learning to enhance decision-making in IoT communication networks for smart farming.
2) Achieving significant improvements in network performance metrics such as latency, reliability, and energy efficiency, surpassing traditional routing protocols.
3) Comprehensive evaluation through simulations and real-world case studies, highlighting the proposed solution's practical applicability and benefits in enhancing smart farming applications.

The remainder of this paper is structured as follows: After this introduction, Sect. 2 presents the background and related work, discussing previous studies and highlighting their limitations in smart farming applications. Section 3 presents the theoretical foundation and detailed description of the proposed Fuzzy Q-Learning routing algorithm. Section 4 discusses the implementation of the proposed solution and provides a comprehensive performance evaluation through simulations and case studies. Finally, Sect. 5 concludes the paper by summarizing the essential findings and discussing potential future research directions.

2 Background and Related Work

Adopting Internet of Things (IoT) technology in agriculture significantly advances modern farming practices. However, the deployment and maintenance of IoT communication networks in smart farming face several notable challenges:

- Connectivity and Infrastructure must be reliable to avoid connectivity gaps in rural areas.
- Scalability is needed to handle the increasing number of devices and data volume.
- Energy Efficiency is crucial since many IoT devices are battery-operated, requiring low-power protocols.
- Effective Data Management is necessary for collecting, processing, storing, and analyzing data.
- Security and Privacy measures are vital to protect sensitive data from cyber threats.
- Interoperability ensures efficient operations across different devices and platforms.
- Finally, Cost and Investment are vital considerations, requiring low-cost solutions and secure funding to support IoT adoption in agriculture.

Challenges in IoT communication networks for smart farming demand improved network technology, energy-efficient design, better data analysis, and complete security solutions. Routing protocol research for these networks is still in its infancy, especially for efficient dynamic routing. This section reviews the recent routing approaches proposed for smart farming IoT networks, considering route discovery, greedy forwarding, position predictive, and energy efficiency as main performance factors.

Several protocols [13–19] have been proposed and implemented in this context, each with unique features and advantages. Some of the commonly used routing protocols include:

In [13], the authors proposed using the RPL protocol for IoT networks in smart farming due to its scalability and energy efficiency, making it suitable for managing communication between battery-powered devices in remote agricultural environments. However, despite these advantages of RPL's applicability to WSNs in general, the actual implementation and deployment are complex. Care must be taken regarding network topology, node mobility (regarding crop growing patterns and maintenance operations), and security. Furthermore, the problems pointed out earlier concerning reliability may also affect performance depending on specific properties of the environment, such as devise density or obstacles.

The FAJIT "FARM-ED" [14] protocol is a notable work in WSNs that relies on fuzzy logic to minimize redundancy and energy consumption by forwarding only required data along the network, thus prolonging the network lifetime. It is suitable for environmental monitoring, precision agriculture, and smart cities. However, critics argued that introducing fuzzy logic might bring complexity and overhead, especially for resource-constrained WSNs.

Authors of "EE-SMART" [15] propose an energy-efficient and scalable multi-hop routing algorithm for IoT-based Smart Farming that minimizes energy consumption to maintain scalability. It makes use of clustering, data aggregation, and dynamic routing. However, there are issues related to clustering and data aggregation overhead and poor performance in a highly dynamic environment. Its real-world performance should be further compared with other existing works.

In the same context, Agro-LORA [16] is a solution to agricultural IoT networks using LoRa technology to improve data transmission. It is suitable for vast agricultural land in remote areas with low power consumption, long-distance communication, scalability, adaptive data transmission, reliability, and cost-effectiveness. Its limitations include lack of bandwidth, network congestion resulting from massive devices, basic security mechanisms, and the requirement to depend on centralized gateways. With the enhanced real-time farm management that it brings about, improving all Internet of Things applications in agriculture is necessary to have more improvements.

The work presented in [17] aims to enhance the security of IoT networks in smart farming by proposing a secure routing protocol. The protocol employs cryptographic techniques and authentication mechanisms that protect against unauthorized access and data breaches, a robust approach to enhancing security. However, implementing such security measures may often introduce additional computational overheads and energy consumption, impacting the performance and lifespan of battery-powered devices in smart farming environments.

As described in [18], the IoT-PSO protocol proposes an energy, reliability, and scalability-optimized routing mechanism in agricultural IoT. In the proposed protocol IoT-PSO, data routing paths are dynamically adjusted considering node residual energy so that the node's power can be evenly consumed, achieving increased network lifetime. Fault tolerance mechanisms and efficient communication strategies are incorporated to provide reliable network operations and scalability features that are suitable even for large-scale operations of agriculture systems. However, real-time adaptability-type scenarios call for further research works so that they can be applied across diverse agriculture environments due to their high computational requirements.

In [19], the investigation into Software-Defined Networking for IoT networks in smart agriculture coined a potential networking paradigm to advance network management and performance. Its centralized control is capable of real-time data and network conditions for spontaneous routing decisions. This attribute is essential in smart agriculture, where systems are adaptive and resources are limited. However, SDN implementation in the IoT network environment presents new challenges that one should face; for instance, a vast amount of data needs a high degree of protection standard thanks to crash polarities from such devices. Still, reliance on the central controller as a single failure point may implicate processing node redundancy in ensuring network functional designation activation.

The reviewed routing protocols show diverse strategies considered to meet the challenges of IoT networks in smart farming. Each protocol will have benefits, be it energy efficient, adaptable, robust, or scalable. The choice should depend on the requirements of applications employed in smart farming, such as the size and expected scale of the network, mobility, and energy constraints.

3 The Proposed Model (FQL-Smart Farming)

The proposed model for efficient predictive and dynamic routing in an IoT communication network within a smart farming environment incorporates fuzzy logic and Q-learning, creating a hybrid Fuzzy Q-learning. This is envisaged to enhance routing efficiency through dynamic adaptation based on network conditions to predict the optimal route using real-time and historical data.

3.1 Model Architecture

The proposed model architecture consists of the following main components: (Fig. 1).

Phase 1: Fuzzy Inference System (FIS).
Phase 2: Q-Learning Framework.
Phase 3: Routing Decision Module.
Phase 4: Performance Monitoring and Feedback Loop.

3.2 Description and Functioning

The details of the proposed model and pseudocode are outlined in the following steps:

3.2.1 Fuzzy Inference System (FIS)

The Fuzzy Inference System is the first layer of the proposed methodology, designed to process the input parameters that influence routing decisions. The key input parameters include:

a. ***Input Parameters:*** The FIS takes several critical network parameters as inputs:

- *Node Energy Levels:* Indicates the remaining battery power of each node.
- *Link Quality:* Measured by Signal-to-Noise Ratio (SNR) and Bit Error Rate (BER) metrics.
- *Traffic Load:* The current load on each node to avoid congestion.
- *Distance to Destination:* Hop count or geographical distance to the destination node.

b. ***Fuzzification:*** The input parameters are converted into fuzzy linguistic variables (e.g., low, medium, high) using membership functions.
c. ***Fuzzy Rules:*** A set of predefined fuzzy rules evaluates the relationships between the fuzzy input variables to determine the suitability of potential next-hop nodes.
d. ***Defuzzification:*** The fuzzy output is converted into a crisp value representing the suitability score for each potential next-hop node.

3.2.2 Q-Learning Framework

The Q-learning framework integrates with the FIS to enable adaptive learning and decision-making. The components of the Q-learning framework include:

a. ***State Representation:*** Each state represents a unique combination of fuzzified input parameters.
b. ***Action Selection:*** The actions involve selecting the next-hop node for data packet transmission. An ε-greedy policy [19] is used to balance exploration and exploitation:

- *Exploration:* Selecting random actions to discover new routes.
- *Exploitation:* Choosing the best-known action based on current Q-values.

c. ***Reward Function:*** The reward function provides feedback on the success of the routing decision. Positive rewards are given for successful data delivery with low energy consumption and high link quality, while negative rewards are assigned for packet loss, high latency, or excessive energy use.
d. ***Q-value Update:*** Q-values are updated using the Bellman equation:

$$Q(s,a) \leftarrow Q(s,a) + \alpha\left[r + \gamma max_{a'} Q\left(s', a'\right) - Q(s,a)\right] \quad (1)$$

In essence, $a\prime$ represents all possible actions in the new state $s\prime$ and $max_{a\prime}Q(s\prime, a\prime)$ is the maximum estimated future reward obtainable from $s\prime$. This term helps update the Q-value for the current state-action pair (s, a) by considering the best possible future rewards, thus guiding the learning process toward optimal actions.

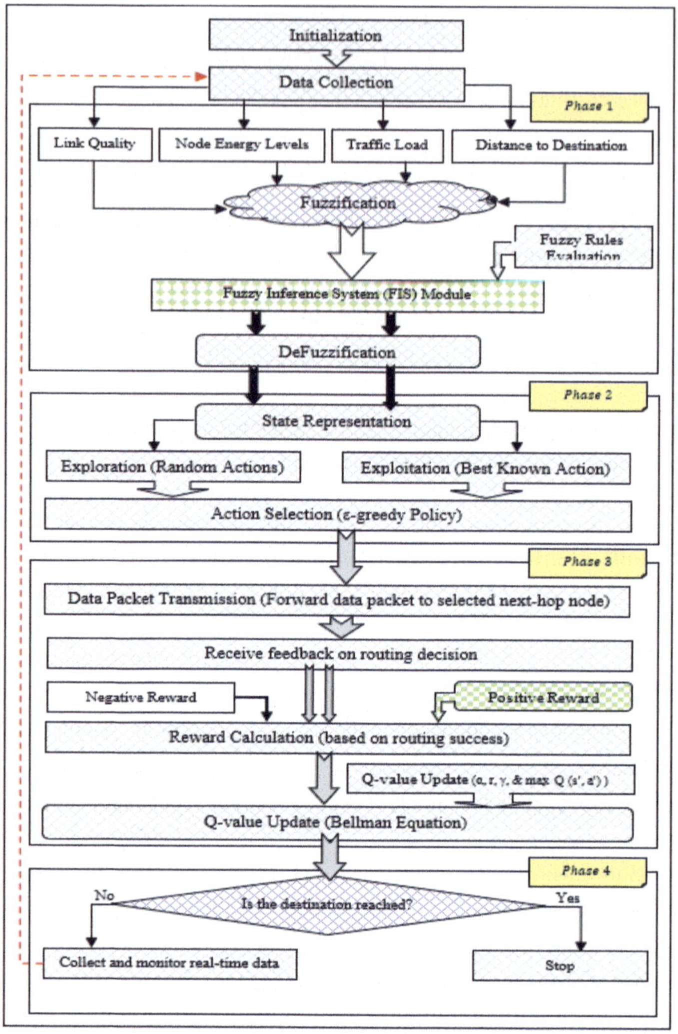

Fig. 1. Architecture of the proposed model (FQL-Smart Farming).

3.2.3 Routing Decision Module

The routing decision module utilizes the outputs from the FIS and the Q-learning framework to make informed routing decisions:

a. ***Next-Hop Selection*:** Based on the suitability scores from the FIS and the Q-values from the Q-learning framework, the module selects the optimal next-hop node for packet transmission.
b. ***Packet Forwarding***: The selected next-hop node receives the data packet, which is repeated until it reaches its destination.

3.2.4 Performance Monitoring and Feedback Loop

The performance of the routing decisions is continuously monitored, and feedback is used to refine the model:

a. ***Data Collection:*** Real-time data on node energy levels, link quality, traffic load, and distances is collected and fed into the FIS and Q-learning framework.
b. ***Feedback Loop:*** The decision to route packets (success or failure) is used to fine-tune fuzzy rules and update Q-values as the network status changes so the model can learn.

4 Simulation Results and Discussions

The proposed Fuzzy Q-Learning model is vividly tested through extensive simulations to authenticate the effectiveness of the proposed model for dynamic routing in IoT communication networks, which is customized particularly for smart farming environments.

The simulations are conducted using the NS-3 (Network Simulator 3) platform [20], a widely used discrete-event network simulator capable of modeling various IoT scenarios.

4.1 Simulation Setup

The simulation environment is meticulously designed to replicate various aspects of smart farming IoT networks to ensure realistic and meaningful results. The key components of the simulation setup include:

- *Network Topology:* The network topology includes static and mobile nodes, representing stationary sensor nodes and mobile robots or vehicles used for data collection.
- *Scenario:* A typical smart agriculture scenario is simulated, consisting of sensor nodes deployed across a farm to monitor various parameters such as soil moisture, temperature, and humidity.
- *Traffic Pattern:* The simulation considers periodic data transmission for regular monitoring and event-driven bursts for detecting anomalies or critical events.
- *Performance Metrics:* In order to evaluate the proposed Fuzzy Q-Learning model, several key performance metrics are used: Energy Consumption, Packet Delivery Ratio (PDR), Average Latency, Network Lifetime, and Throughput.

The following table summarizes the key aspects of the simulation setup: (Table 1).

Table 1. Simulation Parameters.

	PARAMETER	VALUE
	Simulation Tool	NS-3.26
Physical & Mac	Wireless propagation model	Log-Distance/Path loss Exp: 3.0
	Transmission Power	Fixed/0.1 W (20 dBm)
	Data Link Antenna	Omni-directional Antenna
	Frequency Band	ISM Band/2.4 GHz
	MAC Layer Protocol	CSMA/CA
	Sending Capacity	Bit Rate/250 Kbps
Routing	Communication Model	IEEE 802.15.4
	Node Deployment	Random
	Transmission Range	100 m
	Energy Model	Linear Battery Model
	Initial Energy of Nodes	1000 Joules
	Traffic Pattern	CBR (Constant Bit Rate)
	Data Packet Size	50 ytes
	Packet Transmission Interval	1 s
	Queue management	DropTail Queue
Scenario	Simulation Area	1000 m × 1000 m
	Number of Nodes (sensors)	100
	Number of actuators	10
	Simulation Duration	1000 s
	Q-Learning Parameters	$\alpha = 0.1, \gamma = 0.9$, and $\varepsilon = 0.1$
	Mobility Model	Random Waypoint
	Fuzzy Inference System Parameters	Custom-defined

4.2 Comparative Analysis with Recent Protocols in Smart Farming

To evaluate the proposed Fuzzy Q-learning model comprehensively, its performance is compared against several recent and advanced routing protocols designed for smart farming IoT networks. These protocols include FARM-ED [14], EE-SMART [15], Agro-LORA [16], and IoT-PSO [18].

4.3 Results and Discussion

Several detailed schemes and diagrams can be used to visualize the comparative analysis of the proposed Fuzzy Q-learning model against recent protocols. Additionally, specific use cases in smart agriculture will be provided to illustrate the routing protocols' practical application and benefits.

4.3.1 Comparative Analysis Schemes

The comparative analysis of the proposed Fuzzy Q-Learning model against recent routing protocols in smart agriculture reveals several key performance insights.

a. ***Energy Consumption Level:*** Regarding energy consumption, the Fuzzy Q-Learning model (FQL-Smart Farming) demonstrates the lowest energy usage, attributed to its dynamic and adaptive optimization of routing decisions, ensuring minimal energy expenditure per transmission. In contrast, FARM-ED and EE-SMART consume more energy due to aggregation processing and static route adjustments. Agro-LORA takes advantage of the inherently low energy consumption of LoRa technology, but its long-range communication results in higher latency. IoT-PSO shows low energy consumption due to their efficient and secure routing mechanisms. However, they need to match the minimal energy usage of the FQL-Smart Farming model (Fig. 2). The average energy consumption per node in the network can be calculated as:

$$\text{Average Energy Consumption} = \frac{\sum_{i=1}^{N} E_i}{N} \tag{2}$$

where, E_i is the energy consumed by node i, and N is the total number of nodes.

b. ***Packet Delivery Ratio (PDR):*** Regarding packet delivery ratio (PDR), the Fuzzy Q-Learning model achieves a very high PDR, ensuring reliable data transmission even in dynamic network environments. This high reliability stems from its ability to predict and adapt to changing network conditions, selecting the most suitable routes. As illustrated in (Fig. 3), FARM-ED and IoT-PSO also show high PDRs. However, EE-SMART and Agro-LORA exhibit moderate PDRs due to their challenges in highly dynamic environments and long-range signal attenuation. The Packet Delivery Ratio (PDR) is calculated as:

$$PDR = \frac{\textit{Total number of packets received by the destination}}{\textit{Total number of packets sent by the source}} * 100\% \tag{3}$$

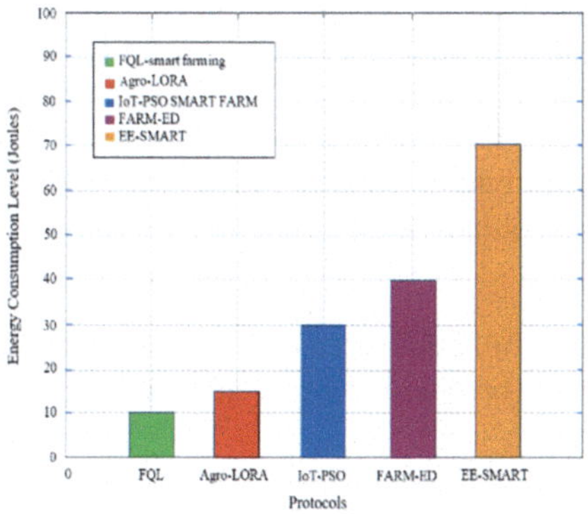

Fig. 2. Energy Consumption Comparison.

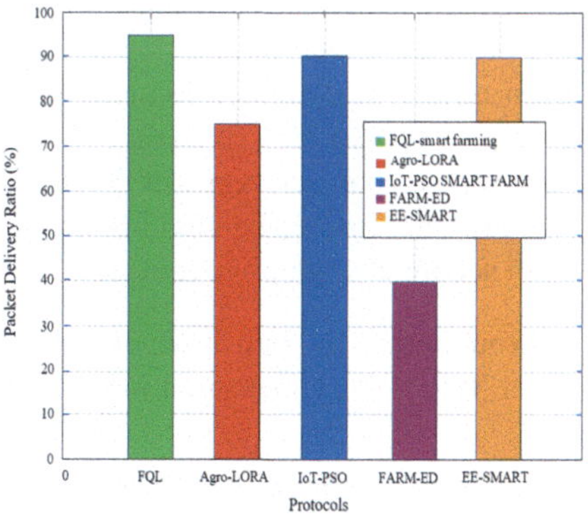

Fig. 3. Packet Delivery Ratio Comparison.

c. ***Average Latency:*** The average latency of the Fuzzy Q-Learning model is the lowest among all compared protocols. Its predictive nature lets it make pre-decisions for selecting delays, minimizing routes, and ensuring timely data delivery. FARM-ED bears average latency due to data aggregation delays, and Agro-LORA witnesses ample time overheads due to the long-range communications and lower rate modulation technique. EE-SMART and IOT-PSO gain average latencies by balancing routing efficiency, overhead computation power consumption, and security measures, as shown in (Fig. 4). The average latency is the meantime taken for a packet to travel from the source to the destination. It can be calculated as:

$$Average\ Latency = \frac{\sum_{I=1}^{P} T_i}{P} \tag{4}$$

where T_i is the time taken for packet i to reach the destination and P is the total number of packets.

d. ***Network Lifetime*:** Regarding network lifetime, the Fuzzy Q-Learning model significantly extends network longevity by efficiently balancing energy consumption across nodes, preventing early node depletion, and maintaining network connectivity for extended periods. While FARM-ED shows a moderate network lifetime due to the energy costs of aggregation processing, EE-SMART and IoT-PSO achieve high network lifetimes with their energy-efficient routing strategies. Agro-LORA boasts a very high network lifetime due to its deficient energy consumption, though it suffers from limitations in throughput and latency (Fig. 5). The network lifetime is often defined as the time until the first node in the network depletes its energy. However, it can also be considered the average time until a significant portion of nodes (e.g., 50%) deplete their energy. If defined by the time to the first node death, it is:

$$Network\ Lifetime = \min\left(T_I\right) \tag{5}$$

e. **Throughput:** Fig. 6 illustrates the throughput performance of various routing protocols in smart farming applications. The model Fuzzy Q-learning has the highest throughput; it refers to a high amount of data being successfully transmitted over the network, which is a significant demand for applications running in smart farming since data collection and monitoring are continuous. Similar to FARM-ED and IoT-PSO, even though it is related to aggregation delays, it could slightly reduce the performance of FARM-ED. Even though EE-SMART has moderate throughput because of static route changes, Agro-LORA has low throughput because LoRa-based technology provides limited bandwidth. Meanwhile, it obtains high throughput by combining both secure and adaptive routing.

Throughput is the rate at which data is successfully delivered over a communication channel. It is usually measured in bits per second (bps) and can be calculated as:

$$Throughput\ (kbps) = \frac{Total\ data\ received\ (kilobits)}{Total\ time\ taken\ (seconds)} \tag{6}$$

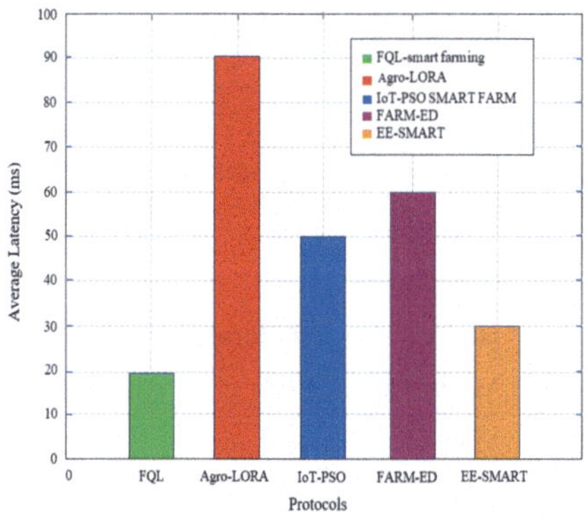

Fig. 4. Average Latency Comparison.

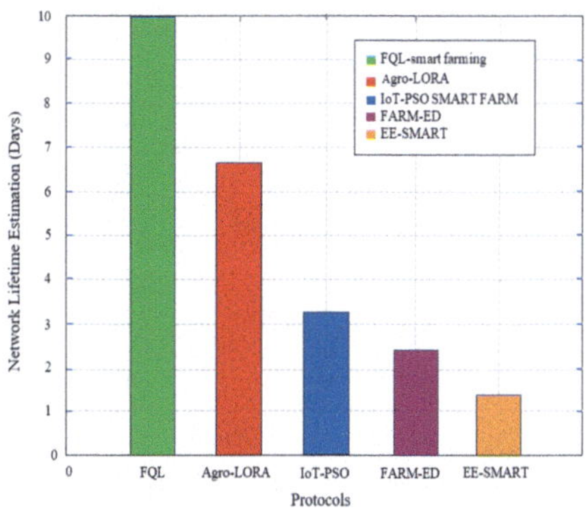

Fig. 5. Network Lifetime Comparison.

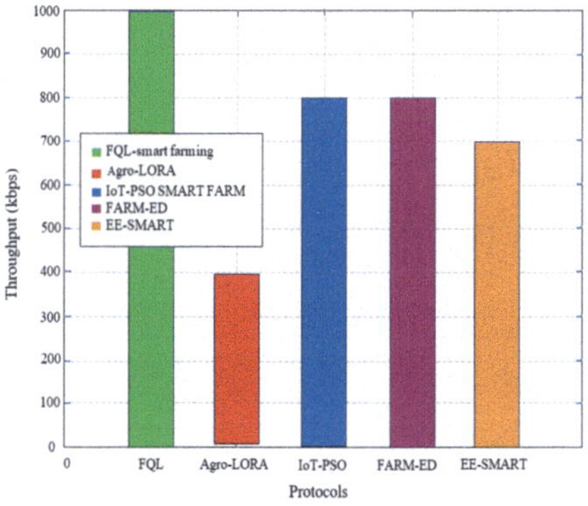

Fig. 6. Throughput Comparison.

4.3.2 Smart Agriculture Use Cases

a. ***Precision Irrigation:*** The Fuzzy Q-Learning model is used in precision irrigation, where soil moisture levels are monitored across different farm zones. It efficiently routes data from numerous sensors to a central system, ensuring timely and reliable data delivery to trigger irrigation systems. Unlike other protocols such as FARM-ED, which offers moderate energy efficiency but may experience delays due to data aggregation (Table 2), or EE-SMART, which provides high energy efficiency but faces challenges with reliability in dynamic conditions, the Fuzzy Q-Learning model adapts to network conditions. It optimizes routes to minimize energy consumption and latency, ensuring real-time soil moisture data is collected and processed efficiently.

b. ***Pest Detection and Control:*** Pest detection and control uses sensors and drones to monitor crop health status and pest infestation. The Fuzzy Q-Learning model (Table 3) is most efficient in routing the data from the sensors and drones in high throughput with low latency for early controlling of pests. It has a very high throughput and PDR with the lowest latency compared to other protocols, e.g., Agro-LORA having low

Table 2. Comparative Performance in Precision Irrigation.

		Performance			
		Energy-Efficiency	*Reliability*	*Adaptability*	*Latency*
Protocols	*FARM-ED*	Moderate	Potential	Low	Moderate
	EE-SMART	High	Challenges	High	Low
	Agro-LORA	Very Low	Higher	Moderate	High
	IoT-PSO	Low	Moderate	High	Moderate
	FQL-Smart Farming	Optimal	Very High	High	Very Low

throughput with higher latency, or IoT-PSO having high adaptability and throughput but moderate latency. Thus, the Fuzzy Q-Learning model can be best suited for fast processing and detecting pests so that appropriate measures can be taken promptly.

Table 3. Comparative Performance in Pest Detection and Control.

		Performance			
		Energy-Efficiency	*Reliability*	*Throughput*	*Latency*
Protocols	*FARM-ED*	High	Potential	High	Moderate
	EE-SMART	High	Challenges	Moderate	Low
	Agro-LORA	Very Low	Higher	Low	High
	IoT-PSO	Low	High	High	Moderate
	FQL-Smart Farming	Optimal	Very High	Very High	Minimal

c. ***Environmental Monitoring:*** The Fuzzy Q-Learning model shines in environmental monitoring, where IoT sensors monitor various parameters like temperature, humidity, and air quality. It extends the network lifetime by balancing energy consumption across nodes, ensuring continuous environmental monitoring, unlike other protocols such as FARM-ED, which provides a moderate network lifetime, or EE-SMART, which offers a high network lifetime with a moderate PDR. The Fuzzy Q-Learning model provides a very high network lifetime, PDR, and efficient energy usage, making it suitable for long-term environmental monitoring in remote areas (Table 4).

Table 4. Comparative Performance in Environmental Monitoring.

		Performance			
		Energy-Efficiency	*Reliability*	*Efficiency*	*Network lifetime*
Protocols	*FARM-ED*	Moderate	Potential	Moderate	Moderate
	EE-SMART	High	Challenges	Moderate	High
	Agro-LORA	Very Low	Higher	Moderate	Very High
	IoT-PSO	Low	High	High	High
	FQL-Smart Farming	Very Low	High	Very High	Very High

The simulation and comparative analysis reveal that Fuzzy Q-Learning is very competitive with recent protocols like FARM-ED, EE-SMART, Agro-LORA, and IoT-PSO in smart agriculture applications. In precision irrigation, low energy consumption and latency-optimized routes help to get efficient real-time data. For pest detection and control of high throughput, minimal latency of the proposed operations helps for quick response. Environmental monitoring of the long life of the network ensures continuous, reliable collection of data from various locations. Overall consideration reveals that this model is much better than others in terms of energy efficiency, adaptability, and reliability in enhancing the IoT communications network in smart farming.

5 Conclusions and Future Work

This study presents a work in which fuzzy logic and Q-learning control the data flow in IoT-based smart agriculture to control network data dissemination. Simulation results show that the FQL-Smart Farming model has a significant edge over existing protocols regarding energy efficiency, Packet Delivery Ratio (PDR), latency, network lifetime, and throughput for different smart agriculture cases. Precision Irrigation outperforms protocols like FARM-ED and EE-SMART regarding route optimization for real-time data processing and efficient energy usage. Pest Detection and Control surpasses Agro-LORA and IoT-PSO in handling high volumes of data with high throughput and low latency. Environmental Monitoring performs better than FARM-ED and EE-SMART in terms of extending network lifetime by balancing energy consumption to ensure continuous monitoring.

The proposed FQL-Smart farming model is a general, robust, adaptive, and efficient solution for IoT communication networks in smart farming. It provides greater efficiency ($\sim$10%) and reliability (7% on average) of IoT networks. Using the Fuzzy Q-learning model to control an IoT communication network in smart farming can lead to precision and sustainable and productive agriculture.

Future research should explore integrating advanced machine learning techniques, conducting scalability studies, implementing the model in real-world systems, investigating security mechanisms, and developing multi-objective optimization strategies. These efforts will refine the Fuzzy Q-Learning model to meet the evolving needs of smart farming, contributing to more efficient, sustainable, and resilient agricultural practices.

References

1. Ojha, T., Misra, S., Raghuwanshi, N.S.: Internet of things for agricultural applications: the state of the art. IEEE Internet Things J. **8**(14), 10973–10997 (2021)
2. Thilakarathne, N.N., Yassin, H., Abu Bakar, M.S., and Emerolylariffion, Pg E.A.: Internet of Things in smart agriculture: challenges, opportunities and future directions. In: Proceedings of the 2021 IEEE Asia-Pacific conference on computer science and data engineering (CSDE), pp. 1–6 (2021)
3. Ahmed, N., De, D., Hussain, I.: Internet of Things (IoT) for smart precision agriculture and farming in rural areas. IEEE Internet Things J. **5**(6), 4890–4899 (2018)
4. Hashmi, A.U.H., Mir, G.U., Sattar, K., Ullah, S.S., Alroobaea, R.: Effects of IoT communication protocols for precision agriculture in outdoor environments. IEEE Access **12**, 46410–46421 (2024)
5. Beghriche, A., Kout, A., Abed, A.: A bio-inspired energy-efficient routing protocol based on penguins search optimization for FANETs. In: Proceedings of the first international conference on computer communications and intelligent systems (I3CIS), pp. 235–240. IEEE (2022)
6. Marietta, J., Chandra Mohan, B.: A review on routing in Internet of Things. Wireless Pers. Commun. **111**, 209–233 (2020)
7. Hassan, R., Qamar, F., Hasan, M.K., Mohd Aman, A.H., Sid Ahmed, A.: Internet of things and its applications: a comprehensive survey. Symmetry **12**(10), 1674 (2020)
8. Lamaazi, H., Benamar, N.: A comprehensive survey on enhancements and limitations of the RPL protocol: a focus on the objective function. Ad Hoc Netw. **96**, 102001 (2020)
9. Tournier, J., Lesueur, F., Le Mouël, F., Guyon, L., Ben-Hassine, H.: A survey of IoT protocols and their security issues through the lens of a generic IoT stack. Internet Things **16**, 100264 (2021)
10. Bajpai, A., Verma, H., Yadav, A.: Optimizing data aggregation and clustering in Internet of Things networks using principal component analysis and Q-learning. Data Sci. Manag. **7**(3), 189–196 (2024)
11. Bhimshetty, S., Ikechukwu, A.V.: Energy-efficient deep Q-network: reinforcement learning for efficient routing protocol in wireless Internet of Things. Indones. J. Electr. Eng. Comput. Sci. **33**(2), 971–980 (2024)
12. Gupta, D., Bajpai, A., Tiwari, N.K., Yadav, S.: Energy-efficient routing optimization for underwater internet of things using hybrid Q-learning and predictive learning approach. Procedia Comput. Sci. **235**, 45–55 (2024)
13. Vasseur, J.P., Agarwal, N., Hui, J., Shelby, Z., Bertrand, P., Chauvenet, C.: RPL: the IP routing protocol designed for low power and lossy networks. Internet Protocol for Smart Objects (IPSO) Alliance (2011)
14. Bhushan, S., Kumar, M., Kumar, P., Stephan, T., Shankar, A., Liu, P.: FAJIT: a fuzzy-based data aggregation technique for energy efficiency in wireless sensor network. J. Ambient. Intell. Humaniz. Comput. **7**, 997–1007 (2021)
15. Bhasker, B., Murali, S.: An Energy-efficient cluster-based data aggregation for agriculture irrigation management system using wireless sensor networks. Sustain. Energy Technol. Assess. **65**, 103771 (2024)
16. Prasad, S.J.S., Thangatamilan, M., Muthusamy, S., Panchal, H.: An efficient LoRa-based smart agriculture management and monitoring system using wireless sensor networks. Int. J. Ambient Energy **43**(2), 1–10 (2021)

17. Dhruva, A.D., Prasad, B., Kamepalli, S., Sakthy, S.S., Kunisetti, S.: An efficient mechanism using IoT and wireless communication for smart farming. Mater. Today Proc. **80**(3), 3691–3696 (2023)
18. Gökçe, B., Koca, Y.B., Aslan, Y.: Particle swarm optimization-based optimal PID control of an agricultural mobile robot. Proc. Bulg. Acad. Sci. **74**(4), 568–575 (2021)
19. Alonso, R.S., Sittón-Candanedo, I., Casado-Vara, R., Prieto, J., and Corchado, J.M.: Deep reinforcement learning for the management of software-defined networks in smart farming. In: Proceedings of the 2020 international conference on omni-layer intelligent systems (COINS) (2020)
20. NS-3 Network Simulator (2016). Available online at: http://www.nsnam.org. Accessed 12 June 2024

An Improved Energy-Efficient MAC Protocol for WSNs

Abdelmalek Bengheni(✉)

Computer Science Department, Math/CS Faculty, LRAIS Laboratory, University of Ibn Khaldoun, Tiaret, Algeria

abdelmalek.bengheni@univ-tiaret.dz, baghanimalik@yahoo.fr

Abstract. Wireless Sensor Networks (WSNs) operating on constrained energy resources present a substantial challenge to researchers. Actually, a lot of ideas have been made to either equip a WSN's sensor nodes with technologies that harvest energy or to preserve the energy that it uses efficiently. This paper presents an Improved Energy-Efficient MAC Protocol for WSNs (IEE-MAC). The functioning of IEE-MAC relies on using transmitter-initiated schemes and low-power listening (LPL) with small preamble messages and based on the value of the collected energy rate, each sensor node in the WSN may decide when to sleep next. In doing so, it lowers the duty cycle more adaptively by utilizing the stored remaining energy, which particularly aids in minimizing latency and energy consumption. According to the simulation findings, our suggested protocol performs better in terms of lowering average latency and power consumption while boosting packet delivery rates as well as extending the lifetime of wireless network compared to the PS-EHWSN scheme.

Keywords: IEE-MAC · WSN · Low Power Listening (LPL) · MAC protocol · Energy Saving · Energy Consumption

1 Introduction

Due to applications relying on wireless sensor networks (WSNs) and their battery-powered devices (i.e. their sensor nodes), energy conservation is a key element in this type of network. The MAC layer in this type of networks manages channel access, scheduling, and buffer access protocols for shared media management and error control while offering efficiency and dependability. It is also essential to managing communications between sensor nodes in order to increase the network's lifetime. However, MAC protocols must first be energy efficient in order to improve latency, and fairness. For example, MAC protocols ought to be tuned for the scenario in which network traffic is minimal or nonexistent. In recent years, a number of studies have attempted to optimize node energy savings through the use of power management (PM) techniques [1] in particular the application of cutting-edge conservation techniques like "Duty-cycling" [2] which is used to preserve energy in WSNs where each of its nodes has the ability to alternate between the sleep and active states. Duty-cycle MAC protocols come in

© The Author(s), under exclusive license to Springer Nature Switzerland AG 2025
M. Hatti (Ed.): IC-AIRES 2024, LNNS 1238, pp. 672–683, 2025.
https://doi.org/10.1007/978-3-031-80301-7_70

two flavors: asynchronous and synchronous, can provide this latest functionality. Synchronous MAC protocols, like SMAC [3], rely on control packets to provide schedule information between nodes, defining their active and sleep periods. However, in order to send or receive data, the schedule information is not communicated by asynchronous MAC technologies. Rather, they do this using beacons or preamble sampling such as RIMAC [4] and XMAC [5] respectively. The incorporation of energy harvesting (EH) technology encouraged researchers to develop self-sustaining WSNs called energy harvesting WSNs (EH-WSNs) [6]. Nodes in these networks have the ability to harvest the energy from outside sources, including mechanical, solar, RF, thermal, and wind energy. This lessens the problem of low battery life that conventional battery-powered WSNs encounter. Moreover, the node may maintain energy neutral operation (ENO) [7], which is the ability to gather energy that is equal to or higher than that which is needed. If not, it consumes battery energy, which may result in a shorter lifetime. Because of this, a duty-cycle-based energy harvesting management (EHM) [8] strategy was developed, in which every node switches to a sleeping state to collect energy and reduce the duration of its activity in an effort to enhance power conservation. In fact, the "duty-cycle" approach is the foundation of the EHM utilized in the majority of MAC protocols of the EH presented, including ODMAC [9] and ERIMAC [10]. However, the majority of the previously proposed protocols avoid employing an estimated approach to modify the duty-cycle, which means that nodes alter its duty-cycle periods disregarding the energy that will be available in the future. As was previously noted, the goal is to reduce the packet mean latency by minimizing the energy consumption resulting from a number of retransmissions for a packet, channel contention, and the packet collision. For the task of reducing the node's active time and, as a result, the duty-cycle is shared by the PM and EHM. This paper presents an Improved Energy-Efficient MAC protocol for WSNs (IEE-MAC), in which every sensor node has an energy collecting device installed. After a channel availability test, communication between the nodes is initiated by transmitting a sequence of brief preambles, in which the exchange is actively engaged by the transmitting node. The issuer-initiated methods established in earlier publications are adopted by the initiator. To enhance the effectiveness of IEE-MAC even further, a method for calculating the nodes' sleep duration is included. This approach will enable the network's nodes to define their sleep durations. Additionally, it enables the nodes to more adaptively modify their duty cycles in response to the energy that is gathered and will eventually become accessible. The duration of the subsequent sleep period is then decided using the energy that is still available. This technique reduces power consumption and latency of all non-target and source nodes by decreasing over-emitting, overhearing, idle listening, and collisions. Furthermore, IEE-MAC's main objective is to decrease mean latency while increasing packet delivery ratio, lowering power consumption, and increasing WSN lifetime utilizing the energy collected.

The rest of the document is organized as follows. Section 2 contains a presentation of related works. The next section describes the IEE-MAC protocol design and the modeling of energy used types by each node in the WSN and latency. The IEE-MAC performance evaluation and comparison covered in Section 4. Section 5 concludes with the findings and recommendations for further study.

2 Related Works

In order to maintain energy neutrality, energy harvesting is a newly developed technique in WSNs that takes into consideration the influence of energy collection rate in compensating for energy consumption [11]. To solve this problem, a many of MAC strategies for EH-WSNs have been created. These schemes take into account the energy harvested while scheduling various tasks, such as duty cycle.

An energy-saving algorithm known as the "Adaptive Duty Cycle Algorithm for Energy Efficiency IEEE 802.15.4 Beacon-Enabled WSN (ADCA)" was introduced by the authors in [12]. Utilizing an IEEE 802.15.4 technology supported by a customized pattern, it automatically sets the duty cycle in exact proportion to the traffic data. The suggested ADCA technique is designed for the WSN's beacon-permitted star network, which enhances the energy efficiency of the system, especially in cases where traffic is highly variable and fluctuates over time.

In order to minimize sensor node sleep latency while preserving the equilibrium of power consumption among them, the authors of [13] presented two dynamic duty cycle scheduling algorithms for power-gathering WSNs. The first one, called Duty-cycle Scheduling based on Residual Energy (DSR), alters the duty-cycle by utilizing the amount of energy that is now left. The subsequent Duty-cycle Scheduling based on the Prospective increase in residual energy (DSP), on the other hand, uses the potential rise in remaining energy as time passes in energy gathering to lower the duty-cycle more aggressively when necessary. To achieve this reduction, the differential between the rates of energy used and energy harvested is computed at the start of every duty-cycle.

The two energy-saving techniques that the authors in [14] recommend for the MAC layer are based on Expansive Decision-making (EDMAC). These schemes are mainly based on the use of a very important component, which is residual energy. The first exponential decision-making MAC protocol Emerging Demand and Supply Requirements (EDDSR), a dynamic duty cycle scheduling-based system, dynamically modifies each sensor node's duty cycle based only on current residual energy. Under the second technique, Dynamic Duty cycle scheduling, which is based on the Prospective energy increase (EDDSP), any sensor node's prospective rise in residual energy over time is forecast. This approach uses exponential decision-making to reduce downtime as likely as possible and achieve greater energy economy.

The authors of [15] proposed a MAC technique that lowers end-to-end network latency by altering a packet's data transmission time in line with estimations of the energy needed for the energy collected, data transfer and the volume of traffic entering the WSN. It accomplishes this by building a system for energy management that keeps track of energy scavenging time and uses it to control network sleep time.

The authors of [16] have developed the ENO based Adaptive Duty-Cycle MAC Protocol for Solar EH-WSNs (ENCOD-MAC), allowing the node to operate as close to the ENO mode as is practical. Improved network performance is achieved by incorporating extra energy collected and controlling their duty cycle according to the level of battery.

The authors of [17] provide a Multi-Threshold Energy strategy for EH-WSNs (MTE-EHWSN), which aims to shorten each sensor node's working duration by lowering its duty-cycle. In MTE-EHWSN, and using a variety of energy thresholds, sensor nodes can reduce its duty-cycle in accordance with its residual energy, which can rise over time in a way that is more profound and dramatic.

In [18], a MAC protocol based on preamble sampling technique for EH-WSNs (PS-EHWSN) is explained. The residual energy of each sensor node within the network may be used to calculate when its next sleep interval is, allowing it to equilibrium its energy consumption and collection over time. This is made possible by the PS-EHWSN system. Furthermore, PS-EHWSN depends on the sender node receiving quick welcome messages and uses Low Power Listening (LPL) technology as its foundation.

For power management in EH-WSNs, the authors of [19] propose a novel energy-saving system called Relay node selection technique and Deep sleep time (RD-EHWSN), which is based on asynchronous duty cycling. It uses a newly proposed energy threshold policy to adjust sensor nodes' duty cycles more deeply and drastically depending on the projected worth of their remaining energy from upcoming presentations of gathered energy. By doing so, the energy consumption of sensor nodes is decreased and the energy usage between them in the network is balanced with the ability to harvest energy. Furthermore, in order to offer load balancing in WSN, RD-EHWSN employs a novel relay node selection procedure in addition to capitalizing on the benefits of transmitter start using the LPL technique with brief preamble messages. Comparing RD-EHWSN with the PSEHWSN system, it increases the WSN lifetime and guarantees that it functions properly when the energy harvesting rate is lower.

3 IEE-MAC Protocol Design, Modeling of Energy used Types and Latency

IEE-MAC employs a transmitter that occupies the wireless channel using a carrier wave whereas the CSMA/CA protocol is used by the receiver to keep an eye on it and prevent collisions. Respectively moment the brief preamble is given; every node in the IEE-MAC WSN that has stored packages that necessity to be forwarded to the sink is monitoring the canal and expecting for a proper acknowledgement (ACK). Sending of data packets begins upon receipt of the appropriate ACK.

3.1 Overview of Basic Communications for IEE-MAC

The following is the basis of IEE-MAC's communication:

At starting up, every node goes into sleep mode. To keep track of network activity, each node switches to listen mode on its own timetable. Transmitter node T determines if the medium is available using the Clear Channel Assessment (CCA) before broadcasting numerous short preamble packets (SP) to surrounding nodes in order to start a preamble session and convey data. It waits for the carrier after every SP packet transmission. After receiving an SP packet and if their back off has passed, the active receiving node R responds to the transmitter in unicast mode by transmitting an ACK (acknowledgement SP) packet; if a channel is not busy, transmitter T will deliver the packet of data to the designated receiver and cease sending the other SP packets and R will respond with an ACK; Or it enters a sleep mode after a certain time length assigned to delivering much SP packets; the transmitter T enters into sleep mode upon receiving the acknowledgement of

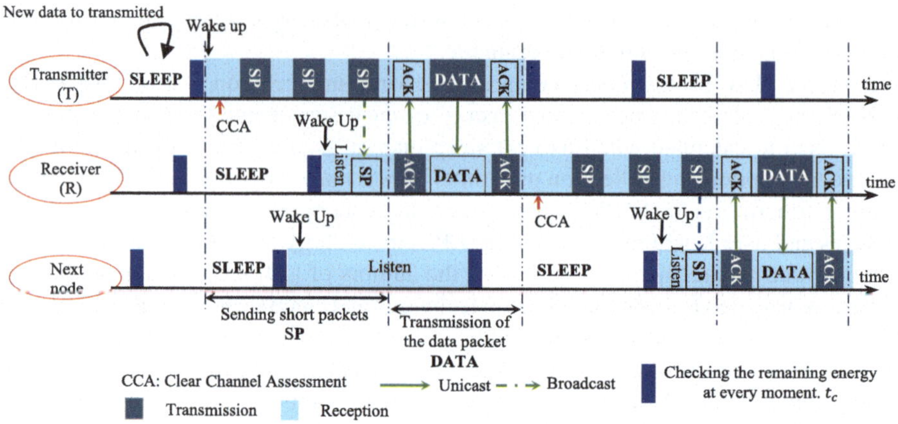

Fig. 1. Adopted procedures for data transfer using the IEE-MAC protocol.

the sent packet of data (ACK). The receiving node then immediately initiates transmitting its data to the following node by following the same paces as the transmitter T before it. Figure 1 represent the essential communication strategy used between a transmitter and a receiver in IEE-MAC scheme.

3.2 Modeling Energy Consumption

We investigate one sort of periodic traffic generation, where nodes create data at a constant periodic interval, and we suppose that our network consists of n nodes. Table 1 lists all of the terms used in the energy and latency model.

IEE-MAC represents an improvement of PS-EHWSN [18], prior to every data packet, SP are transmitted. The amount of energy that node i is expected to consume is:

$$E_{consumed_i} = \left(\sum_{j \in Op} (T_j \times C_j) \right) \times V \tag{1}$$

$Op = \{$carrier sensing (cs), environment sensing (es), Transmission (t_x), Reception (r_x)$\}$, Op: operation.

Every wake-up period in IEE-MAC requires a node to sense the carrier before transmitting a preamble in the time T_{cs} indicated in Eq. (2). In order to assess physical values in the time T_{es} indicated in Eq. (3), it also detects the surroundings.

$$T_{cs} = T_{acs} R_{data} \tag{2}$$

$$T_{es} = T_{asense} R_{es} \tag{3}$$

The total amount of time needed to convey short preambles, packets of data, and acknowledgements (of brief preambles and packets of data) equals the time of transmission of node i. Every time it has data to broadcast, it immediately switches its radio

Table. 1. Terms related to the energy model and network.

Symbol	*Denotation*
C_j (mA)	Current consumption of operation *j*
T_j	Time used by operation *j*
V	Voltage
R_{data}	Rate to TX/RX a packet of data
R_{es}	Rate for sensing the environment
$L_{spreamble}$	Length of short-preamble
L_{packet}	Length of data packet
L_{ack}	Length of acknowledgement
T_{acs}	Average carrier sense time
T_{asense}	Average time for data sampling
$T_{(t/r)xb}$	Time to transmit /receive a byte
T_{sp}	Time to send a short preamble
T_{sw}	Time for radio switching
T_{queu}	Time of queuing
T_{proc}	Time of processing

to listen status so as to get the acknowledgement after sending a brief preamble in the required amount of time T_{sp}. This process must be tracked until the target node acknowledges it. *K* represents the times number a node modifies its radio and transmits a brief preamble in an effort to get an acknowledgement (ack). The receiver changes its radio to send an ack after getting the data payload.

$$L_{spreamble} = T_{sp}/T_{(t/r)xb} \tag{4}$$

$$T_{tx} = R_{data}\left(L_{packet} + 2L_{ack}\right)T_{(t/r)xb} + KT_{sp} + (K+2)T_{sw} \tag{5}$$

It is possible for a node to receive more than one ($\sum$) preamble in one go. However, as soon as it becomes the target node for a particular preamble, it sends the transmitter an acknowledgement, gets the packet, and then replies to the transmitter.

$$T_{rx} = R_{data}\left(L_{packet} + 2L_{ack}\right)T_{(t/r)xb} + \sum T_{sp} + 2T_{sw} \tag{6}$$

3.3 Modeling Remaining Energy

In IEE-MAC, the remaining energy at the instant t_c for node *i* ($E_{rem_i}(t_c)$) is computed using the subsequent equation:

$$E_{rem_i}(t_c) = \left(E_{rem_i}\left(t_c'\right) - E_{consumed_i}\left(t_c - t_c'\right)\right) + E_{harvested_i}\left(t_c - t_c'\right) \text{ with } \left(t_c' < t_c\right) \tag{7}$$

where, $E_{rem_i}(t'_c)$ is the remaining energy at previous check instant t'_c and $E_{harvested_i}(t_c - t'_c)$, $E_{consumed_i}(t_c - t'_c)$ justify the harvesting amount and the consumed amount within same time period $(t_c - t'_c)$ respectively.

3.4 Modeling Latency

The node broadcasts an acknowledgment after receiving each packet of data in the IEE-MAC protocol, which is headed by several preambles. This procedure is run by each node i where each packet is processed and then queued. As a result of reasons (2), (5), and (6), the anticipated delay is as follows:

$$L_{IEE-MAC_i} = T_{queu} + T_{proc} + T_{cs} + T_{tx} + T_{rx}$$

$$= T_{queu} + T_{proc} + T_{cs} + \left(3T_p/2 + 2\left(L_{packet} + L_{ack}\right)t_{(t/r)xb}\right)R_{data} + 3T_{sw} \quad (8)$$

3.5 Dynamicity of Duty Cycle

The duty cycle in IEE-MAC becomes dynamic when the remaining energy $E_{rem_i}(t_c)$ at the t_c moment is less than the energy threshold (E_t) determined by the WSN manager, and this depends on how quickly energy can be gathered. Longer sleep durations, more frequent waking's, and higher packet send and receive rates are generally associated with nodes with superior energy supplies. This strategy also reduces collisions produced by adjacent nodes competing for the channel when they wake up simultaneously. IEE-MAC cannot guarantee that a set of nodes' sleep durations will be the same since it bases its determination of node energy availability deterministically. Therefore, a node with lesser levels of energy selects longer sleep periods in order to gather more energy as feasible and store an elevated level of energy. In the next waking states while ($E_{rem_i}(t_c) \geq E_t$) and by using the equations shown below, one can determine the sleep time T_{slp_i} of node i in an IEE-MAC.

$$T_{slp_i}\left(E_{rem_i}(t_c)\right) = \frac{E_{max} - E_{rem_i}(t_c)}{E_{max} - E_t} \times T_{Ste_i} \quad (9)$$

When: T_{Ste_i} is calculated and $E_t \leq E_{rem_i}(t_c) < (E_{\max} - E_t)$

$$T_{slp_i}\left(E_{rem_i}(t_c)\right) = e^{\log_2\left(\frac{E_{max} - E_{rem_i}(t_c)}{E_{max} - E_t} \times T_{Ste_i}\right)} \quad (10)$$

When: T_{Ste_i} is not calculated (the old computed value is what we utilize) and $E_t \leq E_{rem_i}(t_c) < (E_{\max} - E_t)$

$$T_{slp_i}\left(E_{rem_i}(t_c)\right) = 0 \quad (11)$$

When: $(E_{\max} - E_t) \leq E_{rem_i}(t_c) \leq E_{\max}$

$$\text{With} : T_{Ste_i} = \min\left(T_{Ste_i}, 1\right) \quad (12)$$

where: T_{Ste_i}: Sleep time spent by node i until the energy threshold is reached or exceeded by the remaining energy and Emax represents the maximum energy value of node.

4 IEE-MAC Evaluation and Comparison

Via simulations performed utilizing the OMNeT++ simulator and the MiXiM framework [20], the IEE-MAC protocol was tested against the PS-EHWSN protocol, which uses a dynamic duty cycle comparable to IEE-MAC. We utilized the performance parameters given below for comparing the both protocols. Our findings are derived on a simulation that ran for 1000 simulation seconds, using 10 sensor nodes that are mobile dispersed in an area of 400 m × 400 m to 100 sensor nodes dispersed in an area of 1100 m × 1100 m. All source sensor nodes can use multi-hop transmissions to send their periodic traffic (TrafficParam) to the sink once per 0.1 or 0.5 s; 500 data packets in total are expected to be delivered. For every sensor node, the amount of energy harvested is randomly selected between 0 and 50 mW. The energy harvesting duration is equivalent to 0.1 s. Each sensor node's starting energy value is set to 50 J, with an energy threshold value set to 30 J and a maximum energy value set to 100 J. The following performance metrics were employed.

Average Energy Consumption Per Node: The computation involves dividing the sum of the mean energy consumed by every node in the network by the total nodes number.

Network Lifetime: It's the moment after the original sensor's power is depleted and there is no longer any capacity for the sensor to send data while it is being collected.

Mean Packet Latency: The average of all packets' transit times to the destination node is used.

Packet Delivery Ratio: It represents the proportion of the overall packets number transmitted by all WSN source nodes to the packets number successfully received by the sink.

According to Fig. 2, for varying EH rates, IEEMAC performs better in terms of energy efficiency than PSEHWSN. This is because, in contrast to PSEHWSN, IEEMAC allows each sensor node to dynamically and more adaptively modify its duty cycle according to remaining energy, which may increase over time, as IEEMAC intentionally chooses sleep periods that can be slow or short which are unlikely to be same in the network area of sensor nodes, thus reducing retransmission and collisions, Those are the primary causes of higher energy consumption and more frequent failures when data is being sent inside the network. For instance, IEEMAC beats PSEHWSN in Fig. 2 when the TrafficParam = 0.5 s, saving approximately 15.20% of the energy when there are 10 nodes and 15.46% when there are 100 nodes.

The network lifetime measure shows that for these two protocols, IEEMAC and PSEHWSN, with varying EH rates, the latter rises as the nodes number increases. This is because, based on Eqs. (10 or 11) and (9) in the IEEMAC, respectively, they enable each sensor node to dynamically modify its duty cycle by estimating the duration of its next sleep, which may be long or short. IEEMAC performs better than PSEHWSN since it makes use of a recently suggested technique for estimating the duration of the upcoming sleep period, which allows it to offer load balancing among WSN nodes. Additionally, IEEMAC considers the dynamic adjustment of duty cycles in a more adaptive manner than PS-EHWSN, ensuring that relay nodes never completely run out

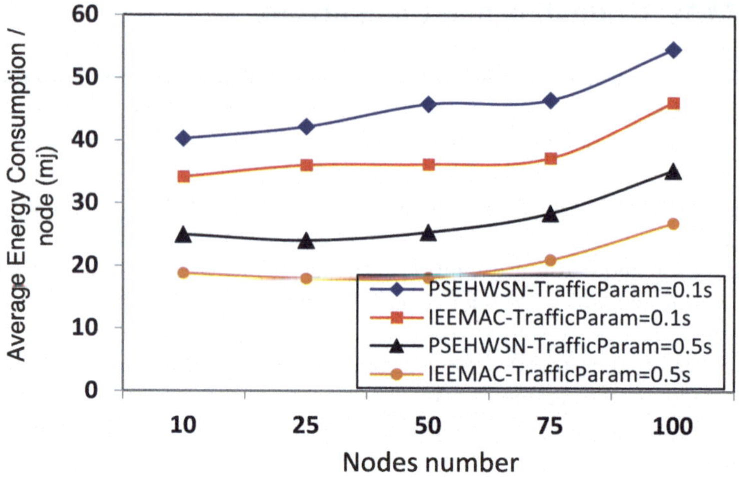

Fig. 2. Average energy consumption per node.

of power after cooperating. Additionally, by shortening the time nodes spend sleeping, it promotes activity on higher power nodes, extending the lifetime of the network. When seen in Fig. 3, IEEMAC enhances network lifetime more than PSEHWSN by 2.77% and 15.61%, respectively, when the nodes number grows from 10 to 100, with the pass parameter set to 0.1 s and among 3.52% and 20.22% with the pass parameter equal to 0.5 s.

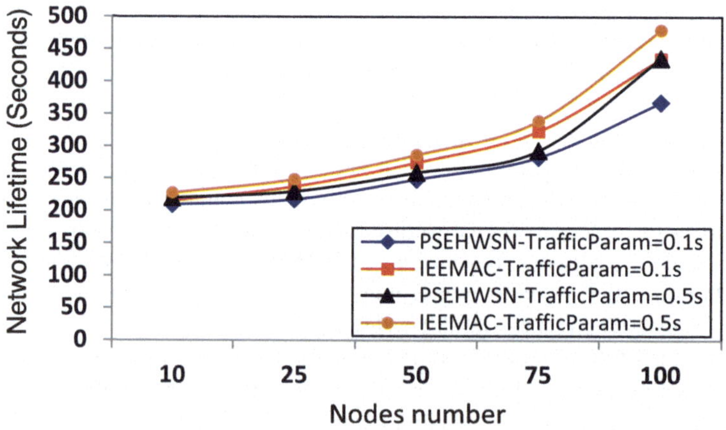

Fig. 3. Network lifetime.

Figure 4 demonstrates that IEEMAC performs better than PSEHWSN as the nodes number rises and that IEEMAC has a lower average latency than PSEHWSN. This is as a result of the new subsequent sleep duration prediction method that the proposed

IEEMAC protocol employs that helps nodes adjust their work cycle more adaptively than PSEHWSN and furthermore to offer load balancing among nodes in WSN, where nodes with greater energy levels are favored to remain active due to their brief sleep period duration; this is determined using Eqs. (10) or (11) if the rate at which remaining energy increases is quite quick in comparison to PSEHWSN. This lowers the average access time and raises the probability that data packets won't be buffered. Compare IEEMAC with PSEHWSN with 100 nodes and 500 sent packets, Fig. 4 demonstrates that IEEMAC achieves a reduced average latency gain of approximately 33.07% when traffic parameter = 0.1 s and 17.15% when traffic parameter = 0.5 s.

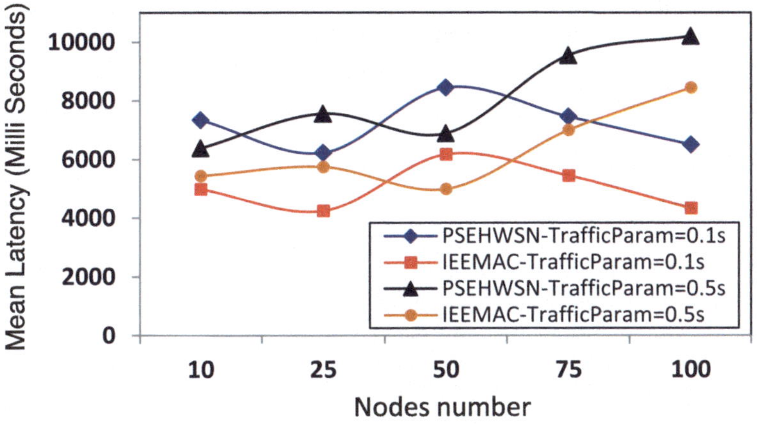

Fig. 4. Mean latency.

Figure 5 illustrates the results, which indicate that the packet delivery rate drops with increasing node count. Furthermore, the findings indicate that IEEMAC can deliver a higher packet delivery rate than PSEHWSN for varying the rates of energy harvesting by employing a novel technique to forecast the subsequent sleep interval according to the rate of energy harvesting, which can vary in length and not be constant across the network's segments where duty cycle adjustments are dynamically adjusted to maximize adaptability for WSN nodes. This reduces packet collision and retransmission rates. If, as seen in Fig. 5, the nodes number is increased from 10 to 50, IEEMAC performs better than PSEHWSN in terms of packet delivery rate, with the difference being among about 19.99% and 35.33% when TrafficParam is adjusted to 0.1 s and among 11.51% and 38.62% when TrafficParam is adjusted to 0.5 s. It is clear from this that employing IEEMAC as opposed to PSEHWSN results in considerable performance increases in the delivery ratio.

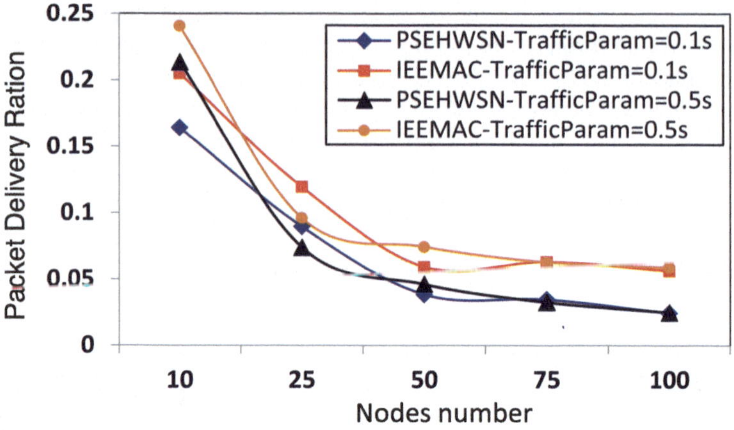

Fig. 5. Packet delivery ratio.

5 Conclusion

The WSN energy aggregation MAC protocol, or IEE-MAC, is a novel EH-based strategy for low-power communication in WSNs and is proposed in this study. IEEMAC broadcasts a series of brief preamble packets using strobed preamble technology in conjunction with a dynamic duty cycle scheduling mechanism. By utilizing the remaining energy level, the suggested protocol forecasts two distinct future sleep durations: long and short, which can be calculated using (9) and ((10) or (11)), respectively. Each node can modify its duty cycle in a more dynamic and adaptive manner based on the total energy function, which will be accessible thereafter. Comparing this idea to the PSEHWSN protocol, network performance is improved. Through a comparison of its efficiency with the PSEHWSN protocol, IEE-MAC controls the energy collected from the sensor node that is used for detection, processing, reception, and transmission. This is illustrated by the outcomes of simulations with OMNeT++ implemented, where the energy efficiency, WSN lifetime duration, mean latency and packet delivery ratio of the IEE-MAC protocol were considerably better than those of the PSEHWSN protocol. Future work to further improve the performance of IEE-MAC will include changing its fixed power threshold to a variable one that can be used by each node and which can be calculated depending on the node's movement and remaining power.

References

1. Chen, F., Guo, L., Chen, C.: A survey on energy management in the wireless sensor networks. IERI Procedia **3**, pp. 60–66 (2012)
2. Bengheni, A., Didi, F., Bambrik, I.: Energy-saving comparison of asynchronous MAC protocols for wireless sensor networks. Int. Conf. on Mathematics and Information Technology (ICMIT), pp. 263–268. Adrar, Algeria (2017)
3. Ye, W., Heidemann, J., Estrin, D.: An energy-efficient MAC protocol for wireless sensor networks. Twenty First Annual Joint Conferences of the IEEE Computer and Communications

Societies. Proceedings.IEEE.INFOCOM, volume 3, pp. 1567–1576. New York, NY, USA (2002)
4. Sun, Y., Gurewitz, O., Johnson, D.B.: RI-MAC: A Receiver Initiated Asynchronous Duty Cycle MAC Protocol for Dynamic Traffic Loads in Wireless Sensor Networks. In: SenSys '08: Proceedings of the 6th ACM Conference on Embedded Networked Sensor Systems, pp. 1–14. Raleigh, NC, USA (2008)
5. Buettner, M., Yee, G.V., Anderson, E., Han, R.: XMAC: A Short Preamble MAC Protocol for Duty-Cycled Wireless Sensor Networks. In: SenSys'06: Proceedings of the 4th international conference on Embedded networked sensor systems, pp. 307–320. New York, NY, USA (2006)
6. Wiliams, A.J., Torquato, M.F., Cameron, I.M., Fahmy, A.A., Sienz, J.: Survey of energy harvesting technologies for wireless sensor networks. IEEE Access **9**, 77493–77510 (2021)
7. Kosunalp, S.: MAC Protocols for Energy Harvesting Wireless Sensor Networks: Survey. ETRI Journal **37**(4), 804–812 (2015)
8. Bengheni, A., Didi, F., Bambrik, I.: August. EEM-EHWSN: Enhanced Energy Management Scheme in Energy Harvesting Wireless Sensor Networks. Wireless Netw **25**, 3029–3046 (2019)
9. Fafoutis, X., Dragoni, N.: ODMAC: An On-Demand MAC Protocol for Energy Harvesting Wireless Sensor Networks. ACM Symp. Performance Evaluation Wireless Ad Hoc, Sensor, Ubiquitous Netw, pp. 49–56. Miami, FL, USA (2011)
10. Nguyen, V.H., Le, D.D., Ji,Y., Duong D.A., Yamada, S.: ERI-MAC: An Energy-Harvested Receiver Initiated MAC Protocol for Wireless Sensor Networks. Int. J. Distrib. Sensor Netw. **10**(5), 1–8 (2014)
11. Ramezani, P., Pakravan, R.M.: Overview of MAC protocols for energy harvesting wireless sensor networks. IEEE 26th Annual International Symposium on Personal, Indoor, and Mobile Radio Communications (PIMRC), pp. 2032–2037. Hong Kong, China (2015)
12. Rasouli, H., Kavian, Y.S., Rashvand, H.F.: ADCA: adaptive duty cycle algorithm for energy efficient IEEE 802.15.4 beacon-enabled wireless sensor networks. IEEE Sensors Journal **14**(11), 3893–3902 (2014)
13. Yoo, H., Shim, M., Kim, D.: Dynamic dutycycle scheduling schemes for energy-harvesting wireless sensor networks. IEEE Communications Letters **16**(2), 202–204 (2012)
14. Varghese, J., Jose, A.L.: Dynamic duty-cycled MAC for wireless sensor networks with energy harvesters. Int. Conf. on Circuits, Communication, Control and Computing, pp. 156–160. Bangalore, India (2014)
15. Park, S.W., Ra, I.H.: Design of MAC protocol for improving energy efficiency and reducing transmission delay in EH-WSN. Korean Institute of Smart Media **8**(2), 21–28 (2019)
16. Sarang, S., Stojanović, G.M., Drieberg, M., Stankovski, S., Jeoti, V.: Energy neutral operation based adaptive duty cycle MAC protocol for solar energy harvesting wireless sensor networks. IEEE 95th Vehicular Technology Conference: (VTC2022-Spring), pp. 1–6. Helsinki, Finland (2022)
17. Bengheni, A.: A multi-threshold energy approach for energy harvesting WSN. 2nd International Conference on New Technologies of Information and Communication (NTIC), pp. 1–5. Mila, Algeria (2022)
18. Bengheni, A., Didi, F., Bambrik, I., Boualem, A.: A preamble sampling scheme based MAC protocol for energy harvesting wireless sensor networks. In: Demigha, O., Djamaa, B., Amamra A. (eds.) Adv. in Computing Syst. and Appl. CSA. Lecture Notes in Networks and Systems, Volume 50. Springer, Cham (2019). 2018
19. Bengheni, A.: Relay node selection scheme and deep sleep period for power management in energy-harvesting wireless sensor networks. Int J Commun Syst. **37**(8), e5742 (2024)
20. MiXiM Documentation. http://mixim.sourceforge.net/. Accessed 10 October 2022

ALTEM: A New Lightweight Trust Evaluation Mechanism for Wireless Sensor Networks

Ahmed Saidi(✉)

Department of Mathematics and Computer Science, Faculty of Exact Sciences, University of Ghardaia, Ghardaia, Algeria
ahmedduc@gmail.com

Abstract. Recently, wireless sensor networks (WSN) have played an important role in various Internet of Things applications such as smart homes, smart grids, smart cities, and intelligent agriculture. Security is a fundamental aspect of WSN. Trust management is a promising security technique used to secure the network for different operations such as routing, localization, and data aggregation. However, most trust schemes fail to consider both direct and indirect trust, leaving the network exposed and susceptible to attacks specifically designed for trust systems. Moreover, these trust models are complicated systems that impact the allocation of resources in sensor nodes. This research presents a novel and efficient trust method that simultaneously assesses both direct and indirect trust. Furthermore, the measurement of indirect trust involves the use of both one-hop and multi-hop trust evaluations. The suggested model obtains high detection rates by employing overstimulation and surpasses other competitor models in performance.

Keywords: Wireless Sensor Networks · Security · Trust management · Direct trust · Indirect trust

1 Introduction

Wireless sensor networks (WSNs) are fascinating fields that offer a solid foundation for the near-term realization of numerous new technologies [1]. WSN is made up of thousands or even more nodes dispersed at random within a monitored area. Every sensor accomplishes two main tasks. It first detects the occurrence of the event, and then, upon detecting any event, it notifies the base station (BS or sink). WSNs are susceptible to various vulnerabilities and attacks [2]. Ensuring security remains an essential feature that has not yet been fully assured. For instance, in the context of health monitoring, a robust level of security is necessary to safeguard patients' data. In military contexts, ensuring the confidentiality and integrity of information is crucial to safeguard the data from potential capture by adversaries.

The Trust Management technique has recently emerged as a security technique capable of identifying malicious nodes and mitigating the harm caused by insider attackers. The phrase "Trust Management" was initially coined by Blaze et al. [3] for the purpose of enhancing security. Multiple studies have been carried out to develop effective

© The Author(s), under exclusive license to Springer Nature Switzerland AG 2025
M. Hatti (Ed.): IC-AIRES 2024, LNNS 1238, pp. 684–695, 2025.
https://doi.org/10.1007/978-3-031-80301-7_71

and lightweight trust models that can handle rogue nodes and safeguard the network from potential risks. A research paper entitled "Reputation-Based Framework for High Integrity Sensor Networks" [4] was published by Ganeriwal et al., the paper proposed A protocol called RFSN, where each node in a network maintains a reputation value for neighboring nodes. From these reputation values, trust values are calculated. The authors aim to ensure data integrity in sensor networks by identifying and excluding unreliable or malicious sensors. Experimental results show that the reputation-based approach offers better results than traditional approaches that do not take sensor reputation into account.

Shaikh et al. [5] proposed GTMS a new group-based trust management approach for clustered wireless sensor networks. The main objective of this approach is to reduce the consumption of resources such as memory and energy associated with trust evaluation in sensor networks. The GTMS approach uses clustering in WSNs, meaning that sensors are grouped in clusters or groups, and trust management is performed at the group level rather than at the individual sensor level. GTMS is designed to be lightweight in terms of resource consumption.

Li et al. [6] suggested LDTS a new lightweight and dependable trust system that employs a clustering algorithm. In a cluster based architecture, the indirect trust of a sensor is evaluated by its CH. In this way, each sensor does not need to maintain feedback from other sensors. LDTS facilitates trust-based decision-making, based on a scheme. This scheme reduces risk and improves the system's efficiency in solving trust calculation problems when direct evidence is insufficient. Evaluations show that LDTS requires less communication overhead than GTMS [5]. Fang et al. [7] proposed an NBBTE algorithm based on the Dempster-Shafer theory to evaluate trust between nodes in wireless sensor networks (WSNs). They used different communication and data characteristics to calculate the direct and indirect trust value using Dempster's revised combination rules to obtain the final trust value of a node. Simulation results show that NBBTE can effectively detect malicious nodes.

Jiang et al. [8] introduced a security approach based on a distributed trust model for WSNs (EDTM). This model calculates direct trust from communication, energy, and data trust. For indirect trust, they estimated the number of recommendation packets received by other sensors in the network. Experimentation was carried out on MATLAB, and the results showed that EDTM is an effective and attack-resistant trust model.

The majority of the proposed techniques greatly contribute to improving the security of WSNs. Upon careful analysis of the advantages and disadvantages of each method, it became evident that the trust system is vulnerable to attacks that exploit both individual and multiple trust factors. This is because these approaches primarily rely on the communication perspective for trust assessment, disregarding other important factors such as data and energy.

Additionally, contemporary trust approaches such as [9] and [10] provide equal weight to all trust indicators, regardless of whether they fall below the threshold for trust. Therefore, despite the absence of a trust factor, a malicious node can still pass the trust test provided it possesses high trust values in the remaining trust factors.

The strategy proposed in [4] proposes to dynamically adjust the attacked trust component's weight. Unfortunately, this strategy takes far too long to discover misbehaviour, particularly in cases where it is minimal or nonexistent. Furthermore, these methodologies do not consider the trustworthiness assessment of the no-neighbor sensor nodes. Certain localization algorithms (like LMAT [12]) or routing protocols (like TPGFPlus [11]) require the sensor node to acquire the trust value of two hops' neighbor nodes in order for it to determine its routing or localization. Consequently, a method for assessing multi-hop trust in WSNs needs to be taken into account. The suggested model surpasses current methods in the following aspects:

- By defending the network against attackers that target one or more trust factors, we apply the rating technique in the behaviour evaluation process. By employing this technique, the amount of positive and negative actions the sensor node produces is used to assess its behaviour. Building a trustworthy relationship between sensor nodes involves these behaviours in conjunction with all the trust criteria, including communication, data, and energy.
- The degree of similarity between the values of the recommendations is measured in order to overcome malicious and incorrect recommendations. Furthermore, through simulated testing, the suggested strategy obtains a high detection rate of malicious nodes. Attacks that target trust factors including erroneous information, sinkholes, black holes, and selfish nodes are prevented from occurring on the network.

The structure of our paper is as follows: Section 3 provides a detailed explanation of the primary functional components of the proposed system. This includes an overview of both the direct and indirect assessment methods, as well as the algorithms associated with them. Section 4 presents the performance and evaluation of the proposed system; finally, Section 5 concludes and introduces future work perspectives.

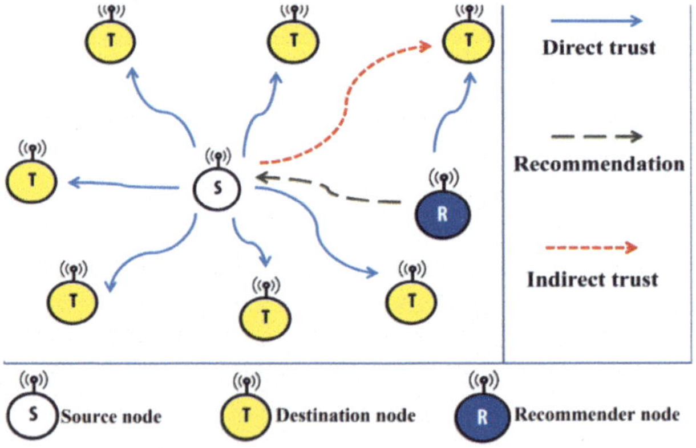

Fig. 1. Network design

2 Overview of the Proposed Approach

Some essential information, including network model and node types, is presented in this section. A summary of our model is also presented, along with the methods for modeling and assessing trust.

2.1 Network Model

We want to detect malicious nodes (bad behaviour) in a wireless sensor network with a trust model based on a distributed architecture in which all nodes are of the same level. Figure 1 illustrates the proposed trust model composed of three node types:

- Source node: represents the node that initiates the evaluation of other nodes.
- Destination node: is the node evaluated by the source node.
- Recommendation node: This node sends recommendations to the other nodes.

2.2 Behaviour Evaluation

If a source node intends to evaluate the behaviour of the destination node, it uses the behaviour evaluation method. The number of positive and negative actions of sensor nodes is determined by Eq. (1), e.g. if a sensor node forwards correctly the received packet or reports a correct data value. It is considered a positive action from the destination node. Similarly, if it drops the packet or is not forwarded to its destination, it is considered a malicious action from the destination node.

$$\left\{ \begin{array}{c} if\ Positive\_action \\ S^{GT} = S^{GT} + 1 \\ else \\ U^{GT} = U^{GT} + 1 \end{array} \right\} \quad (1)$$

Before to the trust assessment, the link quality indicator (LQI) metric is validated. One important measure that is covered by the 802.15.4 standard is the LQI value. According to CC2420, this value goes from zero, which represents the poorest link, to 255, which represents good link quality. The most often used radio that has been suggested in [12] and [13] is CC2420. The lowest allowable level for evaluating trust is set at 210. To prevent misidentifying the benign nodes, the trust test is bypassed if the link quality drops below 210. As a result, the quantity of both positive and negative actions that occur between the source and destination nodes determines the behaviour value. Ultimately, Eq. (2) computes the destination node's behaviour as follows:

$$\boldsymbol{Belif}(i,j) = \left\lceil 10 * \frac{S^{GT}}{S^{GT} + U^{GT}} * \frac{1}{\sqrt{1 + U^{GT}}} \right\rceil \quad (2)$$

The first term represents the proportion of both good and negative interactions. The second term represents the punishment component. A ceiling function is represented by the symbol [], which returns the unsigned integer that is closest to the behaviour value. The behaviour and reputation values are calculated using a range of values (0, 10), where 0 denotes no trust, 5 uncertainty, and 10 total trust. Instead of using an actual number to minimize the memory needed for preserving the trust value, an unsigned integer is utilized, based on the work of [8] and [9].

2.3 Reputation Calculation

When insufficient information about the destination node or the interaction time is very low, the source node seeks secondhand information from neighbour nodes. This type of information indicates the reputation of the destination node in the network. It is calculated by cooperating with neighbour nodes when they are recommended. Every recommender node contributes to the computation by delivering its trust value, and the source node selects the most reliable nodes to be recommenders to obtain a reliable reputation value. However, only a portion of the recommendations are accurate since the sensor nodes are placed in open spaces and can be hacked or reprogrammed to deliver false recommendations in an attempt to undermine the trust system. Therefore, a filtering mechanism is introduced before accepting the recommendation, and it is included in the reputation calculation. There are two cases of reputation calculation:

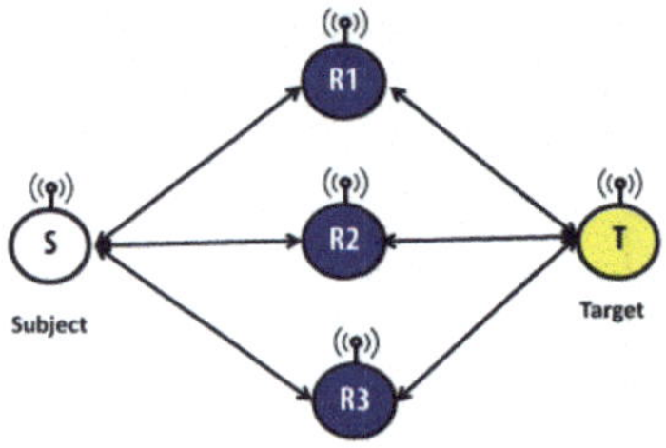

Fig. 2. Example of One hop reputation calculation

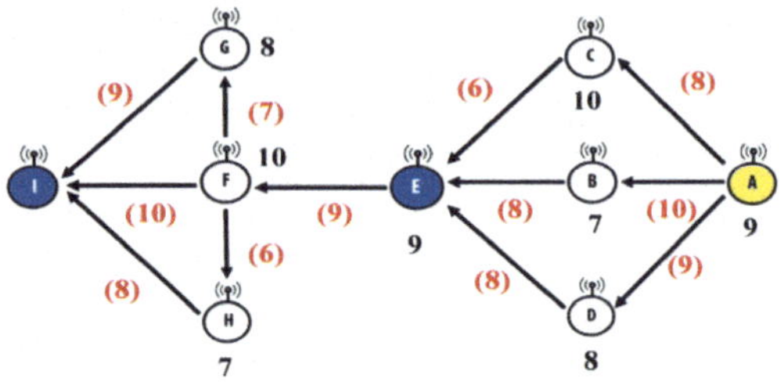

Fig. 3. Example of multi-hops reputation calculation

2.3.1 One-hop Reputation Calculation

In this case, the recommenders are neighbour nodes, and they are connected to both source and destination nodes. Figure 2 illustrates the reputation calculation based on a scenario. The source node is presumed to assess the reputation value of the destination node based on the recommendations it receives from sensor nodes R1, R2, and R3, respectively. Each recommendation received must pass the similarity verification before it is accepted by the source node using Eq. (3).

$$Sim_R = \frac{rec(R_1) * mean\ (rec)}{(rec(R_1))^2 + (mean(rec))^2 - rec(R_1) * mean\ (rec)} mean(rec) = \frac{\sum_{i=1}^{n} rec(i)}{n} \tag{3}$$

It is crucial to note that every recommendation value received and taken into account for the similarity test must be delivered by a recommender node that is already present in the trust list of the source node. Moreover, the mean value is computed using the trust values provided by the reliable nodes listed in the trust list of the source node. This includes the trust value of the source node itself. The mean (rec) represents the average value of all received recommendations, while rec(x) represents the recommendation from node R1.

Equation (3) represents the similarity verification condition. When there is a high similarity between the recommendation received and the mean of all recommendations

(SimR > 0.9), the node is accepted as a recommender. However, if the node sends a faulty recommendation, it is excluded from the list of chosen recommenders.

$$\boldsymbol{Rep(T)} = \frac{\sum_{i=1}^{q} \boldsymbol{rec_{Ri}(T)}}{\boldsymbol{q}} \tag{4}$$

$$W_{Ri} = \frac{\boldsymbol{Rep(Ri)}}{\sum_{i=1}^{q} \boldsymbol{Rep(Ri)}} \tag{5}$$

Equation (4) represents the value of reputation calculation of the destination node, where q represents the number of selected recommenders. Equation (5) represents the calculation of the reputation value of the destination node.

2.3.2 Multi-hop Reputation Calculation

In this case, the destination node is far from the source node. Many recommenders participate in the calculation process and are not necessarily connected to the source or destination node. The recommendations are received from one route or multi-route separately. Different scenarios exist where the source node needs to interact with a destination node that is far from it. This scenario can occur in a data collection if the destination node is a mobile sink or in a localization scenario if the destination node is an anchor node that helps the source node to localize itself. Since WSN is a multi-hop network and trust is transitive, a mechanism of trust propagation is applied in the proposed scheme. In the case of multi-hop reputation calculation, the recommendation arrives from different routes. Thus, three situations can occur:

1) Recommendations arrived from a single route;
2) Recommendations arrived from different routes;
3) Recommendations arrived from overlapping routes;

Figure 3 shows a new scenario to understand the multi-hop reputation calculation. Figure 3 illustrates four types of nodes:

1) Joint node: a node that receives parallel routes;
2) Intermediate node: a node that receives recommendation from a single route;
3) Source and destination nodes;

In Fig. 3, nodes (I and A) are the source and the destination nodes, respectively. Node I and E are joint nodes. Nodes G, F, H, C, B, and D are simple intermediate nodes.

2.3.3 Single Route Calculation

In the case of a single route, the reputation is calculated using Eq. (6), as follows:

$$\boldsymbol{rt_S(S,T)} = \left\lceil \frac{1}{10 * \boldsymbol{length\ (rout)}} * \left\lceil \sum_{i=1}^{q} (\boldsymbol{rec(x)} * \boldsymbol{rep\ (x)}) \right\rceil \right\rceil \tag{6}$$

The example presented in Fig. 3 is used to understand the calculation of a single route. The following route is applied to calculate the single route reputation:

$$A \rightarrow B \rightarrow E \rightarrow F \rightarrow I$$

$$rt_S(A, I) = \frac{1}{10} * [(10 * 9) + (8 * 7) + (9 * 9) + (10 * 10)] = \lceil 8, 175 \rceil = 8$$

2.3.4 Parallel Route Calculation

If the node receives a recommendation from multiple parallel routes, it is considered a joint node because it joins the recommendation received from multiple routes. In this case, an average method is used based on (7):

$$rt_P(S, T) = \frac{rt_{S1}(S, T) + rt_{S2}(S, T) + \ldots + rt_{SK}(S, T)}{K} \tag{7}$$

where, K represents the number of parallel routes. In Fig. 3, the reputation of joint node E is calculated as follows:

$$rt_{S1}(E, A) = E \rightarrow C \rightarrow A \qquad rt_{S1}(E, A) = \frac{1}{20} * [(9 * (8)) + (6 * (10))] = \lceil 6, 6 \rceil = 7$$

$$rt_{S2}(E, A) = E \rightarrow B \rightarrow A \qquad rt_{S2}(E, A) = \frac{1}{20} * [(10 * (9)) + (8 * (7))] = \lceil 7, 3 \rceil = 7$$

$$rt_{S3}(E, A) = E \rightarrow D \rightarrow A \qquad rt_{S3}(E, A) = \frac{1}{20} * [(9 * (9)) + (8 * (8))] = \lceil 7, 25 \rceil = 7$$

$$rt_P(E, A) = \frac{(6, 6) + (7, 3) + (7, 25)}{3} = \lceil 7, 05 \rceil = 7$$

2.3.5 Overlapping Routes Calculation

Overlapping route calculation is a mixture of single and parallel route calculations. A condition is given to the intermediate nodes as described in Eq. (8) to solve this problem:

$$rt_O(A, I) = \begin{cases} AVG(rt_{S1}, rt_{SK}) & \textit{Case of joint node} \\ rt_S & \textit{Otherwise} \end{cases} \tag{8}$$

The average method is used if the intermediate node is a joint node. Otherwise, a single route calculation is applied. The reputation calculation between nodes I and A is assumed in the scenario presented in Fig. 3. The joint method is used for nodes I and E. However, the other nodes use single route calculation. The calculation at node I is as follows:

$$rt_{S1}(I, A) = I \rightarrow G \rightarrow F \rightarrow joint(E) \rightarrow A \qquad rt_{S1}(I, A) = \frac{1}{30} * [(8, 175 * (9)) + (10 * (7)) + (8 * (9))] = 7, 18$$

$$rt_{S2}(I, A) = I \rightarrow F \rightarrow joint(E) \rightarrow A \qquad rt_{S2}(I, A) = \frac{1}{20} * [(8, 175 * (9)) + (10 * (10))] = 8, 67$$

$$rt_{S3}(I, A) = I \rightarrow H \rightarrow F \rightarrow joint(E) \rightarrow A \qquad rt_{S3}(I, A) = \frac{1}{30} * [(8, 175 * (9)) + (10 * (6)) + (7 * (8))] = 6, 319$$

$$rt_P(I, A) = \frac{(7, 18) + (8, 67) + (6, 319)}{3} = \lceil 7, 389 \rceil = 7$$

2.3.6 Assessment of Global Trust

The global trust is established through the integration of the subjective beliefs of the source node with the reputation of the destination node, which is acquired from external sources. During the process of evaluating trust, the source node confirms the presence of the destination node in the trust list. If the target is not there, it initiates the process of evaluating the behaviour. When there are very few interactions between nodes, the source node seeks information about the reputation of the destination node in order to gain more knowledge about it. The behaviour evaluation and reputation information are merged to provide a universal trust value, as described in Eq. (9):

$$\boldsymbol{CT}(\boldsymbol{X}, \boldsymbol{Y}) = \boldsymbol{W}1 * \boldsymbol{Belif}\ (\boldsymbol{X}, \boldsymbol{Y}) + \boldsymbol{W}2 * \boldsymbol{Rep}\ (\boldsymbol{X}) \quad (9)$$

$$\text{where, } W1 + w2 = 1$$

The node's trustworthiness is determined based on the global trust.

3 Simulation

MATLAB served as a simulation tool to conduct the experiment. Various scenarios were created based on the experiment's objective. The experiment was divided into two primary parts. In the first part, the suggested model's accuracy and detection rate were assessed when misbehaving nodes were present. In the second part, we compare the detection rate of our model against EDTM [11] and NBBTE [4]. We chose these models because each is a distributed trust model using one multi-hop trust evaluation method. Moreover, each of them establishes various trust factors to evaluate the trustworthiness of sensor nodes. For all simulated scenarios, we assign equal importance to both direct and indirect trust, with a weight of 0.5. The simulation parameters are presented in Table 1.

Table 1. Simulation parameters

Parameter	Value
Area dimension	500 × 500 m2
Position of the base station	(250 × 250)
Connectivity range	75 m
Node count	400
Initial energy of nodes	0.5 J
Communication energy	0.5 J
E_elec	50 nJ/Bit
E_fs and E_amp	100 PJ/Bit/m2
LQI threshold	210
Initial trust & reputation values	10

3.1 Detection Rate

This experiment investigates the performance of the suggested model in the presence of network misbehaviour nodes. Understanding the classes and types of attacks that are being employed is essential before beginning the evaluation. Attacks can be classified as either internal or external. Conventional security methods, including cryptography and authentication, are effective against external threats but ineffective against internal ones brought on by hacked nodes. Internal attacks are the main focus of this work, particularly those that pertain to nodes' behaviour. The sensor node assesses the actions of other sensor nodes within its connectivity range throughout each of the 50 simulation rounds that are conducted. Figure 4 illustrates a pair of sensor nodes that are undergoing assessment. The sensor node's trust rating falls below the trust barrier as soon as it exhibits malevolent behaviour during cycle 3. The outcome shown in Fig. 4 shows that, even if the sensor node continues to engage in malicious behaviour, its trust value will never rise above the trust threshold, demonstrating the efficacy of our trust mechanism.

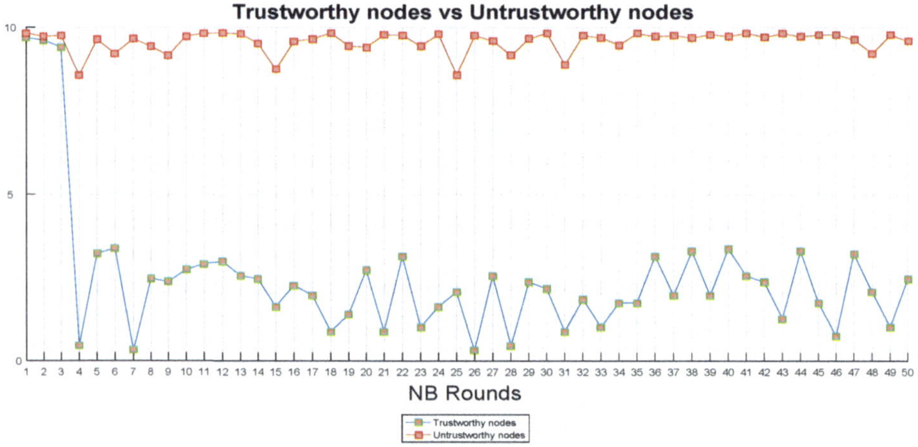

Fig. 4. Trust value of two different sensor nodes

3.2 An Analysis of Trust Models in the Presence of Selfish Nodes

Selfish behaviour refers to a node that does not participate in the network operations to save energy. For example, the node with selfish behaviour drops the received packets, does not participate in forwarding operations, or does not perform sensing and reporting functions. Figure 5 depicts the trust value of a sensor node that starts its selfish behaviour at the beginning of round 6. We first observe that the trust value decreased in all the trust models, which means that all of them can identify this misbehaviour successfully. However, as we can see in the figure, our method identifies the selfish behaviour two rounds before EDTM [11] and three rounds before NBBTE [4] because it includes the punishment factor in the evaluation test. The result of the experiment indicates that our method can deal with selfish behaviour earlier than other models and effectively reduces the damage of misbehaviour nodes.

2) AN ANALYSIS OF TRUST MODELS IN THE PRESENCE OF SELFISH NODES

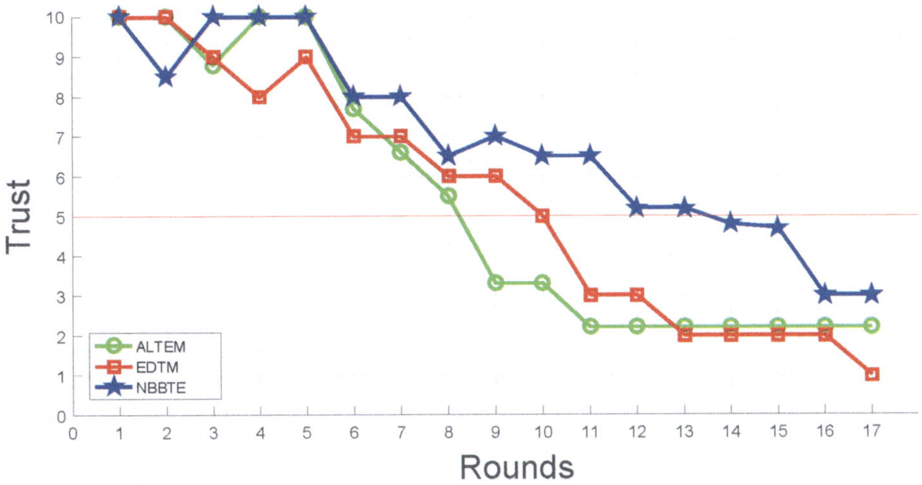

Fig. 5. An analysis of trust models in the presence of selfish nodes

3.3 Comparison of Trust Models Under Different Misbehaviour Rates

In this scenario, we assume different malicious nodes attack the sensor network. Each one of these nodes performs malicious actions, including communication misbehaviour, data misbehaviour, energy misbehaviour, and mobility misbehaviour.

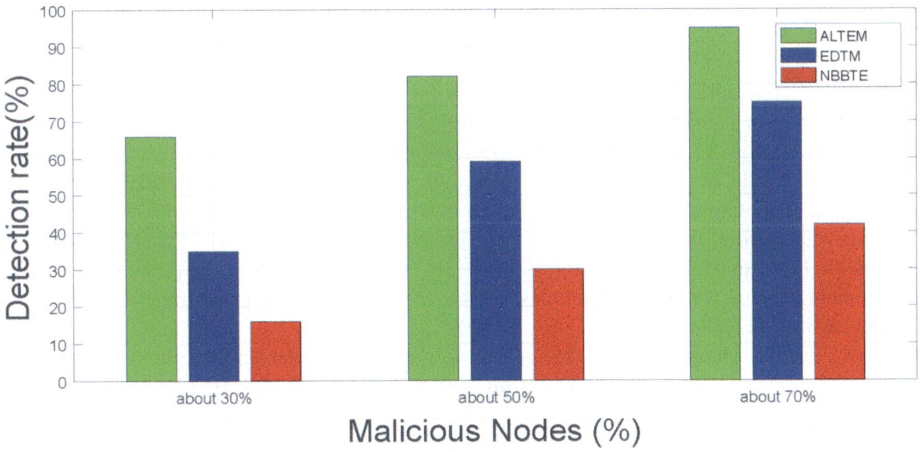

Fig. 6. Comparison of trust models under different misbehaviour rates

Figure 6 illustrates the detection rate of the three models under three different misbehaviour rates of 10%, 30%, and 50%. The result indicates that our method's detection rate is the highest of the three misbehaviour rates. Because our method considers direct and indirect trust types in the evaluation process, it can identify abnormal behaviour effectively. On the other hand, NBBTE [4] has a lower detection rate because it only deals with attacks related to the direct interaction between sensor nodes. However, it is vulnerable to attacks targeting the reputation of sensor nodes, such as bad-mouthing and whitewashing attacks. Unlike NBBTE [4], EDTM [11] is robust against attacks targeting direct and indirect trust types. However, it is vulnerable to badmouthing attacks because it considers dishonest recommendations only in the case of one-hop trust evaluation but not in the case of multi-hop trust evaluation.

4 Conclusion

The limited resources and minimum computational power of nodes in WSNs pose challenges for applying traditional security mechanisms like cryptography and authentication. Trust management is a more suitable approach for WSNs to handle internal attacks. In the evaluation process, this research suggests a novel lightweight trust mechanism that takes into account two different kinds of trust. The one-hop and multi-hop scenarios are taken into account in the suggested model when evaluating trust. The findings show that, in terms of detection rates, our model performs better than the other competing models. Given the encouraging outcomes achieved in the simulation, it is imperative to do a thorough examination of the energy consumption associated with our trust mechanism. Furthermore, it is imperative to consider and solve the memory overhead and message overhead associated with our trust mechanism.

References

1. Rehmani, M.H., Pathan, A.S.K.: Emerging Communication Technologies Based on Wireless Sensor Networks: Current Research and Future Applications. CRC Press (2016)
2. Sarkar, A., Senthil, T.: Cluster head selection for energy efficient and delay-less routing in wireless sensor network. Wireless Network **25**, 303320 (2017)
3. Rathore, H., Jha, S.: Bio-inspired machine learning based wireless sensor network security. In: 2013 World Congress on Nature and Biologically Inspired Computing. IEEE (2013)
4. Han, G., et al.: Crosslayer optimized routing in wireless sensor networks with duty cycle and energy harvesting. Wirel. Commun. Mob. Comput. **15**(16), 1957–1981 (2015)
5. Shaikh, R.A., Jameel, H., d'Auriol, B.J., Lee, H., Lee, S., Song, Y.: Group-based trust management scheme for clustered wireless sensornetworks. IEEE Trans. Parallel Distrib. Syst. **20**, 1698–1712 (2009)
6. Li, X., Feng, Z., Junping, D.: LDTS: A lightweight and dependable trust system for clustered wireless sensor networks. IEEE Trans. Inf. Forensics Secur. **8**, 924–935 (2013)
7. Feng, R., et al.: A trust evaluation algorithm for wireless sensor networks based on node behaviours and ds evidence theory. Sensors **11**(2), 1345–1360 (2011)
8. Jiang, J., et al.: An efficient distributed trust model for wireless sensor networks. IEEE Trans. Parallel Distrib. Syst. **26**(5), 1228–1237 (2014)
9. Wu, X., et al.: BLTM: Beta and LQI Based Trust Model for Wireless Sensor Networks. IEEE Access **7**, 43679–43690 (2019)
10. Zhang, T., Yan, L., Yang, Y.: Trust evaluation method for clustered wireless sensor networks based on cloud model. Wireless Netw. **24**(3), 777–797 (2018)

11. Han, G., et al.: The Insights of Localization through Mobile Anchor Nodes in Wireless Sensor Networks with Irregular Radio. Ksii Transactions on Internet Information Systems **6**(11) (2012)
12. Baccour, N., et al.: Radio link quality estimation in wireless sensor networks: A survey. ACM Transactions on Sensor Networks (TOSN) **8**, 34 (2012)
13. Machado, K., Rosrio, D., Nakamura, E., Abelm, A., Cerqueira, E.: Design of a routing protocol using remaining energy and link quality indicator (REL). In: Proceedings of the 6th Latin America Networking Conference, pp. 33–39. ACM, Quito, Ecuador, New York, NY, USA (2011)

Innovative Materials and Applications

Nanostructured Materials: Synthesis and Application

Henia Fraoucene[1,2](✉), El hadi Khoumeri[2], and Rabea Cheggou[2]

[1] Laboratory of Advanced Technologies of Genie Electrics (LATAGE), Faculty of Electrical and Computer Engineering, Mouloud Mammeri University (UMMTO), BP 17 RP, 15000 Tizi-Ouzou, Algeria
henia.fraoucene@ummto.dz

[2] Laboratory of Innovative Technologies (LTI), National School of Technology, Algiers, Algeria

Abstract. Over the last years, the synthesis of self-organized nanostructures materials has attracted a great scientific and technological interest due to its far-reaching and innumerable applications including renewable energy, sensors, biomedical devices and energy storage. Among these nanostructured materials, self-organized TiO2 nanotubes (TiO2 NTs) and Porous Silicon (PS) layers obtained by electrochimical anodization process of Titanium (Ti) foil and Silicon (Si) wafers respectively, exhibits a high porosity and larger specific area offering an enhancement in their properties. In the present study, we investegate the synthesis process and morphological properties of both structure TiO2 NTs and PS layers through anodic current in fluoride appropriate electrolyte. Thereafter, a brief reminder of the technological applications of these two materials.

Keywords: TiO2 Nanotubes · Si porous · electrochemical anodization · morphological properties · application

1 Introduction

Over time, extensive research has been carried out in the nanomaterials devices due to their advanced properties such as: large surface area, excellent conductivity and high porosity wich controlled by sythesis parameters. Among this nanomaterials, Titanium dioxide Nanotubes (TiO2 NTs) and Porous Silicon (PS) structure exhibed a promosing caracteristics making it advantageous for various applications such as sensors [1], solar cells [2], biomedical devices [3, 4] and energy storage [5], etc. Titanium dioxide Nanotubes (TiO2 NTs) is a n-type semiconductor materials, typically produced through electrochemical etching process of Titanium foil (Ti), a cost-effective fabrication method that ensures high reproducibility of nanostructure TiO2 layers. This method allows precise control over morphological properties of the material [6, 7]. Assefpour-Dezfuly [8] and Zwilling et al. [9] were pioneers in reporting the elaboration of porous TiO2 films. Over the time, significant research has focused on synthesizing TiO2 nanotubes (NTs), resulting in their classification into four generations:

© The Author(s), under exclusive license to Springer Nature Switzerland AG 2025
M. Hatti (Ed.): IC-AIRES 2024, LNNS 1238, pp. 699–707, 2025.
https://doi.org/10.1007/978-3-031-80301-7_72

- The first generation of NTs, with tube lengths of 0.5 μm, was synthesized in hydrofluoric acid [HF] [10].
- The second generation of NTs, reaching up to 05 μm in length, was grown in aqueous electrolytes containing fluorine ions, with pH adjustments [11].
- The third generation featured smooth, longer NTs ranging from 100 to 1000 μm, grown in fluoride organic electrolytes [12].
- The fourth and latest generation was obtained in media free of fluorine (such as HCl, NaCl, H2O2, etc.). These nanotubes exhibit a non-uniform distribution on the surface and grow in irregular agglomerations [13].

Another advenced and suitable material in nanotechnolgy application is the Porous Silicon (PS) structure, which can be tailored through chemical anodization of Silicon (Si) wafers in [HF] electrolyte under an anodic current, allowing precise control over pore size, distribution, and surface properties to suit specific technological needs. PS materials was first discovered by the Uhlirs team in the mid-1950s at Bell Laboratories in USA [14], when there were working on electrolytic shaping of silicon and germanium surface. However, this result did not attract much attention from the scientific community at that time. In 1990, Canham brought renewed interest to PS by reporting significant visible light emission from it when exposed to ultraviolet (UV) light [15]. This breakthrough discovery sparked widespread interest in the material's optical and electronic properties. It is commonly believed that this luminescence arises from quantum- confined structures within the material, which is often referred to as a nanocrystalline film. Until today, PS is widely investigated for various applications in sensory, opto and microelectronics [16], medical and biotechnologies [17], etc.

The architecture of PS layers related through three parameters used in the literature: morphology, geometry and porosity. The morphology describe the properties of the pore; their smooth or branched shapes, perpendicular, titled or random orientation and their interactions. The geometry determines both the average pore diameter and the average spacing between pores; than result in three-size category of the pores: micropores (< 2 nm), mesopores (2–50 nm) and macropores (> 50 nm). The last important parameters is the porosity defined as the ratio of the total volume of pores to the volume of the entire layer [18]. A dramatic porosity effect on the thermal, mechanical and optical properties of PS layers reported by the study of Canham et al [19], governed by the preparation conditions induce a critical parameter for certain application such as drug delivery. In the present work , we investigate the synthesis process to obtained self orginized Titanium dioxide Nanotubes (TiO2 NTs) and high Porous Silicon layers morphology ; under electrochemical anodic etching of Ti fiol and Si wefers in fluoride ethylene glycol organic solution and Hydrofluoric Acid [HF] electrolyte respectively at different synthesised parameters (anodizing voltage and time, HF concentration). Finally, a brief reminder of the technological applications of these two materials were evoqued.

2 Experimental

2.1 Substrats Treatement

Titanium substrates and P- Silicon (P-Si) wafers with a square shape (1.5 cm^2) were degreased by sonication in acetone, methanol and 2-Propanol for 10 minutes respectively, rinsed with pure water and dried in an oven for 05 minutes .

2.2 Elaboration of TIO2 Nanotubes

The anodic anodization method was used to fabricate high order TiO2 NTs, in a both-electrode cell with the Ti substrat as the anode and Pt bar as the cathode at ambient temperature. The electrolyte based with fluoride Ethylene Glycol (EG) organic solutions and distilled water at different voltage (20, 40 and 60 V for 3 h) and time (15, 30, 60 and 120 mn at 60 V). At achieved anodization, the samples were treated respectively in pure water and an oven for 10 minutes.

2.3 Synthesis of Porous Silicon Layers

A *p*-Porous silicon structure was formed on 100 orientation silicon (Si) subtracts with average thickness about 355- 405 μm using electrochemical process at room temperature, for 3 min with anodic current 10 mA/cm^2 using HNO3/H2O electrolyte by varing the HF concentration (35, 20 and 15%). After electrochemical processing, the samples were rinsed respectively, in pure water and ethanol for 10 minutes to remove the surfactant from inside the tranches. Lastly, their dried in an oven.

2.4 Samples Characterization

The surfaces topography of the TiO2 structure and PS layers were investigated by Scanning Electron Micrographs (SEM) using a PHILIPS XL30. The morphological parameters (pore size, average diameter, length) of TiO2 NTs and PS layers were estimated dirctly through SEM images.

The porosity of the synthesized PS layers was calculate from Equation.1 [18, 19], in which the ratio of the mass differences is equivalent to the ratio of the volume:

$$P = \frac{(m_1 - m_2)}{(m_1 - m_3)} = \frac{V_{voids}}{V_{layer}}(\times 100\%) \tag{1}$$

Or, (m1 –m2) define the mass of the material which has been removed to create the pore (m1 –m3) describe the mass of the remaining porous layer.

3 Experimental Results

3.1 Formation of Titanium Dioxide Nanotubes (TIO_2 NTs) at Different Anodization Voltage

The electrochemical anodization process is extensively employed for creating porous and nanotubular TiO2 layers due to its simplicity, cost-effectiveness, and strong mechanical adhesion, as the nanotubular layer forms on the Ti substrate. This allows precise control over the thickness and morphology of the TiO2 films. During the anodic process, the formation of TiO2 NTs involves a competitive interplay between TiO2 formation and the chemical dissolution of this oxide, facilitated by fluoride ions, following relationships 2 and 3:

$$T_i + 2H_2O \rightarrow T_iO_2 + 4H^+ + 4e^- \quad (2)$$

$$T_iO_2 + 6F^- + 4H^+ \rightarrow [T_iF_6]^{2-} + 2H_2O \quad (3)$$

Fig. 1 displays SEM micrographs obtained from Ti samples anodized in fluoride EG for 3 h at : 20 (a), 40 (b) and 60 V (c) and bottom view of TiO2 NTs anodized at 40 V (d).

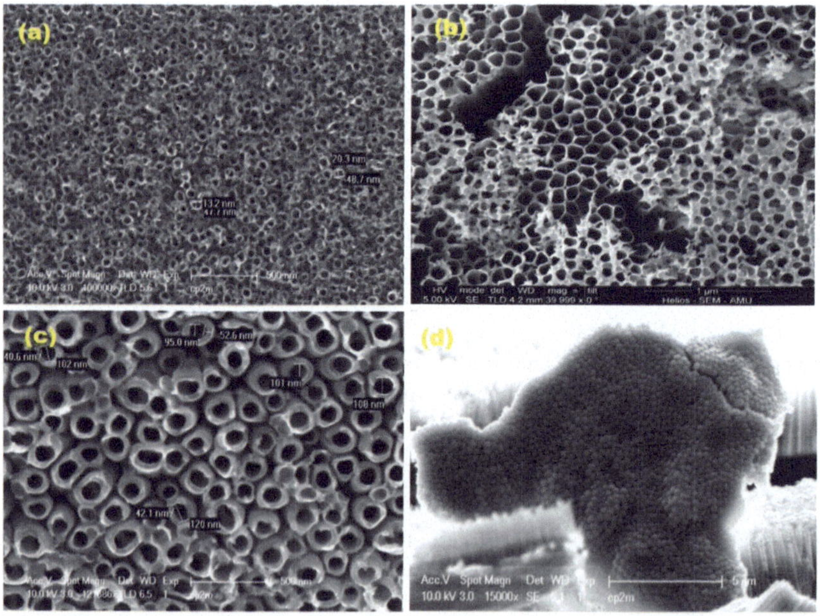

Fig. 1. Topography SEM images of TiO2 NTs obtained by anodizing Ti foil for 3 h in floride EG solution at: 20 (a), 40 (b), 60 V (c) and bottom view of TiO2 NTs anodized at 40 V (d).

The SEM top-view images reveal that varying the anodization voltage allows precise control over the morphology of TiO2 NTs in the sense of diameter, thickness, length, and appearance. Higher anodization voltages result in larger tube diameters, indicating thicker tube walls. At 20 V, non-uniform NT morphology with wall undulations is evident (Fig. 1a). Increasing the voltage to 40 V produces a less ordered structure with enlarged tube diameters (Fig. 1b). A perfectly ordered nanotubular structure with smooth walls is achieved at 60 V, as depicted in Fig. 1c. Additionally, the SEM images confirm that the tubes are open at the top and closed at the bottom (Fig. 1d). Table 1 summarizes the geometric parameters of TiO2 NTs at different anodization voltages. By increasing the anodization voltage from 20 to 60 V, tube diameters increase from 39 to 106 nm, accompanied by reduced inter-tube spacing (x), attributed to intensified surface etching under higher electric fields. Moreover, NTs synthesized at 60 V exhibit the highest periodicity, while those formed at 20 V show the lowest aspect ratio and pore regularity.

Table 1. The geometric parameters of TiO2 NTs anodized at different voltages.

	Inner diameter (Di) (nm)	External diameter (De) (nm)	Thickness (w) (nm)	Spacing between pore (x) (nm)
20 V	39–52	69–75	08–15	16–33
40 V	67–79	87–117	10–19	14–21
60 V	83–106	167–198	42–46	09–19

3.2 Formation of Titanium Dioxide Nanotubes at Different Anodization Time

Various tube geometries can be produced by controlling the anodizing time. Fig. 2 displays SEM micrographs obtained from Ti samples subjected to electrochemical etching in fluoride ethylene glycol at 60 V for different durations: 15 (a), 30 (b), 60 (c), and 120 minutes (d). The top-view SEM pictures reveal that the dimensions (diameter and thickness) and aspect ratio of the NTs can be manipulated by adjusting the anodization time. Initially, a porous TiO2 layer forms at the onset of anodization (Fig. 2a), followed by the localized dissolution and appearance of surface pores over time, as shown in Fig. 2b. After 60 min, a honeycomb-like porous structure develops with an average pore diameter of 80 nm. A regular and well-defined nanotubular structure is observed after 120 min of anodizing, featuring an internal diameter of 96 nm (Fig. 2d). These findings underscore the significant influence of anodization time on tube morphology and growth mechanisms. When oxidation and reduction rates reach equilibrium, the tube thickness stabilizes, maintaining nearly constant values as thermodynamic equilibrium is attained.

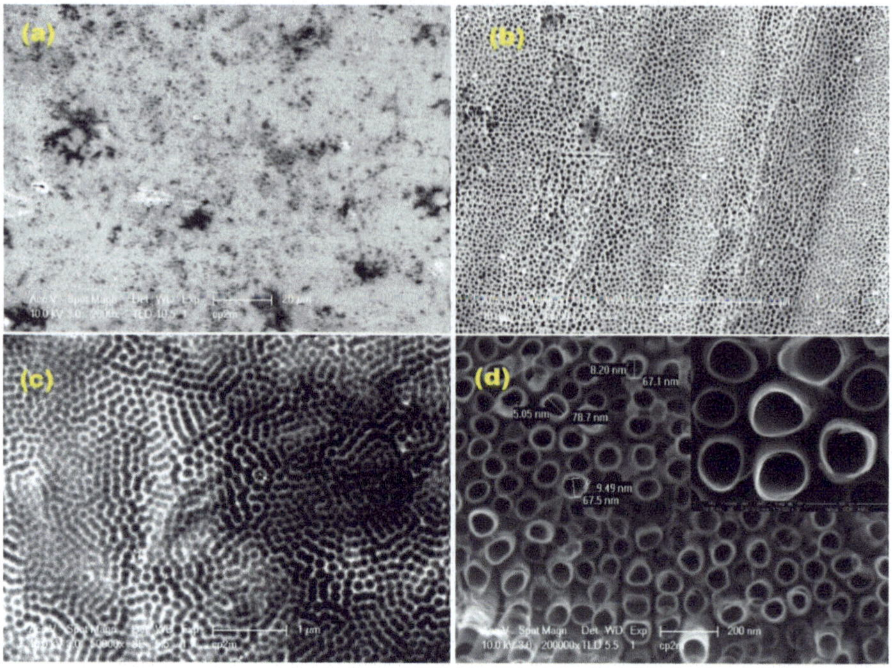

Fig. 2. Surfaciques SEM images of TiO2 NTs obtained by anodizing Ti foil at 60 V in floride EG solution for: (a) 15, (b): 30, (c): 60 and (d) 120 mn.

3.3 Elaboration of Pourous Silicon Layers in a Resistive HF Electrolyte

Fig. 3 displays SEM micrographs of Si wafers anodized for 3 minutes in HNO3/H2O with varying HF concentrations: 35% (a), 20% (b), and 15% (c). These images illustrate the significant influence of HF concentration on morphology, surface topography, pore size, and porosity. Fig. 3a shows well-formed layers of porous silicon (PS) obtained under an applied current of 10 mA in 35% HF electrolyte. As HF concentration decreases, the appearance shifts to a more sponge-like porous structure, evident in the images.

To calculate the geometric parameters of PS layers, measurements such as average pore size and diameter were directly taken from each image. Porosity was calculated using Equa. 2, as mentioned previously. Higher HF concentrations, such as 35%, resulted in lower porosity (30%), as observed. Decreasing HF concentration increased porosity: at 20% HF, porosity reached 40% (Fig. 3b), and at 15% HF, it peaked at 50%, indicating a lower aspect pore ratio (Fig. 3c). Furthermore, HF concentration affects current densities in the solution, thereby influencing the dissolution process on the wafer surface necessary for PS layer formation. Across all samples, average pore sizes ranged from 10–14 nm, with PS layers synthesized in 35% HF concentration showing a higher aspect pore ratio.

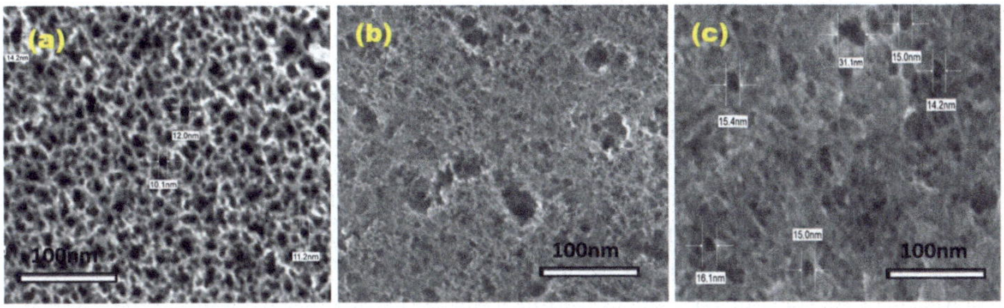

Fig. 3. Top view SEM images of PS layers obtained by anodic etching of Si wafer for 3 mn in HNO3/H2O solution containing: 35 (a), 20 (b) and 15 HF% (c).

3.4 Technological Application of TIO_2 NTs and PS Layers

Titanium dioxide nanotubes (TiO2 NTs) and Porous Silicon (PS) layers are promising materials with applications in microelectronics, sensors, and other fields requiring high surface area structures and efficient absorption capabilities. This section provides a concise overview of their use in energy storage, particularly as anode materials in lithium batteries.

Our experimental study focused on TiO2 NTs synthesis on the Ti alloy Ti-6Al-4V (TA6V) as anode materials for lithium micro-batteries [5]. The investigation revealed excellent electrochemical performance and enhanced stability during lithiation processes, as evidenced by cyclic voltammetry (CV) cycling tests and galvanostatic measurements. Specifically, galvanostatic charge-discharge tests demonstrated a promising capacity of 82 $mAh.g^{-1}$ (34 $\mu Ah.cm^{-2}.\mu m^{-1}$) at a 1C rate over 50 cycles. These findings indicate that TiO2 NTs grown on TA6V alloy possess the potential to serve as competitive anode materials for Li-ion microbatteries.

Porous silicon (PSi) layers are increasingly recognized for their potential in energy storage applications, particularly in the domain of lithium-ion batteries. PSi layers possess a remarkably high surface area per unit volume due to their porous structure, enhancing electrochemical activity by providing abundant active sites for lithium-ion intercalation and deintercalation processes.

Hiroyuki Kawaura et al. [20] synthesized a skeletal porous silicon structure through acid etching of atomized Al–Si alloy powder, demonstrating its efficacy as anode material in lithium-ion batteries. The p-Si19 anode exhibited a finely distributed pore size ranging from 20 to 200 nm, perfect rate characteristics, a reversible discharge capacity of 1607 mAh g^{-1} after 200 cycles at a rate of 0.1 C, with a Coulombic efficiency exceeding 97%, highlighting its robust stability. In result, the utilization of porous silicon layers in energy storage applications shows significant promise, combining high performance, stability, and adaptability. This capability can substantially advance the capabilities of lithium-ion batteries and other energy storage technologies.

4 Conclusion

In this study, Titanium dioxide nanotubes (TiO2 NTs) and Porous Silicon (PS) layers were fabricated through anodic electrochemical etching of titanium foil and silicon wafers, respectively. A well-defined, ordered nanotubular shape was achieved at 60 V for 3 hours in fluoride EG. In contrast, a porous honeycomb-like layer with approximately 80 nm pore diameter was formed after 60 minutes of anodization at 60 V. Furthermore, varying the HF concentration led to porous silicon layers with diverse morphological topographies; increasing HF concentration reduced layer porosity while enhancing the aspect pore ratio.

In conclusion, TiO2 NTs and PS layers represent pivotal materials in energy storage, each offering distinct advantages. These advancements are critical for ongoing developments in energy efficiency and sustainable technologies.

References

1. Jiahui, Z., Haiquan, W., Yahui, C., Junjin, Z., Zhida, G., Yan-Yan, S.: The challenges and opportunities for TiO2 nanostructures in gas sensing. ACS Sensors **9**(4), 1644–1655 (2024)
2. Kim, H., et al.: Structural evolution of anodized TiO2 nanotubes and their solar energy applications. Macromolecular Research, 1–23 (2024)
3. Mabrouk, M., Das, D.B., Salem, Z.A., Beherei, H.H.: Nanomaterials for Biomedical Applications: Production, Characterisations, Recent Trends and Difficulties. Molecules **26**(4), 1077 (2021)
4. McNamara, K., Tofail, S.A.M.: Nanoparticles in biomedical applications. Advances in Physics: X **2**(1), 54–88 (2016)
5. Fraoucene, H., et al.: Optical and electrochemical properties of self-organized TiO2 nanotube arrays from anodized Ti− 6Al− 4V alloy. Front. Chem. **7**, 66 (2019)
6. Yan, J., Zhou, F.: TiO2 nanotubes: structure optimization for solar cells. J. Mater. Chem. **21**(26), 9406–9418 (2011)
7. Giorgi, L., et al.: Titania nanotubes self- assembled by electrochemical anodization: Semiconducting and electrochemical properties. Thin Solid Films **601**, 28–34 (2016)
8. Assefpour-Dezfuly, M., Vlachos, C., Andrews, E.H.: Oxide morphology and adhesive bonding on titanium surfaces. J. Mater. Sci. **19**(11), 3626–3639 (1984)
9. Zwilling, V., Aucouturier, M., Darque-Ceretti, E.: Anodic oxidation of titanium and TA6V alloy in chromic media. An Electrochemical Approach. Electrochimica Acta **45**(6), 921–929 (1999)
10. Zhao, J., Wang, X., Chen, R., Li, L.: Fabrication of titanium oxide nanotube arrays by anodic oxidation. Solid State Commun. **134**(10), 705–710 (2005)
11. Macak, J.M., Taveira, L.V., Tsuchiya, H., Sirotna, K., Macak, J., Schmuki, P.: Influence of different fluoride containing electrolytes on the formation of self-organized titania nanotubes by Ti anodization. J. Electroceram. **16**(1), 29–34 (2006)
12. Paulose, M., et al.: TiO2 nanotube arrays of 1000 μm length by anodization of titanium foil: phenol red diffusion. The Journal of Physical Chemistry C **111**(41), 14992–14997 (2007)
13. Richter, C., Panaitescu, E., Willey, R., Menon, L.: Titania nanotubes prepared by anodization in fluorine-free acids. J. Mater. Res. **22**(6), 1624–1631 (2007)
14. Uhlir, A.: Electrolytic Shaping of Germanium and Silicon. Bell Syst. Tech. J. **35**, 333–347 (1956)

15. Canham, L.T.: Silicon quantum wire array fabrication by electrochemical and chemical dissolution of wafers. Appl. Phys. Lett. **57**, 1046–1048 (1990)
16. Bisi, O., Ossicini, S., Pavesi, L.: Porous silicon: a quantum sponge structure for silicon based optoelectronics. Surf. Sci. Rep. **38**(1–3), 1–126 (2000)
17. Abed, H.R., Zayer, M.Q., Alwan, A.M.: High-Sensitive Detection of E. coli Bacteria Based on Low-Etching Current Density Meso Porous Silicon Electrical Sensor. Brazilian Journal of Physics **54**(5), 165 (2024)
18. Zhang, X.G.: Morphology and formation mechanisms of porous silicon J. Electr. Soc. **151**(1), C69–C80 (2004)
19. Canham, L.: Handbook of porous silicon In: Canham, L. (ed.) Springer Int. Pub., pp. 1–7 (2014)
20. Hiroyuki, K., Ryo, S., Yasuhito, K., Yuji, M.: Scalable Synthesis of Porous Silicon by Acid Etching of Atomized Al–Si Alloy Powder for Lithium-Ion Batteries. ACS Appl. Mater. Interfaces. **15**(29), 34909–34921 (2023)

Optical Characteristics of Photonic Crystal Fiber Splitters: A Study on Refractive Index, Chromatic Dispersion, and Effective Frequency

Ahlem Assia Harrat[1](✉), Mohammed Debbal[1], and Mohammed Chamse Eddine Ouadah[2]

[1] Electronic and Telecommunications Department, Belhadj Bouchaib University, Ain-Temouchent, Algeria
harrat22.ahlem@gmail.com

[2] Telecommunications Department, Mouloud Mammery University, Tizi Ouzou, Algeria

Abstract. The fundamental characteristics of photonic crystal fiber (PCF) and its uses are the main topics of this study. This offers us a better understanding of how PCF came to be and the benefits we enjoy across countless applications. In addition, we can see the outcomes and how they differ from those of traditional fibers. This research examined the many features of a PCF splitter. We will study the characteristics of a splitter Such as the effective index, chromatic dispersion, and the effective frequency.

Keywords: Photonic crystal fibers (PCFs) · chromatic dispersion · refractive index (Neff) · effective frequency (Veff)

1 Introduction

The foundation of photonic crystal fiber, sometimes called holey or microstructure fiber, ise based on a combination of optical fiber and photonic crystal characteristics. Numerous industries, including sensors [1], fiber-optic communications [2], high-power transmission [3], and fiber lasers [4], splitters [5], have found use for PCFs. It is much superior to conventional fibers because it can contain light in hollow-core e and solid-core fibers in various ways and change properties through structural adjustments [6]. Photonic crystal fibers (PCFs) are well known for having superior qualities over traditional optical fibers. According to their light-guiding mechanism, PCFs are divided into two categories [7]. Light is directed by the first high-index guiding PCF using a modified total internal reflection method [8]. Conversely, the second type of fiber guides light using the photonic bandgap (PBG) effect, enabling new features like confining light to a low-index core [9]. When constructing photonic integrated circuits (PICs), one of the most important components is the splitter [10]. Utilizing reliable sources is essential for the PIC's optical chip integration to be successful, compact optical power splitters. Photonic power splitters are an important member of the interconnecting family and one of the basic components of integrated optics [11]. An effective photonic splitter can focus light input, sending

© The Author(s), under exclusive license to Springer Nature Switzerland AG 2025
M. Hatti (Ed.): IC-AIRES 2024, LNNS 1238, pp. 708–715, 2025.
https://doi.org/10.1007/978-3-031-80301-7_73

a particular wavelength to several ports at varying intensities, with the least amount of loss. These inactive gadgets can be used for a variety of tasks, such as distribution of signals [12], feedback systems [13], and equalization of power [14]. Not only is PCF visually appealing, but it also possesses several peculiar qualities, such as single-mode functioning [15], flat dispersion [16], high birefringence [17], a sizable mode area [18], high negative dispersion [19], etc. These qualities enable its use as a practical transmission medium for both short- and long-range communications [20]. Three of these qualities are the ones we will focus on in this study; One of the most crucial optical properties is a material's refractive index, the chromatic dispersion of optical fiber is the frequency dependence of the phase velocity in a transparent medium, and using dispersion compensating fiber is one approach to do this, The V-parameter, which increases without bound as the wavelength lowers, controls the effective frequency, which is the number of bound modes. The main goal of this article is to study the splitter's optical characteristics based on photonic crystal fibers that have been developed by K. Saitoh et al. [21].

2 Theory and Design

PCFs that have a wide range of air hole patterns and arrangements can be used to alter the refractive index difference between the photonic crystal cladding and core in several different ways. This makes PCFs, which have enhanced features and properties not possible in ordinary fibers, useful in a wide range of applications.

A PCF's effective cladding index largely depends on wavelength. This is because the field stays clear of the airholes at short wavelengths. The effective index can be calculated as [22]:

$$n_{eff} = \sqrt{1 + \frac{B_1\lambda^2}{\lambda^2 - C_1} + \frac{B_2\lambda^2}{\lambda^2 - C_2} + \frac{B_3\lambda^2}{\lambda^2 - C_3}} \tag{1}$$

where the Sellmeier coefficients are Bi and Ci (i = 1, 2, and 3) and the working wavelength is λ [23] for the particular content.

To describe the dispersion, the wavelength can be widened or spread out. Chromatic dispersion causes a light pulse containing digitized information to extend and interfere with the other pulse when it travels through a fiber. Dispersion in PCFs may change due to the adaptability in modifying the layout and arrangement of the air holes.

In conventional fibers, dispersion is caused by the waveguide and material dispersion, and due to constraints in parameter adjustment, analog qualities cannot be attained. It can be calculated as [22]:

$$D_W = \frac{-\lambda}{c}\frac{d^2R_e(n_{eff})}{d\lambda^2} \tag{2}$$

The real component of the effective refractive index is denoted by Re() in this equation, where λ stands for wavelength and c for the speed of light.

The functional area is a factor of paramount importance. It was first created as a non-linearity indicator; for non-linear effects to have any significance, a high power density is indicated by a low effective area. In the context of confinement loss, the effective area is critical [24], micro-bending loss [25], macro-bending loss [26], cutting out loss [27], and numerical gap [28], but it can also be connected to the spot-size w through = w2. In general, the above-mentioned losses are higher the larger the effective area. [29].

The formula below can be used to get the effective area. Where is the transverse electric field is the transverse electric field:

$$A_{eff} = \frac{[\iint |E(x, y)dxdy|^2}{\iint |E(x, y)dxdy|^4} \quad (3)$$

A fiber's non-linearity is negatively correlated with its effective mode area. You can figure it out by using [29]:

$$\gamma = \frac{n_2}{cA_{eff}} = \frac{2\pi n_2}{\lambda A_{eff}} \quad (4)$$

n_2 is the coefficient of the nonlinear refractive index. ($3.01*10^{20}$ m^2 / W for Silica).

Photonic crystal fibers exhibit extremely temperature-insensitive acquired birefringence. Because of the non-axisymmetric distribution of the effective refraction index, which is influenced by the size and spatial configuration of the hole, photonic crystal fibers exhibit birefringence.

$$\mathrm{B} = \Delta\mathrm{n} = n_x - n_y \quad (5)$$

The more air-filled pores there are in the core, the more confined the mode becomes. Leakage loss is reduced as a result of it. However, as the modes diverge from the core as the wavelength increases, the loss likewise rises [29].

$$L_c = 8.686 k_o I_m \left[n_{eff} \right] dB/m \quad (6)$$

Novel properties such as endlessly single mode (ESM), a single mode property spanning a wide wavelength range, down to visible regimes, achieved concurrently with yet a reasonably large mode area (LMA), have been developed as a result of research on holey cladding in photonic crystal fibers (PCFs). Helps in determining whether any fiber is single-mode. It is calculable as [29].

$$V_{eff} = 2\pi \frac{\Lambda}{\lambda} \sqrt{n_{core}^2 - n_{eff}^2} \quad (7)$$

V > 2.405: multi-mode;
V < 2.405: single-mode.

The silica core's refractive index, denoted as, is complemented by the effective refractive index, Λ is the pitch, and λ is the wavelength in vacuum. The structure is shown in Fig. 1(a) while Fig. 1(b) presented the index profile.

The PCF design has three cores. The air holes' diameter is denoted by d, while their separation is denoted by Λ, the larger hole diameters d1 and d2, and the smaller hole diameters d3, d4, and d cores = d5.

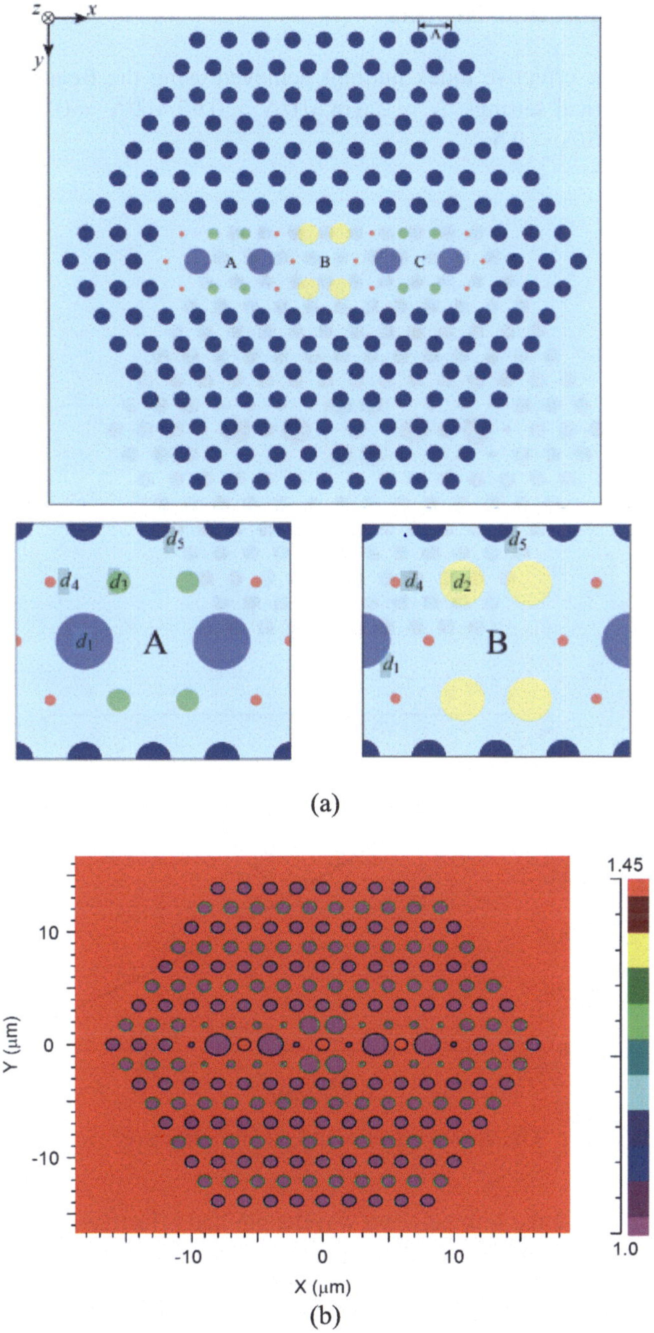

Fig. 1. (a) The design structure. (b) the profile index of the structure.

3 Interpretation and Results

Figure 2 shows the effective index findings achieved using the Beam Prop technique with opto-geometrical settings. $\Lambda = 2$ μm, $d1/\Lambda = 0.95$, $d2/\Lambda = 0.747$, $d3/\Lambda = 0.3$, $d4/\Lambda = 0.2$, and $d5/\Lambda = 0.5$:

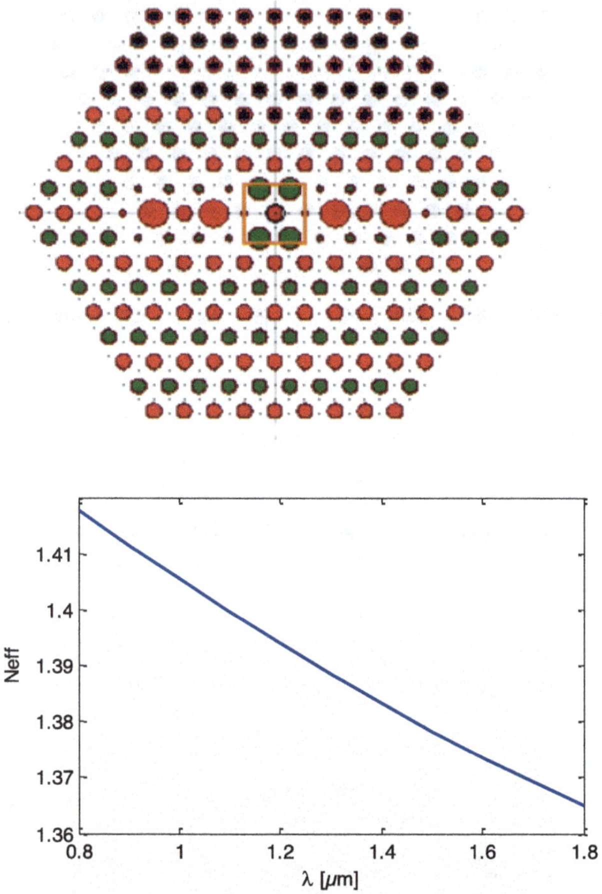

Fig. 2. The effective index of the reference.

We show in Fig. 2 the link between wavelength and effective index. Based on the Figure, it is clear that the effective index varies significantly as a function of wavelength, going from 1.41 to 1.445 (with a silica index of 1.45).

In Fig. 3, from 0.8 to 1.8 μm, we can see that the chromatic dispersion is negative over a broad wavelength range. A highly efficient numerical method must be utilized to determine the PCF modes' dispersion characteristics. The coefficient of dispersion D is defined by the modal effective index's second derivative concerning wavelength. Therefore, it is crucial to estimate the dependence of with accuracy.

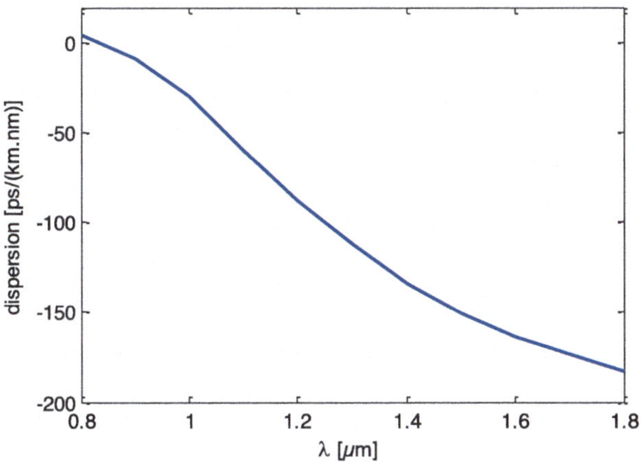

Fig. 3. Chromatic dispersion as a function of wavelength.

After computing, we can find the effective V value for PCF. Figure 4 shows the generated curves that demonstrate differences in the effective V for four different values of d/λ. The cut-off value for Vco is 2.405, with V > 2.405 denoting multi-mode and V < 2.405 denoting single-mode.

Figure 4 depicts curves as monomode fibers only if they never cross this line. On the other hand, curves that are significantly below this line would suggest poorly guided fibers and suffer from significant bend loss. According to the results in Fig. 4, the critical value of the parameter d/ below which this occurs is d/ ≈0.1. We don't have a suitable answer for this splitter.

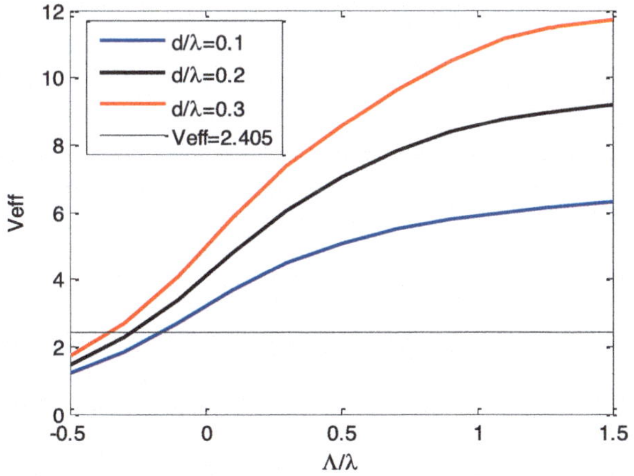

Fig. 4. Effective frequency V_{eff} when $d/\lambda = 0.1, 0.2$ and 0.3

4 Conclusion

In this paper, studies were conducted on the characteristics of a splitter that utilized a photonic crystal fiber. Including its effective index, chromatic dispersion, and effective frequency. The simulation findings make it clear that the negative dispersion photonic crystal fiber described in this study and the recommendations it offers have benefits when used in optical communications applications. So, in this work using the numerical method, we have tried to find a better result in terms of the effective index which depends on the wavelength and effective frequency which helps in determining whether any fiber is single-mode and dispersion.

References

1. Bock, W.J., Jiahua, C., Eftimov, T., Urbanczyk, W.: A photonic crystal fiber sensor for pressure measurements. IEEE Trans. Instrum. Meas. **55**(4), 1119–1123 (2006). https://doi.org/10.1109/TIM.2006.876591
2. Du, F., Lu, Y.-Q., Wu, S.-T.: Electrically tunable liquid-crystal photonic crystal fiber. Appl. Phys. Lett. **85**(12), 2181–2183 (2004)
3. Al-Zahrani, F.A., Ahmed, K.: Novel design of dual guided photonic crystal fiber for large capacity transmission in high-speed optics communications with supporting good quality OAM and LP modes. Alex. Eng. J. **59**(6), 4889–4899 (2020)
4. Li, W., et al.: 151W monolithic diffraction-limited Yb-doped photonic bandgap fiber laser at~ 978nm. Opt. Express **27**(18), 24972–24977 (2019)
5. An, S., et al.: Ultra-short and dual-core photonic crystal fiber polarization splitter composed of metal and gallium arsenide. Optik **226**, 165779 (2021)
6. Sharma, N., Rajawat, N., Agrawal, K.J.I.J.o.S., Research, E.: Photonic Crystal Fiber Characteristics Benefits Numerous Applications **6**, 28–33 (2015)
7. Broeng, J., Mogilevstev, D., Barkou, S.E., Bjarklev, A.J.O.f.t.: Photonic crystal fibers: A new class of optical waveguides **5**(3), 305–330 (1999)
8. Knight, J.C.J.n.: Photonic crystal fibres **424**(6950), 847–851 (2003)
9. Knight, J.C., Broeng, J., Birks, T.A., Russell, P.S.J.J.S.: Photonic band gap guidance in optical fibers **282**(5393), 1476–1478 (1998)
10. Lee, J., et al.: A compact cold-atom interferometer with a high data-rate grating magneto-optical trap and a photonic-integrated-circuit-compatible laser system. Nat. Commun. **13**(1), 1–12 (2022)
11. Li, X., et al.: Compact and low-loss silicon power splitter based on inverse tapers. **38**(20), 4220–4223 (2013)
12. Panda, A., Devi, P.P.: Photonic crystal biosensor for refractive index based cancerous cell detection. Opt. Fiber Technol. **54**, 102123 (2020)
13. Doerr, C.R., Fontaine, N.K., Buhl, L.L.: PDM-DQPSK silicon receiver with integrated monitor and minimum number of controls. IEEE Photonics Technol. Lett. **24**(8), 697–699 (2012)
14. Sun, J., Timurdogan, E., Yaacobi, A., Hosseini, E.S., Watts, M.R.J.N.: Large-scale nanophotonic phased array. **493**(7431), 195–199 (2013)
15. Birks, T.A., Knight, J.C., Mangan, B.J., Russell, P.S.J.J.I.t.o.c.: Photonic crystal fibres: an endless variety **84**(5), 1211–1218 (2001)
16. Li, J., Wang, J., Wang, R., Liu, Y.J.O., Technology, L.: A novel polarization splitter based on dual-core hybrid photonic crystal fibers **43**(4), 795–800 (2011)

17. Knight, J., Birks, T., Cregan, R., Russell, P.S.J., De Sandro, J.J.E.L.: Large mode area photonic crystal fibre **34**(13), 1347–1348 (1998)
18. Habib, M.S., et al.: Broadband dispersion compensation of conventional single mode fibers using microstructure optical fibers **124**(19), 3851–3855 (2013)
19. Lehtonen, M., Genty, G., Ludvigsen, H., Kaivola, M.J.A.p.l.: Supercontinuum generation in a highly birefringent microstructured fiber **82**(14), 2197–2199 (2003)
20. Hasan, M.R., Hasan, M.I., Anower, M.S.J.A.o.: Tellurite glass defect-core spiral photonic crystal fiber with low loss and large negative flattened dispersion over S+ C+ L+ U wavelength bands **54**(32), 9456–9461 (2015)
21. Saitoh, K., Sato, Y., Koshiba, M.: Polarization splitter in three-core photonic crystal fibers. Opt. Express **12**(17), 3940–3946 (2004)
22. Saitoh, K., Koshiba, M., Hasegawa, T., Sasaoka, E.J.O.e.: Chromatic dispersion control in photonic crystal fibers: application to ultra-flattened dispersion **11**(8), 843–852 (2003)
23. Kiroriwal, M., Singal, P.: Design and analysis of ethanol-infiltrated photonic crystal fiber for wavelength conversion. Journal of Optics 1–9 (2022)
24. Kuiri, B., et al.: Design and optimization of photonic crystal fiber with low confinement loss guiding 98 OAM modes in THz band. Opt. Fiber Technol. **68**, 102752 (2022)
25. Zehua, X., et al.: Ultralow bending-loss micro-structured fiber with large-mode-area. Optik **251**, 168440 (2022)
26. Paul, D., Dutta, S.: Exploring Macro-Bending Losses on Photonic Crystal Fiber for Communication Window with Dissimilar Material Composites (2022)
27. Liu, Q., et al.: Research on photonic crystal fiber sensor network based on fast fourier transform bandpass filtering. IEEE Transactions on Instrumentation and Measurement (2023)
28. Xie, Y., Zhang, J., Wang, S., Liu, D.: High-efficiency broadband photonic crystal fiber metalens with a large numerical aperture. Optics Communications **481**, 126524 (2021)
29. Ghatak, A.A., Ghatak, A., Thyagarajan, K., Thyagarajan, K.: An Introduction to Fiber Optics. Cambridge University Press (1998)

The Study of Cooling Systems for Superconducting Coils

Cheriet Rahmouna(✉), Bensaad Burassia, Bouhadjela Fatiha, and Belhenini Soufyane

Smart Structures Laboratory, University of Ain Temouchent Belhadj Bouchaib, Ain Temouchent, Algeria
rahmouna.cheriet@univ-temouchent.edu.dz

Abstract. The storage of electrical energy represents a technological and economic issue that is still in the news. Superconducting coils offer a relevant and promising solution to electrical energy storage [1, 2]. High-performance cryogenic systems are necessary to ensure better operation of SMES [3]. The main objective of this study is to analyze the operation of a superconducting coil cooling system. For this we opted for simulation using the COMSOL calculation code. COMSOL Multiphysics is numerical simulation software based on the finite element method. This software makes it possible to simulate physical phenomena and numerous applications in engineering, and particularly coupled phenomena or multi-physics simulation.

Keywords: Superconductivity · Superconducting Coil · Cooling · Cryogeny

1 Introduction

The main objective of this work is to provide a general overview on the cooling systems of superconducting coils. In order to fully understand the cooling problem of these components, the research undertaken will make it possible to answer the following questions:

- What is superconductivity and what are its applications?
- How can we model and calculate a cryostat using Comsol software?

2 The Objective of the Study

In this study we proposed to reproduce the Dorazio model consisting of analyzing the thermal field in an NMR cryostat," Nuclear Magnetic Resonance [1]. It is a phenomenon which appears when the nuclei of certain atoms, immersed in a uniform magnetic field B0, are exposed to another variable magnetic field Be.

Our main objective is to analyze the thermal field inside a cryostat on COMSOL:

- The development of a cryostat model with particular reference to the type of closure placed at the ends.
- Taking into account the insulation and limit condition to simulate the effect of a "cold head" placed in contact with a very limited part of the cryo-shields.
- Considering tie rods mounted in real time cryostat to keep the cavity empty and to center the magnetic field.

© The Author(s), under exclusive license to Springer Nature Switzerland AG 2025
M. Hatti (Ed.): IC-AIRES 2024, LNNS 1238, pp. 716–721, 2025.
https://doi.org/10.1007/978-3-031-80301-7_74

3 Presentation of the Model

3.1 Simplified Geometry of the Cryostat

At the beginning, a very simplified model of the cryostat was developed, composed of open concentric cylinders spaced apart from the zones (Fig. 1).

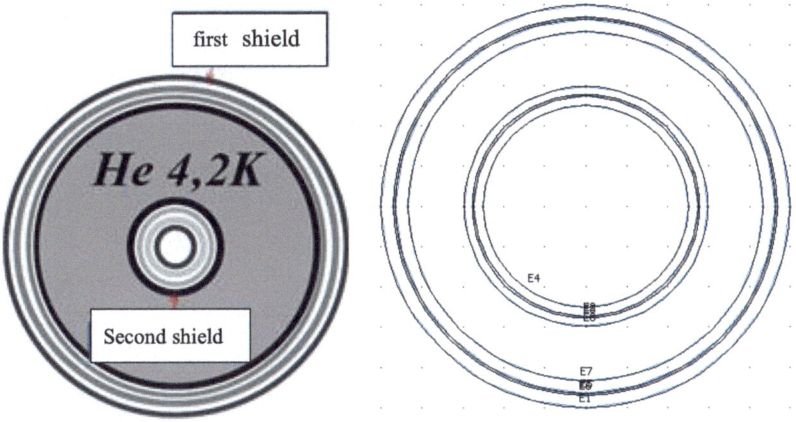

Fig. 1. Description of a cryostat's geometry and a model schematic in 2D.

This model does not take into account the thermal zone near the closures; only the heat exchange in the central part of the cryostat is considered (Fig. 2).

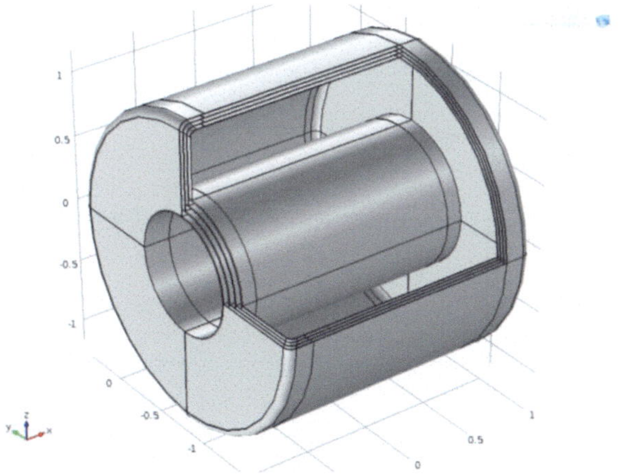

Fig. 2. Simplified geometry of the cryostat (3D).

The thermo-physical properties of the materials [4] making up the various components at operating temperatures are given in table below.

	Hardware	$\rho(kg/m^3)$	Thermal Conductivity [W/mK]	Volume $[m^3]$	Mass [kg]	Specific Heat [J/K.kg]
outer container	304 stainless steel	7999.5	14 {80K}	0.119	951.94	502 {80K}
Radiation shield	6063 -T5 Aluminium	2712.6	205 {20K}	0.155	420.45	850 {20K}
inner tank	304 stainless steel	7999.5	14 {80K}	0.097	263.12	502 {80K}
cool head	6063 -T5 Aluminium	8940	205 {20K}	0.145	1159.92	850 {20K}

3.2 Boundary Condition

The types of boundary conditions describing this problem are:

-Convective heat flows imposed at solid/fluid boundaries:

- Incoming heat flow from the outside to the steel sheets
- Outgoing heat flow from Aluminium foil to Helium,

- Heat exchange by radiation between external surfaces and the environment,
- Heat exchange by radiation between neighbouring surfaces through a vacuum,
- Set temperature.

The amount of heat to be removed by the cooling step through each contact surface is defined as the overall difference between the flow into and out of the shield.

3.3 Initial Conditions

We consider a steady state with the following initial conditions:

- Outside temperature = 293.15 K
- Internal temperature: = 4.2 K

3.4 Meshing

The mesh of the {2D} geometry is shown in Fig. 3:

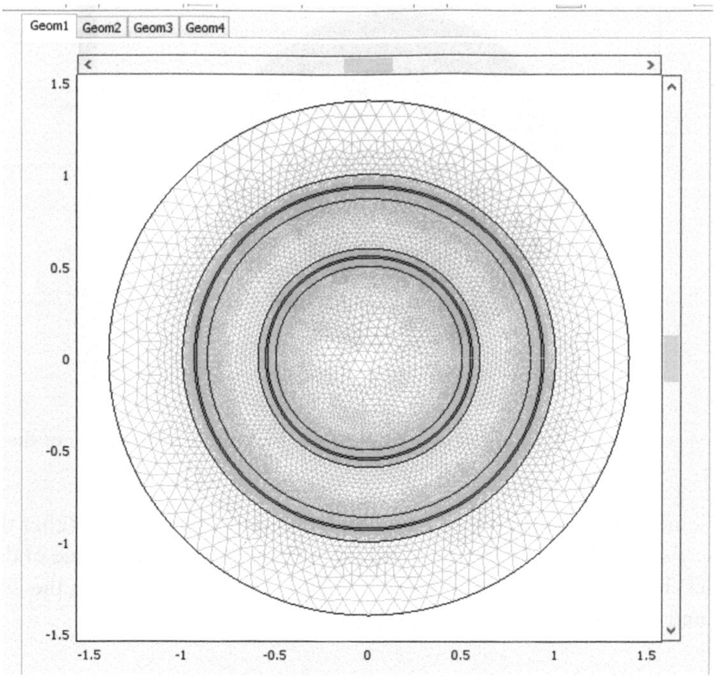

Fig. 3. The (2D) mesh of the model

4 Discussion of Results

Calculating the 3D model is proving slow. Due to time constraints and the cylindrical shape of the device, we propose to present only the 2D model.

Figure 4 shows the temperature field calculated using COMSOL in 2D. The first circular enclosure, with a radius of 14 cm, shows the envelope of air, at T = 293.15 K, containing the simulated component.

Heat exchange by convection and radiation takes place between the air at 293.15 K and the steel enclosure maintained at 80 K. A vacuum is maintained between the two steel and aluminium enclosures. A radiative exchange persists, with the Al envelope maintained at a temperature of 20 K. The helium contained in the aluminium enclosure is maintained at a temperature of 4.2 K [1, 2]; an exchange by convection/radiation takes place between the liquid helium and the aluminium wall.

The same heat exchanges take place through the helium and the two metal shells (Al, Steel) towards the center of the device.

To get a better idea of the thermal field near the walls, we took a reading of the section along the length of the device; the segment in red shows this reading. This segment is used to plot the graph in Fig. 5. The temperature levels and heat exchanges are clearly visible on the curve in this figure. It is very difficult to compare with the fields obtained in D'orazio's work, as the temperature fields in the vicinity of the walls are not provided.

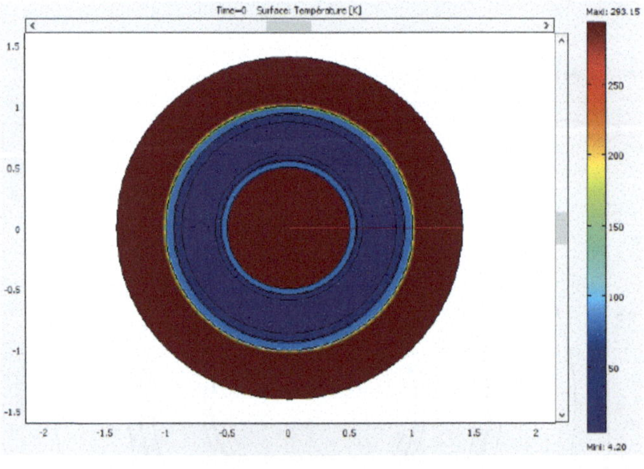

Fig. 4. Measurement of the temperature field over the longitudinal section of the device

In the vicinity of aluminium walls, helium temperatures are much higher than 4.2 K, and there is a risk of rapid evaporation of liquid helium and the appearance of the Quench phenomenon. The same observation was made by D'orazio. Isolation of the device from convective and radiative exchanges is essential.

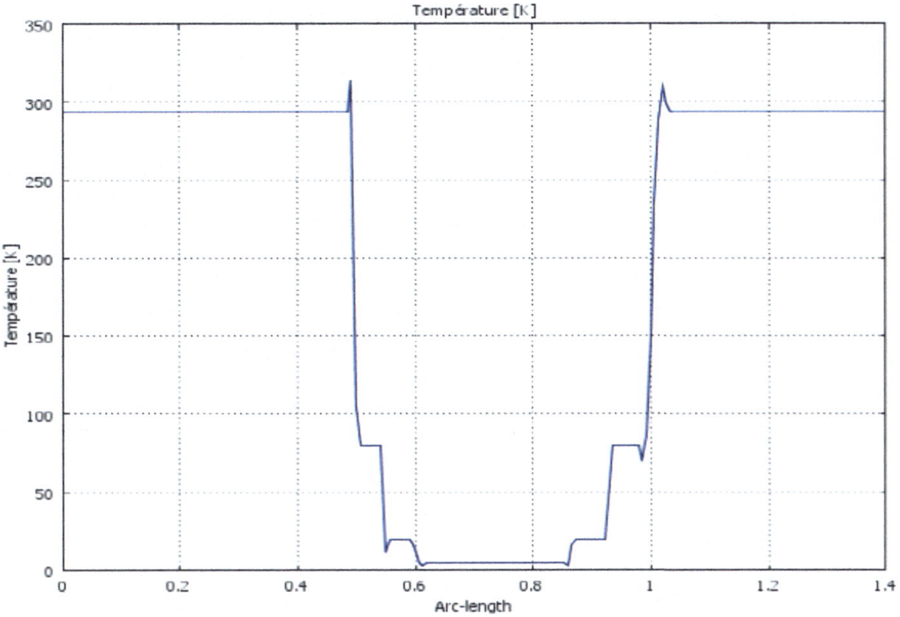

Fig. 5. Plot of temperature distribution using COMSOL

5 Conclusion

A 2D representation of the distribution of thermal fields within the cryostat is obtained. Rapid evaporation of liquid helium near the walls is inevitable under the conditions considered. A more realistic model, with fewer simplifications and including insulation for convective and radiative transfers, could give an idea close to reality and enable us to propose a better solution to the Quench problem.

References

1. Conference comsol for Thermal Field in a NMR Cryostat Annunziata D'Orazio & all D'Orazio A; Agostini C; Fiacco S; University of Sapienza - Rome – Department: Astronautique Ingénierie (2014)
2. Markham, A.H., Netzel, R.G., Dillinger, J.R.: Rev. Sci. Inst. **28**, 382 (1957)
3. Cryostat a température variable entre 1 et 300 K pour direction de neutrons
4. Mechanical properties and constitutive modeling of 6063-T5 aluminum alloy over a wide range of strain rates and temperatures Xin Wang & al

Estimate the Hardness of Deposited Thin Layers Using Fuzzy Logic

Nada Rihane Mechehat[1,2](✉), Djilani Necib[1,2], Imane Rezgui[3], and Mondir Soualah[4]

[1] Department of Mechanical Engineering, Faculty of Technology, University of El Oued, 39000 El Oued, Algeria
mechehat.nada@gmail.com, mechehat-nadaelrihane@univ-eloued.dz
[2] UDERZA Laboratory, University of El Oued, 39000 El Oued, Algeria
[3] Department of Mechanical Engineering, Faculty of Applied Sciences, University of Ouargla, Ouargla, Algeria
[4] Department of Mechanical Engineering, Faculty of Technology, University of Blida1, 9000 Blida, Algeria

Abstract. Thin film coatings are essential in the production of various industrial devices, significantly enhancing material performance through advanced deposition techniques. This study focuses on estimating the hardness of thin titanium nitride (TiN) films, which serve as protective coatings in industrial applications, by utilizing fuzzy logic. The fuzzy logic approach allows for the incorporation of various input parameters and the handling of uncertainties inherent in experimental data. The results indicate a high degree of accuracy, with the fuzzy logic model showing a percentage error of only 2.25% and an impressive precision of 97.75%. This close agreement between experimental outcomes and predicted values underscores the effectiveness of fuzzy logic in predicting material properties. By improving the predictability and reliability of thin film coatings, such methodologies can contribute to advancements in material science and engineering, leading to the development of more durable and efficient industrial devices. The application of fuzzy logic in this context demonstrates its potential as a valuable tool for material property estimation and process optimization in industrial manufacturing.

Keywords: Hardness · thin film · Ti-N · fuzzy logic

1 Introduction

Various studies have investigated methods to enhance thin film properties. Beake et al. [1] introduced a probe testing method for dynamic loading. Sukwisute et al. [2] found that sputtering power affects CrN hardness, peaking at 175 W with values from 6.65 to 9.58 GPa. Zalnezhad et al. [3] used fuzzy logic to predict CrN hardness on AL7075-T6 with 94.664% accuracy. Demas et al. [4] linked surface roughness and contamination to coating hardness measured via nanoindentation. Additional studies by Zalnezhad et al. [5] and Zalnezhad et al. [6] focused on TiN coatings on Al7075- T6 to improve

© The Author(s), under exclusive license to Springer Nature Switzerland AG 2025
M. Hatti (Ed.): IC-AIRES 2024, LNNS 1238, pp. 722–730, 2025.
https://doi.org/10.1007/978-3-031-80301-7_75

wear resistance and hardness. Ghadai et al. [7] optimized thin film hardness using meta-heuristic algorithms. Kaczmarek et al. [8] explored Tb doping effects on TiO2 hardness. Schubert et al. [9] examined substrate hardness and roughness on Al2O3 coatings. Wang et al. [10] investigated CoCrFeNi MEA thin films, while Natrayan et al. [11] enhanced TiAlN films on steel for improved wear resistance and hardness. Lastly, Yazdi et al. [12] applied an MLP neural network to optimize titanium thin film hardness.

Among the applications of digital optimization, neural networks, genetic algorithms and fuzzy logic, etc. We propose fuzzy logic makes it possible to take into account all kinds of qualitative knowledge of designers and operators in the automation of systems.

In this study, we propose to use a fuzzy model based on the Mamdani method; in order to estimate the hardness of thin layers deposited under different working conditions (ion bombardment time, working pressure, chamber pressure, underlayer temperature, roughness and geometry of the coated parts).

2 Procedure

2.1 Fuzzy Model

Fuzzy logic, developed by Lotfi Zadeh in the 1960s, allows for reasoning with uncertain data by working with degrees of truth between 0 and 1, unlike traditional binary logic. This approach is useful for modeling complex systems and is widely applied in decision-making, pattern recognition, and control systems. In this study, we propose estimating the hardness of thin layers through a fuzzy logic-based simulation (Mamdani) using MATLAB software. This method will be based on experimental data compiled in Table 1 from the reference source [12].

2.2 Fuzzy System

In this system, we selected a Mamdani-type fuzzy inference system to analyze the deposition of thin TiN layers. The fuzzy system incorporates six input variables: ion bombardment time (T), working pressure (WP), chamber pressure (CP), underlayer temperature (T), underlayer roughness (Ra), and the geometry of the coated parts (G). The output variable in this model is the hardness of the deposited thin layers (H) (Fig. 1).

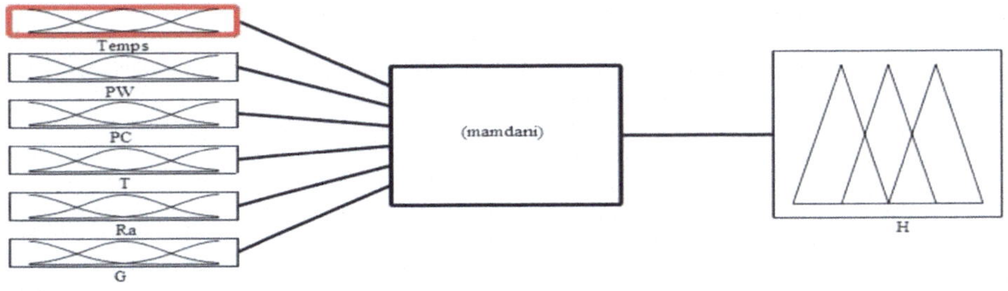

Fig. 1. Fuzzy system for thin TiN films.

2.3 Fuzzy Modeling

- ***Fuzzy variables***

In constructing our model, we selected the following input parameters: ion bombardment time, working pressure, chamber pressure, underlayer temperature, underlayer roughness, and the geometry of the coated parts. The output variable chosen was the hardness of the thin layer. In our study, all variables are categorized into linguistic variables, each corresponding to specific numerical data intervals. For the first and second variable, ion bombardment time and work pressure, the linguistic variables are defined as illustrated in Fig. 2.

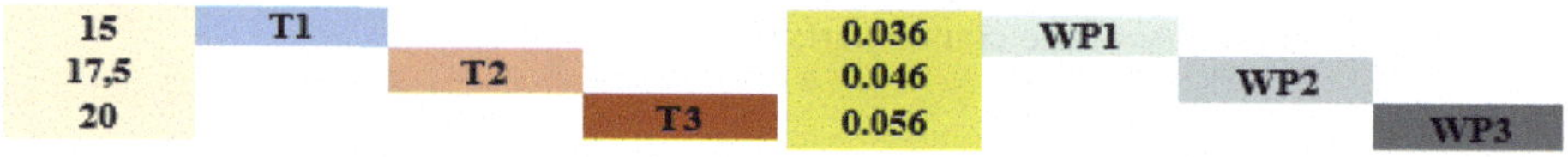

Fig. 2. Linguistic variables for ion bombardment time and work pressure.

The linguistic variables for the third and fourth variables, chamber pressure and underlayer temperature, are illustrated in Fig. 3.

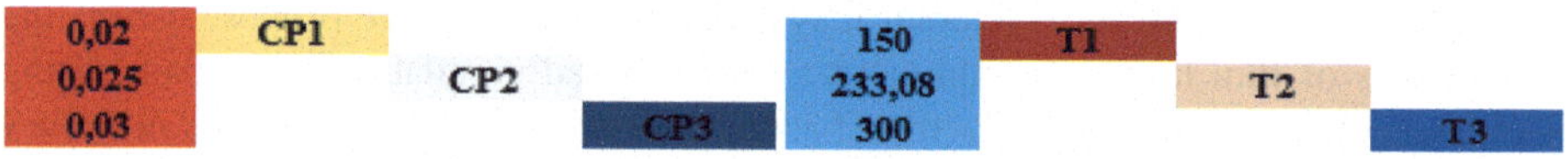

Fig. 3. Linguistic variables for chamber pressure and underlayer temperature.

The linguistic variables for the fifth and sixth variables (roughness and geometry) are chosen as shown in Fig. 4.

Fig. 4. Linguistic variables for roughness and geometry.

The linguistic variables for the output (thin film hardness) are chosen as shown in Fig. 5.

- *THE FUZZY SUBSETS*

Fuzzy subsets were designed to replicate human knowledge representation, improving decision systems by addressing uncertainty. Defined by membership functions, similar to classical set theory, these functions can be triangular or trapezoidal. This study used both types for input and output parameters, shown in the figures (Fig. 6).

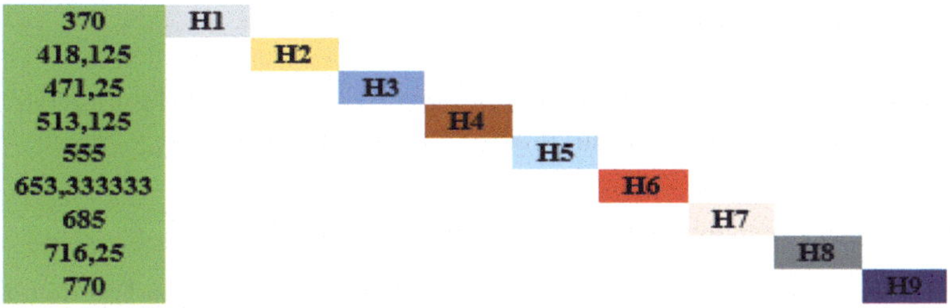

Fig. 5. Linguistic variables for thin film hardness.

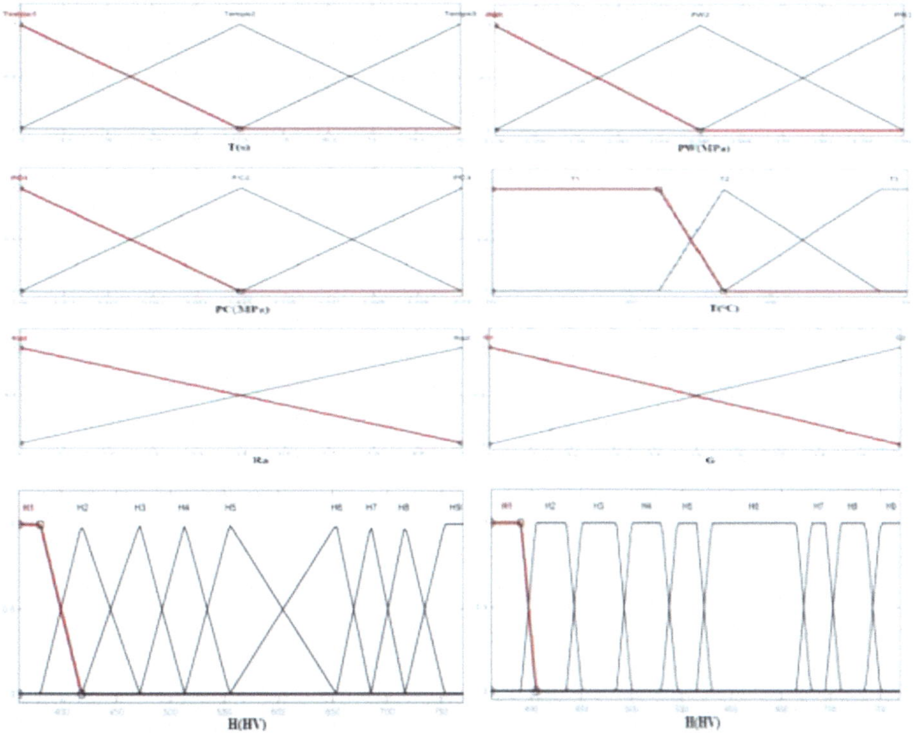

Fig. 6. Triangular and Trapezoidal membership functions.

2.4 The Fuzzy Rules

Sixty-four fuzzy rules were formulated based on the experimental conditions outlined in Table 1. Utilizing the max- min composition method, these fuzzy logic rules generated fuzzy outputs. Each rule follows this structure:

"If T is (linguistic variable), WP is (linguistic variable), CP is (linguistic variable), T (linguistic variable), Ra (linguistic variable) and G (linguistic variable), Then H is (linguistic variable)."

3 Results and Discussions

The final step in fuzzy logic, defuzzification, involves converting the linguistic values produced by the fuzzy controller into precise numerical values.

3.1 Study of the Precision and Error of the Fuzzy System

Errors are quantified by measuring the deviation between observed and predicted values using Eq. 1, with percentage error calculated as the absolute difference divided by the observed value. Accuracy is assessed by comparing predicted values to actual measurements, represented by A in Eq. 2, with N indicating the total datasets. Overall model accuracy is the average of individual accuracies.

$$e_i = \left(\frac{|H_{exp} - H_{prd}|}{H_{exp}}\right) \times 100\% \quad (1)$$

$$A = \frac{1}{N}\sum_{i=1}^{N}\left(1 - \frac{|H_{exp} - H_{prd}|}{H_{exp}}\right) \times 100 \quad (2)$$

Based on the comparison of these functions, we utilized the precision relation (2), yielding the following accuracies:

o Triangular function: A = 97.73% and Trapezoidal function: A = 97.75%

consequently, the results most closely aligned with the experimental data were those obtained using trapezoidal membership functions. Therefore, all results presented in the remainder of this manuscript were derived using trapezoidal membership functions. Table 1 brings together all the results obtained by our fuzzy system.

Table 1. The results obtained by our fuzzy system.

No	Parameters of deposited thin layers						Hardness results			
	T(s)	WP(MPa)	CP(MPa)	T (°C)	RA	G	Hexp (HV)	Hpred (HV)	Errors	Accuracy
1	20	0.056	0.03	300	0	1	470	467	0,6382	99,0493
2	20	0.056	0.03	290	0	1	450	467	3,7777	97,4968
3	17.5	0.046	0.025	280	0	1	435	419	3,6781	95,8753
4	15	0.036	0.02	270	0	1	420	420	0	94,3702
5	20	0.056	0.03	260	0	1	435	444	2,0689	92,8077
6	17.5	0.046	0.025	250	0	1	420	420	0	91,2129
7	15	0.036	0.02	240	0	1	405	420	3,7037	89,6504
8	20	0.056	0.03	230	0	1	420	416	0,9523	88,0301
9	17.5	0.046	0.025	220	0	1	405	399	1,4814	86,4824

(*continued*)

Table 1. (*continued*)

No	Parameters of deposited thin layers						Hardness results			
	T(s)	WP(MPa)	CP(MPa)	T (°C)	RA	G	Hexp (HV)	Hpred (HV)	Errors	Accuracy
10	15	0.036	0.02	210	0	1	390	378	3,0769	84,9431
11	20	0.056	0.03	200	0	1	405	400	1,2345	83,4287
12	17.5	0.046	0.025	190	0	1	390	378	3,0769	81,8854
13	15	0.036	0.02	180	0	1	375	378	0,8	80,3710
14	20	0.056	0.03	170	0	1	390	400	2,5641	78,7960
15	17.5	0.046	0.025	160	0	1	375	378	0,8	77,1935
16	15	0.036	0.02	150	0	1	360	378	5	75,6185
17	20	0.056	0.03	300	1	1	740	737	0,4054	73,9778
18	17.5	0.046	0.025	290	1	1	720	737	2,3611	72,4217
19	15	0.036	0.02	280	1	1	705	707	0,2836	70,8223
20	20	0.056	0.03	270	1	1	690	688	0,2898	69,2553
21	17.5	0.046	0.025	260	1	1	705	719	1,9858	67,6974
22	15	0.036	0.02	250	1	1	690	707	2,4637	66,1038
23	20	0.056	0.03	240	1	1	675	688	1,9259	64,5028
24	17.5	0.046	0.025	230	1	1	690	690	0	62,9103
25	15	0.036	0.02	220	1	1	675	652	3,4074	61,3478
26	20	0.056	0.03	210	1	1	660	623	5,6060	59,8385
27	17.5	0.046	0.025	200	1	1	675	637	5,6296	58,3636
28	15	0.036	0.02	190	1	1	660	623	5,6060	56,8891
29	20	0.056	0.03	180	1	1	645	623	3,4108	55,4141
30	17.5	0.046	0.025	170	1	1	660	637	3,4848	53,9049
31	15	0.036	0.02	160	1	1	645	623	3,4108	52,3969
32	20	0.056	0.03	150	1	1	630	623	1,1111	50,8877
33	17.5	0.046	0.025	300	0	2	565	555	1,7699	49,3425
34	15	0.036	0.02	290	0	2	545	555	1,8348	47,8077
35	20	0.056	0.03	280	0	2	530	515	2,8301	46,2165
36	17.5	0.046	0.025	270	0	2	515	515	0	44,6983
37	15	0.036	0.02	260	0	2	530	532	0,3773	43,1358
38	20	0.056	0.03	250	0	2	515	515	0	41,5674
39	17.5	0.046	0.025	240	0	2	500	515	3	40,0049

(*continued*)

Table 1. (*continued*)

N o	Parameters of deposited thin layers						Hardness results			
	T(s)	WP(MPa)	CP(MPa)	T (°C)	RA	G	Hexp (HV)	Hpred (HV)	Errors	Accuracy
40	15	0.036	0.02	230	0	2	515	507	1,5533	38,3955
41	20	0.056	0.03	220	0	2	500	487	2,6	36,8573
42	17.5	0.046	0.025	210	0	2	485	467	3,7113	35,3354
43	15	0.036	0.02	200	0	2	500	490	2	33,8309
44	20	0.056	0.03	190	0	2	485	467	3,7113	32,2996
45	17.5	0.046	0.025	180	0	2	470	467	0,6382	30,7951
46	15	0.036	0.02	170	0	2	485	490	1,0309	29,2426
47	20	0.056	0.03	160	0	2	470	467	0,6382	27,6640
48	17.5	0.046	0.025	150	0	2	455	467	2,6373	26,1114
49	15	0.036	0.02	300	1	2	770	757	1,6883	24,5077
50	20	0.056	0.03	290	1	2	750	757	0,9333	22,9716
51	17.5	0.046	0.025	280	1	2	735	707	3,8095	21,3945
52	15	0.036	0.02	270	1	2	710	707	0,4225	19,8916
53	20	0.056	0.03	260	1	2	735	735	0	18,3357
54	17.5	0.046	0.025	250	1	2	710	707	0,4225	16,7732
55	15	0.036	0.02	240	1	2	695	708	1,8705	15,2173
56	20	0.056	0.03	230	1	2	710	718	1,1267	13,6255
57	17.5	0.046	0.025	220	1	2	695	654	5,8992	12,0454
58	15	0.036	0.02	210	1	2	680	637	6,3235	10,5751
59	20	0.056	0.03	200	1	2	695	687	1,1510	9,11146
60	17.5	0.046	0.025	190	1	2	680	637	6,3235	7,56695
61	15	0.036	0.02	180	1	2	665	637	4,2105	6,10325
62	20	0.056	0.03	170	1	2	680	687	1,0294	4,60654
63	17.5	0.046	0.025	160	1	2	665	637	4,2105	3,02796
64	15	0.036	0.02	150	1	2	650	637	2	1,53125
Accuracy = 97.75015079%										

3.2 Validation of Results

Figure 7 shows that the fuzzy logic model accurately predicts the hardness of TiN thin films, as evidenced by the close match between experimental and predicted values.

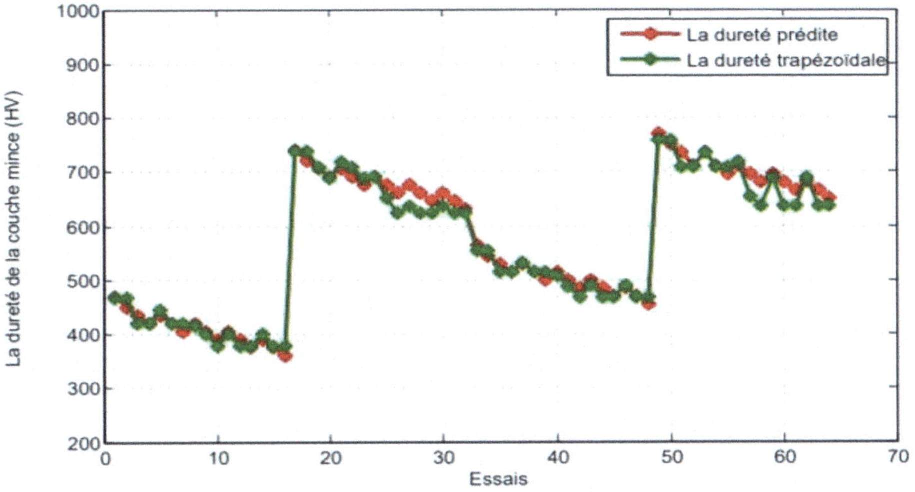

Fig. 7. The curves of the experimental and simulation results of TiN thin films.

4 Conclusion

In conclusion, the fuzzy logic model has demonstrated high reliability and accuracy in predicting the hardness of TiN thin films. The strong agreement between the model's outputs and experimental data supports its effectiveness. As a result, this model can be confidently applied for estimating hardness within a defined parameter range, making it useful for optimizing film properties in practical applications. In this research, we optimized variables like ion bombardment time, chamber pressure, surface roughness, and the geometry of coated parts. Maximum hardness is achieved when ion bombardment time, pressure, and temperature are at their peak values. The prediction accuracy of this model exceeds 97%, highlighting its precision in predicting thin film hardness.

References

1. Beake, B.D., Goodes, S.R., Smith, et J.F.: Micro-impact testing: a new technique for investigating thin film toughness, adhesion, erosive wear resistance, and dynamic hardness. Surface Engineering **17**(3), 187–92 (2001). https://doi.org/10.1179/026708401101517755
2. Sukwisute, P., Sakdanuphab, R., Sakulkalavek, et A.: Hardness and Wear Resistance Improvement of ABS Surface by CrN Thin Film. Materials Today: Proceedings **4**(5), 6553–61 (2017). https://doi.org/10.1016/j.matpr.2017.06.167
3. Zalnezhad, E., Mohammed Sarhan, A.A.D., Hamdi, et M.: Surface Hardness Prediction of CrN Thin Film Coating on AL7075-T6 Alloy Using Fuzzy Logic System. Int. J. Precision Eng. Manufact. **14**(3), 467–73 (2013). https://doi.org/10.1007/s12541-013-0063-5
4. Demas, N.G., Cinta, L.-M., Oyelayo, O.A., Erck, R.A., Shareef, et I.: Measurement of thin-film coating hardness in the presence of contamination and roughness: implications for tribology. Metallurgical and Materials Transactions A **47**(4), 1629–40 (2016). https://doi.org/10.1007/s11661-016-3342-9

5. Zalnezhad, E., Sarhan, A.A.D., Hamdi, et M.: A fuzzy logic based model to predict surface hardness of thin film TiN coating on aerospace AL7075-T6 Alloy. The Int. J. Adv. Manufact. Technol. **68**(14), 415–23 (2013). https://doi.org/10.1007/s00170-013-4738y
6. Ghadai, R.K., Kalita, K., Mondal, S.C., Swain, et B.P.: PECVD process parameter optimization: towards increased hardness of diamond-like carbon thin films. Materials and Manufacturing Processes **33**(16), 1905–13 (2018). https://doi.org/10.1080/10426914.2018.1512114
7. Kaczmarek, D., et al.: Influence of Nanocrystalline Structure and Composition on Hardness of Thin Films Based on TiO2. Open Physics **9**(2), 349–53 (2011). https://doi.org/10.2478/s11534-010-0112-y
8. Zalnezhad, E., Ahmed, A.D., Sarhan, et M.H.: Optimizing the PVD TiN Thin Film Coating's Parameters on Aerospace AL7075-T6 Alloy for Higher Coating Hardness and Adhesion with Better Tribological Properties of the Coating Surface. The Int. J. Adv. Manufact. Technol. **64**(14), 281–90 (2013). https://doi.org/10.1007/s00170-012-4022-6
9. Schubert, M., Hahn, M., Exner, J., Kita, J., Moos, et R.: Effect of Substrate Hardness and Surface Roughness on the Film Formation of Aerosol-Deposited Ceramic Films. Functional Materials Letters **10**(04), 1750045 (2017). https://doi.org/10.1142/S179360471750045X
10. Wang, J.-J., Ouyang, et F.-Y.: Nanotwinned Medium Entropy Alloy CoCrFeNi Thin Films with Ultra-High Hardness: Modifying Residual Stress without Scarifying Hardness through Tuning Substrate Bias. Surface and Coatings Technology **434**, 128191 (2022). https://doi.org/10.1016/j.surfcoat.2022.128191
11. Natrayan, L., et al.: Experimental investigation on mechanical properties of TiAlN thin films deposited by RF magnetron sputtering. Édité par Lakshmipathy R. Journal of Nanomaterials 2021 **1**, 7 (2021). https://doi.org/10.1155/2021/5943486
12. Yazdi, M.R.S., Mahyar Khorasani, A., Faraji, et M.: Optimization of coating variables for hardness of industrial tools by using artificial neural networks. Expert Systems with Applications **38**(10), 12116–27 (2011). https://doi.org/10.1016/j.eswa.2011.03.002

Improved TSR Based MPPT Control Using Hybrid Fuzzy Sliding Mode Control for Wind Turbines

Rania Kadri(✉), Mohamed Abid, and Aissaoui Abdelghani

Department of Electrical Engineering. IRECOM Laboratory, Djillali Liabes University of Sidi-Bel-Abbes, Sidi-Bel-Abbes, Algeria
rania.kadri@univ-sba.dz

Abstract. This study showcases the pivotal role of fuzzy controllers in system control improvement. We applied this approach to enhance the efficiency of MPPT based on an optimal Tip Speed Ratio (TSR) technique for wind turbines. Our study is developed on a wind conversion system in order to produce optimum power (to extract the maximal wind power). The main drawback is that the system is highly nonlinear, and thus a nonlinear control strategy is required. By combining a fuzzy controller with a sliding mode controller via a fuzzy supervisory system, a hybrid controller is proposed. The robustness of the suggested control technique is demonstrated through simulation results, which effectively preserve system stability.

Keywords: wind turbine · maximum power point tracking · tip speed ratio · Chattering phenomenon · sliding mode control · Fuzzy logic control

1 Introduction

Harnessing the natural power of wind, wind turbines stand as iconic symbols of clean energy generation, offering a multitude of environmental, and economic benefits. Aerodynamic wind systems utilizing variable-speed turbines have emerged as key players in the transition to sustainable power generation. By allowing rotor speeds to vary based on wind conditions, these turbines optimize energy capture efficiency. In the pursuit of enhancing the efficiency of wind turbines, optimizing Maximum Power Point Tracking (MPPT) algorithms is crucial. While Sliding Mode Control (SMC) has been widely used for its robustness, its tendency to exhibit chattering phenomena which is induced by a discontinuous term in the control input, which is introduced by the sign function. Various approaches have been put forward to reduce chattering, and one of these involves using fuzzy logic techniques. The steady control action provided by the knowledge base of FLC can help reduce the chattering problem commonly seen in SMC. In this study, a Hybrid Fuzzy Sliding Mode Control (HFSMC) is introduced. This control approach involves combining SMC and FLC through a fuzzy supervisory system, ensuring robust dynamic response [1]. The output of the fuzzy supervisory system is determined based

© The Author(s), under exclusive license to Springer Nature Switzerland AG 2025
M. Hatti (Ed.): IC-AIRES 2024, LNNS 1238, pp. 731–740, 2025.
https://doi.org/10.1007/978-3-031-80301-7_76

on error and its rate of change, combining outputs from SMC and FLC in a manner where SMC dominates during transient state and FLC during steady state [2]. As a result, rapid dynamics response and decreased chattering are attainted. This paper focuses on improving the efficiency of MPPT through enhanced speed control loop performance, aiming to increase the production of generated power and that by using HFSMC. A comparative analysis was conducted to showcase the improvements achieved by employing the new proposed control technique in comparison to conventional sliding mode control SMC.

2 Wind Turbine Modeling

The mathematical model of system with MPPT technique is represented by the block diagram shown in the following figure:

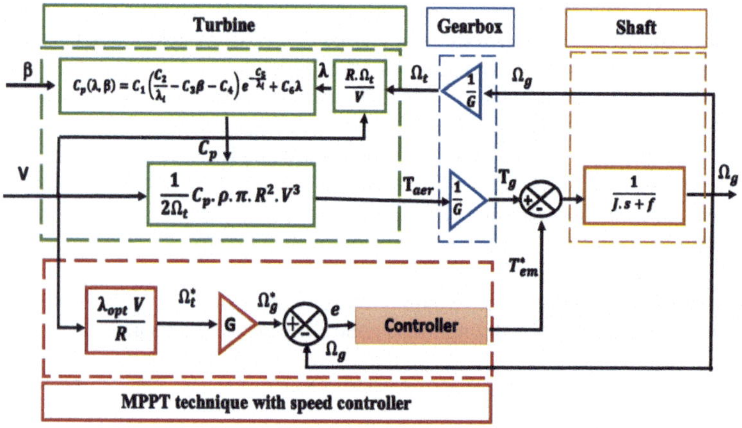

Fig. 1. MPPT technique with speed controller

With:

$$\begin{cases} C_p(\lambda,\beta) = 0.5176\left(\frac{116}{\lambda_i} - 0.4\beta - 5\right)e^{\frac{-21}{\lambda_i}} + 0.0068\lambda \\ \frac{1}{\lambda_i} = \frac{1}{\lambda+0.08\beta} - \frac{0.035}{\beta^3+1} \end{cases} \tag{1}$$

The MPPT control algorithm is crucial for maximizing power generation and optimizing wind energy capture. This allows the turbine to achieve the maximum power coefficient C_{pmax}, resulting in increased power generation and improved wind energy capture [3].

In this study, mechanical speed control is employed to implement the MPPT technique, as illustrated in Fig. 1.This approach involves regulating the torque applied to the turbine shaft to keep it at its reference speed. This process involves utilizing a rotor speed control loop, which requires accurately determining the turbine's speed to optimize power generation [4].

The reference rotational speed of the turbine has defined as follows:

$$\Omega_t^* = \frac{\lambda_{opt} V}{R} \tag{2}$$

The reference generator shaft speed is:

$$\Omega_g^* = G\Omega_t^* \tag{3}$$

In this approach, two control techniques are utilized: conventional sliding mode control and hybrid fuzzy sliding mode control.

3 Hybrid Control Design

3.1 Sliding Mode Control

Sliding mode control is a robust nonlinear control technique that aims to drive the system's states onto a predefined surface, ensuring the states stay on this surface to achieve stable performance. In sliding mode control, there are two principal phases. Initially defining a sliding surface and then the development of a control law that drives the system state towards this surface [5].

The sliding surface is chosen in as follow:

$$S_\Omega = e_\Omega = \Omega_g^* - \Omega_g \tag{4}$$

where Ω_g and Ω_g^* are respectively, the actual and the reference speed.

Sliding mode control law consists of two main command components. The first one is the continuous component or the equivalent command U_{eq} which is obtained by setting the surface derivative to zero, its role being to maintain the system on the sliding surface. The second one is U_N which represents the discontinuous control that guides the system state trajectory towards the sliding surface to achieve convergence.

The general sliding control law is defined as follow:

$$U = U_{eq} + U_N \tag{5}$$

In order to determine the control law of SMC for our system, the use of Eq. (6) is required.

$$J\frac{d\Omega_g}{dt} = \mathrm{T}_g - \mathrm{T}_{em} - f\Omega_g \tag{6}$$

through the derivation of Eq. (4), we get:

$$\dot{S_\Omega} = \dot{\Omega_g^*} - \dot{\Omega_g} \tag{7}$$

By replacing Eq. (6) in Eq. (7):

$$\dot{S_\Omega} = \frac{1}{J}\left(T_{\text{em}} + f\Omega_g - T_{\text{g}}\right) + \dot{\Omega}_g^* \tag{8}$$

By replacing the expression of T_{em} in the previous equation by the equivalent and discrete commands ($T_{em}^{eq} + T_{em}^{n}$), we find:

$$\dot{S_\Omega} = \frac{1}{J}\left(\left(T_{em}^{eq} + T_{em}^{n}\right) + f\Omega_g - T_{\text{g}}\right) + \dot{\Omega}_g^* \tag{9}$$

In this hybrid control, SMC dominates during transient state. The equivalent command T_{em}^{eq} for our SMC in our case is not included because only the FLC is applied during the permanent mode. The control law U_{SM} is determined in a way to ensure that the system state trajectory converges towards the sliding surface. A sufficient condition for achieving this goal is the appropriate selection of the control strategy.

To maintain system stability, Lyapunov function is considered:

$$V = \frac{1}{2}S_\Omega^2 \tag{10}$$

By validating the following condition that ensures an ideal sliding motion in finite time, despite external disturbances or uncertainty [6]:

$$\frac{d}{dt}\left(\frac{1}{2}S_\Omega^2\right) \le -\eta|S| \quad ; \quad \eta > 0 \tag{11}$$

We define the control law as follows:

$$U_{SM} = -K \; sign(S_\Omega) \tag{12}$$

where K is a positive constant.

3.2 Fuzzy Logic Controller

A fuzzy logic controller is a control strategy that employs fuzzy sets to represent uncertainty and imprecision in system variables, allowing for adaptive decision-making and robust performance across varying conditions. FLC develops control rules based on human expertise, employing linguistic rules if-then that address the system's dynamic behavior [7]. For our fuzzy controller, we adopt the speed error "e" and its variation "Δe" as inputs and the change in command ΔU_{FL} as output.

Figure 2 illustrates the membership functions for inputs and output, while Table 1 displays the FLC rules. The terms LP, MP, SP, ZE, SN, MN, LN represent respectively Large-Positive, Medium-Positive, Small-Positive, Zero, Small-Negative, Medium-Negative, Large-Negative.

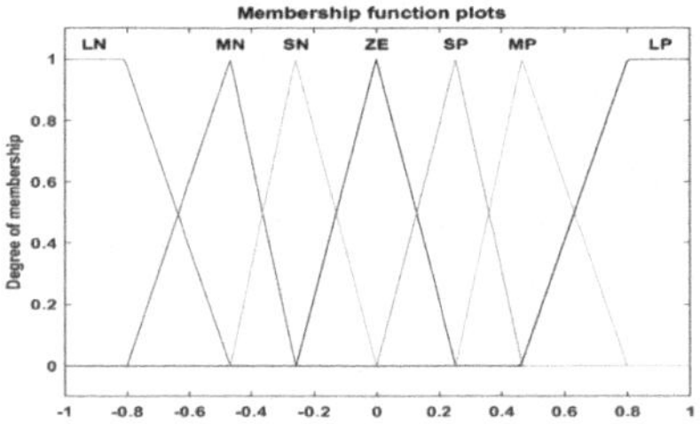

Fig. 2. Membership functions for e, Δe, ΔU_{FL}.

Table 1. Rule bases of FLC controller

Δe \ e	LN	MN	N	ZE	SP	MP	LP
LN	LN	LN	LN	LN	MN	SN	ZE
MN	LN	LN	LN	MN	SN	ZE	SP
SN	LN	LN	MN	SN	ZE	SP	MP
ZE	LN	MN	SN	ZE	SP	MP	LP
SP	MN	SN	ZE	SP	MP	LP	LP
MP	SN	ZE	SP	MP	LP	LP	LP
LP	ZE	SP	MP	LP	LP	LP	LP

3.3 Hybrid Control Implementation

SMC controllers are known for their ability to control systems efficiently and offer quick dynamic responses. However, their main disadvantage is the occurrence of chattering in steady-state. To address this problem, a hybrid controller with a fuzzy supervisory system has been implemented, merging the positive aspects of both SMC and fuzzy controllers. When both the error and its variation are large, the hybrid system applies SMC to quickly adjust the position relative to the reference point. On the other hand, when these values are small, the system switches to fuzzy control, which ensures higher precision near the target position and effectively minimizes the chattering induced by SMC. The shift from one control strategy to another should be smooth and gradual. This transition can be managed by a fuzzy supervisory system that generates the following switching law:

$$\begin{cases} U = \alpha U_{FL} + (1 - \alpha) U_{SM} \\ 0 \leq \alpha \leq 1 \end{cases} \tag{13}$$

where α is the factor generated by the fuzzy supervisory system based on the following set of rules:

- α takes the value "0" when the system is away from the reference trajectory.
- α takes the value “1” when the system is close to the target trajectory.

The proposed structure for the hybrid control system is presented in Fig. 3.

Figures 4 and 5 illustrates the membership functions for inputs and output of the fuzzy supervisory system, while Table 2 displays the fuzzy supervisory rules. The terms Z, M, L represent respectively Zero, Medium, Large and the terms PB, S, B represent Positive-Big, Small, Big.

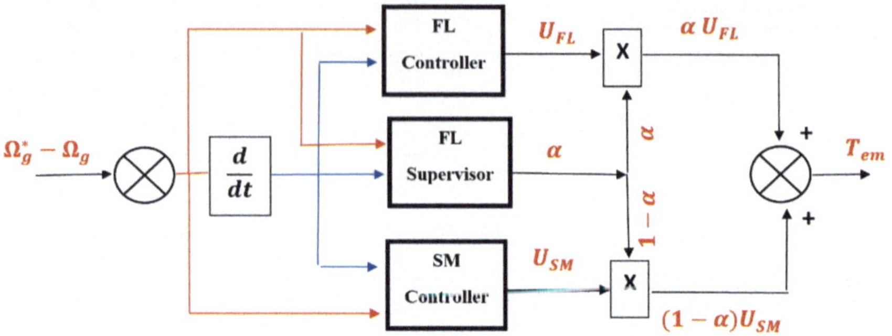

Fig. 3. Hybrid fuzzy sliding mode control structure

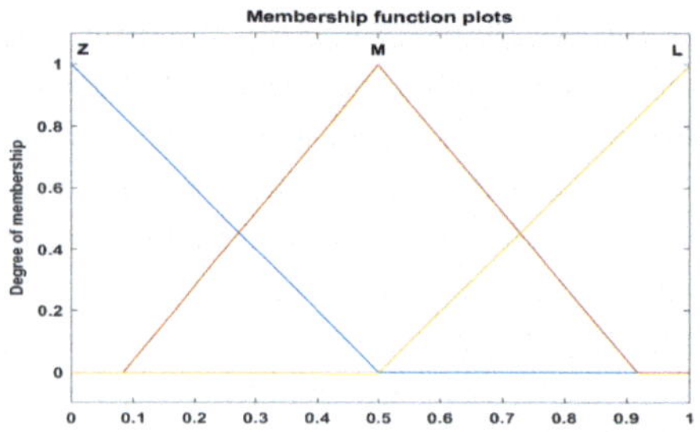

Fig. 4. Membership functions for e and Δe

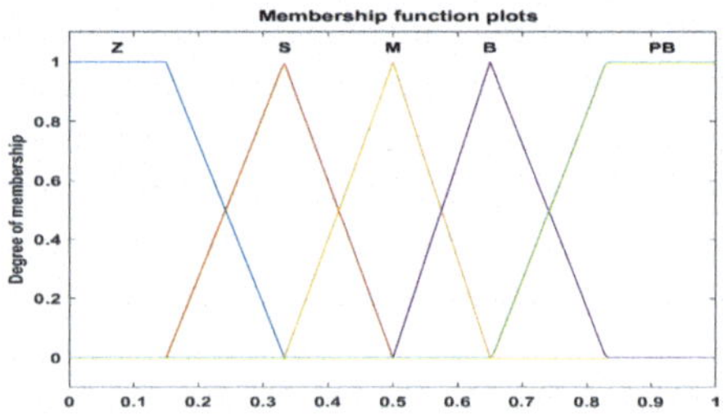

Fig. 5. Membership functions for ΔU_{FL}

Table 2. Rule bases of Fuzzy Supervisor

Δe \ e	**Z**	**M**	**L**
Z	PB	B	M
M	S	S	Z
L	Z	Z	Z

4 Simulation Results

This section focuses on evaluating and comparing the performance of the suggested HFSMC and the conventional SMC, focusing on reference tracking, system stability, and robustness. Simulations for both control approaches are performed using MATLAB/Simulink. The WECS parameters are listed in the Appendix.

In order to validate the effectiveness of the proposed control strategies and evaluate their performance, we carried out robustness and tracking tests. This involved applying the wind profile depicted in Fig. 6, where a change in the wind profile between 0s and 20s followed by a random range of 6 m/s–11.5 m /s.

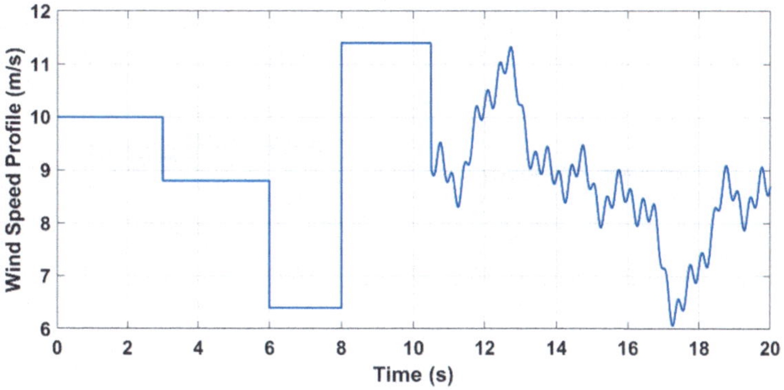

Fig. 6. Wind speed profile

To extract the maximum power generated, we employed the operating conditions cited in [8], where the power coefficient Cp attains its peak at C_{pmax}= 0.48 with a pitch angle $\beta = 0$ and an optimal Tip Speed Ratio (TSR), λ_{opt}= 8.1. As shown in Fig. 7, a comparison of the mechanical speed of the generator shaft under two different control methodologies is presented. We can clearly see that the generator shaft speed converges to a reference speed that adjusts proportionally with the wind speed profile.

The comparison depicted in Fig. 8 shows how the conventional sliding mode controller experiences significant ripples in tracking speed error. In contrast, the HFSMC showcases superior performance, effectively addressing the chattering phenomenon in

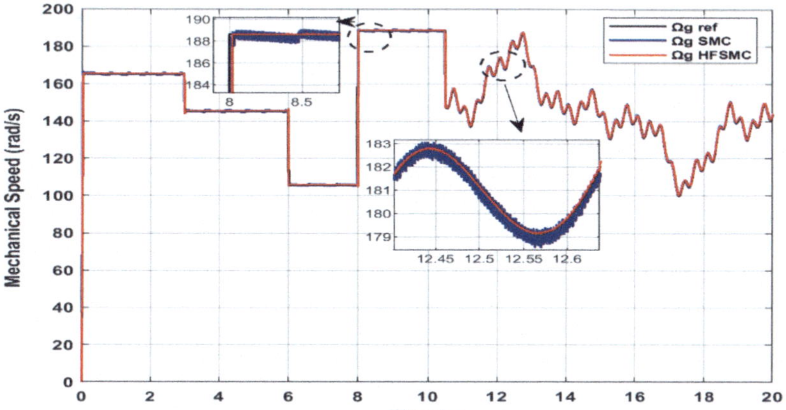

Fig. 7. Generator shaft speed

steady state by using fuzzy logic controller. And ensuring robust dynamic response during transient state in the presence of variations in wind profile using SMC.

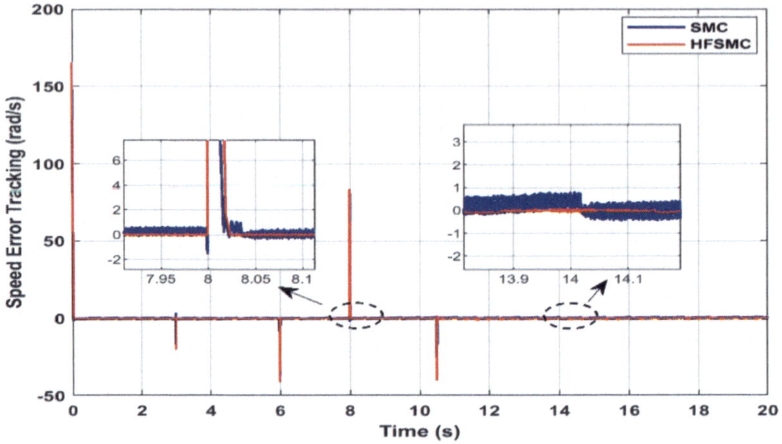

Fig. 8. Generator shaft speed error

In Fig. 9, we observe how the MPPT technique adeptly tracks optimal power points, ensuring that the power coefficient remains around its maximum value, $C_{pmax} \approx 0.48$, even during variations in the wind speed profile.

Figure 10 also reveals that the tip speed ratio is maintained near its optimal value, which is around $\lambda_{opt} \approx 8.1$.

Figure 11 shows the performance of the aerodynamic power, and we can clearly observe that it changes in correspondence with the changes in the wind speed profile.

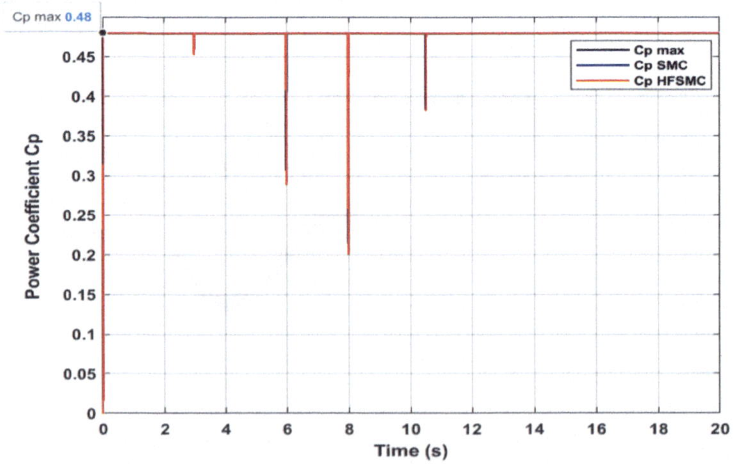

Fig. 9. Power Coefficient

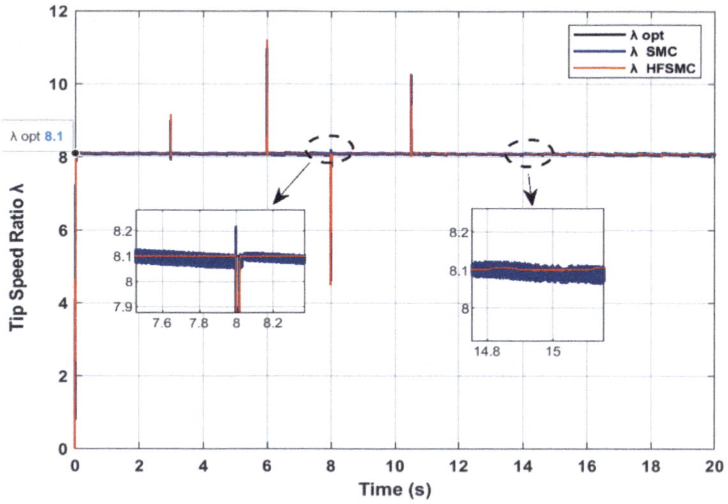

Fig. 10. Tip speed ratio

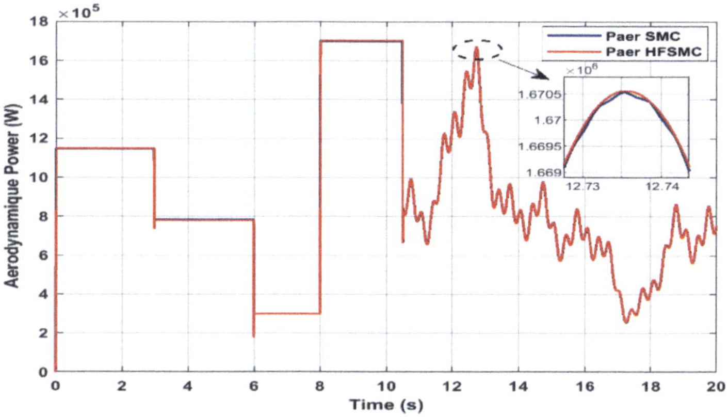

Fig. 11. Aerodynamic power response using SMC, HFSMC controller

5 Conclusion

In order to showcase the importance of fuzzy controllers in *system control enhancement,* a hybrid control approach is introduced to improve the speed regulation in TSR-based MPPT techniques for wind turbines. A comparative study was conducted through simulations to illustrate the enhancements achieved through the adoption of the novel control approach called hybrid fuzzy sliding mode control in comparison to conventional sliding mode control SMC. The results convincingly demonstrate the efficiency of HFSMC. The proposed HFSMC controller stands out as a major advancement over the SMC. It boasts superior qualities including precise reference tracking, high robustness against disturbances, elimination of chattering for smoother performance in steady state by using fuzzy controller which offers higher accuracy near the target position and effectively decrease chattering, and without forgetting faster dynamic response in transient state achieved by using SMC. Consequently, HFSMC establishes itself as the most effective and efficient control strategy for generated power tracking using MPPT techniques for wind turbines.

Appendix

See Table 3.

Table 3 Wind turbine parameters.

Parameters	*Value*
Air density ρ	1.22 kg/m^3
blade radius R	36.6 m
Maximum Power Coefficient C_{pmax}	0.48
Optimal Tip speed ratio λ_{opt}	8.1

References

1. Ahmed, H., Rajoriya, A.: A hybrid of sliding mode control and fuzzy logic control using a fuzzy supervisory switched system for DC motor speed control. Turk. J. Electr. Eng. Comput. Sci. **25**, 1993–2004 (2017). https://doi.org/10.3906/elk-1511-213
2. Kerboua, A., Abid, M.: Hybrid fuzzy sliding mode control of a doubly-fed induction generator in wind turbines
3. Zhang, X., Zhang, Z., Jia, J., Zheng, L.: A maximum power point tracking control method based on rotor speed PDF shape for wind turbines. Appl. Sci. **12**(18), 9108 (2022). https://doi.org/10.3390/app12189108
4. A new fuzzy design for switching gain adaptation of sliding mode controller for a wind energy conversion system using a doubly fed induction generator. Arch. Electr. Eng. (2024). https://doi.org/10.24425/aee.2023.143702
5. Sześciofazowym, P.-Z.: Sliding-mode direct field-oriented control
6. Hamayun, M.T., Edwards, C., Alwi, H.: Integral sliding mode control. In: Hamayun, M.T., Edwards, C., Alwi, H. (eds.) Fault tolerant control schemes using integral sliding modes, vol. 61, pp. 17–37. Springer International Publishing, Cham (2016). https://doi.org/10.1007/978-3-319-32238-4_2
7. Tir, Z., Soufi, Y., Hashemnia, M.N., Malik, O.P., Marouani, K.: Fuzzy logic field oriented control of double star induction motor drive. Electr. Eng. **99**(2), 495–503 (2017). https://doi.org/10.1007/s00202-016-0377-2
8. Nasiri, M., Milimonfared, J., Fathi, S.H.: Modeling, analysis and comparison of TSR and OTC methods for MPPT and power smoothing in permanent magnet synchronous generator-based wind turbines. Energy Convers. Manag. **86**, 892–900 (2014). https://doi.org/10.1016/j.enconman.2014.06.055

Evaluation of the Performance of a Savonius Micro Wind Turbine

Zohra Ouchiha(✉), Hemiche Iddou, Tarek Legouini, and Abdelhakim Hamrit

Faculty of Physics, LMFTA, MDF, USTHB, Bab Ezzouar, Algeria
zouchiha@yahoo.fr

Abstract. This work discusses the importance of harnessing wind energy to generate electrical power, especially in urban areas. The article presents experimental data collected from a prototype Savonius-type wind turbine and demonstrates that the performance can be improved by modifying the design and the structure, in order to better capture the kinetic energy of the wind. The paper also discusses various methods for evaluating the efficiency of wind turbines and highlights several previous studies that have proposed design changes to improve their performance. We conclude by emphasizing the usefulness of experimental data in teaching aerodynamics, particularly with regard to flow regimes and the appearance of vortices. Overall, it provides valuable insight into the design and performance of small-scale wind turbines and highlights their potential to provide a renewable source of energy in urban environment.

Keywords: Rotor Savonius · Wind tunnel · Reynolds number · specific speed · renewable energy

1 Introduction

Savonius vertical axis wind turbine (VAWT) has several advantages over horizontal axis wind turbine (HAWTs), notably: it is noiseless, it operates at low speeds of the wind, etc. However, low power generation is its main disadvantage compared to Darrieus and HAWTs models which are blade lift force dependent. Yadav et al. (2024) examined a possible mix of solar PV and wind energy plant to decarbonize the Biratnagar airport in Nepal. Tree-shaped wind turbines (TSWT) at 58.5 kW and 585 kW were used. They observed low CO_2 emission, when the wind speed operating at 9 m/s or more contributes to the electricity production. Multiple strategies are useful to predict the power coefficient variation with the Tip Speed Ratio. Experimental tests often necessitate resources that may not be consistently available, Mandal et al. (2020). The semi-empirical method which is based on the blade element momentum theory involves the treatment of empirical data using algorithms and also softwares like Blade and Xfoil. However, this approach requires an extensive array of empirical data and fails in accurately predicting wake/blade interactions as achieved by CFD. Hence, researchers lean more towards numerical approaches, Giraldo et al. (2004). Among the numerical techniques solving the Navier-Stokes equations, we find: - The Six Degrees of Freedom method, utilized for wind turbines to anticipate parameters like cut-in, cut-out, and

© The Author(s), under exclusive license to Springer Nature Switzerland AG 2025
M. Hatti (Ed.): IC-AIRES 2024, LNNS 1238, pp. 741–749, 2025.
https://doi.org/10.1007/978-3-031-80301-7_77

nominal power, Wicaksono et al. (2021) - Sliding mesh and Overset mesh, tailored for wind turbine which is examined in wind tunnels, Laws et al. (2020). The effectiveness can be determined either, by a semi-empirical approach, or by the computational fluid dynamics CFD method, Cuevas-Carvajal et al. (2022). Maldonado et al. (2014) used the finite element method and CAD software. They recommended VAWT for the electric power supply in urban areas where the wind blows at 5 m/s. Muneer et al. (2015) studied a three-blade model which is sketched using AutoCAD and a CFD analysis. They found an effect of great drag coefficient for a turbine having a big surface. Wenehenubun et al. (2015) explored the effect of the number of the blades. Best results are obtained with a three blades model corresponding to high λ, and with a four blades device for a low λ. Zemamou et al. (2017) focused on the lift and drag forces effect. In addition, panels have been grafted onto the rotors. Jaohindy (2012) presented a model called ''Turby'' which is sketched with CFD technique for urban use on high buildings of at least six floors, and it is also functional on bridges, dykes, etc. The system operates between 2 m/s and 20 m/s of wind's speed. Kaya et al. (2024) harnessed Tagushi method to study the efficiency of a two bladed Savonius rotor which is involved in the hydrogen production system. The simulation with MATLAB – Simulink model showed that 2.423 kg of hydrogen is produced in a year using a single turbine, while an amount of 57.58 kg is reached in a year with ten turbines. Brusca et al. (2022) examined the performance of a ducted Savonius turbine as a function of overlap ratios. Uniform air velocity at 5 m/s, and several turbine overlap ratios: 1/6, 1/3 and 1/2 have been considered. Saha et al. (2008) stated that the optimum Savonius turbine number of blades is 2 as the concept is simple, low cost and may be widely spread in an urban area. The work hereby is complementary to the analysis of the results obtained on a Savonius rotor, now called "Lhob", and which was carried out as part of a master's thesis Legouini and Hamrit (2016). The objective in this study is to present a high-performance wind device that is simple and efficient to operate for the greatest comfort of its users, and also to make it a practical tool of fairly small dimensions that can be associated with a battery, for example.

2 Aerodynamics of VWAT Model

Relative motion and acting forces on the rotating (anticlockwise motion) blade are sketched in the Fig. 1 for the model of Darrieus, Iddou et al. 2024. The relative wind velocity vector is written as: $\overrightarrow{W} = \overrightarrow{V_\infty} - \overrightarrow{V_{rot}}$.

Where V_{rot} is the rotational velocity of one blade and V_∞ is the wind's free stream velocity. The blades rotate under the resultant, $\overrightarrow{F} = \overrightarrow{F_N} + \overrightarrow{F_T}$. Where F_T (tangential) and F_N (normal) are shown in the Fig. 2. F_T depends on the drag force $\overrightarrow{F_D}$, the lift force $\overrightarrow{F_L}$ and the effective angle of attack α_{eff} where: $\overrightarrow{F_T} = \overrightarrow{F_D}\sin(\alpha_{eff}) - \overrightarrow{F_L}\cos(\alpha_{eff})$.

The torque $\overrightarrow{C}$ generated by a single blade is: $\overrightarrow{C} = \overrightarrow{R} \wedge \overrightarrow{F_T}$ (The radius vector $\overrightarrow{R}$ is perpendicular to $\overrightarrow{F_T}$).

Accordingly, the torque generation is impacted positively by $\overrightarrow{F_D}$ and negatively by $\overrightarrow{F_L}$.

The torque coefficient $C_t(\theta)$ is given by: $C_t(\theta) = \frac{\overrightarrow{C(\theta)}}{^1/_2\rho SRV_\infty^2}$ and the power coefficient Cp is$C_p = \lambda\overline{C_t(\theta)}$.

λ is the tip speed ratio TSR which is representing the ratio of V_{rot} on V_∞:$\lambda = \frac{V_{rot}}{V_\infty}$.

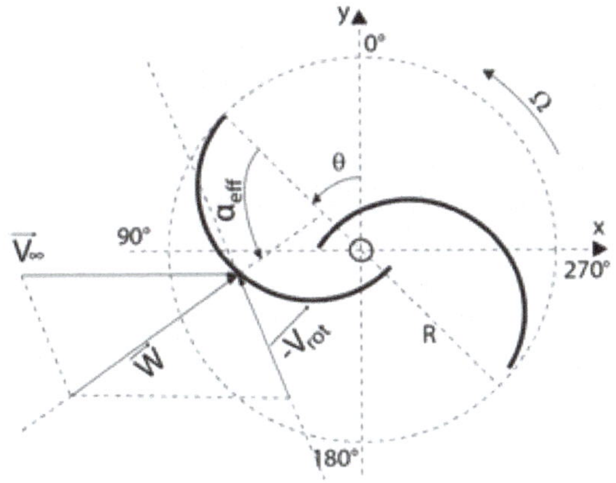

Fig. 1. Relative wind speed W and the angle of attack on a blade element

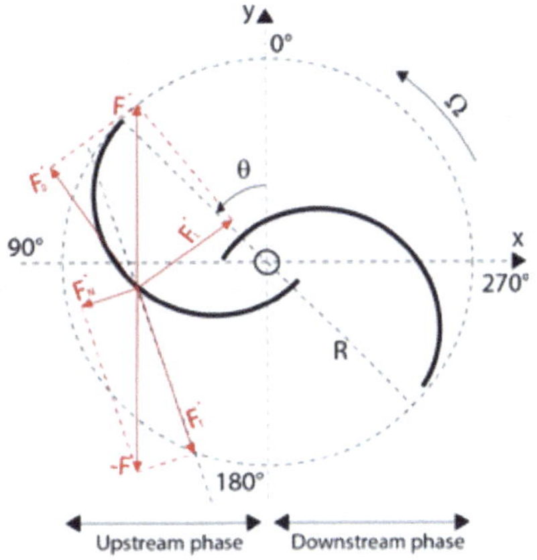

Fig. 2. Aerodynamic forces exerted on the dynamic blade element

3 Experimental Structure

A Savonius model of VAWT is an assemblage of two equally half parts of a cylinder of height H = 30 cm and diameter D = 11.5 cm, which are screwed to a metallic rod. The Fig. 3 depicts an horizontal section of the rotor of S-shaped blades which are settled in reverse and without contact ensuring an aspect ratio a = 3 cm, (Legouini and Hamrit 2016). The wind's action result in a drag force F which is expressed as:

$$F = \frac{1}{2}\rho V_\infty^2 SC_f \tag{1}$$

According to the diagram of the Fig. 3, the coefficient C_f is approximately equal to 1.7 considering both of favorable and unfavorable air currents crossing the rotor. The power P is proportional to the couple C_t of the force F acting on the rotor and its angular speed ω. It is calculated as follow:

$$P = C_t \cdot \omega \tag{2}$$

The magnitude of the torque Ct which is generated by a single blade is given by:

$$||\vec{C}_t|| = C_t = FR \tag{3}$$

F depends on certain parameters such as the air density ($\rho = 1.2175\ kg/m^3$) and its speed V_∞, the drag coefficient C_f and the projected surface S.

The power coefficient C_p is defined as:

$$C_P = \frac{P}{1/2(\rho S V_\infty^3)} \tag{4}$$

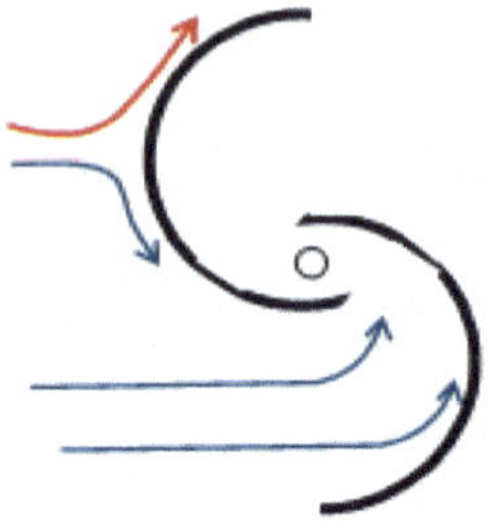

Fig. 3. Diagram showing the air streamlines crossing the rotor

4 Results and Discussion

Series of tests with the device Lhob have been achieved to address appropriately the aerodynamical behavior of the rotor in its surrounding environment. Figure 4 (a) illustrates the variation of V_{rot} versus Reynolds number Re, $R_e = \frac{V_d d}{\nu}$. This parameter depends of the diameter d of the wind tunnel's outlet diaphragm, the wind speed V_d (in meter per second), and the kinematic viscosity ν that is given in meter square per second. Increasing curves are observed whatever the wind turbine is distanced to the wind tunnel, (x = 0.1 m, 0.2 m, 0.3 m, 0.4 m, 0.5 m ou bien 0.6 m). The profiles seem to obey polynomial laws of different degrees varying between 3 and 5. At high wind speeds and therefore for larger Reynolds (Re) values, most of the profiles tend to become linear up to a maximum of the rotational speed at Re = 112,500, and they decrease as x increases from 0.2 m to 0.6 m. At x = 0.1 m and 0.4 m, a decrease in the rotational speed values is observed. The Reynolds number lies between 50,000 and 130,000, at the diaphragm of the wind tunnel outlet, attesting that the flow regime is turbulent at this position. The turbulence increases as we get closer to the wake surrounding the blades. It would also increase measurement errors using the Pitot tube near the rotor. The Fig. 4 (b), represents the smoothing curves of V_{rot} versus Re. At x = 0.2 m, the profile follows a polynomial law of order 5 (Eq. 5), while at x = 0.4 m, the profile tends towards a polynomial of order 4 (Eq. 6). The respective correlation equations are:

$$\begin{aligned} V_{rot} = &- 16.23 + 8.66 \times 10^{-4}(Re) - 1.77 \times 10^{-8}(Re)^2 + 1.76 \times 10^{-13}(Re)^3 \\ &- 7.54 \times 10^{-19}(Re)^4 - 9.66 \times 10^{-25}(Re)^5 \end{aligned} \tag{5}$$

and $V_{rot} = -3.66 + 2.31 \times 10^{-4}(Re) - 5.66 \times 10^{-9}(Re)^2 + 6.36 \times 10^{-14}(Re)^3 - 7.36 \times 10^{-19}(Re)^4$ (6)

The coefficients of the monomials are approximated with negligible margins of error, and also the coefficients of the monomials of the highest degrees, namely: $7.54 \times 10^{-19}(Re)^4$ et 9.66×10^{-25} $(Re)^5$, just like $7.36 \times 10^{-19}(Re)^4$, for x = 0.2 m and x = 0.4 m respectively, give quite low values of V_{rot}. This is consistent with Betz's theory stating that the power coefficient of the wind turbine is variable with respect to the cube power of the wind speed,V_{∞}^3. The Fig. 5 represents the variation of the tip speed ratio TSR as a function of Re. The profile is non-linear and the maximum value does not exceed $\lambda = 0.3$ which is reached at Re $\approx$115000. The graphs of V_{rot} varying with λ (the Fig. 6) can easily fit a polynomial equation. For x = 0.6 m, far from the outlet air section of the wind tunnel there is a notably lower variations. The profiles displayed in the Fig. 7 (a and b) show that V_{rot} and λ are both increasing as the wind power P increases. From zero to nearly 114 W, TSR is approaching 0.29 while the rotational velocity V_{rot} value is limited to 3.69 m/s. V_{rot} increases while P remains constant at the great power values for the positions x = 0.2 m and x = 0.3 m. At x = 0.4 m and x = 0.5 m the wind power do not exceed 88 W. In fact, for low powers both of TSR and Vrot increase two ways: - by ascending with the power then – by an abrupt increase to a maximum value, TSR$\approx$0.29 and $V_{rot} \approx 3.69$ m/s corresponding to the graphs at x = 0.3 m and 0.4 m respectively. Far from the wind source, at x = 0.6 m, V_{rot} increases and do not exceed 2.70 m/s approximately, while the TSR parameter increases sharply at the lowest first values of the power, then remains constant from zero to nearly 38 W. Comparatively, the profiles of TSR and V_{rot} versus $P_0(W)$ of the wind emerging from the wind tunnel outlet do not show reversion directed to the low powers as it is illustrated in the sub-graphs embedding in the Fig. 7 (a and b) respectively. Therefore, the perturbations appearing in the profiles of TSR and V_{rot} in the Fig. 7 (a and b) are attributed to the uncertainties appearing in the operator measurements, added to the wake flow in the rotor vicinity. A decrease of the wind's power along the distance x, is illustrated in the Fig. 8, where the greater values are displayed around x = 0.1 m, 0.2 m. Similar behavior is remarkable for an array of velocities delivered by the wind tunnel according to the graduations: 50 – 60 – 70 – 80 – 90 and 100. However, over the graduation 60, unavoidable perturbations appear in the profiles starting from the position x = 0.3 m. The graphs show irregularities owed to the perturbations close to a strong wake surrounding the rotor. The wind's power represent a branch of polynomial function of degree 3 according to the velocity variations, $P(W) = f(v^3)$ as it is observed in the Fig. 9. Bouanani et al. (2020), reported on the troposkein type and H type of VAWT. Best performances appear for the first type with a 120 W of power and 18 m/s of the wind speed. Our results are comparable with those of Bouanani et al. (2020) in the ascending shape and the power varying between 0 to 120 W when the velocity embeds between 0 to 14.85 m/s. The power and the torque coefficients C_p and $C_t = \frac{C_p}{TSR}$ which are also defined by Yahya et al. (2021) as follow:

$$C_{pth} = \left(\frac{\rho A\left(v_{upw}^2 - v_{dow}^2\right)\left(v_{upw} + v_{dow}\right)}{4}\right) / \left(\frac{\rho A v_{upw}^2}{2}\right).$$

Where v_{upw} and v_{dow} are respectively the velocities in front and behind the rotor, and the ratio of v_{dow} over v_{upw} is denoted$\alpha = \frac{v_{dow}}{v_{upw}}$. Hence, C_p is rewritten as: $C_{p_{th}} = \frac{(1-\alpha^2)(1+\alpha)}{2}$.

The theoretical coefficient $C_{p_{th}}$ as a function of the variable α is plotted in the Fig. 10. An enlargement is also shown as a comparison between $C_{p_{th}}$ and our present work at two velocities corresponding to the graduations 20 and 40 of the wind tunnel. These data are recorded in an intern report of the master student work of Saidj F. under the advisement of Ouchiha Z. Almost all our measurement campaigns using 'Lhob' device led to $C_{pth} \approx 0.1$. In the Fig. 11, the rotational angle θ(t) increases linearly with time for any angular velocity ω. It appears on these graphs that the wind speed increases while its effects on the rotation cycles diminish. This is all the more noteworthy on the last two straight lines which pile up for ω= 422.92 rpm or ω = 426.70 rpm with the same slope. Indeed, the rotations of the rotor do not increase indefinitely as a response to the wind speed increase. The rotor performance is depending of its structure, like the blades shape, the material of the rod itself, etc.

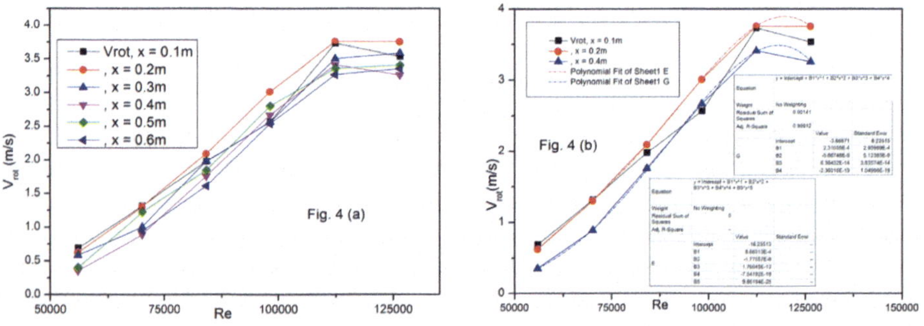

Fig. 4. A) Rotational speed V_{rot} versus Re, (b) Smoothing functions at the positions x = 0.2 m and x = 0.4 m

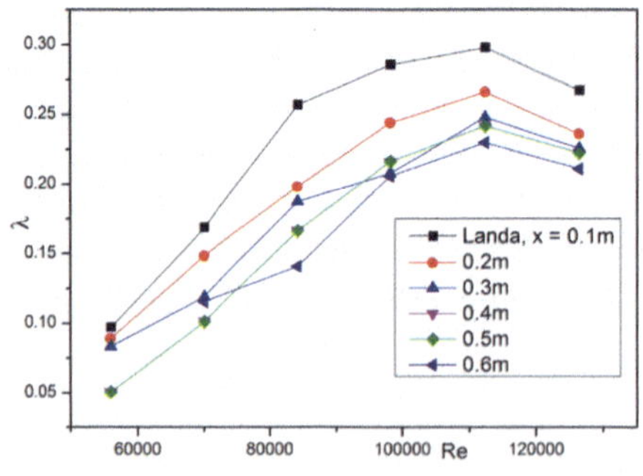

Fig. 5. TSR λ versus Re

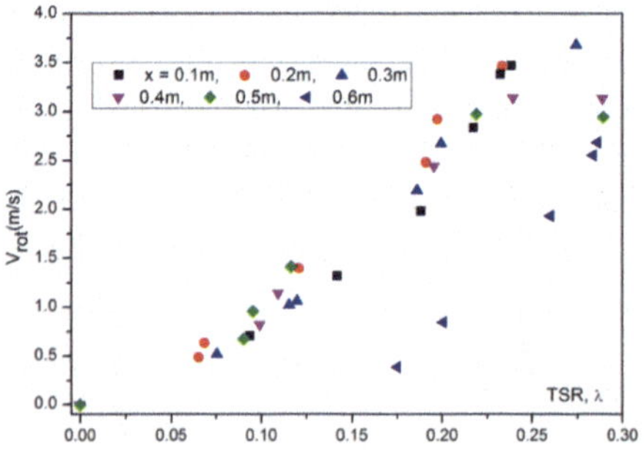

Fig. 6. Rotational velocity versus λ

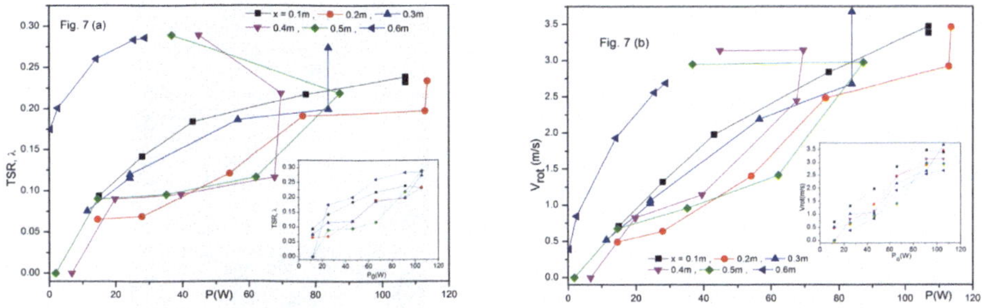

Fig. 7. Variations of (a) TSR λ and (b) V_{rot} with the wind power, P(W)

An amount of energy is dissipated behind the rotor in the third quarter part of the horizontal section view of the Fig. 12, as it is shown by the smoke passing through the rotor. It happens approximately in a region embedded between the angles 270° to 360°. Reinjection of the wasted energy, for example by an additional panel redirecting the air leakage into the wind turbine as it rotates is widely used. However, it may be better to add on the air leakage trajectory a smaller subsidiary rotor, with appropriate dimensions having the ability to work just with the air leaking behind the principal rotor. This pack of two or more rotors would contribute simultaneously to the same VAWT system used for the electricity production. The aerodynamics around a modified NACA 0021 aifoil profile which is created using CST method have been studied by Davendeh and Maghrebi (2023). They found that the optimization of the leading-edge radius reduces the formation of the vortex in the wake flow close to the rotating turbine.

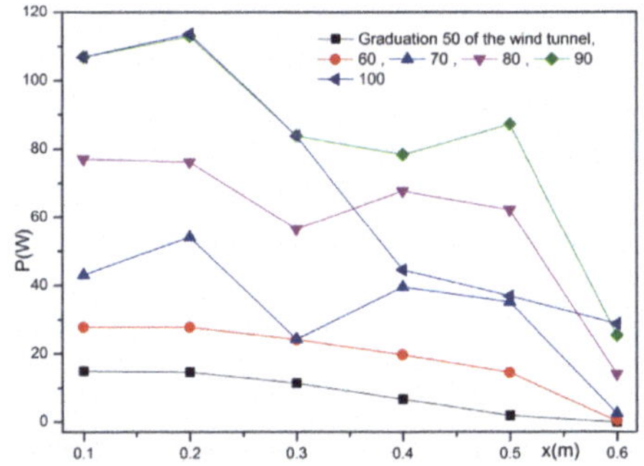

Fig. 8. Wind's power evolution along x

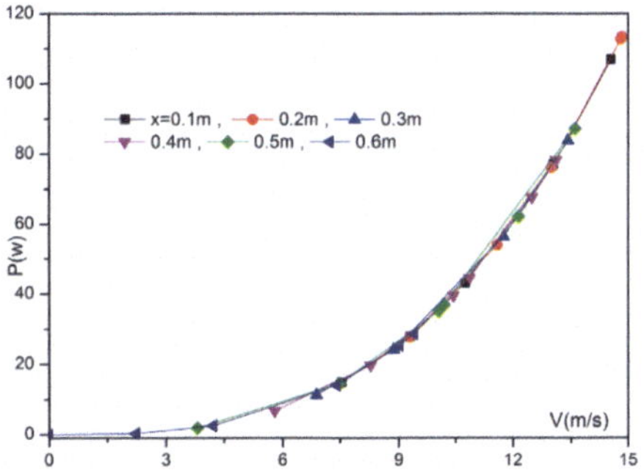

Fig. 9. Wind's power evolution versus the velocity

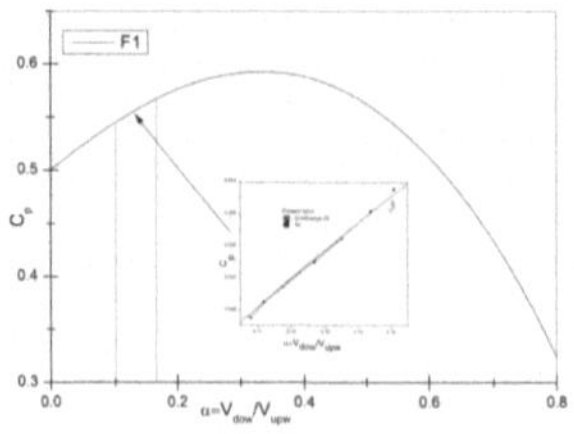

Fig. 10. Theoretical power coefficient, Cp

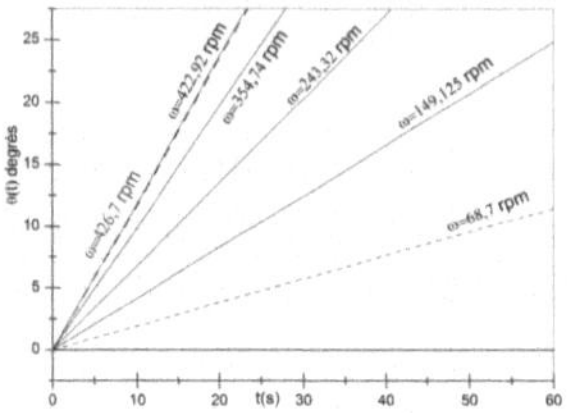

Fig. 11. Variation of the angle of rotation with time

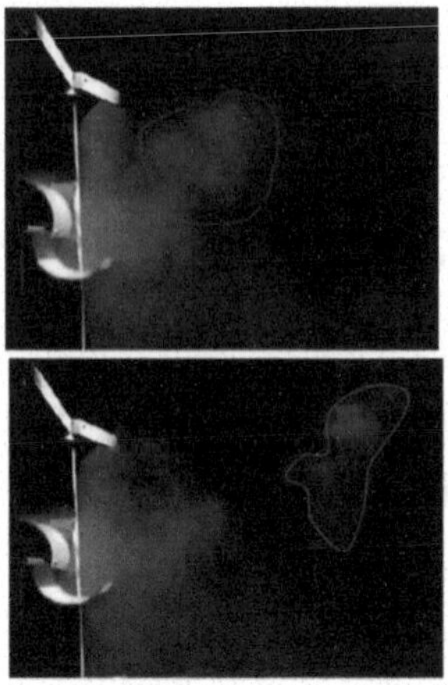

Fig. 12. Two horizontal section views of the cylindrical rotor (The energy dissipation during the rotation is bounded by a blue line) (Color figure online)

5 Conclusion

This work is the synthesis of some important results obtained on a vertical axis wind turbine designed with recycled materials, at the Faculty of Physics (USTHB-Algiers University). The data are to some extent satisfactory and open a door to further promising exploration in this research area. Some experimental results, especially given at the distances 0.3 m and 0.6 m need check-in tests owed to a discard compared to the general behavior of most of the physical parameters such as the pressure, the wind power, etc. Many parameters of this wind turbine can be the subject of a study to produce a maximum power of electricity without environmental nuisance. This model which is a pioneer in the Faculty of Physics, could lead to the development of other perfected models.

References

Bouanani, A., Bouchaour, M., Merad, L.: Modeling and simulation of the vertical axis wind turbine by Qblade Software. Algerian J. Renew. Energy Sustain. Dev. **2**(02), 181–188 (2020)

Brusca, S., Galvagno, A., Lanzafame, R., Mauro, S., Messina, M.: How to increase savonius power coefficient: ducted rotor performance with different overlap ratios. J. Phys. Conf. Ser. **2385**, 012105 (2022)

Cuevas-Carvajal, N., Cortes-Ramirez, J., Norato, J.A., Hernandez, C., Montoya-Vallejo, M.: Effect of geometrical parameters on the performance of conventional Savonius VAWT: a review. Renew. Sustain. Energy Rev. **161**, 112314 (2022)

Davandeh, N., Maghrebi, M.J.: Leading edge radius effects on VAWT performance. J. Appl. Fluid Mech. **16**(9), 1877–1886 (2023)

Giraldo, M., Florez, W., Isaza, C.: Modelling and simulation of a Savonius rotor using the boundary element method. WIT Trans. Modell. Simul. **37** (2004)

Iddou, H., Bouda, N.N., Zereg, K.: Computational fluid dynamics study on the efficiency of straight-bladed vertical axis wind turbine. Int J Thermofluids 22, 100672 (2024). ISSN 2666-2027. https://doi.org/10.1016/j.ijft.2024.100672

Jaohindy, P.: Modélisation des systèmes éoliens verticaux intégrés aux bâtiments: modélisation du couple production. Bâtiment Université de la Réunion (2012)

Kaya, A.F., Morselli, N., Puglia, M., Allesina, G., Pedrazzi, S.: Design optimization of two-blade Savonius wind turbines for hydrogen generation. Renew. Energy **231**, 121028 (2024)

Laws, P., Saini, J.S., Kumar, A., Mitra, S.: Improvement in Savonius wind turbines efficiency by modification of blade designs—a numerical study. J. Energy Res. Technol. **142**(6), 061303 (2020)

Legouini, T., Abdelhkim, H.: Etude et réalisation d'une éolienne à axe vertical. USTHB (2016)

Maldonado, R., et al.: Design, simulation and construction of a Savonius wind rotor for subsidized houses in Mexico. Energy Procedia **57**, 691–697 (2014)

Mandal, A.K., Rana, K.B., Tripathi, B.: Performance improvement of Savonius turbine by design modifications: a review. Int. J. Renew. Energy Technol. **11**(4), 297–318 (2020)

Muneer, A., Khan, M.B., Sarwar, U.B., Khan, Z.A., Badar, M.S.: CFD analysis of a Savonius vertical axis wind turbine. Power Generation System and Renewable Energy Technologies (PGSRET) (2015)

Saha, U., Thotla, S., Maity, D.: Optimum design configuration of Savonius rotor through wind tunnel experiments. J. Wind Eng. Ind. Aerodyn. **96**(8–9), 1359–1375 (2008)

Wenehenubun, F., Saputra, A., Sutanto, H.: An experimental study on the performance of Savonius wind turbines related with the number of blades. Energy Procedia **68**, 297–304 (2015)

Wicaksono, H., Susilo, S.H., Pranoto, B., Fakhruddin, M.: Initial rotation characteristic investigation of a hybrid savonius-darrieus wind turbine using 6 DOF computational fluid dynamics. Mekanika Majalah Ilmiah Mekanika **20**(1), 9–17 (2021)

Yadav, B.K., Yadav, R., Jahangiri, M., Priya, S.S., Bajracharya, T.R., Sudhakarf, K.: Decarbonizing airport using solar and wind farm: a case of Biratnagar, Nepal. e-Prime Adv. Electr. Eng. Electron. Energy **8**, 100583 (2024)

Yahya, W., Ziming, K., Juan, W., Qurashi, M.S., Al-Nehari, M., Salim, E.: Influence of tilt angle and the number of guide vane blades towards the Savonius rotor performance. Energy Rep. **7**, 3317–3327 (2021)

Zakaria, A., Ibrahim, M.: Effect of twist angle on starting capability of a Savonius rotor–CFD analysis. IOP Conf. Ser. Mater. Sci. Eng. **715**, 012014 (2020)

Zemamou, M., Aggour, M., Toumi, A.: Review of savonius wind turbine design and performance. Energy Procedia **141**, 383–388 (2017)

Design and Development of Low-Power Output Vortex Tower Model

A. Bouam(✉), A. L. Cheridi Deghal, A. Dadda Khorsi, A. Hadjam, A. Dahia, N. Koudiah, K. Attari, and L. Bouam

Nuclear Research Centre of Birine, P.O. box 180, Ain-Oussera, Algeria
a.bouam@crnb.dz

Abstract. The realization of the vortex tower prototype was completed with the objective of developing a comprehensive analytical and experimental model to evaluate its performance. This model was intended to assess the performance and power output of the 1/5 scale prototype and to investigate the impact of various ambient conditions on its power output. To eliminate the influence of solar energy on the test bench's performance, several tests were conducted during non-sunny periods. The results indicated that a decrease in ambient temperature accelerates the ascending air, due to differences in flow density, which in turn enhances the equipment's output. Improving the performance of this prototype on a real scale will contribute to the advancement of energy production systems in the country.

Keywords: Vortex tower · Artificial vortex · Ascendant air flow · Thermodynamic cycle

Nomenclatures

Variable	Designation
T	Temperature [°C]
p	Pressure [K Pascal]
U	Velocity [m/s]
t	Time [h]
H	Height [mm]
D	Diameter [mm]
L	Width [mm

Greek variables

φ	Relative humidity [%]
η	Thermal efficiency [–]
Indexes	Designation
Amb	Ambient
Adm	Upstream of the tower (Tower inlet)
C	Chimney
C.Cter	Chimney center

© The Author(s), under exclusive license to Springer Nature Switzerland AG 2025
M. Hatti (Ed.): IC-AIRES 2024, LNNS 1238, pp. 750–757, 2025.
https://doi.org/10.1007/978-3-031-80301-7_78

C.Int	Chimney entrance
C.Out	Chimney outlet
Srce	Hot spring (Source)
Srce.Sf	External surface of the hot source
T	Turbine

1 Introduction

Using a controlled vortex, a Vortex Engine (VE) utilizes the mechanical energy generated during the upward movement of heat by atmospheric convection. By tangentially admitting warm or wet air into a circular arena, a tornado -like whirlwind is formed. The attached convective vortex is created when humid, warm air rises and is rotated by tangential inlets. The convection work is captured by turbines located at the level of the inlet of the chimney (outlet of the convergence chamber), where the speed is highest. The heat source can be hot water or waste heat.

With the exception of replacing the solar chimney's actual tube with a vertical upward vortex, the VE's thermodynamic foundation is the same as that of a solar power plant. It is not necessary to have a solar collector – The solar collector is replaced by a cylindrical tank considered as a hot source. By using less fuel to produce the same amount of energy, the vortex engine helps to mitigate global warming.

Many researchers from around the world have presented various conceptions and configurations for a VE. The idea of using the Vortex air flow to generate electricity was proposed for the first time in 1975 by the Canadian engineer Louis M. Michaud through the publication of his project "Vortex Power Station" in the Bulletin de l'Enérican Meteorological Society [1, 2]. He suggested a method to capture the energy generated by the vortex because when the heat is transported up by convection, it turns into work.

The generated vortex must enter the atmosphere vertically. To reach this ascending movement, several Tour technologies have been proposed and tested. Mamulashvili [3] proposed a combination between the wind process and a solar tower. Wind energy is received by a large-section solar sensor then it is driven to the entrance slots tilted at the base of the chimney. The author indicates that this configuration can generate a swirling air flow quickly inside the chimney and can produce 100 Mwe. In 2005, a generation of convective whirlwind was presented by Takhar and Beg [4]. An electrical energy generator with solar vortex was presented by Dunn [5]. The device consists of a flat solar sensor with slit plates to cancel the harmful effects of the wind and if infrared reflection would increase the effectiveness of the solar sensor.

In order to power generators, the hot air is sucked under the collector and transmitted through a set of moving blades to an air turbine located in the center of the tower. Dunn has proven that it is possible to produce 640 MW of electricity with only 25% efficiency using a square mile of thin steel sheet in a desert. In 2008, Haan et al. [6] presented and built a lab tornado simulator to produce whirlwinds. The simulator uses a rotating forced descending current technique which is generated by two concentric conduits which form a channel with a fan 1.83 m in diameter in the center. The hot air circulates between the canal and the surface of the base and the swirling flow is created by means of fins in the roof of the inner cylinder. The authors indicate that this simulator can produce a Tornado whirlwind with rays ranging from 0.23 to 0.56 m and maximum tangential speeds ranging from 6.9 to 14.5 m/s. Michaud [7] built in 2009 a VGS based on mechanical energy in which is produced via a controlled vortex in the form of a tornado generated at the base of a cylindrical wall which sucks tangentially to produce a convective vortex and acts like a dynamic fireplace.

From this bibliographic study, we offer a thermodynamic and experimental analysis of a Vortex Tower Prototype (VTP), on a laboratory level (1/5), to examine the effect of various ambient conditions on the power output. The measurements of the various parameters along the tower are made by an acquisition system installed for this equipment.

2 Material and Methods

2.1 Description and Installation of the VTP

Small-scale circulations, similar to tornadoes, can occur near any intense surface heat source in the sense that it is a vertical swirling column of air, Fig. 1a. The warm, ascending air column may create a small, detectable cyclonic motion close to the ground if there is sufficient low-level wind shear. Because they occur under favorable weather conditions and have no cloud association, they are not regarded as tornadoes. From these natural phenomena the researchers reflected on the vortex tower. The mechanical energy generated by natural upward convection is captured by the vortex tower through the construction of an artificial vortex. A semi-open roofed cylindrical enclosure at the installation base allows hot air to be admitted tangentially, creating the vortex. The heat source used is geothermal energy or a hot source installed at the base of the tower. Figure 1b illustrates how the airflow enters the lower structure of the tower and passes through the heat source before entering the convergence chamber. When hot air is admitted tangentially to the base of a vertically oriented cylindrical wall, a convective vortex is created that functions as a dynamic stack. Mechanical energy would be extracted from the vortex using a mini-turbine installed at the entrance to the chimney.

Fume could be injected into the tangential inlets to be carried by the air through the chimney in order to visualize the vortex at the outlet. The turbines are powered by the pressure differential between the base of the vortex and the surrounding air. In order to carry out a performance analysis of a vortex tower, a small-scale prototype vortex tower was carried out Fig. 1c. The tower is built and located at the Birine Nuclear Research Centre. A photograph of the system is shown in Fig. 1c.

The geometric parameters of the vortex tower are outlined in Table 1.

2.2 Thermodynamic Analysis

The examination of the thermodynamic process inside the zones, including the environment, the collector, the turbine, and the chimney, is presented at Fig. 2 according to Michaud [7]. Beginning at the manifold inlet, the turbine and stack are successively passed through by the operating fluid, finally it discharges the energy into the environment and flows again back to the collector inlet.

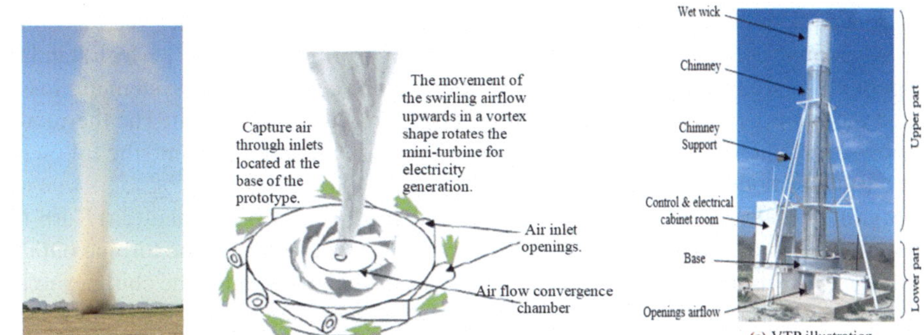

(a) Natural vortex photograph. **(b)** Schematic diagram of the vortex. **(c)** VTP illustration.

Fig. 1. Presentation of the vortex tower and its operating principle (a), (b) and (c).

Table 1. Geometric parameters

Component	Parameter	Values	Component	Parameter	Values
Chimney	Height Interior diameter	7600 mm 720 mm	Heat transfer chamber	Height Diameter	530 mm 2200 mm
Hot Source	Height Diameter 1 Diameter 2	200 mm 1500 mm 300 mm	Air inlet	Number Height Width	08 560 mm 1200 mm
Convergence chamber	Height Diameter	300 mm 1692 mm	Guide vane	Length Width	90 cm 30 m

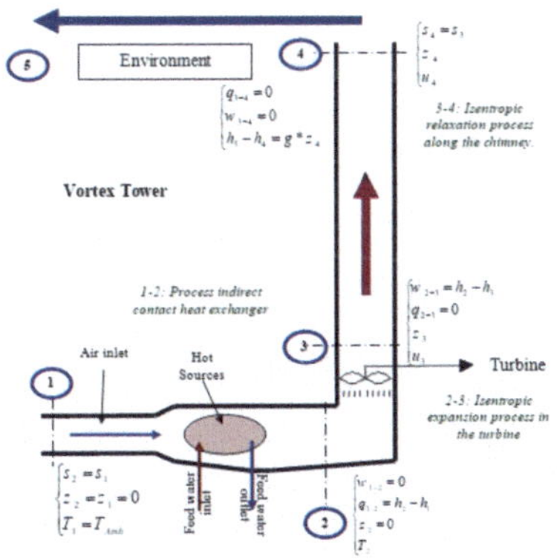

Fig. 2. Diagram of the thermodynamic process of a VTP.

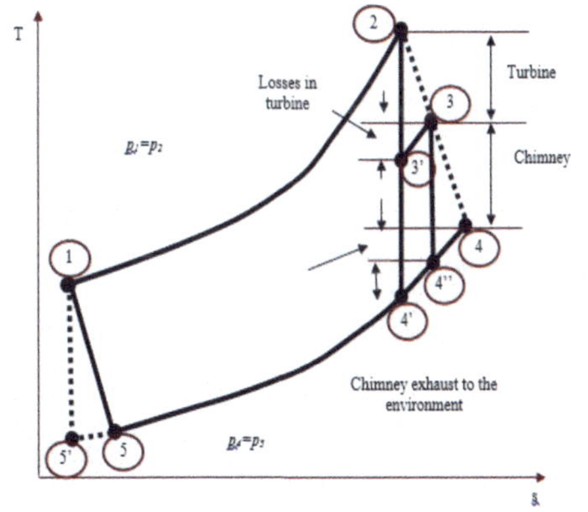

Fig. 3. Temperature diagram-air entropy in the VTP.

Figure 3 shows the schematic circuit of the VTP thermodynamic process and lists the similarities to the gas turbine cycle. Due to these similarities, the analysis is much simplified because it makes use of current gas turbine theory [8]. The goal of the analysis is to determine a correlation between device performance and variables such as chimney height and collector temperature rise. This figure illustrates the perfect standard temperature-entropy diagram of the air in the system, containing all systematic losses except the unimportant macroscopic kinetic energy of the chimney exit. The thermodynamic cycle of the working fluid can be, based on analysis, simplified into the following four basic thermodynamic processes:

Significant air circulation state points are defined as: (1) Air intake (2a) Turn the turbine after it crosses the hot source (2b) Chimney intake, it is also the turbine outlet's condition (3) Chimney outlet (4) The condition of the environment at the chimney exit's size. The work delivered by the turbine of the VTP system can be written as follows [8]:

$$w_{\text{Shaft},i} = h_2 - h_{3'} \tag{1}$$

The energy used to increase the potential energy of the engine fluid when passing through the chimney is:

$$w_{P,i} = h_{3'} - h_{4'} \tag{2}$$

As we have analyzed above, the 5–1 process occurs in the environment, which will not cost a technical work delivered by the system. However, the energy consumed by the 3′-4 process, which does not come from the technical work of the system, will be approximately equal to that of the 5–1 process. Therefore, most of the technical work in the ideal cycle 1-2-3′-4′-5-1 is used by increasing the potential energy of the operating fluid through the fireplace, leaving a minimum proportion to be produced by the axis work by the turbine installed at the lowest part of the chimney. Cycle 1-2-3-4-5-1 represents the real irreversible Brayton cycle for the VTP system, including turbine and chimney losses. The work delivered by the turbine for the VTP system can be written as follows [8, 9]:

$$w_T = h_2 - h_3 = c_p * T_2\left[1 - \left(\frac{p_3}{p_2}\right)^{(k-1)/k}\right] \tag{3}$$

2.3 Thermal Efficiency of the VTP

The thermal efficiency of the VTP can be expressed as follow, [8–10]:

$$\eta = \frac{T_2}{T_2 - T_1}\left[1 - \left(\frac{p_3}{p_2}\right)^{(k-1)/k}\right] \tag{4}$$

or:

$$\eta = \frac{(T_1 + \Delta T)}{\Delta T}\left[1 - \left(1 - \frac{\Delta p_{tur}}{p_1}\right)^{(k-1)/k}\right] \tag{5}$$

Δp_{tur} represents the pressure drop through the turbine:

$$\Delta p_{tur} = p_2 - p_3 = p_1 - p_3 \tag{6}$$

While ΔT represents the increase in the temperature of the work fluid inside the collector:

$$\Delta T = T_2 - T_1 \tag{7}$$

3 Results and Discussions

Figure 4 shows ambient and outlet temperature profiles during the test periods. The ambient temperature of air is measured at the inlet and the outlet air temperature is measured just at the outlet of the chimney.

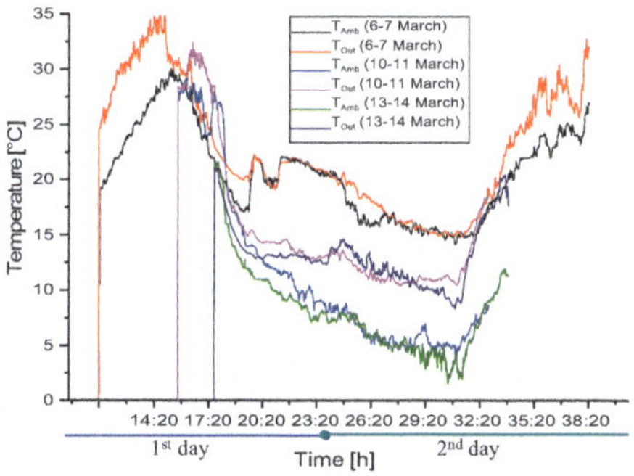

Fig. 4. Temperature profiles during the test periods.

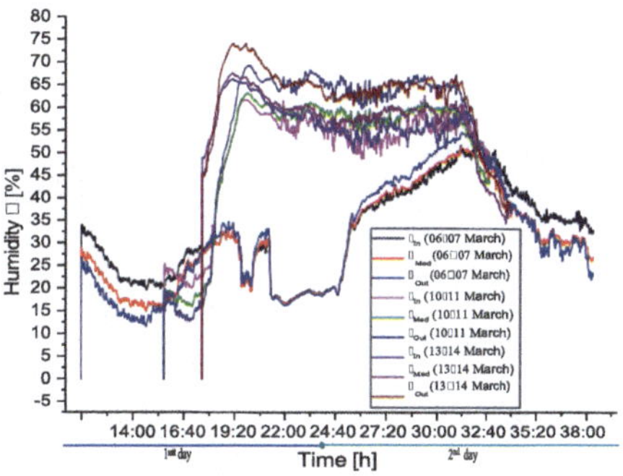

Fig. 5. Humidity profiles during the test periods.

From this figure, we can remark that the temperatures at several locations increase and reach maximums values and then reduce at the afternoon from 18:00 a.m., after that, they start to increase again in the morning of the second day from about 07:00 a.m. In the second test, the temperatures start to decrease at the first time and then they rise in the second day from 07:30 a.m. During the test periods, we note that the air temperature measured at the outlet of the vortex tower prototype almost reached the value of the air ambient temperature. This occurs because heat is removed through outwards movement of air.

The profiles of the humidity measured at different points along the prototype during the experimental periods are presented in Fig. 5. The concerned locations are; the entrance, the middle and the exit of the chimney. It can be seen that the humidity profiles along the chimney are inversely proportional to the prototype temperatures due to the ambient temperature variation during the tests and the hot source where the temperature of the system is generated by this one, which has a significant influence in the humidity evolution. The humidity level increases with the decrease in temperature in parallel.

This evolution is quasi-totally due to the quantity of steam water available in the air circulating inside the tower [1, 11].

During the test periods, the measurement of the output parameters velocity and power was carried out by an instrument of a speed anemometer installed where the kinetic energy is maximum in entrance of the chimney. The velocity anemometer sensor, shown in Fig. 6, includes an integrated dynamo to transform the kinetic energy into mechanical power and finally into electrical energy [12].

We can see that the velocity profile is more crucial in the last test period where the maximum reached velocity is about 6 m/s, while in the first test 5.80 m/s was found. The velocity manometer is driven by a combination of two movements of air; the movement of upward air and the artificial vortex. This movement is due to the phenomenon of natural convection.

Each 1/100 of a second, the air velocity is measured at the chimney inlet and represented in Fig. 7. A density difference in the configuration is caused by the temperature differential (ΔT) between the air inside the tower and the surrounding air. As the ΔT increases, the air density difference also increases. Due to the low density of air inside the tower, cold air with higher density enters the configuration causing the thermosiphon effect.

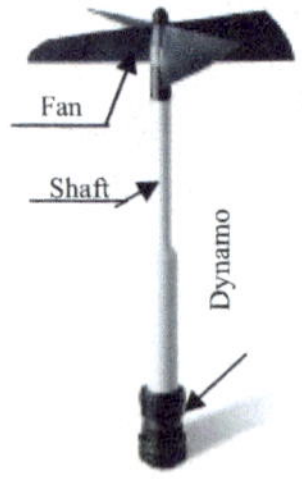

Fig. 6. Velocity anemometer sensor.

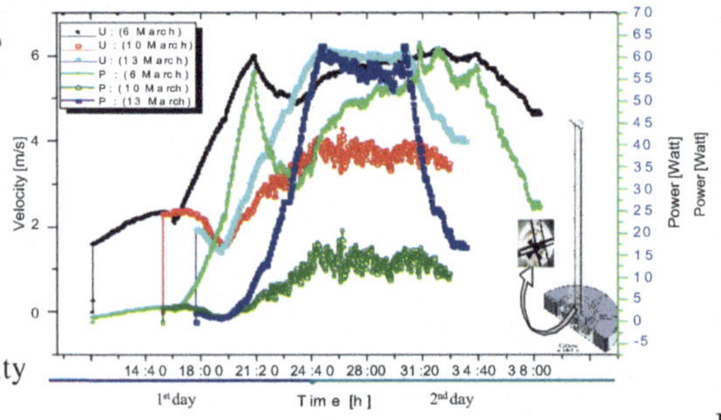

Fig. 7. Velocity and power profiles during the test periods.

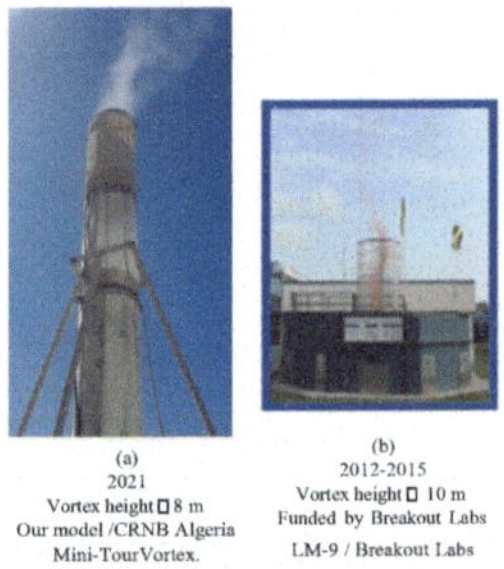

Fig. 8. Vortex generated visualization of our VTP compared to the prototype of Ref. [13].

For validation purpose, a visualization of the vortex behavior was artificially generated by the injection of the fume in the entrance of the VTP, and the simulation of the produced vortex. As seen in Fig. 8, the progression of the fume released from the top of the chimney of our prototype have the same trend than the different flows presented in the work of Ali et al. [13].

4 Conclusion

The project consists to implemented an updraft vortex tower type production plant will be proposed in our country. For this, a model of a VTP was developed and realized as the first step in the initial phase. The model was used to study the possibility of producing low electrical power and to analyze the influence of ambient conditions on the VTP performance generated by an artificial vortex in a chamber comprising bladed zones, where kinetic energy is transferred into mechanical energy. Geothermal energy was generated by a hot spring heated by electrical resistance.

The measurements were recorded by an acquisition system that was developed to collect all connection data every 1/100 s during test periods of 3 days through the spring season.

Although the experiment was conducted during on non-sunny days (at night), the effect of the hot spring was sufficient to produce a temperature difference of 19 °C between the chimney inlet and the ambient temperature. This produced an upward air velocity of 4 m/s in the chimney.

Several results were recorded from these experiments; it was found that the velocity of the upward airflow is inversely proportional to the outlet temperature (ambient temperature). It was also found that the air speed measured at the chimney inlet is inversely proportional to the humidity of the climate. The efficiency of this equipment is inversely proportional to the ambient temperature, which is verified by the Carnot cycle. It was found that the effects of small variations in ambient pressure on the VTP are negligible. In conclusion, ambient temperature and humidity affect the VTP performance.

The siting of this type of vortex power plant combined with universal cooling towers is an essential factor to take into account for future nuclear power plants in countries lacking cooling water and in inland regions with dry climates.

References

1. Michaud, L.M.: Proposal for the use of a controlled tornado-like vortex to capture the mechanical energy produced in the atmosphere from solar energy. Bull. Am. Meteor. Soc. **56**, 530–534 (1975)
2. Michaud, L.M.: On the energy and control of atmospheric vortices. Journal de Recherches Atmosphériques (1977)
3. Mamulashvili, G.: Trailing solar chimney. SU patent, SU1319654 A1 (1987)
4. Takhar, H.S., Beg, O.A.: Mathematical modeling of geophysical vortex flow. Int. J. Fluid Mech. Res. **32**(4), 439–453 (2005)
5. Dunn, MA.: Solar vortex electric power generator. US patent, US6772593B2 August (2004)
6. Haan, F.L., Sarkar, P.P., Gallus, W.A.: Design, construction and performance of a large tornado simulator for wind engineering applications. EngStruct **30**(4), 1146–1159 (2018)
7. Michaud, LM.: The atmospheric vortex engine. In: IEEE Toronto International Conference Toronto. Science and Technology for Humanity (TIC-STH), Canada, pp. 971–975 (2009)
8. Cohen, H., Rogers, G.F.C., Saravanamuttoo, H.I.H.: Gas turbine theory, 4th edn., pp. 37–45. Addison Wesley Longman Limited, England (1996)
9. Gannon, A.J., von Backström, T.W.: Solar chimney cycle analysis with system loss and solar collector performance. J. Solar Energy Eng. **122**, 132–137 (2000)
10. Ming, T.Z., Zheng, Y., Liu, C., Liu, W., Pan, Y.: Simple analysis on thermal performance of solar chimney power generation systems. J. Energy Inst. **83**(1), 6–11 (2010). https://doi.org/10.1179/014426009X12519696923902
11. Varaksin, A.Y., Romash, M.E., Kopeitsev, V.N.: Tornado-like vortices generation due to air turbulent convection14th European Turbulence Conf., France, 1–4 September (2013)
12. Anemometer Young Propeller for vertical measurements. CFT Propeller 27106T, S17100, Wrangelstrasse 100, D-10997 Berlin, Web: www.ammonit.com
13. Ali, A., Ismaeel, et al.: Review and comparative analysis of vortex generation systems for sustainable electric power production. IET Renew. Power Gener.Gener. **11**(13), 1613–1624 (2017)

Ambient Temperature Effect on the Coriolis Force Produced in the Vortex Tower

A. Dadda Khorsi(✉), A. L. Cheridi Deghal, A. Bouam, A. Dahia, A. Hadjam, and A. Kentouche

Nuclear Research Centre of Birine, BP.180 AinOusscra, Djelfa, Algeria
a.dadda@crnb.dz

Abstract. Nuclear power plants could produce electricity using the Vortex Tower. Indeed, the heat from the secondary cooling system of a nuclear reactor can be used as a heat source to produce electricity rather than being released into the environment. An artificial vortex system is intended to increase the kinetic energy of hot air. This kinetic energy is used to produce electrical energy by utilizing a turbine installed just at the entrance to the chimney, where the air velocity is highest. So, the main effect used in Vortex Tower is the chimney force (the Coriolis effect) as a means to take heat out of a hot source.

This study aims to analyse the vortex effects and Coriolis force along the chimney of a prototype vortex tower made at the level of the Nuclear Research Centre of Birine (CRNB) by changing the ambient temperature. The air kinematic field used in this study is calculated from a numerical simulation using the Relap5 code. The results obtained showed that a reduction in the ambient temperature causes an acceleration of the ascending air, which is also due to the difference in flow density as well as a strengthening of the vortex along the chimney.

Keywords: Renewable energy · Vortex tower · Coriolis force · Ambient temperature · Cinematic field · Numerical simulation

1 Introduction

The need for clean energy in the world is becoming important. To this end, the search for new energy sources will be increasingly requested [1, 2]. Some of the energy sources have been identified as renewable energy, solar, geothermal energy or movements in air currents or water waves. Every source has special advantages and application ranges. Solar panels involve the use of rare elements and are expensive, other sources such as geothermal, waterfalls, hydroelectric, wind, etc., require complex assemblies and are not reliable throughout the year [3].The exploitation of the power of tornadoes remains possible, but it has not been carried out in the real case [4]. The atmospheric vortex tower is a mix of the two principles of tornadoes and the solar chimney [5, 6].

The Vortex tower (VT) is a novel technology or electricity production from geothermal energy and wind energy [7]. The kinetic force of hot air rising through a long chimney and heated by hot energy sources is what gives the VT its mechanical power.

© The Author(s), under exclusive license to Springer Nature Switzerland AG 2025
M. Hatti (Ed.): IC-AIRES 2024, LNNS 1238, pp. 758–765, 2025.
https://doi.org/10.1007/978-3-031-80301-7_79

The heat source used is a coaxial cylindrical tank (ring type) full of water fed by electrical resistances.

A group of solar towers that includes wind towers, also referred to as vortex towers, was founded approximately 40 years since by the French architect and industry pioneer Egard Henri Nazare. When weighed against the Nazare project and all of its successors, the aerogenerator towers nevertheless bring considerable novelties, in terms of the amount of forces and natural phenomena employed, the variety of energy sources considered, numerous structural details, the features of the surrounding greenhouses and the energy storage system, and, lastly, by the yield which is much higher than what can be expected from competing projects. The use of a hollow structure in the form of a tower widened at the base and optimized to combine four, or even five natural forces and effects with a view to the massive and permanent production of electrical energy at low cost, without pollution, without consumption of limited natural resources and without being penalized by the irregularity of the wind regime as in the case of wind turbines.

The forces and natural effects resulting during the movement of the ascending air along the chimney are: chimney effect, "force" of Coriolis, Venturi effect (which relates to a convergent diverging chimney). In addition, the wind is likely to provide a back-up, without ever being necessary for the operation of the tower, and it is possible to increase the efficiency and installation profitability of the by utilizing lower temperature calories from NPP, industry, otherwise mostly lost needlessly, incinerators, or geothermal energy.

In the nuclear applications, the nuclear power plant cooling system and the realized VT [8, 9] are coupled, which greatly reduces the amount of cooling water used from rivers or seas, as seen in Fig. 1 [10].

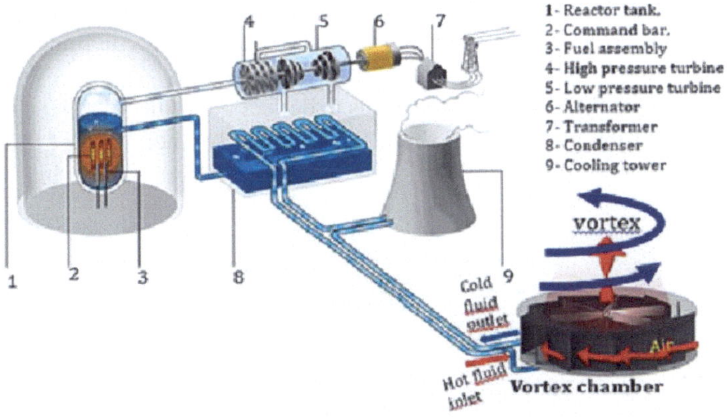

Fig. 1. Schematic diagram of coupling between the VT and the secondary cooling system of nuclear power plant.

This work's goal is to analyse how ambient temperature affects Coriolis and vortex effects created along a vortex tower constructed at the Nuclear Research Centre of Birine (CRNB).The principle used is the chimney effect as a means of extracting heat from a hot source. An implanted vortex system designed to increase the kinetic power of hot air.

This kinetic energy is used to activate a velocity anemometer (mini- turbine) to produce electrical energy.

2 Coriolis Force

TheCoriolisforce is one of the obvious ones. Like all apparent forces (there are only three: the centrifugal force, the force of inertia, and the Coriolis force), it is a fictitious force that has no real existence. This is why physicists prefer to speak of the "Coriolis effect" rather than of force in order to clearly distinguish between effect and cause. The Corioliseffect, moreover predicted by Newton in his treatise published in London in 1687 ("Principles…"), is not due to a force, it is only explained by the spherical form of the earth and its rotational motion. Thus there is no Corioliseffect on the Moon, any more than on any other planet without rotation on itself.

The Coriolisforce is a force acting perpendicular to the direction of air motion, and responsible for rotating the air in a uniformly rotating frame.

The Coriolis acceleration formula is written [11]:

$$\gamma_c = 2 \cdot \Omega \cdot V \cdot \sin \varphi \tag{1}$$

($V = m.s^{-1}$ $\gamma c \equiv m.s^{-2}$)

Where $\Omega = 2\pi/day$ represents the angular velocity of rotation of the Earth (Earth: $\Omega = 7.27 \times 10^{-5}$ $rad.s^{-1}$), φ represents the latitude of the point considered and v represents the linear speed of the particle considered (expressed in ms^{-1}). This relationship shows that, at the equator (center of the earth), the coriolis force is null, and maximum at the poles. This therefore means that, winds direction of the will be essentially given by the pressure gradients at the equator, but will deviate from these predictions for moderate latitudes.

The movement of the Coriolis force tends to deflect the projectile to the right side in the northern part of the globe and to the left side in the southern part of the globe (see Fig. 2).

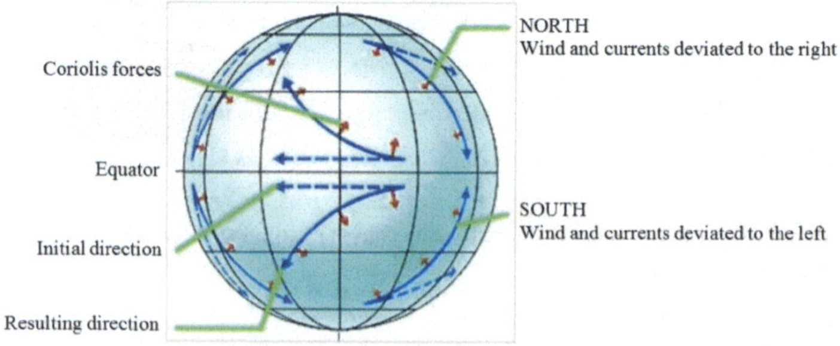

Fig. 2. Deflection of winds and currents by the Coriolis force.

3 Application

The vortex tower is located in Algeria on the Birine site, which is in the northern hemisphere (Fig. 3).

As seen in Fig. 4, the vortex tower prototype is divided into two main components: A lower part where there are eight (08) air inlet openings designed to receive the air flow from the atmosphere and then direct the flow smoothly to the chimney, the convergence chamber (chamber of vortex production), the zone of heat transfer and the hot source. The upper part is the chimney.

The Coriolis force produced in Vortex tower is based on the principle of creating micro tornadoes by the temperature difference between the hot air at ground level and the cool air at altitude.

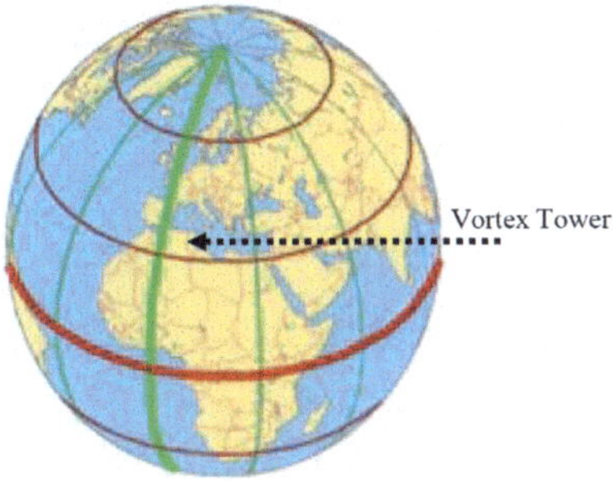

Fig. 3. Our VT's location.

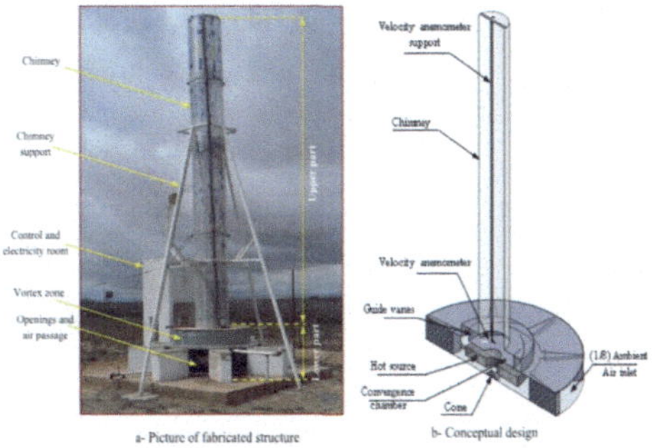

Fig. 4. Prototype of the vortex tower.

- Dimensions of the Vortex Tower
  - Height: 7.600 m
  - Diameter at the base: 3.250 m.
  - Diameter of centrifugal inlets 1.184 m.
  - Internal diameter at the top: 0.710 m.

Volume of the water pool (Hot source) in autonomous operation, much less by combining natural forces and effects with the recovery of industrial cooling calories or from other renewable sources (geothermal).

Through natural convection, the heat is extracted from the tank's upper surface by the airlow. Next, the vortex is produced by introducing hot air tangentially through 08 vanes situated around the tower base's perimeter with pre- rotation.

After that, the air enters with swirling low in the base in order to create a vortex along the chimney. The air circulation in the studied tower from inlet to outlet is illustrated in Fig. 5 [11].

4 Resultats and Interpretations

The air speed values used in this study (Fig. 6) are calculated from a numerical simulation by Relap5.

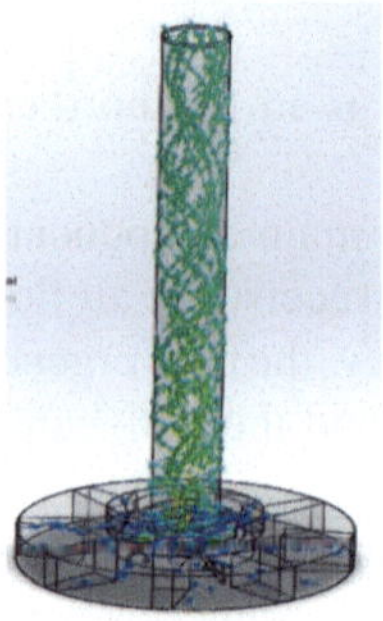

(a) Visualization of 3D simulation of a vortex in the tower

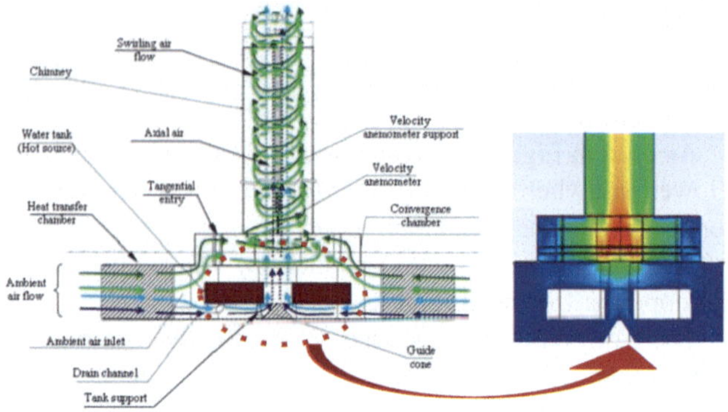

(b) Airlow through the vortex tower model

Fig. 5. Prototype of Vortex Tower

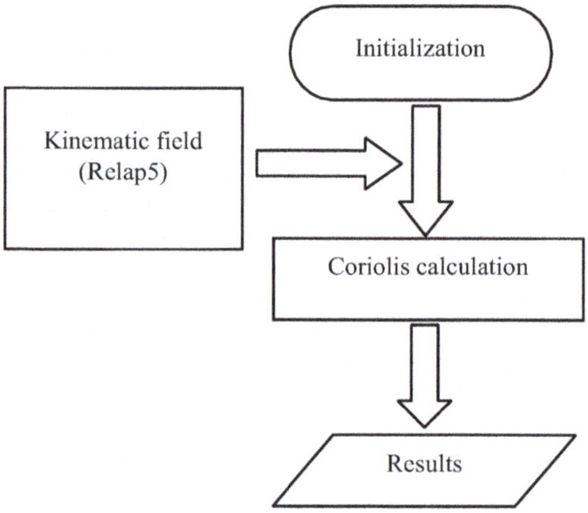

Fig. 6. Flowchart of the calculation.

Table 1 below represents the variation of the Coriolis acceleration as a function of height of the chimney for different ambient temperatures.

Figure 7 represents the evolution of the Coriolis acceleration along the chimney for an altitude of 35°. We notice that there are 2 phases:

- Phase I: 0–0.56 m: we notice that there is a significant acceleration gradient (sudden increase in acceleration in a relatively small interval). This is due to the artificial air vortex in the annular convergence chamber. This leads to the presence of a strong Coriolis effect.

At $z = 0.56$ m, the force reaches the maximum, because the air flow penetrates the chimney, the two components of the speed become negligible.

- Phase II: 0.56 m to the end of the chimney, we note that the evolution of the acceleration is almost stable, because the two components of the air flow velocity are

Table 1. Coriolis acceleration values γ_c as a function of ambient temperature along the chimney.

Z(i) (m)	γ_c(i) [T=20°C]	γ_c(i) [T=25°C]	γ_c(i) [T=30°C]	γ_c(i) [T=35°C]	γ_c(i) [T=40°C]	γ_c(i) [T=45°C]
0.00	2.8090572E-05	2.8090572E-05	2.8090572E-05	2.8090572E-05	2.8090572E-05	2.8090572E-05
0.28	5.5697987E-05	5.5697987E-05	5.5697987E-05	5.5697987E-05	5.5697987E-05	5.5697987E-05
0.56	1.5882129E-03	1.5801508E-03	1.5705158E-03	1.5565266E-03	1.5036602E-03	1.3551455E-03
0.863	2.9410829E-04	2.9301277E-04	2.9250715E-04	2.9208578E-04	2.9183296E-04	2.9177676E-04
1.623	2.7801521E-04	2.7698147E-04	2.7650391E-04	2.7610504E-04	2.7585783E-04	2.7581572E-04
2.383	2.7803768E-04	2.7700391E-04	2.7652641E-04	2.7612751E-04	2.7588030E-04	2.7584098E-04
3.143	2.7806297E-04	2.7702920E-04	2.7655167E-04	2.7615280E-04	2.7590559E-04	2.7586345E-04
3.903	2.7808541E-04	2.7705170E-04	2.7657414E-04	2.7617527E-04	2.7592809E-04	2.7588874E-04
4.663	2.7811070E-04	2.7707699E-04	2.7659943E-04	2.7620056E-04	2.7595335E-04	2.7591121E-04
5.423	2.7813317E-04	2.7709946E-04	2.7662190E-04	2.7622303E-04	2.7597582E-04	2.7593650E-04
6.183	2.7815846E-04	2.7712472E-04	2.7664719E-04	2.7624829E-04	2.7600111E-04	2.7595897E-04
6.943	2.7818093E-04	2.7714722E-04	2.7666966E-04	2.7627079E-04	2.7602358E-04	2.7598426E-04
7.703	2.7820622E-04	2.7717248E-04	2.7669495E-04	2.7629323E-04	2.7604884E-04	2.7600670E-04
8.463	2.7065826E-04	2.6965264E-04	2.6918913E-04	2.6879870E-04	2.6855990E-04	2.6852058E-04

Note: the line in green; the Coriolis acceleration reaches the maximum.

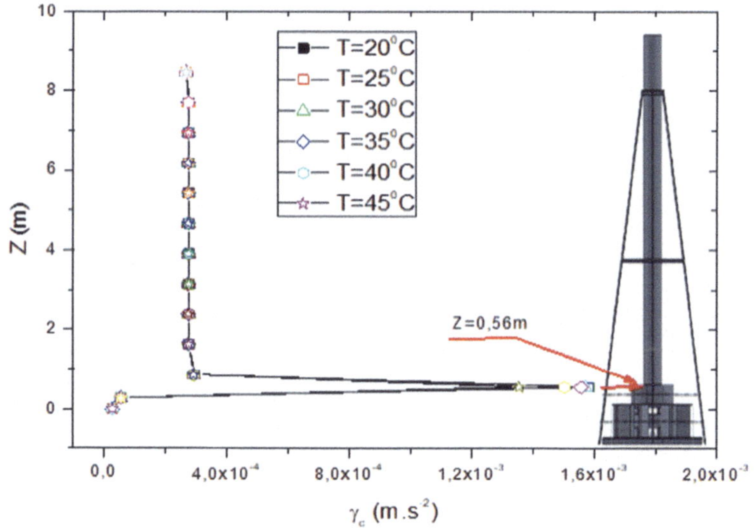

Fig. 7. Evolution of the Coriolis acceleration as a function of ambianttemperatur along the chimney.

negligible compared to that of the axial chimney effect important compared to the Coriolis effect.

Table 2 and Fig. 8 show the maximum Coriolis acceleration values for each ambient temperature.

From Fig. 8, we notice that when the ambient temperature is low, the Coriolis effect becomes significant. This explains that the decrease in temperature leads to an acceleration of the air density difference as well as a strengthening of the cyclone along the chimney.

Table 2. Maximum values of Coriolis acceleration as a function of ambient temperature

Ambiante temperature(°C)	20	25	30	35	40	45
γ_c (m.s^{-2}) (E−03)	1.5882129	1.5801508	1.5705158	1.5565266	1.5036602	1.3551455

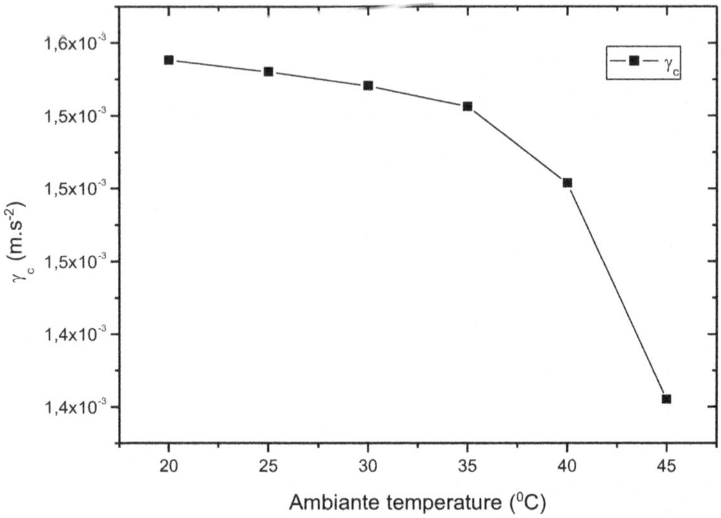

Fig. 8. Evolution of the Maximum Coriolis acceleration as a function of ambient temperature.

We can conclude that even though gravity has an acceleration of 9.81 m/s^2, the γ_c is roughly 1000 times smaller.

Therefore, for our vortex tower, the rotation is ensured in the positive direction even if the acceleration due to the Coriolis force is very low.

5 Conclusion

In this work, a study of Coriolis and vortex effects along the chimney of a prototype of a vortex tower, installed at the Birine Nuclear Research Center (CRNB), was carried out. The analysis consists of varying the ambient temperature during the tests. The kinematic field used in this study are calculated from a numerical simulation using the Relap5 code.

The results show that the Coriolis acceleration is significant when the air flow begins to enter the chimney, where the kinetic energy is transformed into mechanical energy, then to electrical energy using a turbine which comprises a dynamo. Furthermore, the results obtained indicate that when the ambient temperature decreases, the Coriolis effect becomes significant. A reduction in ambient temperature causes an acceleration of the rising air, which is also due to the difference in flux density as well as a strengthening of the vortex along the chimney. Also, we can conclude that, for our vortex tower, the rotation is ensured in the positive direction even if the acceleration due to the Coriolis force is very low.

We deduce that the primary factor influencing the vortex's strength is the temperature difference between the ambient air and the air entering the chimney, and, therefore, the power provided by the vortex tower. This parametric study was conducted and is a good starting point for future designs.

References

1. Michaud, L.M.: On the energy and control of atmospheric vortices. Journal de RecherchesAtmospheriques (1977)
2. Dhahri, A., Omri, A., Orfi, J.: Numerical Study of a solar chimney power plant. Res. J. Appl. Sci. Eng. Technol. **8**(8), 1953–1965 (2014)
3. Djezzarmahfoud, T.: "Numerical simulation of natural convection in a solar chimney (2003)
4. Church, C.R., Snow, J.T., Baker, G.L., Agee, E.M.: Characteristics of tornado like vortices as a function of swirl ratio: a laboratory investigation. J. Atmospheric Sci. **36**, 1755–1776 (1979)
5. Louis. M.M." p Eng The Atmospheric Vortex Engine. AVE tec Energy Corporation (2009)
6. Koonsrisuk, A., Lorente, S., Bejan, A.: Constructal solar chimney coniguration. Int. J. Heat Mass Transf. **53**(1), 327–333 (2010). https://doi.org/10.1016/j.ijheatmasstransfer.2009.09.026
7. Zhou, X., Wang, F., Ochieng, R.M.: A review of solar chimney power technology. Renew. Sustain. Energy Rev. **14**(8), 2315–2338 (2010)
8. Deghal Cheridi, A.L., Bouam, A., Dadda, A., Dahia, A., Koudiah, N., Attari, K.: Realization and experimentation of a novel vortex cooling tower. Energ. Conversion Manag. **270**, 116197 (2022). https://doi.org/10.1016/j.enconman.2022.116197
9. Hadjam, A., et al.: Numerical analysis of a small-scale novel vortex tower integrated with heat source for Nuclear Application. Nucl. Eng. Des. **414**, 112660 (2023). https://doi.org/10.1016/j.nucengdes.2023.112660
10. Coriolis, G.: Sur les équations du mouvement relatif des systèmes de corps. J. Éc. Polytech. Paris, Cahier XXIV, Tome **XV**, 142–154 (1835)
11. DeghalCheridi, A.L., et al.: Numerical investigation of a novel cooling vortex tower using Relap5 computer code. Nuclear Eng. Design 391, 111730 (2022). https://doi.org/10.1016/j.nucengdes.2022.111730

An Experimental Study of a Photovoltaic Stand-Alone House System in Adrar

Abdeldjalil Dahbi[1,2(✉)], Abderrahmane Khelfaoui[1], Miloud Benmedjahed[1], Ahmed Bouraiou[1], Tidjar Boudjema[1], Abdeldjalil Slimani[1], Nouar Aoun[1], and Messaoud Hamouda[2]

[1] Unité de Recherche en Energies Renouvelables en Milieu Saharien (URERMS), Centre de Développement des Energies Renouvelables (CDER), 01000 Adrar, Algeria
Dahbi_j@yahoo.fr

[2] Laboratory of Sustainable Development and Computing, (L.D.D.I), University of Adrar Department of Electrical Engineering, Adrar University, 1000 Adrar, Algeria

Abstract. This paper presents an analysis and experimental study of a photovoltaic stand-alone house system in desert area (Adrar, South West of Algeria). This system is proposed to be installed in remote area for farmer, companies…etc. It is composed from photovoltaic panels, cables, converters, battery and load. All these elements have been presented. The real UDTS50 PV panel was simulated according to Adrar weather conditions. After that, the experimental results were presented. The obtained results proved a good sizing and operation of the system which encourages to install photovoltaic systems in Saharan regions, especially in remote houses.

Keywords: Photovoltaic Stand-alone House System (PSHS) · UDTS50 · Simulation · Sensors

1 Introduction

Nowadays, with the current environmental issues, the depletion and the high cost of fossil fuels, the production of electrical energy poses a serious difficulty. Many farmers have agriculture in remote areas, where they do need electrical energy for many applications such as water pumping, irrigation and house needs. However, there is not a near electrical grid to link their houses or it is very costly. Face to these problems, the renewable energy and especially photovoltaic energy is used to fix them [1]. In this context, a Photovoltaic Stand-alone House System (PSHS) is proposed and studied. This system is proposed to be installed on the house roof in order to benefit from its area by installing photovoltaic panels. Moreover, for obtaining shadow and electrical energy. This later can be used for different applications at home such as lighting, which is the subject of this paper [2] According to estimations from the literature and measures, Algeria has a great solar potential that encourages us to benefit from photovoltaic energy for a long period especially in remote area [3–5]. However, prior to implementation, a photovoltaic

© The Author(s), under exclusive license to Springer Nature Switzerland AG 2025
M. Hatti (Ed.): IC-AIRES 2024, LNNS 1238, pp. 766–773, 2025.
https://doi.org/10.1007/978-3-031-80301-7_80

system's design and engineering become crucial to ensure a proper operation and effective exploitation of the system [6–11]. In this work, the PSHS has been studied to be installed in Adrar which has a good photovoltaic potential such as Saharan regions [12, 13], Fig. 1.

This paper is organized as following: in the second section; the description of the photovoltaic stand-alone house system components. After that, the simulation of the photovoltaic panels (UDTS50) is given. Then, the experimental results of PSHS system are discussed. Then, the calculation results is presented in the last section.

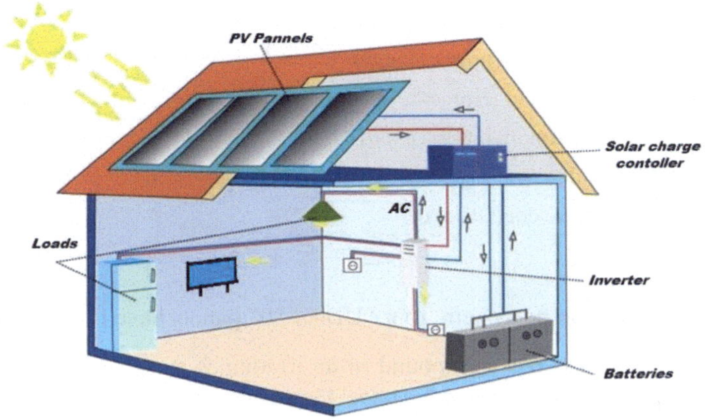

Fig. 1. Photovoltaic stand-alone house system

2 Description of the Photovoltaic Stand-Alone House System

The stand-alone house system is devoted for remote areas. It is equipped by an autonomous system that powers the house without any need to an external electrical grid power. The main components of the studied system are: PV panels, a converter that controls the charge, batteries, loads and sensors to measure the produced energy.

A. ***Photovoltaic panels:*** They convert the received sun irradiation into electrical power. If one module is not enough to supply the load, several photovoltaic modules should be coupled in series and in parallel or both of them in order to increase the voltage and the current to be suitable for those of the inverter, battery and the load.

The characteristics of the used UDTS 50 photovoltaic panels are listed in Table 1

Table 1. Photovoltaic panel characteristics:

PV module parameters	
Maximum power (*Pmax*)	50 W
Open circuit voltage (*Voc*)	21.6 V
Voltage at maximum power (*Vmpp*)	17.5 V
Short circuit current (*Isc*)	3.18 A
Current at maximum power (*Impp*)	2.9 A
Voc Temperature coefficient of	−0.33
Isc Temperature coefficient of	0.273

In the studied system, four "UDTS 50" panels have been used, Fig. 2.

B. ***Sensors:*** it is compound of an irradiance sensor and thermocouple (Type-K) to measure the photovoltaic panel temperature and ambient temperature; Figs. 3 and 4.
C. ***Acquisition system:*** in this experiment a (Fluke Hydra 2635 A) data logger was used. It operates as a digital multimeter, but it can save all measurements and electrical parameters in function of the time (photovoltaic array voltage and current, solar irradiation, panel temperature, ambient temperature, voltage and current of load); Fig. 5.
D. ***Storage battery:*** Since the irradiation and temperature are random sources and they effect directly the photovoltaic energy production, the storage system is very important to supply loads with a stable energy and avoid the intermittence problem habitually caused by the renewable energies.

Fig. 2. UDTS50 Photovoltaic panels

Fig. 3. Pyranometer

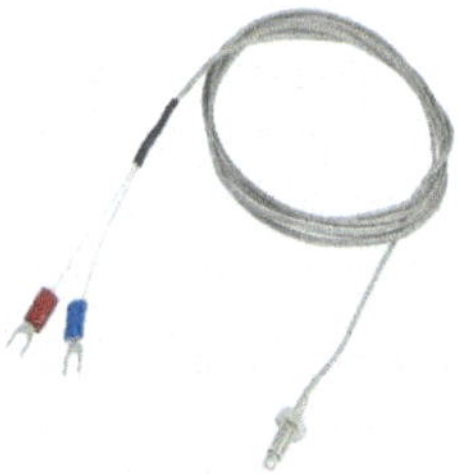

Fig. 4. Thermocouple

When the produced energy by PV is enough, batteries store the electrical energy in a chemical form, and then they generate it when the load need enough energy, such as lighting at night. The used battery is a gel electrolyte type, with 12 V and 195AH as it is shown in Fig. 6.

E. ***Solar charge controller:*** The role of the solar charge controller is to manage the energy flow from the photovoltaic panel into the battery and the load. Moreover, it controls the battery state of charge (SOC), and protects the battery from overcharging and deep discharging. Thus, it prolongs the battery life. The used solar charge controller is compatible with 12 V of the battery systems. This device can support an input voltages ranging from 0 V to 50 V, and it provides a 12 V as an output voltage; Fig. 7
F. ***Load:*** The loads are all electrical devices supplied by the photovoltaic system.

Fig. 5. Fluke Hydra 2635A data logger

Fig. 6. Storage battery

Fig. 7. Solar charge controller

Figures 8 and 9 present the photovoltaic system and the panels on the roof of the mobile house.

Fig. 8. Inside photovoltaic system with load

Fig. 9. Photovoltaic system with a mobile home

3 Results and Analysis

The section below depicts the simulation and experimental behavior results of the photovoltaic used panel (UDTS50) across various cases and under real weather conditions in Adrar. Figures 10.A and B present the characteristics of UDTS50 in Standard Test Conditions (STC), 1000 W/m^2 and 25 °C.

It is seen that there is a good agreement between the real characteristics of the panel presented in Table 1 and the obtained results (*Voc, Isc, Pmax, Vmpp, Impp*), which proves the validity of the simulation model.

According to the simulation results, Fig. 10.C shows that the current value is directly proportional with irradiation, but it has not a noticed influence on the voltage.

Figure 10D shows that the voltage and temperature values are inversely proportional. However, the current has a little change. As a result, the maximum power value is effected by the temperature and irradiation because of the variation of voltage and current. The obtained power is inversely proportional with the temperature, but it is proportional with the irradiation. However, the irradiation effect is more significant. The power reached only 46,3 W at 45 °C, which decreases the total harvested power in hot weather.

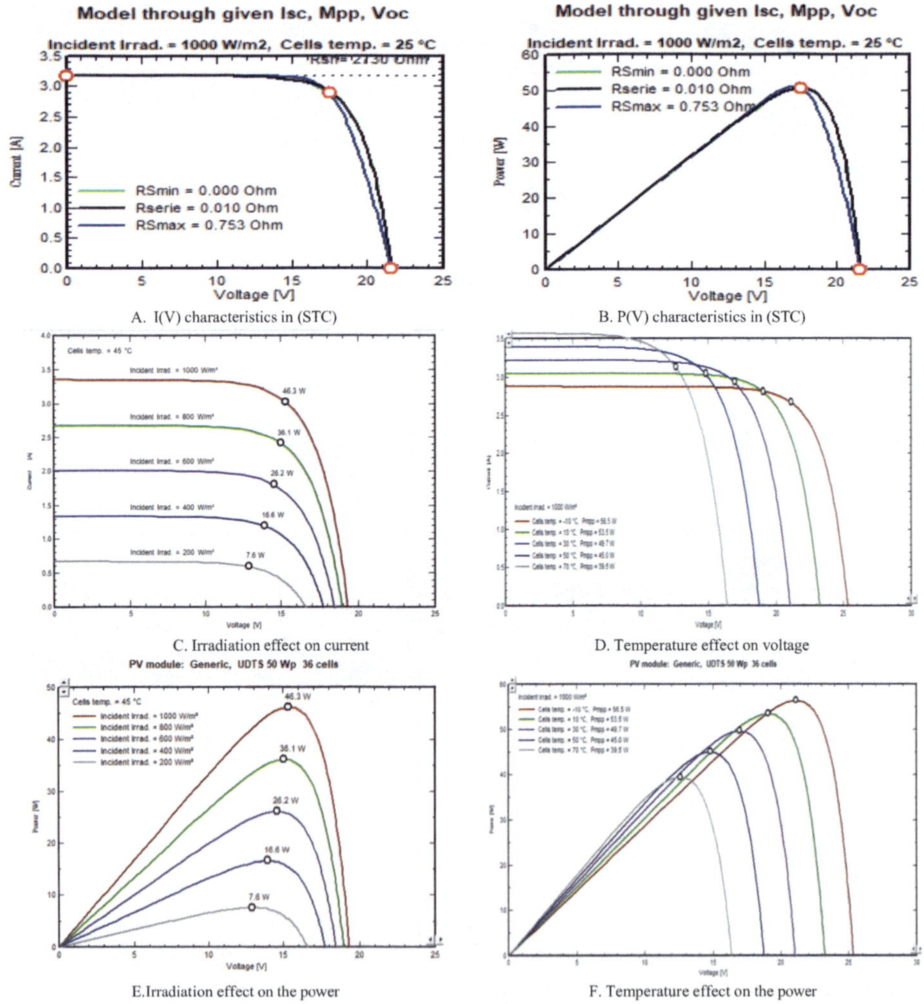

A. I(V) characteristics in (STC)

B. P(V) characteristics in (STC)

C. Irradiation effect on current

D. Temperature effect on voltage

E.Irradiation effect on the power

F. Temperature effect on the power

Fig. 10. UDTS 50 photovoltaic panel characteristics

The next section is devoted for the experimental results of the PV house system using UDTS50 in Adrar Saharan region. After installing the photovoltaic system in the mobile house, the measurement were done on the lighting system. An example of the obtained results on 14/03/2023 is presented below.

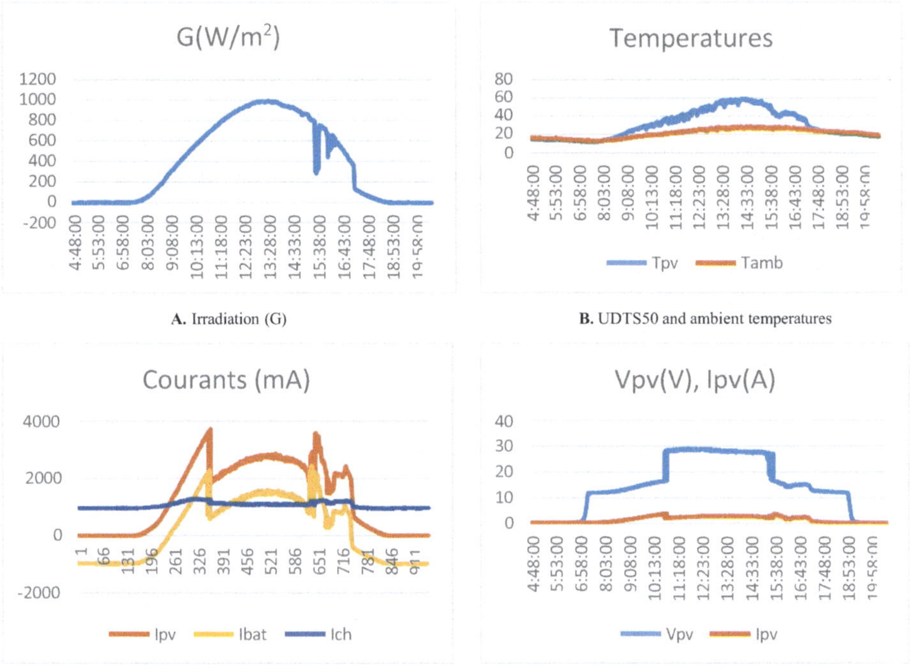

A. Irradiation (G)

B. UDTS50 and ambient temperatures

C. Currents of the PV system

D.PV panels voltage and current

Fig. 11. Experimental results of the mobile house lighting photovoltaic system

Figure 11.A shows the real variation of the irradiation during the given day. It increases from zero in the sun rise at 07:00 till the maximal value (1000 W/m^2) after the solar real time (13:00). After that, it decreases till the sun set at (19:00). It can be deduced that the irradiation in Adrar is very high with a long delay, which produce more energy. It is seen in Fig. 11.B that the photovoltaic panel temperature does not equal the ambient temperature, but both of them are proportional to the irradiation. The ambient temperature increases till 26 °C at 13:47, but the PV temperature increases till 58 °C, which decreases the PV efficiency. In Fig. 11.C, after sun rise, the current and voltage of a the photovoltaic array increase until reaching peak values of 3.5 A and 28 V, at 10:54. Then, the voltage remains fixed at its maximal value, while the current decreases rapidly to 2 A to protect the battery. After that, the current increases according to the changes of solar radiation, but the voltage maintains its value stability until 15:45. After that point, the voltage and current decrease gradually with the solar radiation until zero at sun sets. It can be seen clearly in Fig. 11.C that during the day the PV current equals to the sum of battery and load currents. The load current remains stable in all cases even at night, it is always ensured either from the PV panels or from the battery or both of them. This proves the good sizing of the system. This system can be powered by a low cost weather station and Artificial Intelligence to forecaste the electrical produced energy and detect system faults [14]

4 Conclusion

In this work, a study of photovoltaic lighting system of a mobile stand-alone house in Adrar (Algeria) was achieved. The elements system were presented. Then, the real UDTS50 PV panels was simulated according to Adrar weathe conditions. After that, the experimental results were presented. The obtained results proved a good sizing and operation of the system.

It is deduced that the photovoltaic power can be acceptable even in Saharan medium, because the irradiation effect is stronger that the temperature effect. Moreover, the desert has a big irradiation intensity with long days durations. This encourage installing PV systems in these region especially in remote houses.

The Artificial Intelligence can be applied in this system in order to forecast the energy production and detect faults.

References

1. Rezaei, M., Ghanbari, M.: Optimization of sizing and placement of photovoltaic (PV) system in distribution networks considering power variations of pv and consumers using dynamic particle swarm optimization algorithm (DPSO)'. Indian J. Fundam. Appl. Life Sci. **5**(S1), 3321–3327 (2015)
2. Coonick, C.: BRE National Solar Centre and David Gance, A Technical Guide to Multifunctional Solar Car Parks. BRE National Solar Centre (2018)
3. Boussaid, M., Belghachi, A., Agroui, K., Djarfour, N.: Mathematical models of photovoltaic modules degradation in desert environment. AIMS Energy **7**(2), 127–140 (2019)
4. Boussaid, M., Belghachi, A., Agroui, K., et al.: Solar cell degradation under open circuit condition in out-doors- in desert region. Results Phys. **6**, 837–842 (2016)
5. Boussaid, M., Belghachi, A., Agroui, K.: Contribution to the degradation modeling of a polycrystalline photovoltaic cell under the effect of stochastic thermal cycles of a desert environment. Int. J. Control, Energy Electric. Eng. 6, 66–72 (2018)
6. Chaib, M., et al.: 'Modeling, simulation and analysis of the input climat parameter effect on the photovoltaic panel. In: The Second International Conference on Energy Transition and Security (2ICET&S'2023), 12–14 December Adrar. IEEE proc 2023. https://doi.org/10.1109/ICETS60996.2023.10410824.
7. Kamran, Z., Saif, U.I., Waqar, U., et al.: An overview of transformerless inverters for grid connected photovoltaic system. In: IEEE Proc, 978-1-5386-7939 (2018).
8. Unité de recherche en énergies renouvelables en milieu saharien, 2024, URERMS, CDER, 01000, Adrar, Algeria.
9. Ghada Bel Hadj Ali, 27 Octobre 2014 'Les installations Photovoltaïques Raccordées au réseau', Report GIZ, Tunis.
10. Dahbi, A., Nait-Said, N., Nait-Said, M.-S.: A novel combined MPPT-pitch angle control for wide range variable speed wind turbine based on neural network. Int. J. Hydrogen Energy **41**(22), 9427–9442 (2016)

11. Dahbi, A., Hachemi, M., Nait-Said, N., Nait-Said, M.-S.: Realization and control of a wind turbine connected to the grid by using PMSG. Energy Conv. Manag. **84**, 346–353 (2014)
12. Benmedjahed, M., et al.: Adrar initial state investigation for the use of renewable energies in irrigation systems. Int. J. Sustainable Dev. Plann. **18**(6), 1671–1680 (2023). https://doi.org/10.18280/ijsdp.180603
13. Abdeldjalil, D., et al.: Analysis study and simulation of a real mono crystalline photovoltaic plant in desert area. In: The Second International Conference on Energy Transition and Security (2ICET&S'2023), Université d'Adrar. IEEE (2023). https://doi.org/10.1109/ICETS60996.2023.10410722 https://ieeexplore.ieee.org/abstract/document/10410722.
14. Dahbi, A., et al.: Study and realization of a weather station. In: The Fifth International Conference on Electrical Engineering and Control Applications (ICEECA'2022), Khenchla, Algérie, Springer (2024). https://doi.org/10.1007/978-981-97-4776-4_49.

Evaluation of Wind Energy Potential in the Dhaya Region, Sidi Bel Abbès

Fatima Zohra Bochra Amrani(✉), Boutlilis Fatima, and Benaissa Bekkouche

Faculty of Science and Technology, Department of Electrical Engineering, University of Mostaganem, Mostaganem, Algeria
amranibouchra98@gmail.com

Abstract. This study aims to assess the wind potential of the Dhaya region located in the southern part of Sidi Bel Abbès, Northwest Algeria, using three years of collected data (2020–2021–2022). Statistical analysis of the data confirms the significant wind potential in the study area, with an average wind speed of 6.5 m/s at a height of 10 m, a power density of 383 w/m^2, and Weibull parameters A = 6.9 m/s, K = 1.65 from the South and South-South-West (SSW) direction. In terms of temporal analysis, it was observed that the nighttime conditions at the site are relatively calm, with windier periods occurring in June (8.13 m/s) and November (10.3 m/s). Consequently, based on these positive findings, we have opted to establish a 12 MW wind farm in this region utilizing the Vestas V66 2 MW turbine technology, known for its efficiency and reliability in similar wind conditions. The wind resources in this selected region are highly conducive to electricity generation highlighting the potential for sustainable energy development in the Dhaya region.

Keywords: Wind potential · wind speed · power density · Weibull distribution · Dhaya

1 Introduction

Wind energy is increasingly recognized as the leading SUSTAINABLE energy option to replace coal, oil, gas, and nuclear power sources [1]. In Algeria, where electricity production heavily relies on fossil fuels, there is considerable solar and wind energy potential, particularly in the southern regions due to the country's geographical positioning. Algeria is currently implementing a significant renewable energy initiative aiming to generate 22,000 MW of clean electricity by 2030, with 5,010 MW coming from wind power [2]. The initial exploration of Algeria's wind potential was pioneered by Said (1984) [3], followed by subsequent studies from Bensaid (1985) [4], Abdeladim et al. (1996) [5], and Merzouk (2000) [6]. More recent research, such as that by Boudia & Guerri (2015) [7] and Abdeslame et al. (2017) [8], has leveraged up-to-date meteorological data and extensive measurement points compared to earlier investigations.

Previous studies have shown a lack of assessment regarding wind potential in Algeria, particularly with no evaluations conducted at the Dhaya site. The current study aims to fill this gap by employing a scientifically rigorous approach using internationally recognized

© The Author(s), under exclusive license to Springer Nature Switzerland AG 2025
M. Hatti (Ed.): IC-AIRES 2024, LNNS 1238, pp. 774–782, 2025.
https://doi.org/10.1007/978-3-031-80301-7_81

standards to assess wind potential in the Dhaya region in southwestern Sidi Bel Abbès within the Tell Atlas Mountain range in northwestern Algeria (see Fig. 1). The selection of Dhaya as the study site was motivated by compelling wind measurements recorded at the meteorological station over several months of the year. This assessment employs WindPro software, using recent wind speed and direction data collected by the National Meteorological Office (ONM) network from 2020 to 2022. The WindPro tool determines the average wind speed and power density at the site, enabling the identification of optimal locations for potential wind farm projects that align with Algeria's renewable energy development objectives.

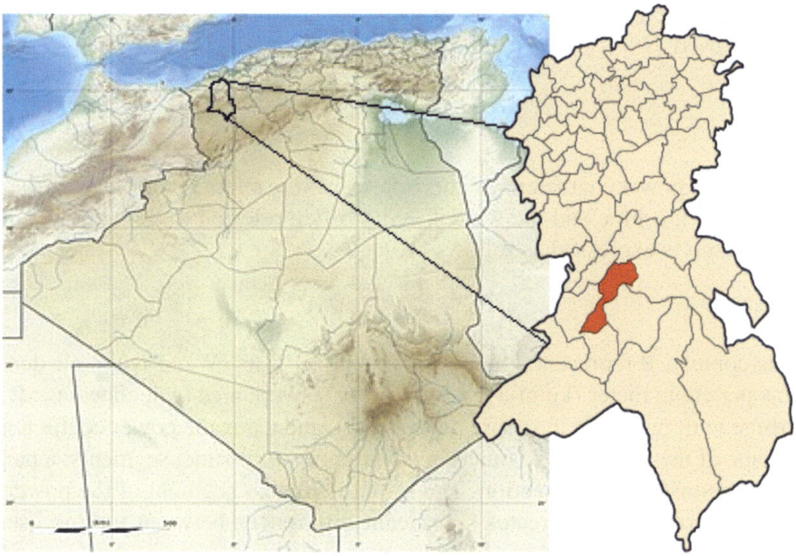

Fig. 1. Geographical location of Dhaya region Sidi Bel Abbès (Source: google map).

2 Material and Methods

In our analysis of wind potential in the case study region of Dhaya, we employed WindPro software for simulation purposes. The methodology employed in this study to identify an optimal location for a wind farm involves four key steps. Given the critical importance of wind speed in wind power plant construction, the initial step involves a detailed analysis of wind data. The next phase involves using WindPro software to estimate wind climatology, based on three-hourly wind speed and direction data collected at a height of 10 m above ground level (a.g.l) for the selected site. In the third stage, topographical data for each site is entered into WindPro to create a wind atlas that accurately reflects the wind field characteristics of the study areas. Finally, the fourth and final step focuses on decision-making for site selection and wind farm planning. This includes determining the appropriate turbine types and quantities, selecting installation sites, and estimating the energy production potential for the chosen location.

The software applies the Weibull probability distribution function to describe variations in wind speed and assess the available wind power density at the selected site. The Weibull distribution is widely utilized in this context due to its broad applicability and ease of parameter determination. It is particularly effective in modeling wind speed patterns over given periods, offering a reliable match to experimental data and closely approximating many observed wind speed distributions [9].

The Weibull probability distribution function is typically expressed as:

$$p(V) = \frac{k}{A}\left(\frac{V}{A}\right)^{k-1} \exp\left[-\left(\frac{V}{A}\right)^{K}\right] \tag{1}$$

To calculate the mean wind speed $\overline{v}$ using the Weibull distribution, the following expression is used:

$$\overline{v} = \frac{1}{n}\left[\sum_{i=1}^{n} v_i\right] \tag{2}$$

The turbine's power characteristic curve describes the relationship between the power delivered by the machine and the wind speed V, often referred to as the power density. This relationship is expressed as follows [10]:

$$p(V) = \frac{1}{2}\rho S c_p(V) V^3 \tag{3}$$

In this context, P represents the power output in watts (W), ρ is the air density in kilograms per cubic meter (kg·m-3), S is the rotor's swept area (calculated as πR2 for a wind turbine with radius R) in square meters (m^2), and Cp is the power coefficient. The overall shape of the power curve function consists of two distinct segments: a parabolic section and a constant linear section. The initial nonlinear segment of the power curve is particularly crucial, as it creates significant differences between turbine users and manufacturers [11].

3 Results and Discussions

This section outlines the key findings and provides an in-depth discussion of the results from the wind resource assessment in the Dhaya region. It includes an analysis of mean wind speed, predominant wind direction, duration, and availability. Additionally, the assessment evaluated the resulting power density, key factors for wind farm design, and a detailed analysis of potential energy output. This thorough examination highlights the feasibility and potential for sustainable wind energy generation at the identified optimal location.

4 Wind Speed Variation

Wind speed is a key factor in assessing the viability of wind turbines for power generation. Evaluating wind speed variability, including daily, monthly, seasonal, and annual fluctuations, is vital for forecasting energy production levels over various time frames [12]. Figure 2 depicts the monthly mean wind speeds recorded at Dhaya region at 10-m hub heights above ground level (a.g.l).

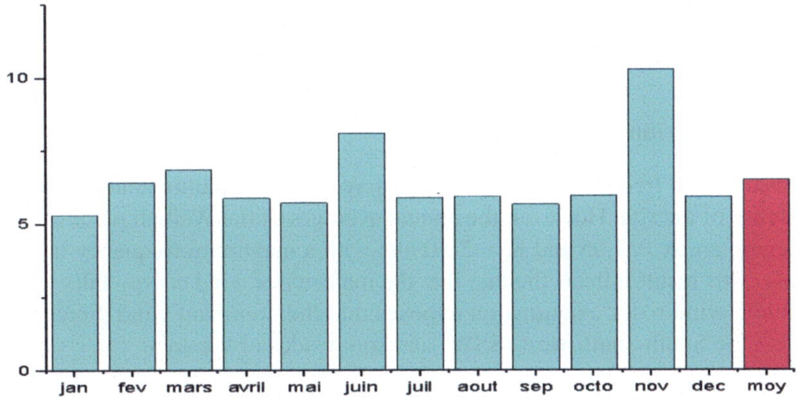

Fig. 2. Monthly average wind speed fluctuations at 10 m a.g.l for the Dhaya region.

Monthly wind speed variations demonstrate consistent wind speed patterns across the site. The analysis reveals higher wind speeds during the year, with notable peaks in March (6.88 m/s), June (8.13 m/s), and November (10.3 m/s) (Fig. 2). These seasonal fluctuations reflect transitional periods between winter and summer extremes, characterized by less stable weather patterns and changing temperature and pressure gradients [13].

The wind power density is categorized into seven classes based on wind speed and annual wind power density, as detailed in Table 1:

Table 1. Wind Power classification at 10 m a.g.l [9]

Wind Power Class	Mean wind speed (m/s)	Wind Power Density (W/m^2)
1	0–4.4	0–100
2	4.4–5.1	100–150
3	5.1–5.6	150–200
4	5.6–6.0	200–250
5	6.0–6.4	250–300
6	6.4–7.0	300–400
7	7.0–9.4	400–1000

As depicted in Fig. 2 Dhaya exhibits a high annual mean wind speed at 6.51 m/s revealing different wind speed and annual wind power density classifications, placing it in wind class 6. Dayal et al. (2021) [9] indicate that a Wind Power Class of 3 or higher is typically suitable for utility-scale wind power projects, consistent with similar studies in Algeria. For instance, El Bayadh features a mean wind speed of 4.73 m/s and a power density of 157.49 W/m^2 while Djelfa is documented to have a mean wind speed of 3.88 m/s and a power density of 81.57 W/m^2 [14]. Therefore, these findings clearly demonstrate that the Dhaya site possesses superior wind resources compared to some other sites and holds significant potential for wind farm development.

4.1 Weibull Distribution and Wind Rose

Figure 3 shows the frequency density curve of Weibull distribution and Fig. 4 displays the wind rose of the site. However, the annual averages of the Weibull parameter for the Dhaya region are A = 1,95 and K = 7,40 m/s with a maximum frequency in the order of 5.5 m/s. This result reflects the fact that the majority of wind energy falls within this range. While wind roses examination exposes that the dominant wind direction of the site is from the South-south-west (SSW) and south sides of the area.

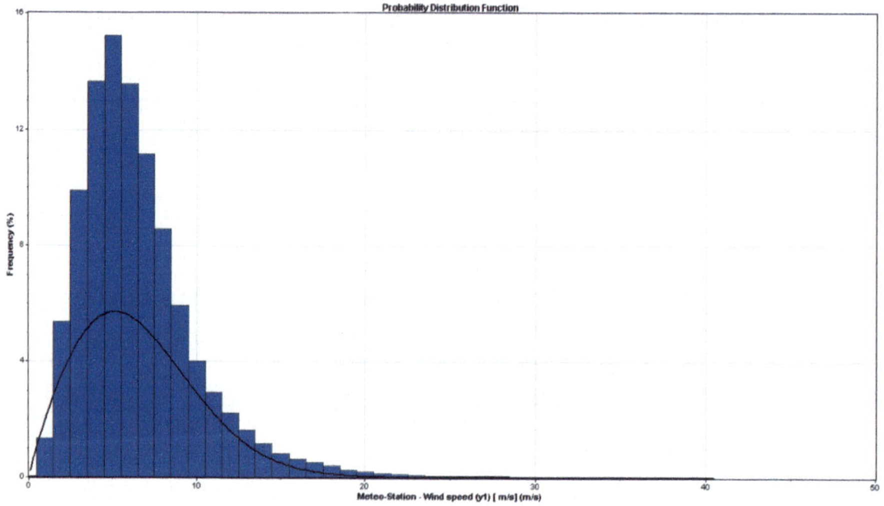

Fig. 3. Annual frequencies of wind speed for the Dhaya region fitted by Weibull distribution.

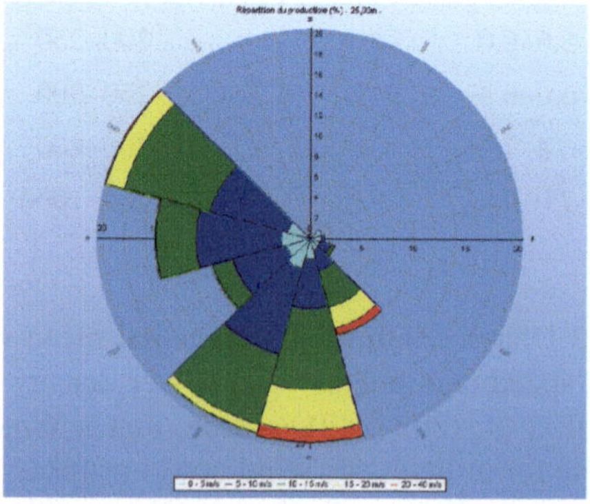

Fig. 4. Annual Wind rose for the site.

4.2 Seasonal Weibull Distribution Parameters

Table 2 provides the Weibull shape factor K and scale factor A for each season as well as for the entire year at the studied location. The shape factor K helps predict wind stability and distribution, while the scale factor A is used to estimate wind strength at the location. The measurements' shape and scale factors are between 1.90 and 2.56, and the speeds are between 6.49 in summer and 7.52 m/s in autumn at 10 m hub heights, respectively.

Table 2. Seasonal Weibull parameters.

Site	Parameters	Winter	Spring	Summer	Autumn	Annual
Dhaya	A m/s	6.57	7.05	6.49	7.52	7.40
	K	2.15	2.27	2.56	1.90	1.95

The analysis of seasonal variations in shape and scale factors indicates that wind speeds are more consistent during summer compared to Winter and Autumn across the studied site. These results align with expected seasonal weather patterns, with summer generally characterized by stronger and more consistent winds, while Winter and Autumn tend to have more variable and less intense wind conditions.

Comparing our obtained values with those from other studies conducted in the country, we observe that the shape factor in the Ksar-Chellala region ranges between 1.5 and 1.24 [15]. Additionally, in Tiaret and Mecheria, the shape factor is reported as 1.48 and 2.02, respectively [16, 17], while in Djelfa and El Bayadh, the shape factor is 1.71 and 1.62, respectively [14] This comparison highlights the presence of large, stable winds at high speeds in the Dhaya region, contrasting with smaller, less stable winds at lower speeds observed in other locations.

4.3 Wind Farm Design and Energy Output

Building on the favorable results from the initial analysis of the Dhaya site, the second phase involves designing a wind farm with a 12 MW capacity and evaluating its feasibility using the Vestas turbine, which has a power rating of 2 MW and Fig. 5 displays its power curve. The choice of the wind turbine was made by comparing three wind turbine technologies (Bonus 2 MW, NEG-Micon 2000–500/72, Vestas V66 2 MW) using WindPro software, and the most suitable technology for the creation of a 12 MW wind farm was selected.

In this study, we chose a one-row configuration with a spaced gap between two cascaded wind turbines on the first row to minimize wake impact on the second-row wind turbines, as shown in Fig. 6, while Tables 3 and 4 show the simulation results of the installation.

The simulation findings not only highlight the considerable potential but also offer compelling evidence of the wind arm's efficiency. The projected gross annual energy production is estimated at 44.25 GWh, with a net energy output of 45.82 GWh, indicating

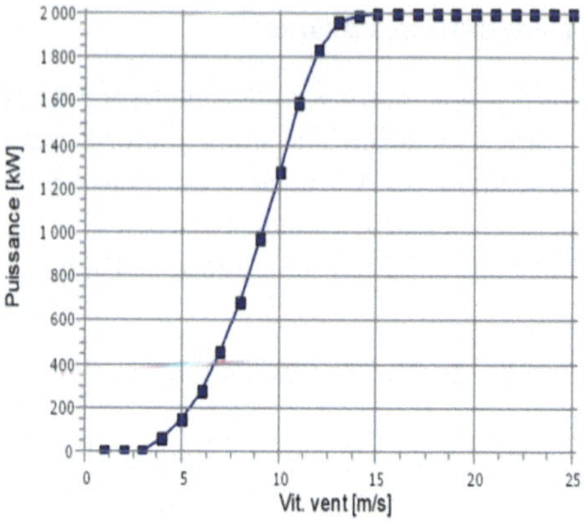

Fig. 5. Vestas V66 2 MW power curve

Fig. 6. Turbines' placement for the 12 MW wind farm.

Table 3. Result parameters produced by the farm.

Turbine	Annual production (MWh/a)	Wake effect (%)	Wind speed (m/s)
01	7 327,2	4,2	7,15
02	7 299,0	4,5	7,15
03	7 308,9	4,4	7,15
04	7 332,9	4,1	7,15
05	7 388,0	3,6	7,15
06	7 598,0	0,6	7,15

a minimal loss of 3.5% due to a minor wake effect. Furthermore, the average wind speed at the wind turbine hub is an impressive 7.15 m/s, yielding a power density of 383 W/m^2.

These findings provide additional support for the installation of a wind farm and highlight the region's favorable conditions for wind energy projects.

Table 4. Simulation results for the total annual production of the wind farm.

Variable	Total
Total gross AEP [MWh]	44 254,0
Total net AEP [MWh]	45 872,3
Proportional wake loss [%]	3.5
Mean speed [m/s]	6.7
Power density [W/m^2]	383

5 Conclusion

In this study, we assessed the wind resource at a newly identified location in north-western algeria within the tell atlas mountain range known as el dhaya, which has not been previously studied. Additionally, the annual wind speed and power density at the el dhaya site qualify it as a class 6 wind power rating (6.7 m/s, 383 w/m^2). This classification highlights the site's strong wind energy potential and its appropriateness for establishing a well-developed and efficient wind energy project. Moreover, this research has highlighted the significance of the windpro tool, which facilitated the evaluation of wind potential in this area more efficiently compared to other tools like the aiolos tool.

References

1. Solarin, S.A., Bello, M.O.: Wind energy and sustainable electricity generation: evidence from Germany. Environ., Dev. Sustain. **24**(7), 9185–9198 (2021). https://doi.org/10.1007/s10668-021-01818-x
2. Stambouli, A.B., Khiat, Z., Flazi, S., Kitamura, Y.: A review on the renewable energy development in Algeria: current perspective, energy scenario and sustainability issues. Renew. Sustain. Energy Rev. **16**(7), 4445–4460 (2012). https://doi.org/10.1016/j.rser.2012.04.031
3. Said, I.A.: Energy in the Arab world. Energy (1984). https://doi.org/10.1016/0360-5442(84)90109-9
4. Bensaid, H.: The Algerian programme on wind energy. In: Proceeding of WEAC, pp. 21–27 (1985)
5. Abdeladim, K., Romeo, R., Magrì, S.: Wind mapping of a region in the north-east of Algeria. Renew. Energy **9**(1–4), 789–793 (1996). https://doi.org/10.1016/0960-1481(96)88401-1
6. Merzouk, N.K.: Wind energy potential of Algeria. Renew. Energy **21**(3–4), 553–562 (2000). https://doi.org/10.1016/s0960-1481(00)00090-2
7. Boudia, S.M., Guerri, O.: Investigation of wind power potential at Oran, northwest of Algeria. Energy Convers. Manage. **105**, 81–92 (2015). https://doi.org/10.1016/j.enconman.2015.07.055

8. Abdeslame, D., Merzouk, N.K., Mekhtoub, S., Abbas, M., Dehmas, M.: Estimation of power generation capacities of a wind farms installed in windy sites in Algerian high plateaus. Renew. Energy **103**, 630–640 (2017). https://doi.org/10.1016/j.renene.2016.10.07
9. Dayal, K.K., Cater, J., Kingan, M.J., Bellon, G., Sharma, R.N.: Wind resource assessment and energy potential of selected locations in Fiji. Renew. Energy **172**, 219–237 (2021). https://doi.org/10.1016/j.renene.2021.03.034
10. Cardona-Mancilla, C., del Río, J.S., Chica-Arrieta, E., Hincapié-Zuluaga, D.: Turbinas hidrocinéticas de eje horizontal: una revisión de la literatura. Tecnol. cienc. agua **09**(3), 180–197 (2018). https://doi.org/10.24850/j-tyca-2018-03-08
11. Carrillo, C., Montaño, A.O., Cidras, J., Díaz-Dorado, E.: Review of power curve modelling for wind turbines. Renew. Sustain. Energy Rev. **21**, 572–581 (2013). https://doi.org/10.1016/j.rser.2013.01.012
12. Abdelrahman, M.A., Abdel-Hamid, R.H., Abo Adma, M.A., Daowd, M.: Techno-economic analysis to develop the first wind farm in the Egyptian western desert at Elkharga Oasis. Clean Energy **6**(1), 211–225 (2022). https://doi.org/10.1093/ce/zkac006
13. Wooten, R.D.: Statistical analysis of the relationship between wind speed, pressure and temperature. J. Appl. Sci. **11**(15), 2712–2722 (2011). https://doi.org/10.3923/jas.2011.2712.2722
14. Merzouk, N.K., Merzouk, M.: Efficiency of three wind turbines installed on high plains region of Algeria. Procedia Eng. **33**, 450–457 (2012). https://doi.org/10.1016/j.proeng.2012.01.1224
15. Louassa, S., Guerri, O., Merzouk, N.K., Merzouk, M.: Wind resources estimation and performance evaluation of two wind farms in an Algerian arid zone. In: Proceedings of 2016 International Renewable and Sustainable Energy Conference, 2017. IRSEC 2016, pp. 444–449 https://doi.org/10.1109/IRSEC.2016.7983943
16. Belabes, B., Youcefi, A., Guerri, O., Djamai, M., Kaabeche, A.: Evaluation of wind energy potential and estimation of cost using wind energy turbines for electricity generation in north of Algeria. Renew. Sustainable Energ. Rev. **51**, 1245–1255 (2015). https://doi.org/10.1016/j.rser.2015.07.043
17. Amrani, F.Z.B., Marih, S., Missoum, I., Boutlilis, F., Bekkouche, B.: Site suitability analysis of wind energy resources in different regions of Algeria's southwestern highland. Int. J. Renew. Energ. Dev. **13**(1), 62–70 (2024). https://doi.org/10.14710/ijred.2024.56476

A Systematic Review of AI-Powered Generative Design to Improve Efficiency and Thermal Comfort in Early-Stage Residential Design

Ibtissam Chekkal(✉) and Djamel Mestoul

Architecture and Environment Laboratory (LAE), Polytechnic School of Architecture and Urban Planning, EPAU, Algiers, Algeria
{i.chekkal,d.mestoul}@epau-alger.edu.dz

Abstract. The generative design for energy efficiency and thermal comfort in residential buildings is a growing field with increasing research interest. The review identified diverse generative design methods such as EPSAP, machine learning, and rule-based approaches, often integrated with advanced software to optimize design variables such as window characteristics and insulation. Studies have achieved significant improvements in energy efficiency (up to a 13% reduction) and occupant comfort (up to a 38% increase), while highlighting the importance of location-specific design and the impact of urban context on building performance. Additionally, the review found success in combining natural ventilation with conventional air conditioning for substantial energy savings (18% to 40%). The advancements in generative design in optimizing energy efficiency and thermal comfort go beyond mere building performance. They signal a significant shift in contemporary architecture – the growing importance of embedding sustainability principles directly within the design process itself.

Keywords: Generative design · energy efficiency · thermal comfort · residential buildings · early-stage design

1 Introduction

The design of a building transcends aesthetics and functionality, playing a critical role in user comfort and achieving energy efficiency from the outset. The global energy demand is heavily influenced by the building sector, accounting for 37.02% of final energy consumption, with residential buildings alone responsible for 20.96% [1]. Indoor thermal comfort, once considered a luxury, has become a desired amenity, driving the widespread use of Heating, Ventilation, and Air Conditioning (HVAC) systems. The HVAC systems contribute to 32% of the overall energy consumption in the residential sector [2]. Therefore, maximizing energy efficiency in building design is paramount [2–5] to enhance the sustainable utilization of energy. Optimizing building operations is a vital step in the pursuit of global carbon neutrality [5]. Building energy efficiency relies heavily on incorporating passive design strategies, like natural ventilation and shading, from the very beginning of the design process [2, 6]. Optimizing these elements during this early

© The Author(s), under exclusive license to Springer Nature Switzerland AG 2025
M. Hatti (Ed.): IC-AIRES 2024, LNNS 1238, pp. 783–791, 2025.
https://doi.org/10.1007/978-3-031-80301-7_82

stage minimizes the need for energy-guzzling mechanical systems later and avoids the high costs associated with retrofitting a building after construction [7]. A new and exciting tool emerges to address this challenge: generative design. This iterative process acts as a proactive explorer, automatically considering a vast array of design options based on specific criteria. By leveraging this computational power, architects can delve into a much wider range of possibilities than manual methods allow, potentially uncovering innovative and highly energy-efficient solutions they might have otherwise overlooked [8, 9]. While existing reviews have explored the utilization of generative design in product design and even urban design energy efficiency, a significant gap remains when it comes to its use in residential buildings [10, 11]. This study aims to bridge this gap by conducting a systematic review using the PRISMA methodology. This review will carefully analyze existing literature to understand how generative design can be harnessed to improve both energy efficiency and thermal comfort specifically in residential buildings. This foundational information will support future research endeavors, ensuring efficiency and rigor in studies. This review seeks to address the following questions: Q1: What is the current state and extent of research exploring generative design's application to enhancing energy efficiency and thermal comfort in the initial stages of residential building design? Q2: How are generative methods applied with optimized design variables to enhance both thermal comfort and energy efficiency in residential settings during the early design stage?

2 Methodology

Transparent and systematic reviews, following PRISMA guidelines and incorporating meta-analysis techniques, are utilized to improve the reliability and reproducibility of findings. This approach also effectively reduces bias, thus addressing the research gap [12].

2.1 Database Search

Scopus and Google Scholar served as the primary data sources for this study. Scopus, the largest database collects abstracts, references, and works from renowned authors around the world. Google Scholar plays a vital role in minimizing publication bias by encompassing both academic gray or unconventional literature, thus being the search engine of choice for many researchers [13].

2.2 Data Collection

The process of research conforms to the PRISMA guidelines, which detail four crucial stages for reporting systematic literature reviews aligned with the specified research objectives (Fig. 1).

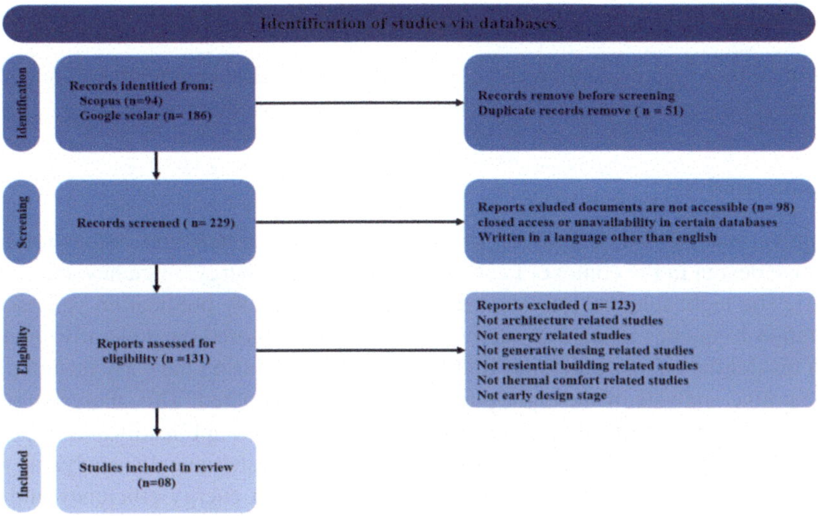

Fig. 1. PRISMA-based flowchart for systematic literature review.

2.3 Identification Phase

The documents collected from Scopus were selected using the option (title + abstract + keywords) with the research keywords (generative) AND (design) AND (building) AND (energy) AND (efficiency). A total of 94 documents were retrieved from the database spanning the period from 2004 to 2024. The research criteria in Google Scholar involved the terms ("generative design") AND ("residential building") AND ("energy efficiency") within the document's title and abstract. A total of 186 publications were identified between 2023 and 2024.

2.4 Screening Phase

A total of 280 publications, spanning from 2004 to 2024, were identified during the initial document search and will undergo the screening process. An initial assessment indicated that the documents were suitable for further analysis. Subsequent screening resulted in the exclusion of 51 publications due to factors such as restricted access, absence from accessible databases, or non-English language.

2.5 Eligibility Phase

Following the screening process, 131 publications advanced to a rigorous evaluation to ensure the review's reliability and impartiality. By assessing their alignment with research objectives focused on generative design, architecture, energy efficiency, thermal comfort, residential buildings, and early design stages, 123 publications were determined to be outside the study's scope and consequently excluded.

2.6 Including Phase

Following a thorough evaluation and collaborative discussion among all authors, a final selection of 08 publications was incorporated into this systematic review.

2.7 Data Analysis

Descriptive analysis was employed to examine research trends and output related to generative design in the context of early-stage residential energy efficiency and thermal comfort. The eight selected publications were categorized by publication year, source, and citation frequency using a deductive approach. To explore how generative methods impact residential energy efficiency and thermal comfort during the early design phase, a comparative analysis of the eight publications was conducted. This analysis focused on categorizing generative design methods, particularly their applications in improving energy efficiency in early-stage residential design. By systematically analyzing these methods, commonalities in enhancing residential building energy efficiency and their potential to achieve specific energy performance targets were identified.

3 Result

This section presents the findings of our systematic literature review, organized into two key areas. The first part provides a descriptive overview of research on generative design and energy efficiency in residential buildings, including publication trends over time, sources, and citation impact. The second part conducts a comparative analysis to investigate the diverse applications of generative design methods in enhancing energy efficiency and thermal comfort within residential buildings. This analysis also pinpoints the design parameters optimized in this domain.

3.1 Publications on Generative Design for Enhancing Energy Efficiency and Thermal Comfort During the Design Stage in Residential Building

From 2018 to 2024, 08 articles were selected for review (Table 1). The number of publications steadily increased, with a significant surge in recent years, peaking at four articles in 2023 (Fig. 2). These articles were published in various journals, including "Sustainability," "Building Engineering," and "Building and Environment," among others. The high citation counts of the articles highlight their significant impact on the field of generative design and energy efficiency in residential buildings.

Table 1. A list of sources included in the systematic literature review.

Article Title	Authors	Year	Puplishing source	Citation
"The potential impact of low thermal transmittance construction on the European design guidelines of residential buildings"	Eugénio et al. [14]	2018	Preprint submitted to Energy and Buildings	31
"Thermal transmittance effect on energy consumption of Mediterranean buildings with different thermal mass"	Eugénio et al. [15]	2019	Applied energy	73
"Performance-based design of multi-story buildings for a sustainable urban environment: a case study"	Eugénio et al. [16]	2019	Renewable and Sustainable Energy Reviews	31
"A Deep Learning Approach toward Energy-Effective Residential Building Floor Plan Generation"	Da Wan et al. [17]	2022	sustainability	11
"Climate change and ideal thermal transmittance of residential buildings in Iran"	Eugénio et al. [18]	2023	Building Engineering	19
"Prediction of Cooling Load of Tropical Buildings with Machine Learning"	Gebrail et al. [19]	2023	sustainability	08
"Generative design of floor plans of multi-unit residential buildings based on consumer satisfaction and energy performance"	Ting-Kwei W and Weina D. [20]	2023	Developments in the Built Environment	04
"Sustainability through Optimal Design of Buildings for Natural Ventilation using Updated Comfort and Occupancy Models"	Jihoon et al. [21]	2023	Building and Environment	00

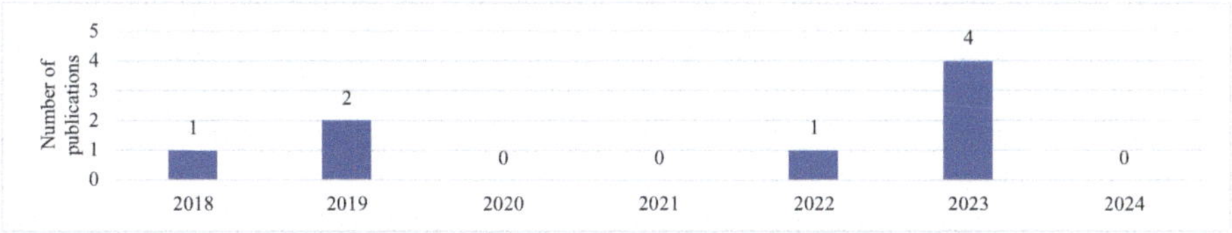

Fig. 2. Annual Publication Count or Publications by Year.

3.2 Generative Design Approaches for Energy Efficiency and Thermal Comfort During the Design Stage in Residential Buildings

This literature review highlights the diverse forms and complexities of generative design methods, emphasizing their role in achieving efficient design iterations and exploring wide solution spaces for optimization [20]. Generative design empowers designers to concentrate on the creative process by utilizing advanced software platforms such as

Rhino and Grasshopper or Revit and Dynamo [22]. EPSAP is notable for creating diverse building layouts (Fig. 3) in multi-unit projects [23]. Rule-based generative design uses simple algorithms governed by specific rules [24], while evolutionary design employs various computation techniques to handle complex problems with numerous variables and solutions [25, 26]. Machine learning and AI enhance decision-making by extracting insights from design data, with notable methodologies including GANs [27, 28]. Several studies employed diverse methodologies: three used EPSAP (Eugénio et al. for generative design and simulation [14, 21]; for automated layout design in sustainable urban design [16]). Two studies mentioned SD-GAN (Da Wan et al. for energy-efficient floor plans [17]; Eugénio et al. for space optimization [21]). Monte Carlo was used in two studies (Gebrail et al. for cooling load prediction and PCA for data analysis [19]; Eugénio et al. for data synthesis [21]). Ting-Kwei W and Weina D. used K-means clustering, PCA for data analysis, the NSGA-II algorithm, and NSFDSS for optimal solutions [20]. Jihoon et al. utilized the WFC algorithm for generating building geometries [21]. Common tools included EnergyPlus (Eugénio et al., 2019; Jihoon et al., 2023), BIM applications (Gebrail et al., 2023), Dynamo Revit (Ting-Kwei W and Weina D., 2023), and Grasshopper with Ladybug and Honeybee (Jihoon et al., 2023).

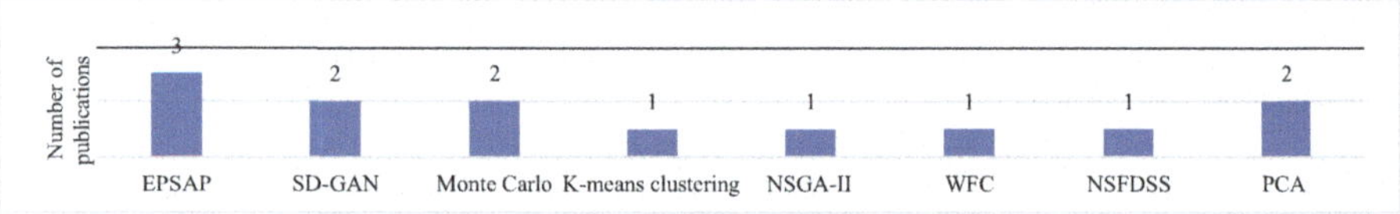

Fig. 3. Generative design methods used.

By integrating advanced methods and software tools to optimize specific design variables (Fig. 4) such as floor area, window characteristics, insulation materials, and building orientation, the studies achieved significant improvements in energy efficiency, occupant comfort, and overall building performance. A building's thermal performance is determined by how effectively it retains or releases heat, influenced primarily by its envelope's ability to insulate, focusing on factors specifically transmittance (U-value) and the building's external envelope [13, 17]. Each climatic location has an ideal range of U-values that minimizes energy demand while making the effect of building geometry less significant [14].

Eugénio et al. (2019) Despite having identical apartment layouts on each floor, the study revealed that the building's location significantly influences its energy performance. This resulted in a performance gap of 9% to 12% between the most and least energy-efficient floors. The most energy-efficient apartments were at intermediate levels [16]. Da Wan et al. (2022) showed that the SD-GAN model, trained on energy efficiency, generated plans with 7.47% to 13.38% lower annual energy consumption compared to originals [17]. Gebrail et al. (2023) achieved high accuracy with the histogram gradient boosting regression algorithm and stacking implementations, reaching an R2 score of 0.9949 [19]. Wang T.K. and Duan W. (2023) used generative design for residential floor plans, resulting in A notable increase of up to 38.78% in consumer satisfaction was accompanied by an energy performance enhancement of up to 11.18% [20]. Jihoon

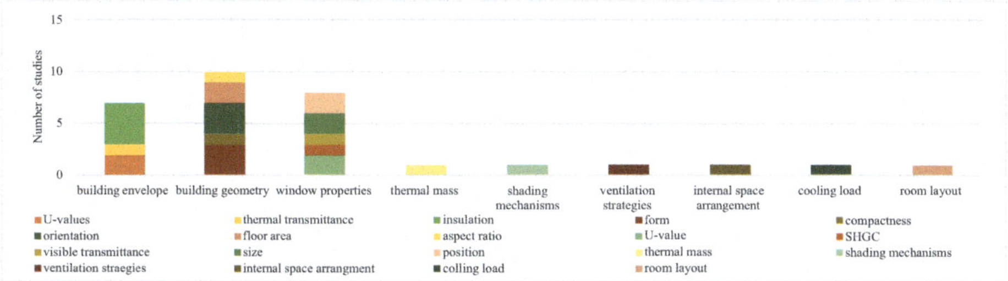

Fig. 4. The design variables optimized.

et al. found that NV and multi-mode (MM) cooling strategies reduced energy consumption compared to conventional HVAC, with MM mode decreasing Energy Use Intensity (EUI) by 18% to 40% across various climate zones [21].

4 Discussion

The generative design for energy savings and thermal comfort in residential buildings is a growing field with increasing research interest since 2018. The systematic review by Phattranis et al. (2024) mentions that evolutionary algorithms have traditionally been the dominant force in generative design. However, parametric search strategies have gained significant traction in recent years [23]. The review identified diverse generative design methods such as EPSAP, machine learning, and rule-based approaches, often integrated with advanced software to optimize design variables such as window characteristics and insulation. Studies have achieved significant improvements in energy efficiency (up to a 13% reduction) and occupant comfort (up to a 38% increase) [17], while highlighting the importance of location-specific design and the impact of urban context on building performance. Additionally, the review found success in combining natural ventilation with conventional air conditioning for substantial energy savings (18% to 40%) [21]. Limitations include the relatively small number of studies. Future research directions include conducting real-world case studies for optimization in creating a more sustainable built environment.

5 Conclusions

This systematic review uncovered generative design as a powerful tool for creating energy-efficient and comfortable residential buildings. The review highlighted how generative design uses advanced software and methods like EPSAP, machine learning, and rule-based systems. These methods allow for fine-tuning important design elements like windows and insulation. Studies consistently showed significant improvements, with buildings achieving up to 13% less energy use and occupants experiencing up to 38% more comfort. Additionally, combining natural ventilation with traditional air conditioning offered substantial energy savings, ranging from 18% to 40%. However, the review also identified a need for more real-world applications to confirm these findings in various city settings and climates. By addressing this limitation, generative design has the potential to significantly contribute to sustainable building practices and achieving global energy efficiency goals.

References

1. « Final energy consumption of buildings relative to other sectors, 2022 – Charts – Data & Statistics », IEA. Consulté le: 15 juillet 2024. [En ligne]. Disponible sur: https://www.iea.org/data-and-statistics/charts/final-energy-consumption-of-buildings-relative-to-other-sectors-2022
2. González-Torres, M., Pérez-Lombard, L., Coronel, J.F., Maestre, I.R., Yan, D.: Review on buildings energy information: trends, end-uses, fuels and drivers. Energy Rep. **8**, 626–637 (2022). https://doi.org/10.1016/j.egyr.2021.11.280
3. Emissions Gap Report 2022: The closing window : climate crisis calls for rapid transformation of societies. Nairobi: United Nations Environment Programme (2022)
4. Rodrigues, E., Fernandes, M.S., Gomes, Á., Gaspar, A.R., Costa, J.J.: Performance-based design of multi-story buildings for a sustainable urban environment: a case study. Renew. Sustainable Energy Rev. **113**, 109243 (2019). https://doi.org/10.1016/j.rser.2019.109243
5. Xiang, X., et al.: Historical decarbonization of global commercial building operations in the 21st century. Appl. Energy **322**, 119401 (2022). https://doi.org/10.1016/j.apenergy.2022.119401
6. Chel, A., Kaushik, G.: Renewable energy technologies for sustainable development of energy efficient building. Alexandria Eng. J. **57**(2), 655–669 (2018). https://doi.org/10.1016/j.aej.2017.02.027
7. Bragança, L., Vieira, S.M., Andrade, J.B.: Early stage design decisions: the way to achieve sustainable buildings at lower costs. Sci. World J. **2014**, 1–8 (2014). https://doi.org/10.1155/2014/365364
8. Ma, W., et al.: Generative design in building information modelling (BIM): approaches and requirements. Sensors **21**(16), 5439 (2021). https://doi.org/10.3390/s21165439
9. Lee, J., Cho, W., Kang, D., Lee, J.: Simplified methods for generative design that combine evaluation techniques for automated conceptual building design. Appl. Sci. **13**(23), 12856 (2023). https://doi.org/10.3390/app132312856
10. Bekdaş, G., et al.: Prediction of Cooling Load of Tropical Buildings with Machine Learning. Sustainability **15**(11), 9061 (2023)
11. Suphavarophas, P., Wongmahasiri, R., Keonil, N., Bunyarittikit, S.: A systematic review of applications of generative design methods for energy efficiency in buildings. Buildings **14**(5), 1311 (2024). https://doi.org/10.3390/buildings14051311
12. Bekkouche, S.M.E.A., et al.: Thermal resistances of local building materials and their effect upon theinterior temperatures case of a building located in Ghardaïa region. Construct. Build. Mater. **52**, 59–70 (2014)
13. Shamseer, L., et al.: Preferred reporting items for systematic review and meta-analysis protocols (PRISMA-P) 2015: elaboration and explanation. BMJ **349**(jan02 1), g7647–g7647 (2015). https://doi.org/10.1136/bmj.g7647
14. Rodrigues, E., Fernandes, M.S., Soares, N., Gomes, Á., Gaspar, A.R., Costa, J.J.: The potential impact of low thermal transmittance construction on the European design guidelines of residential buildings. Energy Build. **178**, 379–390 (2018). https://doi.org/10.1016/j.enbuild.2018.08.009
15. Rodrigues, E., Fernandes, M.S., Rodrigues Gaspar, A., Gomes, Á., Costa, J.J.: Thermal transmittance effect on energy consumption of Mediterranean buildings with different thermal mass. Appl. Energy **252**, 113437 (2019). https://doi.org/10.1016/j.apenergy.2019.113437

16. Rodrigues, E., Fernandes, M.S., Gomes, Á., Gaspar, A.R., Costa, J.J.: Performance-based design of multi-story buildings for a sustainable urban environment: a case study. Renew. Sustainable Energy Rev. **113**, 109243 (2019). https://doi.org/10.1016/j.rser.2019.109243
17. Wan, D., Zhao, X., Lu, W., Li, P., Shi, X., Fukuda, H.: A deep learning approach toward energy-effective residential building floor plan generation. Sustainability **14**(13), 8074 (2022). https://doi.org/10.3390/su14138074
18. Rodrigues, E., Fereidani, N.A., Fernandes, M.S., Gaspar, A.R.: Climate change and ideal thermal transmittance of residential buildings in Iran. J. Build. Eng. **74**, 106919 (2023). https://doi.org/10.1016/j.jobe.2023.106919
19. Bekdaş, G., Aydın, Y., Isıkdağ, Ü., Sadeghifam, A.N., Kim, S., Geem, Z.W.: Prediction of cooling load of tropical buildings with machine learning. Sustainability **15**(11), 9061 (2023). https://doi.org/10.3390/su15119061
20. Wang, T.-K., Duan, W.: Generative design of floor plans of multi-unit residential buildings based on consumer satisfaction and energy performance. Dev. Built Environ. **16**, 100238 (2023). https://doi.org/10.1016/j.dibe.2023.100238
21. Chung, J., Shahmansouri, N., Goldstein, R., Stoddart, J., Locke, J.: Sustainability through Optimal Design of Buildings for Natural Ventilation using Updated Comfort and Occupancy Models. arXiv, 11 décembre 2023. https://doi.org/10.48550/arXiv.2312.06498
22. Holzer, D., Hough, R., Burry, M.: Parametric design and structural optimisation for early design exploration. Int. J. Architectural Comput. **5**(4), 625–643 (2007). https://doi.org/10.1260/147807707783600780
23. «A-Systematic-Review-of-Applications-of-Generative-Design-Methods-for-Energy-Efficiency-in-Buildings_2024_Multidisciplinary-Digital-Publishing-Institute-MDPI (1).pdf »
24. Gilibert, P., Mesnil, R., Baverel, O.: Rule-based generative design of translational and rotational interlocking assemblies. Autom. Construct. **135**, 104142 (2022). https://doi.org/10.1016/j.autcon.2022.104142
25. Frazer, J.H., Tang, M.X., Jian, S.: Towards a Generative System for Intelligent Design Support, présenté à CAADRIA 1999, Shanghai, China, pp. 285–294 (1999). https://doi.org/10.52842/conf.caadria.1999.285
26. Gradišar, L., Klinc, R., Turk, Ž, Dolenc, M.: Generative design methodology and framework exploiting designer-algorithm synergies. Buildings **12**(12), 2194 (2022). https://doi.org/10.3390/buildings12122194
27. Zheng, H., Yuan, P.F.: A generative architectural and urban design method through artificial neural networks. Build. Env. **205**, 108178 (2021). https://doi.org/10.1016/j.buildenv.2021.108178
28. Oh, S., Jung, Y., Kim, S., Lee, I., Kang, N.: Deep generative design: integration of topology optimization and generative models. J. Mech. Des. **141**(11), 111405 (2019). https://doi.org/10.1115/1.4044229

Author Index

A
Abadlia, Issam 51
Abbadi, A. 133
Abbadi, Amel 354
Abdelghani, A. 99
Abdelghani, Aissaoui 731
Abdelghani, Choucha 179
Abdelhak, Achouri 27
Abdelkader, Boutadara 437
Abdelkarim, Ammar 372
Abderrezak, Aibeche 453
Abdoune, Fateh 51
Abed, Amina 656
Abedelkader, Hadidi 437
Abid, Mohamed 731
Adardour, Haroun Errachid 634
Addaoui, N. 646
Afaf, Smaani Anissa 573
Aibeche, Abderrezak 464
Ait Hammouda, C. 152
Akroum, Hamza 235
Alti, Adel 573
Amira, Slimani 372
Ammar, A. 285
Ammari, Chouaib 207
Ammi, M. 141
Amor, Bourek 372
Amrani, Fatima Zohra Bochra 774
Annane, Boubakeur 573
Annane, Boubekeur 656
Aoun, Nouar 766
Arezki, S. 99
Atia, Abdelmalek 405
Attari, K. 750

B
Bahloul, Derradji 3
Barazane, L. 554
Bechekir, S. 60
Bedri, Mohammed Amine 563
Beghriche, Abdesselem 656
Bekkouche, Benaissa 774
Belabed, M. 60
Belarbi, EL-Habib 263
Belarbi, S. 554
Belhachat, F. 594
Belkacem, Z. 397
Bellaoui, Mebrouk 43
Benabdellah, Abdelkader 263
Benazzouz, Djamel 32
Benchettara, A. 423
Bendaho, H. 60
Bendiab, Gueltoum 342
Bendib, A. 152
Bengheni, Abdelmalek 672
Bengourina, Mohamed Rida 51
Benhabyles, L. 416
Benmansour, J. E. 161
Benmansour, Khelifa 78
Benmbarek, M. 475
Benmedjahed, Miloud 217, 766
Benmir, Nora 235
Bennia, R. 594
Benrabah, Abdeldjabar 78
Berrabah, S. E. 423
Bouam, A. 226, 750, 758
Bouam, L. 750
Boudaoud, Ali 330
Boudiaf, A. 321
Boudjema, Tidjar 766
Boudoucha, A. 141
Boudour, M. 99
Bougandoura, Djihane 86
Bouguenna, Elouahab 170
Boukenoui, R. 152
Boumesaid, Chaima 342
Bouraiou, Ahmed 766
Bourek, Yacine 207
Bousmaha, I. S. 60
Boutaghane, Said 563
Boutana, Yacine 304
Bouttout, A. 504

© The Editor(s) (if applicable) and The Author(s), under exclusive license to Springer Nature Switzerland AG 2025

M. Hatti (Ed.): IC-AIRES 2024, LNNS 1238, pp. 793–796, 2025.
https://doi.org/10.1007/978-3-031-80301-7

Bouzerdoum, M. 312
Bouzgou, Hassen 428, 544, 584
Bouziane, Abdelghani 525
Bouziane, Mohammed 525
Bradai, R. 152
Bradai, Rafik 71, 116
Brahami, M. 60
Brahimi, A. 60
Brik, A. 141
Burassia, Bensaad 716

C
Chaker, Yassine 263
Chaouch, W. Braham 397
Cheggaga, N. 293
Cheggaga, Nawal 364
Cheggou, Rabea 699
Chekkal, Ibtissam 783
Cheridi, A. L. Deghal 226
Cherifi, Dalila 330, 563

D
Dahbi, Abdeldjalil 217, 766
Dahia, A. 226, 750, 758
Dahli, Khadidja 364
Daif, M. 475
Daouli, Benhaoua Larbi 269
Debbal, Mohammed 708
Debdab, Mansour 263
Deghal, A. L. Cheridi 750, 758
Derradji, L. 504
Djaber, A. 397
Djedjiga, Boukhlef 437
Djellal, A. 187
Djilali, Abdelkadir Belhadj 12, 18
Doghman, Moahamed. Z. 235
Doghmane, M. Z. 244
Doghmane, Mohamed Zinelabidine 443, 464
Douifi, N. 133
Drir, N. 554

F
Fatiha, Bouhadjela 716
Fatima, Belhi 437
Fatima, Boutlilis 774
Ferroudji, Fateh 43
Fihakhir, Amine 196
Foullanine, M. 397
Foullanine, Meriama 276
Four, I. 646
Fraoucene, Henia 699

G
Gama, Amor 544
Gougam, Fawzi 32
Guenfaf, Lakhdar 483
Guerbouz, Abdesselam 196
Guerfa, Yousra 342
Gueymard, Christian 584
Guezam, Abdelhak 263

H
Hachemi, Ahmed T. 124
Haddad, Sofiane 304
Hadibi, Samira 443
Hadjam, A. 226, 750, 758
Halloub, Saad 263
Hamid, Abdi 27
Hamidia, F. 133
Hamidia, Fethia 354
Hamouda, Messaoud 766
Hamrit, Abdelhakim 741
Haouzi, Amel 634
Harrat, Ahlem Assia 708
Hasni, Mourad 483
Hassaine, Linda 51
Hassina, Ghodbane 3
Hattabi, Intissar 71, 116
Hatti, Mustapha 263, 615
Hfaifa, Ahmed 196
Hiber, Houssam Eddine Nadir 254

I
Iddou, Hemiche 741
Idir, A. 244
Imene, Bensalah 573
Iourzikene, Zouhir 32

K
Kada, Hartani 383
Kadri, Rania 731
Kameche, Samir 634
Kebir, A.-H. 293
Kellil, N. 312, 321
Kentouche, A. 226, 758
Khaldi, Fouad 3
Kheldoun, A. 285, 475

Kheldoun, Aissa 71, 116
Khelfaoui, Abderrahmane 217, 766
Khellaf, L. 187
Khenissa, M. I. 423
Kherbachi, A. 152
Khettab, Soufian 71, 116
Khorsi, A. Dadda 226, 750, 758
Khouane, B. 161
Khoucha, Farid 78
Khoudir, Kakouche 372
Khoumeri, El hadi 699
Kidouche, M. 244
Kidouche, Madjid 443, 464
Korzane, Hicham 78
Kouba, N. 141
Koudiah, N. 750
Kourdi, Z. 646

L
Ladaci, Samir 170
Lahcen, Lakhdari 536
Lahdeb, M. 124
Lakehal, Abderrahim 573
Lakhdar, Chaib 179
Larbes, C. 594
Larbes, Cherif 86
Larbi, A. A. 397
Legouini, Tarek 741
Limam, A. 504
Louchene, Houssem Eddine 584
Loumani, A. 397
Loumani, Akil 276

M
Madjid, Kidouche 453
Mameri, Abdelbaki 405
Mammeri, Souhila 623
Mana, Hassan 269
Manal, Khaldi 437
Mansour Bekhtache, R. 646
Mansouri, Khaled 405
Matouk, M. A. 141
Mawloud, Tellaa 453
Mayache, H. 187
Mechehat, Nada Rihane 722
Mechri, H. 285
Meddah, S. 244
Mediani, A. 397
Mediani, Ahmed 276
Medjber, Ahmed 354
Megaache, Salah Eddine 544
Mekkaoui, Abdelkader 493
Mekroud, Sifeddine 330
Mellit, A. 312, 321
Meriem, Hammou 383
meriem, mami 573
Merrouche, Walid 170
Merzouk, Imad 196
Messini, El Mouatez Billah 207
Mestoul, Djamel 783
Meziane, Assia 405
Miloua, Hadj 254
Mizat, L. 124
Mohamed, Habbab 536
Morsli, A. 133
Morsli, Abdelkader 354
Moussaoui, C. 312, 321

N
Naima, Khatir 525
Necib, Djilani 722
Norediene, Aouadj 383

O
Ouadah, Mohammed Chamse Eddine 708
Oubbati, Y. 124
Ouchefoun, Zoulikha 483
Ouchiha, Zohra 741
Oukaci, S. 504
Ould Amara, Said 623

R
Rached, I. 161
Rahmouna, Cheriet 716
Rahrah, Khaoula 428
Rebbah, Mohammed 493
Rezgui, Imane 722
Rouibah, N. 504

S
Saber, M. 416
Safaa, Madani 354
Saidi, Ahmed 684
Saihi, Lakhdar 43
Selam, Karim 364
Sheta, A. 187
Skender Mohamed, Redha 354
Skender, M. R. 133

Slimani, Abdeldjalil 766
Smail, Omar 493
Soualah, Mondir 722
Soufyane, Belhenini 716
Soukkou, Ammar 304

T
Tadjer, S. A. 244
Taibi, Soufiane 428
Taleb, Rachid 12, 18
Tchoketch-Kebir, S. 293
Tellaa, Mawloud 464
Tidjar, Boudjemaa 217
Tigani, C. 397
Titri, Sabrina 86
Toualbia, Asma 12, 18

Y
Yassmine, Bouziane 383
Yazid, Mohand 623

Z
Zahra, Zouak Fatma 179
Zemouli, A. S. 161
Zerari, Naima 544
Zermane, Hanane 516
Zine, Saadi 437
Zinelabidine, Doghmane Mohamed 453

The manufacturer's authorised representative in the EU is Springer Nature Customer Service Centre GmbH, Europaplatz 3, 69115 Heidelberg, Germany. If you have any concerns regarding our products, please contact ProductSafety@springernature.com

Printed and bound by CPI Group (UK) Ltd, Croydon, CR0 4YY
01/10/2026
02225957-0018